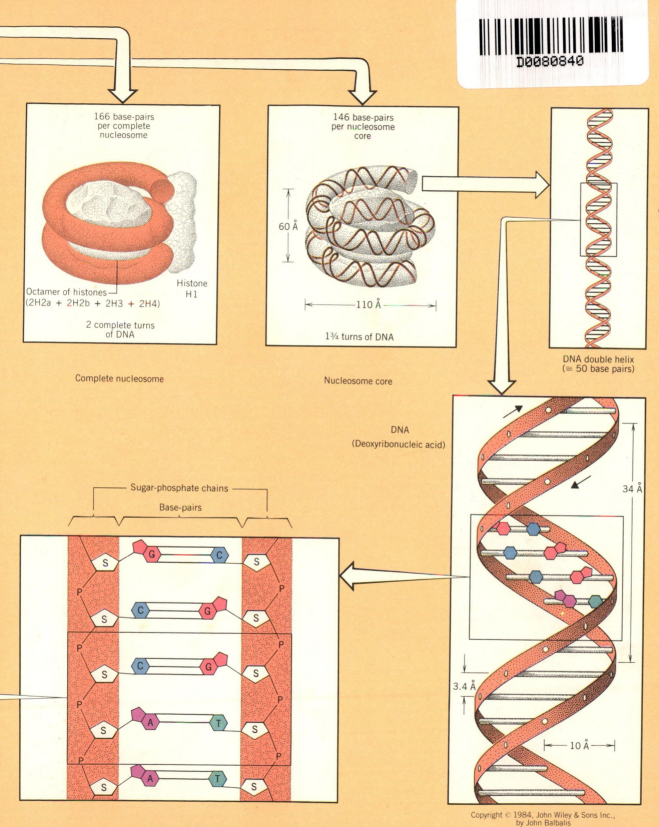

166 base-pairs
per complete
nucleosome

Octamer of histones
(2H2a + 2H2b + 2H3 + 2H4)

Histone
H1

2 complete turns
of DNA

Complete nucleosome

146 base-pairs
per nucleosome
core

60 Å

110 Å

1¾ turns of DNA

Nucleosome core

DNA double helix
(≅ 50 base pairs)

DNA
(Deoxyribonucleic acid)

34 Å

3.4 Å

10 Å

Sugar-phosphate chains

Base-pairs

S G   C S
P   P
S C   G S
P   P
S C   G S
P   P
S A   T S
P   P
S A   T S

PRINCIPLES OF GENETICS

PRINCIPLES OF GENETICS
SEVENTH EDITION

ELDON J. GARDNER
Utah State University

D. PETER SNUSTAD
University of Minnesota

JOHN WILEY & SONS
New York
Chichester
Brisbane
Toronto
Singapore

Production Supervised by Linda Indig
Photo Researched by Kathy Bendo
Illustrations by John Balbalis with the assistance
of the Wiley Illustration Department.
Cover: Painting by Kathleen Borowick
Text Design: Edward A. Butler
Manuscript editor: Genevieve Danser

Library of Congress Cataloging in Publication Data

Gardner, Eldon John, 1909–
    Principles of genetics.

    Includes bibliographies and index.
    1. Genetics. I. Snustad, D. Peter. II. Title.
[DNLM: 1. Genetics. QH 430 G226p]

QH430.G37 1984   575.1   83-21798
ISBN 0-471-87610-0

Printed in the United States of America

10 9 8 7 6

PREFACE

This revision is a continuing attempt to provide a general genetics text that is current, readable, and challenging for students; it is flexible in that it can serve as a textbook for university quarter, semester, or two-quarter courses. It is a beginning-level text, focused on the basic principles of genetics, but with sufficient rigor to satisfy the needs of biology majors. This edition has been reorganized somewhat and extensively rewritten. In making the book more concise, basic aspects of genetics have been preserved and extended in depth. Coverage of molecular genetics has again been expanded in this edition.

The textbook begins with a concise introduction, giving historical and functional perspectives, as well as a perspective of the principles covered in the 16 chapters that follow. Chapter 2 provides a unified statement of Mendelian genetics followed in Chapter 3 by a description of the cellular mechanics that explain Mendelian principles, and in Chapter 4 by an outline of sex determination and sex linkage. The biochemical nature of genes, their mode of replication, and the organization of DNA in chromosomes are introduced in Chapter 5. These early discussions of cellular and molecular biology provide background for the consideration of crossing over, bacterial recombination, mutation, genetic fine structure, and gene regulation in Chapters 6 to 11.

Linkage, chromosome mapping, tetrad analysis, gene conversion, and the molecular basis of crossing over are integrated in Chapter 6. Recombination mechanisms unique to bacteria — conjugation, transduction, and transformation — are covered in Chapter 7, along with plasmids, episomes, and insertion sequences. Chapter 8 covers the genetic control of metabolism, with emphasis on the one gene — one polypeptide concept, transcription, translation, and the genetic code. Chapter 9 describes the nature and molecular basis of spontaneous and induced mutation. Repair mechanisms, the consequences of defective repair pathways, and the correlation between mutagenicity and carcinogenicity are also discussed. The structure of genes and complex loci are covered in Chapter 10. Evidence for noncoding intervening sequences or introns within genes of eukaryotes is presented, along with a discussion of the recombinant DNA and gene-cloning, restriction enzyme mapping, and DNA sequencing techniques used to detect these sequences. In Chapter 11, we have attempted to provide a concise description of what is known about the regulation of gene expression in both prokaryotic and eukaryotic systems. Oncogenes, proto-oncogenes, and the genetic control of antibody structure are also covered in Chapter 11.

Chromosomal structural aberrations and chromosomal numerical changes with their mechanisms and consequences are the topics of Chapter 12 and 13. Most extranuclear inheritance (Chapter 14) is attributed to evolutionary symbiosis that has brought free living bacterial cells with their own DNA into intimate association with eukaryotic organisms. Chapter 15 and 16 are devoted to quantitative inheritance and population genetics and evolution. Current aspects and the status of behavior genetics are outlined in Chapter 17. Examples from human genetics are given throughout the book. A glossary of terms, answers to all problems in the book, and a combined author and subject index complete the book.

Our thanks to students, teachers, and colleagues who have suggested improvements for the book. We especially thank Rayla Temin, R. H. Whalen, Linda Kosturko, Clint Magill, Peter Wejksnora, Charles Rodell, Hugh Stanley, Larry Puckett, Robert M. Fineman, Irene Uchida, Scott R. Woodward, John R. Simmons, and Franklin D.

Enfield who have made valuable contributions. We also thank the following reviewers for their valuable suggestions and comments: Judith Van Houten, The University of Vermont; Robert Petters, Pennsylvania State University; Darrell S. English, Northern Arizona University; Romesh C. Mehra, Indiana University at South Bend; Glenn Wolfe, Kansas University; Paul J. Homsher, Old Dominion University; Paul A. Roberts, Oregon State University; Asim Esem, Virginia Polytechnic Institute and State University; Howard Laten, Loyola University. We also gratefully acknowledge the assistance of Frederick C. Corey, editor, John Balbalis and staff of the Wiley Illustration Department, Kathy Bendo, photograph editor, Genevieve Danser, manuscript editor, and Linda Indig, production supervisor.

Credits for illustrations, tables, and quotations from other individuals or publications are given with the individual item.

Eldon J. Gardner
D. Peter Snustad

CONTENTS

PRINCIPLES OF GENETICS

ONE

INTRODUCTION

This book deals with the **principles** of genetics. The objective is to present the basic foundation and building blocks on which the science is established. We begin with the principles uncovered by Gregor Mendel; the first procedure is mating of experimental organisms, from which transmission genetics was discovered. With complex living organisms, the mechanism for the Mendelian principles of segregation and independent assortment is demonstrated in reproductive cells. Chromosomes are associated with particular roles in the inheritance process. Thus, cells and their chromosomes are the basis of most living processes.

Combinations in chromosomes account for the behavior of the genes they carry. Genes have an organization within chromosomes that can be changed and thus provide variation in the traits of organisms. Mutations create variations in the genes themselves. Genes replicate themselves or their changed selves and reproduce chromosomes, cells, and organisms of their own kind. This is all possible because of the properties of the chemical (DNA) from which the genes are composed. DNA also directs the synthesis of proteins (enzymes, antigens) within cells. Molecular genetics has opened the way to recombinant DNA, cloning (asexual reproduction of identical progeny by a single cell or other biological unit), and other applications of genetic engineering. Genetics has gone commercial with bacterial cells, making human insulin, interferon (which has antiviral and possibly antitumor activity), and other valuable life-saving products.

Chromosomes undergo structural and numerical changes, causing more variations in the organisms they influence. A small amount of DNA does not join the central chromosomal DNA but remains in some cell organelles, such as mitochondria, and directs limited activity in living systems. Some chromosomal genes work together, each making a small contribution to height, weight, or intelligence. These multiple genes undergo Mendelian segregation and have practical significance in quantitative traits such as milk, meat, and wool production.

Genes not only have a basic role in the origin and life of individual organisms, but they also, through variations in gene frequencies, cause changes in populations. Such changes are brought about by natural selection, mutation, migration, random genetic drift, meiotic drive, and molecular drive. These factors control the unfolding of changes or evolution within populations. Finally, the behavior of members of populations is at least partly under genetic control.

A SCIENCE OF POTENTIALS

Genetics is a science of potentials. It deals with the transfer of biological information from cell to cell, from parents to offspring, and thus from generation to generation. Geneticists are concerned with the whys and hows of these transfers, which are the basis for certain differences and similarities that are recognized in groups of living organisms. Genetics deals with the physical and chemical nature of the information itself. What is the source of genetic variation? How are differences distributed in populations? Not all variation among living things, however, is inherited. Environmental and developmental factors are also significant and therefore of interest to the geneticist.

Long before humans began to wonder about genetic mechanisms, they were already operating effectively in nature. How and why were such mechanisms discovered? Populations of plants and animals are now known to have built-in potentials for constancy and change that are dependent on genetics. Change that becomes established through these mechanisms over long periods of time in a population of living things is called **evolution.**

Many potential changes have been accomplished by human intervention in genetic mechanisms that now accrue to benefit human beings. Wild animals and plants have become domesticated. By selective breeding, domesticated organisms have been made to serve human society increasingly better than did their wild, unselected counterparts. Improved quantity and quality of milk, eggs, meat, wool, maize, wheat, rice, cotton, and many other sources of food, fiber, and shelter attest to the successes of human interventions.

HISTORICAL PERSPECTIVES

BIRTH OF A SCIENCE

Gregor Mendel (1822–1884) is appropriately called the "father of genetics." His precedent-setting experiments with garden peas *(Pisum sativum)*, published in 1866, were conducted in the limited space of a monastery garden (Fig. 1.1) while he was also employed as a substitute schoolteacher. The conclusions that he drew from his elegant investigations constitute the foundation of today's science of genetics. Why was Mendel so successful in discovering basic principles of genetics?

Mendel was not the first to perform hybridization experiments, but he was one of the first to consider the results in terms of single traits. Sageret in 1826 had studied the inheritance of contrasting traits. Others of Mendel's predecessors had considered whole organisms, which incorporate a nebulous complex of traits; thus, they could observe only that similarities and differences occurred among parents and progeny. They missed the significance of individual differences. Employing the scientific method, Mendel designed the necessary experiments, counted and classified the peas resulting from his crosses, compared the proportions with mathematical models, and formulated a hypothesis for these differences. Although Mendel devised a precise mathematical pattern for the transmission of hereditary units, he had no concept of the biological mechanism involved. Nevertheless, on the basis of his preliminary experiments and hypotheses, he predicted and subsequently verified his predictions with the results of later crosses.

In 1900, Mendel's paper was discovered simultaneously by three botanists: Hugo de Vries (Fig. 1.2) in Holland, known for his mutation theory and studies on the evening primrose and maize; Carl Correns in Germany, who investigated maize, peas, and beans; and Eric von Tschermak-Seysenegg in

FIGURE 1.1

Garden at Altbrünn, Austria, Monastery garden where Mendel's experiments on garden peas were conducted. (Courtesy of Professor Jaroslav Křiženecký.)

FIGURE 1.2
Hugo de Vries (1848–1935), Dutch botanist who discovered Mendel's principles in 1900. He was a pioneer in plant genetics. (Portrait by Everett Thorpe.)

FIGURE 1.3
William Bateson (1861–1926), English biologist who repeated and supported Mendel's principles and introduced Mendelian genetics to the English-speaking world. (Portrait by Everett Thorpe.)

Austria, who worked with several plants, including garden peas. Each of these investigators obtained evidence for Mendel's principles from his own independent studies. They all found Mendel's report while searching the literature for related work and cited it in their own publications. William Bateson (Fig. 1.3), an Englishman, gave this developing science the name "genetics" in 1905. He coined the term from a Greek word meaning "to generate."

CONCEPT OF THE GENE

In addition to naming the science, Bateson actively promoted Mendel's view of paired genes. He used the word "allelomorph," shortened to "allele," to identify members of pairs that control different alternative traits. Also during the early 1900s, a Frenchman, Lucien Cuénot, showed that genes controlled fur color in the mouse; an American, W. E. Castle, related genes to sex and to fur color and pattern in mammals; and a Dane, W. L. Johannsen (Fig. 1.4), studied the influence of heredity and environment in plants. Johannsen began using the word "gene" from the last syllable of Darwin's term "pangene." The gene concept, however, had been implicit in Mendel's visualization of a physical element or factor (Anlage) that acts as the foundation for development of a trait. These men and their peers were able to build on the basic principles of cytology, which were established between 1865 (when Mendel's work was completed) and 1900 (when it was discovered). Why were Mendel's important discoveries not recognized for such a long time (35 years) after the studies were completed and reported?

FIGURE 1.4
W. L. Johannsen (1857–1927), Danish botanist who recognized the significance of "pure lines," used the term "gene" to identify the unit of inheritance. (Portrait by Everett Thorpe.)

FIGURE 1.5
Theodor Boveri (1862–1915), German biologist who made basic contributions to the mechanics of cell division and recognized the chromosome as the carrier of genes. (Portrait by Everett Thorpe.)

CHROMOSOME THEORY

Wilhelm Roux had postulated as early as 1883 that chromosomes within the nucleus of the cell were the bearers of hereditary factors. The only model he was able to devise that would account for his observed genetic results was a row of lined-up objects duplicated exactly. To explain the mechanics of gene transmission from cell to cell, he therefore suggested that nuclei must have invisible structures held in rows or chains that duplicated themselves when the cell divided. Constituents of the nucleus that seemed best designed to carry genes and fill these requirements were **chromosomes.** Experiments of T. Boveri (Fig. 1.5) and W. S. Sutton in 1902 brought confirming evidence that a gene is part of a chromosome. The theory of the gene as a discrete unit of a chromosome was developed by

T. H. Morgan (Fig. 6.1) and his associates from studies on the fruit fly, *Drosophila melanogaster.* H. J. Muller later promoted the merger of the two sciences that had contributed most to the chromosome theory — cytology (the study of cells) and genetics — as **cytogenetics.**

CHEMICAL NATURE OF THE GENE

In the 1930s, G. W. Beadle, B. Ephrussi, E. L. Tatum, J. B. S. Haldane, and others provided a basis for understanding the functional properties of genes and suggested functional extensions of the classical gene concept. The gene was at first characterized as an indivisible unit of structure, unit of mutation, and unit of function, with all three of these attributes considered equivalent. Investigators then recalled that a physician, A. E. Garrod, had indicated in 1902 that genes in humans function through **enzymes.** Geneticists of the 1940s following Garrod's lead sought an ideal experimental system for investigating functional aspects of genes. Prokaryotes (organisms lacking well-defined nuclei and not undergoing meiosis; that is, bacteria and blue-green algae) were chosen for experimental material even though eukaryotes (organisms made up of cells with true nuclei bounded by envelopes and undergoing meiosis) had more practical significance for geneticists.

Early triumphs were the identification of macromolecules carrying genetic information in a bacterium by O. T. Avery and associates and in a virus by A. Hershey and M. Chase. The Avery et al. experiments showed that a chemical, deoxyribonucleic acid **(DNA),** could bring about genetic change (transformation) in a pneumococcus bacterium; Hershey and Chase demonstrated that the nucleic acid component (DNA) and not the protein is the genetic material carried by the bacteriophage. H. Fraenkel-Conrat and B. Singer showed that ribonucleic acid **(RNA)** is the genetic material in tobacco mosaic virus. Thus, RNA performs the functions in some viruses that DNA performs in other organisms. J. D. Watson and F. H. C. Crick worked out the double helix structure of the chemical DNA. The central problem of genetics was thus resolved with the discovery that DNA is the genetic material. Genetic mechanisms could now be formulated in biochemical terms. How do units (genes) of DNA control specific traits in organisms, and how do assemblages of DNA units carried in fertilized eggs provide "blueprints" for development of entire organisms?

FUNCTIONAL PERSPECTIVE

Genes accomplish their function (1) through replication that results in more units like themselves and (2) through transcription and translation, whereby proteins that function as determiners in the metabolism of the cell are synthesized. Although genes are usually stable, they are susceptible to occasional change or mutation, which provides altered forms of genes **(alleles).** Mendel first postulated the existence of genes from their end effects, as expressed in altered characteristics. Now genes have been defined chemically and are known for what they do in directing the formation of traits through the specificity of protein enzymes. Thus, DNA carries the specifications for growth, differentiation, and functioning of cells in the organism.

In the animal body, the same set of genes is present in virtually all nucleated cells. Different genes, however, become active at different times during development. But how does this selective activation occur? At times of activity, information contained in a gene is known to be decoded by processes of transcription and translation to produce **proteins.** Proteins that are enzymes catalyze cellular biochemical reactions. But how does each cell or group of cells become activated at the proper time to take its place in the blueprint of a coordinated sequence to form an organism? Contents of a single fertilized cell include the information needed for the development of a finely adjusted organism — for example, a human being destined at one time to contain at least a million billion (10^{15}) cells. It is difficult to imagine how a complex organism is assembled; however, metabolic disorders often provide unique opportunities to combine genetic concepts with tools of biochemistry and to reconstruct some steps in normal development. At a different level of organization, the DNA in a species population is the evolutionary storehouse, carrying the information for that species.

SUMMARY

Genetics deals with the inherent mechanisms that control constancy and change in living organisms. The science of genetics was born with the discovery of Mendelian principles at the turn of the twentieth century. It has taken a prominent place among the biological sciences with (1) the concept of the gene, (2) chromosome theory, and (3) the discovery that the chemical DNA is genetic material. A few basic principles represent the core of the science. Applications have developed through plant and animal breeding and through our increased understanding of the mechanisms of living systems.

REFERENCES

Dunn, L. C. 1965. *A short history of genetics.* McGraw-Hill Book Co., New York.

Gardner, E. J. 1983. *Human heredity.* John Wiley and Sons, New York.

Iltis, H. 1932. *Life of Mendel.* (Trans. E. and C. Paul.) W. W. Norton and Co., New York.

Sturtevant, A. H. 1965. *A history of genetics.* Harper & Row, New York.

TWO

MENDELIAN GENETICS

Mendel chose the garden pea as his experimental organism because it is an annual plant with well-defined characteristics, and it can be grown and crossed easily. Moreover, garden peas have perfect flowers containing both female and male (pollen-producing) parts, and they are ordinarily self-fertilized. Pollen from another plant can be experimentally introduced to the stigma, but cross-pollination is rare without human intervention.

Mendel was fortunate in choosing a diploid plant with only two sets of chromosomes. If he had chosen a polyploid organism with more than two sets, he would not have obtained simple, understandable results. Through many generations of natural self-fertilization, garden peas had developed into pure lines. A single alteration in a trait was therefore demonstrated by a visible difference between varieties. Furthermore, in the seven pairs of contrasting traits Mendel chose to study, one form was dominant over a well-defined, contrasting alternative. Vines were either tall or dwarf; unripe pods were green or yellow and inflated or constricted between the seeds; flowers were either distributed along the stem (axial) or bunched at the top (terminal); nutritive parts of ripe seeds were green or yellow; the outer surface of the seed was smooth or deeply wrinkled; and the seed coats were either white or gray. Flower color was positively correlated with this last trait. Seeds with white seed coats were produced by plants that had white flowers, and those with gray seed coats came from plants that had violet flowers. Much of Mendel's success in his first experiments may be attributed to his good judgment in making crosses, as far as possible, between parents that differed in only one trait. When this was not feasible, he considered only one trait at a time.

MENDEL'S EXPERIMENTS

Crosses were made with great care when the peas were in blossom. To prevent self-fertilization in "test" flowers, anthers were removed from those chosen to be seed parents before their pollen-receiving parts were fully mature. Pollen from the designated pollen parent was transferred at the appropriate time to the stigma of the seed-parent flower. Seeds were allowed to mature on the vines. With a trait such as seed color, classification could be made immediately; but before traits such as plant size could be classified, the seeds had to be planted in the next season and the plants raised to maturity. Hybridization experiments were carried through several generations, and backcrosses were made between hybrids and pure-parent varieties. Mendel **visualized** clearly each problem to be solved and **designed** his crosses to that end. He observed that weather, soil, and moisture conditions affected the growth characteristics of the peas, but heredity was the main factor under the conditions of his experiments. In a given environment, tall plants were 6 to 7 feet high, whereas dwarfs measured from 9 to 18 inches. No dwarfs ever turned into tall plants, and no tall plants became dwarfs.

PRINCIPLE OF SEGREGATION

In one experiment, Mendel crossed tall and dwarf varieties of garden peas. All offspring in the first (F_1) generation (F symbolized *filial* from the Latin, meaning "progeny") were tall. The dwarf trait had disappeared in the F_1 progeny. When tall hybrid plants were self-fertilized and progeny (second, or F_2 generation) were classified, some were tall and some were dwarfs. Careful classification of plants showed that when large numbers were considered, about three-fourths were tall and one-fourth were dwarfs. To be exact, an F_2 of 1064 consisted of 787 tall plants and 277 dwarfs, a nearly perfect $\frac{3}{4} : \frac{1}{4}$ **ratio.**

The experiment could have been concluded at this point; but to test his hypothesis that independent factors (genes) were responsible for observed hereditary patterns, Mendel predicted what would occur in the F_3 generation and planted F_2 seeds to test this prediction. On the basis of his hypothesis, he predicted that about one-third of the F_2 tall plants would produce only tall F_3 progeny, whereas two-thirds would produce both tall and short progeny. The F_2 short plants were expected to produce all short F_3 progeny. Expected results were obtained: tall F_2 plants produced both tall and short in the proportion of about two tall to one short, and the short F_2 plants produced only short progeny.

In other crosses, the remaining six of the originally selected seven pairs of contrasting traits were studied. One member of each pair dominated the other in the same way as tall dominated dwarf; this member Mendel identified as dominant in contrast to the other (recessive) member. Mendel's conclusions were based on his concept of **unit characters,** which was in marked contrast to the prevailing belief in a blending inheritance. On the basis of good experimental evidence, he visualized the **physical elements** as occurring in pairs of **alleles** (different forms of a given gene). In garden peas, for example, a gene for height has two alleles, one for tall and one for dwarf. The allele for tall behaves as a dominant, whereas that for dwarf is recessive. Similarly, the gene for seed coat color has two alleles, a dominant for yellow and a recessive for green. During meiosis, the members of each pair of alleles separate into different sex cells or gametes; thus, they occur in different offspring. Mendel called this separating or segregation process the "splitting of hybrids." In these experiments, Mendel had used the scientific method very well indeed.

The significant deduction from Mendel's results was that the separation of pairs of determiners resulted in a "purity of gametes." The concept of segregation, identified as Mendel's principle, can be phrased in this way: **The paired genes (allelic pairs) separate from one another and are distributed to different sex cells.**

SYMBOLS AND TERMINOLOGY

Symbolism is not the important part of genetics and no universal notation has been established. But symbols are useful for providing a language to describe important aspects of genetics. Mendel used letters of the alphabet as symbols for genes. A capital letter signified a dominant and a lowercase letter a recessive member of a pair of alleles. Mendel considered factors as abstract units, any one of which could be symbolized by A or B or some other letter. Consider that many genes are known and that several may be involved in a single series of experiments. To avoid confusion as to which gene is indicated, appropriate letter symbols are chosen to represent particular genes. The **mutant trait** that deviates from the "wild type" is usually chosen as the basis for the symbol. This trait usually is produced by the recessive allele, because most mutations occur as recessives. The dominant allele usually produces a functional product. A mutational change removes all or part of this product and leaves the recessive to come to expression only in the absence of a dominant allele. For example, the mutant vermilion is a recessive for eye color in fruit flies. As compared with the dominant wild type, the color production is diminished and the gene symbol is v. A few mutant alleles, such as the one for wrinkled wings in the fruit fly, however, are dominant and are therefore represented by a capital letter (e.g., W).

If the mutant is to provide the symbol, the history of the organism under investigation must be known well enough to suggest which member of the contrasting pair (e.g., tall or dwarf in peas) is the mutant trait. World collections of pea species show no dwarfs in natural populations. Dwarf peas occur only in certain cultivated stocks that have been developed. Dwarf is probably the recessive mutant (d), and tall is the allele for "wild type" (D). With the aid of these symbols, Mendel's experiment (Fig. 2.1) may be reconstructed in steps. **Parents (P),** with either of the two alleles (DD or dd), are represented as follows:[1]

Tall parent   Dwarf parent

$$DD \qquad \times \qquad dd$$

Segregation, the separation of pairs, occurs during formation of mature reproductive cells or **gametes.** Each gamete produced by the tall parent carries only one D allele, and each gamete from the dwarf parent carries only one d allele. Therefore, the fertilized egg **(zygote),** which results from fusion of the male and female gametes, must have one allele of each kind (Dd). Because the D was always present and dominant, the F_1 plants (first-generation progeny) were all tall. When the F_1 tall (Dd) plants were self-fertilized, half the gametes carried the D allele and half the d allele. Results of self-ferti-

[1] In diagrams of crosses, the female or seed parent is usually written first.

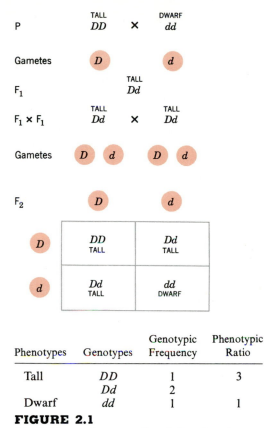

FIGURE 2.1

Mendel's cross between tall and dwarf garden peas and summary of phenotypic and genotypic results.

Phenotypes	Genotypes	Genotypic Frequency	Phenotypic Ratio
Tall	*DD*	1	3
	Dd	2	
Dwarf	*dd*	1	1

lizing the F_1 indicated to Mendel that alleles were entirely separate from each other.

Gene symbols represented in pairs designate zygotes and individual plants or animals that have arisen from zygotes. Members of pairs of alleles are represented separately to designate mature germ cells or gametes, either eggs or sperm. Circles or brackets placed around gamete symbols indicate mature germ cells, as distinguished from plants or animals. A female and a male gamete combine in fertilization to produce a zygote. Zygotes or individual organisms carrying two units of one allele (*DD* or *dd*) are **homozygous,** and those with two alleles (*Dd*) are **heterozygous.** Two other useful terms, **phenotype** and **genotype,** refer to visible expression of trait and actual gene constitution, respectively. Letter symbols are used to represent genotypes.

When the F_1 (*Dd*) plants from Mendel's experiments were crossed back to the dwarf (*dd*) variety, half the progeny were tall and half dwarf, as illustrated in Fig. 2.2. This demonstrated more conclusively the principle of segregation, but the separation of alleles could be detected only in the parent (*Dd*) that produced two kinds of gametes (*D*) and (*d*). The dwarf parent (*dd*) could produce only one kind of gamete.

As the genetics of a particular species advanced, the 26 letters in the alphabet were soon depleted and more symbols were needed. *Drosophila* (fruit fly) geneticists met this limitation by adding a second letter and a third and fourth, when necessary, taken from the name of the mutant phenotype. Another technical advance from *Drosophila* geneticists is the + to symbolize wild-type alleles of recognized mutant genes. For example, *Cy* is the dominant allele for curly wings, and + is the allele for wild-type wings. The lowercase *b* is assigned to the recessive allele for black body color, and + the dominant allele for wild-type gray body. If doubt can exist in the meaning of a given +, the symbol is added as a superscript ($+^w$ or w^+). Thus, w^+ symbolizes the wild-type allele for red, and *w* the allele for white eyes in *Drosophila*. Another useful device initiated by *Drosophila* geneticists was the separation of alleles by one or two crossbars or slash marks to indicate chromosomes. A heterozygous pair of alleles at the *w* locus, for example, was symbolized

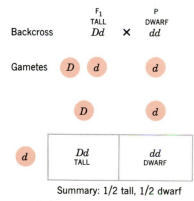

Summary: 1/2 tall, 1/2 dwarf

FIGURE 2.2

Backcross between F_1 tall garden pea and the dwarf parent variety from the cross illustrated in Fig. 2.1.

w^+/w. In summary, the various ways of denoting a pair of alleles are: $Aa, A/a, A//a, +a, +/a, \overset{\pm}{a}$, and $\overset{\pm}{\underset{\cdot}{a}}$. Gene symbols are printed in italics.

Hybrids are offspring from a cross between two genetically unlike individuals (e.g., $AA \times aa \rightarrow Aa$). A monohybrid is heterozygous for one pair of alleles (e.g., Aa). By extension, crosses (e.g., $AA \times aa$) involving parents that differ with respect to one pair of alleles are called "monohybrid crosses."

MONOHYBRID CROSSES

Monohybrid crosses are basic to Mendelian genetics. Pertinent information about genetic segregation as it occurs in monohybrid combinations is summarized in Table 2.1 and discussed in the following pages. Such crosses may occur in all major groups of sexually reproducing organisms. Dominance is the major form of interaction between alleles, because the dominant allele is usually the one that produces a functional product while the recessive allele does not. Therefore, the normal phenotype is produced if the dominant allele is present.

DOMINANCE

If the phenotype of allele A is dominant, AA and Aa individuals are alike phenotypically. In the heterozygous (Aa) condition, allele a is completely masked and the trait is recessive. Phenotypic recognition of dominance may be influenced by factors in the internal and external environment; thus, dominance is not caused by a single allele alone. In practice, however, phenotypes attributable to single allele substitutions are called dominants, and those requiring homozygous combinations for expression are called recessives. Dominants are easier to detect than recessives, because they are expressed when paired with either kind of allele. Criteria for identifying dominant, defect-transmitting alleles from human pedigree studies are summarized as follows: (1) The trait is transmitted by a parent to about half of his or her children. (This assumes that the parent is heterozygous, which is usually the case, because most dominant defective alleles in humans seem to be homozygous lethal.) If each family includes three or four children, the trait usually occurs in every generation. (2) Persons who do not express the trait do not carry the allele and therefore do not transmit it to their children. Each of these criteria is a simple consequence of the definition of dominance. This pattern of dominant inheritance assumes complete expression of dominant alleles in each individual in whom the allele is carried. Some alleles do not have full expression, as discussed in Chapter 11 under penetrance and expressivity.

Dominant inheritance can be illustrated by results of a study of dentinogenesis imperfecta (opales-

TABLE 2.1 Crosses Involving One Allelic Pair. Expected Gametes, Genotypic Frequencies of Progeny, and Phenotypic Ratios for Dominance and Intermediate Inheritance Are Given for the Different Combinations

MATING COMBINATIONS	GAMETES FIRST PARENT	GAMETES SECOND PARENT	GENOTYPIC FREQUENCIES	PHENOTYPIC RATIO WHEN DOMINANCE IS COMPLETE	PHENOTYPIC RATIO WHEN DOMINANCE IS INTERMEDIATE
$AA \times aa$	A	a	all Aa	all dom.	all int.
$Aa \times Aa$	$\frac{1}{2}A\ \frac{1}{2}a$	$\frac{1}{2}A\ \frac{1}{2}a$	$\frac{1}{4}AA, \frac{1}{2}Aa, \frac{1}{4}aa$	$3:1$	$1:2:1$
$Aa \times AA$	$\frac{1}{2}A\ \frac{1}{2}a$	A	$\frac{1}{2}AA, \frac{1}{2}Aa$	all dom.	$1:1$
$Aa \times aa$	$\frac{1}{2}A\ \frac{1}{2}a$	a	$\frac{1}{2}Aa, \frac{1}{2}aa$	$1:1$	$1:1$

cent dentin) in a family group (Fig. 2.3). Among the descendants of II-I, 16 people have the condition. Fifteen brothers and sisters of those 16 have normal teeth. Thus, about half the children who have one parent with dentinogenesis imperfecta express the trait. This is expected from matings between heterozygous people with dentinogenesis imperfecta and people who are homozygous for the recessive allele for normal white teeth. All the 16 people with the condition had an affected parent. Among the descendants of family members with normal teeth, no opalescent teeth have occurred. Thus, the two criteria for identifying dominant alleles were met. Radiographs of affected teeth (Fig. 2.4) showed that the central pulp cavities of most teeth were filled with dentin. Histological studies confirmed that dentin, which is completely covered in normal white teeth, could be seen through the deficient enamel,

and this gave the teeth their opalescent appearance.

From this pattern of dominant inheritance and the probability factors involved, it can be predicted that in future generations those who have dentinogenesis imperfecta (presumably heterozygous) and marry individuals with normal teeth might expect about half of their children to have the disorder. Individuals who do not express the trait will not transmit it.

RECESSIVENESS

Recessive alleles are expressed only in homozygous (*aa*) individuals. A population of crossbreeding organisms usually includes all three genotypes (*AA*, *Aa*, and *aa*), but it has more heterozygous (*Aa*) **carriers** than homozygous (*aa*) individuals who express the trait. Carriers are not detectable phenotypically, but recessive alleles can be identified ex-

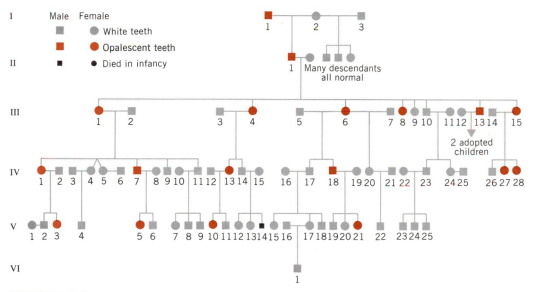

FIGURE 2.3

Pedigree chart showing the distribution of opalescent dentin in a family group. Generations are identified with Roman numerals in vertical sequence at left. Horizontal lines of squares and circles show individuals in each generation, squares for males and circles for females. A solid line connecting a male and female horizontally represents a marriage. Descending vertical lines illustrate progeny in order of age for each sibship from left to right. Horizontal numbers identify members of generations along with their spouses. Shaded squares and circles represent individuals expressing a particular trait (i.e., opalescent dentin, shaded; no shading, normal teeth). (From Gardner, The Journal of Heredity, *42, 289–290, 1951.)*

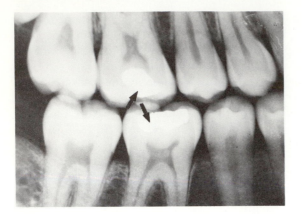

(a)

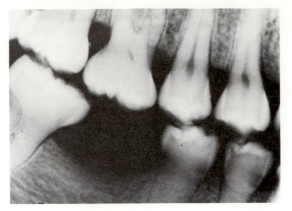

(b)

FIGURE 2.4

Dental radiographs of normal teeth compared with those of a patient with dentinogenesis imperfecta (opalescent dentin). (a) Normal teeth showing normal enamel, dentin and patent pulp chambers, and root canals. Restorations are present in both first molars (arrows). The teeth are normal in color on clinical examination. (b) Opalescent teeth. The enamel is normal, but the pulp chambers and root canals in most teeth are obliterated with abnormal dentin. There is an increased constriction at the junction between the crowns and roots of the molars. (Courtesy of Professor L. S. Levin, Department of Otolaryngology. The John Hopkins University.)

perimentally with a testcross. Many recessives represent nonfunctional enzymes.

In humans, the influence of recessive alleles can be detected from pedigree studies using the following criteria: (1) The trait is usually detected among sibs (offspring of the same parents), but not in their parents or other relatives; (2) on the average, one-fourth of the sibs are affected. These criteria are to be expected on the basis of the definition of a recessive. The allele (*c*) associated with albinism, for example, is a relatively rare recessive. Albino people (*cc*) are characterized by a marked deficiency or complete absence of pigment in the skin, hair, and iris of the eyes. A mating between two normal people, both of whom are carriers for the allele (*c*) for albinism, is illustrated in Fig. 2.5.

In one Caucasian family of six (Fig. 2.6), two albinos occurred. The parents were within the normal range of pigment for Caucasian people, but they were carriers (*Cc*) and both passed on a recessive allele (*c*) to each of the two albino sons (III-5, -13, Fig. 2.7). III-5 married a normally pigmented woman (III-4), who must be a carrier, and they have

five albino and two normal children. For an infrequent allele, this occurrence is rare, indeed, if the parents (III-4 and III-5) are not related. This example represents an exception to the first criterion for detecting recessives from pedigrees. The cross is the

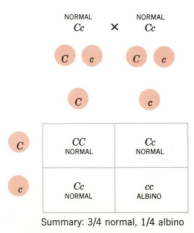

Summary: 3/4 normal, 1/4 albino

FIGURE 2.5

Cross between two normally pigmented people, both of whom were carriers of the gene (c) for albinism.

FIGURE 2.6
*Family consisting of father, mother
(front, center) and six sons, two of
whom are albino. (Courtesy of McKay
Kunz and Charles Kunz.)*

testcross type (*Cc* × *cc*) and is similar to the one reconstructed in Fig. 2.2, but for a different trait. About half the children of III-4 and III-5 would be expected to express the albino trait, whereas the other half would be normal phenotypically but still carriers (*Cc*).

CODOMINANCE
When both alleles of a pair are fully expressed in a heterozygote, they are called **codominants**. Such alleles act in distinctive ways. In general, the two products are the same with respect to function but different in exact amino acid sequence. In humans, the ABO blood group antigens are a good example. Allele I^A for A-type blood is codominant with its allele I^B for B-type blood. The heterozygote (I^AI^B) expresses the characteristics of both A and B antigens (AB-type blood). Since the two alleles control different protein products, a mating between a homozygous A-type person (I^AI^A) and a homozygous B-type person (I^BI^B) would result in all heterozygous (I^AI^B) offspring. Mating between heterozygotes

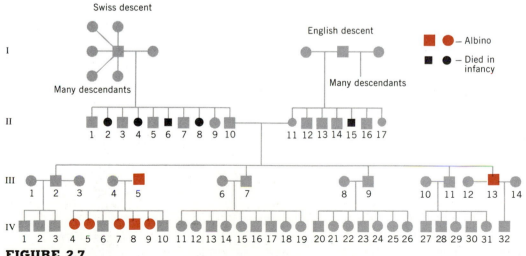

FIGURE 2.7
Pedigree of a family group in which albinism has occurred.

($I^A I^B \times I^A I^B$) would result in a ratio of 1 A-type ($I^A I^A$) : 2 AB-type ($I^A I^B$) : 1 B-type ($I^B I^B$). A phenotypic ratio of 1 : 2 : 1 has thus replaced the 3 : 1 ratio, because the alleles are codominant. In the ABO blood group, a third allele (i) occurs, and the homozygous arrangement (ii) produces O-type blood. This is a good example of triple allelism in humans.

SEMIDOMINANCE

In the absence of complete dominance, every genotype has a distinguishable phenotype. **Semidominant** alleles may produce the same product but in lesser quantity as compared with the dominant allele. Semidominance is sometimes used for incomplete inheritance. In the heterozygous condition, the total product is intermediate between that of the dominant and recessive alleles. In snapdragons, for example, heterozygotes for color alleles have pink flowers in contrast to red and white for the dominant and recessive homozygotes, respectively. The phenotypic ratio for the monohybrid cross then becomes 1 : 2 : 1 instead of 3 : 1, as it does for dominant alleles.

LETHALS

Genes may affect viability as well as the visible traits of an organism. Appropriate experiments have shown that animals carrying certain genes are disadvantaged through impaired biochemical as well as physical functioning. White-eyed and vestigial-winged *Drosophila,* for example, have lower viabilities than the wild type. Detrimental physiological effects are apparently associated with the genes involved (w and vg, respectively). Some other genes have no effect on the appearance of the fly, but do influence viability in some way. Some genes have such serious effects that the organism is unable to live. These are called **lethal genes.** Obviously, if the lethal effect is dominant and immediate in expression, all individuals carrying the gene will die and the gene will be lost. Some dominant lethals, however, have a delayed effect so that the organism lives for a time. Recessive lethals carried in the heterozygous condition have no effect but may come to expression when matings between carriers occur.

The dominant gene (C) in chickens is responsi-

ble for profound developmental changes that result in aberrant forms called "creepers," and the homozygous genotype (CC) is lethal. These birds have short, crooked legs and are of little value except as novelties. When two creepers were mated, a ratio of 2 creepers to 1 normal instead of 3 : 1 appeared, as illustrated in Fig. 2.8. This is a characteristic ratio for all crosses involving lethals. In this particular case, the CC class is missing. All creepers that lived could be shown by testcrosses to be heterozygous (Cc). When a creeper was mated with a normal chicken, the expected backcross result of 1 creeper to 1 normal was obtained (Fig. 2.9). How can investigators determine whether they are working with a dominant for a phenotype (e.g., creeper) and a recessive for lethality? If the numbers of progeny from matings between affected individuals (e.g., creepers) are large enough, statistical analysis might distinguish between a 2 : 1 and 3 : 1 ratio. Testcross results could indicate whether progeny from crosses

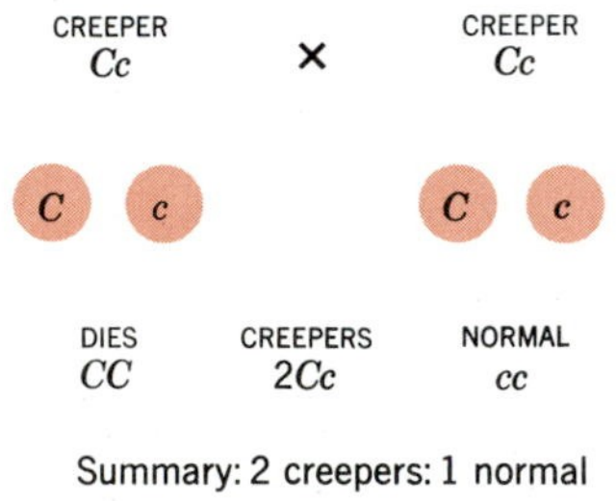

CREEPER   *Cc*   ×   CREEPER   *Cc*

| *C* | *c* | | *C* | *c* |

DIES *CC*   CREEPERS *2Cc*   NORMAL *cc*

Summary: 2 creepers: 1 normal

FIGURE 2.8

Cross between two creeper chickens. The 2 : 1 ratio replaces the 3 : 1 because the homozygous (CC) embryos die.

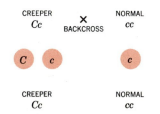

CREEPER *Cc*   × BACKCROSS   NORMAL *cc*

| *C* | *c* | | *c* |

CREEPER *Cc*   NORMAL *cc*

Summary: 1 creeper: 1 normal

FIGURE 2.9

Cross between a creeper and a normal chicken. Expected ratio is 1 creeper to 1 not-creeper (normal).

between affected individuals are heterozygous or homozygous. In some animals, such as chickens, it is possible to observe dead or dying embryos at the appropriate time in incubating eggs. The allele C is expressed in heterozygous arrangement (Cc) in creeper chickens as a dominant. As a lethal, C is a recessive in the sense that two doses of C are required for lethality.

PRINCIPLE OF INDEPENDENT ASSORTMENT

Mendel also crossed plants that differed in two pairs of alleles (Fig. 2.10). In this cross, designed to clarify the relation of different pairs of alleles, he crossed plants having round, yellow seeds with plants having wrinkled, green seeds. The F_1 progeny from such a cross between homozygous parents are hybrids (heterozygotes) for two gene pairs. The F_1 progeny ($GgWw$) are dihybrids, and, by extension, the $GGWW \times ggww$ cross is a **dihybrid cross.** Alleles for both round and yellow were known from previous studies to be dominant over their respective alleles, producing wrinkled and green seeds.

All the F_1 seeds resulting from the cross were round and yellow, as expected. When the F_1 hybrids were allowed to self-fertilize, four F_2 phenotypes were observed in a definite pattern. From a total of 556 seeds, the following distribution was obtained: 315 round, yellow; 108 round, green; 101 wrinkled, yellow; and 32 wrinkled, green. These results closely fit a ratio of $9:3:3:1$ (i.e., $\frac{315}{556} \cong \frac{9}{16}$, $\frac{108}{556} \cong \frac{3}{16}$, $\frac{101}{556} \cong \frac{3}{16}$, and $\frac{32}{556} \cong \frac{1}{16}$). Mendel recognized this as the result of **two monohybrid** crosses, each expected to result in a $3:1$ ratio, operating together. The product of the two monohybrid ratios $(3:1)^2$ or $(3+1)^2$ was equal to the **dihybrid ratio** $(3+1)^2 = (9+3+3+1)$, thus conforming to the **law of probability,** called the "product rule," which states: **The chance of two or more independent events occurring together is the *product* of the chances of their separate occurrences.**

The results were those expected from the assortment of two independent pairs of alleles, each showing dominance of one member. Not only did the members of each pair of alleles segregate, but

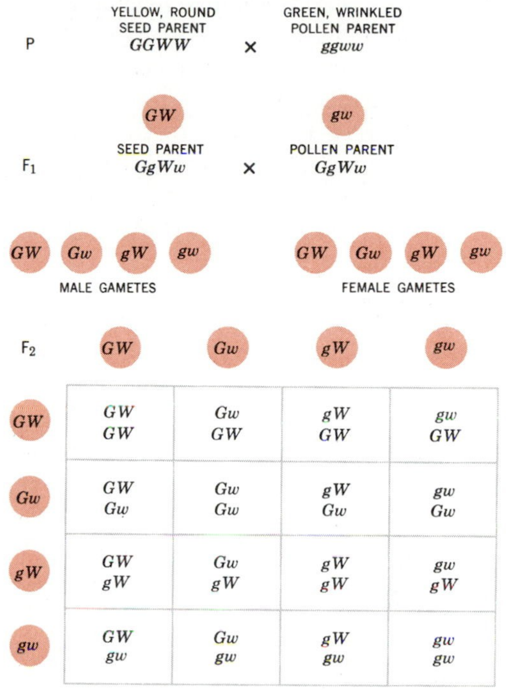

Phenotypes	Genotypes	Genotypic Frequency	Phenotypic Ratio
Yellow, round	$GGWW$	1	9
	$GGWw$	2	
	$GgWW$	2	
	$GgWw$	4	
Yellow, wrinkled	$GGww$	1	3
	$Ggww$	2	
Green, round	$ggWW$	1	3
	$ggWw$	2	
Green, wrinkled	$ggww$	1	1

FIGURE 2.10

Diagram and summary of a cross between a variety of garden peas with yellow, round seeds and a variety with green, wrinkled seeds. The $F_1 \times F_1$ represented illustrates a dihybrid cross.

the allelic pairs of different genes behaved independently with respect to each other. Mendel therefore drew another conclusion: Members of different

pairs of alleles **assort independently into gametes.**
This concept of independent assortment of different pairs of alleles is designated as his second principle. It is a simple corollary of meiosis (Chapter 3).
Mendel's two principles were set forth in a paper
entitled, "Experiments in Plant Hybridization,"
which was read before the Brünn Natural History
Society in 1865 and published in the proceedings of
that society in 1866.

Mendel's principle of independent assortment
has a practical application in plant and animal
breeding. Desirable traits carried in different varieties can be combined and maintained in a single
type. A variety of barley resistant to rust, for example, was needed in a rust-infested area in the United
States. The best available rust-resistant variety,
however, like most barley varieties, had hulls on the
seeds and did not thresh well. Another variety had
no hulls and threshed out clean like wheat, but had
poor rust resistance. These two varieties were combined by appropriate crosses, and a valuable new
strain with rust resistance and no hulls was obtained.

DIHYBRID RATIOS

The basic mechanics of genetics were postulated
and later established from particular ratios, such as
$3:1$ and $9:3:3:1$. Ratios of this kind merely represent the grouping expected when particular conditions are met. Common patterns such as the
$9:3:3:1$ ratio may serve as models for analyzing
results of experiments. When such a ratio is obtained from a cross in which parental genotypes are
not known, the geneticist may postulate that two
independent pairs of alleles are involved, and that
one member of each pair behaves as a dominant
over its allele. Mendel's dihybrid cross between
plants with round, yellow seeds and those with wrinkled, green seeds is represented diagrammatically in
Fig. 2.10 as a pattern applicable in analyzing other
crosses.

When the F_1 plants were selfed (i.e., pollen and
eggs from the same plant were united), four kinds of
gametes were produced by the male parts and four
by the female parts of the F_1. At the top of the
checkerboard (Punnett square, Fig. 2.10), the four
kinds of gametes from the seed parent are shown.
The four possible gametes from the pollen parent

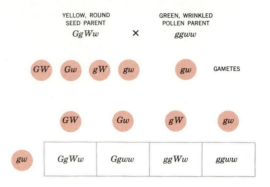

Phenotypes	Genotypes	Genotypic Frequency	Phenotypic Ratio
Yellow, round	$GgWw$	1	1
Yellow, wrinkled	$Ggww$	1	1
Green, round	$ggWw$	1	1
Green, wrinkled	$ggww$	1	1

FIGURE 2.11
*Diagram and summary illustrating a method for solving a
backcross-type problem involving two gene pairs. This cross is
between an F_1 garden pea with yellow, round seeds and the fully
recessive parent type with green, wrinkled seeds.*

are represented in a vertical row at the left. This
Punnett square is merely a geometrical device that
helps in visualizing all possible combinations of male
and female gametes. It is valuable as a learning
exercise, but we will shortly turn to methods that
are less time-consuming. Letter symbols in the 16
squares represent combinations of independent
genes brought together by the fusion of gametes.
When these are collected according to the phenotypes represented, the $9:3:3:1$ ratio becomes apparent. The completed summary chart (Fig. 2.10)
illustrates the F_2 results of the cross in tabular form.

The $1:1:1:1$ ratio is expected from a dihybrid
backcross to the recessive parent, that is, a cross
between an F_1 that carries two heterozygous pairs of

alleles and a parent type with the full recessive combination for these two genes. This cross is illustrated in Fig. 2.11. A cross of this type is used in practical breeding programs to determine the genotype of an individual that may carry a recessive allele, the expression of which could be obscured by a dominant allele. The cross with the double recessive genotype for testing purposes is called a **testcross,** to distinguish it from a **backcross,** which is the cross of the F_1 with either of the parent types.

TRIHYBRID RATIOS

Virtually all cross-fertilizing plants or animals differ from other members of their species in more than one or two pairs of alleles. Therefore, matings in natural breeding populations usually produce new combinations of many genes. Genetic analysis of such crosses is complicated. In many cases, however, complex combinations can be simplified by resolving them into monohybrid crosses or by using formulas devised to handle several traits in the same

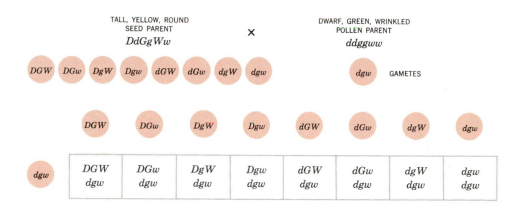

Phenotypes	Genotypes	Genotypic Frequency	Phenotypic Ratio
Tall, yellow, round	DdGgWw	1	1
Tall, yellow, wrinkled	DdGgww	1	1
Tall, green, round	DdggWw	1	1
Tall, green, wrinkled	Ddggww	1	1
Dwarf, yellow round	ddGgWw	1	1
Dwarf, yellow wrinkled	ddGgww	1	1
Dwarf, green, round	ddggWw	1	1
Dwarf, green, wrinkled	ddggww	1	1

FIGURE 2.12

Method for solving backcross-type problems involving three gene pairs. This cross is between an F_1 garden pea with tall vines, yellow and round seeds, and the fully recessive parental type with dwarf vines, green and wrinkled seeds.

problem. A cross between homozygous parents that differ in three gene pairs (i.e., producing trihybrids) is a combination of three single-pair crosses operating together. Thus, $(AA \times aa)$ $(BB \times bb)$ $(CC \times cc)$ could be combined in the same cross as $AABBCC \times aabbcc$.

What results might be expected (1) in the F_1, (2) in the backcross to the fully recessive parent, and (3) in the F_2 from a cross between two varieties of garden peas differing in three traits? A diagram of a cross in which the seed parent is homozygous for the genes producing a tall vine and yellow, round seeds ($DDGGWW$) and the pollen parent is homozygous for a dwarf vine and green, wrinkled seeds ($ddggww$) can answer that question. The three traits represented in the seed parent are known from previous experiments to depend on dominant genes. The first-generation cross may be illustrated as follows:

$$DDGGWW \quad \times \quad ddggww \quad \textbf{P}$$
$$(DGW) \qquad\qquad (dgw) \quad \textbf{Gametes}$$
$$DdGgWw \qquad\qquad \textbf{F}_1$$

When the F_1 plants are crossed with the full recessive type, $DdGgWw \times ddggww$, eight kinds of gametes (DGW, DGw, DgW, Dgw, dGW, dGw, dgW, and dgw) are produced by the F_1 parent and only one kind, dgw, by the full recessive parent. As a result of fertilization, eight kinds of peas are expected in equal proportion. Thus, the trihybrid backcross genotypic and phenotypic ratio of $1:1:1:1:1:1:1:1$ is explained by the fertilization of eight different kinds of gametes from F_1 by the one kind of gamete from the fully recessive parents. The sequence involved in this backcross and the summarized results are illustrated in Fig. 2.12.

When the F_1 plants are selfed (that is, $DdGgWw \times DdGgWw$), eight kinds of gametes (DGW, DGw, DgW, Dgw, dGW, dGw, dgW, and dgw) are produced from both the male and female parts. These gametes represent all combinations. If the $F_1 \times F_1$ cross were represented by a Punnett square, 64 (8^2) squares would be required with a phenotypic ratio of $27:9:9:9:3:3:3:1$ as a result. In the next section, we look at less time-consuming methods for determining the results of complex combinations.

FORKED-LINE METHOD FOR GENETIC PROBLEMS

A method for bringing the combinations of a trihybrid cross together may be illustrated as follows: First, visualize the trihybrid cross as three monohybrid crosses—that is, $Dd \times Dd$, $Gg \times Gg$, and $Ww \times Ww$—operating together. If one member of each pair is dominant, a $3:1$ ratio would be predicted from each monohybrid cross. Since the three pairs are independent, each monohybrid segregant may occur with any combination possible from any other pair of alleles. The combinations therefore can be systematically arranged together. The $3:1$ ratio from $Dd \times Dd$ may be combined with the $3:1$ ratios from each of the other two monohybrid crosses, $Gg \times Gg$ and $Ww \times Ww$, as shown in Fig. 2.13.

Usually, the genotypes as well as the phenotypes are necessary for the complete solution of such a problem. The same forked-line system may be employed to represent and combine genotypes expected from monohybrid crosses. From each monohybrid cross in the example, a genotypic frequency of $1:2:1$ may be predicted. The three monohybrid units may be combined as shown in Fig. 2.14. The forked-line system is merely another device for analyzing crosses.

MATHEMATICAL METHOD FOR GENETIC PROBLEMS

This introduction to combinations suggests a third way to anticipate the results from crosses that involve independent pairs of alleles. A mathematical manipulation provides a way to arrive at the product of the combinations without drawing them out mechanically. As an example, consider the crosses made by Toyama between two varieties of the silk moth, *Bombyx mori*. In one variety, the caterpillars were striped and the cocoons were yellow; and in the other variety, the caterpillars were unstriped and the cocoons were white. From previous crosses, striped was known to be dominant over unstriped and yellow over white. What proportions might be expected in the F_2?

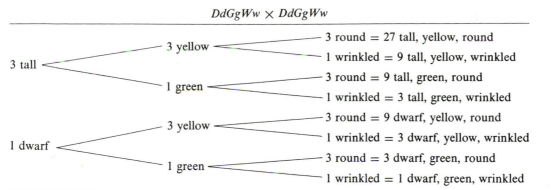

FIGURE 2.13

Forked-line method for solving genetic problems in which independent assortment is involved. Genotypes of parents are given, but only phenotypes of progeny are listed in this example.

If the striped and unstriped alleles are considered separately, three-fourths of the F_2 progeny are expected to be striped and one-fourth unstriped. Similarly, three-fourths are expected to be yellow and one-fourth white. The phenotypes and their proportions are summarized in Table 2.2. Toyama obtained results from actual crosses that satisfied the predictions.

The expected F_2 result from a trihybrid cross that involves independent assortment and dominance of one allele in each pair is the product of three pairs, such as

$$(A:a)(B:b)(C:c).$$

Written algebraically,

$$(3 + 1)(3 + 1)(3 + 1)$$

or $(3 + 1)^3$, which expands to

$$27 + 9 + 9 + 9 + 3 + 3 + 3 + 1$$

or

$$27:9:9:9:3:3:3:1.$$

A cross with four gene pairs under the same conditions would result in $(3 + 1)^4$, five gene pairs $(3 + 1)^5$, and so on. Numbers of gametes, genotypes, and phenotypes expected from different numbers of heterozygous pairs of genes are sum-

TABLE 2.2 Expectations and Results from Crosses Between Striped, Yellow and Unstriped, White Moths[a]

PHENOTYPES	GENOTYPES	PROPORTIONS		F_2 EXPECTATIONS	F_2 OBSERVED
Striped, yellow	S-Y-	$\frac{3}{4} \times \frac{3}{4}$		$\frac{9}{16} \times 11,322 = 6,368.6$	6,385
Striped, white	S-yy	$\frac{3}{4} \times \frac{1}{4}$		$\frac{3}{16} \times 11,322 = 2,122.9$	2,147
Unstriped, yellow	ssY-	$\frac{1}{4} \times \frac{3}{4}$		$\frac{3}{16} \times 11,322 = 2,122.9$	2,099
Unstriped, white	ssyy	$\frac{1}{4} \times \frac{1}{4}$		$\frac{1}{16} \times 11,322 = 707.6$	691
Total				11,322.0	11,322

[a] Data from Toyama.

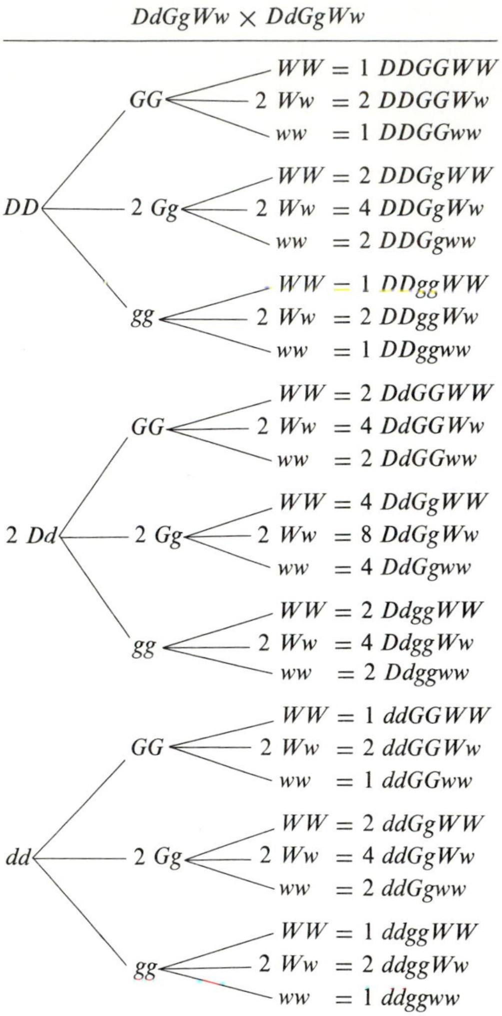

$DdGgWw \times DdGgWw$

(a)

```
                    ┌── WW = 1 DDGGWW
            GG ─────┼── 2 Ww = 2 DDGGWw
                    └── ww = 1 DDGGww
                    ┌── WW = 2 DDGgWW
    DD ──── 2 Gg ───┼── 2 Ww = 4 DDGgWw
                    └── ww = 2 DDGgww
                    ┌── WW = 1 DDggWW
            gg ─────┼── 2 Ww = 2 DDggWw
                    └── ww = 1 DDggww

                    ┌── WW = 2 DdGGWW
            GG ─────┼── 2 Ww = 4 DdGGWw
                    └── ww = 2 DdGGww
                    ┌── WW = 4 DdGgWW
 2 Dd ─── 2 Gg ────┼── 2 Ww = 8 DdGgWw
                    └── ww = 4 DdGgww
                    ┌── WW = 2 DdggWW
            gg ─────┼── 2 Ww = 4 DdggWw
                    └── ww = 2 Ddggww

                    ┌── WW = 1 ddGGWW
            GG ─────┼── 2 Ww = 2 ddGGWw
                    └── ww = 1 ddGGww
                    ┌── WW = 2 ddGgWW
    dd ──── 2 Gg ───┼── 2 Ww = 4 ddGgWw
                    └── ww = 2 ddGgww
                    ┌── WW = 1 ddggWW
            gg ─────┼── 2 Ww = 2 ddggWw
                    └── ww = 1 ddggww
```

(b)

Phenotypes	Genotypes	Genotypic Frequency	Phenotypic Ratio
Tall, yellow, round	DDGGWW	1	27
	DDGGWw	2	
	DDGgWW	2	
	DDGgWw	4	
	DdGGWW	2	
	DdGGWw	4	
	DdGgWW	4	
	DdGgWw	8	
Tall, yellow, wrinkled	DDGGww	1	9
	DDGgww	2	
	DdGGww	2	
	DdGgww	4	
Tall, green, round	DDggWW	1	9
	DDggWw	2	
	DdggWW	2	
	DdggWw	4	
Tall, green, wrinkled	DDggww	1	3
	Ddggww	2	
Dwarf, yellow, round	ddGGWW	1	9
	ddGGWw	2	
	ddGgWW	2	
	ddGgWw	4	
Dwarf, yellow, wrinkled	ddGGww	1	3
	ddGgww	2	
Dwarf, green, round	ddggWW	1	3
	ddggWw	2	
Dwarf, green, wrinkled	ddggww	1	1

FIGURE 2.14

(a) *Cross between two F_1 garden peas of the genotype* DdGgWw. *The forked-line method is employed and the genotypes are illustrated. These results represent the F_2 of a cross similar to those obtained from the* Punnett square *method, which involves 64 squares.* (b) *Summary of F_2 from trihybrid cross, resulting in a $27:9:9:9:3:3:3:1$ phenotypic ratio.*

marized in Table 2.3. It will be observed that the number of kinds of **gametes** is a multiple of 2, that is 2^n; the number of F_2 **genotypes** is a multiple of 3, that is 3^n; and the number of **phenotypes is 2^n when dominance is present.** This pattern forms the basis for predicting results when any number of independent pairs of alleles is involved in the production of hybrids.

TABLE 2.3   Relations among Pairs of Independent Alleles, Gametes, F_2 Genotypes, and F_2 Phenotypes When Dominance Is Present

NUMBER OF HETEROZYGOUS PAIRS	NUMBER OF KINDS OF GAMETES	NUMBER OF F_2 GENOTYPES	NUMBER OF F_2 PHENOTYPES
1	2	3	2
2	4	9	4
3	8	27	8
4	16	81	16
10	1024	59,049	1024
n	2^n	3^n	2^n

GENE INTERACTION

In the preceding discussion, the emphasis has been placed on the fact that the genes studied by Mendel were segregating independently of each other. What has not yet been mentioned is that these genes must also have been **functioning** (exerting their effects on the phenotype) **independently** of one another.

If each gene were expressing itself in a separate test tube, it would be reasonable to expect all genes to be functionally independent. But these genes are not in separate test tubes; they are all located in the same nuclei of the same cells. Thus, it should not be surprising that the expression of an allele of one gene will sometimes alter the expression of one or more of the alleles of a second (nonallelic) gene.

A classic example of gene interaction based on the results of crosses between different breeds of chickens was reported in the early part of this century by William Bateson and his associate, R. C. Punnett (after whom the Punnett square was named). Bateson began to confirm and extend Mendel's work immediately after its discovery in 1900 and soon became a pioneer in transmission genetics. He had chickens in his research coops and sweet peas in his garden for immediate use in genetic investigations.

Domestic breeds of chickens have different comb shapes (Fig. 2.15). Wyandottes have a characteristic type of comb called "rose," whereas brahmas have a "pea" comb. Leghorns have "single" combs. The investigators crossed wyandottes and brahmas, and all F_1 chickens had walnut combs, a phenotype not expressed in either parent. When the F_1 chickens were mated among themselves and large F_2 populations were produced, a familiar dihybrid ratio, $9:3:3:1$, was recognized, but the phenotypes representing two of the four classes were different from those expressed in the parents. About $\frac{9}{16}$ of the F_2 birds were walnut, $\frac{3}{16}$ were rose, $\frac{3}{16}$ were pea, and $\frac{1}{16}$ had single combs. Neither single comb nor walnut was expressed in the original parent lines. These two phenotypes were explained as the result of gene product interaction. Results based on a total of 16 indicated that two different allelic pairs were involved; one pair was introduced by the rose-comb parent and one by the pea-comb parent. A gene for rose and a gene for pea would interact and produce walnut, as in the F_1.

Analysis of the F_2 results and appropriate testcrosses indicated that the $\frac{9}{16}$ class, with the two dominant genes ($R\text{-}P\text{-}$), was walnut, like the F_1 chickens. The $\frac{1}{16}$ class, representing the full recessive combination ($rrpp$), was characterized as single combs. The two $\frac{3}{16}$ (rose and pea) classes were $R\text{-}pp$ and $rrP\text{-}$. It was then determined that the homozy-

(a)

(b)

(c)

(d)

FIGURE 2.15

Comb types characteristic of different breeds of chickens: (a) *rose, wyandottes;* (b) *pea, brahmas;* (c) *walnut, hybrid cross between chickens with rose and pea combs; and* (d) *single, leghorns. (Courtesy of R. G. Somes, Jr.)*

gous genotype of the rose-combed parent (wyandotte) was *RRpp* and of the pea-combed parent (brahma), *rrPP*. Although the usual 9:3:3:1 ratio was obtained, the result from this cross was unusual in two important respects: (1) The F$_1$ progeny differed from those of the parent; that is, none was rose or pea, but all were walnut. (2) Two phenotypes (walnut and single) not expressed in the original parents appeared in the F$_2$.

Genes *R* and *P* were nonallelic with respect to each other but each was dominant over its allele (i.e., *R* over *r* and *P* over *p*). When *R* and *P* were together, as in the F$_1$ (*RrPp*), the two different products interacted to produce walnut comb. The two nonallelic genes *R* and *P* acted independently in different ways, similar to the ways in which codominant alleles act.

EPISTASIS

The functional interaction of different (nonallelic) genes is called **epistasis** (Greek, "standing upon"). Any gene that masks the expression of another, nonallelic gene is **epistatic** to that gene. **Epistasis should not be confused with dominance. Epistasis is the interaction between different genes (nonalleles). Dominance is the interaction between different alleles of the same gene.**

Metabolic processes in living organisms take place via sequences of enzyme-catalyzed reactions. Each step of each metabolic pathway thus requires the activity of a specific enzyme. Each enzyme, in turn, is the product of a specific gene.

(Some enzymes contain two or more different subunits, each specified by a particular gene. Thus, some enzymes are coded for by two or more different genes. This somewhat more complex picture will be developed further in Chapters 8–10. For now, the assumption that each enzyme is coded for by one specific gene is adequate.)

Consider the following simple two-step pathway:

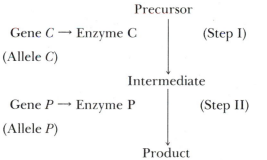

If the recessive alleles *c* and *p* of the two genes specify inactive forms of the two enzymes, organisms that are homozygous recessive for either gene (*cc* or *pp*) will not be able to synthesize the product. Organisms of genotype *cc* will be blocked at Step I; those of genotype *pp* will be blocked at Step II.

Suppose that the product of this pathway is an anthocyanin pigment responsible for colored flowers in a plant such as the sweet pea, *Lathyrus odoratus.* If the precursor and intermediate in this pathway are colorless compounds, only plants that carry at least one copy of the dominant allele of each gene (*C-P-*) will have colored flowers.

Clearly, genes controlling the synthesis of enzymes involved at different steps of the same metabolic pathway will not be functionally independent. Moreover, as a result, such genes will not yield the classical Mendelian segregation ratios. Instead, they will yield modified phenotypic segregation ratios such as those listed in Table 2.4.

In sweet peas, the anthocyanin pigment resulting in colored flowers is synthesized by a pathway like that described above. In a classical study, Bateson and Punnett crossed two different white-flowered varieties of sweet peas. The F_1 plants all had purple flowers. When the F_1 plants were intercrossed, they produced F_2 progeny of which $\frac{9}{16}$ had purple flowers and $\frac{7}{16}$ had white flowers (Fig. 2.16).

In Bateson and Punnett's experiment, one vari-

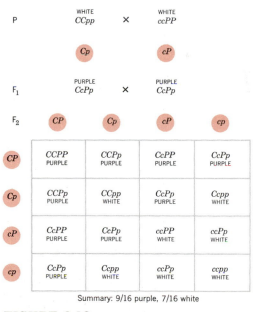

Summary: 9/16 purple, 7/16 white

FIGURE 2.16

Cross between two white varieties of sweet peas from which an F_2 ratio of 9:7 was obtained.

TABLE 2.4   Epistatic and Dominance Interactions among Nonalleles and Alleles Resulting in Modifications of the 9:3:3:1 Mendelian Ratios from Crosses, *AaBb* × *AaBb*, Each Pair Assorting Independently. Genotypes Expected from This Cross Are as Follows:

$$
AA \begin{cases} BB \rightarrow 1\ AA\ BB \\ 2Bb \rightarrow 2\ AA\ Bb \\ bb \rightarrow 1\ AA\ bb \end{cases}
$$

$$
2Aa \begin{cases} BB \rightarrow 2\ Aa\ BB \\ 2Bb \rightarrow 4\ Aa\ Bb \\ bb \rightarrow 2\ Aa\ bb \end{cases} \quad \text{or for phenotypic classes} \quad \begin{array}{l} 9\ A\text{-}B\text{-} \\ 3\ A\text{-}bb \\ 3\ aaB\text{-} \\ 1\ aabb \end{array}
$$

$$
aa \begin{cases} BB \rightarrow 1\ aa\ BB \\ 2Bb \rightarrow 2\ aa\ Bb \\ bb \rightarrow 1\ aa\ bb \end{cases}
$$

GENOTYPES

INTERACTIONS		*AABB* *AABb* *AaBB* *AaBb* *AAbb* *Aabb* *aaBB* *aaBb* *aabb*					EXAMPLE	PHENOTYPES
EPISTASIS	DOMINANCE	PHENOTYPIC CLASSES						
(classical ratio)	Complete for *A,B*	9	3	3	1		Garden peas seed color and surface	9 yellow round, 3 yellow wrinkled, 3 green round, 1 green wrinkled
None	Complete for *A*, incomplete for *B*	× − × − − × − − −	3	○● ●	1		Guinea pigs, hair length and color	3 short yellow (×), 6 short cream (−), 3 short white, 1 long yellow (○), 2 long cream (●), 1 long white
aa epistatic to *B,b*	Complete for *A,B*	9	3	4			Mice coat pattern and color	9 agouti, 3 colored, 4 white

ety of sweet peas was of genotype *ccPP* and was blocked at one of the steps in the synthesis of the anthocyanin pigment. The other variety, of genotype *CCpp*, was blocked at another step in pigment synthesis. The F₁ plants were all of genotype *CcPp*. Since in this case *C* and *P* are dominant to their alleles *c* and *p*, respectively, all of the F₁ plants synthesized the anthocyanin pigment and thus had colored flowers. In the F₂ generation, $\frac{9}{16}$ of the progeny would be expected to carry at least one dominant allele of each gene (*C-P-*) and have colored flowers. The other $\frac{7}{16}$ of the progeny should be homozygous recessive for one or the other or both of the genes ($\frac{3}{16}$ *ccP-* + $\frac{3}{16}$ *C-pp* + $\frac{1}{16}$ *ccpp*) and have white flowers (Fig. 2.16).

Other modified dihybrid ratios are known to result from different types of epistasis. For example, if two different (nonallelic) genes code for enzymes that are able to catalyze the same reaction, for example,

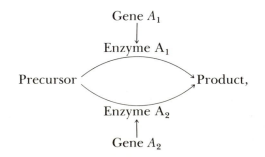

Precursor → Product,

TABLE 2.4   (continued)

INTERACTIONS		GENOTYPES AABB AABb AaBB AaBb AAbb Aabb aaBB aaBb aabb	EXAMPLE	PHENOTYPES
EPISTASIS	DOMINANCE	PHENOTYPIC CLASSES	EXAMPLE	PHENOTYPES
A epistatic to B,b	Complete for A,B	——— 12 ——— 3 ¦ 1	Summer squash fruit color	12 white, 3 yellow, 1 green
A epistatic to B,b; bb epistatic to A,a	Complete for A,B	— 13 ——— 3 ¦ (1)	Chickens color	13 white, 3 colored
aa epistatic to B,b; bb epistatic to A,a	Complete for A,B	9 — 7 ———	Yellow daisy color	9 purple, 7 yellow center
A epistatic to B,b; B epistatic to A,a	Complete for A,B	——— 15 ——— 1	Shepherd's purse seed capsules	15 triangular, 1 ovoid
aa epistatic to B,b; bb epistatic to A,a; bb epistatic to aa	Complete for A, incomplete for B	× − × − ¦ 4 ¦ 3 ¦ (1) − × − −	Flour beetle color	6 sooty (−), 3 red (×), 3 jet, 4 black

one can easily see how a $15:1$ F_2 dihybrid phenotypic segregation ratio might result. If the recessive alleles, a_1 and a_2, of these two genes produce inactive forms of the two enzymes, a cross of $a_1a_1A_2A_2 \times A_1A_1a_2a_2$ would yield a 15 $A_{1 \text{ or } 2}$ --- : 1 $a_1a_1a_2a_2$ F_2 phenotypic segregation ratio. Such $15:1$ ratios have been observed in crosses involving genes controlling kernel color in wheat and seed capsule shape in shepherd's purse.

In other cases of epistasis, the product of one gene may inhibit or otherwise prevent the expression of the product of another gene. If the inhibitor is the product of a dominant allele, a $12:3:1$ F_2 dihybrid ratio is produced (Fig. 2.17). If it is the product of a recessive allele, a $9:3:4$ F_2 dihybrid ratio will result. Still other kinds of interactions between gene products can produce other modified ratios (Table 2.4).

It should be emphasized that **genes do not act in isolation.** The final phenotype of an organism is the result of the action of, and the interaction between, a large number of genes. In addition, all of these genes are influenced to a greater or lesser extent by environmental factors. The interaction of many genes and their interaction with the environment in controlling many of the most important components of phenotype are discussed further in Chapter 15.

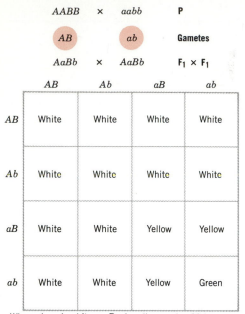

Where *A---* is white, *aaB-* is yellow, and *aabb* is green.

FIGURE 2.17

Example of epistasis in fruit color of summer squash. Gene A *is epistatic to* B *and* b*. Please compare Fig. 2.10 for dominance, where* G *is dominant over allele* g *and* W *is dominant over allele* w*.*

PROBABILITY IN MENDELIAN INHERITANCE

Probability is the ratio of a specified event to total events. Limits of probability are from 0 if an event never occurs to 1 if it always occurs. Mendel had studied probability in his mathematics courses and recognized the 3 : 1 ratio as a particular mathematical relation that suggested a model for the mechanism of **segregation.** If two alleles, one dominant and one recessive, could segregate freely many times in succession, or if several similar pairs were behaving in this way at one time, the expected summation would be about three of the dominant expressions to one recessive. The analysis, therefore, was based on a mathematical relation with which Mendel was familiar, but its application and the concept of segregation were Mendel's own contributions.

The laws of probability apply to the genetic mechanism as well as to other processes in which uncertainty exists. In the F_2 results of Mendel's cross between tall and dwarf garden peas, one-fourth were dwarf, one-fourth homozygous tall, and one-half heterozygous tall. A similar result might be obtained from a simple experiment in tossing coins. A coin that is tossed freely is equally likely to fall heads or tails. If one coin is tossed 100 times, it would be expected to fall heads about 50 times and tails about 50 times. When two coins are tossed together, each behaves independently and falls either heads or tails. From 100 trials, about 25 heads–heads, 50 heads–tails (actually 25 heads–tails and 25 tails–heads), and 25 tails–tails would be expected. A ratio of 1 : 2 : 1 or 1 : 1 : 1 : 1 could be interpreted, depending on the classification of combinations. This result, which parallels Mendelian segregation, is merely the chance occurrence of independent events. The experimenter would usually not obtain exactly 25 heads–heads, 50 heads–tails, and 25 tails–tails. It would be surprising if precisely those results were obtained very often. The ratio represents only an average of expected results when independent events occur.

How frequently can the various combinations be expected to occur in succession? The law of probability applied for the solution of this problem is stated as follows: **If two or more events are independent, the chance that they will occur together is the product of their separate probabilities.** When a single coin is tossed repeatedly, the chance of heads occurring twice in succession would be $\frac{1}{2} \times \frac{1}{2} = \frac{1}{4}$. The chance of three such occurrences would be $(\frac{1}{2})^3$ or $\frac{1}{8}$, and of four, $(\frac{1}{2})^4$ or $\frac{1}{16}$. When two coins are tossed together, the tails–tails combination is expected in one-fourth of the trials $(\frac{1}{2} \times \frac{1}{2} = \frac{1}{4})$. The chance of occurrence of two such tails–tails combinations for two coins in succession would be $(\frac{1}{4})^2$ or $\frac{1}{16}$.

FITTING RESULTS FROM CROSSES TO HYPOTHESES

The **"goodness of fit"** of the numerical result obtained from an actual cross or other experiment, relative to predicted results based on a particular mechanism and a perfect genetic segregation, is of vital concern. The geneticist must know how much the experimental result can differ from the hypothetical or calculated figure and still be regarded as statistically close to expectation. In evaluating the results of crosses and determining which modes of inheritance are involved, how much deviation is permissible without casting some doubt as to whether the data agree with a given hypothesis? Too much deviation would surely make investigators question their hypotheses or discard them entirely. Where should the line be drawn? Unfortunately, there is no precise answer to this question. The best the geneticist can do is to determine the likelihood of the deviation occurring by chance and use statistical inference to decide whether a particular result supports a given hypothesis. These numerical data are the only means of evaluating goodness of fit of an experimental result as compared with a particular expectation.

CHI-SQUARE

The chi-square (χ^2) test is a valuable tool that aids the investigator in determining goodness of fit. The test takes into account the size of the **sample** and the **deviations** from the expected ratio. It not only can be used for samples of different sizes, but can be adapted to ratios with different numbers of classes, such as those of monohybrid crosses with two classes and those of dihybrid crosses with four classes. Essentially, the chi-square test is a mechanism by which deviations from a hypothetical ratio are reduced to a single value based on the size of the sample. This allows the investigator to determine the probability that a given sum of deviations will occur by chance. Expected values are obtained from the total size of the sample. If the hypothesis is that a 1:1 ratio results from a cross, the total is divided into two equal parts. For any other expected ratio, the total is divided into appropriate proportions.

A formula for χ^2, designed for a sample consisting of two classes (i.e., 1:1 or 3:1 ratios), is symbolized as follows:

$$\chi^2 = \frac{(O_1 - E_1)^2}{E_1} + \frac{(O_2 - E_2)^2}{E_2}$$

where O_1 is the experimentally observed number

for the first class, and E_1 is the expected number for the same class derived from the ratio; O_2 is the observed for the second class and E_2 the expected. When each of these deviations is squared $[(O - E)^2]$ and divided by the expected value (E) for that class, the resulting fractions can be added (Σ) to give a single χ^2 value. The formula can be symbolized as follows:

$$\chi^2 = \Sigma \frac{(O - E)^2}{E}$$

where $(O - E)$ is the deviation between each observed and expected class value, E the expected value in the respective class, and the Greek letter Σ the summation sign. If the deviations of expected from observed events are small, χ^2 approaches 0 and the fit is good; if the deviations are large, χ^2 is increased and the fit is poor. As an example, calculate χ^2 for the two arbitrary samples, 15:35 and 240:260 on the basis of a 1:1 hypothesis. This example will illustrate how the χ^2 relates the size of the deviation to the size of the sample. For the 15:35 result, with a total of 50, the expected (E) value for each class is 15 + 35 ÷ 2, or 25. The deviations ($O - E$) on either side of E are 10; that is, 25 − 15 = 10 and 25 − 35 = −10. For the larger sample, 240:260, the expected (E) for each class is 250. The deviations ($O - E$) are 250 − 240 = 10, and 250 − 260 = −10.

$$(1) \quad \chi^2 = \Sigma \frac{(O - E)^2}{E} = \frac{(10)^2}{25} + \frac{(-10)^2}{25}$$

$$= \frac{200}{25} = 8$$

$$(2) \quad \chi^2 = \Sigma \frac{(O - E)^2}{E} = \frac{(10)^2}{250} + \frac{(-10)^2}{250}$$

$$= \frac{200}{250} = .8$$

The χ^2 value of 8 for the smaller sample is considerably greater than that of .8 for the larger sample, even though the actual deviations in the two examples are the same. Thus, the χ^2 value is related to sample size as well as to the variability within the sample.

When more than two groups are classified from the sample (e.g., $1:2:1$ or $9:3:3:1$ ratios), each class is included in the summation, which is χ^2. It should be emphasized that the χ^2 formula is based on actual numerical frequencies and totals, not on percentages that always total 100. When data are reduced to percentages, the factor of sample size is eliminated.

The next step is to interpret the χ^2 value in terms of probability. In any experimental procedure dealing with quantitative data, some variation, called "experimental error," can be attributed to chance alone. It is important to determine whether observed deviations from a hypothesis are significantly different from the experimental error. For interpreting χ^2 values, the number of classes on which a χ^2 is based must be considered. The value for a two-class distribution includes only two squared deviations, whereas that for a distribution with more classes has more squared deviations. It is therefore necessary to consider the number of classes that contribute to a given χ^2 in evaluating the "goodness of fit." The effect of the number of independent classes is included in the mathematical concept as **degrees of freedom.** As an analogy, a person has two socks, for his two feet, but in placing the socks he has only one degree of freedom. If he places a sock on one foot, the other sock must go on the other foot. When the total number of objects or classes is fixed, and all except one have been placed, the one remaining is not free but must fill a particular vacant niche. In general, therefore, the number of degrees of freedom is one less than the number of classes. Two-class ratios (e.g., $1:1$ or $3:1$) have one degree of freedom; three-class ratios ($1:2:1$) have two degrees of freedom; four-class ratios ($9:3:3:1$) have three degrees of freedom; and so on.

When χ^2 and the degrees of freedom have been determined, Table 2.5 may be consulted for the probability (P) value. Locate the figure representing the number of degrees of freedom at the left, read across horizontally, and find the figures nearest the χ^2 value in the body of the table; then read the P values directly above on the top line. The χ^2 of 8, calculated for the first example, is not on the table. The highest value on the line for one degree of freedom is 6.635, which has a P of .01. This indicates that the probability for obtaining, by chance, deviations as great as or greater than those of $\chi^2 = 8$ would be less than 1 percent. Experimental error is expected, but the difference here is much greater than that expected by chance. Therefore, the fit of these results to a $1:1$ ratio is not good. Another hypothesis (other than the $1:1$ ratio) might be considered for these data.

In the second example, $\chi^2 = .8$ falls between .455 and 1.642, or between the P values of .50 and .20. The probability of obtaining a deviation as great as, or greater than, $\chi^2 = .8$ is between .20 and .50. This probability value indicates that, if numerous independent repetitions of an ideal experiment involving two independent events were conducted, chance deviations as large as, or larger than, those observed here (± 10 corresponding to $\chi^2 = .8$) would be expected to occur in $20-50$ percent of the trials. Such a deviation could be explained readily by chance. The data fit the $1:1$ ratio hypothesis very well.

A hypothesis is never proved or disproved by a P value. The results of an experiment are evaluated by the investigator as acceptable or unacceptable with respect to the hypothesis. The 5 percent point (.05) on the table is usually chosen as an arbitrary standard for determining the significance or goodness of fit. Probability at this point is 1 in 20 that a true hypothesis will be rejected. Sometimes the 1 percent point (.01) is used as a level of significance. At this level, there is a smaller probability (.01) that a true hypothesis will be rejected, but a correspondingly greater chance that a false hypothesis will be accepted.

It should be emphasized that these are arbitrary points, and judgment is required in making interpretations. In any event, the P value represents the probability that a deviation as great as, or greater than, that obtained from the experiment will occur by chance alone. If the P is small, it is concluded that the deviations are not due entirely to chance, and the hypothesis is rejected. If the P is greater than the predetermined level (e.g., .05), the data conform well enough to the hypothesis and the hypothesis is accepted.

TABLE 2.5   Table of Chi-Square (χ^2)[a]

DEGREES OF FREEDOM	$P = .99$	.95	.80	.50	.20	.05	.01
1	.000157	.00393	.0642	.455	1.642	3.841	6.635
2	.020	.103	.446	1.386	3.219	5.991	9.210
3	.115	.352	1.005	2.366	4.642	7.815	11.345
4	.297	.711	1.649	3.357	5.989	9.488	13.277
5	.554	1.145	2.343	4.351	7.289	11.070	15.086
6	.872	1.635	3.070	5.348	8.558	12.592	16.812
7	1.239	2.167	3.822	6.346	9.803	14.067	18.475
8	1.646	2.733	4.594	7.344	11.030	15.507	20.090
9	2.088	3.325	5.380	8.343	12.242	16.919	21.666
10	2.558	3.940	6.179	9.342	13.442	18.307	23.209
15	5.229	7.261	10.307	14.339	19.311	24.996	30.578
20	8.260	10.851	14.578	19.337	25.038	31.410	37.566
25	11.524	14.611	18.940	24.337	30.675	37.652	44.314
30	14.953	18.493	23.364	29.336	36.250	43.773	50.892

[a] Selected data from Fisher and Yates, *Statistical Tables for Biological, Agricultural and Medical Research*, Oliver and Boyd, Ltd., London, 1943.

INDEPENDENT ASSORTMENT AND PROBABILITY

Probability must be considered in explaining the Mendelian principle of independent combinations or independent assortment as well as that of segregation. It was through Mendel's understanding of the mathematical laws of combinations that he was able to recognize and interpret the dihybrid ratio of 9 : 3 : 3 : 1 as a multiple of the 3 : 1 monohybrid ratio. If, for example, the 3 : 1 ratio is changed to an algebraic expression, $3 + 1$, the product of the expected results of two monohybrid crosses is $(3 + 1)^2 = 9 + 3 + 3 + 1$. Because the terms represent separate classes, they are not grouped together, but the product can be converted back to a ratio: 9 : 3 : 3 : 1. This is an example of the **binomial expansion** of $(a + b)^n$, in this case $(A + a)^2$, where $A = 3$ and $a = 1$ or as phenotypes $3A\text{-} + aa$.

Using the F_2 results of the cross between peas with round, yellow seeds and those with wrinkled, green seeds, Mendel tested the mathematical relation between the monohybrid and dihybrid cross. He observed that about $\frac{3}{4}$ $\left(\frac{423}{556}\right)$ of the F_2 seeds were round and $\frac{1}{4}$ $\left(\frac{133}{556}\right)$ were wrinkled. Similarly, seeds from about $\frac{3}{4}$ $\left(\frac{416}{556}\right)$ were yellow, and those from $\frac{1}{4}$ $\left(\frac{140}{556}\right)$ were green. This observed proportion provided a cross-check for the hypothesis of **independence**.

When the two characters were considered together, the results conformed to the mathematical model expected for two independent events occurring together.

On the basis of the law of probability, Mendel predicted that $\frac{9}{16}$ $\left(\frac{3}{4} \times \frac{3}{4}\right)$ of the F_2 would be round, yellow; $\frac{3}{16}$ $\left(\frac{3}{4} \times \frac{1}{4}\right)$ round, green; $\frac{3}{16}$ $\left(\frac{1}{4} \times \frac{3}{4}\right)$ wrinkled, yellow; and $\frac{1}{16}$ $\left(\frac{1}{4} \times \frac{1}{4}\right)$ wrinkled, green. The results that Mendel actually obtained (315 : 108 : 101 : 32) resembled very closely the calculated ratio of 9 : 3 : 3 : 1, based on the hypothesis of complete independence of the genes influencing the shape and color of the seeds. When the χ^2 test for goodness of fit between the actual and the expected result is applied to these figures, the probability (P) of finding deviations as great as, or greater than, those obtained by Mendel is between .80 and .95.

$$\chi^2 = \sum \frac{(O - E)^2}{E} = \frac{(2.25)^2}{312.75} + \frac{(3.75)^2}{104.25} + \frac{(-3.25)^2}{104.25}$$
$$+ \frac{(-2.75)^2}{34.75}$$
$$= .016 + .135 + .101 + .218$$
$$= .470$$
$$P = .80 - .95$$

The actual results fit very closely with those

expected on the basis of the hypothesis of independent combinations. In fact, it is so close that the probability is less than .20 that further data would fit the hypothesis this well. Investigators are as concerned about results that are very close to a calculated expectation as about those with large deviations. A very close fit may be obtained occasionally by chance; but when the observed data fit "too well," the investigator may have intentionally or unintentionally biased the results.

Mendel presented some experimental results that had poor agreement with expectation. In one experiment from which he expected a 3 : 1 ratio, he obtained 43 round and only 2 wrinkled seeds. From another experiment with the same expectation, he obtained 32 yellow and 1 green seeds. He explained these results on the basis of fluctuations due to chance, and in the second case there was also some difficulty in distinguishing between yellow and green seeds. This would add a factor of classification in addition to chance. Fisher has calculated Mendel's results as a whole and has shown the chance of getting such good agreement to be only .0007. Thus, the examples given as 43 : 2 and 32 : 1 are atypical.

From his results, Mendel was able to predict the numbers of genotypes to be expected when more than two pairs of alleles were involved in the cross.

EXPANSION OF A BINOMIAL

Various combinations in groups of a given size representing a particular ratio can be calculated by the binomial expansion of $(p + q)^n$, where p and q represent the probabilities of occurrence of two alternative events (e.g., $p =$ the probability of a boy and $q =$ the probability of a girl) and n is the size of the group involved. How many boys and how many girls would be expected in randomly selected families of 2, 3, 4, 5, or more? Combinations for families of a given size may be calculated by the binomial expansion of $(p + q)^n$. Thus, for two-child families,

$$(p + q)^2 = p^2 + 2pq + q^2$$
$$p = q = \tfrac{1}{2}$$
$$p^2 = \text{families of 2 boys} = \tfrac{1}{4}$$
$$2pq = \text{families of 1 boy and 1 girl} = \tfrac{1}{2}$$
$$q^2 = \text{families of 2 girls} = \tfrac{1}{4}$$

Among families of 2 children, $\tfrac{1}{4}$ would be expected to be composed of all boys, $\tfrac{1}{2}$ of one boy and one girl, and $\tfrac{1}{4}$ of all girls. For three-child families,

$$(p + q)^3 = p^3 + 3p^2q + 3pq^2 + q^3$$
$$p^3 = \text{families of 3 } \male = (\tfrac{1}{2})^3 = \tfrac{1}{8}$$
$$3p^2q = \text{families of 2 } \male, 1 \female = 3(\tfrac{1}{2})^2(\tfrac{1}{2}) = \tfrac{3}{8}$$
$$3pq^2 = \text{families of 1 } \male, 2 \female = 3(\tfrac{1}{2})(\tfrac{1}{2})^2 = \tfrac{3}{8}$$
$$q^3 = \text{families of 3 } \female = (\tfrac{1}{2})^3 = \tfrac{1}{8}$$

($\male$ is the male symbol, representing the shield and spear of Mars, the Roman war god; $\female$ is the female symbol, representing the mirror of Venus, the Roman goddess of love.) The binomial expansion includes all possible combinations of the two events. In three-child families, for example, there are eight combinations.

If the chance of a particular birth order is included in the problem, the probability of each sequence is $\tfrac{1}{8}$. If, on the other hand, only the total number of boys and girls is considered, the probability of all boys is $\tfrac{1}{8}$, 2 boys and 1 girl $\tfrac{3}{8}$, 1 boy and 2 girls $\tfrac{3}{8}$, and all girls $\tfrac{1}{8}$. The expected distributions of males and females in families of one to five children are summarized in Table 2.6.

The probability for each combination can be determined from the binomial coefficient for this combination, as compared with all possible combinations. In sibships of 5, for example, the following proportions of boys and girls would be expected: $\tfrac{1}{32}$, all boys; $\tfrac{5}{32}$, 4 boys and 1 girl; $\tfrac{10}{32}$, 3 boys and 2 girls; $\tfrac{10}{32}$, 2 boys and 3 girls; $\tfrac{5}{32}$, 1 boy and 4 girls; and $\tfrac{1}{32}$, all girls.

When the probability of only a certain combination in a given size group is required, factorials may be employed. These are products of factors derived from functions by successively increasing or decreasing by a constant, usually 1. For example, factorial 4 (4!) is the product of $4 \times 3 \times 2 \times 1$, or 24. [Factorial 0 (0!) = 1 and 1! = 1 by definition.] The probability for a particular combination may be calculated from a formula based on the $x + 1$th term in the binomial expansion of

$$(p + q)^n : P = \frac{n!}{x!(n - x)!} p^x q^{(n-x)}$$

where $n!$ is the product of the integers making up the

TABLE 2.6   Distribution of Boys and Girls in Families

NUMBER OF CHILDREN IN FAMILY	$(p + q)^n$	DISTRIBUTION $p = \male$, $q = \female$
1	$(\frac{1}{2} + \frac{1}{2})^1$	$\frac{1}{2}(1\ \male) + \frac{1}{2}(1\ \female)$
2	$(\frac{1}{2} + \frac{1}{2})^2$	$\frac{1}{4}(2\ \male) + \frac{1}{2}(1\ \male:1\ \female) + \frac{1}{4}(2\ \female)$
3	$(\frac{1}{2} + \frac{1}{2})^3$	$\frac{1}{8}(3\ \male) + \frac{3}{8}(2\ \male:1\ \female) + \frac{3}{8}(1\ \male:2\ \female) + \frac{1}{8}(3\ \female)$
4	$(\frac{1}{2} + \frac{1}{2})^4$	$\frac{1}{16}(4\ \male) + \frac{4}{16}(3\ \male:1\ \female) + \frac{6}{16}(2\ \male:2\ \female) + \frac{4}{16}(1\ \male:3\ \female) + \frac{1}{16}(4\female)$
5	$(\frac{1}{2} + \frac{1}{2})^5$	$\frac{1}{32}(5\ \male) + \frac{5}{32}(4\ \male:1\ \female) + \frac{10}{32}(3\ \male:2\ \female) + \frac{10}{32}(2\ \male:3\ \female) + \frac{5}{32}(1\ \male:4\ \female) + \frac{1}{32}(5\ \female)$

total size of the group (n = the total number in the group); $x!$, the product of the integers for the class with probability p; and $(n - x)!$, the product of the integers for the class with the probability q. The symbol p represents the probability for one occurrence (e.g., a boy), and q is the probability for the other (e.g., a girl). If, for example, 6 babies are born in a given hospital on the same day, what is the probability that 2 will be boys and 4 will be girls? For this problem, assume that $p = q = \frac{1}{2}$. Substituting,

$$P = \frac{n!}{x!(n - x)!}\, p^x q^{(n-x)}$$

$$= \frac{6!}{2!4!}\left(\frac{1}{2}\right)^2\left(\frac{1}{2}\right)^4$$

$$= \frac{6 \times 5 \times 4 \times 3 \times 2 \times 1}{2 \times 1(4 \times 3 \times 2 \times 1)}\left(\frac{1}{4}\right)\left(\frac{1}{16}\right)$$

$$= 15 \times \frac{1}{4} \times \frac{1}{16}$$

$$= \frac{15}{64}$$

The probability of 2 boys and 4 girls in groups of 6 is $\frac{15}{64}$.

For examples of boys and girls in families of different sizes, the probability values p and q were equal ($p = q = \frac{1}{2}$). The binomial distribution can be applied for other values of p and q. If, for example, the trait being considered is albinism in a human family, and the parents are known to be heterozygous (Cc), the probability for a normally pigmented child (p) would be $\frac{3}{4}$, and the probability for an albino child (q) would be $\frac{1}{4}$. In families of 4, what is the probability that 2 will be normally pigmented and 2 will be albino? Substituting,

$$P = \frac{n!}{x!(n - x)!}\, p^x q^{(n-x)}$$

$$= \frac{4!}{2!(2)!}\left(\frac{3}{4}\right)^2\left(\frac{1}{4}\right)^2$$

$$= \frac{4 \times 3 \times 2 \times 1}{2 \times 1(2 \times 1)}\left(\frac{3}{4}\right)^2\left(\frac{1}{4}\right)^2$$

$$= \frac{54}{256} = \frac{27}{128}$$

The probability of 2 normally pigmented and 2 albino children in families of 4 children from heterozygous (Cc) parents would be $\frac{27}{128}$. Other values could be substituted into the binomial expansion for families of given size and the probability for various combinations could be calculated.

PROBABILITY IN PEDIGREE ANALYSIS

Practical applications of the Mendelian principles and the laws of probability are made by human geneticists and, in some instances, by animal breeders in analyzing pedigrees. Traits with a simple pattern of inheritance may sometimes be traced accurately enough to justify prediction concerning the likelihood of their expression in future children, if related individuals or those with a family history of such traits marry. The first step in such an analysis is to determine whether the trait in question is behaving as a dominant or a recessive. Although human traits are often affected by many genes, many are associated with the differential action of certain specific genes and have been identified as dominants or recessives in family groups. McKusick lists 2811 traits associated with specific genes (*Mendelian Inheritance in Man*, 1978). A confusing feature of this type of analysis is that some phenotypes (for instance, deafness) may behave as dominant in some families and recessive in others. Obviously, several different gene substitutions can result in deafness.

Recessive genes are difficult to follow because they may remain hidden by their dominant alleles generation after generation. Carriers in the population usually cannot be identified until an affected child is born. Traits dependent on recessive genes sometimes appear unexpectedly in families having no visible history of such traits. Recessives are expressed more frequently in families in which the father and the mother are more closely related than parents in the general population. The likelihood of similar genes being present is enhanced when the parents have descended from a common ancestor.

In the absence of data to indicate which individuals are carriers, the geneticist may resort to probability as the best available tool for determining the likelihood of expression of a given recessive gene in a certain family. If no expression has occurred in the history of the family, an estimate indicating the frequency of the gene in the general population may be used as a basis of probability. If the trait has appeared in the family, more precise calculations are possible. Probability is then based on the family history, which may be recorded in a pedigree chart.

The use of probability in human pedigree analysis is illustrated in Fig. 2.18. The trait, adherent ear lobes (Fig. 2.19), dependent on a recessive gene, appeared only once in the known history of the family, as indicated by the single darkened circle in Fig. 2.18. No information other than that shown on the chart is available. Unless there is evidence to the contrary, it may be assumed (to avoid dealing with small probabilities) that those individuals who have married into the family are homozygous for the dominant genes and do not carry the gene in question.

The first step is to identify the genotypes of as many individual family members as possible from the information given. The woman (II-3) in whom the trait is expressed must be homozygous (*aa*) for the recessive gene. Each of her parents (I-1 and I-2) who did not express the trait but contributed an *a* gene to their daughter (II-3) must carry the hetero-

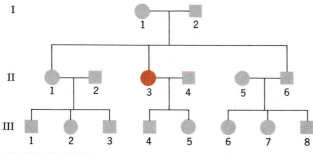

FIGURE 2.18

Pedigree chart to illustrate probability in pedigree analysis. In this family group, a trait (adherent ear lobes), dependent on a recessive gene, has appeared in one individual (identified by the darkened circle). Three basic steps may be followed in all pedigree analysis: (1) What is the chance that one parent is a carrier of the gene in question? (2) What is the chance that the other parent is a carrier for the same gene? (3) What is the chance that a child of these two parents (genotypes) could express the trait involved? The product of the separate probabilities is the chance that a particular future child will express the trait.

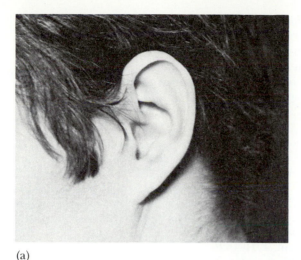

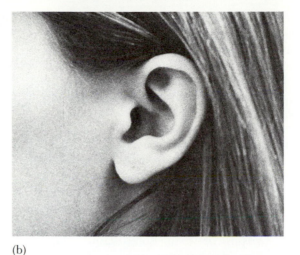

(a)

(b)

FIGURE 2.19

Adherent ear lobes compared with free ear lobes. (a) *Adherent;* (b) *free.* [(a) *Photo by Teri Leigh:* (b) *photo by Kathy Bendo.*]

zygous genotype (*Aa*). The sister (II-1) and brother (II-6) must be *AA* or *Aa*. Obviously, they are not *aa* because they do not express the trait. There is no way to determine whether each of these individuals is *AA* or *Aa*. Therefore, the probability that each individual is a carrier (*Aa*) represents the best information available. From the parent cross (*Aa* × *Aa*), the probability for the occurrence of *Aa* in any child with free earlobes is $\frac{2}{3}$ and the probability for the occurrence of *AA* is $\frac{1}{3}$. In the absence of more definite information, II-1 and II-6 may be considered *Aa* with $\frac{2}{3}$ probability. The children of II-1 and II-6 have a $\frac{1}{2}$ chance of being carriers for the gene (*a*) if their parent is a carrier. Therefore, the probability that III-1, III-2, III-3, III-6, III-7, or III-8 is a carrier is $\frac{2}{3} \times \frac{1}{2} = \frac{1}{3}$. The children of II-3 (III-4 and III-5) must be carriers (probability = 1).

The problem may be carried a step further by calculating the likelihood for an expression (*aa*) of the trait in the first child, resulting from a marriage between two of the cousins represented in generation III. The mating III-1 × III-5 will serve as an example. The probability of being a carrier (*Aa*) is $\frac{1}{3}$ for III-1 and 1 for III-5. Both could be carriers and yet avoid an expression of the trait in their family.

Therefore, another probability must be included, that of parents with genotypes *Aa* having an *aa* child, (*Aa* × *Aa* = 1*AA*, 2*Aa*, 1*aa*), which is $\frac{1}{4}$. The probability for an expression of the trait in the child of the individuals indicated is

$$\tfrac{1}{3} \times 1 \times \tfrac{1}{4} = \tfrac{1}{12}$$

The chance of each future child expressing the trait is also 1 in 12.

If the first child should express the trait, the probability that a second child would also express the trait would be $\frac{1}{4}$, because evidence would then be available to indicate that the genotypes of III-1 and III-5 were both *Aa*. One element of uncertainty or probability would thus be eliminated. At best, probability is a poor substitute for certainty. It is employed in analyses only when definite information is not available.

MODERN EVALUATIONS OF MENDEL'S CONCLUSIONS

New interpretations are inevitable in scientific disciplines as additional data accumulate. Mendel considered a single gene to be responsible for a single trait. It is now known that many genes are involved in the production of some traits, although single gene substitutions can influence basic biochemical reactions and thus be responsible for alternative end products. Furthermore, it is the **genes** and not the traits that are **inherited.** Genes behave as separate units, whereas traits may result from complex interactions involving many genes.

Complete dominance was indicated in all seven allelic pairs that Mendel reported. It was natural, therefore, for him to consider dominance an inherent property of genes. When sweet peas and snapdragons were studied, shortly after the discovery of Mendel's paper, intermediate traits were observed in hybrids. Crosses between homozygous snapdragons with red flowers and those with white flowers resulted in F_1 progeny with pink flowers. Heterozygotes could thus be distinguished phenotypically from both parents. Dominance has now been shown to be influenced by factors in the external, internal (hormonal), and genetic environment. Thus, Mendel's view of dominance as a fundamental inherent property of the allele alone is no longer tenable for all cases. Dominance of some genes may eventually be explained on the basis of modifier genes that are present in the genetic environment. In other cases, dominance may depend on the quantity or activity of enzymes that are gene-controlled.

The most important concepts that Mendel inferred from his experiments were (1) **segregation,** the process through which alleles separate and produce haploid gametes, and (2) **independent assortment** of different pairs of alleles. These principles are the basic foundation of Mendelian heredity.

SUMMARY

Mendelian genetics is based on the transmission of chemical units (genes) from parents to progeny and thus from generation to generation. The mechanism of transmission includes (1) segregation, the separation of pairs of alleles into different gametes, and (2) independent assortment, the independent segregation of members of different pairs of alleles, as demonstrated in dihybrid crosses. Hereditary mechanisms operate in all plants and animals. Prob-

ability is involved in genetic mechanisms and must be recognized in predicting the transmission and expression of both dominant and recessive alleles. Gene-product interactions such as epistasis modify phenotypes and Mendelian ratios.

REFERENCES

Carlson, E. A. 1973. *The gene: a critical history,* 2nd ed. W. B. Saunders, Philadelphia.

Mendel, G. 1866. "Versuch über pflanzenhybriden." (Available in the original German in *J. Hered.* 42:1–47. English translation under the title "Experiments in plant hybridization.") Harvard University Press, Cambridge, Mass.

Morgan, T. H. 1926. *The theory of the gene.* Yale University Press, New Haven, Conn.

Novitski, E., and S. Blixt. 1978. "Mendel, linkage, and synteny." *BioScience* 28:34–35.

PROBLEMS AND QUESTIONS

2.1. On the basis of Mendel's hypothesis and observations, predict the results from the following crosses in garden peas: (a) a tall (dominant and homozygous) variety crossed with a dwarf variety; (b) the progeny of (a) selfed; (c) the progeny from (a) crossed with the original tall parent; (d) the progeny from (a) crossed with the original dwarf-parent variety.

2.2. Mendel crossed pea plants that produced round seeds with those that produced wrinkled seeds. From a total of 7324 F_2 seeds, 5474 were round and 1850 were wrinkled. Using the symbols W and w for genes, (a) symbolize the original P cross; (b) the gametes; and (c) F_1 progeny. (d) Represent a cross between two F_1 plants (or one selfed); (e) symbolize the gametes; and (f) summarize the expected F_2 results under the headings: phenotypes, genotypes, genotypic frequency, and phenotypic ratio.

2.3. The French biologist Cuénot crossed wild, gray-colored mice with white (albino) mice. In the first generation, all were gray. From many litters, he obtained in the F_2 198 gray and 72 white mice. (a) Propose a hypothesis to explain these results. (b) On the basis of the hypothesis, diagram the cross and compare the observed results with those expected.

2.4. A woman has a rare abnormality of the eyelids called ptosis, which makes it impossible for her to open her eyes completely. The condition has been found to depend on a single dominant gene (*P*). The woman's father had ptosis, but her mother had normal eyelids. Her father's mother had normal eyelids. (a) What are the probable genotypes of the woman, her father, and her mother? (b) What proportion of her children will be expected to have ptosis if she marries a man with normal eyelids?

2.5. In pigeons, the checkered pattern is dependent on a dominant gene *C* and a plain exterior on the recessive allele *c*. Red color is controlled by a dominant gene *B* and brown by the recessive allele *b*. Diagram completely a cross between homozygous checkered red birds and plain brown birds. Summarize the expected F_2 results.

2.6. In mice, the gene (*C*) for colored fur is dominant over its allele (*c*) for white. The gene for normal behavior (*V*) is dominant over the one for waltzing (*v*). Give the probable genotypes of the parent mice (each pair was mated repeatedly and produced the following results): (a) colored, normal mice mated with white, normal mice produced 29 colored normal and 10 colored waltzers; (b) colored normal mated with colored normal produced 38 colored normal; 15 colored waltzers; 11 white normal; and 4 white waltzers; (c) colored normal mated with white waltzer produced 8 colored normal; 7 colored waltzers; 9 white normal; and 6 white waltzers.

2.7. In rabbits, black fur is dependent on a dominant gene (*B*) and brown on the recessive allele (*b*). Normal length fur is determined by a dominant gene (*R*) and short (rex) by the recessive allele (*r*). (a) Diagram and summarize the F_1 and F_2 results of a cross between a homozygous black rabbit with normal length fur and a brown rex rabbit. (b) What proportion of the normal black F_2 rabbits from the above cross may be expected to be homozygous for both gene pairs? (c) Diagram and summarize a testcross between the F_1 and the fully recessive brown rex parent.

2.8. In shorthorn cattle, the gene (*R*) for red coat is not dominant over that for white (*R'*). The heterozygous combination (*RR'*) produces roan. A breeder has white, red, and roan cows and bulls. What phenotypes might be expected from the following matings, and in what proportions:
(a) red × red   (d) roan × roan
(b) red × roan   (e) roan × white
(c) red × white   (f) white × white
(g) Would it be easier to establish a true-breeding (homogeneous for color) herd of red or a true-breeding herd of roan shorthorns? Explain.

2.9. Albinism in humans is controlled by a recessive gene (*c*). From marriages between normally pigmented people known to be carriers (*Cc*) and albinos (*cc*): (a) What proportion of the children would be expected to be albinos? (b) What is the chance that any pregnancy would result in an albino child? (c) What is the chance in a family of three that one would be normal and two albino?

2.10. If both partners were known to be carriers (*Cc*) for albinism, what is the chance of the following combinations in families of four: (a) all four normal; (b) three normal and one albino; (c) two normal and two albino; (d) one normal and three albino?

2.11. In *Drosophila*, a dominant gene (*D*) for a phenotype called "dichaete" alters the bristles and also makes the wings remain extended from the body while the fly is at rest. It is homozygous lethal. (a) Diagram a cross between two dichaete (*Dd*) flies and summarize the expected results. (b) Diagram a cross between dichaete and wild type and summarize the expected results.

2.12. In humans, two abnormal conditions, cataracts in the eyes and excessive fragility in the bones, seem to depend on separate dominant genes located in different chromosomes. A man with cataracts and normal bones, whose father had normal eyes, married a woman free from cataracts but with fragile bones. Her father had normal bones. What is the probability that their first child will (a) be free from both abnormalities; (b) have cataracts but not fragile bones; (c) have fragile bones but not cataracts; (d) have both cataracts and fragile bones?

2.13. The inheritance pattern represented by colored squares and circles (symbolizing the same trait in different families—see the accompanying figure) may be assumed to depend on a single autosomal dominant or a single autosomal recessive gene. (a) Indicate which is the most likely mode of inheritance for the trait. (b) Based on your answer to (a), symbolize the probable genotype for each individual in each of the four pedigrees.

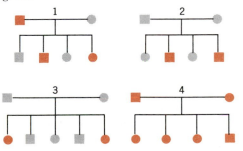

2.14. In garden peas, the genes for tall vine (D), yellow seed (G), and round seed (W) are dominant over their respective alleles for dwarf (d), green (g), and wrinkled (w). (a) Symbolize a cross between a homozygous tall, green, round plant and a dwarf, yellow, wrinkled plant. Represent the gametes possible from each parent and the F_1. (b) Symbolize a cross between two F_1 plants. Complete this cross by making use of the forked-line method and summarize the expected phenotypes. (c) Using the forked-line method, diagram a cross between the F_1 and the dwarf, green, wrinkled parent. Summarize the results for phenotypes, genotypes, genotypic frequency, and phenotypic ratio.

2.15. How many different kinds of F_1 gametes, F_2 genotypes, and F_2 phenotypes would be expected from: (a) $AA \times aa$; (b) $AABB \times aabb$; (b) $AABBCC \times aabbcc$? (d) What general formula can be applied for F_1 gametes, F_2 genotypes, and F_2 phenotypes?

2.16. The shape and color of radishes are controlled by two independent pairs of alleles that show no dominance; each genotype is distinguishable phenotypically. The color may be red (RR), purple ($R'R$), or white ($R'R'$), and the shape may be long (LL), oval ($L'L$), or round ($L'L'$). Using the Punnett square method, diagram a cross between red, long ($RRLL$) and white, round ($R'R'L'L'$) radishes and summarize the F_2 results under the headings phenotypes, genotypes, genotypic frequency, and phenotypic ratio.

2.17. In poultry, the genes for rose comb (R) and pea comb (P) together produce walnut comb. The alternative alleles of both in a homozygous condition (i.e., $rrpp$) produce a single comb. From information concerning interactions of these genes given in the chapter, determine the phenotypes and proportions expected from the following crosses: (a) $RRPp \times rrPp$; (b) $rrPP \times RrPp$; (c) $RrPp \times Rrpp$; and (d) $Rrpp \times rrpp$.

2.18. Rose-comb chickens mated with walnut-comb chickens produce 15 walnut, 14 rose, 5 pea, and 6 single-comb chicks. Determine the probable genotypes of the parents.

2.19. White fruit color in summer squash is dependent on a dominant allele (W), and colored fruit on the recessive allele (w). In the presence of ww and a dominant gene (G), the color is yellow, but when G is absent (i.e., gg), the color is green. Give the F_2 phenotypes and proportions that are expected from crossing a white-fruited ($WWGG$) with a green-fruited ($wwgg$) plant.

2.20. The white leghorn breed of chickens is known to carry in homozygous conditions a color allele (C) and a dominant inhibitor (I) that prevents the action of C. The white wyandotte ($iicc$) has neither the inhibitor nor the color gene. Give the F_2 phenotypes and proportions expected from crossing a white leghorn ($IICC$) with a white wyandotte ($iicc$).

2.21. What phenotypic ratio would be expected from a testcross of F_1 and a full recessive ($AaBb \times aabb$) if the F_2, resulting from $F_1 \times F_1$ (i.e., $AaBb \times AaBb$), were as follows: (a) $13:3$; (b) $15:1$; (c) $9:3:4$; (d) $12:3:1$; (e) $1:2:1:2:4:2:1:2:1$?

2.22. In the F_2 generation of a certain tomato experiment, 3629 fruits were red and 1175 were yellow. A $3:1$ ratio was expected. (a) Are the discrepancies between the observed and expected ratios significant? (b) In the same experiment, 671 plants with green foliage and 569 with yellow were counted. This was a backcross and the hypothetical ratio was $1:1$. Test with χ^2 and explain.

2.23. The following are some of Mendel's results, with the hypotheses to which they were fitted. Test each for goodness of fit and indicate whether each is significantly different from the hypothesis.

CROSS	RESULTS	HYPOTHESIS
(a) Round × wrinkled seed (F_2)	5474:1850	3:1
(b) Violet × white flower (F_2)	705:224	3:1
(c) Green × yellow pod (F_2)	428:152	3:1
(d) Round yellow (F_1) × wrinkled green	31:26:27:26	1:1:1:1
(e) Round yellow (F_1) × wrinkled green	24:25:22:27	1:1:1:1

2.24. When four coins are tossed together in a series: (a) what proportion of the total results will be in the class of four heads; (b) four tails; (c) three heads and one tail; (d) three tails and one head; and (e) two heads and two tails?

2.25. If four babies are born at a given hospital on the same day: (a) What is the chance that two will be boys and two girls? (b) What is the chance that all four will be girls? (c) What combination of boys and girls among four babies is most likely to occur? Why? (d) If a certain family has four girls, what is the chance that the fifth child will be a girl?

2.26. What is the probability in families of six of: (a) one boy and five girls; (b) three boys and three girls; and (c) all six girls?

2.27. The trait represented by colored squares and circles in the following pedigree chart is inherited through a single dominant gene. Calculate the probability of the trait appearing in the offspring if the following cousins should marry: (a) III-1 × III-3; (b) III-2 × III-4.

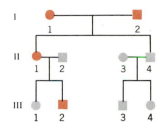

2.28. In the family pedigree shown in the following chart, an abnormal trait is inherited as a simple recessive. Unless there is evidence to the contrary, assume that the individuals who have married into this family do not carry the recessive gene for the trait. Colored squares and circles represent expressions of the trait (see accompanying figure). Calculate the probability of the trait appearing in a given offspring if the following cousins and second cousins in generations III and IV should marry: (a) III-1 × III-12; (b) III-4 × III-14; (c) III-6 × III-13; (d) IV-1 × IV-2.

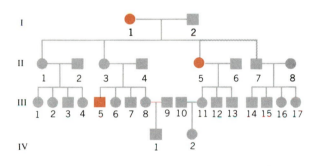

2.29. Phenylketonuria in humans is caused by a recessive allele p. If both partners are known to be carriers (Pp), what is the chance in the following combinations with five children that (a) all are normal; (b) four are normal and one is affected; (c) three are normal and two are affected; (d) two are normal and three are affected; (e) one is normal and four are affected; (f) all are affected?

2.30. Consider the following hypothetical scheme of inheritance of coat color in an extinct mammal. Gene A controls the conversion of a white pigment P_0 to a gray pigment P_1; the dominant allele A produces the enzyme necessary for this conversion but the recessive allele a produces an enzyme with no activity. Gene B controls the conversion of the gray pigment P_1 to a black pigment P_2; the dominant allele B produces the active enzyme that catalyzes the $P_1 \rightarrow P_2$ reaction. The recessive allele b produces a defective enzyme with no activity. The dominant allele C of a third gene produces a product that completely inhibits the activity of the enzyme produced by gene A; that is, it prevents the reaction $P_0 \rightarrow P_1$. Allele c produces a defective product that does not inhibit the reaction $P_0 \rightarrow P_1$. Genes A, B, and C assort independently, and no other genes are involved. In the F_2 of a cross, $AAbbCC \times aaBBcc$, what is the expected phenotypic segregation ratio?

2.31. What F_2 phenotypic segregation ratio would be expected for the cross described in Problem 2.30 if the dominant allele, C, of the third gene produced a product that completely inhibited the activity of the enzyme produced by gene B, that is, prevented the reaction $P_1 \rightarrow P_2$, *rather than* inhibiting the activity of the product of gene A ($P_0 \rightarrow P_1$)?

2.32. Consider the following genetically controlled biosynthetic pathway in a hypothetical plant:

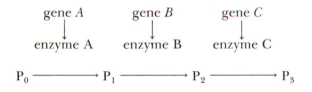

Assume that gene A controls the conversion of a white pigment, P_0, to another white pigment, P_1; the dominant allele A codes for the enzyme necessary to catalyze this conversion, but the recessive allele a codes for a defective enzyme (with no activity). Gene B controls the conversion of the white pigment, P_1, to a pink pigment, P_2; again, the dominant allele, B, produces the enzyme necessary for the $P_1 \rightarrow P_2$ conversion, but the recessive allele b produces an inactive product. The dominant allele, C, of a third gene codes for an enzyme that catalyzes the conversion of the pink pigment, P_2, to a red pigment, P_3; its recessive allele, c, produces an altered enzyme with no activity.

The dominant allele, D, of a fourth gene produces a gene product that completely inhibits the activity of enzyme C; that is, it blocks the reaction $P_2 \rightarrow P_3$. Its recessive allele, d, produces a defective gene product that does *not* block the $P_2 \rightarrow P_3$ reaction.

Assume that flower color is determined solely by these four genes and that they assort independently (i.e., are located on four different chromosomes). In the F_2 of a cross between plants of genotype $AAbbCCDD$ and plants of genotype $aaBBccdd$, what proportion of the plants would be expected to have: (a) red flowers? (b) pink flowers? (c) white flowers?

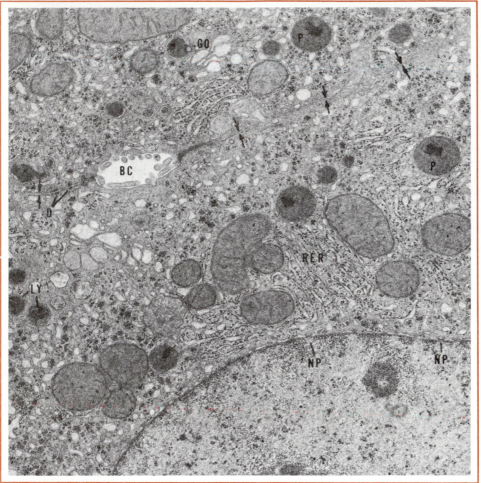

THREE

Electron photomicrograph showing portions of two adjacent liver parenchyma cells: opposed arrows identify plasma membranes of the two cells, BC, bile canaliculus, NP, nuclear pores, RER, rough endoreticulum, GO, Golgi bodies, D, desmosome, P, peroxisome. (Courtesy of Hugh P. Stanley, Utah State University.)

CELL MECHANICS

In the photomicrograph (Fig. 3.1) of a generalized liver cell, structural characteristics are illustrated. Although cells within a single plant or animal vary widely in structure and function, they all represent units of living material and have important properties in common. Both plant and animal cells, for example, are essentially alike in terms of having genes and chromosomes. Chromosomes are involved in the division cycles of individual cells (Fig. 3.2) as well as in the basic reproduction processes of the entire organism (whether animal or plant).

Mitotic cell division is the process through which single cells reproduce themselves and multicellular organisms grow. When cells divide, each resultant part is a complete, although at first relatively small, cell. Immediately following division, the newly formed daughter cells grow rapidly, soon reaching the size of the original cell. Mitotic cell division is actually **duplication** or multiplication

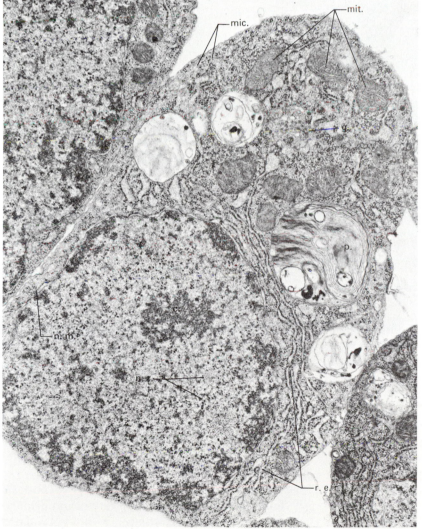

FIGURE 3.1

Electron photomicrograph of a cell showing structural characteristics: mic., microtubules; mit., mitochondria; g., golgi; p., phagolysosome; r.e.r., rough endoplasmic reticula; n.r., nuclear ribosomes; c., chromatin; n.m., nuclear membrane; n., nucleus of another cell. (Courtesy J. J. Bozzola, Microbiology, S.I.U., Carbondale.)

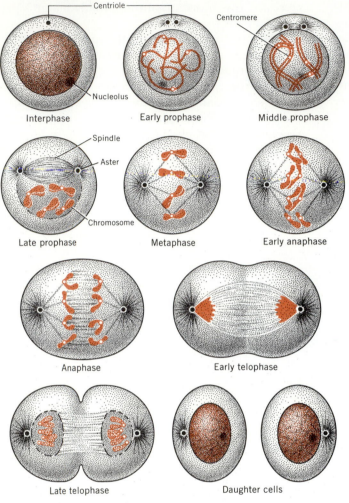

Interphase
Early prophase
Middle prophase
Centriole
Centromere
Nucleolus
Spindle
Aster
Chromosome
Late prophase
Metaphase
Early anaphase
Anaphase
Early telophase
Late telophase
Daughter cells

FIGURE 3.2

Schematic diagram of mitosis in an animal cell with four chromosomes.

rather than division in the usual sense. In unicellular animals, however, cell duplication equates with reproduction, since two new individuals are formed from the original parent. Each daughter cell receives the genetic information carried by the mother cell.

In higher organisms, growth occurs through mitotic cell division with subsequent enlargement and differentiation of the cells produced. In reaching adulthood, the number of cells increases in all multicellular organisms. Each human being, for example, begins as a single fertilized cell or zygote, which develops eventually into a person with more

than a million billion cells. The number of cells then remains approximately constant. Cell division in human skin is rapid throughout life, but this is replacement division and does not increase cell number. In some organisms, such as the adult fruit fly, there is little cell division and little change in cell number.

The complex body of a multicellular organism eventually contains a variety of specialized cells. Epithelial cells have a relatively short life span; replacements must be made continuously. Those in the lining of the respiratory, digestive, and urinary tracts, for example, are replaced within a few days.

Some gland cells have a life span of only a few hours. Cells in the human central nervous system, on the other hand, once established do not divide. Loss of cells is not compensated for by mitotic cell division, and the number diminishes through the mature life of the individual.

CELL CYCLE

Growth requires an increase in cell mass, a duplication of the genetic material, and a division assuring that each daughter cell receives an equal complement of the genetic material to ensure perpetuation of the cell line. These steps occur in an ordered progression during the cell life span or cycle (Fig. 3.3). Initially, a diploid cell (containing $2n$ chromosomes) undergoes a period of growth and increase in cell mass. This is called G_1. For a cell requiring 24 hours for the complete cycle, the G_1 stage would take the first 10 hours. This stage is devoted to cell growth and chemical preparation for DNA synthesis. At a defined time, duplication of the genetic material commences. During this synthesis (S) phase

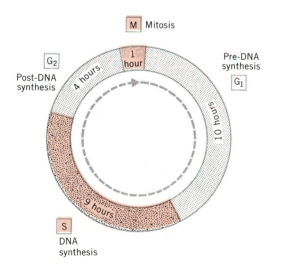

FIGURE 3.3
Diagrammatic illustration of the stages in the cell cycle of a "typical" mammalian cell growing in tissue culture with a generation time of 24 hours. (After L. E. Hood, J. H. Wilson, and W. B. Wood, Molecular Biology of Eucaryotic Cells, *W. A. Benjamin, Inc.)*

of nine hours, every chromosome is duplicated. These duplicated structures, called sister chromatid pairs, each contain two identical chromosome copies. After the completion of chromosome replication, the cell enters a second growth phase called G_2. This post-DNA synthesis stage requires four hours, continuing until the onset of mitosis (M), which requires one hour. During mitosis the sister chromatid pairs separate, one sister going to each of the two daughter cells (Fig. 3.2).

MITOSIS

Details of cell reproduction were elucidated in animal cells in the latter part of the nineteenth century by Walther Flemming, and in plant cells by Edward Strasburger and several other investigators. Two interrelated processes were found to be involved: (1) **mitosis,** the nuclear division, and (2) **cytokinesis,** the changes in the cytoplasm that include division of the cell proper. Fig. 3.2 illustrates these processes as they occur in animal cells.

The names **interphase** (between divisions), **prophase, metaphase, anaphase,** and **telophase** have been associated with different stages of the continuous mitotic cycle for convenience in describing the changes that occur. The prophase and telophase stages of mitosis are usually long and involved, whereas metaphase and anaphase are commonly brief.

First indications of approaching mitosis in animal cells are observed in the cytoplasm of the interphase cell. A region of differentiated cytoplasm containing the centriole begins to divide. The centriole is a reproducing organelle in the cytoplasm that initiates division and organizes the mitotic apparatus consisting of (1) asters and (2) spindle, microtubules called spindle fibers. In early prophase, the daughter centrioles move apart. Thin, uncoiled, replicated sister chromatids become coiled, shortened, and discrete. The number of coils decreases with a concomitant increase in diameter of each coil (see diagram inside front cover).

By late prophase, the two chromatids of each chromosome are held together at a constricted area called the primary constriction, where the **centromere** or spindle fiber attachment region is located.

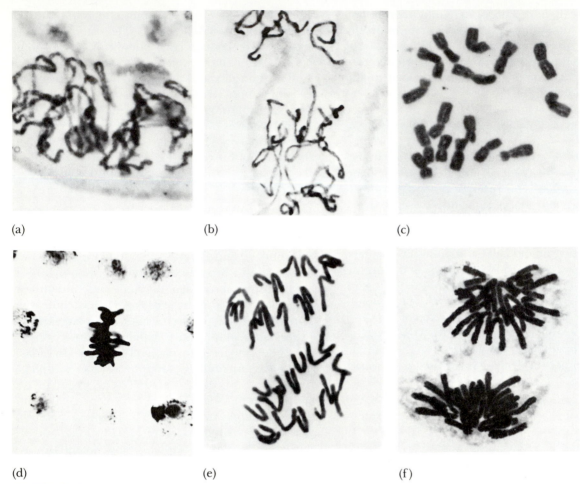

(a)

(b)

(c)

(d)

(e)

(f)

FIGURE 3.4

Photomicrographs representing major stages in the mitotic sequence of the onion, Allium cepa, *root tip. This species has 16 chromosomes. (a) early prophase; (b) middle prophase; (c) metaphase, polar view; (d) metaphase, side view; (e) anaphase showing separate chromosomes; (f) telophase. All six photographs are made to the same scale, ×800. (Courtesy of W. S. Boyle.)*

Centrioles have now spread to opposite poles and continuous spindle fibers have connected the poles. The nuclear membrane and nucleolus, which is the site of ribosomal RNA genes, disappear. A spindle fiber becomes attached to each chromatid at the centromere. Each of two sister chromatids becomes attached to a different pole of the spindle, but the centromeres remain together. At metaphase, discrete chromatid pairs take places in the center or equatorial plate. Metaphase chromatids are tightly coiled and discrete, thus facilitating counts and gross structural comparisons. Arms of sister chromatids are extended from the centromeres, but the chromatids are held together by the centromeres until the beginning of anaphase.

When anaphase separation occurs, each chromatid has its own centromere and is a chromosome. Anaphase chromosomes elongate somewhat by relaxation of the tight metaphase coiling and move to the respective poles of the spindle. Duplication and separation of chromatids fulfill the requirements of Roux's models (Chapter 1). Mitosis ensures that each daughter cell has the same genetic information as the mother cell.

During telophase, a nuclear membrane is reconstructed around each daughter nucleus and the nucleolus reappears in a particular place on a chromosome where the ribosomal RNA genes are located. This location corresponds with a secondary constriction in some chromosomes. The cytoplasmic part of the cell then divides. Animal cells, with their flexible outer layers, accomplish this by a constriction that converges from the two sides and eventually separates the two daughter cells. The surface around the equator pushes in toward the center and pinches the cell into two parts. Plant cells, with their rigid walls, form a partition or cell plate between the daughter cells. After the middle lamella (cell plate) is formed, walls of cellulose are deposited on either side.

As indicated above, each cell division (nuclear and cytoplasmic phases) is a continuous process from the time a cell first shows evidence of beginning to divide until the two daughter cells are completely formed. Nuclear and cytoplasmic phases are distinct but coordinated processes. The entire procedure ordinarily requires a few hours to several days, with variations dependent on the type of organism and environmental conditions. The actual mitotic sequence is illustrated with photomicrographs of dividing onion cells in Fig. 3.4.

MEIOSIS

Meiosis is a process in sexual reproduction through which the chromosome number of diploid ($2n$) germ cells is reduced to half (n) in formation of mature reproductive cells or gametes. This reduction is accomplished by two successive cell divisions during which the chromosomes are duplicated only once. Preceding the first of the two divisions, the homologous chromosomes come together in pairs and duplicate. These paired chromosomes are separated, that is, pulled apart in the first meiotic cell division. In the second meiotic cell division, the two chromatids of each chromosome separate, making a

total of four separate mature sex cells each capable of fertilization. Through fertilization the diploid ($2n$) chromosome number is restored. An organism characteristic of a species may then be developed largely by mitotic cell division.

Yeast cells, *Saccharomyces cerevisiae*, provide a simple uncomplicated example of meiosis. They are single-celled organisms reproducing asexually by budding. But their vegetative cells may act as sexual cells. The two meiotic divisions occur within a closed sac, the ascus, which holds all four products of a single meiotic event, called ascospores, together (Fig. 3.5).

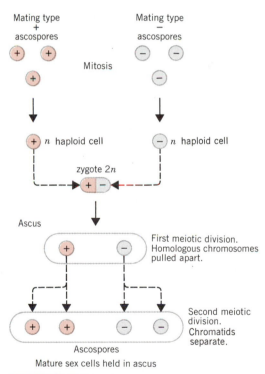

FIGURE 3.5

Mating types of yeast illustrating mitosis and meiosis in their simplest form. Haploid cells multiply by mitosis (budding). Cells of different mating type fuse to produce a diploid zygote, which may then undergo meiosis. Homologous chromosomes synapse and replicate to produce sister chromatids, and then the homologous chromosomes are pulled into separate nuclei during the first meiotic division. The two progeny cells divide in the second meiotic division giving rise to four haploid cells called ascospores, which are held in the ascus.

GAMETE FORMATION IN ANIMALS

Virtually all normal cells can reproduce themselves. The sex or germ cells, however, can initiate reproduction of the entire organism. A special sequence, **gametogenesis,** the formation of haploid (reduced chromosome number) female and male gametes, results in the development of mature sex cells. Gametogenesis includes meiosis (from the Greek, "to reduce"). During meiosis, the chromosome number is changed from the diploid or $2n$ number, characteristic of body cells and premature germ cells, to the haploid or n number, which is characteristic of mature germ cells. Gametogenesis also includes differentiation of eggs and sperm, a process necessary for their functioning. Eggs of animals usually accumulate nutrient materials that sustain the developing embryo for a brief period; sperm of most animal species each develop a flagellum for independent motility.

SPERMATOGENESIS

Sperm originate in the male reproductive organs, or testes, through a sequence called spermatogenesis (Fig. 3.6, left). Considered grossly, the process consists of growth in cell size, two successive cell divisions, and a metamorphosis of the resulting cells from spherical static bodies to elongated motile sperm. Spermatogenesis is initiated in diploid $(2n)$, or unreduced, germ cells called **spermatogonia** that enlarge and become **primary spermatocytes.** These spermatocytes then undergo the first meiotic division, with each producing two **secondary spermatocytes.** In turn, each secondary spermatocyte undergoes the second meiotic division to produce two **spermatids.** Each spermatid then changes shape, develops a flagellum, and becomes a mature sperm. While the cells are dividing **twice,** the chromosomes duplicate themselves only **once.** Reduction in chromosome number from $2n$ to n is accomplished in the first meiotic division. Units of two chromatids resulting from the reduction division are the sister chromatids whose separate DNA arose from earlier replication. They separate in the second meiotic division, each entering a different mature sperm.

Cells of the salamander *Amphiuma means tridactylum* Cuvier (the three-toed *Amphiuma*) are unusu-

ally large, and so the chromosomes are conducive to studies of mitosis and meiosis. A series of photographs (Fig. 3.7; see also Fig. 3.8) showing chromosomes from this salamander illustrate the actual appearance of stages in the meiotic sequence of spermatogenesis. The sequence of developmental changes in the meiotic prophase is long and involved. The cell in which the **prophase** occurs, called the primary spermatocyte, enlarges during the entire prophase. Five major prophase stages are distinguishable in the continuous transition from spermatogonium to metaphase of the first meiotic division. The first stage (Fig. 3.7*a*) is called the **leptotene** stage (the noun form, leptonema). Chromosomes ($2n = 28$), which appear as single threads, represent the maternal and paternal chromosomes received by the individual (male) from the gametes of his parents. Replication of genetic information occurred in the preceding interphase.

The pairing process, called "synapsis" (from the Greek, "conjunction" or "union"), brings together maternal and paternal members of the same pair of chromosomes (Fig. 3.7*b*). Corresponding segments of particular maternal and paternal chromosomes come together along their length in zipperlike fashion. When synapsis is finished, the number of visible structures is half that in leptonema, and the visible bodies in the nucleus are now homologous chromosome pairs rather than single chromosomes. This is the **zygotene** stage (zygonema).

Following zygonema, homologous chromosomes can be observed side by side. This is the **pachytene** stage (pachynema; Fig. 3.7*c*). Exchanges between homologs and chiasma formation occur in this stage. The exchange event takes place first and appears visually as a **chiasma.** At the ends of bivalents, four chromatids can be seen in some preparations. These have resulted from the duplication of each homolog forming sister chromatids. A ribbonlike structure called a **synaptinemal complex** can be seen with the electron microscope between the synapsed homologs, apparently binding the chromatids along their length and thus facilitating chromatid exchange and genetic **crossing over** (Chapter 6). Chromatids continue to shorten and thicken, and the four chromatids in a group (bivalent) be-

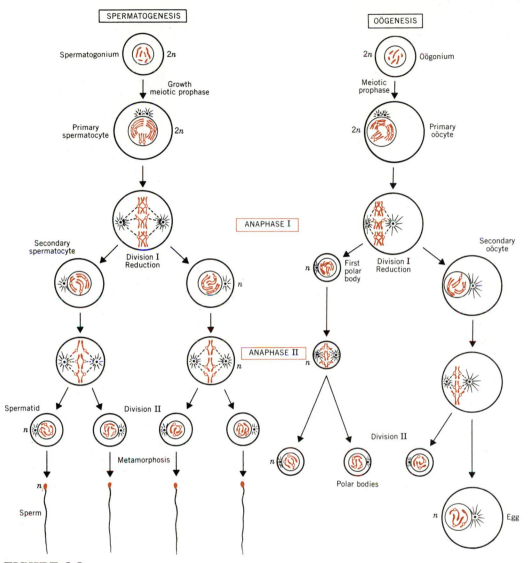

FIGURE 3.6

Meiotic sequence in a male and female animal. (Left) Process of spermatogenesis resulting in the formation of four sperm; (right) oögenesis resulting in the formation of one egg and three polar bodies.

come more apparent. This is the **diplotene** stage (diplonema). Paired chromatids appear to repel each other, causing the strands to separate longitudinally in some areas and to form loops (Fig. 3.7*d*).

In diplonema (Figs. 3.7*d* and 3.7*e*), the centromere in each chromosome is not split, and the longitudinal separation of the chromosomes is incomplete. Bivalents are held together at various places along their length because of interchanges between chromatids (chiasmata). From one to several chiasmata may be observed in favorable preparations, depending on the length of the bivalent. Each

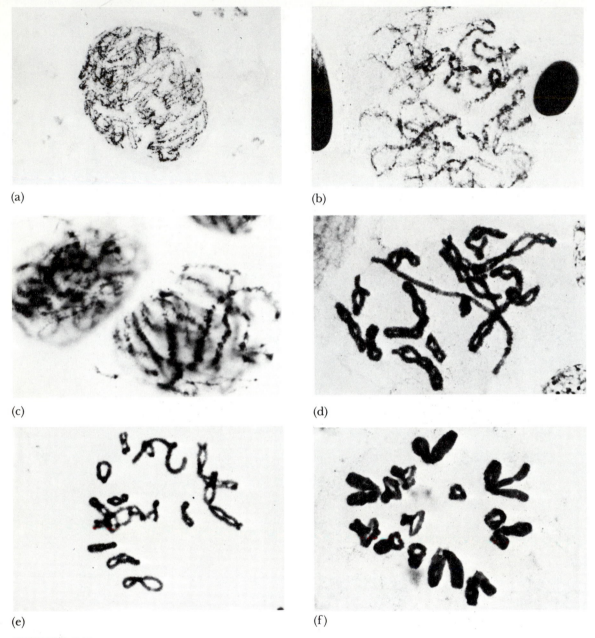

FIGURE 3.7

Stages in the meiotic sequence of spermatogenesis in Amphiuma means tridactylum *Cuvier. (a) Leptotene-zygotene in spermatocyte with single filaments and some pairing visible. Feulgen stained; ×360. (b) Detail of zygotene spermatocyte showing paired strands. Feulgen stained; ×680. (c) Early pachytene stage showing some thin filaments that may represent asynaptic regions. Feulgen stained; ×1200. (d) Diplotene spermatocyte showing long arms of bivalent (in center) and chiasmata in other bivalents; ×680.*

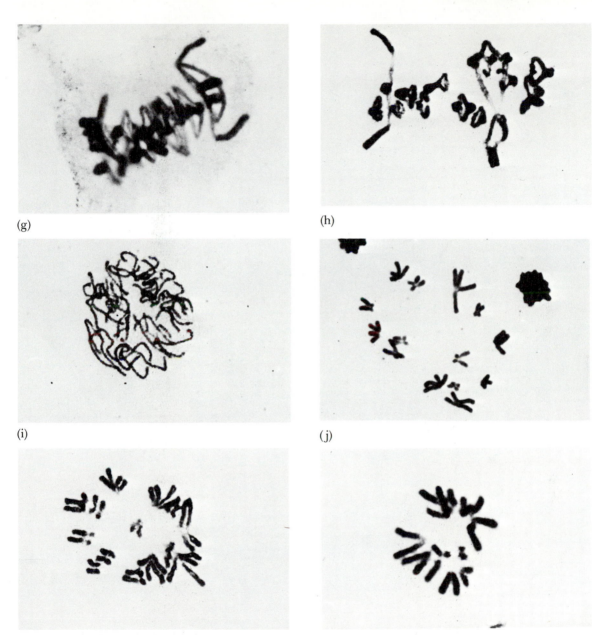

(g)

(h)

(i)

(j)

(k)

(l)

(e) *Late diplotene with 14 bivalents;* ×680. (f) *Diakinesis stage showing 14 distinct bivalents and a lack of chiasmata;* ×680. (g and h) *Metaphase I;* ×735. (i) *Prophase II with identifiable reduced* (n) *chromosomes, each with 2 chromatids;* ×820. (j) *Metaphase II;* ×740. (k) *Anaphase II showing early separation of daughter chromosomes;* ×820. (l) *The haploid complement of 14 chromosomes after anaphase II separation;* ×1120. (Courtesy of G. M. Donnelly, A. H. Sparrow, and R. F. Smith. Brookhaven National Laboratory.)

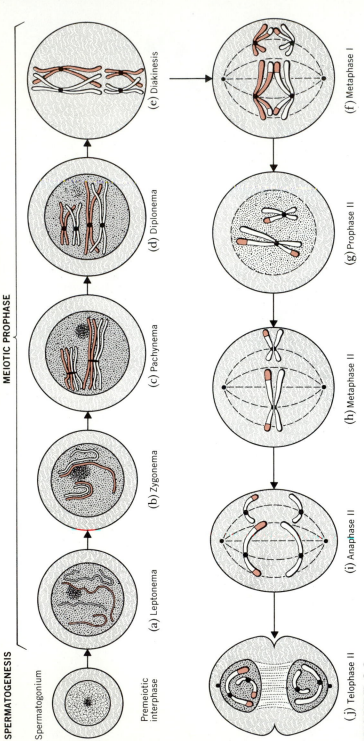

SPERMATOGENESIS

MEIOTIC PROPHASE

Spermatogonium

Premeiotic
interphase

(a) Leptonema

(b) Zygonema

(c) Pachynema

(d) Diplonema

(e) Diakinesis

(f) Metaphase I

(g) Prophase II

(h) Metaphase II

(i) Anaphase II

(j) Telophase II

FIGURE 3.8

Diagrams illustrating chromosome changes that were shown in Fig. 3.7, photographs a to l. Stages in meiosis in an animal with two pairs of chromosomes, one member of each pair having come from the mother (white) and one from the father (colored). (a–e) Meiotic prophase stages parallel with a to e in Fig. 3.7; (f) metaphase I parallels with (f) in Fig. 3.7, showing the results of random assortment; (g–j) second meiotic division stages parallel with g to l in Fig. 3.7, showing the results of segregation of sister chromatids.

chiasma observable at this stage apparently represents an exchange that occurred between nonsister chromatids during the pachytene stage. In late diplonema (Fig. 3.7*e*) when the homologs repel each other, it is the chiasmata that hold the homologs together in the bivalent. The place where a chiasma appears, however, is not necessarily where the exchange actually occurred, because chiasmata tend to slip toward the ends of the bivalents and thus to become terminalized as the meiotic prophase continues. Genetic implications of chiasmata (crossovers) are discussed in Chapter 6.

Shortening of the bivalent continues through the next stage, **diakinesis** (Fig. 3.7*f*), resulting in discrete units (of four chromatids), which, in favorable preparations, can be counted and found to represent half the 2*n* chromosome number of the salamander, or 14, as expected following the pairing process. As the meiotic prophase is completed, the bivalents become angular or oval in appearance and take their places in the equatorial plane, forming the equatorial plate of meiotic **metaphase I** (Figs. 3.7*g* and 3.7*h*).

The first of the two cellular divisions in the meiotic sequence separates the homologous chromosomes (nonsister chromatids) that paired during the zygotene stage. In the anaphase of this so-called reduction division, the original maternal and paternal chromosomes (each composed of one centromere and two chromatids) separate. Thus, the number of chromosomes in each resultant cell is *reduced* from the original 2*n* to the *n* number. These are shown in **prophase II** (Fig. 3.7*i*) and **metaphase II** (Fig. 3.7*j*). After the first meiotic division, the cells are called secondary spermatocytes.

During the secondary spermatocyte division, the centromere of each bipartite chromosome parts, separating each new chromosome with its own centromere (Fig. 3.7*k*). Each chromosome then moves to a pole of the spindle. The chromosome number in a spermatid is haploid (*n*), the same as it is in the secondary spermatocyte. Chromosomes of spermatids, however, are unipartite (Fig. 3.7*l*), whereas those of the secondary spermatocyte are bipartite, being composed of two chromatids. Each **spermatid** nucleus has a **single set of nonhomologous chro-**

mosomes. The second division is a mitotic-type division (called the equational division), because it separates the duplicated (sister) chromatids. The division is different from the reductional division in which the homologous chromosomes that came together in synapsis separate (Fig. 3.8).

The terminal process in spermatogenesis is a complicated differentiation called **spermiogenesis.** Through a progressive sequence of changes, each of the comparatively large, spherical, nonmotile spermatids is metamorphosed into a small, elongated, motile sperm composed typically of three parts: head, middle piece, and tail. In most animal species, spermatids begin differentiation by secretion of the apical body or acrosome, division of the centriole into two, and production of the flagellum by one centriole. Sloughing off of cytoplasm diminishes the overall size, as the developing sperm changes from a spherical to an oval and finally to an elongate shape. The nucleus moves to one edge of the cell, becoming elongated and increasingly compact. The acrosome, which is produced by the Golgi apparatus, takes its place around the anterior end of the sperm head. It contains lytic enzymes that facilitate sperm entry into ova (Fig. 3.9).

The middle piece contains the centriole, which lies next to the nucleus, and the proximal part of the axial filament, which continues in the tail. The mitochondria of the spermatid become concentrated around the axial filament in the middle piece. The tail of the sperm is composed of two parts: the outer sheath, which is cytoplasmic in origin, and the axial filament inside the sheath, which extends from the base of the head to the posterior end of the tail. Much of the cytoplasm of the spermatid is not used in formation of the sperm and is reabsorbed (taken up by Sertoli cells that presumably have nutritive and endocrine functions).

In most sexually mature male animals, spermatogenesis is constantly or periodically occurring in the testes, and many millions of sperm are produced. Insects generally require only a few days to complete their cycle of spermatogenesis, but in mammals the cycle extends over weeks or months. In mature human males, spermatogenesis occurs in the seminiferous tubules of the testes. Spermato-

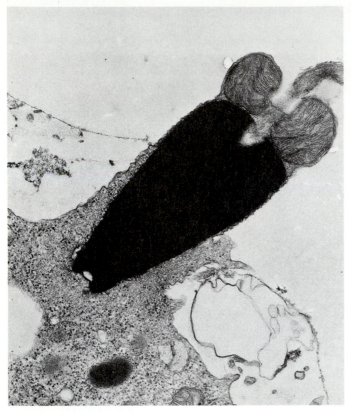

FIGURE 3.9

Sea urchin sperm entering egg. Magnification × 19,800. The egg cytoplasm bulges up around the sperm head forming a fertilization cone through which the sperm enters. (Courtesy of Everett Anderson, Harvard Medical School, Boston. From J. Cell Biol. *37:514, 1968.)*

gonia, undergoing mitotic division and continuing the population of stem cells, can be observed in cross sections of tubules at the periphery. The spermatogonia that appear to be in the innermost cell layer of the periphery enlarge and form primary spermatocytes with 23 bivalents. These spermatocytes undergo the two mechanically interwoven meiotic divisions in rapid succession and produce spermatids, which develop into sperm with single sets of 23 chromosomes. It should be noted that the sperm cells that develop in the testes of the male descend from original or primordial cells that migrated to the testes during early embryogenesis. The developmental time from primitive spermatogonia to mature human sperm is about 74 days.

OÖGENESIS

The process of gamete formation in the female (oögenesis, the origin of the egg) is also illustrated in Fig. 3.6 (right). Oögenesis is essentially the same as

spermatogenesis as far as nuclear division is concerned, but aspects of the cytoplasm are quite different. Much more **nutrient material** is accumulated during oögenesis than during spermatogenesis. This is particularly true of oviparous animals — those that lay eggs that hatch outside the body of the mother. These animals must provide yolk material for the nourishment of the developing embryo outside the mother's body. Even in viviparous animals — which retain and nourish the young inside the body of the mother — a considerable amount of nutrient material accumulates. Because of the accumulated nutrient materials, an egg is usually considerably larger than a sperm of the same species (Fig. 3.10).

In addition, the cells that result from divisions in oögenesis are of unequal size. Nutrient material in the primary oöcyte is not divided equally into four cells that result from the meiotic sequence. One large cell (much larger in proportion than that

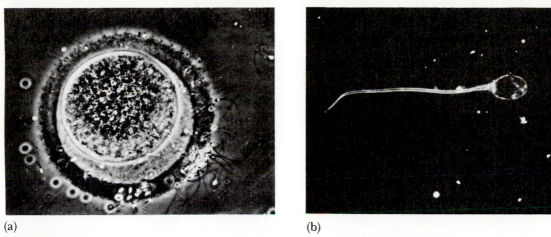

(a)   (b)

FIGURE 3.10

Human egg and sperm. (a) Egg with its surrounding layer of supporting cells as it is being penetrated by sperm. Several sperm are shown at lower right, but only one will penetrate the egg. (b) Single human sperm greatly enlarged. [(a) Courtesy of The American Museum of Natural History; (b) L. V. Bergman and Associates.]

shown in Fig. 3.6, right, for each division) retains essentially all the yolk, while the other cells, called "polar bodies," get very little. First and second **polar bodies** receive the same chromosome complements as the secondary oöcytes and ova from the respective divisions, but they do not become functional sex cells. During differentiation, special egg membranes are formed and the nucleus is generally reduced in size. The cytoplasm contains materials required for cell activity and differentiation.

In some animal species, oögenesis proceeds rapidly and continuously in sexually mature females, and numerous eggs are produced. Usually, these eggs complete the second meiotic division and become mature before encountering sperm. In many other animals, including mammals, the meiotic divisions are not completed until after sperm entry. In the human female, for example, oögenesis begins before birth. Oögonia located in the follicles of cortical tissue in the fetal ovary begin to differentiate into primary oöcytes at about the third month of intrauterine development. At the time of birth of the female infant, all the primary oöcytes are in the prophase of the first meiotic division. They remain in "suspended prophase" for many years until sex-

ual maturity is reached. Then, as the ovarian follicles mature, the meiotic prophase is resumed. The first meiotic division for each developing egg is completed shortly before the time of ovulation for that egg. One cell becomes a secondary oöcyte and the other a polar body.

The second meiotic division is in progress when the developing egg is extruded from the ovary and passes into a Fallopian tube. This division is not completed, however, until after penetration by the sperm, which usually occurs in a Fallopian tube. Sperm entry is a random process in that any available sperm may fuse with any mature egg. If penetration by a sperm is accomplished, the **secondary oöcyte** divides and forms a **mature ovum** with a pronucleus containing a single set of 23 maternal chromosomes. The other cell resulting from this division is a second polar body not capable of further development. The sperm head forms a pronucleus with 23 paternal chromosomes. After the two pronuclei fuse, the resultant zygote, with $(2n)$ 46 chromosomes, begins mitotic division or **first cleavage,** which produces the two-cell stage of a beginning embryo. If the fertilized egg should undergo a mitotic division, identical twins could be produced.

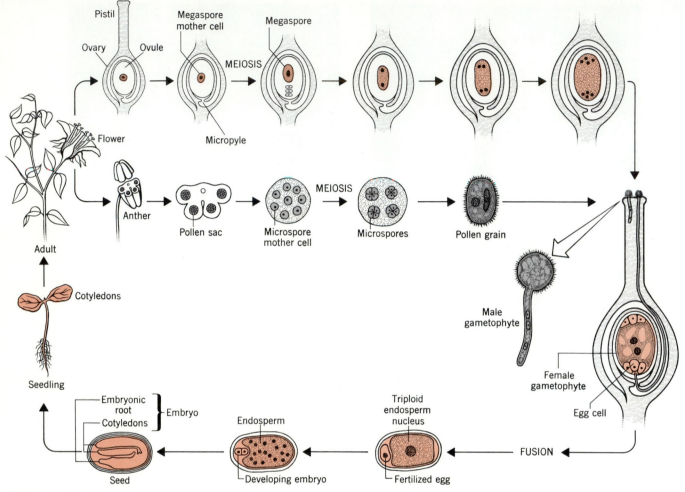

FIGURE 3.11

The life cycle of a seed plant. These enlarged diagrams show fertilization and seed formation.

When two eggs are produced at the same time and fertilized by separate sperm, fraternal twins, no more alike genetically than other sibs, may develop together in the same uterus.

SPORE FORMATION IN PLANTS

Gamete formation in plants, like that in animals, requires a reduction in chromosome number from $2n$ to n. The meiotic process itself is similar to that in animals, but the life cycle of plants is somewhat more complicated (Fig. 3.11). Gamete formation usually does not follow meiosis directly. Sporogenesis in plants involves the formation of spores rather than sex cells. Spores produce gametophytes, and gametes come from gametophytes in which the chromosome number is already reduced. An **alternation of generations** between a haploid and a diploid phase in virtually all plants can separate the reduction of chromosome number from fertiliza-

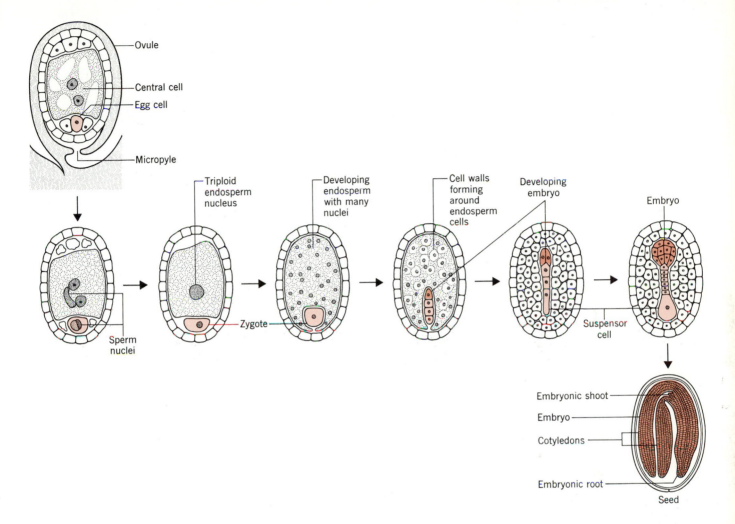

Ovule

Central cell

Egg cell

Micropyle

Sperm nuclei

Triploid endosperm nucleus

Zygote

Developing endosperm with many nuclei

Cell walls forming around endosperm cells

Developing embryo

Embryo

Suspensor cell

Embryonic shoot

Embryo

Cotyledons

Embryonic root

Seed

tion in plant reproduction. Diploid plants, called **sporophytes,** undergo meiosis to produce spores with reduced chromosome number. Spores develop into haploid **gametophytes,** which ultimately produce gametes capable of fertilization. Zygotes resulting from **fertilization** develop into sporophytes, completing the cycle. Sporophyte and gametophyte phases vary in length and importance in different plants.

PLANT FERTILIZATION
In the male cycle (Fig. 3.11), three nuclei arise from two mitotic divisions: the tube nucleus and two generative nuclei. This is the gametophyte phase.

The generative nuclei are carried through the micropyle into the embryo sac and accomplish the **double fertilization** process characteristic of the higher plants. One male nucleus fuses with the egg nucleus and gives rise to the $2n$ zygote, which divides repeatedly to form the **embryo** of the seed. The second male nucleus unites with the two polar nuclei to form the $3n$ nutrient **endosperm** of the seed.

The process of double fertilization introduces genetic material from the pollen parent into the endosperm tissue as well as the embryo. Therefore, it might be expected that both maternal and paternal inheritance would be represented. This hereditary influence of the pollen parent genes on the endosperm is called xenia. When, for example, maize from a variety normally bearing white kernels is pollinated with a yellow-kernel variety, the endosperm of the hybrid kernels is yellow. The dominant gene for yellow from the pollen comes to expression in the endosperm in the same manner as expected for embryonic tissue.

The diploid number of chromosomes is restored in the fertilized cell that gives rise to the plant embryo. Thus, through fertilization, the genetic contributions from each parent are combined. Subsequent continuous mitotic division of cells, and in plants the growth of individual cells, produces a new individual, representative of the species to which the parent belongs.

MEIOSIS AND MENDEL'S PRINCIPLES

Mendelian segregation and independent assortment are simple **corollaries of meiosis.** The physical bases of these two principles are inherent in the chromosome mechanism of meiosis. By observing microscopically distinguishable pairs of chromosomes during a reduction division, the maternal and paternal chromosomes can be followed. This was accomplished by E. E. Carothers in studies of grasshoppers. The three distinguishable pairs of maternal and paternal chromosomes are symbolized $AA'BB'CC'$, and the various observed arrangements of the division spindle are illustrated in Fig. 3.12. The independence of the reduction division

explains Mendel's second principle. A chance distribution of maternal and paternal chromosomes occurs in such a way that each has an equal chance of facing one pole or the other on the equatorial plate of the reduction division. When chromosomes become established on the equatorial plate, each goes to the pole it faces and eventually becomes a part of the daughter cell formed around that pole of the spindle.

To illustrate this concept, consider the phenotypes vine height, seed color, and shape that are observed in F_1 peas (Chapter 2). Genes for three contrasting phenotypes (E. Novitski and S. Blixt, 1978) are located on separate homologous chromosome pairs (chromosomes 4, 1, and 7, respectively). Gene D in the example is located on the maternal member of one pair (chromosome 4), whereas its allele, d, is in the corresponding position of the homologous (paternal) chromosome 4. Fig. 3.13a shows, diagrammatically, the six chromosomes representing three homologous pairs in unpaired condition before synapsis. Maternal and paternal members of each pair carrying their respective alleles are present in the same nucleus. At synapsis in meiosis, each chromosome finds its mate, and alleles come together in corresponding positions, as illustrated in Fig. 3.13b. A particular maternal chromosome is equally likely to face one or the other pole of the spindle.

Zygotes resulting from fusion of haploid (n) gametes are diploid ($2n$) and therefore carry pairs of alleles. Alleles are alternate forms of a given gene,

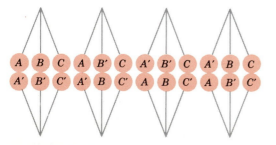

FIGURE 3.12

Arrangements of maternal and paternal chromosomes that could be identified in Carothers' microscope studies of grasshopper chromosomes.

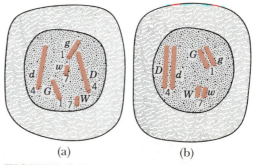

(a)   (b)

FIGURE 3.13

Cells showing three chromosome pairs: (a) *before synapsis has occurred and* (b) *after synapsis has occurred.*

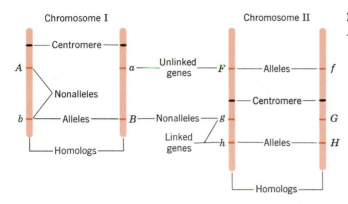

FIGURE 3.14
Alleles and nonalleles, and linked and unlinked genes.

located at the same chromosome locus, which may substitute for one another. Body cells of higher organisms originate from the zygote through continued cell duplication and differentiation. Like the zygote, body cells arising from the zygote carry two genomes included in two sets of homologous chromosomes. Two pairs of homologous chromosomes illustrating alleles and linked and unlinked nonalleles are shown in Fig. 3.14. The zygote from which these two pairs were taken would carry the diploid number characteristic of the species.

In animals and plants with a large number of chromosomes, an almost infinite number of possible combinations may be expected. In organisms with 23 pairs of chromosomes, for example, the probability that a gamete produced by an individual in the population will have any specific combination of chromosomes is $(\frac{1}{2})^{23}$, which is in the order of one in eight million. This calculation is an underestimate due to the possibility of any crossing over, which is another source of variability (see Chapter 6). Further increased numbers of gene combinations are possible in zygotes that result from **random fertilization.** Much of the variation observed in natural populations can therefore be explained on the basis of the recombination of chromosomes and genes already present in the breeding population.

SUMMARY

Genes are organized into linear strands in chromosomes. In nondividing cells, the genes remain intact

within the nuclei. Genes replicate in interphase of both mitosis and meiosis. Each of the two replicated chains becomes a part of a new daughter cell. Most normal living eukaryotic cells can reproduce themselves by mitosis. Germ cells produce mature sex cells with reduced (*n*) numbers of chromosomes through meiosis. Through fertilization, a male and a female sex cell initiate reproduction of the entire organism. In development of a new organism, the 2*n* zygote replicates its genes and divides. This process continues and results in the numerous cells that make up the organism. Chromosome mechanisms of the germ cells in meiosis provide the biological basis for the Mendelian principles of segregation and independent assortment. Mitosis provides each daughter cell with the same genetic information as the mother cell. Meiosis reduces reproductive cells to haploidy, so that the chromosome number remains constant from generation to generation.

REFERENCES

Alberts, B., D. Bray, J. Lewis, M. Raff, K. Roberts, and J. D. Watson. 1983. *Molecular biology of the cell.* Garland Publishing, Inc., New York.

Dalton, A. J., and F. Haguenau. 1968. *The nucleus.* Academic Press, New York.

De Robertis, E. D. P., and E. M. F. De Robertis, Jr. 1980. *Cell and molecular biology,* 7th ed. Saunders College/ Holt, Rinehart and Winston, Philadelphia.

Donnelly, G. M., and A. H. Sparrow. 1965. "Mitotic and meiotic chromosomes of *Amphiuma.*" *J. Hered.* 56:91–98.

Du Praw, E. J. 1968. *Cell and molecular biology.* Academic Press, New York.

Dyson, R. D. 1975. *Essentials of cell biology.* Allyn and Bacon, Boston.

Mittwoch, U. 1973. *Genetics and sex differentiation.* Academic Press, New York.

Novitski, E., and S. Blixt. 1978. "Mendel, linkage, and synteny." *BioScience* 28:34–35.

Swanson, C. P., and P. Webster. 1977. *The cell.* Prentice-Hall, Englewood Cliffs, N. J.

Voeller, B. R. 1968. *The chromosome theory of inheritance.* Appleton-Century-Crofts, New York.

Zimmerman, A., and A. Forer, eds. 1981. *Mitosis/cytokinesis.* Academic Press, New York.

PROBLEMS AND QUESTIONS

3.1. Mark the true statements with a plus (+) and the false with a zero (0). (a) Skin cells and gametes of the same animal contain the same number of chromosomes. (b) Any chromosome may pair with any other chromosome in the same cell in meiosis. (c) The gametes of an animal may contain more maternal chromosomes than its body cells contain. (d) Of 10 chromosomes in a mature sperm cell, 5 are always maternal. (e) Of 22 chromosomes in a primary oöcyte, 15 may be paternal. (f) Homologous parts of two chromosomes lie opposite one another in pairing. (g) A sperm has half as many postmitotic chromosomes as a spermatogonium of the same animal.

3.2. In each somatic cell of a particular animal species, there are 46 postmitotic chromosomes. How many should there be in a (a) mature egg, (b) first polar body, (c) sperm, (d) spermatid, (e) primary spermatocyte, (f) brain cell, (g) secondary oöcyte, (h) spermatogonium?

3.3. If spermatogenesis is normal and all cells survive, how many sperm will result from (a) 50 primary spermatocytes, (b) 50 spermatids?

3.4. In humans, a type of myopia (an eye abnormality) is dependent on a dominant gene (*M*). Represent diagrammatically (on the chromosomes) a cross between a woman with myopia who is heterozygous (*Mm*) and a normal man (*mm*). Show the kinds of gametes that each parent could produce and summarize the expected results from the cross.

3.5. Beginning with the myopic woman in Problem 3.4, diagram the oögenesis process that produces the egg involved in the production of a child with myopia. Label all stages.

3.6. In what ways is cell division similar and different in animals and plants?

3.7. How does meiosis differ from mitosis? Consider differences in mechanism as well as end results.

3.8. How does gamete formation in higher plants differ from that in higher animals with reference to (a) gross mechanism and (b) chromosome mechanism?

3.9. How is double fertilization accomplished in plants, and what is the fate of the egg and the endosperm nucleus?

3.10. In humans, an abnormality of the large intestine called intestinal polyposis is dependent on a dominant gene *A*, and a nervous disorder called Huntington's chorea is determined by a dominant gene *H*. A man carrying the gene *A* (*Aahh*) married a woman carrying the gene *H* (*aaHh*). Assume that *A* and *H* are on nonhomologous chromosomes. Diagram the cross and indicate the proportions of the children that might be expected to have each abnormality, neither, or both.

3.11. Beginning with the oögonium in the woman described in Problem 3.10, diagram the steps in the process of oögenesis necessary for formation of the egg that produced an *H* child. Label all stages.

3.12. Diagram completely the process of spermatogenesis involved in the production of the sperm in Problem 3.10 necessary for the production of an *A* child. Label all stages.

3.13. A man produces the following kinds of sperm in equal proportions: *AB, Ab, aB,* and *ab.* What is his genotype with reference to the genes specified?

3.14. Would greater variability be expected among asexually reproducing organisms, self-fertilizing organisms, or bisexual organisms? Explain.

3.15. If biopsies were taken from follicle tissues of the human ovary at the following developmental periods, what stages in the process of oögenesis might be observed: (a) fifth month of intrauterine development, (b) at birth, (c) 10 years of age, (d) 17 years of age?

FOUR

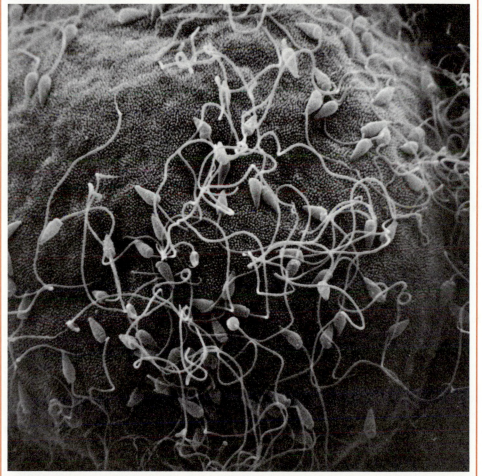

SEX DETERMINATION AND SEX LINKAGE

The first investigations relating chromosomes to sex determination were carried out at the turn of the century. H. Henking, a German biologist, discovered in 1891 that a particular nuclear structure could be traced throughout spermatogenesis of certain insects. Half the sperm received this structure and half did not. Henking did not speculate on the significance of this body, but merely identified it as the "X body" and showed that sperm differed in its presence or absence. In 1902, these observations were verified and extended by C. E. McClung, who made cytological observations on many different species of grasshoppers and demonstrated that the somatic cells in the female grasshopper carry a different chromosome number than do corresponding cells in the male. He followed the X body in spermatogenesis but did not succeed in tracing the oögenesis of the female grasshopper. McClung associated the X body with sex determination, but erroneously considered it to be peculiar to males. Had he been able to follow oögenesis, his interpretation would undoubtedly have been different.

Contributions to basic knowledge about sex determination were made in the early part of the century by E. B. Wilson and his associates. Wilson reported extensive cytological investigations on several different insects, notably from the genus *Protenor*, an uncommon group of insects closely related to the box elder bug. In these insects, different numbers of chromosomes were observed in germ cells of the two sexes. He succeeded in following oögenesis as well as spermatogenesis and found that unreduced cells of the male carried 13 chromosomes and those of the female carried 14. Some male gametes were found to carry 6 chromosomes, whereas others from the same individual carried 7. The female gametes all had 7. Eggs fertilized with 6-chromosome sperm produced males, and those fertilized with 7-chromosome sperm produced females.

MECHANISMS OF SEX DETERMINATION

The X body of Henking thus was found to be a chromosome that determined sex. It was identified in several insects and became known as the sex or X chromosome. All eggs of these insects carried an X chromosome, but it was included in only half of the cells forming sperm (Fig. 4.1). All sperm, however, had the usual complement of other chromosomes (**autosomes**). Eggs fertilized by sperm containing the X chromosome produced zygotes with **two X chromosomes,** which became **females.** Eggs receiving sperm without an X chromosome produced zygotes with **one X,** which became **males.**

Wilson observed another chromosome arrangement in the milkweed bug, *Lygaeus turcicus.* In this insect, the same number of chromosomes was present in the cells of both sexes. The homolog of the X, however, was distinctly smaller and was called the Y chromosome. XX zygotes became females and XY zygotes became males. This was called the XY type of sex determination. As evidence was accumulated from a wider variety of animals, the XY mechanism was found to be more prevalent than the XO. The XY type is now considered characteristic in most higher animals and occurs in some plants (e.g., *Melandrium album*).

We also find the XY pattern in humans; the human X chromosome is considerably longer than the Y. The total complement of human chromosomes includes 44 autosomes: XX in the female and XY in the male (Fig. 4.2). Eggs produced by the female in oögenesis have the usual complement of autosomes (22) plus an X chromosome. Sperm from the male have the same autosomal number and either an X or a Y. Eggs fertilized with Y chromosome sperm result in zygotes that develop into males; those fertilized with X chromosome sperm develop into females. Segregation of the XY pair and random fertilization thus explain, at least superficially, why some individuals develop into females and some into males, and why about half of the members of each population of higher animals are males and half are females.

Historically, the association of the most conspicuous phenotype (i.e., sex) with a particular chromosome greatly strengthened the hypothesis that genes are in chromosomes. This idea originally had been postulated largely because of the parallel observed between the behavior of chromosomes in

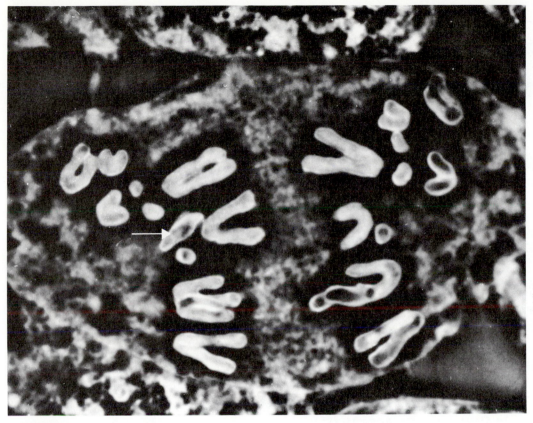

FIGURE 4.1

Photomicrograph of chromosomes of the grasshopper in the anaphase of the first division of spermatogenesis. Twelve chromosomes are at the left of the equatorial plane and 11 are at the right. The difference is the single X chromosome (arrow). Zygotes receiving X chromosomes from the sperm will become females, and those receiving sperm with no X chromosomes will become males. (Courtesy A. M. Winchester, University of Northern Colorado.)

meiosis and the Mendelian behavior of genes. Control of sex determination by a particular chromosome provided tangible evidence that **genes are in chromosomes.**

Experiments with insects formed a basis for speculation and experimentation concerning the sex-determining mechanism in higher forms. Because invertebrate hormones are not functionally comparable with steroid hormones in birds and mammals, such animals as chickens and mice were primarily used in experimental work on secondary

sex characteristics (those characteristics that distinguish the two sexes, but have no direct role in reproduction) and hormonal influences on phenotypes.

BALANCE CONCEPT OF SEX DETERMINATION

Soon after sex chromosomes were identified, it became obvious that sex determination was more complicated than preliminary observations had indicated. A more intricate mechanism than segrega-

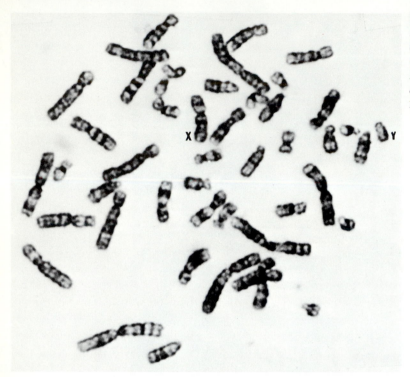

FIGURE 4.2
Micrograph showing the chromosomes of a young boy. The chromosomes were prepared with modified Giemsa (G-banding)-staining technique; X and Y chromosomes as well as the autosomes are identified by this technique. (Courtesy of S. W. Rogers, Utah State University.)

TABLE 4.1 Ratio of X Chromosomes to Autosomes and Corresponding Sex Type in *Drosophila melanogaster*[a]

X CHROMOSOMES (X) AND SETS OF AUTOSOMES (A)		RATIO X/A	SEX
1X	2A	0.5	Male
2X	2A	1.0	Female
3X	2A	1.5	Metafemale
4X	3A	1.33	Metafemale
4X	4A	1.0	Tetraploid female
3X	3A	1.0	Triploid female
3X	4A	0.75	Intersex
2X	3A	0.67	Intersex
2X	4A	0.5	Tetraploid male
1X	3A	0.33	Metamale

[a] After Bridges.

FIGURE 4.3

Calvin B. Bridges (1889–1938), American cytogeneticist who pioneered in research on chromosomes of Drosophila melanogaster. *He prepared maps of the giant salivary chromosomes of Drosophila. (Courtesy of Everett Thorpe.)*

tion of a single pair of chromosomes was in evidence. The investigations on *Drosophila* by C. B. Bridges (Fig. 4.3) showed that female determiners were located in the X chromosomes and male determiners were in the autosomes. No specific loci were identified, and the present evidence suggests that many chromosome areas are involved. It was thus shown that sex-determining genes are carried in certain chromosomes in *Drosophila,* and that all individuals carry genes for both sexes. The **genetic balance theory** of sex determination was devised as a more detailed explanation of the mechanics of sex determination.

XO or XY chromosome segregation was interpreted as a means of tipping the balance between maleness and femaleness, whereas more deep-seated processes were involved in actual determination. Bridges experimentally produced various combinations of X chromosomes and autosomes in *Drosophila* and deduced from comparisons that one X chromosome (X) and two sets of autosomes (A) had a ratio in terms of sex-determining capacity of 0.5. This combination of 1X and 2A resulted in a male (Table 4.1); 2X and 2A $= \frac{2}{2} = 1$, which produced a female.

The first irregular chromosome arrangement from Bridges' experiments resulted from **nondisjunction,** the failure of paired chromosomes to separate in anaphase. X chromosomes, which ordinarily came together in pairs in the meiotic prophase of oögenesis and separated to the poles in anaphase, remained together and went to the same pole. As a result, some female gametes received two X chromosomes and some received no X chromosomes (Fig. 4.4). Following fertilization by sperm from wild-type males (AAXY), all zygotes had $2n$ autosomes (2A), but some received two Xs from the mother and an X from the father (3X). The ratio of $3:2$ resulted in flies, called metafemales, that were highly inviable. The XXY flies (2X/2A) from the same mating were normal females in appearance; XO (1X/2A) males were sterile, while those with a Y but no X chromosomes did not survive. These results indicated that, in *Drosophila,* the Y chromosome is not involved in sex determination but it does control male fertility. In XO *Drosophila melanogaster,* sperm develop but they are nonmotile, whereas in XO *D. hydei* sperm do not develop.

Flies produced experimentally with 4X/3A were also metafemales. Those with 4X/4A and also those with 3X/3A, both with an X/A ratio = 1, were females. The combinations 3X/4A = 0.75 and 2X/3A = 0.67 were intermediate in characteristics between males and females and were called "intersexes." Combinations of 2X/4A = 0.5 were males and those of 1X/3A = 0.33 were metamales.

No other animals or plants have been investigated with equal thoroughness, but evidence suggests that some such balance is involved in many organisms. **Intersexes** can be produced experimentally in some animals by upsetting this balance during developmental stages.

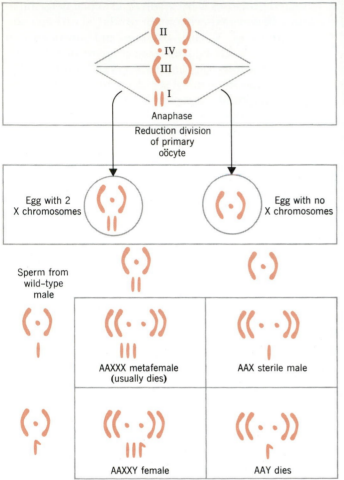

Anaphase
Reduction division
of primary
oöcyte

Egg with 2
X chromosomes

Egg with no
X chromosomes

Sperm from
wild–type
male

AAXXX metafemale
(usually dies)

AAX sterile male

AAXXY female

AAY dies

FIGURE 4.4

Nondisjunction in Drosophila, *and the zygotes that result from fertilization by wild-type males (AAXY). (The AAXXY) females and (AAX) males were the exceptional flies in Bridges' experiment. In the primary oöcyte, autosomes II and III are represented by pairs of bent rods; autosomes IV, which are small and take their places in the center of the equatorial plate, are represented by a pair of dots. The X chromosomes (I) are symbolized as short rods; the Y chromosome introduced by the sperm is illustrated by a rod with a hook or short arm. (After Bridges.)*

Y CHROMOSOME IN SEX DETERMINATION

The XX is one sex and the XY is the other. Usually, XX is female and XY male. Such females are homogametic (all gametes of one type; namely, with an X) and males heterogametic (two types of gametes, those with an X and those with a Y). Some organisms, such as birds, moths, and some fish, are the reverse—female heterogametic (XY, sometimes designated as ZW) and male homogametic (XX, sometimes designated ZZ).

In *Drosophila*, the Y does not determine sex; instead, the number of Xs relative to the number of sets of autosomes determines sex. In contrast, the **Y** determines sex in mammals, *Axolotl* (an amphibian genus), and *Melandrium* (a plant genus). Human sex determination is XY type, with female XX and male XY. When chromosome constitution is irregular (Chapter 13), however, X chromosomes alone in any number (e.g., XXX or XXXX) are associated with females. A Y is required for maleness. The Y chromosome induces development of the undifferentiated gonadal medulla into a testis, whereas an XX chromosomal complement induces the undifferentiated gonadal cortex to develop ovaries. Even in the presence of three or more X chromosomes, a single Y chromosome is usually sufficient to produce testes and male characteristics.

In *Melandrium album*, which follows the XY mode of sex determination, H. E. Warmke, M. Westergaard, and others have shown that sex is determined by a balance between male-determining genes in the Y and female-determining genes in the X and in the autosomes. In this plant, which is normally dioecious, XY individuals are staminate (i.e., pollen bearing) and XX plants are pistillate (egg bearing). The Y chromosome is the largest and most conspicuous member of the complement.

Experimental procedures using spontaneous fragmentations have resulted in mapping of major sections of the *Melandrium* Y chromosome (Fig. 4.5). Three distinct regions influencing sex determination and male fertility have been localized on the differential part of the Y chromosome (which does not have a homologous part on the X). Region I suppresses femaleness. In the absence of this region,

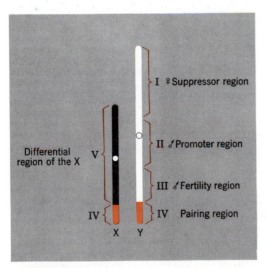

FIGURE 4.5

Sex chromosomes in Melandrium. *Regions I, II, and III of the Y chromosome do not have homologous segments in the X, and thus they make up the differential portion of the Y. Regions IV are homologous in the X and Y and are pairing regions at meiosis. Region V is the differential portion of the X chromosome. When I is lost from a Y chromosome, a bisexual plant is produced. When II is lost, a female plant is produced. If Region III is absent, male-sterile plants with abortive anthers appear. (After Westergaard,* Hereditas *34:257–279, 1948.)*

plants are bisexual; that is, they express both male and female characteristics. Region II promotes male development. When this region (with or without Region I) is missing, a female plant is produced. Region III carries male fertility genes; loss of this region results in male sterility. A part of the Y chromosome is homologous with a part of the X, but the major part of the X is differential with no structural counterpart in the Y. Although most female-determining genetic material is in the X chromosome, Westergaard found that autosomes were also involved in female determination.

X CHROMOSOME SEGMENT IN HYMENOPTERA

More involved mechanisms for sex determination have been described in the insect order, Hymenoptera, which includes ants, bees, wasps, and sawflies. In several species of Hymenoptera, **males develop parthenogenetically** (from unfertilized eggs) and have a haploid chromosome number (16 in the drone honeybee *Apis mellifera*). The **queen** honeybee and **workers,** which arise from **fertilized eggs,** carry the diploid chromosome number (32). In the parasitoid wasp *Bracon hebetor* (formerly *Habrobracon juglandis*), females are diploid with 20 chromosomes, and males are haploid with 10 chromosomes. Females originate from fertilized eggs and males from unfertilized eggs.

Results of experiments by Whiting showed that homozygous, heterozygous, or hemizygous (gene present in single dose) status of certain chromosome segments controls sex determination. Female determination depends on heterozygosity for part of a chromosome. Therefore, xaxb, xaxc, or xbxc are all female. Haploid individuals (xa, xb, or xc) cannot be heterozygous (but are hemizygous) and are therefore male. Whiting showed that sex determination depends on this region of the chromosome, and not on diploidy versus haploidy per se, by producing homozygous diploid males (xaxa, xbxb, or xcxc).

SINGLE GENES AND SEX DETERMINATION

Sex determination in some organisms is influenced by the differential action of a single gene. Maize, for example, is **monoecious** (both sexes in the same plant), having staminate flowers in the tassel and pistillate flowers in the developing ear. A substitution of two single gene pairs makes the difference between monoecious and **dioecious** (separate sexes) plants. The allele for barren plant (*ba*), when homozygous, makes the stalk staminate by eliminating silks and ears, whereas the allele for tassel seed (*ts*), when homozygous, transforms the tassel into a pistillate structure that produces no pollen. A plant of the genotype *ba*/*ba ts*/*ts* lacks silks on the stalk, but has a transformed tassel and is therefore only pistillate (female). A plant with *ba*/*ba ts*$^+$/*ts*$^+$ is only staminate (male). These data suggest how monoecious plants could become dioecious, and vice versa, by the alteration (mutation or recombination) of just two genes: in this case, *ba*$^+$ to *ba* and *ts*$^+$ to *ts*.

EXTERNAL ENVIRONMENT AND SEX DETERMINATION

In some lower animals, sex determination is nongenetic and depends on factors in the external environment. Males and females have similar genotypes, but stimuli from environmental sources initiate development toward one sex or the other. Males of the marine worm *Bonellia*, for example, are small and degenerate and live within the reproductive tract of the larger female (Fig. 4.6). All organs of his body are degenerate except those of the reproductive system. F. Baltzer found that any young worm reared from a single isolated egg became a female. If he released newly hatched worms into water containing mature females, however, some young worms were attracted to females and became attached to the female proboscis. These were transformed into males and eventually migrated to the female reproductive tract where they became parasitic. Genetic determiners for both sexes are apparently present in young worms. Extracts made from the female proboscis influence young worms toward maleness.

GYNANDROMORPHS

Abnormal chromosomal behavior in insects can result in **sexual mosaics** called gynandromorphs. Some parts of the animal express female character-

istics while other parts express those of the male. Some gynandromorphs in *Drosophila* are bilateral intersexes (Fig. 4.7), with male color pattern, body shape, and sex comb on one half of the body and female characteristics on the other half. Both male and female gonads and genitalia may be present.

Bilateral gynandromorphs have been explained on the basis of an irregularity in mitosis at the first cleavage of the zygote. Infrequently, a chromosome lags in division and does not arrive at the pole in time to be included in the reconstructed nucleus. When one of the **X chromosomes** of an XX (female) zygote **lags** in the spindle, one daughter nucleus receives only one X chromosome, while the other receives XX, as illustrated in Fig. 4.8. A mosaic pattern is thus established. One nucleus in the two-nuclei stage would be XX (female) and one would be XO (male). If the cleavage plane is oriented so that one daughter nucleus goes toward the right, that nucleus will give rise to all cells that make up the right half of the adult body and the other gives rise to the left half. If the same chromosome loss occurred at a later cell division, smaller proportions of the adult body would be male. Position and size of the mosaic sector are determined, therefore, by place and time of the division abnormality. In *Bracon hebetor*, gynandromorphs may occur in the anterior–posterior plane, giving rise to wasps with such peculiar arrangements as male heads and female abdomens and female heads with male abdomens.

SEX DIFFERENTIATION

SEX CHROMATIN BODIES

M. L. Barr observed **chromatin bodies** in the nerve cells of female cats that were not present in cells of the male. Barr and others reported a consistent cytological difference between the nuclear contents of human male and female cells in several tissues, including epithelium of buccal mucosa (lining inside of the mouth). With appropriate staining techniques, a small chromatin body could be identified in the nucleus of female cells. This **Barr body** was related to the number of X chromosomes and was

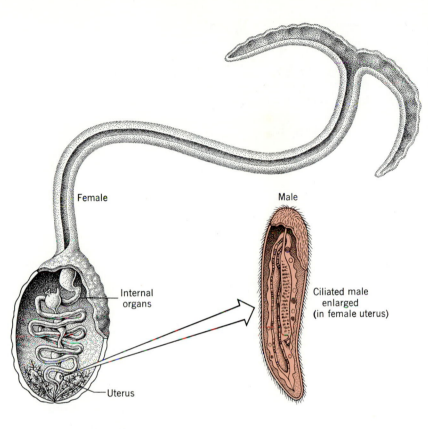

FIGURE 4.6
Female and male of the marine worm
Bonellia viridis. *The male is shown in
the uterus of the female and is greatly
enlarged at the right to show details of
internal structure. (Redrawn from Dobzhansky,* Evolution, Genetics and
Man, *1955, John Wiley & Sons.)*

not observed in normal male cells. Nuclei from cells of the two sexes are illustrated in Fig. 4.9. With this technique, the sex of human embryos can be distinguished at early stages of development. More recent studies have shown that most if not all tissues of female mice have some cells with Barr bodies. This cellular characteristic generally seems to apply to mammals.

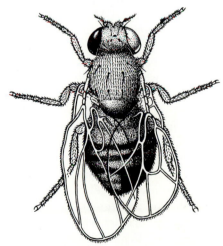

FIGURE 4.7
Bilateral gynandromorph in Drosophila, *showing male characteristics including sex comb on the right foreleg. The white eye on the right is the expression of the sex-linked gene* w *when carried in one (hemizygous) X chromosome. (After Morgan.)*

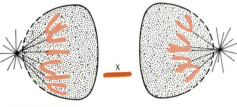

FIGURE 4.8
A lagging X chromosome in the first cleavage of Drosophila *illustrating the origin of a bilateral gynandromorph. (After Morgan and Bridges.)*

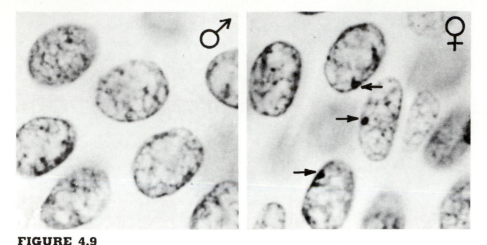

FIGURE 4.9
Cells of human epidermis illustrating the sex chromatin bodies or Barr bodies. (Right)
Epidermal cells of a female showing sex chromatin bodies (see arrows). (Left) Epidermal
cells of a male showing no sex chromatin bodies. (Courtesy of M. L. Barr.)

Sex chromatin bodies not only distinguish normal female cells from male cells; they are also useful for diagnosing various kinds of sex chromosome abnormalities in people. Those who have two or more X chromosomes have **one less chromatin body than the number of X chromosomes present.** Cells of abnormal females with only one X chromosome (Turner's syndrome) have no sex chromatin bodies, and cells of males with two X chromosomes and one Y chromosome (Klinefelter's syndrome) have one sex chromatin body (Chapter 13). Abnormal females with three X chromosomes have two sex chromatin bodies in their cell nuclei.

DOSAGE COMPENSATION

For many years, geneticists have observed that, in some cases, females homozygous for genes in the X chromosomes do not express a trait more markedly than do hemizygous males. Thus, there must be a mechanism of "dosage compensation" through which the effective dosage of genes of the two sexes is made equal or nearly so. Several investigators arrived almost simultaneously at a hypothesis, explaining that the sex chromatin body and dosage compensation in mammals is determined by the inactivation of one X chromosome in the normal

female. The hypothesis was named after Mary F. Lyon (Fig. 4.10), who first formulated it in detail from cytological observations and genetic studies on mouse coat color. Female mice, heterozygous for certain coat-color genes, show a mottled effect unlike the homozygote and the intermediate color pattern. The fur pattern (Fig. 4.11) is a **mosaic** made up of **randomly arranged patches** of the two colors. Normal male mice and abnormal XO female mice never have the mottled phenotype.

The Lyon hypothesis was based on the observation that the number of sex chromatin bodies in interphase cells of adult females is one less than the number of X chromosomes observed in metaphase preparations. It is therefore proposed that the chromatin body is a heterochromatinized X chromosome. On this premise, it must be assumed that only one X chromosome is required for cellular metabolism in the female cell. Any additional Xs become heteropyknotic (i.e., stain more densely than other chromosomes) and are genetically inactive. Thus, the mammalian X chromosome has the capability of being heterochromatic (see Chapter 5) in some cells and euchromatic in others.

Which X chromosome becomes inactive is a matter of chance; but once an X has become inacti-

FIGURE 4.10

Mary F. Lyon, British geneticist who with others formulated the Lyon hypothesis to explain the sex chromatin body and gene dosage compensation in mammals. The hypothesis is based on the inactivation of one X chromosome in the normal female. (Courtesy of Godfrey Argent.)

FIGURE 4.11

Female mouse heterozygous for an X-linked gene that controls the coat color tortoise-shell. The mosaic phenotype of such females provided one of the first lines of evidence for the Lyon hypothesis. (From M. W. Thompson. Reproduced by permission of the National Council of Canada from the Canadian Journal of Genetics and Cytology, *volume 7, 1965.)*

vated, all cells arising from that cell will keep the same inactive X chromosome (a sex chromatin body). In the mouse, the inactivation apparently occurs early in development; and in human embryos, sex chromatin bodies have been observed by the sixteenth day of gestation. Some human traits could therefore be influenced by both X chromosomes during the first 16 days. Later (after the sixteenth day), only one X is functional in a body cell. The female is mosaic with some parts having the alternate allele expressed. X chromosome inactivation occurs only when at least two X chromosomes are present. When several X chromosomes are in the same nucleus, all but one are inactivated. The number of sex chromatin bodies after the period of inactivation is, therefore, one less than the number of X chromosomes in the original cell.

The Lyon hypothesis explains certain genetic consequences of genes in mammals: (1) the dosage compensation for females with two X chromosomes that regulates enzyme activity to the level of males with only one X chromosome and (2) the variability of expression in heterozygous females because of the random inactivation of one or the other X chromosome.

Support for the hypothesis that, in mammalian females, dosage compensation occurs through inactivation of all but one X chromosome, thus producing functional equivalence in sexes, came from studies of glucose-6-phosphate dehydrogenase (G-6-PD). **Enzymatic activity** was shown to be about equal in the two sexes and at the level of the male with one X chromosome. Two alleles of the G-6-PD locus, which produce electrophoretically distinct enzymes (F and S), were tested with heterozygous females. Isolated cells from several biopsies of skin from the same heterozygous person were cloned. Each clone contained either the F or S form of the enzyme but never both. This supports the premise that an X chromosome received from either mother or father (of the female being tested) was inactive. After inactivation has occurred, a clone of cells is uniform for the expression of only one allele.

HORMONES AND SEX DIFFERENTIATION

The hormonal system that regulates the internal or physiological environment of the organism does not directly influence the fundamental process of sex determination. It is important, however, to the development of the **secondary sex characteristics.** Sex hormones of higher animals are synthesized by the ovaries, testes, and adrenal glands with stimulation from pituitary gland hormones. The adrenals produce steroids that are chemically related to those of the gonads and that also influence the secondary sex characteristics. Ovaries and testes each have a dual function; they are responsible for the production of the gametes, the eggs and sperm, as well as the sex hormones. These hormones influence the development of secondary sexual characteristics such as physiological differences (rate of metabolism, blood pressure, heart beat, and respiration), bone structure, voice, breast development, and hair.

SEX-INFLUENCED DOMINANCE

Dominance of alleles may differ in heterozygotes of the two sexes. This phenomenon is called "sex-influenced dominance." Gene products of heterozygotes in the two sexes may be influenced differentially by sex hormones. Autosomal genes responsible for horns in some breeds of sheep, for example, may behave differently in the presence of the male and female **sex hormones.** The results are equally well explained by secondary effects due to differences in programing. More than a single pair of genes is involved in the production of horns; but assuming all other genes to be homozygous, the example can be treated as if only a single pair were involved. Among Dorset sheep, both sexes are

horned, and the gene for the horned condition is homozygous (h^+h^+). In Suffolk sheep, neither sex is horned, and the genotype is hh. Among the heterozygous F_1 progeny from crosses between these two breeds, horned males and hornless females are produced. Because both sexes are genotypically alike (h^+h), the gene must behave as a dominant in males and as a recessive in females; that is, only one allele is required for an expression in the male, but the allele must be homozygous for expression in the female.

When F_1 hybrids are mated, a ratio of 3 horned to 1 hornless is produced among F_2 males, whereas a ratio of 3 hornless to 1 horned is observed among the F_2 females. Genotypes and phenotypes of the two sexes are summarized in Table 4.2. The only departure from the usual pattern is concerned with the heterozygous (h^+h) genotype. This genotype in the male sheep results in the horned condition, but females with the same genotype are hornless. Dominance of the gene is apparently influenced by sex hormones.

TABLE 4.2   Expression of h Alleles in Sex-influenced Inheritance

GENOTYPES	MALES	FEMALES
h^+h^+	Horned	Horned
h^+h	Horned	Hornless
hh	Hornless	Hornless

Some human traits, such as a certain type of white forelock, absence of upper lateral incisor teeth, a particular type of enlargement of the terminal joints of the fingers, and premature pattern baldness (Fig. 4.12), have been reported to follow this mode of inheritance (Fig. 4.13).

SEX-LIMITED GENE EXPRESSION

One sex may be uniform in expression of a particular trait and yet transfer genes that produce a different phenotype in offspring of the other sex. This is called "sex-limited gene expression." **Sex hormones** are apparently **limiting factors** in the expressions of some genes. More basic factors are also

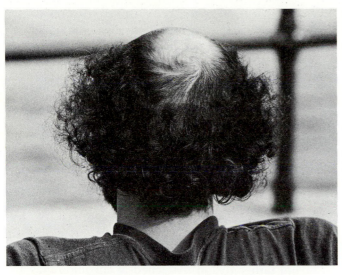

FIGURE 4.12

Young man with premature balding beginning on the crown of his head. (Ira Kirschenbaum/Stock, Boston.)

involved with sex-limited characteristics. Milk production in mammals, for example, is limited to females, but certain bulls are in great demand among dairy breeders and artificial insemination associations because their mothers and daughters have increased milk-production records.

SEX LINKAGE

When parallelism was discovered between the X chromosome cycle and sex determination, it was generally assumed that genes other than sex determiners were also located in the X chromosome. Indeed, many mutant genes such as the Xg^a blood alleles in humans have been identified in the X chromosome. Because of their location in the same chromosome as sex determiners, they are said to be "sex-linked." The first extensive experimental evidence for sex linkage in a particular species came in 1910 with the discovery by T. H. Morgan of a white-eye mutant in *Drosophila.* A gene had undergone a change that resulted in a phenotypic alteration. This change expressed itself as white eyes rather than the normal red eyes. The white-eyed male first discovered was mated with a red-eyed female. The F_1 flies were all red-eyed, but the F_2 included both red and white in the proportion of about 3 red to 1 white. All white-eyed flies in the F_2 generation, however, were males. About half of the F_2 males had white eyes and half had red, but all females had red eyes. In this experiment, the recessive allele was expressed only in males. Morgan arrived at an explanation by associating this **gene** with the **X chromosome,** as illustrated in Fig. 4.14.

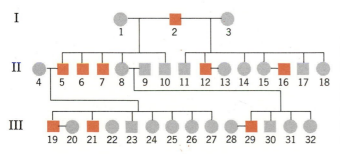

FIGURE 4.13

Pedigree showing the incidence of premature baldness in a family group. The men represented by the darkened squares were bald-headed before they reached the age of 35. Those symbolized by light squares were over 35 when the study was made and all had thick hair. No women in this family group expressed the trait.

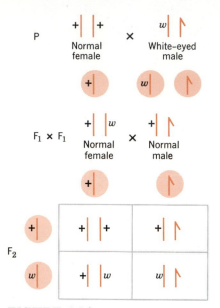

FIGURE 4.14

A cross illustrating sex linkage in Drosophila. *This cross is between a red-eyed female and a white-eyed male. Alleles are identified with their chromosomes. The* ⌐ *represents a Y chromosome.*

Because the male fly had only one X chromosome and a Y chromosome that lacked most genes of the X, it was postulated that a hemizygous allele for white eyes was expressed. Furthermore, the mutant allele present in the X chromosome of the original white-eyed male was passed on to his daughters (he transmitted a Y chromosome to his sons). All the daughters therefore were carriers for the allele. The F_2 **hemizygous males** obtained their X chromosomes from their heterozygous mothers. Half received the w^+ allele and developed red eyes and half received the w allele and developed white eyes. The equal proportion of red-eyed and white-eyed F_2 males was thus explained on the basis of segregation of X chromosomes from the F_1 mothers.

Could white-eyed females occur? On the basis of his hypothesis that the gene was carried in the X chromosome, Morgan predicted that a female of genotype ww could be produced and would have white eyes. This was tested experimentally with crosses between males with white eyes and F_1 (ww^+)

females with red eyes. From these crosses, half the females as well as half the males had white eyes, as predicted. Later studies identified many other genes of *Drosophila* on the X chromosome.

Most sex-linked genes in male heterogametic animals are in the X chromosome. Some animals, however, may carry a few genes in the Y chromosome that have visible effects. Such "Y-linked" or **holandric** genes would be transmitted directly from father to son and never appear in females. Since all sex-linked genes are on the same chromosome, they are linked with each other as well as with sex-determining genes.

Further *Drosophila* studies with attached-X chromosomes ($\widehat{XX}$) showed that X chromosomes with X-linked genes can be transmitted directly from father to son. Attached-X chromosomes are aberrant compound X chromosomes possessing a fused centromere. When attached-X females are mated with normal XY males, the $\widehat{XX}$ female gametes fertilized with Y male gametes result in $\widehat{XX}$Y females (the Y is not a male sex determiner in *Drosophila*). Male X chromosome gametes that are fertilized with female gametes lacking an X chromosome result in XO-sterile males (see Fig. 4.4) that express the X-linked genes of the father.

In the absence of dominant alleles, a recessive allele, such as the one (w) responsible for white eyes in *Drosophila,* can express itself. The crisscross pattern of inheritance, which is characteristic of sex-linked genes, means that traits appearing in males are transmitted (unexpressed) through their daughters to the males in the next generation, where they are expressed. Cytological studies on the nature and behavior of chromosomes have supported the genetic interpretation of sex linkage. In *D. melanogaster,* the X and Y chromosomes can readily be identified by their appearance. The X is rod-shaped with the centromere near one end, whereas the Y is hooklike, having a long and a short arm.

Although the Y chromosome is composed of heterochromatin and essentially is devoid of genes, one small part of the short arm in *D. melanogaster* has a homologous section in the X. The allele (bb) for bobbed bristles in the X chromosome, or its wild-type allele (bb^+), may occur in the short arm of the Y chromosome. Genes such as *bb* with a locus in the Y chromosome as well as in a homologous part of the X are said to be incompletely sex-linked. The *Drosophila* Y chromosome also contains at least four fertility factors beside the bobbed region.

Like the Y chromosome in *Drosophila,* supernumerary chromosomes in some species of animals and plants (e.g., mealy bugs and maize) are composed mostly of **heterochromatin.** Sometimes the heterochromatin regions of different chromosomes in the same cell coalesce and form an amorphous "chromocenter." This occurs in the giant salivary gland chromosomes of *D. melanogaster,* as shown in Fig. 4.15. The usual metaphase configuration observed in other cells of this species is shown in Fig. 4.15*e* with heterochromatin shaded. Giant chromosomes from a cell of a male larva are shown in Fig. 4.15*a*. The heterochromatin near the centromere of each autosome and the X chromosome, along with that making up nearly all the Y chromosome, compose the chromocenter (Fig. 4.15*c*).

In summary, the Y chromosome is mostly heterochromatin and has only a few genes. The X chromosome, with many genes, and the Y with virtually none establishes the characteristic pattern for sex linkage. The mother, with two X chromosomes, transmits one X to each gamete. Zygotes that receive a Y chromosome from the father develop into males. These hemizygous male progeny express the sex-linked genes from the mother. Zygotes that receive two X chromosomes, one from each parent, develop into females. The expression of sex-linked traits in females follows the same pattern as autosomal traits, with the recessive phenotype appearing only in homozygotes.

SEX LINKAGE IN HUMANS

The inheritance pattern associated with sex linkage is so obvious that it is ideal for genetic studies. This was, in fact, the first pattern of inheritance to be recorded for humans. Early Greek philosophers noticed that some human traits tended to skip a generation. An inherited characteristic was observed in a father but not in any of his children, either male or female, and then it would reappear in males of the next generation. This distinctive crisscross pattern,

from father through daughter to grandson, replacing the usual pattern for the F_1 and F_2 generations, is now interpreted as evidence of sex linkage. Since humans are not subjected to experimental procedures, the characteristic inheritance pattern in family groups that can be illustrated in pedigree charts (Fig. 4.16) is the standard means of detecting sex-linked genes.

Sex linkage has now been indicated for more than 200 traits in humans, including, in addition to those already mentioned, such important and distinctive traits as optic atrophy (degeneration of the optic nerve), juvenile glaucoma (hardening of the eyeball), myopia (nearsightedness), defective iris, epidermal cysts, distichiasis (double eyelashes), white occipital lock of hair, mitral stenosis (abnormality of mitral valve in the heart), and one form of mental retardation (Fig. 4.17). Some of these traits have alternative forms that are dependent on autosomal genes.

Although the pattern now associated with sex linkage was observed in human pedigrees many years ago, the understanding of the genetic mechanism was a direct consequence of Morgan's experimental work with the white-eyed mutant in *Drosophila.* The explanation given for sex-linked inheritance in *Drosophila* applies equally to traits in humans that are associated with genes in the X chromosome.

Several kinds of defective color vision have now been identified, and the genetic mechanisms are more complex than at first suspected. Here, only the protan defect (Fig. 4.16) will be considered. It will be treated as a single sex-linked recessive allele, without reference to other genes which influence color vision. A man defective in green color vision has a single recessive allele (g) in his X chromosome. Since the Y chromosome carries no color vision locus, the single allele is expressed, causing color blindness. If the wife of this man is homozygous for the dominant allele (g^+) for normal color vision (Fig. 4.16), all their daughters will receive an X chromosome from the mother carrying g^+ and a g from their father and will be heterozygous. Sons, with only one X chromosome, will have only one allele (g^+) from the mother and will be free from green

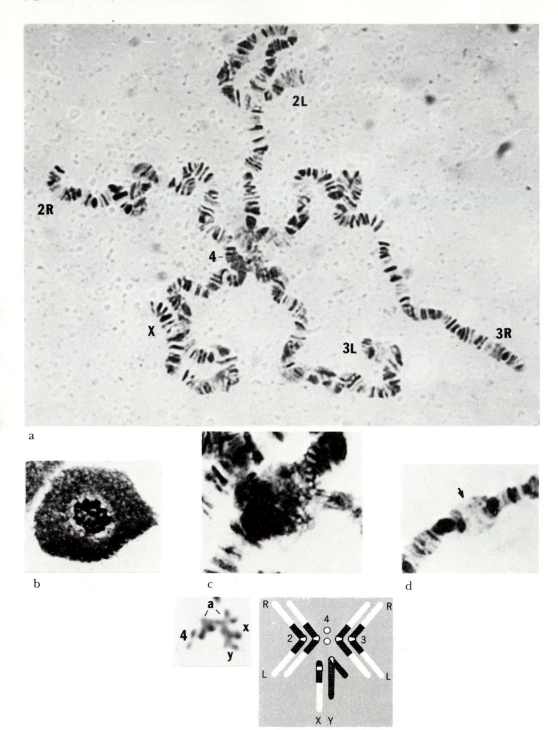

a

b

c

d

e

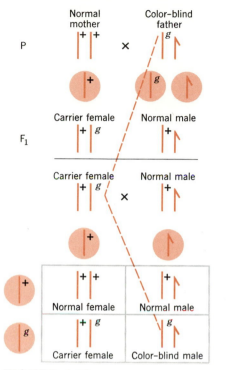

FIGURE 4.16

Crisscross inheritance from father to daughter to grandson (of original father) that is characteristic of a sex-linked gene. Genes are shown on the chromosomes illustrating a cross between a woman with normal vision and a green (protan) color vision defective man. The symbol g represents the sex-linked recessive gene for green color defective vision.

defective color vision. The Y chromosome, carrying no allele for this trait, will be contributed by the father only to his sons. In the next generation, about half of the sons of the carrier females will be normal and half will be color defective because the X chromosome carrying g^+ will segregate to about half of the heterozygous mother's gametes, and the other half will carry g. If the father's are normal (g^+), half of the daughters of carrier mothers will be carriers, not expressing the trait.

Segregation of X chromosomes and expression of single recessive alleles in hemizygous males explain the higher incidence of green color defective males than females. About 5 percent of Caucasian men in the United States, but less than 1 percent of the Caucasian women, are green color defective, since females must receive two recessive alleles. Color-defective people occur in all human populations. The gene frequency, however, varies among those of different ancestral groups. Only about 4 percent of Negro men in the United States are reported to be green color defective.

The criteria for identifying sex-linked (X-linked) recessive traits from pedigree studies may be summarized as follows: (1) the trait occurs much more frequently in males than in females, (2) traits are transmitted from an affected man through his daughters to half his grandsons, (3) an X-linked allele is never transmitted directly from father to son, and (4) all affected females have an affected father and a carrier or affected mother. Criteria for identifying sex-linked genes are consequences of the definition of sex-linked.

FIGURE 4.15

Chromosomes of Drosophila melanogaster. (a) *Polytene chromosomes from salivary gland of mature larva with chromosomes and chromosome arms identified ($\times 400$),* (b) *Cell from salivary gland of mature larva. Note polytene chromosomes condensed in nucleus ($\times 300$).* (c) *Chromocenter of polytene chromosomes; same as center area of chromosome spread above, but stained here to show chromocenter ($\times 1100$).* (d) *Salivary chromosome puff marked by arrow ($\times 1100$). (Left,* e) *Photograph of metaphase chromosomes from neuroblast cell of male larval brain:* a *represents the large autosomes, 4 represents the small fourth chromosomes dividing,* X *the X chromosome,* Y *the Y chromosome. (Right) Diagram of idealized metaphase somatic chromosomes, with chromosomes and chromosome arms identified. (Courtesy of S. W. Rogers.)*

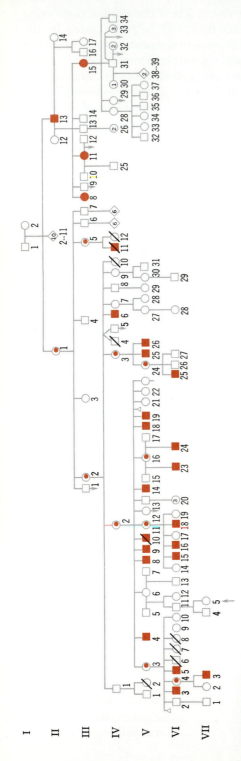

FIGURE 4.17

Pedigree of a family group in which a particular kind of mental retardation follows the pattern of recessive sex-linked inheritance. Darkened squares and circles represent mentally retarded males and females, respectively. Circles with dark centers represent carrier females, about half of whose sons are mentally retarded and half of whose daughters are carriers of the mental retardation gene. Original parents, I-1 and I-2, are unknown, but I-2 is presumed to have been a carrier for mental retardation. From Lehrke, R. G. Clinical Studies of x-linked mental retardation, in Bergsma, D., (ed.) x-linked mental retardation and verbal disability. Miami: Symposia Specialists for the National Foundation March of Dimes BD: OASX(1): 41–70, 1974.

For dominant X-linked traits, such as the rare blood type *Xg^a*, males expressing the trait would be expected to transmit it to all of their daughters but none of their sons. Heterozygous females would transmit the trait to half of their children of either sex. If a female expressing the trait is homozygous, all her children will inherit it. Sex-linked dominant inheritance cannot be distinguished from autosomal inheritance in the progeny of females expressing the trait, but only in the progeny of affected males.

MALES EXPRESS SEX-LINKED RECESSIVE GENES

Congenital hyperuricemia (**Lesch-Nyhan syndrome**), characterized by excess production of uric acid, is inherited through a **sex-linked recessive gene.** This means that the mother contributes the X chromosome with the defective gene to a male zygote. Half of the male children of carrier mothers may be expected to inherit the disease. These are deficient for the enzyme hypoxanthine-guanine phosphoribosyltransferase (**HGPRT**). This enzyme is involved in nucleotide synthesis. Infants who receive the gene appear normal at birth and for several months, but may show symptoms of excessive uric acid in the urine, such as orange "sand" (uric acid crystals). By about 10 months of age, they may become abnormally irritable and lose motor control. Weak and flabby muscles prevent the child from sitting, walking, or speaking normally. By the second year of life, the nervous condition has progressed to a degree that self-mutilation occurs, manifested by lip-biting, finger-chewing, teeth-grinding, and marked swinging of the arms. Death, which is usually secondary to severe renal and neurological damage, usually occurs within a few years, but some victims live into their 20s. Affected features as well as sex can be detected by amniocentesis.

Juvenile muscular dystrophy also depends on a **sex-linked recessive gene.** If the mother is known to be a carrier for this gene, either from her pedigree or through tests that are available, about half of her male children are expected to be affected. Male fetuses can be identified by a chromosome study. Juvenile muscular dystrophy afflicts males, usually before they reach teen age, with muscular deterioration that progresses rapidly during the early teen years. Muscles of the legs and shoulders become stiff, and the children usually become paralyzed and crippled during their middle or late teens. Virtually all die before age 21. All female children born to a carrier mother are expected to be normal, since the possibility for their being homozygous for a sex-linked recessive gene is virtually nonexistent.

Another severe disease following the pattern of **X-linked recessive** inheritance is the **Hunter syndrome.** It is characterized by mental retardation, coarse features, hirsutism (abnormal hairiness), and a characteristic facial appearance that includes a broad bridge of the nose and a large protruding tongue. Symptoms appear in early childhood. A chemical means of diagnosing this condition is being developed. Certain constituents in the amniotic fluid indicate the presence of this disease, which is associated with an abnormal processing of mucopolysaccharides synthesized in early pregnancy. Mucopolysaccharides also accumulate in skin cells of persons who are homozygous for the gene for Hunter syndrome. When amniotic or skin cells are grown in culture and stained with *o*-toluidine blue, any mucopolysaccharide cell inclusions will be stained pink. It is thus possible to identify a heterozygous carrier of the gene as well as an affected fetus.

The pattern of mucopolysaccharide (containing an amino sugar as well as uronic acid units) metabolism by Hunter cells is so strikingly different from the normal that it can be used along with chromosome analysis for sex determination in prenatal diagnosis, a situation in which clinical observation is obviously impossible. Of the many cell types originally present in amniotic fluid, fetal fibroblasts are the only ones to multiply in culture. Like fibroblasts from skin biopsies, they show an excessive accumulation of mucopolysaccharide or stainable cell inclusions if the fetus is affected with the Hunter syndrome.

Y CHROMOSOME LINKAGE IN HUMANS

Published pedigrees have indicated that the human Y chromosome has distinctive genes controlling holandric inheritance that are transmitted exclusively through the male line. Pedigree analysis for transmission from father to son provides the only evidence for holandric genes. Although pedigrees have been interpreted to show the holandric pattern for hairy pinna of the ear, other pedigrees clearly show autosomal inheritance for that trait. A histocompatibility gene (*H-Y*), however, has recently been mapped on the short arm of the human Y chromosome. Few Y-linked genes have been found in experimental animals. The pedigree pattern is so obvious that if there were many Y-linked genes, more would have been discovered. The conclusion is that very few genes are located on the Y chromosome.

SUMMARY

Investigations of sex determination have shown that the embryo is bipotential, having the ability to develop into either sex. Determination for one sex or the other is usually accomplished by a balance between genetic factors for maleness and those for femaleness. Several different combinations involving chromosomes, genes, cytoplasm, and hormones have been associated with this balance, particularly in the secondary sex characteristics. Hormones influence the expressions of some genes. Sex mosaics (gynandromorphs), resulting from irregularity in an early cell cleavage of insects, have female characteristics in one part and male characteristics in another part of the body. Sex chromatin bodies that result from the inactivation of one X chromosome in normal female mammals are useful for determining the genetic sex of abnormal fetuses and intersex individuals. Cells of normal males have no sex chromatin bodies. The number of sex chromatin bodies in individuals with more than two X chromosomes is one less than the number of X chromosomes. Genes other than sex determiners are also located in sex chromosomes. They behave according to the segregation pattern of these chromosomes and are sex-linked.

REFERENCES

Barr, M. L. 1960. "Sexual dimorphism in interphase nuclei." *Am. J. Hum. Genet.* 12:118–127.

Bridges, C. B. 1925. "Sex in relation to chromosomes and genes." *Am. Nat.* 59:127–137.

Koo, G. C., S. S. Wachtel, K. Krupen-Brown, L. R. Mittl, W. R. Breg, M. Genel, I. M. Rosenthal, D. S. Borgaonkar, D. A. Miller, R. Tantravahl, R. R. Schreck, B. F. Erlanger, and O. J. Miller. 1977. "Mapping the locus of the H-Y gene on the human Y chromosome." *Science* 198:940–942.

Lyon, M. F. 1962. "Sex chromatin and gene action in the mammalian X-chromosome." *Am. J. Hum. Genet.* 14:135–148.

Mittwoch, U. 1973. *Genetics and sex differentiation.* Academic Press, New York.

Moore, K. L. 1966. *The sex chromatin.* W. B. Saunders Co., Philadelphia.

Morgan, T. H., and C. B. Bridges. 1916. *Sex-linked inheritance in Drosophila.* Carnegie Inst. Wash. Publ. 237:1–22.

Wachtel, S. S. 1977. "H-Y antigen and the genetics of sex determination." *Science* 198:797–799.

Whiting, P. W. 1945. "The evolution of male haploidy." *Quart. Rev. Biol.* 20:231–260.

PROBLEMS AND QUESTIONS

4.1. What difference exists between male- and female-determining sperm in animals with heterogametic males?

4.2. In line with Bridges' genetic balance theory for sex determination, what is the expected sex of individuals with each of the following chromosome arrangements: (a) 4X4A, (b) 3X4A, (c) 2X3A, (d) 1X3A, (e) 2X2A, and (f) 1X2A?

4.3. List the expected results in terms of sex and intersex combinations from a cross between a triploid (3*n*) female fly, with two X chromosomes attached and one free, and a normal diploid male. (Assume that the cross is successful and the gametes of the female will carry one or two whole sets of autosomes.)

4.4. In plants of the genus *Melandrium,* which sex will be determined by the following chromosome arrangements: (a) XY, (b) XX, (c) XY with Region I removed, and (d) XY with Region II removed?

4.5. What sex is expected for individuals of the following genotypes in *Bracon hebetor:* (a) xb, (b) xaxb, (c) xcxc, and (d) xbxc?

4.6. How could maize plants, which are ordinarily monoecious, give rise to plants that are dioecious?

4.7. How many sex chromatin bodies are expected to occur in cell nuclei with each of the following chromosome arrangements: (a) XY, (b) XX, (c) XXY, (d) XXX, (e) XXXX, and (f) XYY?

4.8. In sheep, the gene h^+ for the horned condition is dominant in males and recessive in females. If a hornless ram were mated to a horned ewe, what is the chance that an (a) F_2 male sheep will be horned? (b) F_2 female sheep will be horned?

4.9. In chickens, the gene h, which distinguishes hen feathering from cock feathering, is sex-limited. Males may be hen-feathered or cock-feathered, but females are always hen-feathered. If a cock-feathered male (hh) were mated to a homozygous (h^+h^+) hen-feathered female, what patterns of feathering might be expected among the (a) male F_2 and (b) female F_2 progeny?

4.10. In a particular species of grasshoppers, two pairs of autosomes are heteromorphic; that is, they can be distinguished by microscopic observation. In one pair, one homolog is rod-shaped and the other has a small hook at the end. One member of the other pair has a knob on one end. List all distinguishable combinations, with reference to these two pairs, that can be found in the sperm.

4.11. In *Drosophila,* the recessive gene (bb) for bobbed bristles is located in the X chromosome. The Y chromosome of *Drosophila* carries a homologous section in which bb or its allele, bb^+, may be located. Give the genotypes and phenotypes of the offspring from the following crosses: (a) $X^{bb}X^{bb} \times X^{bb}Y^{bb+}$, (b) $X^{bb}X^{bb} \times X^{bb+}Y^{bb}$, (c) $X^{bb+}X^{bb} \times X^{bb+}Y^{bb}$, (d) $X^{bb+}X^{bb} \times X^{bb}Y^{bb+}$.

4.12. Consider two different sex-linked traits, one recessive and one dominant, with equal effect on viability and equal frequency in the same population. If males are heterogametic, would the recessive trait or the dominant trait be expressed more frequently in (a) males or (b) females?

4.13. If a white-eyed male fruit fly should occur in a culture of red-eyed flies, how could the investigator obtain evidence to answer the following questions: (a) Is a mutant gene or an environmental change responsible for the new phenotype? (b) If a mutation has occurred, is it sex-linked? (c) Can white-eyed females occur?

4.14. The gene (w) for white eyes in *D. melanogaster* is recessive and sex-linked; males are heterogametic. (a) Symbolize on the chromosomes the genotype of a white-eyed male, red-eyed male, red-eyed female (two genotypes), and white-eyed female. (b) Diagram on the chromosomes a cross between a homozygous red-eyed female and a white-eyed male. Carry through the F_2 and summarize the expected sex and eye-color phenotypes. (c) Diagram on the chromosomes and give the expected phenotypes from a cross between F_1 females and (i) a white-eyed male and (ii) a red-eyed male.

4.15. In humans, green defective color vision results from the sex-linked recessive gene (g) and normal vision from its allele (g^+). A man (a) and woman (b), both of normal vision, have the following three children, all of whom are married to people with normal vision: a color-defective son (c) who has a daughter of normal vision (f), a daughter of normal vision (d) who has one color-defective son (g) and two normal sons (h), and a daughter of normal vision (e), who has six normal sons (i). Give the probable genotypes of all the individuals in the family (a to i).

4.16. If a mother carried the sex-linked gene for protan defective color vision and the father was normal, would their sons or daughters be defective in color vision?

4.17. If a father and son are both defective in green color vision, is it likely that the son inherited the trait from his father?

4.18. Diagram on the chromosomes a cross between a normal woman whose father was defective in green color vision and a color-defective man. Summarize the expected results for sex and eye condition.

4.19. In humans, the gene (h) for hemophilia is sex-linked and recessive to the gene (h^+) for normal clotting. Diagram on the chromosomes the genotypes of the parents of the following crosses and summarize the expected phenotypic ratios resulting from the crosses: (a) hemophiliac woman $\times$ normal man, (b) normal (heterozygous) woman $\times$ hemophiliac man, (c) normal (homozygous) woman $\times$ hemophiliac man.

4.20. A normal woman, whose father had hemophilia, married a normal man. What is the chance of hemophilia in their children?

4.21. The $Xg^{(a+)}$ allele is dominant and X-linked. If a woman heterozygous for this gene [$Xg^{(a+)}Xg^{(a-)}$] married a man carrying the allele $Xg^{(a-)}$, what is the probability that (a) each daughter and (b) each son will receive the $Xg^{(a+)}$ gene?

FIVE

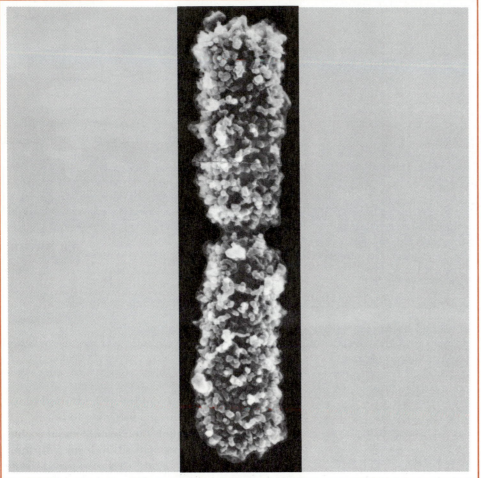

Scanning electron micrograph of a metacentric Chinese hamster metaphase chromosome isolated by nitrogen cavitation using the method of Wray and Stubblefield (*Exp. Cell Res.* 59:469–478, 1970). The chromosome was stabilized using 2 percent aqueous uranyl acetate, dehydrated using a graded series of acetone solutions, critical-point dried, and sputter-coated with gold-palladium prior to observation using a JEOL JEM-100CX Scanning-transmission electron microscope. The chromatids are characterized by looped and coiled 25 nm chromatin fibers. Bar, 0.5 μm. (Micrograph courtesy of Drs. Susanne M. Gollin and Wayne Wray, Department of Cell Biology, Baylor College of Medicine, Houston, Texas. Copyright © by S. M. Gollin and W. Wray, 1983.)

GENETIC MATERIAL: PROPERTIES AND REPLICATION

Genes were first detected and analyzed by Mendel, and subsequently many other scientists, by following their patterns of transmission from generation to generation (Chapter 2). These studies, while greatly elucidating the nature of inheritance in living organisms, provided no insight into the structure or molecular composition of genes.

Subsequent studies established the precise correlation between the patterns of transmission of genes from generation to generation (segregation and independent assortment) and the behavior of chromosomes during sexual reproduction, specifically the reduction division of meiosis and fertilization (Chapter 3). These and related experiments provided strong early evidence that genes are usually located on chromosomes. Thus, in posing questions about the chemical basis of heredity, scientists began by probing the biochemical composition of chromosomes.

Whatever its chemical composition, it was clear even in Mendel's time that the genetic material had to fulfill two key requirements.

1. The **genotype function** or **replication.** The genetic material must be capable of storing genetic information and transmitting this information faithfully from parents to progeny, generation after generation (although, as we will see in Chapter 9, the genetic material does undergo occasional heritable changes called **mutations**).
2. The **phenotype function** or **gene expression.** The genetic material must control the development of the phenotype of the organism, be it a virus, a bacterium, a plant, or an animal such as a human being. That is, the genetic material must dictate the growth and differentiation of the organism from the single-celled zygote to the mature adult. To control this complex process, the genetic material must not only express itself accurately, but each gene must act at the proper time and place to guarantee that the liver is made up of liver cells, the nervous system of nerve cells, and so on (see Chapters 8 and 11).

Chromosomes are composed of two types of large organic molecules (macromolecules) called **proteins** and **nucleic acids.** The nucleic acids are of two types: **deoxyribonucleic acid (DNA)** and **ribonucleic acid (RNA).** For many years, there was considerable disagreement among scientists as to which of these three macromolecules carries the genetic information. During the 1940s and early 1950s, several elegant experiments were carried out which clearly established that the genetic information resides in the nucleic acids rather than in proteins. More specifically, these experiments showed that the genetic information resides in DNA. (In a few simple viruses, however, RNA carries the genetic information; these particular viruses contain no DNA.)

DNA, THE GENETIC MATERIAL

Several lines of indirect evidence have long suggested that DNA contains the genetic information of living organisms. Most important, results obtained using several different experimental procedures showed that most of the DNA is located in the chromosomes, whereas RNA and proteins are also abundant in the cytoplasm. Moreover, a precise correlation exists between the amount of DNA per cell and the number of sets of chromosomes per cell. That is, most somatic cells of diploid organisms, for example, contain exactly twice the amount of DNA as the haploid germ cells or gametes of the same species. Finally, the molecular composition of the DNA in all of the different cells of an organism is the same (with rare exceptions), whereas the composition of RNA and proteins varies both qualitatively and quantitatively from one cell type to another. While these correlations strongly suggest that DNA is the genetic material, they by no means prove it. Fortunately, direct evidence has established that the genetic information is encoded in DNA.

TRANSFORMATION IN PNEUMOCOCCUS

The first direct evidence showing that the genetic material is DNA rather than protein or RNA was published by O. T. Avery, C. M. MacLeod, and M. McCarty in 1944. They demonstrated that the component of the cell responsible for the phenomenon of transformation in the bacterium *Diplococcus pneumoniae* (pneumococcus) is DNA. Transformation is a mode of recombination (exchange or transfer of genetic information between organisms or from one organism to another) occurring in several, but not all, species of bacteria. It does not involve direct contact between the bacterial cells or mediation by any vector such as a virus (see Chapter 7, pp. 200–202).

The phenomenon of transformation was discovered by Frederick Griffith in 1928. It should be emphasized that although Griffith's experiments demonstrated the occurrence of transformation in pneumococcus and thus set the stage for the work of Avery, MacLeod, and McCarty, they provided no evidence that DNA was involved in any way.

Pneumococci, like all other living organisms, exhibit genetic variability that can be recognized by the existence of different phenotypes (Table 5.1). The two phenotypic characteristics of importance in Griffith's demonstration of transformation were (1) the presence or absence of a surrounding polysaccharide (complex sugar polymer) capsule and (2) the type of capsule, that is, the specific molecular composition of the polysaccharides present in the capsules. When grown on appropriate media (such as blood agar) in petri dishes, pneumococci with a capsule form large, smooth colonies and are thus designated Type S. Such encapsulated pneumococci are quite pathogenic to most mammals (e.g., causing pneumonia in humans). These virulent (disease causing) Type S pneumococci mutate to a nonvirulent (or nonpathogenic) form which has no polysaccharide capsule (at a frequency of about one cell in 10^7). Such nonencapsulated, nonvirulent pneumococci form small, rough-surfaced colonies when grown on blood agar medium and are thus designated Type R (Table 5.1). (The polysaccharide capsule is required for virulence since it protects the bacterial cell against phagocytosis by leukocytes.) When a capsule is present, it may be of several different antigenic types (type II, III, etc.), depending on the specific molecular composition of the polysaccharides and, of course, ultimately on the genotype of the cell.

The different capsule types can be identified immunologically. If Type II cells are injected into the bloodstream of rabbits, the immune system of the rabbits will produce antibodies (a specific set of large proteins whose function is to protect the organism against foreign substances such as macromolecules, viruses, and bacteria) that react specifically with Type II cells. Such Type II antibodies will agglutinate Type II pneumococci but not Type III pneumococci, and vice versa.

Griffith's unexpected discovery was that if he injected heat-killed Type IIIS pneumococci (virulent when alive) plus live Type IIR pneumococci (nonvirulent) into mice, many of the mice succumbed to pneumonia, and live Type IIIS cells were recovered from the carcasses (Fig. 5.1). When mice were injected with heat-killed Type IIIS pneumococci alone (Fig. 5.1, top), none of the mice died. The observed virulence was therefore not due to a few Type IIIS cells that survived the heat treatment. It is critical to note that the live virulent pneumococci recovered from the carcasses were of polysaccharide Type III, since it is known that nonencapsulated Type R cells can mutate back to virulent encapsulated Type S cells. When such a mutation

TABLE 5.1   Characteristics of *Diplococcus pneumoniae* Strains When Grown on Blood Agar Medium

| TYPE | COLONY MORPHOLOGY | | CAPSULE | VIRULENCE | REACTION WITH ANTISERUM PREPARED AGAINST: | |
	APPEARANCE	SIZE			TYPE IIS	TYPE IIIS
IIR[a]	Rough	Small	Absent	Nonvirulent	None	None
IIS	Smooth	Large	Present	Virulent	Agglutination	None
IIIR[a]	Rough	Small	Absent	Nonvirulent	None	None
IIIS	Smooth	Large	Present	Virulent	None	Agglutination

[a] Although Type R cells are nonencapsulated, they carry genes that would direct the synthesis of a specific kind (antigenic Type II or III) of capsule if the block in capsule formation were not present. When Type R cells mutate back to encapsulated Type S cells, the capsule Type (II or III) is determined by these genes. Thus, R cells derived from Type IIS cells are designated Type IIR. When these Type IIR cells mutate back to encapsulated Type S cells, the capsules are of Type II.

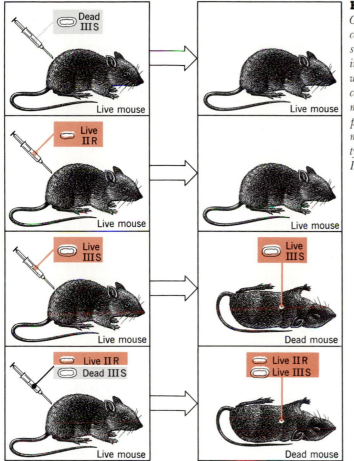

FIGURE 5.1

*Griffith's demonstration of **transformation** in pneumococcus. When heat-killed encapsulated (designated* **S** *for
smooth colony formation) type III pneumococci were injected
into mice, the mice did not develop pneumonia. Similarly,
when living nonencapsulated (designated* **R** *for rough
colony formation) type II cells were injected into mice, the
mice showed no ill effects. Injection of living type IIIS
pneumococci resulted in severe pneumonia and the death of
many of the mice. Surprisingly, the injection of heat-killed
type IIIS cells (virulent* if alive) *together with living type
IIR cells (nonvirulent) caused the death of many of the mice.*

occurs in a Type IIR cell, however, the resulting cell
will be Type IIS, not Type IIIS. Thus, the **"transformation" of nonvirulent Type IIR cells to virulent Type IIIS cells cannot be explained by mutation, rather some component of the dead Type
IIIS cells (the "transforming principle") must
convert living Type IIR cells to Type IIIS.**

Subsequent experiments showed that the phenomenon described by Griffith, now called **transformation,** was not mediated in any way by a living
host. The same phenomenon occurred in the test
tube when live Type IIR cells were grown in the
presence of dead Type IIIS cells or extracts of Type
IIIS cells. Since it was clearly shown that the new
phenotype, Type IIIS, was hereditary, that is, was
due to a permanent inherited change in the geno-

type of the cells, the demonstration of transformation neatly set the stage for determining the chemical basis of heredity in pneumococcus. What
remained was to determine what component of the
cell extract was responsible for transformation.

PROOF THAT THE "TRANSFORMING PRINCIPLE" IS DNA

The "transforming principle" was shown to be
DNA in 1944 when Avery, MacLeod, and McCarty
published the results of a set of extensive and laborious experiments. They showed that **if highly purified DNA from Type IIIS pneumococci was
present with Type IIR pneumococci, some of the
pneumococci were transformed to Type IIIS** (Fig.
5.2). But how could one be sure that the DNA was

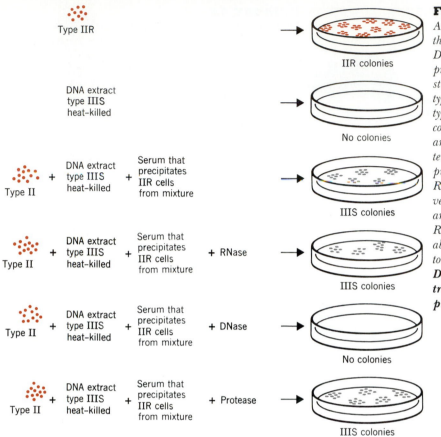

FIGURE 5.2
*Avery, MacLeod, and McCarty's proof that the "transforming principle" is DNA. Transformation of type IIR pneumococci to type IIIS could be demonstrated using highly purified DNA from type IIIS cells as well as using heat-killed type IIIS cells. Proof that the active component was DNA and not small amounts of contaminating RNA or protein was accomplished by treating the purified DNA with the enzymes DNase, RNase, and trypsin (a protease), which very specifically degrade DNA, RNA, and protein, respectively. Treatment with RNase or protease had **no effect** on the ability of the purified DNA preparation to transform type IIR cells to type IIIS.* **DNase treatment destroyed the transforming activity of the DNA preparation.**

really pure? Proving the complete purity of any macromolecular substance is extremely difficult. Maybe the DNA preparation contained a few molecules of protein and these contaminating proteins were responsible for the observed transformation. The most definitive experiments in Avery, MacLeod, and McCarty's "proof" that DNA was the transforming principle involved the use of enzymes (proteins that catalyze specific metabolic reactions) that degrade DNA, RNA, or protein. In separate experiments, highly purified DNA from Type IIIS cells was treated with (1) deoxyribonuclease ("DNase"; which degrades DNA), (2) ribonuclease ("RNase"; which degrades RNA), or (3) proteases (which degrade proteins) and then tested for its ability to transform Type IIR cells to Type IIIS. Only DNase had any effect on the transforming activity of the DNA preparation; it totally elimi-

nated all transforming activity (Fig. 5.2).

Although the molecular mechanism by which transformation occurred remained to be worked out in subsequent investigations, the results obtained by Avery and co-workers clearly established that the genetic information in pneumococcus was present in DNA. We now know that the segment of the DNA in the chromosome of pneumococcus that carries the genetic information specifying the synthesis of a Type III capsule is physically integrated into the chromosome of the Type IIR recipient cell by a specific recombination process occurring during transformation (see Chapter 7).

THE "HERSHEY-CHASE EXPERIMENT"

Additional direct evidence indicating that DNA is the genetic material was published in 1952 by A. D. Hershey (1969 Nobel Prize winner) and M. Chase.

These experimenters showed that the genetic information of a particular bacterial virus (bacteriophage T2) was present in DNA. Their results, although probably less definitive than the results of Avery, MacLeod, and McCarty, had a great impact on the acceptance by scientists of DNA as the genetic material. This large impact undoubtedly was the result of the elegant simplicity of the so-called "Hershey-Chase experiment."

Viruses are the smallest living organisms; they are living at least in the sense that their reproduction is controlled by genetic information stored in nucleic acids via the same processes as in cellular organisms. Viruses, however, are acellular obligate parasites that can reproduce only in appropriate host cells. Their reproduction is totally dependent on the metabolic machinery (ribosomes, energy-generating systems, etc.) of the host. Viruses have been extremely useful in studying many genetic processes because of their simple structure and chemical composition (many contain only proteins and nucleic acids) and their very rapid reproduction (15–20 minutes for some bacterial viruses under optimal conditions).

Bacteriophage T2, which infects the common colon bacillus *Escherichia coli,* is composed of about 50 percent DNA and about 50 percent protein (Fig. 5.3). Experiments prior to 1952 had shown that all bacteriophage T2 reproduction takes place within *E. coli* cells. Therefore, when Hershey and Chase showed that the DNA of the virus particle entered the cell, whereas most of the protein of the virus remained adsorbed to the outside of the cell, this strongly implied that the genetic information necessary for viral reproduction was present in DNA. The basis of the Hershey-Chase experiment is that **DNA contains phosphorus but no sulfur, whereas proteins contain sulfur but no phosphorus.** Thus, Hershey and Chase were able to specifically label either (1) the phage DNA by growth in a medium containing the radioactive isotope of phosphorus, ^{32}P, in place of the normal isotope, ^{31}P, or (2) the phage protein coats by growth in a medium containing radioactive sulfur, ^{35}S, in place of the normal isotope, ^{32}S (Fig. 5.3). When T2 phage particles labeled with ^{35}S were mixed with *E. coli* cells for a few minutes and were then subjected to shearing forces by placing the infected cells in a Waring blender, it was found that most of the radioactivity (and thus the proteins) could be removed from the cells without affecting progeny phage production. When T2 phage in which the DNA was labeled with ^{32}P were used, however, essentially all of the radioactivity was found inside the cells; that is, it was not subject to removal by shearing in a blender. The sheared-off phage coats were separated from the infected cells by low-speed centrifugation, which pellets (sediments) cells while leaving phage particles suspended. These results indicated that the DNA of the virus enters the host cell, whereas the protein coat remains outside the cell. Since progeny viruses are produced inside the cell, Hershey and Chase's results indicated that the genetic information directing the synthesis of both the DNA molecules and the protein coats of the progeny viruses must be present in the parental DNA. Moreover, the progeny particles were shown to contain some of the ^{32}P, but none of the ^{35}S of the parental phage.

However, the Hershey-Chase experiment did not provide unambiguous proof that the genetic material of phage T2 is DNA. A significant amount of ^{35}S (and thus protein) was found to be injected into the host cells with the DNA. Thus, one could always argue that this small fraction of the phage proteins contained the genetic information. More recently, however, it has been possible to develop conditions in which protoplasts (cells with the walls removed) of *E. coli* can be infected with pure phage DNA. Normal infective progeny phage are produced in these experiments, called **transfection** experiments, proving that the genetic material of such bacterial viruses is DNA.

RNA AS GENETIC MATERIAL IN SMALL VIRUSES

As more and more viruses were identified and studied, it became clear that many of them contain RNA and proteins, but no DNA. In all cases so far studied, it is clear that these "RNA viruses" store their genetic information in nucleic acids rather than in proteins just like all other organisms, although in these viruses the nucleic acid is RNA. One of the

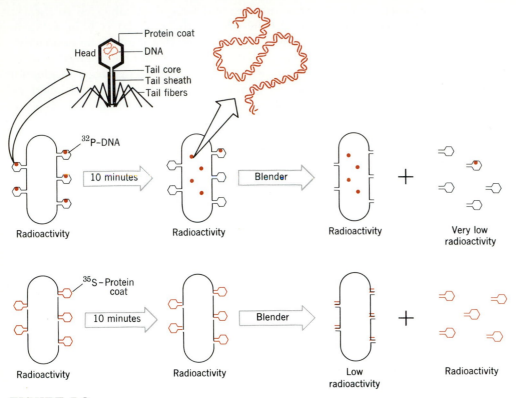

FIGURE 5.3

The "Hershey-Chase experiment": evidence that DNA is the genetic material in bacterio-phage T2. E. coli cells were infected with ^{32}P*-labeled phage (DNA-labeled), and after being allowed time for infection, they were agitated in a blender which sheared off the phage coats. The phage coats and the infected cells were then separated by centrifuga-tion. Radioactivity was measured in the cell pellet (the sediment) and in the phage coat suspension. Most of the radioactivity was found in the cells. When the same experiment was done using phage with* ^{35}S*-labeled proteins, the results were very different. Most of the radioactivity was found in the suspension of phage coats; very little entered the host cells. Since phage reproduction (both DNA synthesis and new-coat protein synthesis) occurs inside the infected cells, and since only the phage DNA enters the host cell, the DNA, not the protein, must carry the genetic information. (Based on R. Sager and F. J. Ryan,* Cell Heredity, *John Wiley & Sons, 1961.)*

first experiments that established RNA as the genetic material in RNA viruses was the so-called reconstitution experiment of H. Fraenkel-Conrat and B. Singer, published in 1957. Fraenkel-Conrat and Singer's simple, but definitive, experiment was done with tobacco mosaic virus (TMV), a small virus composed of a single molecule of RNA encapsu-lated in a protein coat. Different strains of TMV can be identified on the basis of differences in the chemi-cal composition of their protein coats.

By using the appropriate chemical treatments, one can separate the protein coats of TMV from the RNA. Moreover, this process is reversible; by mix-ing the proteins and the RNA under appropriate

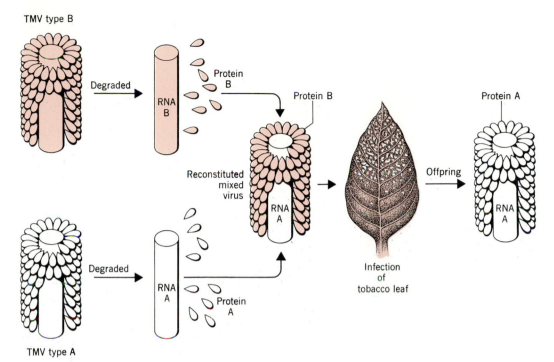

TMV type B

Degraded

RNA B

Protein B

Protein B

Reconstituted mixed virus

RNA A

Infection of tobacco leaf

Offpring

Protein A

RNA A

Degraded

RNA A

Protein A

TMV type A

FIGURE 5.4

Proof that the genetic material of tobacco mosaic virus (TMV) is RNA, not protein. The RNA molecules and the protein coats of two different strains (A and B) of TMV were separated biochemically. The RNA of strain A was then mixed with the protein coats of strain B under conditions where complete, infective virus particles are reconstituted. When the reconstituted viruses were rubbed onto live tobacco leaves, the progeny viruses were phenotypically and genotypically identical to strain A from which the RNA was obtained and unlike strain B from which the protein was obtained. When the reconstituted viruses contained RNA of type B and protein of type A, the progeny were of type B. (After H. Fraenkel-Conrat and B. Singer, Biochim. et Biophys. Acta *24:540, 1957.)*

conditions, "reconstitution" will occur, yielding complete, infective TMV particles. Fraenkel-Conrat and Singer took two different strains of TMV, separated the RNAs from the protein coats, and reconstituted "mixed" viruses by mixing the proteins of one strain with the RNA of the second strain, and vice versa. When these mixed viruses were used to infect tobacco leaves, the progeny viruses produced were always found to be phenotypically and genotypically identical to the parent strain from which the RNA had been obtained (Fig. 5.4). Thus, the genetic information of TMV is stored in RNA, not protein.

DNA STRUCTURE

The genetic information of all living organisms, except the RNA viruses, is therefore apparently stored in DNA. What, then, is the structure of DNA, and in what form is the genetic information stored? What features of the structure of DNA allow for the transmission of genetic information from generation to generation?

Nucleic acids, first called "nuclein" because they were isolated from cell nuclei by F. Miescher in 1869, are macromolecules composed of repeating subunits called **nucleotides.** Each nucleotide is composed of (1) a phosphate group, (2) a five-carbon

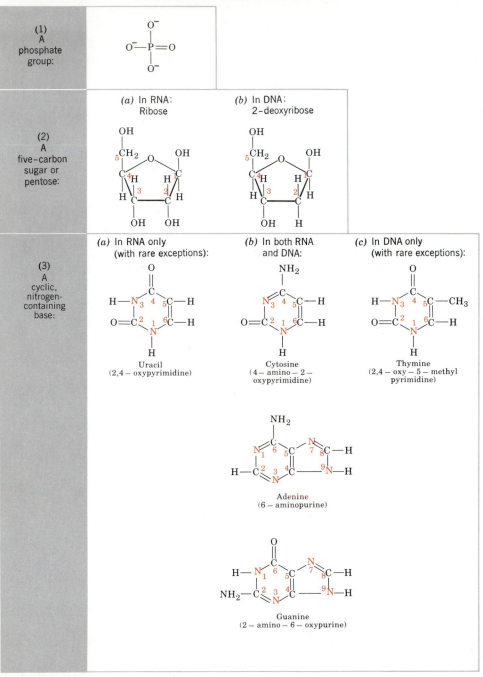

Nucleic Acids Are Composed of Repeating Subunits Called Nucleotides. Each Nucleotide Is Composed of Three Units.

FIGURE 5.5

Structural formulas for the constituents of nucleic acids. When the pentoses are present in nucleosides, nucleotides, or nucleic acids, the five carbons are numbered 1', 2', 3', 4', and 5', respectively, to distinguish them from the carbons of the bases.

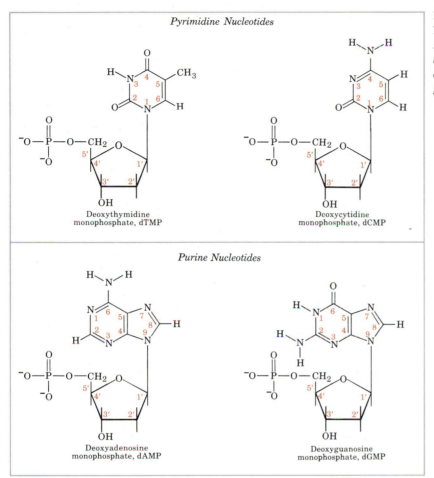

Pyrimidine Nucleotides

Deoxythymidine
monophosphate, dTMP

Deoxycytidine
monophosphate, dCMP

Purine Nucleotides

Deoxyadenosine
monophosphate, dAMP

Deoxyguanosine
monophosphate, dGMP

FIGURE 5.6

*The four common deoxyribonucleotides of DNA. RNA contains similar **ribo**nucleotides, which contain the pyrimidines uracil and cytosine and the purines adenine and guanine.*

sugar (or pentose), and (3) a cyclic nitrogen-containing compound called a base (Fig. 5.5). In DNA, the sugar is 2-deoxyribose (thus the name **deoxyribonucleic acid**); in RNA, the sugar is ribose (thus **ribo**nucleic acid). There are four different bases commonly found in DNA, **adenine, guanine, thymine,** and **cytosine.** RNA also usually contains adenine, guanine, and cytosine, but has a different base, **uracil,** in place of thymine. Adenine and guanine are double-ring bases called **purines;** cytosine, thymine, and uracil are single-ring bases called **pyrimidines** (Fig. 5.5). Both DNA and RNA, therefore, contain four different subunits or nucleotides, two purine nucleotides, and two pyrimidine nucleotides (Fig. 5.6). RNA usually exists as a single-stranded

polymer that is composed of a long sequence of nucleotides. DNA, however, has one very important additional level of organization; it is usually a double-stranded molecule.

THE WATSON AND CRICK DNA DOUBLE HELIX

The correct structure of DNA was first deduced by J. D. Watson (Fig. 5.7) and F. H. C. Crick in 1953. Their **double helix** model of DNA structure was based on two major kinds of evidence.

1. When the composition of DNA from many different organisms was analyzed by E. Chargaff and

FIGURE 5.7

J. D. Watson, American investigator in biochemical genetics. Along with the British investigators F. H. C. Crick and M. H. F. Wilkins, Dr. Watson won the Nobel Prize in physiology and medicine in 1962. The prize was awarded for his contribution to our knowledge and understanding of the chemical nature of the gene. The progress of this team up to 1953 is summarized by the Watson-Crick model, which displays the structure of the DNA molecule. (Courtesy of J. D. Watson.)

colleagues, it was observed that the **concentration of thymine was always equal to the concentration of adenine and the concentration of cytosine was always equal to the concentration of guanine.** This strongly suggested that thymine and adenine as well as cytosine and guanine were present in DNA with some fixed interrelationship. Of course, it also necessitated that the total concentration of pyrimidines (thymine plus cytosine) always equal the total concentration of purines (adenine plus guanine; see Table 5.2). However, the (thymine + adenine)/(cytosine + guanine) ratio was found to vary widely in DNAs of different species (Table 5.2).

2. When X rays are focused through isolated macromolecules or crystals of purified molecules, the X rays are deflected by the atoms of the molecules in specific patterns, called diffraction patterns, which provide information about the organization of the components of the molecules. These X-ray diffraction patterns can be recorded on X-ray sensitive film just as one photographs patterns of light with a

TABLE 5.2 Base Composition of DNA from Various Organisms[a]

SPECIES	% ADENINE	% GUANINE	% CYTOSINE	% THYMINE	MOLAR RATIOS $\dfrac{A+G}{T+C}$	$\dfrac{A+T}{G+C}$
I. **Viruses**						
Bacteriophage λ	26.0	23.8	24.3	25.8	0.99	1.08
Bacteriophage T2	32.6	18.1	16.6	32.6	1.03	1.88
Herpes simplex	13.8	37.7	35.6	12.8	1.06	0.36
Pseudorabies	13.2	37.0	36.3	13.5	1.00	0.36
Vaccinia	31.5	18.0	19.0	31.5	0.98	1.70
II. **Bacteria**						
Escherichia coli	26.0	24.9	25.2	23.9	1.04	1.00
Diplococcus pneumoniae	29.8	20.5	18.0	31.6	1.02	1.59
Micrococcus lysodeikticus	14.4	37.3	34.6	13.7	1.07	0.39
Ramibacterium ramosum	35.1	14.9	15.2	34.8	1.00	2.32
III. **Fungi**						
Neurospora crassa	23.0	27.1	26.6	23.3	1.00	0.86
Aspergillus niger	25.0	25.1	25.0	24.9	1.00	1.00
Saccharomyces cerevisiae	31.7	18.3	17.4	32.6	1.00	1.80
IV. **Higher Eukaryotes**						
Zea mays (corn)	25.6	24.5	24.6	25.3	1.00	1.04
Nicotiana tabacum (tobacco)	29.3	23.5	16.5	30.7	1.12	1.50
Arachis hypogaea (peanut)	32.1	17.6	18.0	32.2	0.99	1.80
Drosophila melanogaster	30.7	19.6	20.2	29.4	1.01	1.51
Bombyx mori (silkworm)	30.7	18.9	19.4	31.1	0.98	1.61
Rana pipiens (frog)	26.3	23.5	23.8	26.4	0.99	1.11
Homo sapiens (human)						
liver	30.3	19.5	19.9	30.3	0.99	1.53
thymus	29.8	20.2	18.2	31.8	1.02	1.60
spermatozoa	30.5	19.9	20.6	28.9	1.02	1.47

[a] Selected data from H. Sober (ed.), *Handbook of Biochemistry, Selected Data for Molecular Biology,* 2nd ed. The Chemical Rubber Co., Cleveland, 1970.

FIGURE 5.8

One of the X-ray diffraction photographs of DNA that led to the double helix model of DNA structure. An X-ray crystallographer can recognize the central cross-shaped pattern as indicative of a helical structure. The heavy dark patterns (top and bottom) indicate that the bases are stacked perpendicular to the axis of the molecule with a periodicity of 3.4 Å. (Courtesy M. H. F. Wilkins, Biophysics Department, King's College, London.)

camera and light-sensitive film. Watson and Crick had available X-ray crystallographic data on DNA structure from the studies of M. H. F. Wilkins, R. Franklin, and their co-workers (Fig. 5.8). These data indicated that DNA was a highly ordered, multiple-stranded structure with repeating substructures spaced every 3.4 angstroms [1 angstrom (Å) = 10^{-8} cm] along the axis of the molecule.

On the basis of Chargaff's chemical data, Wilkins and Franklin's X-ray diffraction data, and inferences drawn from model building, Watson and Crick proposed that DNA exists as a **double helix** in which the two polynucleotide chains are coiled about one another in a spiral (Fig. 5.9). Each polynucleotide chain consists of a sequence of nucleotides linked together by **phosphodiester bonds,** joining adjacent deoxyribose moieties (Table 5.3; Fig. 5.10). The two polynucleotide strands are held together in their helical configuration by **hydrogen bonding** (Table 5.3) between bases in opposing strands, the resulting base-pairs being stacked between the two chains perpendicular to the axis of

the molecule like the steps of a spiral staircase (Fig. 5.9). The base-pairing is specific; **adenine is always paired with thymine, and guanine is always paired with cytosine** (Fig. 5.11). Thus, all base-pairs consist of one purine and one pyrimidine. The specificity of base-pairing results from the hydrogen-bonding capacities of the bases in their normal configurations. In their most common structural configurations, adenine and thymine form two hydrogen bonds, and guanine and cytosine form three hydrogen bonds (Fig. 5.11). Analogous hydrogen bonding between cytosine and adenine, for example, is not possible except when they exist in their rare structural states (see Chapter 9).

Once the sequence of bases in one strand of a DNA double helix is known, the sequence of bases in the other strand is also known because of the specific base-pairing. The two strands of a DNA double helix are thus said to be **complementary (not identical);** it is this property, complementarity of the two strands, that makes DNA uniquely suited to store and transmit genetic information (see the following section on the replication of DNA).

The base-pairs in DNA are stacked 3.4 Å apart with 10 base-pairs per turn (360°) of the double helix (Fig. 5.9). The sugar-phosphate backbones of the two complementary strands are **antiparallel;** that is, they have **opposite chemical polarity** (Fig. 5.10). As one moves unidirectionally along a DNA double helix, the phosphodiester bonds in one strand go from a 3′ carbon of one nucleotide to a 5′ carbon of the adjacent nucleotide, while those in the complementary strand go from a 5′ carbon to a 3′ carbon. This opposite polarity of the complementary strands is very important in considering the mechanism of replication of DNA.

The high degree of stability of DNA double helices results in part from the large number of hydrogen bonds between the base-pairs (even though each hydrogen bond by itself is quite weak, much weaker than a covalent bond) and in part from the hydrophobic bonding (or "stacking forces") between the stacked base-pairs (Table 5.3; Fig. 5.9). The planar sides of the base-pairs are relatively nonpolar and thus tend to be water insoluble ("hydrophobic"). This hydrophobic core of stacked

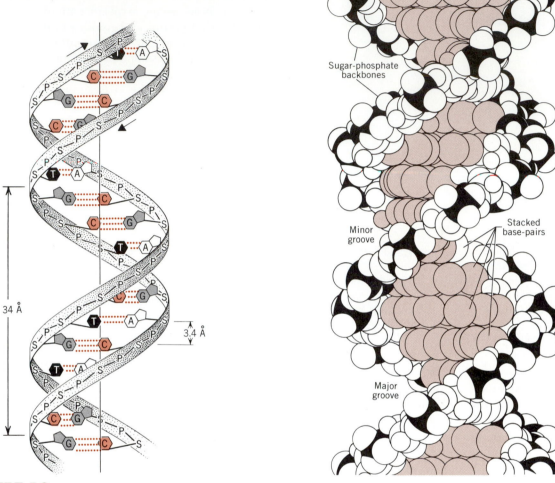

FIGURE 5.9

Diagram (left) and space-filling model (right) of the Watson-Crick double helix model of the structure of DNA. A, T, G, and C represent adenine, thymine, guanine, and cytosine, respectively. S and P represent sugar (2-deoxyribose) and phosphate groups. (The space-filling model is based on a diagram by M. Feughelman et al., Nature 175:834, 1955.)

base-pairs contributes considerable stability to DNA molecules present in the aqueous protoplasms of living cells.

CONFORMATIONAL FLEXIBILITY OF DNA MOLECULES

The vast majority of the DNA molecules present in the aqueous protoplasms of living cells almost cer-

tainly exist in the Watson-Crick double-helix form described above (Fig. 5.9). This is the **B-form** of DNA. The B-form is the conformation that DNA takes under physiological conditions (in aqueous solutions containing low concentrations of salts). However, DNA is not a static, invariant molecule. To the contrary, **DNA molecules exhibit a considerable amount of conformational flexibility.**

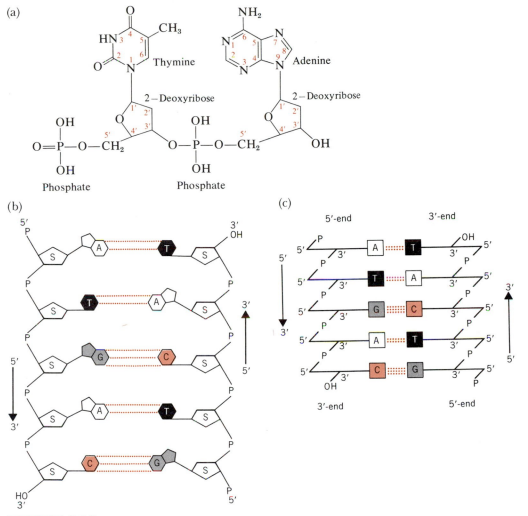

FIGURE 5.10

Molecular structure of DNA showing the sugar-phosphate backbones of the polynucleotide chains and their antiparallel nature (opposite chemical polarity). (a) A deoxythymidylate-deoxyadenylate dinucleotide, showing the numbering system used for nucleotides and the phosphodiester linkage between the 3' and 5' carbons of adjacent nucleotides. Note the 5' to 3' chemical polarity going from left to right. (b) Molecular structure and (c) schematic representation of a short segment of a DNA molecule, emphasizing the opposite polarity of the complementary strands.

The structures of DNA molecules change as a function of their environment. The exact conformation of a given DNA molecule or segment of a DNA molecule will depend on the nature of the molecules with which it is interacting. In fact, intracellular B-form DNA appears to have an average of 10.4 nucleotide-pairs per turn, rather than precisely 10 as shown in Fig. 5.9. In high concentrations of salts or in a dehydrated state, DNA exists in the **A-form**, which has 11 nucleotide-pairs per turn. It is

TABLE 5.3 Chemical Bonds Important in DNA Structure

1. *Covalent bonds*
   Strong chemical bonds formed by sharing of electrons between atoms.
   (a) In bases and sugars

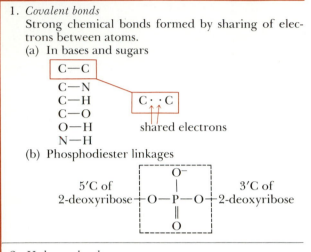

   shared electrons

   (b) Phosphodiester linkages

$$\begin{array}{ccc} & O^- & \\ 5'C \text{ of} & | & 3'C \text{ of} \\ 2\text{-deoxyribose} - O - P - O - & 2\text{-deoxyribose} \\ & \| & \\ & O & \end{array}$$

2. *Hydrogen bonds*
   A weak bond between an electronegative atom and a hydrogen atom (electropositive) that is covalently linked to a second electronegative atom.

3. *Hydrophobic "bonds"*
   The association of nonpolar groups with each other when present in aqueous solutions because of their insolubility in water.

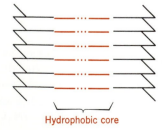

   Water molecules are very polar (δ^- O and δ^+ Hs). Compounds which are similarly polar are very soluble in water ("hydrophilic"). Compounds which are nonpolar (no charged groups) are very insoluble in water ("hydrophobic").
   The stacked base-pairs provide a hydrophobic core.

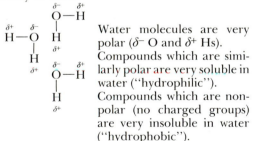

   Hydrophobic core

Hydrogen-bonding in A-T and G-C base-pairs.

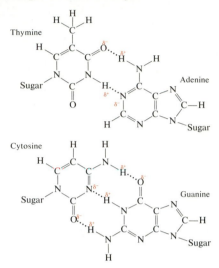

Lack of hydrogen-bonding potential between A and C, G and T.

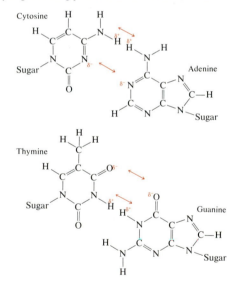

FIGURE 5.11

Base-pairing in DNA, adenine with thymine and guanine with cytosine, based on hydrogen-bonding between appropriately spaced, negatively charged =O and ≡N groups and positively charged —Hs. Note that the similar apposition of adenine and cytosine or guanine and thymine results in the juxtaposition of identically charged (+ or −) groups at two of the three sites of potential hydrogen bonding. Thus, adenine is not normally found base-paired with cytosine (nor guanine with thymine) in DNA.

very unlikely that DNA molecules ever exist in the A-form *in vivo*. This structure is of interest, however, because it is the conformation of DNA-RNA heteroduplexes (double helices containing a DNA strand base-paired with a complementary RNA strand) or RNA-RNA duplexes *in vivo*.

Recently, certain DNA sequences have been shown to exist in a unique left-handed, double helical form called **Z-DNA** (Z for the zigzagged path of the sugar-phosphate backbones of the structure). The helices of A- and B-form DNA are wound in a right-handed manner. Moreover, specific segments of DNA molecules can undergo conformational shifts from the B-form to the Z-form and vice versa. Very possibly, certain regulatory proteins may bind only to the Z-form (or B-form) of a DNA sequence and cause it to shift to the B-form (or Z-form) (see Chapter 11, pp. 407–410). In any case, one must remember that the structure of DNA is not invariant and that structural variations in DNA molecules may play important biological roles.

SEMICONSERVATIVE REPLICATION OF DNA

Living organisms perpetuate their kind through reproduction. This may be simple duplication (cell fission) as in bacteria or complex modes of sexual reproduction as in higher plants and animals. In all cases, however, reproduction entails the faithful transmission of the genetic information of the parents to the progeny. Since the genetic information is stored in DNA, **the replication of DNA is central to all of biology.**

When Watson and Crick proposed the double helix structure of DNA with its complementary base-pairing, they immediately recognized that the base-pairing specificity could provide the basis for a simple (superficially simple, at least) mechanism for DNA duplication. If the two complementary strands of a double helix separated (by breaking the hydrogen bonds of each base-pair), each parental strand could direct the synthesis of a new complementary strand because of the specific base-pairing requirements (Fig. 5.12). That is, each parental strand could serve as a **template** for a new comple-

mentary strand. Adenine, for example, in the parent strand would serve as a template via its hydrogen-bonding potential for the incorporation of thymine in the nascent complementary strand. This mechanism of DNA replication is called **semiconservative** replication, since each of the complementary strands of the parental double helix is conserved (or the double helix is "half-conserved") during the process.

In considering possible mechanisms of DNA replication, three different hypothetical modes are apparent (Fig. 5.13). In addition to (1) semiconservative replication, one can propose (2) "conservative" replication in which the parental double helix remains intact (is totally conserved) and somehow directs the synthesis of a "progeny" double helix composed of two newly synthesized strands, and (3) "dispersive" replication in which segments of parental strands and progeny or nascent strands become interspersed through some kind of a fragmentation, synthesis, and rejoining process.

THE "MESELSON-STAHL EXPERIMENT"

The results of the first critical test of Watson and Crick's proposal that DNA replicates semiconservatively were published in 1958 by M. S. Meselson and F. W. Stahl. Their results showed that the chromosome (now known to contain a single Watson-Crick double helix of DNA) of the common colon bacillus *Escherichia coli* replicated semiconservatively.

Meselson and Stahl grew *E. coli* cells for many generations in a medium in which the heavy isotope of nitrogen, ^{15}N, had been substituted for the normal, light isotope, ^{14}N. The purine and pyrimidine bases in DNA contain nitrogen; thus, the DNA of cells grown on medium containing ^{15}N will have a greater density (weight per unit volume) than the DNA of cells grown on medium containing ^{14}N. Since molecules of different densities can be separated by a procedure called **equilibrium density-gradient centrifugation,** Meselson and Stahl were able to distinguish between the three possible modes of DNA replication by following the changes in the density of DNA of cells grown on ^{15}N-medium and then transferred to ^{14}N-medium for various periods of time (so-called **density transfer experiments**).

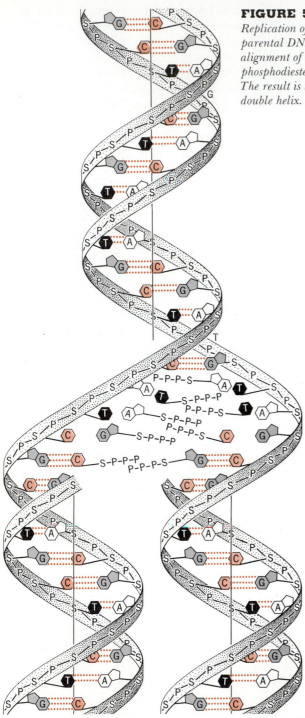

FIGURE 5.12

Replication of DNA as proposed by Watson and Crick. As the double-stranded parental DNA molecule unwinds, the separated strands serve as templates for the alignment of incoming nucleoside triphosphates, which are then linked by phosphodiester bonds to form new strands complementary to the parental strands. The result is the production of two progeny double helices identical to the parental double helix.

The density of most DNAs is about the same as the density of concentrated solutions of heavy salts such as cesium chloride (CsCl). For example, the density of 6 M CsCl is about 1.7 g/cm³. *E. coli* DNA containing ¹⁴N has a density of 1.710 g/cm³. Substitution of ¹⁵N for ¹⁴N increases the density of *E. coli* DNA to 1.724 g/cm³.

When a heavy salt solution such as 6 M CsCl is centrifuged at very high speeds (30,000–50,000 revolutions per minute) for 48–72 hours, an **equilibrium-density gradient** is formed (Fig. 5.14). The centrifugal force caused by spinning the solution at high speeds sediments the salt toward the bottom of the tube. Diffusion, on the other hand, results in movement of salt molecules back toward the top (low salt concentration) of the tube. After a sufficient period of high-speed centrifugation, an equilibrium between sedimentation and diffusion is reached, at which time a linear gradient of increasing density exists from the top of the tube to the bottom of the tube (Fig. 5.14). If DNA is present in such a gradient, it will move to a position where the density of the salt solution is equal to its own density. Thus, if a mixture of *E. coli* DNA containing ¹⁵N ("heavy" DNA) and *E. coli* DNA containing ¹⁴N ("light" DNA) is subjected to CsCl equilibrium density-gradient centrifugation, the DNA molecules will separate into two "bands," one containing "heavy" DNA and one containing "light" DNA (Fig. 5.14).

Meselson and Stahl took cells that had been growing in medium containing ¹⁵N for several generations (and thus contained "heavy" DNA), washed them to remove the ¹⁵N-containing medium, and transferred them to medium containing ¹⁴N. After allowing the cells to grow in the presence of ¹⁴N for varying periods of time, the DNA was extracted and analyzed in CsCl equilibrium-density

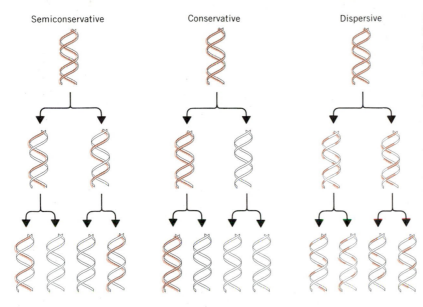

Semiconservative   Conservative   Dispersive

FIGURE 5.13

Three theoretical modes of DNA replication: (1) semiconservative, (2) conservative, and (3) dispersive. During **semiconservative DNA replication** *(left), each single strand of the parental double helix is conserved, acting as a template for the synthesis of a new complementary strand. During* **conservative DNA replication** *(center), the entire parental double helix is conserved and directs the synthesis of a new double helix. During* **dispersive DNA replication** *(right), both strands of the parental double helix are fragmented and parental and newly synthesized segments of DNA become interspersed in each of the strands of the two resulting DNA double helices.*

gradients. The results of their experiment (Fig. 5.15) are only consistent with semiconservative replication, excluding both conservative and dispersive models of DNA synthesis. All of the DNA isolated from cells after one generation of growth in medium containing ^{14}N had a density halfway between the densities of "heavy" DNA and "light" DNA. This intermediate density is usually referred to as "hybrid" density. After two generations of growth in medium containing ^{14}N, half of the DNA was of "hybrid" density and half was "light." These results are precisely those predicted by Watson and Crick semiconservative replication (Fig. 5.15). One generation of semiconservative replication of a parental double helix containing ^{15}N in medium containing only ^{14}N would produce two progeny double helices both of which had ^{15}N in one strand (the "old" strand) and ^{14}N in the other strand (the "new" strand). Such molecules would be of "hybrid" density.

Conservative replication would not produce any DNA molecules with "hybrid" density; after one generation of conservative replication of "heavy" DNA in "light" medium, half of the DNA would still be "heavy" and the other half would be "light." If replication were dispersive, Meselson and Stahl would have observed a shift of the DNA from "heavy" toward "light" each generation (i.e., "half-

heavy" or "hybrid" after one generation, "quarter-heavy" after two generations, etc.). Meselson and Stahl's results are clearly inconsistent with either of these possibilities.

Subsequent studies have verified Meselson and Stahl's conclusion that DNA replication is semiconservative, and have extended it to many other organisms, including higher plants and animals.

AUTORADIOGRAPHY OF REPLICATING BACTERIAL CHROMOSOMES

The visualization of replicating chromosomes was first accomplished by J. Cairns in 1963 using the technique called **autoradiography.** Autoradiography is a method for detecting and localizing radioactive isotopes in cytological preparations or macromolecules by exposure to a photographic emulsion that is sensitive to low-energy radiation. Autoradiography is particularly useful in studying DNA metabolism because **DNA can be specifically labeled by growing cells on ^3H-thymidine,** the tritiated deoxyribonucleoside of thymine. Thymidine is incorporated exclusively into DNA; it is not present in any other major component of the cell.

Cairns grew *E. coli* cells in medium containing ^3H-thymidine for varying periods of time, lysed the cells very gently so as not to break the chromosomes (long DNA molecules are very shear sensitive), and

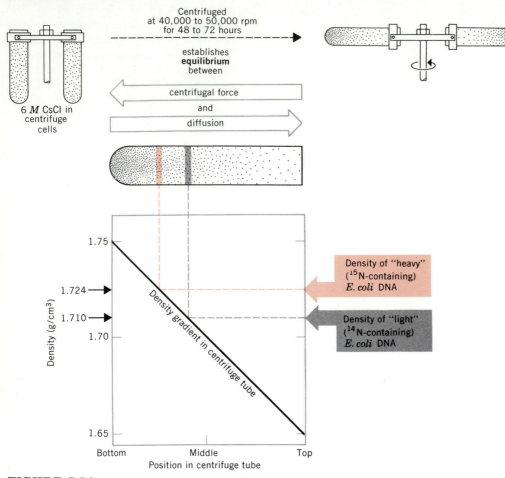

FIGURE 5.14

The use of cesium chloride (CsCl) in density-gradient centrifugation to separate DNAs of different densities. This procedure was used by Meselson and Stahl to demonstrate that the chromosomes *of* E. coli *replicate semiconservatively (see Fig. 5.15). The density of 6 M CsCl is about 1.7 g/cm³. If such a solution is centrifuged at very high speeds for a long enough period of time, a* **density gradient** *will be formed because of the* **equilibrium** *between (1) sedimentation of the CsCl to the bottom of the centrifuge tube as a result of the centrifugal force and (2) diffusion of the CsCl toward the top of the tube. The densities of most naturally occurring nucleic acids fall within the range covered by such gradients. CsCl density gradients have thus been very useful in the study of nucleic acids.*

carefully collected the chromosomes on membrane filters. These filters were affixed to glass slides, coated with emulsion sensitive to β-particles (the low-energy electrons emitted during decay of tritium), and stored in the dark for a period of time to allow sufficient radioactive decays. The autoradiographs observed when the films were developed (Fig. 5.16) showed that the chromosomes of *E. coli* are circular structures that exist as θ-shaped intermediates during replication. These autoradio-

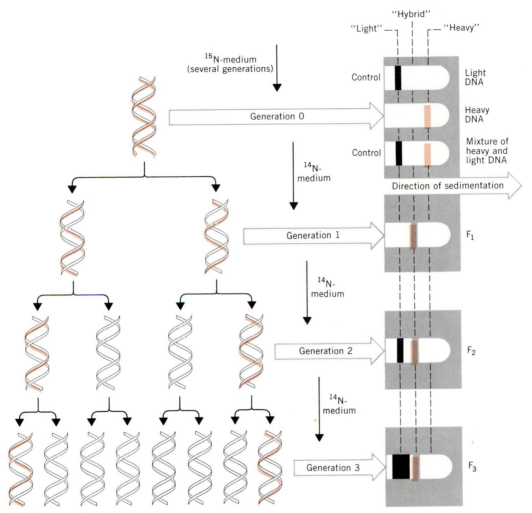

FIGURE 5.15

Results (right) and interpretation (left) of the Meselson and Stahl experiment which demonstrated that the E. coli *chromosomes replicate semiconservatively. The DNA of* E. coli *was density-labeled by growing cells for several generations in medium containing the heavy isotope of nitrogen, ^{15}N. The cells were then transferred to medium containing the normal isotope of nitrogen, ^{14}N, and allowed to grow for varying periods of time (one generation, two generations, etc.). The transfer of the density label (^{15}N) from parental DNA molecules to progeny DNA molecules was followed by the extraction of DNA from cells grown for varying periods in ^{14}N and the analysis of the DNA in CsCl equilibrium density gradients (see Fig. 5.14). DNA that contains* ^{15}N *in both strands ("heavy" DNA) forms a band in the CsCl density gradient at a higher density position than DNA that contains ^{14}N in both strands ("light" DNA). After one generation of growth in ^{14}N-medium, the DNA bands at an intermediate ("hybrid") density. Such "hybrid" DNA contains ^{15}N in one strand and ^{14}N in the other strand. After two generations of growth in ^{14}N-medium, half of the DNA bands at the "hybrid" density and half bands at the "light" density. These results are precisely those predicted by semiconservative replication. (After Meselson and Stahl,* Proc. Nat. Acad. Sci., U.S. *44:671, 1958.)*

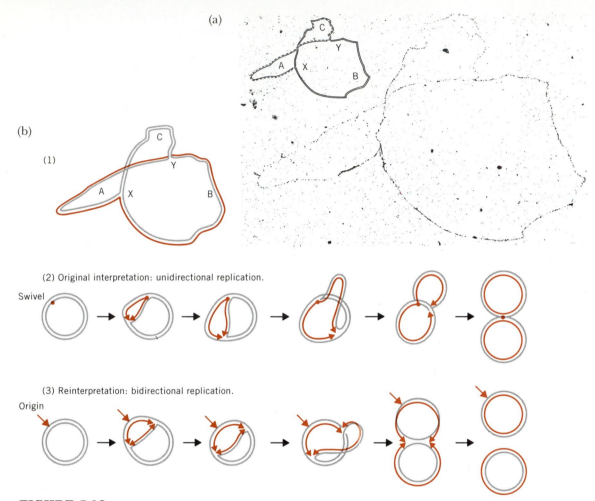

(a)

(b)

(1)

(2) Original interpretation: unidirectional replication.

Swivel

(3) Reinterpretation: bidirectional replication.

Origin

FIGURE 5.16

Depiction of a replicating chromosome of E. coli *by autoradiography.* (a) *Cairns' autoradiograph and an interpretative drawing (upper left inset) of a θ-shaped replicating chromosome from a cell that had been grown for two generations in the presence of ³H-thymidine. Loops A and B have completed a second replication in ³H-thymidine; section C remains to be replicated the second time. Cairns' drawing (upper left inset) shows radioactive strands of DNA as solid lines and nonradioactive strands as dashed lines. Note that in the autoradiograph loop B, with two radioactive strands, exhibits about twice the grain density of loop A, with only one radioactive strand. (J. Cairns, Imperial Cancer Research Fund, London, England. Reproduced with permission from* Cold Spring Harbor Symp. Quant. Biol. *28:43,* *1963. Copyright © 1963 by Cold Spring Harbor Laboratory.)* (b) *Simplified diagram of Cairns' autoradiograph illustrating only the events of the replication in progress (1) and analogous diagrams illustrating the two possible interpretations of Cairns' results, namely, unidirectional replication (2) and bidirectional replication (3). In these diagrams, the gray lines represent the two strands of the parental DNA double helix, and the colored lines represent the nascent (or daughter) strands. Originally, one of the two branch points (X and Y) was believed to be a replication fork and the other a terminus containing a "swivel," which served as an axis of rotation for unwinding the double helix (2). Replication has subsequently been shown to be bidirectional in* E. coli. *Thus, both X and Y are replication forks (3).*

graphs further indicated that the unwinding of the two complementary parental strands (which is necessary for their separation) and their semiconservative replication occur simultaneously or are closely coupled. Since the parental double helix must rotate 360° to unwind each gyre of the helix, this necessitates the existence of some kind of "swivel" in the chromosome. Present evidence suggests that a transient single-strand break (cleavage of one phosphodiester bond in one strand of the double helix) provides an axis of rotation to allow unwinding.

Cairns' interpretation of the autoradiographs was that semiconservative replication started at a site on the chromosome, which he called the "origin," and proceeded sequentially and unidirectionally around the circular structure (Fig. 5.16). Subsequent evidence has shown his original interpretation to be incorrect on one point: replication actually proceeds **bidirectionally,** not unidirectionally. Each Y-shaped structure is a replication fork, and the two replication forks move in opposite directions sequentially around the circular chromosome (Fig. 5.16).

UNIQUE ORIGINS AND BIDIRECTIONAL REPLICATION

Cairns' results provided no information as to whether the origin of replication (site at which replication is initiated) is a unique site or occurs at random on the chromosome. Moreover, his results did not allow him to differentiate between uni- and bidirectional replication. We now have direct evidence showing that replication in *E. coli* and several other organisms proceeds bidirectionally from a unique origin. These features of DNA replication can be illustrated most simply and convincingly by experiments with some of the small bacterial viruses.

Bacteriophage lambda (phage λ) is, like phage T2, a virus that grows in *E. coli.* It has a small chromosome consisting of a single linear molecule of DNA only 17.5 μ long. The phage λ chromosome is unique in that it has a single-stranded region, 12 base-pairs long, at the 5′-end of each complementary strand (Fig. 5.17). These single-stranded ends, called "cohesive" or "sticky" ends, are complementary to each other. The cohesive ends of a lambda chromosome can thus base-pair to form a hydrogen-bonded circular structure. One of the first events to occur after a lambda chromosome is injected into a host cell is its conversion to a covalently closed circular molecule (Fig. 5.17). This conversion from the hydrogen-bonded circular form to the covalently closed circular form is catalyzed by **polynucleotide ligase,** a very important enzyme that seals single-strand breaks in DNA double helices. (Polynucleotide ligase is required in most, if not all, organisms for DNA replication, DNA repair, and recombination between DNA molecules.) The lambda chromosome replicates in its circular form via θ-shaped intermediates (Fig. 5.18), as does the *E. coli* chromosome.

The feature of the lambda chromosome that facilitated the demonstration of bidirectional replication was its differentiation into regions containing high concentrations of adenine and thymine ("A-T rich" regions) and regions with large amounts of guanine and cytosine ("G-C rich" regions). In particular, it contains a few segments with very high A-T content ("A-T rich clusters"). These A-T rich clusters were used as physical markers by M. Schnös and R. B. Inman to demonstrate, using a technique called "denaturation mapping," that replication of the lambda chromosome is initiated at a unique origin and proceeds bidirectionally rather than unidirectionally.

When DNA molecules are exposed to high temperature (100°C) or high pH (pH 11.4), the hydrogen and hydrophobic bonds that hold the complementary strands together in the double helix configuration are broken, and the two strands separate. This process is called **denaturation.** Because A-T base-pairs are held together by only two hydrogen bonds, compared with three hydrogen bonds in G-C base-pairs, A-T rich molecules denature more easily (at lower pH or temperature) than G-C rich molecules. When lambda chromosomes are exposed to pH 11.05 for 10 minutes under the appropriate conditions, the A-T rich clusters denature to form "denaturation bubbles," which are detectable by electron microscopy, while the G-C rich regions

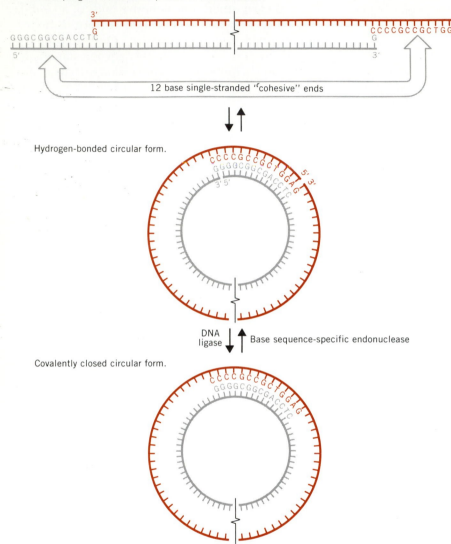

Linear phage λ chromosome present in mature virions

12 base single-stranded "cohesive" ends

Hydrogen-bonded circular form.

DNA ligase   Base sequence-specific endonuclease

Covalently closed circular form.

FIGURE 5.17

Interconversion of the linear lambda chromosome with its complementary "cohesive" ends, the hydrogen-bonded circular lambda chromosome, and the covalently closed circular lambda chromosome. The linear form of the chromosome would appear to be an adaptation to facilitate its injection from the phage head through the small opening in the phage tail into the host cell during infection. Prior to replicating in the host cell, the chromosome is converted to the covalently closed circular form. Only the ends of the chromosome of the mature virion are shown; the jagged vertical line ($\lessgtr$) indicates that the central portion of the chromosome is not shown. The entire lambda chromosome is about 4.5×10^4 nucleotide-pairs long.

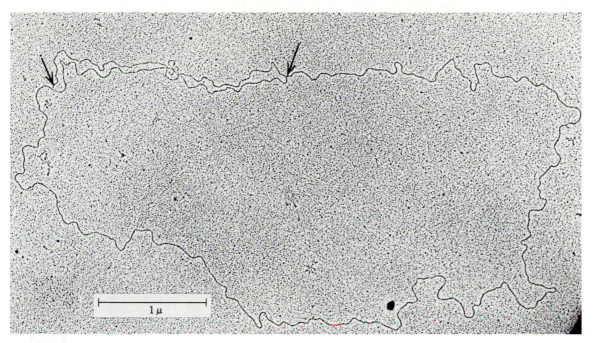

FIGURE 5.18

Electron micrograph of a θ-form replicative intermediate of the chromosome of bacterio-phage lambda. The two replication forks are indicated by the arrows. [Reproduced with permission from M. Schnös and R. B. Inman, J. Mol. Biol. *51:61–73, 1970. Copyright © 1970 by Academic Press, Inc. (London), Ltd.]*

remain in the duplex state (Fig. 5.19). These denaturation bubbles can be used as physical markers whether the lambda chromosome is in its mature linear form, its circular form, or its θ-shaped replicative intermediates. By examining the positions of the branch points (Y-shaped structures) relative to the positions of the denaturation bubbles in a large number of θ-shaped replicative intermediates, Schnös and Inman demonstrated that both branch points are replication forks that move in opposite directions around the circular chromosome. A summary of their results is shown in Fig. 5.20. A schematic illustrating the rationale of this procedure is shown in Fig. 5.21.

Bidirectional replication from a fixed origin has also been demonstrated for several organisms with chromosomes that replicate as linear structures. Replication of the chromosome of phage T7, another small coliphage, begins at a unique site near

one end to form a so-called "eye" structure (⟨○⟩) (Fig. 5.22a) and then proceeds bidirectionally until one fork reaches the nearest end. Replication of the "Y-shaped" structure (Fig. 5.22b) continues until the second fork reaches the other end of the molecule, producing two progeny chromosomes.

Replication of DNA molecules in the chromosomes of eukaryotes is also bidirectional (see pp. 136–141). However, bidirectional replication is not universal. The chromosome of coliphage P2, which like the lambda chromosome is circular during replication, replicates unidirectionally from a unique origin.

In eukaryotes, DNA molecules contain multiple origins of replication (see pp. 139–140). Multiple origins are clearly required to allow the very large DNA molecules in the chromosomes of eukaryotes to complete replication within the observed cell division times. In *Drosophila melanogaster*,

(a) "A-T rich" denaturation sites in linear λ chromosome.

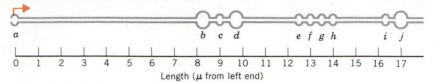

(b) "A-T rich" denaturation sites in circular form of λ chromosome.

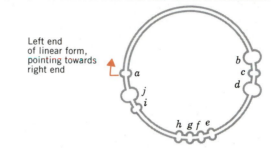

Left end
of linear form,
pointing towards
right end

(c) "A-T rich" denaturation sites in θ-form replicative intermediate.

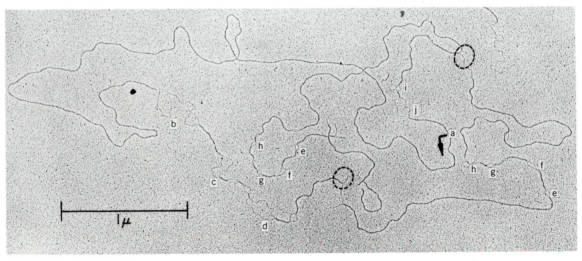

Interpretation:

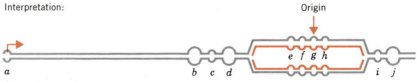

FIGURE 5.19

The use of "A-T rich" denaturation sites as physical markers on the phage λ chromosome to prove that replication is bidirectional rather than unidirectional. (a) The locations of the denaturation bubbles are seen in the linear form on the 17.5 μ long lambda chromosome after treatment at pH 11.05 for 10 minutes, temperature 25°C. Their use as physical markers on the circular form of the lambda chromosome is shown in (b). An electron micrograph of a partially denatured, θ-shaped replicating λ chromosome is

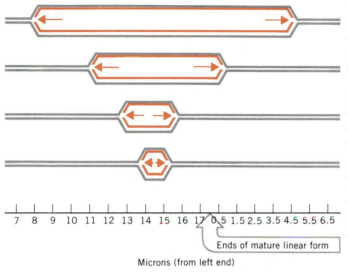

7 8 9 10 11 12 13 14 15 16 17 0.5 1.5 2.5 3.5 4.5 5.5 6.5

Ends of mature linear form

Microns (from left end)

FIGURE 5.20

Diagrammatic summary of the results of electron microscope denaturation mapping of a population of phage lambda replicative intermediates. The circular replication form of the lambda chromosome is diagramed in linear form by breaking the circle near the terminus of replication (not at the cohesive ends of the chromosome of the mature virion). The origin of replication is located at about 14.3 μ from the left end of the chromosome (in its mature linear form). Four chromosomes are shown at different stages of replication. Since both branch points occur at variable positions relative to the physical markers (denaturation bubbles, see Figs. 5.19 and 5.21), both are clearly replication forks that move in opposite directions around the circular chromosome during replication. (Based on the data of M. Schnös and R. B. Inman, J. Mol. Biol. 51:61, 1970.)

for example, the giant DNA molecules of the largest chromosomes contain about 6.5×10^7 nucleotide pairs (see pp. 124–129). The rate of DNA replication in *Drosophila* is about 2600 nucleotide pairs per minute at 25°C. A single replication fork would therefore take about $17\frac{1}{3}$ days to traverse one such giant DNA molecule. With two replication forks moving bidirectionally from a central origin, such a DNA molecule could be replicated in just over $8\frac{1}{2}$ days. The chromosomes of *Drosophila* embryos replicate within 3 to 4 minutes, and the nuclei divide once every 9 to 10 minutes during the early cleavage divisions. Complete replication of the DNA of the largest chromosomes within $3\frac{1}{2}$ minutes would require over 7000 replication forks distributed at equal intervals along the molecules.

The number of active origins varies at different stages of development and in different cell types. In *Drosophila*, there are about 10 times as many active origins per nuclei during early cleavage divisions as in dividing cells of adults. Unfortunately, we still do not know what determines how many or which origins are active in different cells and at different stages of development. Whether origins of replication in eukaryotes and other prokaryotes are always unique sites is also unknown. It is clear that secondary, normally inactive origins exist in some organisms. If the primary origin of replication of phage T7 is deleted (physically removed from the chromosome), a second unique site on the T7 chromosome takes over this function. Why this secondary origin is inactive in the presence of the primary origin is not known.

DNA POLYMERASES AND *IN VITRO* DNA SYNTHESIS

Much has been learned about the molecular mechanisms involved in biological processes by fractionat-

shown in (c), *with an interpretation shown in linear form below it. The bent arrow identifies the small denaturation bubble at the left end of the linear chromosome and points toward the right end. Ten of the most obvious denaturation bubbles are labeled a to j, to aid in orientation. The two replication forks are circled. [The micrograph shown in* (c) *is reproduced with permission from M. Schnös and R. B. Inman,* J. Mol. Biol. *51:61–73, 1970. Copyright © 1970 by Academic Press, Inc. (London), Ltd.]*

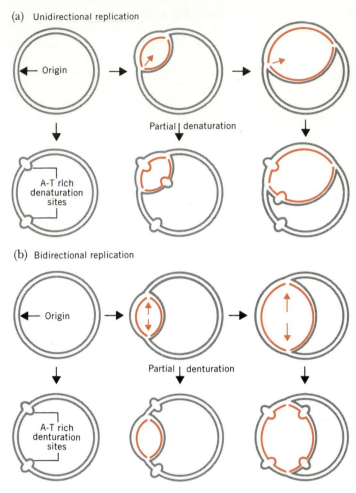

(a) Unidirectional replication

Origin

Partial denaturation

A-T rich denaturation sites

(b) Bidirectional replication

Origin

Partial denturation

A-T rich denturation sites

FIGURE 5.21

The use of electron microscope denaturation mapping to distinguish between (a) *unidirectional and* (b) *bidirectional modes of chromosome replication. Treatment of chromosomes at pH 11.05 for 10 minutes at 25°C under appropriate conditions will denature only regions containing high concentrations of A-T base-pairs. These "denaturation bubbles" can be used as physical markers for specific sites on the chromosome. By examining the positions of the replication forks relative to these markers in a* **population** *of replicating chromosomes, one can distinguish between uni- and bi-directional replication.*

ing cells into their various organelles, macromolecules, and other components, and reconstituting systems in the test tube, so-called *in vitro* systems, capable of carrying out particular metabolic events. Such *in vitro* systems can be dissected biochemically much more easily than *in vivo* systems. Clearly, the information obtained from studies on *in vitro* systems has been invaluable. One should never assume, however, that a phenomenon demonstrated *in vitro* occurs *in vivo*. Such an extrapolation should be made only when independent evidence from *in vivo* studies validate the *in vitro* studies.

The *in vitro* synthesis of DNA was first accomplished by Arthur Kornberg and his co-workers in 1957. Kornberg, who received the Nobel Prize in 1959 for this work, isolated an enzyme from *E. coli*

(initially called **DNA polymerase** or **"Kornberg enzyme"**; now known as DNA polymerase I) that catalyzes the covalent addition of nucleotides to preexisting DNA chains. The enzyme requires the 5'-triphosphates of each of the four deoxyribonucleosides: deoxyadenosine triphosphate (dATP), deoxythymidine triphosphate (dTTP; often written as TTP in the past because "ribo" derivatives of thymine had not been identified), deoxyguanosine triphosphate (dGTP), and deoxycytidine triphosphate (dCTP). The enzyme is active only in the presence of Mg^{++} ions and preexisting DNA. This DNA must provide two essential components, one serving a **primer** function and the other a **template** function (Fig. 5.23). The overall reaction catalyzed by DNA polymerase I is shown in Fig. 5.24; a dia-

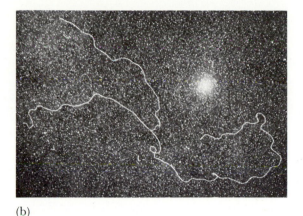

(a) (b)

FIGURE 5.22

Electron micrographs of bacteriophage T7 chromosomes in early stages of replication. The phage T7 chromosome, unlike the E. coli *and phage lambda chromosomes, replicates as a linear structure. Its origin of replication is located 17 percent of the length of the chromosome from one end (the left end of the chromosomes shown). The chromosome in (a) illustrates the "eye" form (⟨⟩) characteristic of early stages in the replication of linear DNA molecules. Replication proceeds bidirectionally until the fork moving in a leftward direction reaches the left end of the molecule, yielding a Y-shaped structure such as that shown in (b). (From J. Wolfson, D. Dressler, and M. Magazin,* Proc. Natl. Acad. Sci., U.S. *69:499, 1972. Original micrographs courtesy of D. Dressler.)*

gram of the functional sites of the enzyme is shown in Fig. 5.25.

(1) Primer DNA. DNA polymerase I cannot initiate the synthesis of DNA chains *de novo*. It has an **absolute requirement for a free 3′-hydroxyl on a preexisting DNA chain.** DNA polymerase I catalyzes the formation of a phosphodiester bridge be-

tween the 3′-OH at the end of the **primer** DNA chain and the 5′-phosphate of the incoming deoxyribonucleotide. The direction of synthesis is thus always 5′ ⟶ 3′ (Fig. 5.26).

(2) Template DNA. DNA polymerase I does not contain sequence specificity. That is, the enzyme requires a DNA **template** chain (Fig. 5.23)

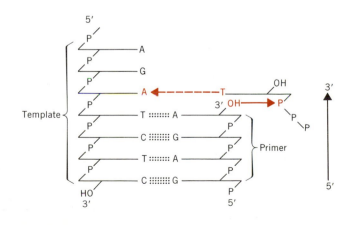

FIGURE 5.23

*Template and primer requirements of DNA polymerases. DNA polymerase requires a **primer** strand (shown on the right) with a free 3′-hydroxyl. It is the primer strand that is covalently extended by the addition of nucleotides (such as dTMP, from the incoming precursor dTTP shown). In addition, DNA polymerases require a **template** strand (shown on the left). It is the template strand that determines the base sequence of the strand being synthesized; the new strand will be complementary to the template strand.*

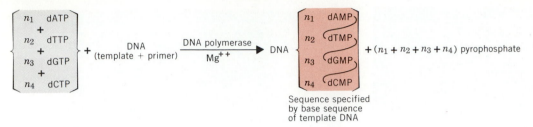

FIGURE 5-24

Overall reaction for DNA polymerase-catalyzed synthesis of DNA, in vitro. *There are requirements for template and primer (see Fig. 5.23), Mg^{++} ions, and the four common deoxyribonucleoside triphosphates (dATP, dTTP, dGTP, and dCTP). The values n_1, n_2, n_3, and n_4 are the moles of dATP, dTTP, dGTP, and dCTP consumed and the moles of dAMP, dTMP, dGMP, and dCMP incorporated during polymerization, respectively. These values will depend on the base composition of the template DNA used in the reaction. The sequence of the nucleotides in the product will also be determined by the base sequence of the template DNA.*

whose base sequence dictates, via the DNA base-pairing requirements, the synthesis of a complementary base sequence in the strand being synthesized.

Since Kornberg's discovery and extensive pioneering work with DNA polymerase I of *E. coli,* a large number of DNA polymerases have been isolated and characterized from many different organisms. Three different DNA polymerases (I, II, and III) have been identified and studied in *E. coli* and *B. subtilis.* Similarly, three DNA polymerases, called α, β, and γ, have been identified in several different eukaryotes, and, recently, a fourth DNA polymerase, named δ, has been isolated from calf thymus and rabbit bone marrow. Thus, there are at least four different DNA polymerases in eukaryotes.

The precise functions of some of the polymerases are still not clear. However, in *E. coli* and *B. subtilis,* DNA polymerase III, rather than DNA polymerase I as first believed, is the major replicative enzyme. Some of the strongest evidence for this has come from studies of mutants, so-called *polA* mutants, that are deficient in DNA polymerase I. These *polA* mutants of *E. coli* replicate their DNA at normal rates, but are defective in their capacity to repair damage to DNA caused, for example, by ultraviolet irradiation. This and other evidence sug-

gests that a major function of DNA polymerase I is DNA repair. Still other evidence indicates that DNA polymerase I is responsible for the excision of the RNA primers used in the initiation of DNA synthesis (see pp. 111–113). The function(s) carried out by DNA polymerase II is uncertain, although it can function in DNA repair in the absence of DNA polymerases I and III. DNA Polymerase III, however, plays an essential role in DNA replication, because in mutant strains growing under conditions where no functional polymerase III is synthesized, DNA synthesis stops.

In eukaryotes, DNA polymerase α is responsible for the replication of chromosomal DNA. DNA polymerases α and β are both localized in nuclei of cells. DNA polymerase γ is located in the mitochondria in animals and in the chloroplasts in plants. Thus, DNA polymerase γ is probably responsible for the replication of the chromosomes of these organelles (see Chapter 14). Of course, one should keep in mind that these organelles may contain other, yet-to-be-identified DNA polymerases.

Most of the prokaryotic DNA polymerases studied so far not only exhibit the $5' \rightarrow 3'$ polymerase activity discussed above, but also have a $3' \rightarrow 5'$ **exonuclease** activity. (An **exonuclease** is an enzyme that degrades nucleic acids from the ends, as opposed to an **endonuclease** that degrades nucleic

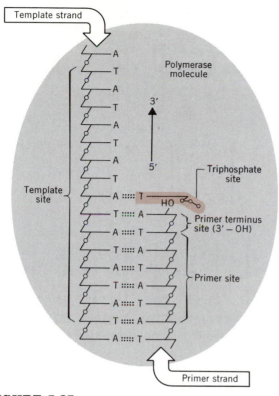

FIGURE 5.25

The interaction of DNA polymerase I with the template and primer strands of DNA and the incoming nucleoside triphosphate. (From A. Kornberg, Science *163:1410–1418, 1969. Copyright © 1969 by the American Association for the Advancement of Science.)*

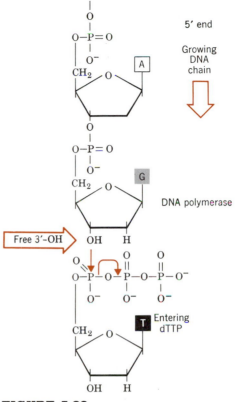

FIGURE 5.26

Covalent extension of DNA chains in the 5′→3′ direction as catalyzed by all known DNA polymerases. The existing chain terminates at the 3′-end with the nucleotide deoxyguanylate (or deoxyguanosine-5′-phosphate). The diagram shows the DNA polymerase-catalyzed addition of deoxythymidine monophosphate [from the precursor deoxythymidine triphosphate (dTTP)] to the 3′-end of the chain with the release of pyrophosphate (P_2O_7).

acids by making internal cuts.) Both activities, polymerase and exonuclease, are present in the same protein macromolecule. The 3′ → 5′ exonuclease activity catalyzes the removal of nucleotides, one by one, from the 3′-ends of polynucleotide chains. Some polymerases, such as DNA polymerase I of *E. coli*, also have **5′ → 3′ exonuclease** activity. When present, 5′ → 3′ exonuclease activity is found at a site on the protein molecule distinct from the active site catalyzing the 5′ → 3′ polymerase reaction and the 3′ → 5′ exonuclease reaction. Both of these polymerase-associated exonuclease activities play important roles in DNA metabolism.

The 3′ → 5′ exonuclease activity of DNA polymerases carries out a **critical "proofreading" or**

"editing" function that is necessary for the high degree of fidelity characteristic of DNA replication. When presented with a template-primer DNA that has a terminal mismatch (an unpaired or incorrectly paired base or sequence of bases at the 3′-end of the primer), the 3′ → 5′ exonuclease of the polymerase clips off the unpaired base or bases (Fig. 5.27). When an appropriately base-paired terminus results, the polymerase begins resynthesis by adding nucleotides to the 3′-end of the primer strand and

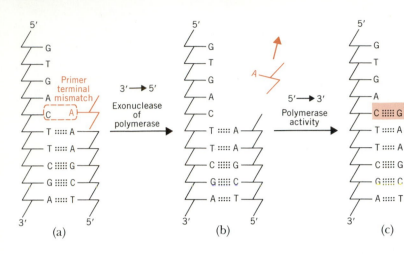

(a)   (b)   (c)

FIGURE 5.27

"Proofreading" by the 3′→5′ exonuclease activity of DNA polymerases during DNA replication. If DNA polymerase is presented with a template and primer containing a 3′ primer terminal mismatch (a), it will not catalyze covalent extension ("polymerization"). Instead, the 3′→5′ exonuclease activity, an integral part of many DNA polymerases, will cleave off the terminal mismatched nucleotide (b). Then, presented with a correctly base-paired primer terminus, DNA polymerase will catalyze 5′→3′ covalent extension of the primer strand (c).

continues until the template is exhausted. This proofreading function of the 3′ → 5′ exonuclease, built into DNA polymerases, is very important. DNA replication must be extremely accurate. A tolerable mistake level could probably not be achieved without such a proofreading mechanism. Thus, it is somewhat surprising that of the eukaryotic DNA polymerases studied to date, only DNA polymerase δ possesses the 3′ → 5′ exonuclease proofreading activity.

The 5′ → 3′ exonuclease activity of many prokaryotic DNA polymerases is also very important. It functions in the removal of segments of DNA damaged by ultraviolet light irradiation and other agents (see Chapter 9). The 5′ → 3′ exonuclease of polymerases, such as *E. coli* DNA polymerase I, may also function in the removal of RNA primers from DNA (see pp. 111–113). This 5′ → 3′ exonuclease activity has not been found in any of the eukaryotic DNA polymerases. In eukaryotes, the functions of the 5′ → 3′ exonuclease activity associated with DNA polymerases in prokaryotes must be carried out by a separate enzyme.

THE "GROWING-POINT PARADOX" AND DISCONTINUOUS DNA SYNTHESIS

Studies of replicating DNA molecules by autoradiography and electron microscopy indicate that the two progeny strands being synthesized at each replicating fork are being extended in the same overall direction, at least on the macromolecular level. Since the complementary strands of a double helix

have opposite polarity, this means that synthesis is occurring at the 5′-end of one strand (or 3′ → 5′) and the 3′-end of the other strand (5′ → 3′). As discussed in the preceding section, however, all known polymerases have an absolute requirement for a free 3′-hydroxyl; they only carry out 5′ → 3′ synthesis. This paradox existed for many years during which biochemists searched in vain for new polymerases that could carry out 3′ → 5′ synthesis. No such polymerase has yet been found. Instead, strong evidence has accumulated indicating that **all synthesis occurs in the 5′ → 3′ direction.** The resolution of the paradox resulted from the demonstration that the synthesis of at least one DNA strand is **discontinuous.**

Autoradiography and electron microscopy show that the two nascent DNA strands being synthesized at each replicating fork are being extended in the **same direction at the macromolecular level.** Since the complementary strands of a DNA double helix have opposite chemical polarity, this means that one strand is being extended in an overall 5′ → 3′ direction and the other strand is being extended in an overall 3′ → 5′ direction (Fig. 5.28, top). At the **molecular level,** however, synthesis is actually occurring in **opposite directions** (Fig. 5.28, bottom). At the molecular level, both new strands are being synthesized in the 5′ → 3′ direction. The strand being extended in the overall 3′ → 5′ direction grows by the synthesis of short segments (synthesized 5′ → 3′), and the subsequent joining of these short segments by polynucleotide ligase. The evidence for this **discontinuous** mode of DNA replication has come from studies in which intermediates in DNA synthesis were radioactively labeled by growth of cells for very short periods of time in medium containing ³H-thymidine ("pulse-labeling"). When *E. coli* cells were pulse-labeled for 15 seconds, for example, all of the label was found in small pieces, 1000–2000 nucleotides long. These small pieces or segments of DNA, often called "Okazaki fragments" after R. Okazaki, who first identified them, are smaller, about 100–200 nucleotides long, in eukaryotes. When longer pulse-labeling periods are used, more of the label is recovered in large DNA molecules—probably the size of molecules containing all of the DNA present in intact chromosomes. In short pulse-labeling periods, the radioactivity present in short DNA "fragments" becomes incorporated in chromosome-sized DNA molecules during subsequent growth of the cells on medium containing nonradioactive thymidine. This is important because it indicates that the "Okazaki fragments" are true intermediates in DNA synthesis rather than some kind of metabolic by-product. It now seems quite clear that DNA synthesis is continuous for the strand growing in the overall 5′ → 3′ direction (sometimes called the "leading" strand) and is discontinuous for the strand growing in the overall 3′ → 5′ direction (sometimes called the "lagging" strand) as is shown in Fig. 5.28.

INITIATION AND THE "PRIMER PROBLEM"

As has been emphasized earlier, all known DNA polymerases have an absolute requirement for a free 3′-OH on a DNA primer plus an appropriate DNA template strand for activity. Thus, **no known DNA polymerase can initiate the synthesis of a new strand of DNA.** Since the synthesis of each "Okazaki fragment" requires an initiation event, an efficient mechanism of chain initiation is essential for ongoing DNA replication. RNA polymerase, a complex enzyme that catalyzes the synthesis of RNA molecules from DNA templates, has long been known to be capable of initiating the synthesis of new RNA chains at specific sites on the DNA. When this occurs, an RNA-DNA hybrid is formed in which the nascent RNA is hydrogen-bonded to the

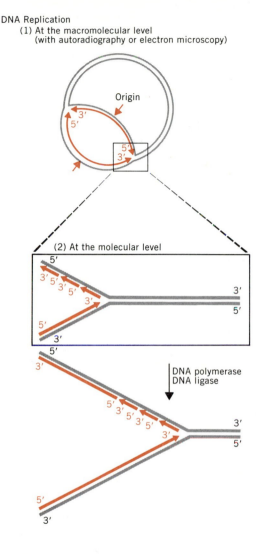

FIGURE 5.28

Discontinuous DNA synthesis. (1) Relatively low-resolution techniques such as autoradiography and electron microscopy show that both nascent DNA chains are extended in the same overall direction at each replication fork. Since the two chains have opposite polarity, the overall or macromolecular direction of extension must be 5′→3′ on one chain and 3′→5′ on the other chain. Both are being extended from left to right, for example, at the framed replication fork. (2) Higher resolution biochemical techniques such as pulse-labeling and density-gradient analysis show that replication is actually discontinuous, at least for the chain being extended in the overall 3′→5′ direction. Short fragments are synthesized in the 5′→3′ direction and subsequently are joined by polynucleotide ligase ("DNA ligase").

DNA template. Since DNA polymerases are capable of extending polynucleotide chains containing an RNA primer with a free 3'-OH, scientists in several laboratories began testing the idea that DNA synthesis is initiated by RNA primers. There is now a considerable amount of evidence supporting the proposal that DNA synthesis is "primed" by short segments of RNA, which are later removed by a 5' → 3' exonuclease and replaced by DNA prior to covalent sealing by polynucleotide ligase (Fig. 5.29). In *E. coli*, the RNA primers appear to be excised by the 5' → 3' exonuclease activity of DNA polymerase I. This probably occurs simultaneously with the synthesis of new DNA strands (replacing the excised RNA primer strands) by the 5' → 3' polymerase activity of this enzyme (Fig. 5.29).

The synthesis of the RNA primers is catalyzed by enzymes called **primases,** which have properties quite distinct from those of the RNA polymerases. The *E. coli* primase is the product of the *dnaG* gene. In prokaryotes, the RNA primers are 10–60 nucleotides in length. In eukaryotes they are quite short, about 10 nucleotides long. The use of RNA primers is almost certainly the most common mechanism used to initiate DNA synthesis. Nevertheless, certain viruses appear to have evolved quite different mechanisms for the initiation of DNA synthesis (see Kornberg, *DNA Replication*).

PHAGE ΦX174 AND "ROLLING CIRCLE" REPLICATION

Bacteriophage ΦX174 is representative of a group of small viruses, both bacterial and eukaryotic, that store their genetic information in a single-stranded, circular molecule of DNA. When these viruses infect a host cell, *E. coli* in the case of ΦX174, the single-stranded viral DNA [called the **"positive"** (+) strand] is converted to a double helical form [called the **"replicative form"** (RF)] by the synthesis of a complementary **"negative"** (−) strand. This double-stranded parental RF then replicates several times to produce a population of progeny RF molecules (double-stranded), which in turn replicate asymmetrically to produce a large population of progeny viral (+) strands. The progeny viral strands are then incorporated into protein coats to complete the reproductive cycle. The replication of the ΦX174 chromosome can thus be divided into three stages: (1) parental (+) strand → parental RF, (2) parental RF → progeny RFs, and (3) progeny RFs → progeny (+) strands (Fig. 5.30). In the last two stages, DNA synthesis occurs by a different mechanism called **"rolling circle" replication.**

Most of the features of "rolling circle" replication are the same as those discussed above for replication via the more common θ, "eye," and Y-shaped structures. In this case, however, the replicative structure is a circular DNA molecule with a single-stranded tail (Fig. 5.31).

Rolling circle replication is initiated when the sequence-specific endonuclease activity of the phage ΦX174 gene *A* protein cleaves the positive strand of the parental RF at the origin of the replication (Fig. 5.30). This endonuclease activity is site-specific; it **cuts the ΦX174 chromosome at only one site, the origin of replication.** It produces 3'-OH and 5'-phosphate termini at the site of the cut in the (+) strand; the (−) strand remains intact. The 5'-end of the (+) strand is unwound and "peeled off," while the intact (−) strand rotates about its axis (thus the name "rolling circle"). This yields the circle with its tail (Figs. 5.30 and 5.31). As initially proposed by W. Gilbert and D. Dressler, the rolling circle model of DNA replication included a specific enzyme, called a "transferase," which attached the 5'-end of the (+) strand to a specific site on the cell membrane. While most, if not all, replicating chromosomes are attached to the membrane, little is known about the specific nature of such attachments. In any case, membrane attachment is not an essential feature of rolling circle replication. As the circle rotates and the 5'-end is displaced, DNA polymerase catalyzes covalent extension at the other (3'-OH) end.

During parental RF to progeny RF replication, the nascent positive strands are used as templates for the discontinuous synthesis of complementary negative strands. In some cases, the synthesis of the complementary strand may occur discontinuously on the single-stranded tail before synthesis of the first strand has been completed. In such cases, a double-stranded tail will be produced. The switch from double-stranded RF DNA synthesis to single-stranded viral (+) DNA synthesis occurs when specific proteins of the viral coat are produced in the

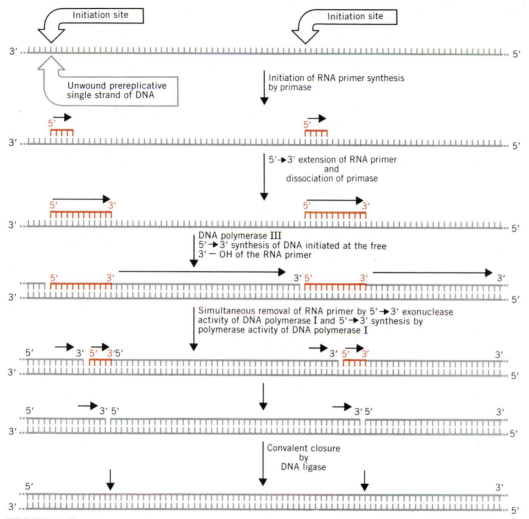

FIGURE 5.29

Schematic illustration of the initiation of DNA synthesis via RNA primers. A short RNA strand is synthesized to provide a 3'-OH primer for DNA synthesis. This RNA primer is subsequently removed and replaced with DNA by the dual 5'→3'

exonuclease and the 5'→3' polymerase activities built into DNA polymerase I. DNA ligase then covalently closes the nascent DNA chain, catalyzing the formation of phosphodiester linkages between adjacent 3'-hydroxyls and 5'-phosphates.

cell. Rolling circle replication continues, but as the viral strand is displaced, these coat proteins bind to it and prevent the synthesis of complementary (−) strands (Fig. 5.30).

The phage ΦX174 gene *A* protein is a key protein in ΦX174 replication. It possesses a remarkable set of activities. (1) Gene *A* protein possesses a site-specific endonuclease activity that cleaves the

positive strand at the origin. (2) Gene *A* protein then maintains the energy of the cleaved phosphodiester linkage by means of a covalent attachment of the 5'-phosphoryl terminus to itself. (3) Gene *A* protein remains bound to the 5'-terminus of the positive strand and to the replication fork while the fork traverses the complete circular minus-strand template. (4) When a complete positive strand has been

Stage I: Viral positive strand ⟶ parental RF

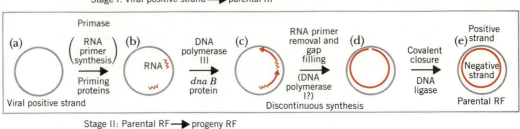

Stage II: Parental RF ⟶ progeny RF

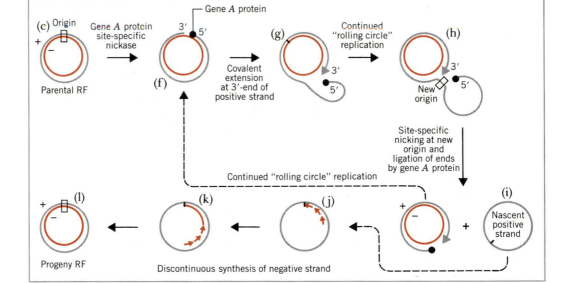

Stage III: Progeny RF ⟶ progeny positive strands

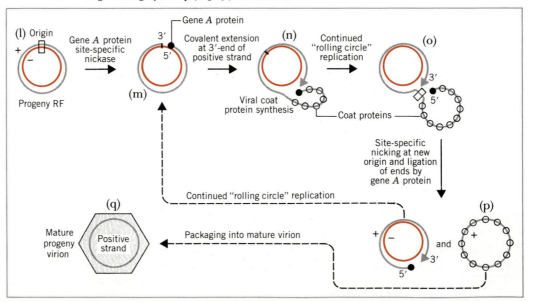

FIGURE 5.30

Three stages in the replication of the single-stranded DNA of bacteriophage ΦX174. *(Top,* a–e) ***Stage I: conversion of the single-stranded chromosome* (a) *to a double-stranded parental replicative (RF) form* (e).** (b) *Synthesis of the complementary "negative"* (−) *strand is initiated by the synthesis of a short RNA primer. This reaction is catalyzed by primase and requires the activity of a complex of at least six different priming proteins; this complex is sometimes called the "primosome."* (c) *DNA polymerase III next catalyzes the covalent addition of deoxyribonucleotides to the 3′-end of the RNA primer. Synthesis of the complementary negative strand then takes place discontinuously (see Figs. 5.28 and 5.29) until the "positive"* (+) *strand template is exhausted. The primosome appears to travel around the circular template strand, pausing to initiate the synthesis of each new "Okazaki fragment."* (d) *Excision of the RNA primers and gap filling appear to be catalyzed by DNA polymerase I, as shown in Fig. 5.29. Polynucleotide ligase ("DNA ligase") then catalyzes the formation of a covalent linkage between the adjacent 3′-OH and 5′-PO₄ groups, to produce the closed, double-stranded parental RF* (e). *(Center,* e–l) ***Stage II: "rolling circle" replication of the parental RF* (e) *to produce a population of progeny RFs* (l).** (f) *The positive strand of the parental RF is cut at the origin by the site-specific endonuclease ("nickase") activity of the ΦX174 gene A protein. The gene A protein nicks the parental RF only at the origin; it will not cut most other DNA molecules at all.* (f–h) *During the nicking event, the gene A protein becomes covalently attached to the 5′-phosphoryl group of the positive strand. It remains linked to*

the 5′-terminus until a complete progeny positive strand has been synthesized. (g) *The 5′-end of the positive strand is displaced from the negative strand and deoxyribonucleotides are added to the free 3′-OH as the circle (maintained by the intact negative strand) rotates about its axis.* (g–h) *The gene A protein remains bound at the replication fork as it travels around the circular negative strand.* (h–i) *Once a new positive strand origin has been synthesized, the gene A protein cleaves the nascent origin and simultaneously ligates the 3′- and 5′-termini to produce a covalently closed circular positive strand.* (i–l) *Synthesis of the complementary negative strand then takes place discontinuously as in stage I using the nascent positive strand as template. The "Okazaki fragments" are initiated by RNA primers in this stage also; however, these RNA primers are not shown in the diagram.* (i–f–l) *The parental RF continues to replicate by the rolling circle mode until a population of about 60 progeny RFs are produced. (Bottom,* l–q) ***Stage III: synthesis of single-stranded progeny chromosomes.*** (l–p) *Rolling circle replication of progeny RFs occurs just as for parental RFs in stage II, except that negative strands are not synthesized. Instead, the positive strands are packaged in progeny virions.* (n–p) *The switch from RF synthesis (stage II) to progeny positive strand synthesis (stage III) results from the binding of newly synthesized viral coat proteins to the nascent positive strand, preventing it from serving as a template for negative strand synthesis.* (q) *Maturation of the progeny virion completes the phage ΦX174 life cycle. Approximately 500 progeny virions are produced per infected cell.*

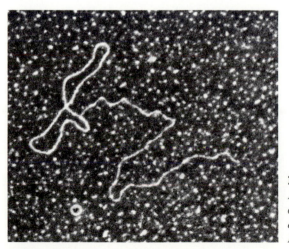

FIGURE 5.31

Electron micrograph of a "rolling circle" intermediate in the replication of the DNA of bacteriophage ΦX174. A single-stranded tail is seen extending from a double-stranded, circular replication form (RF). (From K. Koths and D. Dressler, Proc. Natl. Acad. Sci. 75:605, 1978.)

synthesized, gene A protein cleaves the new origin, ligates the 3′-hydroxyl and 5′-phosphoryl termini, and once again becomes covalently linked to the newly generated 5′-positive-strand terminus. This cycle of gene A protein activities is repeated until a population of progeny RFs (stage II) or progeny positive strands (stage III) are produced.

To date, evidence for rolling circle replication has been found for (1) single-stranded DNA viruses like ΦX174, (2) the replication associated with chromosome transfer during "mating" (conjugation) in bacteria (see Chapter 7), and (3) the replication of small extrachromosomal DNA molecules carrying clusters of rRNA genes during oögenesis in amphibians (see Chapter 11).

PROKARYOTE NUCLEOID STRUCTURE

Most textbooks in biology have until very recently presented a very erroneous picture of the chromosomes of prokaryotes. They have characterized prokaryotic chromosomes as "naked molecules of DNA," in contrast to eukaryotic chromosomes with their associated proteins and complex morphology. This misconception has resulted, at least in part, because (1) the pictures of prokaryotic chromosomes most often published have been autoradiographs and electron micrographs of isolated DNA molecules, **not metabolically active or functional chromosomes,** while (2) the most common pictures of eukaryotic chromosomes have been of highly condensed meiotic or mitotic chromosomes—again, **metabolically inactive chromosomal states.** We now know that functional bacterial chromosomes or "nucleoids" ("nucleoids" rather than "nuclei" since they are not bounded by a nuclear membrane) bear little resemblance to the structures seen in Cairns' autoradiographs just as the metabolically active interphase chromosomes of eukaryotes have little morphological resemblance to mitotic or meiotic metaphase chromosomes.

The contour length of the circular DNA molecule of E. coli is about 1100μ. The E. coli cell has a diameter of only $1-2 \mu$. Clearly, then, the chromosome must exist in a highly folded or coiled configu-

ration within the cell. When the E. coli chromosome is isolated by very gentle procedures in the absence of ionic detergents (commonly used to lyse cells) and is kept in the presence of a high concentration of cations such as polyamines (small basic or positively charged proteins) or $1 M$ salt to neutralize the negatively charged phosphate groups of DNA, the chromosome remains in a highly condensed state comparable in size to the nucleoid *in vivo*. This structure, called the **"folded genome,"** is apparently the functional state of the E. coli chromosome. In its folded genome state, the single DNA molecule of E. coli is arranged into about 50 loops or domains, each of which is highly twisted or **"supercoiled"** (much like a tightly coiled telephone cord). This structure is dependent on RNA and protein, both of which are components of the folded genome (Fig. 5.32). The folded genome can be relaxed by treatment with either deoxyribonuclease (DNase) or ribonuclease (RNase).

Supercoiling of DNA is an important feature of all chromosomes, from those of the smallest viruses to those of eukaryotes. It occurs whenever the DNA is either underwound ("negative supercoils") or overwound ("positive supercoils"). If one takes a covalently closed, circular double helix of DNA, breaks one strand, and rotates one of the ends that is produced for 360° around the complementary strand, while holding the other free end fixed, one supercoil will be introduced into the molecule (Fig. 5.33). If the free end is rotated in the same direction as the DNA double helix is wound (right-handed), a positive supercoil will be produced. If the free end is rotated in the opposite direction (left-handed), a negative supercoil will result. While this is probably the simplest way to visualize the phenomenon of supercoiling in DNA, it is **not** the mechanism used by enzymes to introduce supercoils into DNA (see Fig. 5.34).

Many biological functions of chromosomes can be carried out only when the participating DNA molecules are negatively supercoiled. For example, the phage ΦX174 gene A protein will nick only the ΦX174 replicative form (RF) and initiate replication (see the preceding section) when the RF is in its negatively supercoiled form. All bacterial chromosomes studied to date appear to be negatively super-

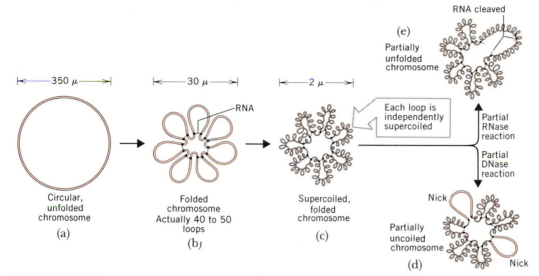

FIGURE 5.32

(a) *Structure of the chromosome of* E. coli *when it is isolated as a completely unfolded, "relaxed" (no supercoils), covalently closed, circular molecule of DNA, much like the structure observed in Cairns' autoradiographs (Fig. 5.16), but here it is nonreplicating. (b) The chromosome segregated into seven loops by RNA "connectors." There are actually 40 – 50 such loops per chromosome in its* in vivo *folded genome form; only seven are shown in the diagram for the sake of simplicity. (c) Each loop is independently supercoiled. The introduction of single-strand* "nicks" (d), *by treatment of the folded genomes with an endonuclease, will relax the DNA only in the nicked loops; all other loops will remain supercoiled. Destruction of the RNA "connectors" (e) by ribonuclease (RNase) will unfold the folded genome partially by eliminating the segregation of the DNA molecule into 40 – 50 loops. However, RNase will not affect supercoiling. (After D. E. Pettijohn and R. Hecht.* Cold Spring Harbor Symp. Quant. Biol. *38:31, 1973.)*

coiled in their functional states. When the basic proteins are carefully dissociated from the DNA of the chromosomes of *Drosophila melanogaster,* the DNA is found to contain the same amount of negative supercoiling as is found in the folded genomes of *E. coli.*

It is very likely that negative supercoiling is universally involved in certain of the biological functions of DNA molecules. Considerable evidence suggests that supercoiling is involved in recombination, gene expression, and the regulation of gene expression. In addition, negative supercoiling is almost certainly required for the replication of most, if not all, DNA molecules. An enzyme called **DNA gyrase,** which catalyzes the formation of negative supercoils in DNA, has now been isolated from several different organisms, both prokaryotes and eukaryotes. The DNA gyrase from *E. coli* has been the most extensively studied. Its activity is inhibited by the drugs novobiocin and nalidixic acid, two potent inhibitors of DNA synthesis in bacteria. This clearly indicates that DNA gyrase activity is required for DNA replication.

Although the exact role (or roles) of DNA gyrase in DNA replication is not yet established, an obvious possibility is that the introduction of negative supercoils may aid in the unwinding of the strands of the double helix. Mechanistically, DNA gyrase is a most interesting enzyme. *In vitro,* it can tie knots in DNA molecules or join two circular molecules to produce interlocking rings. DNA gyrase is now known **not** to introduce negative supercoils into DNA molecules by cleaving one strand and rotating one of the ends as originally proposed (see the visual definition of supercoiling in Fig. 5.33). Rather, it produces negative supercoils two at a time by cleaving both strands of one segment of a DNA molecule, passing another segment of the molecule

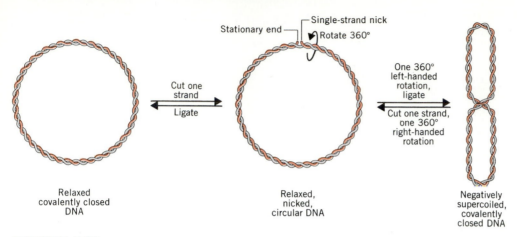

FIGURE 5.33

Visual definition of negative supercoiling. If one strand of a covalently closed circular molecule of DNA (or a linear molecule with fixed ends) is cleaved and one of the resulting free ends is rotated around the intact complementary strand while the other end remains stationary, supercoils will be produced. One supercoil, as shown on the right, will be produced for each 360° rotation of the free end. If the direction of rotation is the same as *the direction in which the DNA double helix is wound (right-handed), a positive supercoil will be produced. If the direction of rotation is left-handed, a negative supercoil will be produced.* **Note that this diagram is for definition purposes only. This is NOT the mechanism used by enzymes such as DNA gyrase to introduce negative supercoils into DNA molecules. That process is illustrated in Fig. 5.34.**

through the temporary gap, and then rejoining the ends (Fig. 5.34). During each catalytic event, DNA gyrase "holds on" to the cut ends as the other DNA molecule or segment of the DNA molecule is passed through the transient gap, so that it can efficiently

rejoin the ends to complete the process. DNA gyrase "holds on" by means of covalent linkage of the transient ends of the cleaved DNA molecule to itself, in the same manner as the phage ΦX174 gene *A* protein (see the preceding section).

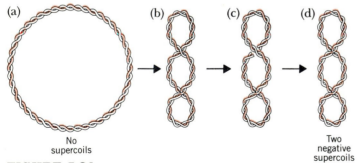

FIGURE 5.34

Mechanism of DNA gyrase-catalyzed negative supercoiling. A relaxed (no supercoils) DNA molecule (a) is wrapped back on itself in a right-handed manner (b). The underlying strand is then transiently cleaved and the overlaying strand is passed through the gap (c). The transient break is resealed yielding two negative supercoils in the molecule (d). The molecule is probably *"wrapped back" on itself by winding it around a portion of the enzyme. The energy of the cleaved phosphodiester bond is conserved by means of transient covalent linkages of the ends produced to the enzyme itself. This energy is then apparently used to reseal the cut. (After P. O. Brown and N. R. Cozzarelli,* **Science** *206:1081–1083, 1979.)*

EUKARYOTE CHROMOSOME STRUCTURE

Much of our information about the structure and mechanisms of DNA replication has come from studies of prokaryotes. The reason for this is that prokaryotes are less complex, both genetically and biochemically, than eukaryotes. Prokaryotes are monoploid (mono = one); they have only one set of genes (one copy of the genome). Most higher animals and many higher plants, by contrast, are diploid, having two complete sets of genes, one from each parent (see Chapters 2 and 3). Many higher plants are polyploid (poly = many), that is, carry several copies of the genome (see Chapter 13). As discussed earlier in this chapter, **the genetic information of most prokaryotes is stored in a single chromosome, which in turn contains a single molecule of nucleic acid** (either RNA or DNA). The smallest known RNA viruses have only three genes. In fact, the complete nucleotide sequences of several of these viruses are now known. (The nucleotide sequence of the coat gene of bacteriophage MS2 is shown in Fig. 8.24. The phage MS2 genome is 3569 nucleotides long and contains four genes.) The smallest known DNA viruses have only 9–11 genes. Again, the complete nucleotide sequences are known in a few cases. (Part of the nucleotide sequence of the phage ΦX174 chromosome, which is 5387 nucleotides long and contains 10 genes, is shown in Fig. 10.28.) The largest DNA viruses, like bacteriophage T2 and the animal pox viruses, contain about 150 genes. Bacteria like *E. coli* have 3000–4000 genes, most of which are present in a single molecule of DNA.

Eukaryotes have from 2 to 10 times as many genes as *E. coli* (see Chapters 9 and 10) but have orders of magnitude more DNA (Fig. 5.35). One of the most challenging problems being studied by geneticists today is the question of the function(s) of this "excess" DNA (DNA not required for structural genes that code for proteins; see Chapter 8). This question will be discussed later in this chapter and again in Chapters 8–10.

Not only do eukaryotes contain many times the amount of DNA of prokaryotes, but this DNA is packaged in several chromosomes, and each chromosome is present in two (diploids) or more

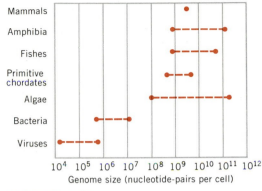

FIGURE 5.35

Increasing genome size with increasing evolutionary or developmental complexity. The minimum and maximum genome sizes observed in each group are given as nucleotide-pairs of DNA per haploid cell. (From "Repeated Segments of DNA" by R. J. Britten and D. E. Kohne, Sci. Amer. 222(4):24–31, 1970. Copyright © 1970 by Scientific American, Inc. All rights reserved.)

(polyploids) copies. Recall that the chromosome of *E. coli* has a contour length of 1100 μ or about 1 mm. Now consider that the haploid chromosome complement, or genome, of the human contains about 1000 mm of DNA (or about 2000 mm per diploid cell). This meter of DNA is, of course, subdivided among 23 chromosomes of variable size and shape, each chromosome containing from 15 to 85 mm of DNA. Until recently, we had little information as to how this DNA was arranged in the chromosomes. Is there one molecule of DNA per chromosome as in prokaryotes, or are there many molecules of DNA per chromosome? If many, how are they arranged relative to each other? How does the 85 mm (85,000 μ) of DNA in the largest human chromosome get condensed into a mitotic metaphase structure that is about 0.5 μ in diameter and 10 μ long? What is the structure of the metabolically active interphase chromosomes?

Interphase chromosomes are not visible with the light microscope; moreover, electron microscopy of thin sections cut through eukaryotic nuclei has provided essentially no information about their structure. Recently, however, chemical analysis,

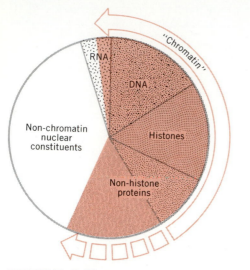

FIGURE 5.36

Eukaryotic chromatin composition as a fraction of total nuclear content. Estimates of nonhistone protein content are variable depending on the isolation procedures and cell types used. (Reprinted with permission from K. E. Van Holde and I. Isenberg, Accounts of Chemical Research 8:327–335, 1975. Copyright © 1975 by the American Chemical Society.)

electron microscopy, and X-ray diffraction studies on isolated **chromatin** (the complex of the DNA, chromosomal proteins, and other chromosome con-

stituents isolated from nuclei) have provided a solid framework for a rapidly emerging picture of chromosome structure in eukaryotes.

CHEMICAL COMPOSITION OF EUKARYOTIC CHROMOSOMES

When chromatin is isolated from interphase nuclei, the individual chromosomes are not recognizable. Instead, one observes an irregular aggregate of nucleoprotein. Chemical analysis of isolated chromatin shows that it consists primarily of DNA and proteins with lesser amounts of RNA (Fig. 5.36). The proteins are of two major classes: (1) basic proteins (positively charged at neutral pH) called **histones** and (2) a heterogeneous, largely acidic (negatively charged at neutral pH) group of proteins collectively referred to as **nonhistone chromosomal proteins.**

Histones play a major structural role in chromatin. They are present in the chromatin of all higher eukaryotes in amounts equivalent to the amounts of DNA (weight/weight). The histones of all higher plants and animals consist of five different major proteins. These five major histones, called H1, H2a, H2b, H3, and H4 (Fig. 5.37), are present in all cell types (except some sperm, where they are replaced by another class of small basic proteins called **protamines**).

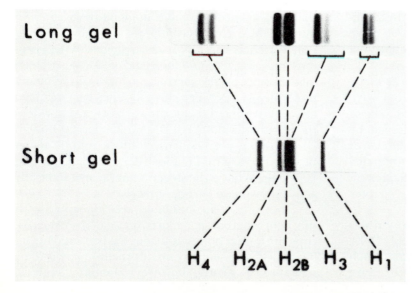

FIGURE 5.37

Acrylamide gel electropherograms showing the five types of histone found in the chromatin of all higher eukaryotes. Note that the long gel (top) shows heterogeneity of histones H1, H3, and H4. The histones shown are from calf chromatin. (Reprinted with permission from S. Panyim and R. Chalkley, Biochemistry 8:3972–3979, 1969. Copyright © 1969 by the American Chemical Society. Original photograph courtesy of Roger Chalkley.)

The five histones are present in molar ratios of approximately 1 H1 : 2 H2a : 2 H2b : 2 H3 : 2 H4. They are specifically complexed with DNA to produce the basic structural subunits of chromatin, small (approximately 110 Å in diameter by 60 Å high) ellipsoidal "beads" called **nucleosomes** (see the following section). The histones have been highly conserved during evolution, four of the five types of histone being very similar in all higher eukaryotes.

Proteins, like nucleic acids, are large macromolecules composed of a large number of small subunits covalently linked together into long polymers (see Chapter 8). In the case of proteins, the subunits are called amino acids, of which 20 different species make up all proteins. Most of the amino acids are neutral, that is, have no charge at neutral pH. A few, however, are basic and a few are acidic. The histones are basic because they contain 20–30 percent arginine and lysine, two positively charged amino acids (Fig. 5.38). The exposed $-NH_3^+$ groups of arginine and lysine allow histones to act as polycations. This is important in their interaction with DNA, which is polyanionic due to the negatively charged phosphate groups.

The remarkable constancy of histones H2a, H2b, H3, and H4 in all cell types of an organism and

even between widely divergent species is consistent with the idea that they are important in chromatin structure ("DNA packaging") and are only nonspecifically involved in the regulation of gene expression.

The nonhistone protein fraction of chromatin, on the other hand, consists of a large number of very heterogeneous proteins (Fig. 5.39). Moreover, the composition of the nonhistone chromosomal protein fraction varies widely among different cell types of the same organism. They are thus likely candidates for roles in the regulation of expression of specific genes or sets of genes (see Chapter 11).

NUCLEOSOME STRUCTURE

When isolated chromatin is examined by electron microscopy, it is found to consist of a series of ellipsoidal "beads" (about 110 Å in diameter and 60 Å high) joined by thin threads (Fig. 5.40). Further evidence for a regular, periodic packaging of DNA has come from studies on the digestion of chromatin with various nucleases. These studies indicated that segments of DNA of 146 nucleotide-pairs in length were somehow protected from degradation by certain nucleases. Moreover, partial digestion of chromatin with these nucleases yielded fragments of DNA in a set of discrete sizes that were integral multiples of the smallest sized fragment (Fig. 5.41). These results are neatly explained if chromatin has a repeating structure, supposedly the "bead" seen by electron microscopy, within which the DNA is packaged in a nuclease-resistant form (Fig. 5.42). According to this model, the **"interbead" threads of DNA** or **linkers** are susceptible to nuclease attack.

After partial nuclease digestion of chromatin, an approximately 200 nucleotide-pair length of DNA is found associated with each nucleosome (produced by a cleavage in each linker region). After extensive nuclease digestion, a 146 nucleotide-pair-long segment of DNA remains present in each nucleosome. This nuclease-resistant structure is called the **nucleosome core.** Its structure is essentially invariant in all eukaryotes, consisting of a 146-nucleotide-pair length of DNA and two molecules each of histones H2a, H2b, H3, and H4. Physical studies (X-ray diffraction and similar analyses) of nucleosome-core crystals have shown that the DNA

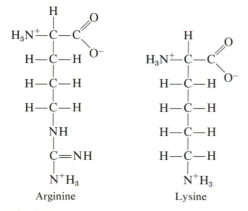

Arginine   Lysine

FIGURE 5.38

Structures of the basic (positively charged at neutral pH) amino acids arginine and lysine. These two amino acids, which make up 20–30 percent of the amino acids of histones, are responsible for the polycationic nature of histones and facilitate their interaction with polyanionic nucleic acids.

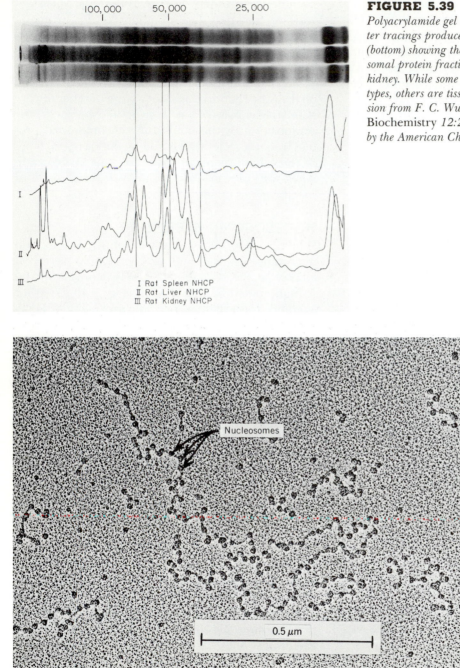

I Rat Spleen NHCP
II Rat Liver NHCP
III Rat Kidney NHCP

FIGURE 5.39

Polyacrylamide gel electropherograms (top) and densitometer tracings produced by photometric scans of the gels (bottom) showing the complexity of the nonhistone chromosomal protein fractions from rat spleen, rat liver, and rat kidney. While some nonhistones are present in all three cell types, others are tissue-specific. (Reproduced with permission from F. C. Wu, S. C. R. Elgin, and L. E. Hood, Biochemistry 12:2792–2797, 1973. Copyright © 1973 by the American Chemical Society.)

FIGURE 5.40

Electron micrograph of rat liver chromatin showing the "beads-on-a-string" nucleosome substructure. (From F. Thoma and T. Koller, Cell 12:101–107, 1977. Copyright © MIT; published by The MIT Press.)

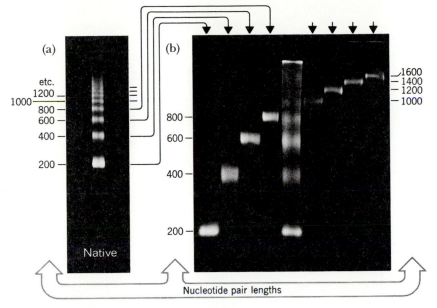

FIGURE 5.41

Demonstration of the repeating subunit (nucleosome) structure of rat liver chromatin by partial nuclease digestion. Rat liver chromatin was partially digested with micrococcal nuclease (an endonuclease). The DNA fragments produced were then extracted from the chromatin and separated by a technique called polyacrylamide gel electrophoresis. This procedure separates DNA molecules on the basis of size. Each DNA fragment preparation is placed on the top of a tube of polyacrylamide that is permeated with a buffer solution; then a high-voltage field is applied across the tube. The negatively charged DNA fragments move through the polyacrylamide gel toward the anode (the positive electrode). The polyacrylamide acts as a molecular sieve; thus, small fragments move faster than larger fragments with the same charge per unit of mass. (a) Total native (double-stranded) DNA fragments. (b) Each fragment band of the gel shown in (a) was cut out and electrophoresed on a separate gel, this time under denaturing conditions (conditions which cause the complementary strands of the double helices to separate). The central gel shown in (b) contained the same total chromatin, nuclease-digested extract as the gel shown in (a).

Note that the fragments fall into a set of discrete sizes where the larger fragment sizes are integral multiples of the size of the smallest fragments (i.e., 200, 400, 600, 800, etc., nucleotide pairs). Consider how these discrete fragment sizes might result from endonuclease cuts of the DNA in the "exposed" linker or internucleosome regions of chromatin (Fig. 5.42). (Data from M. Noll, J. O. Thomas, and R. D. Kornberg, Science 187:1203–1206, 1975. Copyright © 1975 by the American Association for the Advancement of Science.)

is wound as $1\frac{3}{4}$ turns of a superhelix around the outside of the histone octamer (Fig. 5.43a).

The complete chromatin subunit consists of the nucleosome core, the linker DNA, an average of one molecule of histone H1, and the associated nonhistone chromosomal proteins. However, note that it has not been firmly established that histone H1 is evenly distributed, one molecule per nucleosome or linker, in chromatin. The size of the linker DNA varies from species to species and from one cell type to another. Linkers as short as 8 nucleotide pairs and as long as 114 nucleotide pairs have been reported. Some evidence suggests that the "complete nucleosome" (as opposed to the nucleosome core) contains one molecule of histone H1, which stabilizes two full turns of DNA superhelix (a 166-nucleotide-pair length of DNA) on the surface of the histone octamer (Fig. 5.43b). Other evidence

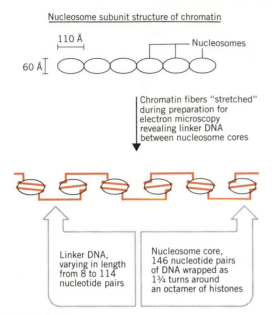

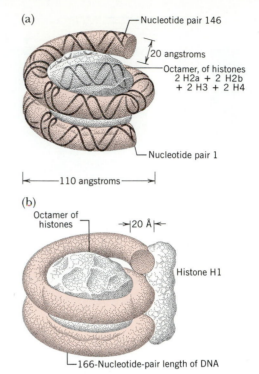

FIGURE 5.42

Nucleosome structure of chromatin. Chromatin of eukaryotes is composed of repeating subunits called nucleosomes. Each nucleosome core contains 146 nucleotide pairs of DNA wrapped around an octamer (two molecules each of histones H2a, H2b, H3, and H4) in a relatively nuclease-resistant complex. These nucleosome cores are joined by nuclease-sensitive segments of DNA called linkers. The linkers are of variable size depending on species and cell type. One molecule of histone H1 is somehow associated with each linker or nucleosome. Thus, the complete repeating unit consists of about 200 nucleotide pairs of DNA and nine histone molecules. In vivo, the linker DNA is probably condensed (coiled?) between closely juxtaposed nucleosome cores (top). The "beads-on-a-string" structure of chromatin (bottom) seen in electron micrographs (see Fig. 5.40) probably results when the fibers are stretched during preparation of the chromatin for electron microscopy. Histone H1 is probably not present in this latter structure.

FIGURE 5.43

Structure of the nucleosome core (a) and the proposed role of histone H1 in stabilizing two complete turns of DNA superhelix around the octamer of histones (b). The nucleosome core consists of a 146 nucleotide-pair-long segment of DNA wound on the surface of a somewhat cylindrical octamer of histones (two each of H2a, H2b, H3, and H4) to yield a roughly ellipsoidal structure. The exact shapes of the histone octamer and histone H1 are not known. The path of the DNA molecule on the surface of the histone octamer has been firmly established. (After R. D. Kornberg and A. Klug, Sci. Amer. *244(2):52–64, 1981.)*

indicates that histone H1 is involved in the coiling or folding of the nucleosome fiber to form a 300 Å chromatin fiber (see Fig. 5.49) and may be involved in other higher levels of organization of chromatin.

Clearly, the basic structural component of eukaryotic chromatin is the nucleosome. But are the structures of all nucleosomes the same? How does the replication fork move past the nucleosome during replication? What role(s), if any, does nucleosome structure play in gene expression and the regulation of gene expression? The structure of nucleosomes in genetically active regions of chromatin is known to differ from that of nucleosomes in genetically inactive regions. But what are the details of this structure–function relationship? Present and future studies on the fine structure of nucleosomes will undoubtedly prove very informative with regard to these and other questions.

ONE DNA MOLECULE PER CHROMOSOME

How is the 1–20 cm (10^4 to 2×10^5 μ) of DNA, which is present in an average eukaryotic chromosome, arranged in the highly condensed mitotic and meiotic structures that are seen with the light microscope (Chapter 3)? Are there many DNA molecules that run parallel throughout the chromosome (the "multineme" or "multistrand" model), or is there just one DNA double helix extending from one end of the chromosome to the other end (the "unineme" or "single-strand" model)? (Note that "strand" here refers to the DNA double helix, not the single-strand of DNA.) Are there many DNA molecules joined end-to-end or arranged in some other fashion in the chromosome, or does one giant, continuous molecule of DNA extend from one end to the other in a highly coiled and folded form? The evidence supporting the unineme model of chromosome structure is now overwhelming. In addition, solid evidence presently supports the concept of chromosome-sized DNA molecules. That is, each chromosome appears to contain a single, giant molecule of DNA.

Some of the strongest evidence supporting the unineme model of chromosome structure has come from studies on the large, so-called "lampbrush" chromosomes, which are present during prophase I of oögenesis in many vertebrates, particularly amphibians. Lampbrush chromosomes are up to $800\,\mu$ long; they thus provide very favorable material for cytological studies. The homologous chromosomes are paired, and each has duplicated to produce two chromatids at the lampbrush stage. Each lampbrush chromosome contains a central axial region, where the two chromatids are highly condensed, and numerous pairs of lateral loops (Figs. 5.44 and 5.45). The loops are transcriptionally active regions of single chromatids. The integrity of both the central axis and the lateral loops is dependent on DNA. Treatment with DNase fragments both the axis and the loops. Treatment with RNase or proteases removes surrounding matrix material, but does not destroy the continuity of either the axis or the loops. Electron microscopy of RNase- and protease-

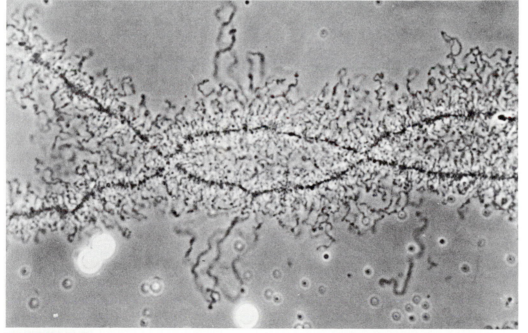

FIGURE 5.44

Phase contrast micrograph of the giant (400–800 μ long) lampbrush chromosomes in the oöcytes of the newt, Triturus viridescens. *The homologous chromosomes of the lampbrush bivalent shown are joined at two chiasmata. Recall that each chromosome has already replicated and thus consists of two chromatids. Note the numerous lateral loops of various lengths. (Courtesy of J. G. Gall.)*

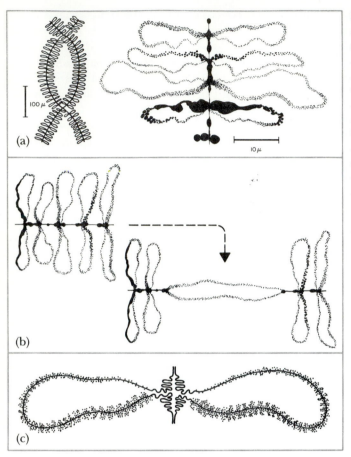

FIGURE 5.45
Schematic representation of the structure of the giant lampbrush chromosomes of amphibian oöcytes. (Left, a) Diagrammatic view of a lampbrush bivalent (compare Fig. 5.44). (Right, a) Semidiagrammatic view of the central chromosome axis with pairs of morphologically identical loops. Each chromosome has already duplicated in interphase. Each homologous chromosome thus contains two chromatids. (b) A portion of a lampbrush chromosome that has been stretched during preparation, revealing the continuity of the loops and the central axis. (c) A single pair of loops displaying the continuity of the single DNA molecule that extends through the loop and axial region of each chromatid. The matrix is primarily nascent RNA that is synthesized on the extended DNA in the loop regions. (From J. G. Gall, "Mutation," Brookhaven Symp. in Biology 8:18, 1955.)

treated lampbrush chromosomes reveals a central filament of just over 20 Å in diameter in the lateral loops. Since each loop is a segment of one chromatid, and since the diameter of a DNA double helix is 20 Å, these lampbrush chromosomes must be unineme structures (Fig. 5.45). This conclusion is also supported by studies on the kinetics of nuclease digestion of lampbrush chromosomes. That is, the kinetics observed are those expected if a single double helix of DNA is the central filament in the loops. The axial region then contains two DNA molecules, one from each of the two tightly paired chromatids.

Lampbrush chromosomes are meiotic or "germ line" chromosomes. Their structure is thus particularly relevant to an understanding of genetic phenomena. "Nongerm line" or somatic cell chromosomes may have different structures. Some are,

in fact, known to be multineme structures (see Chapter 12, pp. 435–437).

The question of whether the unineme chromosomes of eukaryotes contain a single large molecule of DNA or many smaller molecules linked end-to-end has proven very difficult to answer. A centimeter-long molecule of DNA has a length-to-width (diameter) ratio of 5 million to 1. Such a structure is very shear-sensitive. If such a DNA molecule is in solution in a test tube, the slightest vibration will break the molecule into many fragments. For this and other reasons, accurate estimates of the molecular sizes of eukaryotic DNAs cannot be obtained by conventional biochemical methodology. Recently, biophysicists have used a technique called **viscoelastometry** (a procedure for analyzing viscosity parameters of molecules in solution) to estimate the sizes of DNA molecules from eukaryotic chromosomes. For example, the largest DNA molecules in *Drosophila melanogaster* cells were estimated to have a weight of 4.1×10^{10} daltons by viscoelastometry. (A **dalton** is the unit of weight equal to the weight of a single hydrogen atom. It is the most frequently used unit in dealing with sizes of macromolecules like DNA.) The largest chromosomes of *Drosophila* have been shown to contain about 4.3×10^{10} daltons of DNA (total, whether one molecule or many) by direct microspectrophotometric analysis of metaphase chromosomes. Thus, the viscoelastometric estimate of the size of the largest DNA molecules in *Drosophila* nuclei correlates almost exactly with the total amount of DNA present in the largest chromosomes.

Autoradiographs of DNA molecules from *Drosophila,* obtained by very gentle cell lysis procedures, also support the concept of chromosome-sized DNA. The largest molecules observed so far have a contour length of 1.2 cm (Fig. 5.46). While not as large as the viscoelastometric estimates of the largest molecules, such molecules correspond to a weight in the range of 2.4 to 3.2×10^{10} daltons, $\frac{2}{3}$ to $\frac{3}{4}$ the size of the largest chromosome-sized DNA molecules.

How is a DNA molecule, 20 Å in diameter and 1.2×10^8 Å long, packaged into a structure such as a metaphase chromosome (Chapter 3)? Electron micrographs of isolated metaphase chromosomes show masses of tightly coiled or folded lumpy fibers (Fig. 5.47). These fibers have an average diameter of 300 Å. When the structures seen by light and electron microscopy during earlier stages of meiosis are compared, it becomes clear that the light microscope simply detects regions where these 300 Å fibers are tightly packed or condensed (Fig. 5.48).

What is the substructure of the 300 Å fiber seen

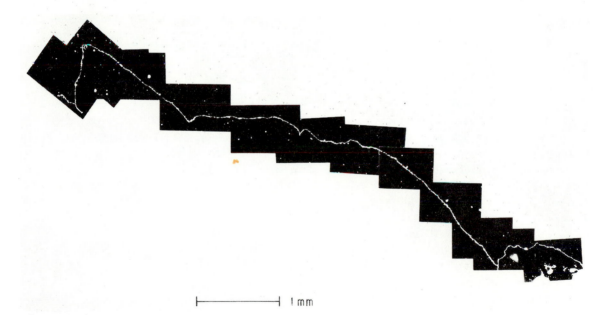

|———————————| 1 mm

FIGURE 5.46

Autoradiograph of a giant DNA molecule from Drosophila melanogaster. *Contour length = 1.2 cm. Such a molecule would have a molecular weight of about 2.8×10^{10}, or about two-thirds of the DNA content of the largest chromosome of* D. melanogaster. *(From R. Kavenoff, L. C. Klotz, and B. H. Zimm,* Cold Spring Harbor Symp. Quant. Biol. *38:1–8, 1973. Copyright © 1973 by Cold Spring Harbor Laboratory. Original photograph courtesy of R. Kavenoff.)*

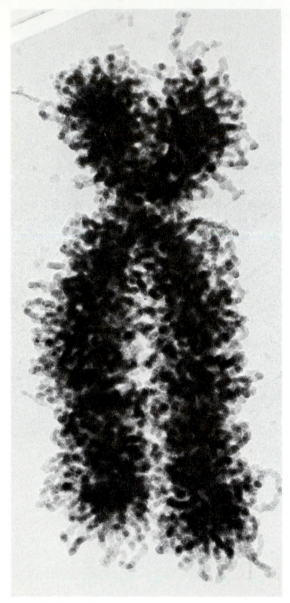

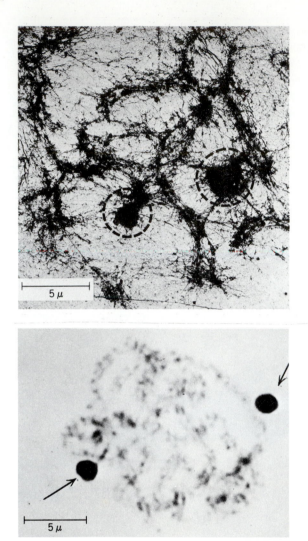

FIGURE 5.47

Electron micrograph of a human metaphase chromosome. Each of the two chromatids is composed of a densely packed mass of fibers. The average diameter of the individul fibers is 300 Å. (From E. DuPraw, DNA and Chromosomes, *Holt, Rinehart and Winston, New York, 1970. Original photograph courtesy of E. DuPraw.)*

FIGURE 5.48

Comparison of the electron (top) and light (bottom) microscope images of meiotic prophase (pachytene) chromosomes of the male milkweed bug, Oncopeltus fasciatus. *The heterochromatic sex univalents are circled on the electron micrograph and marked by arrows on the light micrograph. Note that the images seen with the light microscope correspond to regions where the 300 Å fibers are present at high density in the electron micrograph. The more dispersed fibers are not visible with the light microscope. (From S. L. Wolfe and B. John,* Chromosoma *17:85, 1965.)*

in mitotic and meiotic chromosomes? Although we do not have a firm answer to this question, we now know that the DNA is wound as a supercoil about a histone octamer to yield the roughly 100 Å in diameter nucleosome. Under some conditions, the nucleosomes pack together without a detectable linker to yield a 100 Å fiber. If this 100 Å fiber is, in turn, wound in a higher order supercoil (a "super-supercoil" or solenoid), a 300 Å fiber can easily be generated (Fig. 5.49).

How these 300 Å fibers are further condensed into the observed metaphase structure is not clear.

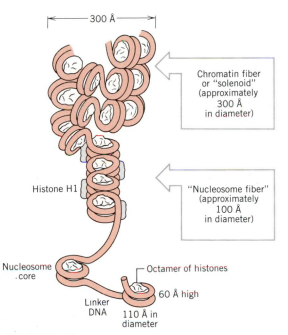

FIGURE 5.49

"Solenoid" model for the structure of the basic 300 Å fiber of meiotic and mitotic chromosomes of eukaryotes. Two levels of supercoiling are proposed; the first level of supercoiling is firmly established by experimental results. (1) The DNA double helix is wound in a negatively supercoiled form in the nucleosome (bottom, see Fig. 5.43). If the nucleosomes are positioned in close juxtaposition to each other with the linker regions coiled between them, a 100–110 Å "nucleosome fiber" is produced (middle). Coiling of this 100 Å fiber can give rise to a 300 Å fiber or solenoid (top). Histone H1 is believed to be involved in the formation and stabilization of these fibers. (Based in part on the results of F. Thoma et al., J. Cell Biol. *83:403–427, 1979.)*

However, there is evidence that the metaphase structure is not dependent on histones. Electron micrographs of isolated metaphase chromosomes from which the histones have been removed reveal a central core or "scaffold," which is surrounded by a huge pool or "halo" of DNA (Fig. 5.50). Note the absence of any apparent ends of DNA molecules in the micrograph, again supporting the concept of chromosome-sized DNA molecules.

At least three levels of condensation are thus required to package the 10^3–10^5 μ of DNA in a eukaryotic chromosome into a metaphase structure a few μ long. (1) The first involves packaging DNA as a supercoil into approximately 100 Å in diameter nucleosomes. This clearly involves an octamer of histone molecules, two each of histones H2a, H2b, H3, and H4. (2) The second may involve a second level of supercoiling to produce the 300 Å fiber characteristic of mitotic and meiotic chromosomes. The involvement of histone H1 has been postulated in the supercoiling of the 100 Å nucleosome fiber to produce the 300 Å fiber. (3) Finally, nonhistone chromosomal proteins may be involved in condensing the 300 Å fibers into the tightly packed metaphase chromosomes. No precise pattern or order is yet apparent for this third level of packaging.

EUCHROMATIN AND HETEROCHROMATIN

When chromosomes are stained by various procedures, such as the Feulgen reaction which is specific for DNA, and are examined by light microscopy, some regions of the chromosomes are observed to stain very darkly, while other regions stain only lightly. When examined by electron microscopy, the intensely staining chromatin, called **heterochromatin,** is seen to consist of densely packed chromatin fibers (300 Å in diameter). The lightly staining chromatin, called **euchromatin,** is composed of less tightly packed 300 Å fibers. Heterochromatin can often be shown to remain highly condensed throughout the cell cycle, whereas euchromatin is not visible with the light microscope during interphase.

Genetic analyses indicate that heterochromatin

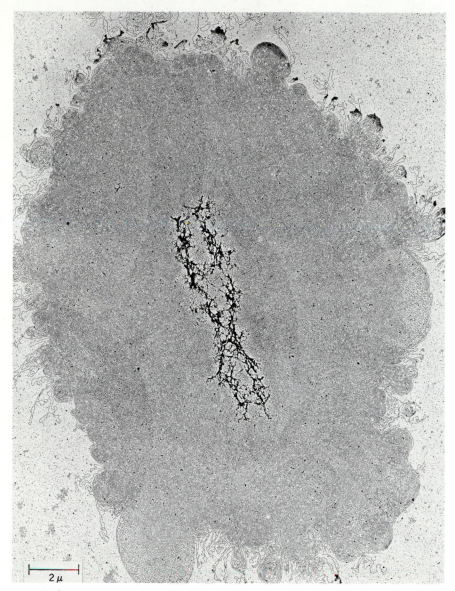

2 μ

FIGURE 5.50

Electron micrograph of a human metaphase chromosome from which the histones have been removed. The entire histone-depleted chromosome has been spread on a monolayer of protein (cytochrome C), which is floating on an aqueous surface; the chromosome is then picked up on a grid for electron microscopy. The chromosome consists of a central, dense "scaffold" or core surrounded by a huge halo of DNA. Because of the relatively low magnification, individual DNA molecules can be seen only near the periphery. Note the absence or rarity of ends of DNA molecules in the chromosome. (From J. R. Paulson and U. K. Laemmli, Cell 12:817–828, 1977. Copyright © 1977 MIT; published by The MIT Press. Original photograph courtesy of U. K. Laemmli.)

is largely genetically inactive. Most of the genes of eukaryotes that have been extensively characterized are located in euchromatic regions of the chromosomes. A structure–function correlation is thus evident: the highly condensed chromatin tends to be genetically inactive; the less condensed chromatin tends to be genetically active (see Chapters 8 and 11).

REPETITIVE DNA AND SEQUENCE ORGANIZATION

The chromosomes of prokaryotes contain DNA molecules with unique (nonrepeated) base-pair sequences. That is, each gene (consider a gene to be a linear sequence of a few hundred to a few thousand base-pairs; see Chapter 10) is present only once in the genome. (The genes for rRNA molecules are an exception.) If prokaryotic chromosomes are broken into many short fragments, each fragment will contain a different sequence of base-pairs. The chromosomes of eukaryotes are much more complex in this respect. Certain base sequences are repeated many times in the haploid chromosome complement, sometimes as many as a million times. DNA containing such repeated sequences, called **repetitive DNA,** often represents a major component (20–50 percent) of the eukaryotic genome.

The first evidence for repetitive DNA came from density-gradient analysis (see Fig. 5.14) of eukaryotic DNA. When the DNA of a prokaryote, such as *E. coli,* is isolated, fragmented, and centrifuged to equilibrium in a cesium chloride (CsCl) density gradient, the DNA usually forms a single band in the gradient. For *E. coli,* this band will form at a position where the CsCl density is equal to the density of DNA containing about 50 percent A-T and 50 percent G-C base-pairs. DNA density increases with increasing G-C content. The extra hydrogen bond in a G-C base-pair is believed to result in a tighter association between the bases and thus a higher density than for A-T base-pairs. On the other hand, CsCl-density-gradient analysis of DNA from eukaryotes usually reveals the presence of one large band of DNA (usually called **"mainband" DNA**) and one to several small bands. These small bands of DNA are called **satellite bands,** and the

DNA in these bands is often referred to as **satellite DNA.** *Drosophila virilis* DNA, for example, contains three distinct satellite DNAs (Fig. 5.51). When isolated and analyzed, each satellite DNA was found to contain a repeating sequence of seven base-pairs. One satellite repeat sequence is 5′-ACAAACT-3′ (one strand; the other strand will have the complementary sequence). A second satellite DNA has a 5′-ATAAACT-3′ repeat; the third satellite repeat sequence is 5′-ACAAATT-3′. They thus differ from each other at only two positions. In three related species of crabs, a satellite DNA is present that contains 97 percent A-T base-pairs. Some satellite DNAs in other eukaryotes have longer repetitive sequences.

The chromosomal locations of several satellite DNAs have been determined by a technique called *in situ* **hybridization.** The complementary strands of DNA molecules are separated by heat or alkali denaturation. Conditions can then be reversed by lowering the temperature or lowering the pH, and the separated strands will **renature** or **reanneal** to reform base-paired double helices, a process called **DNA renaturation.** DNA renaturation is a specific type of **nucleic acid hybridization,** the formation of hydrogen-bonded double helices by single-stranded

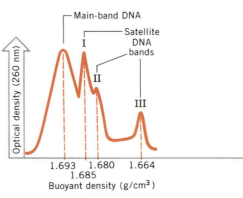

FIGURE 5.51

The three satellite DNAs of Drosophila virilis. *DNA from* D. virilis *embryos was centrifuged to equilibrium in 6 M CsCl. The DNA in the bands was quantitated by measuring the absorption of ultraviolet light (260 nm). (Courtesy of M. Blumenfeld, Department of Genetics and Cell Biology, University of Minnesota.)*

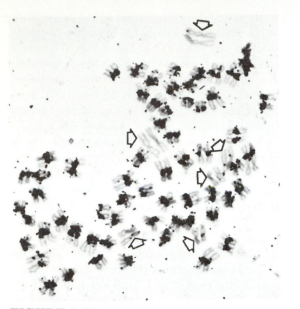

FIGURE 5.52

Autoradiograph showing the hybridization of ³H-labeled RNA transcripts of the DNA satellite HS-β of the kangaroo rat (D. ordii) to metaphase chromosomes. These chromosomes have undergone endoreduplication, a chromosome duplication without separation of the daughter chromosomes. Thus, each chromosome is represented by a pair of closely associated metaphase chromosomes (four chromatids). Most of the chromosomes show radioactivity localized at or near the centromere. Six endoreduplicated chromosomes (three endoreduplicated pairs of homologs) are unlabeled (arrows). Thus, the HS-β satellite DNA sequence is apparently present in the centromeric regions of all but three of the chromosomes of the kangaroo rat. Magnification ×37,400. (Courtesy of D. M. Prescott and Chromosoma 42:205, 1973.)

chromosome squash preparations. After washing out the nonhybridized radioactive material, the locations of the satellite DNA sequences in chromosomes are determined by autoradiography (Fig. 5.52). All satellite DNAs characterized to date are located in heterochromatic regions of chromosomes. In several cases, the satellite DNA sequences are found in heterochromatin in the centromere regions of chromosomes. In other cases, the satellite sequences are located in heterochromatic telomeres (chromosome termini). Satellite DNA sequences are usually found **not** to be transcribed (first step in gene expression; see Chapter 8).

It should be emphasized that a repetitive DNA sequence will be identified as satellite DNA only if the sequence has a base composition sufficiently different from that of main-band DNA; only then is it recognizable as a distinct band in a density gradient. Many repetitive DNA sequences cannot, therefore, be identified by this procedure.

A much more complete picture of the frequency and complexity of repetitive DNA sequences in eukaryotes has resulted from studies of DNA renaturation kinetics. Consider a long DNA molecule with no repeated sequences (such as the DNA of a prokaryote). If such a molecule is sheared into fragments of a particular length, say 400 nucleotide pairs, is denatured, and is then allowed to renature under appropriate conditions, the rate of renaturation will depend on (1) the concentration of DNA in solution and (2) the complexity of the DNA, that is, the number of different 400 base-pair fragments. Consider a particular fragment composed of two complementary strands, *a* and *a'*. Reassociation

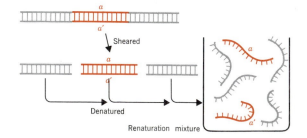

of *a* and *a'* will require a specific collision between these two single strands. Collisions between *a* or *a'* and any other single strand will not lead to hybridization. For a given concentration, the larger the

molecules containing complementary base sequences. Hybridization will occur between complementary single strands regardless of their source. If both participating strands are DNA, the process is called **DNA hybridization.** If one strand is DNA and the complementary strand is RNA, the process is called **DNA-RNA hybridization.**

In situ hybridization in the case of the satellite DNAs usually involves annealing single strands of isolated radioactive satellite DNA, or complementary RNA sequences synthesized using satellite DNA as template, directly to denatured DNA in

DNA molecule, and thus the more nonidentical 400 base-pair fragments, the slower the reassociation reaction will be, because a smaller proportion of the random collisions will be between complementary single strands such as a and a' (Fig. 5.53).

Note that every reassociation event, like a with a', will require a collision between **two** single strands that are **present in the renaturation mixture in equal concentration.** Since the reaction requires the interaction of **two** equally frequent molecules, the rate of renaturation should be a function of the square of the concentration of single strands (so-called second-order or **bimolecular** reaction kinetics).

Or

$$\frac{-dC}{dt} = kC^2$$

(or $\dfrac{-dC}{C^2} = kdt$, rearranged for integration)

where C = the concentration of single-stranded DNA in moles of nucleotide per liter

t = time in seconds

and k = a second-order rate constant in liter per mole seconds

Literally, this equation states that the change (decrease) in concentration of single-stranded DNA $(-dC)$ with time (dt) is equal to the proportionality constant (k) times the square of the concentration of single-stranded DNA.

Integration of the above equation from the initial conditions ($t = 0$ seconds and $C = C_0$, where C_0 equals the concentration of single-stranded DNA at $t = 0$) yields

$$\frac{1}{C} - \frac{1}{C_0} = kt$$

or, rearranged,

$$\frac{C}{C_0} = \frac{1}{1 + kC_0 t}$$

This equation states that the fraction of input single-stranded DNA remaining in a renaturation reaction mixture (C/C_0) at any given time (t) is a function of the initial concentration (C_0) times

elapsed time (t), or $C_0 t$. It is thus convenient to present data on hybridization kinetics in a plot of C/C_0 versus $C_0 t$. These so-called $C_0 t$ (pronounced "caught") curves have provided a great deal of information about the types of repetitive DNA in eukaryotic genomes.

Consider a DNA molecule containing a 400 base-pair sequence that is repeated (present twice).

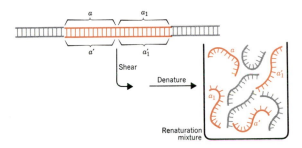

Now a and a_1 are identical single strands, and a' and a_1' are identical complementary strands. Reassociation of each repeated sequence will take only **half** as long as in the previous example, where each fragment contained a unique base sequence. Now reassociation will result from a collision of a with either a' or a_1' (likewise for a_1 and either a' or a_1'). Thus, the time required for reassociation of a particular DNA sequence is inversely proportional to the number of times that sequence is present in the genome. Clearly, highly repetitive DNA sequences will renature very rapidly.

The proportion of the DNA that has renatured at any time can be quantitated in several ways. One method is simply to treat samples taken at various times with a nuclease that is specific for single-stranded nucleic acids. The DNA remaining in the samples after digestion (all double-stranded) can then be quantitated by direct chemical analysis.

Careful renaturation kinetic experiments show not only what proportion of a genome consists of a particular class of repetitive DNA, but also how many copies of that particular sequence (or class of sequences) are present. Such studies have now been done on many eukaryotes. They reveal a surprising amount of diversity, both in the proportion of the genome consisting of repetitive DNA and in the types of repetitive DNA present (Fig. 5.54). These studies of renaturation kinetics indicate that eu-

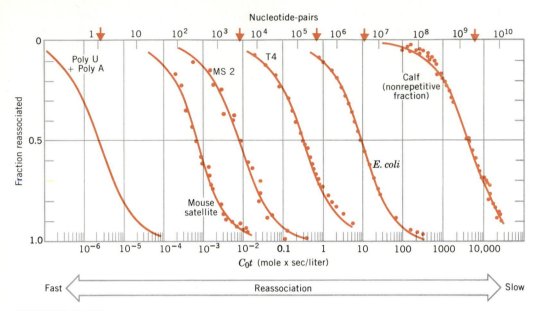

FIGURE 5.53

Renaturation kinetics as a function of genome size (complexity in this case is equivalent to total genome size). The complexity or size of the DNAs in terms of nonrepeating nucleotide-pair sequences, which are plotted along the top, should be read from the mid-points (50 percent reassociation) of the reassociation curves. The fraction reassociated (C/C_0) is plotted as a function of C_0t [the product of the initial concentration (C_0) in moles per liter, and time (t) in seconds]. "Poly U + poly A" is a double helical structure in which one strand contains only U and the other strand contains only A. All base-pairs are thus A:U, and every fragment has the same base sequence. The complexity of such a
duplex is one (or the repeating sequence is equivalent to one A:U base-pair). "Poly U + poly A" thus exhibits the fastest bimolecular reassociation possible. The very complex nonrepetitive DNA from calf chromosomes, by contrast, shows very slow reassociation kinetics. DNAs of intermediate complexities show intermediate rates of reassociation. A plot of genome size (excluding repetitive sequences) versus renaturation rate is, in fact, linear. (From R. J. Britten and D. E. Kohne, Science 161:529–540, 1968. Copyright © 1968 by the American Association for the Advancement of Science.)

karyotic genomes characteristically contain (1) a large fraction (40–80 percent) of **unique** or **single-copy** DNA sequences, (2) a fraction of **middle-repetitive** DNA sequences present in 2 to 10^5 copies, and (3) a fraction of **highly repetitive** DNA sequences present in greater than 10^5 copies (Fig. 5.55). The middle-repetitive class appears to be quite heterogeneous in some eukaryotes; it contains many different sequences with different degrees of reiteration. In *Drosophila melanogaster*, for example, 12 percent of the DNA contains middle-repetitive DNA with an average reiteration frequency of 70.

The highly repetitive DNA contains both satellite and nonsatellite DNA sequences.

When the sequence organization of eukaryotic genomes is studied — by combining the techniques of density-gradient analysis, hybridization kinetic analysis, biochemical analyses, and electron microscopy — a pattern emerges. Much of the genome consists of **middle-repetitive sequences interspersed with single-copy sequences** (Fig. 5.56). In toads, sea urchins, and humans, the sequences are quite short. The middle-repetitive sequences average 300 nucleotide pairs in length; the single-copy

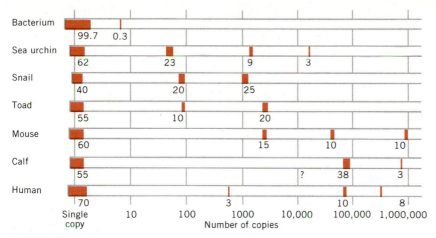

FIGURE 5.54

Patterns of repetitive DNA sequences in a few representative organisms. The fractions of the total DNA of the organism that are present in single-copy sequences and in each known repetition class (represented by shaded bands) are indicated by the widths of the bands and the numbers below the bands. In certain cases, some of the DNA sequences in the genome are not well characterized; thus, the fractions represented do not total 100 percent. The position of the band indicates the number of times that the specific sequence is present in the genome. For example, 25 percent of the DNA of the snail consists of a sequence present in about 1000 copies. (From "Repeated Segments of DNA," by R. J. Britten and D. E. Kohne, Sci. Amer. *222(4):24–31, 1970. Copyright © 1970 by Scientific American, Inc. All rights reserved.)*

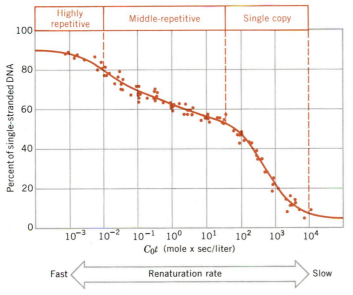

FIGURE 5.55

The complex renaturation kinetics of human DNA illustrating the presence of highly repetitive sequences and middle-repetitive sequences as well as unique or single-copy sequences. (Data from C. W. Schmid and P. O. Deininger, Cell *6:345–358, 1975. Copyright © 1975 MIT; published by The MIT Press.)*

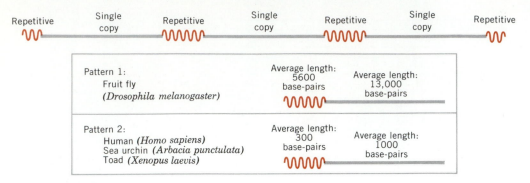

FIGURE 5.56

Patterns of interspersion of single-copy and middle-repetitive DNA sequences that make up the major components of eukaryotic genomes. The significance of such interspersion patterns is not yet clear. However, most structural genes (genes coding for proteins) are present in a single copy in the genome. Many geneticists believe that middle-repetitive sequences are involved in the regulation of gene expression.

sequences are about 800–1200 nucleotide pairs long. *Drosophila melanogaster* DNA also exhibits interspersion of middle-repetitive and single-copy sequences, but the sequences are much longer (5600 and 13,000 nucleotide pairs, respectively).

What are the functions of the different kinds of sequences in eukaryotes? Most of the structural genes (genes coding for proteins or RNA molecules; see Chapter 8) are single-copy sequences. The genes for histones, rRNA molecules, and ribosomal proteins, all gene products that are present in the cell in large quantities, are exceptions; these genes are **redundant** (present in middle-repetitive DNA). Many geneticists believe that most other middle-repetitive sequences are involved in the regulation of gene expression (see Chapter 11). Their interspersion with single-copy sequences, and thus their location adjacent to structural genes, is certainly consistent with a regulatory role. Recently, several intermediate-repetitive sequences of *Drosophila melanogaster* have been shown to be capable of moving from one location in a chromosome to another location in the same chromosome or even to a different chromosome. These intermediate-repetitive DNA sequences have been called **"nomadic" sequences** or **transposable elements,** because of their ability to migrate from one position to another in the genome. These "nomadic" middle-repetitive DNA sequences are clearly analogous to the well-charac-

terized transposable elements of prokaryotes (see Chapter 7, pp. 221–226). The **transposability** of these intermediate-repetitive sequences has led to much speculation as to their possible involvement in the regulation of gene expression during development and their almost certain involvement in the evolution of eukaryotic genomes. The actual function(s) of these "nomadic" sequences is still unknown. However, they are known to be responsible for a surprisingly large number of the mutations of yeast and *Drosophila*.

The function of highly repetitive DNA — most, if not all, of which is located in genetically inactive heterochromatic regions of chromosomes — is also completely unknown. Postulated functions for highly repetitive DNA include: (1) structural or organizational roles in chromosomes, (2) involvement in chromosome pairing during meiosis, (3) involvement in crossing over or recombination (see Chapter 6), and (4) "protection" of important structural genes, like histone, rRNA, or ribosomal protein genes. The validity of these postulated roles remains to be determined.

REPLICATION OF EUKARYOTIC CHROMOSOMES

The duplication of the chromosomes of eukaryotes involves not only the replication of their giant DNA molecules, but also the synthesis of the associated histones and nonhistone chromosomal proteins. The packaging of DNA and histones into nucleosomes supposedly is the final stage in chromosome duplication. Thus, some features of the replication of eukaryotic chromosomes are clearly unique to eukaryotes. At the molecular level, however, the replication of DNA in eukaryotes appears to involve many of the same enzymes and mechanisms as in prokaryotes. The eukaryotic DNA polymerases have the same absolute requirements for template and primer as prokaryote polymerases. DNA replication is semiconservative and is discontinuous, at least for the strand synthesized in the overall $3' \rightarrow 5'$ direction. However, the Okazaki fragments produced by discontinuous synthesis are only 100–200 nucleotides long in eukaryotes. This may be related to the fact that DNA synthesis is slower (about $1\,\mu$ of DNA per minute) in eukaryotes than in prokaryotes (about $30\,\mu/\text{min}$). The slower rate of replication, in turn, may result from the need to disassemble or conformationally rearrange the nucleosomes ahead of the advancing replication forks.

THE CELL CYCLE

DNA synthesis occurs continuously in most prokaryotic organisms growing under optimal conditions. That is, DNA synthesis occurs from the time a new cell is formed by cell fission until the time that cell divides again. Eukaryotic organisms, on the other hand, exhibit a more complex cell cycle. The important chromosome condensation and segregation events of mitosis (M phase; see Fig. 3.3) occur within a short period (often about one hour) of the total cell cycle. Most DNA synthesis occurs during interphase over a period (designated S for synthesis) of several hours. In many eukaryotic cells, there is an interval called G_1 (for "first gap") after the completion of mitosis and prior to the initiation of DNA synthesis. Another interval, called G_2, follows the completion of DNA synthesis and precedes mitosis. Interphase can thus be divided into three stages, G_1, S, and G_2. G_1, S, and G_2 are periods of intense metabolic activity and growth of cells; most cellular metabolism stops during mitosis. The sequence $G_1 \rightarrow S \rightarrow G_2 \rightarrow M \rightarrow G_1$ is known as the **cell cycle.**

The durations of the various stages, particularly G_1 and G_2, are highly variable, depending on the cell type and species involved.

SEMICONSERVATIVE CHROMOSOME REPLICATION

Eukaryotic chromosomes, like their simpler prokaryotic counterparts, replicate in a semiconservative manner. This was first demonstrated in 1957 by J. H. Taylor, P. Woods, and W. Hughes for root-tip cells of the broad bean, *Vicia faba*. Taylor and his colleagues labeled *V. faba* chromosomes by growing root tips for eight hours (less than one cell cycle) in medium containing radioactive (^{3}H) thymidine. The root tips were then removed from the radioactive medium, washed, and transferred to nonradioactive medium containing **colchicine.** Colchicine binds to microtubules and prevents the formation of functional spindle fibers. As a result, daughter chromosomes do not undergo their normal anaphase separation. Thus, the number of chromosomes per nucleus will double once per cell cycle in the presence of colchicine. This allowed Taylor and his colleagues to determine how many DNA duplications each cell had undergone **subsequent** to the incorporation of radioactive thymidine. At the first metaphase in colchicine (**c-metaphase),** nuclei will contain 12 pairs of chromatids (still joined at the centromeres). At the second c-metaphase, nuclei will contain 24 pairs, and so on. The distributions of radioactive DNA at the first and second c-metaphases were determined by autoradiography (Figs. 5.57a and 5.57b). Both chromatids of each pair were similarly labeled at the first c-metaphase. At the second c-metaphase, however, only one of the chromatids of each pair was radioactive. These results are precisely predicted by the semiconservative replication of DNA, given one DNA molecule per chromosome (Fig. 5.57c). At the time, however, Taylor and his colleagues could only conclude that the chromosomal DNA segregated in a semiconservative manner during each cell cycle. Analogous experiments have been carried out with several other eukaryotes and, in all cases, the results indicate that replication is semiconservative.

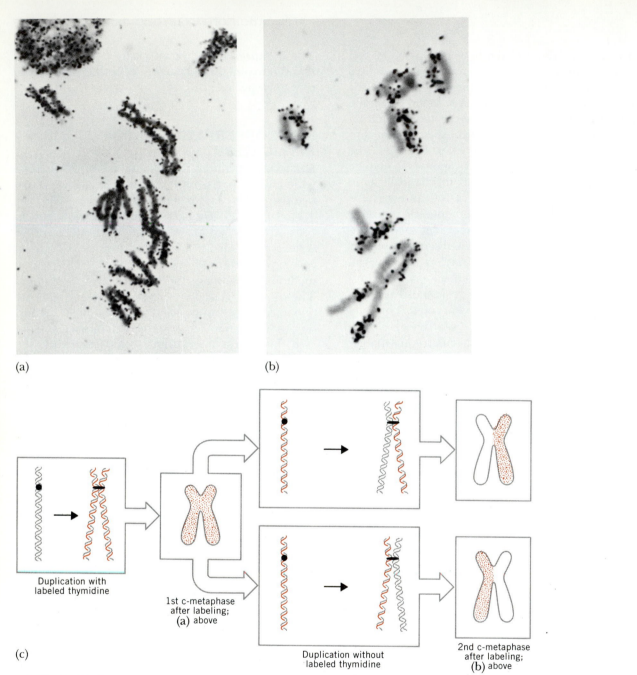

(a)

(b)

(c)

Duplication with
labeled thymidine

1st c-metaphase
after labeling;
(a) above

Duplication without
labeled thymidine

2nd c-metaphase
after labeling;
(b) above

FIGURE 5.57

(Top) Autoradiographs of Vicia faba *chromosomes at the first
metaphase after replication in the presence of* ³H-thymidine (a)
*and at the second metaphase after an additional replication in
nonradioactive medium* (b). (c) *Diagrammatic interpretation of
the results shown in* a *and* b. *Radioactive single strands of DNA
are shown in color. Radioactive chromatids at metaphase are
indicated by colored dots. Colchicine has been used to inhibit
spindle fiber formation and thus the anaphase separation of
sister chromatids. Under these "c-metaphase" conditions, separation of sister centromeres is delayed. (From J. H. Taylor, "The
Replication and Organization of DNA in Chromosomes,"* Molecular Genetics, *Part I, J. H. Taylor (ed.), Academic Press,
New York, 1963.)*

MULTIPLE "REPLICONS" PER CHROMOSOME

When eukaryotic DNAs are pulse-labeled with ^{3}H-thymidine and the DNAs are extracted from the chromosomes and autoradiographed, tandem arrays of radioactivity are observed, suggesting that single macromolecules of DNA contain multiple origins of replication (Fig. 5.58a). Moreover, when the pulse-labeling period is followed by a short period of growth in nonradioactive medium ("pulse-

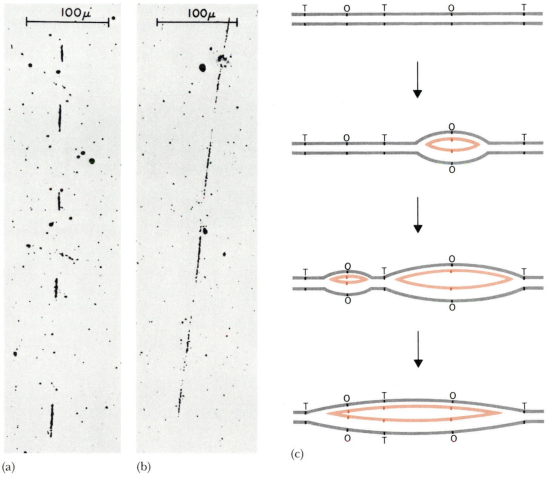

(a)　　　　　(b)　　　　　(c)

FIGURE 5.58

Bidirectional replication of the multiple replicons of DNA molecules of mammalian chromosomes. (a) An autoradiograph of a segment of a DNA molecule from a Chinese hamster cell that has been pulse-labeled with ^{3}H-thymidine. The tandem arrays of radioactivity indicate that replication was occurring at five distinct sites in this segment of the DNA molecule. (b) An autoradiograph of a segment of a DNA molecule from a cell that was pulse-labeled with tritiated thymidine and then allowed to grow for a period in nonradioactive medium (pulse-chase). The decreasing grain densities from the centers to both ends of the tracks *indicate that two replication forks were moving in opposite directions. (c) A schematic interpretation of the autoradiographs. Replication begins at multiple origins (O) of the DNA molecules and proceeds bidirectionally toward termini (T). Initiation does not necessarily occur simultaneously at all origins. [Reproduced with permission from J. A. Huberman and A. D. Riggs, J. Mol. Biol. 32:327–341, 1968. Copyright © 1968 by Academic Press, Inc. (London) Ltd. Original photographs courtesy of J. A. Huberman.]*

chase'' experiments), the tandem arrays contain central regions of high grain density with ''tails'' of decreasing grain density **at each end** (Fig. 5.58*b*). This indicates that replication in eukaryotes is bidirectional, as it is in most prokaryotes. The tails of decreasing grain density result from the gradual dilution of the intracellular pools of ^{3}H-thymidine by cold (^{1}H) thymidine as the replication forks move bidirectionally from central origins toward replica-

tion termini (Fig. 5.58*c*). A segment of the chromosome whose replication is under the control of an origin and two termini has been called a **replicon.** In prokaryotes, the entire chromosome is usually one replicon.

The existence of many units of replication (or replicons) per eukaryotic chromosome, each with its own origin, has been verified by electron microscopy (Fig. 5.59). Some evidence indicates that repli-

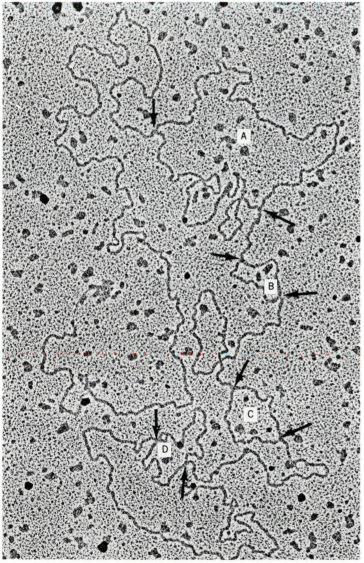

FIGURE 5.59

Electron micrograph of a DNA molecule in **Drosophila melanogaster** *showing multiple sites of replication. Four eye-shaped (◁▷) replication structures (labeled A – D) are present in the segment of the DNA molecule shown. (From D. R. Wolstenholme,* Chromosoma *43:1, 1973. Original photograph courtesy of D. R. Wolstenholme.)*

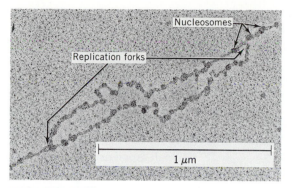

FIGURE 5.60

Electron micrograph of replicating chromatin from blastoderm embryos of Drosophila melanogaster. *Note the organization of the chromatin into nucleosomes both ahead of the replication forks ("prereplicative chromatin") and behind the replication forks ("postreplicative chromatin"). If the nucleosomes are disassembled to allow the replication forks to pass, as would seem necessary, they are clearly reassembled quite rapidly after DNA replication has occurred. Recall that replication is bidirectional; thus each branch point is a replication fork. (From S. L. McKnight and O. L. Miller, Jr.,* Cell *12:795–804, 1977. Copyright © 1977 MIT; published by The MIT Press. Original photograph courtesy of O. L. Miller, Jr.)*

cation may be initiated at more sites during the very rapid cell divisions of embryogenesis than it is during later stages of development. If so, it would be very interesting to know what determines which origins are operational at a given time or in a particular type of cell! One point that is clear is that the newly replicated DNA of eukaryotes is packaged into nucleosomes very rapidly. Electron microscopy of replicating chromatin reveals the presence of nucleosomes with apparently normal spacing on both sides of the replicating forks (Fig. 5.60).

THE COMPLEX "REPLICATION APPARATUS"

When Watson and Crick worked out the double helix structure of DNA, they immediately recognized that the complementary nature of the two strands provided a simple basis for the faithful duplication of genetic material. Meselson and Stahl's demonstration of the semiconservative replication of the *E. coli* chromosome solidified the concept that the two strands of the double helix unwind and serve as templates for the synthesis of complementary strands. Thus, a parental double helix directs the synthesis of two identical progeny double helices. Kornberg's isolation of an enzyme, DNA polymerase I, capable of synthesizing DNA *in vitro*, appeared to provide the final link in what was thought to be an elegantly simple mechanism for the replication of the genetic material—but such was not the case. Twenty years later, scientists are still trying to work out the details of the mechanism of DNA replication.

DNA replication is complex. It is carried out by a **multienzyme complex,** often called the **replication apparatus** or the **replisome.** In eukaryotes, the components of the replication machinery are just beginning to be identified. Even in prokaryotes, DNA replication requires many different proteins. *E. coli* DNA replication requires at least two dozen different gene products. Many of these gene products have been purified and their roles in DNA replication studied *in vitro*. Figure 5.61 shows the involvement of some of these *E. coli* proteins in DNA replication; it is intended to illustrate the complexity of the replication process rather than to detail the roles of the individual gene products.

Several of the components essential for DNA replication have only been identified genetically. That is, *E. coli* strains carrying mutations (heritable changes in the genetic material; see Chapter 9) that result in the inability to replicate DNA under certain conditions (usually high temperature) have been identified. When these mutations were characterized genetically (see Chapters 6 and 7), they were found to identify a set of genes (designated *dnaA*, *dnaB*, etc.) whose products are required for DNA synthesis *in vivo*. The products of a few of these genes are known. For example, *dnaE*, *dnaN*, *dnaX*, and *dnaZ* code for four of the seven subunits (polypeptides) of the complete DNA polymerase III enzyme, and *dnaG* specifies the primase. Other replication enzymes, like polynucleotide ligase, were first identified biochemically.

Three different types of proteins appear to

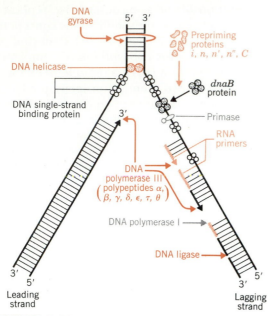

FIGURE 5.61

Complexity of the E. coli *replication apparatus. Only those proteins that have been purified (or partially purified) and studied in vitro are shown. Other gene-products, such as the products of genes* dnaJ, dnaK, dnaL, dnaP, *and* dnaT, *are known to be required for replication. However, these gene products have not yet been identified. (After A. Kornberg, DNA Replication, Freeman, 1980.)*

a result of the fact that the complexes are held together by relatively weak protein-protein interactions, which are disrupted during the isolation procedures. In addition, replication complexes may be membrane-bound and require membrane structures for their assembly. There is considerable evidence that replication forks are associated with the cell membrane in prokaryotes and with the nuclear envelope in eukaryotes.

For excellent, more detailed accounts of replication and the components of the replication apparatus, the reader is referred to Kornberg's *DNA Replication* and *1982 Supplement to DNA Replication* and to Lewin's *Genes.*

SUMMARY

The genetic information of all living organisms, except for certain viruses, is stored in **DNA. RNA** is the genetic material of some viruses. DNA usually has a **double helical structure** in which the two strands are held together by **hydrogen bonding** between the two bases of each nucleotide pair and **hydrophobic bonding** in the core of "stacked" base-pairs. The **base-pairing is specific: adenine always pairs with thymine, and guanine always pairs with cytosine.** The two strands of a DNA double helix are thus **complementary;** once the sequence of bases in one strand is established, the sequence of bases in the other (complementary) strand is fixed.

Replication of DNA is semiconservative in both prokaryotes and eukaryotes. During replication the two complementary strands unwind, and each single strand serves as a template directing the synthesis of a new complementary strand. The net result of replication is thus two progeny DNA molecules identical to the parental double helix. Replication often begins at **unique origins** and proceeds **bidirectionally** from these origins. Prokaryotic chromosomes usually contain a single primary origin. Eukaryotic chromosomes, on the other hand, contain multiple origins of replication.

The two complementary strands of a DNA double helix have **opposite** chemical polarity, one 3′ to 5′, the other 5′ to 3′. All known **DNA polymerases,** both prokaryotic and eukaryotic, have an

contribute to unwinding the strands of double helices. (1) **DNA single-strand binding proteins** bind tightly to single-stranded DNA and, in so doing, aid unwinding and help stabilize the extended single-stranded templates needed for polymerization. (2) **DNA unwinding proteins** or **DNA helicases** are directly involved in catalyzing the unwinding of the double helices. (3) Finally, **DNA gyrases,** which catalyze the formation of negative supercoils in DNA (see Fig. 5.34), appear to be essential for the unwinding process.

Hopefully, the exact functions of the many gene products involved in replication will be worked out during the next few years. Attempts to isolate intact, functional replisomes, however, have been largely unsuccessful. Reconstitution of subcomplexes of replication apparatuses from purified proteins has been more successful. This is undoubtedly absolute requirement for a free **3'-hydroxyl primer.** They only catalyze covalent extension in the **5' → 3' direction.** Thus, synthesis of the strand growing in the overall 3' → 5' direction is **discontinuous.** Short segments are synthesized in the 5' → 3' direction and then covalently joined by **polynucleotide ligase.** No known DNA polymerase can catalyze the initiation of new DNA chains. Instead, nascent DNA chains are initiated by short **RNA primers.**

The chromosomes of prokaryotes exist as highly compact **"folded genomes"** *in vivo.* The DNA molecules are segregated into many independently **negatively supercoiled loops by RNA-containing cross-links.**

Each eukaryotic chromosome appears to contain one giant molecule of DNA. In interphase chromatin, this DNA is held in a **negatively supercoiled** configuration by **histone** packaging. Interphase chromatin has a "beads-on-a-string" structure. Each "bead" or **nucleosome** is an ellipsoid-like structure about 110 Å in diameter and 60 Å high. Each nucleosome contains a highly conserved **core** consisting of a **146 nucleotide-pair-long segment of DNA wrapped around an octamer of histones.** Each octamer contains two molecules each of the histones H2a, H2b, H3, and H4. During mitosis and meiosis, the chromatin is further condensed, possibly by a second level of supercoiling, into **300 Å diameter fibers.** At metaphase, the highly coiled or folded 300 Å fibers are organized about a "scaffold" that is composed of nonhistone chromosomal proteins.

Eukaryotic chromosomes contain some DNA sequences that are repeated many times in the genome. DNA containing repeated sequences is called **repetitive DNA.** DNA containing sequences that are repeated more than 10^5 times is called **highly repetitive DNA.** Highly repetitive sequences are usually located in **heterochromatin,** chromatin that remains condensed during interphase and is genetically inactive. Heterochromatin is found at specific sites in chromosomes, often in centromeric or telomeric regions of chromosomes. **Euchromatin** is not visible during interphase with the light microscope; it contains most, if not all, of the structural genes. Euchromatin consists of both **unique or** single-copy base sequences and base sequences repeated 2 to 10^5 times (**middle-repetitive sequences**). In all eukaryotes studied to date, a major fraction of the **genome contains single-copy sequences of DNA interspersed with middle-repetitive sequences.**

Most structural genes are presumed to correspond to the single-copy DNA sequences; several are known to be single-copy sequences. A few structural genes are redundant (middle-repetitive sequences). There is much speculation that some middle-repetitive sequences have regulatory functions.

Chromosome replication is accomplished by a **complex replication apparatus** composed of many different proteins and enzymes. At least three different proteins are involved in unwinding the complementary strands of the DNA double helix. Several proteins are involved in initiating the synthesis of new strands. Others are involved in discontinuous synthesis of DNA per se. The **3' → 5' exonuclease activity built into prokaryotic DNA polymerases provides a proofreading** function that is essential for accurate DNA replication.

REFERENCES

Avery, O. T., C. M. MacLeod, and M. McCarty. 1944. "Studies on the chemical nature of the substance inducing transformation in pneumococcal types." *J. Expl. Med.* 79:137–158.

Britten, R. J., and D. E. Kohne. 1970. "Repeated segments of DNA." *Sci. Amer.* 222(4):24–31.

Cairns, J. 1966. "The bacterial chromosome." *Sci. Amer.* 214(1):36–44.

Cantor, C. R. 1981. "DNA choreography." *Cell* 25:293–295.

Cozzarelli, N. R. 1980. "DNA gyrase and the supercoiling of DNA." *Science* 207:953–960.

Freifelder, D. 1978. *The DNA molecule: structure and properties.* W. H. Freeman, San Francisco.

Gilbert, W., and D. Dressler. 1968. "DNA replication: the rolling circle model." *Cold Spring Harbor Symp. Quant. Biol.* 33:473–484.

Hershey, A. D., and M. Chase, 1952. "Independent functions of viral protein and nucleic acid in growth of bacteriophage." *J. Gen. Physiol.* 36:39–56. (Reprinted in G. S. Stent, *Papers on bacterial viruses,* 2nd ed. Little, Brown, Boston.)

Huberman, J. A., and A. D. Riggs. 1968. "On the mechanism of DNA replication in mammalian chromosomes." *J. Mol. Biol.* 32:327–341.

Igo-Kemenes, T., W. Hörz, and H. G. Zachau. 1982. "Chromatin." *Ann. Rev. Biochem.* 51:89–121.

Jelinek, W. R., and C. W. Schmid. 1982. "Repetitive sequences in eukaryotic DNA and their expression." *Ann. Rev. Biochem.* 51:813–844.

Kornberg, A. 1980. *DNA replication.* W. H. Freeman, San Francisco.

Kornberg, A. 1982. *1982 Supplement to DNA replication.* W. H. Freeman, San Francisco.

Kornberg, R. D., and A. Klug. 1981. "The nucleosome." *Sci. Amer.* 244(2):52–64.

Lehninger, A. L. 1982. *Principles of biochemistry.* Worth, New York.

Lewin, B. 1974. *Gene expression-1, bacterial genomes.* John Wiley & Sons, New York.

——— 1980. *Gene expression-2, eucaryotic chromosomes,* 2nd ed. John Wiley & Sons, New York.

——— 1977. *Gene expression-3, plasmids and phages.* John Wiley & Sons, New York.

——— 1983. *Genes.* John Wiley & Sons, New York.

Loeb, L. A., and T. A. Kunkel. 1982. "Fidelity of DNA synthesis." *Ann. Rev. Biochem.* 51:429–457.

Meselson, M. S., and F. W. Stahl. 1958. "The replication of DNA in *Escherichia coli.*" *Proc. Natl. Acad. Sci. U.S.* 44:671–682.

Schnös, M., and R. B. Inman. 1970. "Position of branch points in replicating λ DNA." *J. Mol. Biol.* 51:61–73.

Taylor, J. H., P. S. Woods, and W. L. Hughes. 1957. "The organization and duplication of chromosomes as revealed by autoradiographic studies using tritium-labeled thymidine." *Proc. Natl. Acad. Sci., U.S.* 43:122–128.

Wang, J. C. 1982. "DNA topoisomerases." *Sci. Amer.* 247(1):94–109.

Watson, J. D., and F. H. C. Crick. 1953. "A structure for deoxyribose nucleic acid." *Nature* 171:737–738.

Worcel, A., and E. Burgi. 1972. "On the structure of the folded chromosome of *Escherichia coli.*" *J. Mol. Biol.* 71:127–147.

Zimmerman, S. B. 1982. "The three-dimensional structure of DNA." *Ann. Rev. Biochem.* 51:395–427.

PROBLEMS AND QUESTIONS

5.1. (a) How did the transformation experiments of Griffith differ from those of Avery and his associates? (b) What was the significant contribution of each? (c) Why was Griffith's work not evidence for DNA as the genetic material whereas the experiments of Avery et al. provided direct proof that DNA carried the genetic information?

5.2. A cell-free extract is prepared from Type IIIS pneumococcal cells. What effect will treatment of this extract with (a) protease, (b) RNase, (c) DNase have on its subsequent capacity to transform recipient Type IIR cells to Type IIIS? Why?

5.3. How could it be demonstrated that the mixing of heat-killed Type III pneumococcus with live Type II resulted in a transfer of genetic material from Type III to Type II rather than a restoration of viability to Type III by Type II?

5.4. What is the macromolecular composition of a bacterial virus or bacteriophage such as phage T2?

5.5. What chemical properties do DNA and proteins possess that allow researchers to specifically label one or the other of these macromolecules with a radioactive isotope?

5.6. (a) What was the objective of the experiment carried out by Hershey and Chase? (b) How was the objective accomplished? (c) What is the significance of this experiment?

5.7. (a) What background material did Watson and Crick have available for developing a model of DNA? (b) What was their contribution to the building of the model?

5.8. (a) Why was a double helix chosen for the basic pattern of the molecule? (b) Why were hydrogen bonds placed in the model to connect the bases?

5.9. (a) If a virus particle contains double-stranded DNA with 200,000 base-pairs, how many nucleotides would be present? (b) How many complete spirals would occur on each strand? (c) How many atoms of phosphorus would be present? (d) What would be the length of the DNA configuration in the virus?

5.10. If one strand or helix on the Watson-Crick model has a base sequence of 5′-GTCATGAC-3′, what would be the base sequence of the complementary DNA strand?

5.11. What are the differences between DNA and RNA?

5.12. DNA was extracted from cells of *Staphylococcus afermentans* and analyzed for base composition. It was found that 37 percent of the bases are cytosine. Using this information, is it possible to predict what percentage of the bases are adenine? If so, what percentage? If not, why not?

5.13. RNA was extracted from TMV (tobacco mosaic virus) particles and found to contain 20 percent cytosine (20 percent of the bases were cytosine). Using this information, is it possible to predict what percentage of the bases in TMV are adenine? If so, what percentage? If not, why not?

5.14. Indicate whether each of the following statements about the structure of DNA is true or false. (Each letter is used to refer to the concentration of that base in DNA.) (a) A + T = G + C. (b) A = G; C = T. (c) A/T = C/G. (d) T/A = C/G. (e) A + G = C + T. (f) G/C = 1. (g) A = T within each single strand. (h) Hydrogen bonding provides stability to the double helix in aqueous cytoplasms. (i) Hydrophobic bonding provides stability to the double helix in aqueous cytoplasms. (j) When separated, the two strands of a double helix are identical. (k) Once the base sequence of one strand of a DNA double helix is known, the base sequence of the second strand can be deduced. (l) The structure of a DNA double helix is fully conserved during its replication. (m) Each nucleotide pair contains two phosphate groups, two deoxyribose molecules, and two bases.

5.15. *E. coli* cells are grown for many generations in a medium in which the only available nitrogen is the heavy isotope ^{15}N. They are then transferred to a medium containing ^{14}N as the only source of nitrogen. (a) What distribution of ^{15}N and ^{14}N would be expected in the DNA molecules of cells that had grown for one generation in the ^{14}N-containing medium **assuming** that DNA replication was (i) conservative, (ii) semiconservative, or (iii) dispersive? (b) What distribution would be expected after two generations of growth in the ^{14}N-containing medium **assuming** (i) conservative, (ii) semiconservative, or (iii) dispersive replication?

5.16. A culture of bacteria is grown for many generations in a medium in which the only available nitrogen is the heavy isotope (^{15}N). The culture is then switched to a medium containing only ^{14}N for one generation of growth; it is then returned to a ^{15}N-containing medium for one final generation of growth. If the DNA from these bacteria is isolated and centrifuged in a CsCl equilibrium density gradient, how would you predict the DNA would band in the gradient?

5.17. The Boston straggler is an imaginary plant with a diploid chromosome number of 4. Boston straggler cells are easily grown in suspended cell cultures. ^{3}H-thymidine was added to the culture medium in which a G_1-stage cell of this plant was growing. After one cell generation of growth in ^{3}H-thymidine-containing medium, colchicine was added to the culture medium.

The medium now contained both ^{3}H-thymidine and colchicine. After two "generations" of growth in ^{3}H-thymidine-containing medium (the second "generation" occurring in the presence of colchicine as well), the two progeny cells (each now containing eight chromosomes) were transferred to culture medium containing nonradioactive thymidine (^{1}H-thymidine) **plus colchicine.** Note that a "generation" in the presence of colchicine consists of a normal cell cycle's chromosomal duplication, but no cell division. The two progeny cells were allowed to continue to grow, proceeding through the "cell cycle," until **each cell** contained a set of metaphase chromosomes that looked as follows.

If autoradiography were carried out on the above metaphase chromosomes (4 large plus 4 small), what pattern of radioactivity (as indicated by silver grains on the autoradiograph) would be expected? (Assume no recombination between DNA molecules.)

5.18. Suppose that the experiment described in Problem 5.17 were carried out again, except this time replacing the ^{3}H-thymidine with nonradioactive thymidine at the same time that the colchicine was added (after one cell generation of growth in ^{3}H-thymidine-containing medium). The cells were then maintained in colchicine plus nonradioactive thymidine until the metaphase shown in Problem 5.17 occurred. What would the autoradiographs of these chromosomes look like?

5.19. DNA polymerase I of *E. coli* is a single polypeptide of molecular weight 109,000. (a) What enzymatic activities other than polymerase activity does this polypeptide possess? (b) What are the *in vivo* functions of these activities? (c) Are these activities of major importance to an *E. coli* cell? Why?

5.20. A DNA template-primer with the structure

$$3' \ \textcircled{P} - T\ G\ C\ G\ A\ A\ T\ T\ A\ G\ C\ G\ A\ C\ A\ T - \textcircled{P}\ 5'$$
$$5' \ \textcircled{P} - A\ T\ C\ G\ G\ T\ A\ C\ G\ A\ C\ G\ C\ T\ T\ A\ A\ C - OH\ 3'$$

(where $\textcircled{P}$ = a phosphate group) is placed in an *in vitro* DNA synthesis system (Mg^{++}, an excess of the four deoxyribonucleoside triphosphates, etc.) containing a mutant form of *E. coli* DNA polymerase I which lacks

$5' \rightarrow 3'$ exonuclease activity. The $5' \rightarrow 3'$ polymerase and $3' \rightarrow 5'$ exonuclease activities of this aberrant enzyme are identical to those of normal *E. coli* DNA polymerase I. It simply has *no* $5' \rightarrow 3'$ exonuclease activity. (a) What will be the structure of the final product? (b) What will be the first step in the reaction sequence?

5.21. The satellite DNAs of *Drosophila virilis* can be isolated, essentially free of main-band DNA, by density gradient centrifugation. If these satellite DNAs are sheared into approximately 40 nucleotide-pair-long fragments and are analyzed in denaturation-renaturation experiments, how would you expect their hybridization kinetics to compare with the renaturation kinetics observed using similarly sheared main-band DNA under the same conditions? Why?

5.22. It has been demonstrated experimentally that most highly repetitive DNA sequences in the chromosomes of eukaryotes are *not* transcribed. What does this indicate about the function of highly repetitive DNA?

5.23. The available evidence indicates that each eukaryotic chromosome (excluding polytene chromosomes) contains a single giant molecule of DNA. What different levels of organization of this DNA molecule are apparent in chromosomes of eukaryotes at various times during the cell cycle?

5.24. How might continuous and discontinuous modes of DNA replication be distinguished experimentally?

5.25. Suppose that the DNA of cells (growing in a cell culture) in an unknown eukaryotic species was labeled for a short period of time by the addition of ^{3}H-thymidine to the medium. Next, assume that the label was removed and the cells were resuspended in nonradioactive medium. After a short period of growth in nonradioactive medium, the DNA was extracted from these cells, diluted, gently layered on filters, and autoradiographed. If autoradiographs of the following type:

———·· · · ———··· · · ———·· · · ·

were observed, what would this indicate about the nature of DNA replication in these cells? Why?

5.26. A diploid nucleus of *Drosophila melanogaster* contains about 2×10^8 nucleotide pairs. Assume (1) that all of the nuclear DNA is packaged in nucleosomes and (2) that an average internucleosome linker size is 60 nucleotide pairs. How many nucleosomes would be present in a diploid nucleus of *D. melanogaster*? How many molecules of histone H2a, H2b, H3, and H4 would be required?

5.27. Identify the proteins that are involved in DNA replication in *E. coli* and list their known or putative functions.

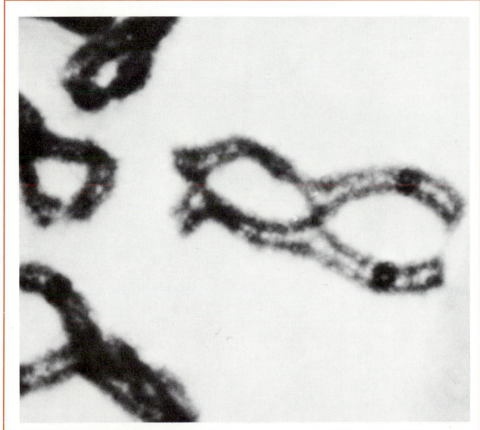

SIX

Photomicrograph of pairs of homologous chromosomes of the salamander *Oedipina poelzi,* showing *chiasmata* where the chromatids of the homologous chromosomes appear to exchange pairing partners. The tetrads of four chromatids can easily be recognized at this late diplotene stage of meiosis. (J. Kezer, "Meiosis in Salamander Spermatocytes," in *The Mechanics of Inheritance* by F. W. Stahl, copyright © 1964, p. 109. Reprinted by permission of Prentice-Hall, Inc., Englewood Cliffs, N.J.)

LINKAGE, CROSSING OVER, AND CHROMOSOME MAPPING

In his classic paper, Mendel reported the results of crosses involving alleles of seven genes that controlled seven different phenotypic characteristics of the garden pea (Chapter 2). The observed segregation of the two alleles of each gene was the basis of Mendel's first law—the law of segregation. When Mendel crossed pea plants differing in two phenotypic characteristics, so-called dihybrid crosses, the segregation of the alleles of the second gene occurred independently of the segregation of the alleles of the first gene. This provided the basis of Mendel's second law—the law of independent assortment.

The garden pea, *Pisum sativum*, has seven pairs of chromosomes. Like all other organisms, it has many more genes than chromosomes. In fact, in all organisms except the simplest viruses, there are orders of magnitude more genes than chromosomes. We don't know exactly how many genes the garden pea has, but it seems safe to conclude that the correct number is over 10,000. Indeed, the correct number might be considerably higher. Clearly, then, there must be hundreds to thousands of genes per chromosome.

Mendel's original paper detailed the results of his dihybrid cross between pea plants with yellow, round seeds and pea plants with green, wrinkled seeds. The genes controlling these two traits, yellow versus green seed color and round versus wrinkled seed shape, are now known to be located on chromosomes 1 and 7, respectively, of the garden pea. Had these traits been controlled by genes located near one another on the same chromosome, Mendel would not have observed the independent assortment of the alleles of these genes, and, thus, would not have been able to deduce his second law from the results of this cross. Combinations of **genetic markers** (alleles of each of the segregating genes) that are present on a particular chromosome of the parents tend to remain together on that chromosome in the progeny. That is, such genetic markers behave as though they are partially **linked.**

Clearly, **each chromosome must contain many genes, and these genes would not be expected to assort independently since the basis of independent assortment is (usually) the independent segregation of the different pairs of homologous chromosomes during the reductional division of meiosis** (Chapter 3). Genes that are located far apart on the same chromosome will also assort independently for reasons that will be explained later in this chapter.

As mentioned above, Mendel studied the segregation of the alleles of seven different genes. On the basis of Mendel's descriptions of the seven pairs of alternative phenotypic traits and of present knowledge of the chromosomal locations of known genes, S. Blixt has concluded that the genes studied by Mendel are located on four of the seven pairs of chromosomes in the garden pea. Two of the genes are on chromosome 1, and three are on chromosome 4. Why, then, did Mendel not detect the linked segregation of the alleles of pairs of genes in his dihybrid and trihybrid crosses?

For the two genes on chromosome 1 and for two of the three combinations of the three genes on chromosome 4, the answer is that the genes are located so far apart on the chromosome that linked segregation is not observed or expected. Only in the case of the genes controlling plant height (tall versus dwarf) and pod shape (inflated versus constricted) is linked segregation of the alleles expected. Mendel states in his classic paper, after presenting the data for dihybrid and trihybrid crosses involving the segregation of the alleles of the genes controlling seed shape, seed endosperm color, and flower color, that "further experiments were made with a smaller number of experimental plants in which the remaining characters by twos and threes were united as hybrids: all yielded approximately the same results" (from W. Bateson's translation of Mendel's paper, reprinted in *Classic Papers in Genetics*, J. A. Peters, ed.). This statement clearly implies that Mendel carried out all possible di- and trihybrid crosses involving the seven genes that he studied. However, H. Lamprecht has investigated this question and has concluded that Mendel probably did not study the simultaneous segregation of the alleles of the genes controlling plant height and pod shape.

LINKAGE

Consider a dihybrid cross, a cross in which two pairs of alleles are segregating. For example, let us look at a cross in plants and, for simplicity, assume: (1) that both genes are autosomal, (2) that one allele of each gene is completely dominant to the other allele, and (3) that no epistasis (gene interaction) is involved. In Chapter 2, such a cross might have been diagramed as follows:

Parental genotypes: $AABB$ × $aabb$

Parental gametes: AB ——— ab

Progeny or F_1 genotype: $AaBb$

Next, suppose that the F_1 plants are testcrossed, that is, crossed to fully homozygous recessive plants, as follows:

	F_1 Plant	**Testcross Plant**
Genotypes:	$AaBb$ ×	$aabb$
Gamete types:	?	ab

What types and frequencies of offspring would we expect from such a testcross? Clearly, all the gametes produced by the doubly homozygous recessive testcross parent will be ab in genotype. But what types of gametes will the F_1 parent ($AaBb$) produce, and in what frequencies? We find that this question does not have a simple answer!

Up to this point, we have considered only crosses where the two genes involved were located on different (nonhomologous) chromosomes and thus showed independent assortment. Such crosses might be more specifically diagramed as follows:

to indicate that the two loci involved are on different chromosomes. The F_1 plants would then be:

From Chapter 2, we know that such an F_1 plant would produce gametes of type:

in equal proportions (1 : 1 : 1 : 1). Thus, the expected frequencies of testcross progeny will be:

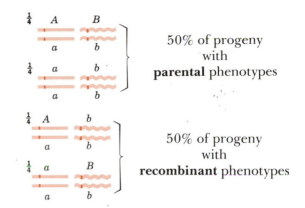

(Note that "recombinant phenotypes" include all new combinations of the inherited phenotypic traits, that is, all combinations except those of the parents.)

Now, suppose that gene C is located on the same chromosome as gene A. A dihybrid cross in which alleles of genes A and C are segregating might be diagramed:

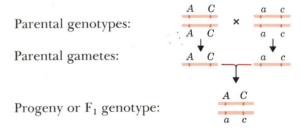

A testcross of the F_1 plants would be:

What progeny types and frequencies might we expect from this testcross? One's first guess might be that the F_1 plants would produce gametes of only two types:

If so, genes A and C would be **completely linked,** and all gametes would carry the **parental combina-**

tions of markers. **No recombinant gametes** would be produced, in sharp contrast to the **50 percent recombinant gametes characteristic of independent assortment.** However, complete linkage seldom if ever occurs. The actual results of such testcrosses involving two genes located on the same chromosome are usually somewhere between the results expected for complete linkage and the results expected for independent assortment. **At some frequency less than 50 percent, recombinant gametes are formed.** Again, an examination of the testcross progeny will show this directly:

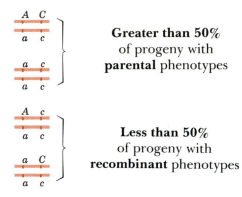

Greater than 50% of progeny with **parental** phenotypes

Less than 50% of progeny with **recombinant** phenotypes

Gametes that contain recombinant combinations of genetic markers located on the same chromosome are produced by a phenomenon called **crossing over,** which will be discussed in detail in the following section.

Genetic markers located on the same chromosome thus tend to remain together during sexual reproduction. That is, they do not exhibit independent assortment. Such genetic markers are said to be **linked,** and the phenomenon, or transmission pattern, of linked genes is called **linkage.**

Strictly speaking, genetic markers are said to be **linked whenever over 50 percent of the gametes produced contain parental combinations** of the markers and **less than 50 percent of the gametes contain recombinant combinations** of the markers. There are rare cases when genes located on different chromosomes exhibit linkage and fairly common cases when genes located on the same chromosome assort independently. Usually,

however, when linkage is observed, it results because the genes involved are located close together on the same chromosome.

The effects of linkage were first evident in the results of a dihybrid cross in sweet peas that were reported by W. Bateson and R. C. Punnett in 1906. However, Bateson and Punnett did not interpret their results in terms of the behavior of genes located on the same chromosome. T. H. Morgan (Fig. 6.1) was the first to relate linkage to the segregation of homologous chromosomes and the occurrence of crossing over between homologous chromosomes during meiosis. Morgan's interpretation of linkage was published in 1911 in a paper in which he reported the results of crosses involving linked genes in the fruitfly *Drosophila melanogaster.* Many of our current concepts about linkage, crossing over, and chromosome mapping have evolved from the work of Morgan and two of his students, C. B. Bridges and A. H. Sturtevant.

The effects of linkage can be illustrated by considering the results of two crosses involving pairs of alleles of two genes located on the second chromosome of *D. melanogaster.* One gene affects body

FIGURE 6.1
Thomas Hunt Morgan, American geneticist and embryologist. Dr. Morgan was Nobel Laureate in biology and medicine in 1933. (Culver Pictures.)

color. Flies homozygous for its recessive mutant allele b have black bodies. The presence of its dominant wild-type allele b^+ results in gray bodies. The second gene affects the phenotype of the wing. In the homozygous recessive mutant form, it results in "vestigial" or short, dumpy wings. The symbol for this mutant allele is vg; however, we will take the liberty of substituting the symbol s for short wings, and use s^+ for its dominant wild-type allele, which results in long wings. First, consider a cross between homozygous flies with long wings and gray bodies (wild-type flies) and homozygous flies with short wings and black bodies. This cross will produce heterozygous F_1 flies with long wings and gray bodies, as diagrammed in Cross I. Of the progeny of this testcross, 82 percent exhibited one or the other (41 percent each) of the parental combinations of traits. The other 18 percent of the progeny had new or recombinant combinations.

Next, consider a different cross, one between homozygous flies with long wings and black bodies and homozygous flies with short wings and gray bodies, as shown in Cross II. Again, in Cross II, 82 percent of the testcross progeny have parental phenotypes (phenotypes identical to one or the other of the original parents), and 18 percent have new or recombinant phenotypes. Although the F_1 flies have the same phenotype (long wings, gray bodies) in both crosses, the testcross progeny of the F_1 female flies contain very different frequencies of the four phenotypic classes in the two cases. (The reason for using F_1 females and not F_1 males in the testcross is that crossing over occurs infrequently in wild-type male *Drosophila*; see p. 183.) For example, 41 percent of the testcross progeny in Cross I are wild-type (have long wings and gray bodies); in Cross II, only 9 percent are wild-type. Clearly, the allelic forms of the two genes that are present together on

Cross I

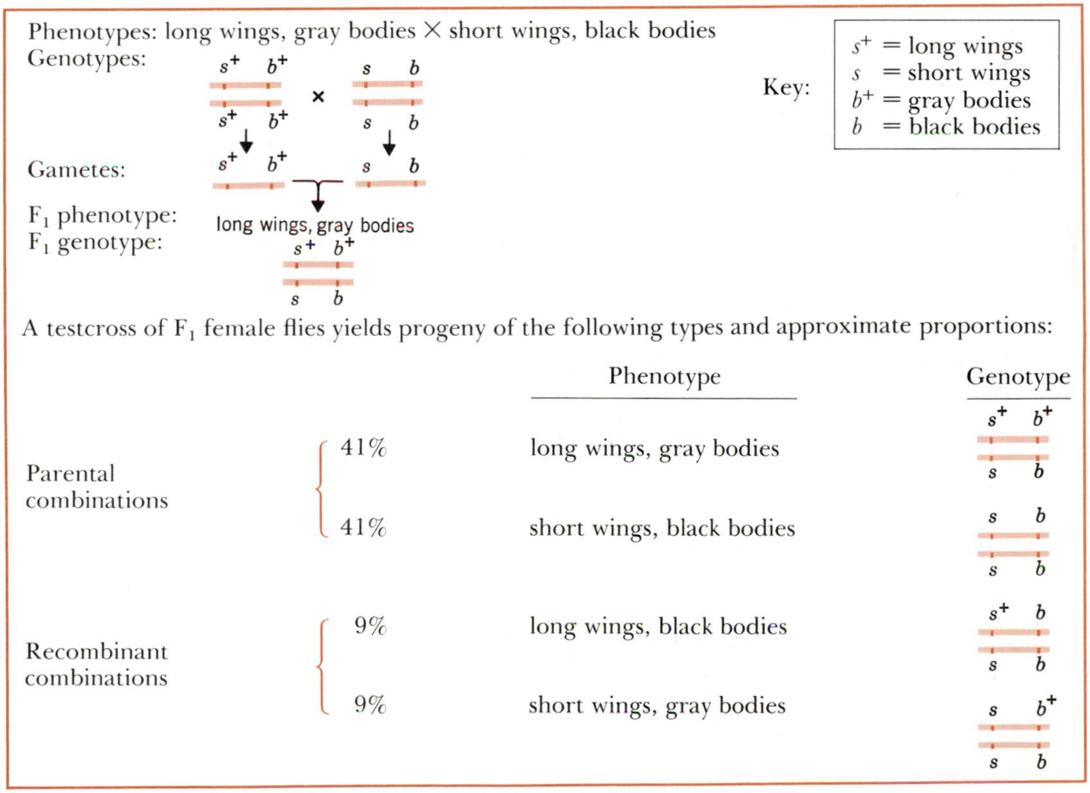

Phenotypes: long wings, gray bodies × short wings, black bodies
Genotypes:

Key:
s^+ = long wings
s = short wings
b^+ = gray bodies
b = black bodies

Gametes:

F_1 phenotype: long wings, gray bodies
F_1 genotype:

A testcross of F_1 female flies yields progeny of the following types and approximate proportions:

		Phenotype	Genotype
Parental combinations	41%	long wings, gray bodies	
	41%	short wings, black bodies	
Recombinant combinations	9%	long wings, black bodies	
	9%	short wings, gray bodies	

Cross II

Phenotypes: long wings, black bodies × short wings, gray bodies
Genotypes:

s^+ b s b^+
s^+ b s b^+

Key:
s^+	= long wings
s	= short wings
b^+	= gray bodies
b	= black bodies

Gametes: s^+ b s b^+

F_1 phenotype: long wings, gray bodies
F_1 genotype:
s^+ b
s b^+

A testcross of F_1 female flies produces progeny of the following types and proportions:

		Phenotype	Genotype
Parental combinations	41%	long wings, black bodies	s^+ b / s b
	41%	short wings, gray bodies	s b^+ / s b
Recombinant combinations	9%	long wings, gray bodies	s^+ b^+ / s b
	9%	short wings, black bodies	s b / s b

the homologous chromosomes of the parents tend to remain together on the chromosomes of the progeny.

In Cross I, the F_1 flies carried the wild-type forms ($s^+ b^+$) of the two genes on one homolog and the mutant forms (s b) on the other homolog. The genotype of a heterozygote of this type is frequently written as

$$s^+ \quad b^+ \,/\, s \quad b$$

or simply $s^+ b^+ / s b$. This arrangement of mutant and wild-type forms of two genes in a heterozygote is called the **coupling** or *cis*-**configuration** or state. The alternative arrangement, illustrated in Cross II [where each homolog contains one mutant gene and one wild-type gene ($s^+ b/s b^+$)], is called the **repulsion** or *trans*-**configuration** or state. (If you have taken organic chemistry, you may recall that *cis* and *trans* represent substitutions or side-groups on the

same (*cis*) and on *opposite* (*trans*) sides of a carbon-carbon double bond; these usages may help in remembering the analogous usages of *cis* and *trans* in genetics.)

An important feature of the two crosses described above is that—regardless of whether the cross was done in *cis* (coupling; Cross I) or in *trans* (repulsion; Cross II)—the F_1 female flies produced gametes 18 percent of which carried recombinant combinations of the genetic markers. How are the recombinant chromosomes in these gametes formed, and why is their frequency the same in the two crosses?

CROSSING OVER

Recombination of genes on the same chromosome is accomplished by **crossing over,** a process by which parts of homologous chromosomes are interchanged. The importance of crossing over in evolution should be quite apparent. Crossing over and independent assortment are the most important mechanisms for the generation of new combinations of genes. Natural selection then acts to preserve those combinations that produce organisms with maximum fitness, that is, maximum probability of perpetuation of the genotype.

We can summarize the most important features of the concept of crossing over as follows.

1. The location of a gene on a chromosome is called a **locus** (plural **loci**). The loci of the genes on a chromosome are arranged in a linear sequence. Sometimes the term locus is used to refer to the location of a set of contiguous genes with related functions.

2. The two alleles of a gene in a heterozygote occupy corresponding positions in the homologous chromosomes. That is, allele *A* occupies the same position in homolog 1 that allele *a* occupies in homolog 2.

3. Crossing over involves the breakage of each of two homologous chromosomes (actually chromatids) and the exchange of parts.

4. Crossing over occurs during the **synapsis of the homologous chromosomes in prophase I** (zygotene and pachytene) of meiosis. Since chromosome replication occurs during interphase, meiotic crossing over occurs in the **postreplication tetrad stage,** that is, after each chromosome has doubled such

that **four chromatids are present for each pair of homologous chromosomes.** Crossing over that involves sister chromatids (the two chromatids of one homolog) occurs, but it is seldom detectable genetically since the sister chromatids are usually identical.

5. Chromosomes with recombinant combinations of linked genes are formed by the occurrence of crossing over in the region between the two loci.

6. The probability that crossing over will occur between two loci increases with increasing distance between the two loci on the chromosome.

Some of these features of crossing over are illustrated in Fig. 6.2. The molecular mechanism by which crossing over occurs is discussed on pp. 182–188 of this chapter.

CYTOLOGICAL BASIS OF CROSSING OVER

Morgan first proposed crossing over to explain the formation of recombinant combinations of genes that were shown to be linked by genetic data. He hypothesized that this linkage was the result of the location of these genes on the same chromosome. If crossing over occurs as diagramed in Fig. 6.2, one might expect to be able to observe it (or the consequences of it) cytologically. In fact, cross-shaped structures, in which two of the four chromatids of homologous chromosome pairs appear to exchange

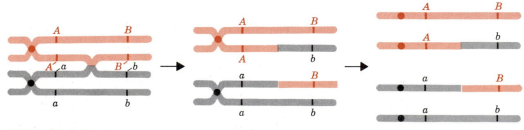

FIGURE 6.2

Diagram illustrating the occurrence of crossing over between two loci. Note that the crossover involves only two of the four chromatids of the pair of homologous chromosomes. These two chromatids interchange corresponding segments by a breakage and exchange mechanism. Also notice that of the four products of this meiotic event only two contain recombinant combinations of the alleles of the two genes. The other two daughter chromosomes (top and bottom, right) carry parental combinations of the alleles of the genes.

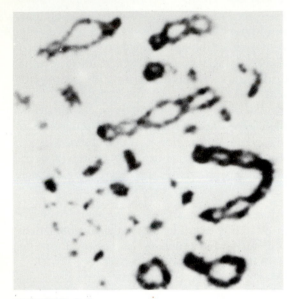

FIGURE 6.3

Meiotic prophase I chromosomes of the fowl (Gallus domes-ticus), *showing multiple chiasmata in the large bivalents. The preparation is from the testis of a White Leghorn male. All domestic fowl and the ancestral Red Jungle Fowl contain 76 pairs of autosomes plus a pair of sex chromosomes (ZZ homoga-metic males and ZW heterogametic females). The chromosomes of fowl exhibit a vast variation in size, from a few pairs of very large chromosomes to many pairs of very small chromosomes. (Photograph courtesy of R. N. Shoffner, Cytogenetics Laboratory, Department of Animal Science, University of Minnesota.)*

pairing partners, are readily detected in cytological studies of meiosis in many organisms (Fig. 6.3). These cross-shaped structures were first detected in amphibians and called **chiasmata** (singular **chiasma**) by F. Janssens (see Chapter 3). Studies in which chiasma frequencies were correlated with re-combination frequencies have shown a direct rela-tionship between crossing over and chiasma.

Direct cytological evidence that homologous chromosomes exchange parts during crossing over was first obtained in 1931 by C. Stern (working with *Drosophila*) and by H. B. Creighton and B. McClin-tock (working with maize). Normally, the two chro-mosomes of any homologous pair are morphologi-cally indistinguishable. Stern, Creighton, and

McClintock, however, identified homologs that were morphologically distinguishable; that is, they were **not entirely homologous.** The chromosome pairs studied by these workers were homologous along most of their length, such that they paired and segregated normally during meiosis. The homologs differed, however, at their ends, having distinct morphological features that could be recognized by microscopy.

Stern studied two X chromosomes that differed from the normal X chromosome of *Drosophila*. One X chromosome had part of a Y chromosome (that had broken off its normal location) attached to one end. (Such chromosomal rearrangements are well-known; they are called translocations; see Chapter 12.) The second X chromosome was shorter than normal; a piece had been broken off and translo-cated to another chromosome (in this case, to chro-mosome IV). Crosses were done to produce female flies heterozygous for these two morphologically distinguishable X chromosomes. These female flies were also heterozygous for alleles of two genes that are located on the X chromosome. One gene affects eye shape; the other eye color. The partially domi-nant mutant allele *B* results in bar-shaped eyes; its wild-type allele B^+ produces round eyes when ho-mozygous. The mutant allele *car* of the second gene results in carnation-colored eyes; its dominant wild-type allele car^+ yields red eyes. The females used in Stern's study carried the allelic pairs in the *cis*-con-figuration, as shown in the following figure. Stern

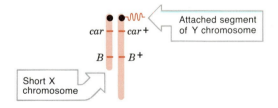

crossed such heterozygous females with males hav-ing carnation-colored, normal-shaped eyes (*car* B^+/Y males). This cross and the results Stern ob-tained are diagrammed in Fig. 6.4. Stern determined the genotypes of the progeny and asked whether the progeny with recombinant genotypes carried X chromosomes with recombinant combinations of the morphological markers. He observed that the

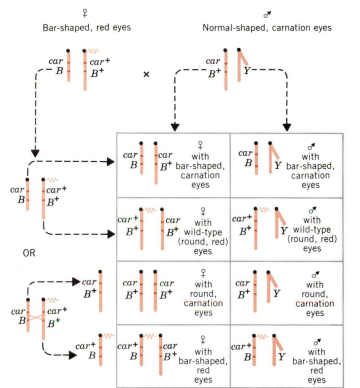

FIGURE 6.4

C. Stern's classic experiment demonstrating that crossing over involves the interchange of parts of homologous chromosomes. Details of the experiment are described in the text. The ✕ *(bottom left) indicates the occurrence of crossing over between the* **car** *and* **B** *loci. The centromeres on the chromosomes are represented by circles. The two X chromosomes in the female parent (with bar-shaped, red eyes) are morphologically distinguishable. The short one (left) is missing the distal end; its homolog has a piece of a Y chromosome attached to the centromere end (shown as a horizontal zigzagged line).*

combinations of morphological markers present on the X chromosomes of each recombinant progeny were precisely those predicted (from the arrangement of the markers in the heterozygous female parent) if crossing over involved the breakage and exchange of parts of homologous chromosomes. Recombinant male progeny with bar-shaped red eyes (car^+B/Y), for example, were found to carry the short X chromosome, but now with the translocated piece of the Y chromosome at one end. Male flies with normal-shaped carnation eyes ($car\ B^+/Y$) contained long X chromosomes, but without the attached piece of the Y chromosome.

Creighton and McClintock also demonstrated the correlation between crossing over and the exchange of parts of homologous chromosomes. They analyzed crosses involving two loci on chromosome 9 of maize. One gene controlled kernel color (*C*, colored; *c*, colorless); the other controlled the type of carbohydrates in the kernel (*Wx*, starchy; *wx*, waxy). They made use of a chromosome with a densely staining "knob" at one end and an extra (translocated) piece of chromosome at the other end (as shown below).

When a testcross of the *trans*-heterozygote was made, the *C Wx* and *c wx* recombinant progeny were found to carry a chromosome with recombinant combinations of the morphological markers, as shown in the following diagram (plus, of course, a normal chromosome from the testcross parent).

These two experiments, by Stern and by Creighton and McClintock, are true classics in genetics. They provided confirmation of Morgan's hypothesis that crossing over involves the interchange of parts of homologous chromosomes; they also provided strong evidence that indicated that genes are indeed located on chromosomes.

CROSSING OVER OCCURS IN THE POSTREPLICATION TETRAD STAGE

An important feature of crossing over as illustrated in Fig. 6.2 is that crossing over occurs after chromosome duplication, when each homologous chromosome is represented by two chromatids. Each pair of synapsed homologs is called a **tetrad,** because it consists of **four chromatids.** A genetically detectable crossover occurs between any two nonsister chromatids of a homologous pair of chromosomes. Note that **crossing over** refers to the **exchange of segments of chromosomes** (the process), whereas **recombination** implies the **formation of new combinations of genes** (one possible outcome of the process). Thus **crossing over occurs in completely homozygous organisms, but new combinations can be formed only in organisms that are heterozygous at two or more loci.**

Proof that crossing over occurs after duplication of the chromosomes rather than prior to their duplication is most easily obtained by studying fungi of the class known as Ascomycetes. Of the Ascomycetes, the bread mold *Neurospora crassa* has been of particular importance in genetic studies. Five features of *N. crassa* make it highly suited for certain types of genetic analysis. (1) After two haploid nuclei from cells of two different mating types fuse to form a diploid fusion nucleus (much like fertilization in higher organisms), meiosis occurs, just as it does in higher plants and animals. (2) The haploid products of meiosis, called **ascospores,** are maintained in linear order within an elongate, tubelike structure called an **ascus** (plural, **asci**). Thus each ascus contains **all four products of a single meiotic event.** Moreover, all of the ascospores in each ascus can usually be recovered and analyzed genetically. (3) These haploid ascospores germinate and grow to produce multicellular **mycelia** (singular, **mycelium**), and all of these cells are also haploid. The

genotype of each product of meiosis can thus be determined without carrying out testcrosses or other genetic manipulations. Because of the haploid state of the mycelium, the presence of recessive markers is never masked by dominant alleles. (4) *N. crassa* can be grown on a simple synthetic medium containing only inorganic salts, a carbohydrate (such as sucrose), and one other organic compound (the vitamin biotin). (5) *N. crassa* reproduces asexually as well as sexually, facilitating the maintenance of strains of particular genotypes. The sexual and asexual life cycles of *N. crassa* are illustrated in Fig. 6.5.

In most organisms, such as *Drosophila* and maize, one *cannot* recover and analyze the genotypes of all four haploid products of a **single meiotic event.** Instead, one is forced to perform testcrosses, for example, and to examine a **population of randomly selected products of meiosis.** The genotypes of the progeny in these random samples are then used to deduce what has occurred during meiosis. In the case of *N. crassa,* however, one can isolate and determine the genotypes of all four products of a single meiotic event. The ability to analyze all four products of individual meiotic events has provided geneticists with several kinds of useful information (e.g., see pp. 170–176). Data from crosses in which the genotypes of all of the products of meiosis have been determined are called **tetrad data.**

An analysis of tetrad data from crosses in which pairs of alleles of two genes located on the same chromosome were segregating readily shows that crossing over occurs after replication, in the "four-strand" or tetrad stage, rather than before replication, in the "two-strand" (per homologous chromosome pair) stage. If crossing over occurred prior to replication (or chromosome duplication), all of the products of a meiotic event in which a crossover occurred between the two loci would have recombinant combinations of the genetic markers (Fig. 6.6a). On the other hand, if crossing over occurs after replication, in the tetrad stage, only two of the four products of each such meiotic event will be recombinant (Fig. 6.6b). The other two products will have parental combinations of the segregating genetic markers. Tetrad data clearly show that the

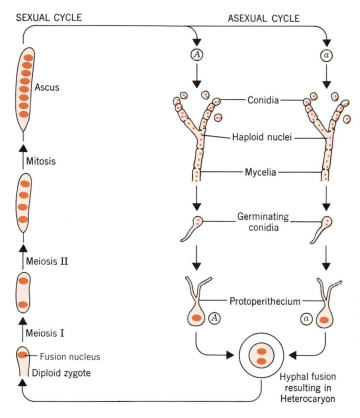

SEXUAL CYCLE   ASEXUAL CYCLE

Ascus

Mitosis

Meiosis II

Meiosis I

Fusion nucleus
Diploid zygote

Conidia

Haploid nuclei

Mycelia

Germinating conidia

Protoperithecium

Hyphal fusion resulting in Heterocaryon

FIGURE 6.5

Sexual and asexual reproduction in the bread mold
Neurospora crassa. *During sexual reproduction, the
products of meiosis are maintained in a tubelike structure
called an ascus. Each ascus contains four pairs of
ascospores, with the members of each pair being identical
twins, as a result of the mitotic (equational) division
following meiosis. Asexual reproduction occurs by mitotic
divisions of haploid cells to form spores called conidia.*
Hyphal fusion *can also occur between mycelia. If the fusion occurs between cells with nuclei that are genetically
identical, the resulting cells are called* **homokaryons**
*("same nuclei"). If the nuclei are of two different genotypes,
the resulting cells are called* **heterokaryons** *("different
nuclei"). Note that during the formation of homokaryons
and heterokaryons, the nuclei do* **not** *fuse.*

second alternative is correct. Tetrads produced by
meiotic events in which a single crossover has occurred between the loci involved give rise to asci
containing 50 percent genotypically recombinant
ascospores and 50 percent genotypically parental
ascospores (Fig. 6.6*b*). Tetrads containing all recombinant ascospores are rare, indicating that they
are formed by meiotic events in which two crossovers have occurred (see pp. 170–176).

CHROMOSOME MAPPING

Earlier in this chapter, we considered the results of
crosses in which pairs of alleles of two genes located
on the same chromosome were segregating. Such
genes were shown to be **linked** in that **over 50
percent of the gametes** produced by F_1 double
heterozygotes contained **parental combinations** of
the genetic markers and **less than 50 percent of the**

gametes contained **recombinant combinations** of
the markers. This result was shown to contrast with
the predictions of independent assortment, namely,
50 percent parental and 50 percent recombinant
combinations. Subsequently, we showed that recombinant combinations of linked genes are produced by crossing over, a process during which the
breakage and exchange of parts of homologous
chromosomes occurs.

When the above kinds of analysis have been
used to investigate the linkage relationships of a
large number of genes of any species, the genes have
always been found to occur in a distinct set of **linkage groups,** with each linkage group corresponding
to one of the pairs of homologous chromosomes in
the genome of that species. In *D. melanogaster,* for
example, there are 4 linkage groups, corresponding
to the 4 pairs of chromosomes (Fig. 6.7). Maize has
10 linkage groups and 10 pairs of chromosomes
(Fig. 6.8). The mouse has 20 pairs of chromosomes

(a)   If crossing over occurs before chromosome replication:

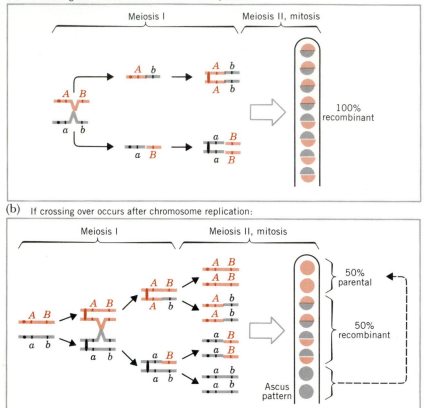

(b)   If crossing over occurs after chromosome replication:

FIGURE 6.6

The Neurospora *asci patterns predicted to result from a single crossover between two loci (**A**, **a** and **B**, **b**) if crossing over occurs (a) prior to replication (when only the two homologous chromosomes are present) or (b) after replication (in the four-chromatid or tetrad stage). Recall (Fig. 6.5) that in* Neurospora *the four haploid nuclei produced by meiosis immediately undergo one mitotic division to produce eight ascospores (or four pairs of ascospore "twins") per ascus as shown. The chromosomes, chromatids, and ascospores containing the chromosomes from one parent are shown in color; those from the other parent are shown in gray. The centromeres are represented by circles. Analyses of tetrad data from many crosses carried out in many different laboratories have shown conclusively that crossing over occurs in the postreplication tetrad stage as diagramed in* b.

and 20 linkage groups, and so on. The linkage groups of humans can**not** be investigated by the standard procedures used with other species. However, new methodologies, especially somatic cell hybridization (see pp. 176–182) and chromosome-banding techniques, have facilitated the construction of linkage maps for each of the 23 pairs of human chromosomes.

   An important feature of all linkage maps is their **linearity.** That is, all genes in a given linkage group can be shown to map in a linear array. [A circle or ring exhibits linearity; it is simply a closed line (no ends). Most importantly, it has no branches. The linkage maps of prokaryotes (see Chapters 7 and 10) and of eukaryotic organelles (mitochondria and chloroplasts, see Chapter 14) are usually circular.] In Chapter 5, we presented evidence indicating that the genetic information is stored in DNA (RNA in some cases) in the form of the specific sequences of base-pairs. Since a sequence of base-pairs in DNA is linear, it is appropriate that linkage maps are also linear.

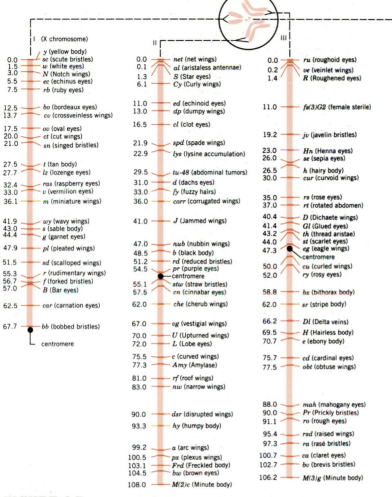

FIGURE 6.7

Abbreviated linkage map of Drosophila melanogaster. *The correlation of each linkage group with one of the four pairs of chromosomes (circular inset) is shown by the dashed lines. Map positions are given on the left of each linkage map; the genetic symbols used to represent the mutant allele(s) of each gene are shown on the right of each linkage map. The phenotypic description commonly used for each mutant allele is given in parentheses after the gene symbol. The map positions are those given by D. L. Lindsley and E. H. Grell,* Genetic Variations of Drosophila melanogaster, *Carnegie Institution of Washington Publication 627, 1968, except where updated by R. C. King, pp. 625–652, in* Handbook of Genetics, *Vol. 3,* Invertebrates of Genetics Interest, *Plenum Press, New York, 1975.*

TWO-FACTOR CROSSES

Recombinant combinations of the alleles of two linked genes are produced by crossing over in the interval between the two segregating loci. The rationale behind genetic mapping is that the probability of a crossover occurring between two loci is a

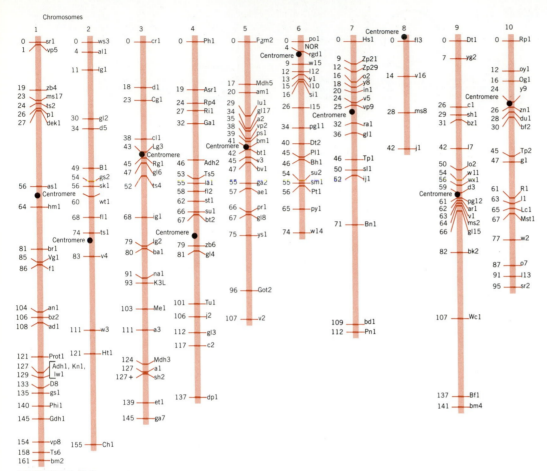

FIGURE 6.8

Linkage map of maize. Descriptions of phenotypes caused by mutant alleles of a few genes are given below. Designations are in the form: chromosome no. – map position, gene symbol, name of mutant allele, mutant phenotype. 1-0, sr1, *striate-1, white striations on leaves; 1-56,* as1, *asynaptic-1, failure of the meiotic prophase chromosomes to synapse; 2-30,* gl2, *glossy-2 cuticle wax altered such that leaf surface appears glossy; 2-121,* Ht1, *Helminthosporium turcicum resistance, resistance to the northern leaf blight caused by this fungus; 3-0,* cr1, *crinkly-1, leaves broad and crinkled; 3-139,* et1, *etched-1, pitted or scarred endosperm; 4-24,* Rp4, *rust resistance-4, resistance to the fungus* Puccinia sorghi; *4-101,* Tu1, *tunicate-1, kernels enclosed in long glumes; 5-20,* am1, *ameiotic-1, meiosis fails; 6-3,* NOR, *nucleolus organizer, tandem repeats of genes coding for rRNA; 6-74,* w14, *white-14, white seedlings; 7-16,* o2, *opaque-2, endosperm starch soft and opaque, has high lysine content; 8-28,* ms8, *male sterile-8, anthers not exserted; 9-26,* c1, *colorless-1, nonpigmented aleurone; 9-56,* wx1, *waxy-1, no amylose in endosperm or pollen; 10-16,* Og1, *old gold-1, variable bright yellow stripes on leaf blades. [Based on the map of Coe, Hoisington, and Neuffer, 1982. For additional details and descriptions of the mutant phenotypes for all of the genes shown, see E. H. Coe, Jr., D. A. Hoisington, and M. G. Neuffer, 1982. Linkage map of corn (maize)* (Zea mays L.), *p. 377–393 in* Genetic Maps, *Vol. 2, S. J. O'Brien, ed., National Cancer Institute, Frederick, Maryland.]*

function of the length of the interval separating the loci. Intuitively, this seems very reasonable. Consider, for example, the following two crosses:

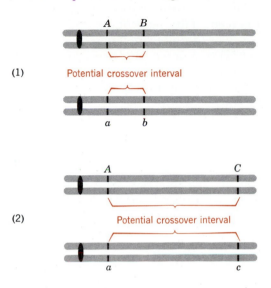

(1)

Potential crossover interval

(2)

Potential crossover interval

All three genes, *A, B,* and *C,* are on the same chromosome. The *A* and *B* loci are close together, whereas *A* and *C* are quite far apart. A crossover occurring anywhere within the long interval between the *A* locus and the *C* locus will produce recombinant combinations (*A c* and *a C*) of the two pairs of alleles segregating in Cross 2. Recombinants (*A b* and *a B*) will be produced in Cross 1 only when a crossover occurs within the short interval between the *A* locus and the *B* locus. It seems very reasonable, therefore, to expect more recombinants to be produced in Cross 2 than in Cross 1.

On the basis of the above type of logic, one of Morgan's students, A. H. Sturtevant (while still an undergraduate), suggested that the frequency of recombinant gametes produced be used as an index of the distance between two loci on a chromosome. From Sturtevant's suggestion, linkage maps of the type shown in Figs. 6.7 and 6.8 have evolved. **Linkage maps** are made **quantitative** by **defining one map unit as the distance that yields 1 percent recombinant chromosomes or gametes.** Thus, the genes *b* (black body) and *vg* (short or vestigial wings) that yielded **18 percent recombinant** testcross progeny — whether the F_1 dihybrid (double

heterozygote) carried the markers in the *cis* (coupling)-configuration or in the *trans* (repulsion)-configuration (see pp. 150–152) — are about **18 map units** apart on linkage Group II of the genetic map of *D. melanogaster* (Fig. 6.7).

It is very important not to confuse the frequency of **crossing over, the event occurring in meiotic tetrads,** with the frequency of **crossover** or **recombinant chromosomes, the products of crossing over. Linkage map distances are determined by the frequencies of crossover or recombinant chromosomes.** As is illustrated in Fig 6.2, **each meiotic crossing over event yields two crossover chromosomes** (two recombinant chromosomes if the interval within which crossing over occurred is flanked by heterozygous loci). Thus, if a single crossover occurs between two loci in 100 percent of the tetrads, only 50 percent of the progeny chromosomes will be recombinant (i.e., the recombination frequency will be 50 percent).

If one **assumes** that the probability of a crossover occurring between two loci is directly proportional to the distance between the two loci, that is,

$$\text{Probability of crossover} = K \text{ (distance)}$$

where *K* is a proportionality constant, then one would **predict** that **map distances would be additive.** This property of **additivity** can be illustrated by the following example. **If loci *P* and *Q* are linked** and are **8 map units apart** (yield 8 percent recombinant gametes), and loci ***P* and *R* are linked** and are **3 map units apart,** then loci ***Q* and *R* are also linked** and are **either (1) 5 map units apart or (2) 11 map units apart.** That is, **additivity** can be achieved only by the following two linkage arrangements:

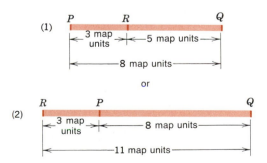

The assumption that the probability of crossing over is directly proportional to distance is an oversimplification. The assumption is reasonably accurate, however, at least as a first approximation, for linkage distances up to 10–20 map units. As distances become large, significant deviations from additivity are observed. These deviations are of the following nature. Consider, for example, that X and

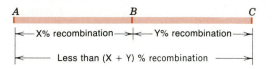

Y are both 15; that is, two-factor crosses, in which alleles of either genes *A* and *B* or genes *B* and *C* are segregating, would produce 15 percent recombinant gametes. Additivity would predict that 30 percent recombinant gametes would be produced in two-factor crosses involving genes *A* and *C*. In actual crosses analogous to the gene *A*-gene *C* cross, recombinant gametes are produced at frequencies significantly lower than 30 percent.

The reason for this systematic deviation from additivity as distances become large is that, when loci are far apart, double crossovers occur at significant frequencies. Consider any two homologous, nonsister chromatids that have recombinant combinations of genetic markers as a result of a crossover occurring between the two marked loci. A second crossover involving these two chromatids in the interval between the two loci will, in effect, cancel out the first crossover. Although two crossovers

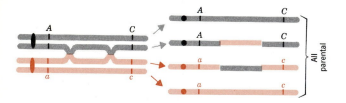

have occurred, they will not be counted when the data for such a two-factor cross are recorded. For this reason, two-factor crosses involving loci that are far apart (10–20 map units or more) will underestimate true linkage distances.

MAXIMUM FREQUENCY OF RECOMBINATION IS 50 PERCENT

The **maximum frequency of recombination** that can result from crossing over between linked genes is **50 percent.** Fifty percent is also, of course, the frequency of recombination observed for genes on different chromosomes that are assorting independently. Thus, recombination frequencies never exceed 50 percent. For two-factor crosses, recombination frequencies are observed to asymptotically approach 50 percent as the **additive map distance** between the loci involved increases (Fig. 6.9). The additive map distance is the summation of the map distances between genetic markers spanning the shortest marked intervals. The additive map distance between genes *A* and *Z* in Fig. 6.9 is 130 map units, the sum of the distances between *A* and *B*, *B* and *C*, *C* and *D*, . . . , *X* and *Y*, and *Y* and *Z*. This deviation from additivity results from the increasing number of multiple crossovers as distances become large. All multiple crossovers with an **even number of exchanges** between two segregating loci will yield parental combinations of the genetic markers and thus go undetected. When intervals between loci are small, such multiple crossovers will be rare and have a negligible effect on additivity relationships.

The maximum frequency of recombination for genes located on the same chromosome is 50 percent—the same frequency as observed for genes located on different chromosomes and thus assorting independently. The reason for the 50 percent maximum frequency can be illustrated as follows. Assume that a single crossover occurs between two loci in every meiotic event. In reality, that will not happen, since crossovers are random events such that no crossovers will occur in some meiotic events, double crossovers will occur in others, and so on. But assume that it does happen. The resulting recombination frequency will be 50 percent because each crossover involves only two of the four chromatids, and the gametes carrying the two noncrossover chromatids will have parental combinations of the genetic markers (as shown in Fig. 6.2).

Now consider an analogous hypothetical situation. Assume that two crossovers always occur be-

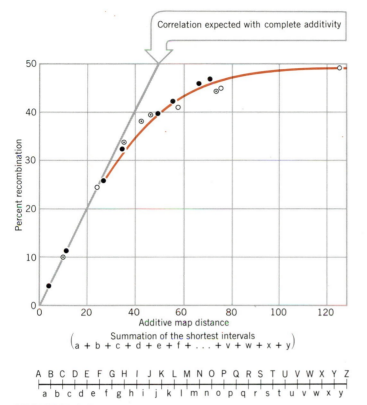

FIGURE 6.9

Frequency of recombination in two-factor crosses as a function of the additive map distance separating two loci. The data shown are those summarized by Perkins, Genetics *47:1253–1274, 1962, for* Drosophila (●), Neurospora (⊙), *and maize* (○). *The frequency of recombination asymptotically approaches 50 percent, the value characteristic of unlinked genes, as the additive distance between two loci increases. At map distances in excess of about 20 map units, the deviation from linearity (predicted if recombination frequencies and thus map distances were completely additive) becomes quite large; this results from the inability to detect double crossovers (or any even number of multiple exchanges) in two-factor crosses. The additive map distance is the summation of the map distances obtained from two-factor crosses in which pairs of alleles of loci spanning the shortest intervals are segregating. Thus, the additive map distance from* A *to* Z *is 130 map units, the summation of the distance from* A *to* B (a *map units), the distance from* B *to* C (b *map units), and so on. Even though genes* A *and* Z *are 130*

map units apart, less than 50 percent of the gametes produced by an A Z/a z *heterozygote will be recombinant (*A z *and* a Z). *Many linkage maps are over 100 map units long. In Fig. 6.7 the linkage map corresponding to chromosome II of* D. melanogaster *is 108 map units long. The linkage map corresponding to chromosome 1 of maize (Fig. 6.8) is over 160 map units long.*

This relationship between recombination frequencies and additive map distances can be described mathematically by so-called **mapping functions.** *The first mapping function was developed by J. B. S. Haldane in 1919; it is often referred to as the "Haldane function." Since then, many different mapping functions have been devised, some of which are applicable to only one or a few related species. These mapping functions are useful in allowing one to convert recombination frequencies to physical distances (usually in terms of nucleotide pairs) with fair accuracy. The reader is referred to Stahl,* Genetic Recombination, *1979, for an excellent discussion of mapping functions.*

tween two loci. Again, the recombination frequency will be 50 percent. As is illustrated in Fig. 6.10, double crossovers can occur in three different ways.

(1) **Two-strand double crossovers** occur when both crossovers involve the same two chromatids. (2) **Three-strand double crossovers** are those in which

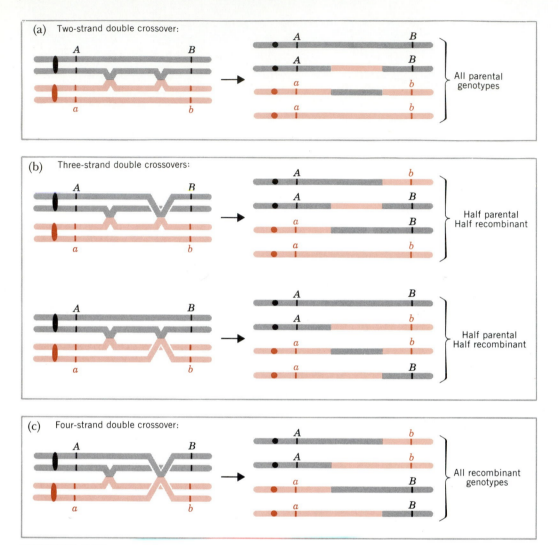

FIGURE 6.10

Diagram illustrating the consequences of (a) *two-strand,* (b) *three-strand, and* (c) *four-strand double crossovers between two loci. All heterozygotes are in the* **cis** *(coupling)-configuration. The chromatids of one homolog are shown in gray; those of the other homolog are shown in color. The circles represent the centromeres. All of the gametes resulting from meiotic events in which two-strand double crossovers have occurred between the two loci have parental combinations of the segregating pairs of alleles. Those resulting from meiotic events in which three-strand doubles have occurred are half parental and half recombinant in genotype. Only meiotic events in which four-strand double crossovers have occurred yield all gametes with recombinant genotypes. Three-strand double crossovers can occur in twice as many ways as two-strand or four-strand double crossovers. Given the 1:2:1 ratio of two-strand to three-strand to four-strand double crossovers, the total double crossover gamete population will be 50 percent parental and 50 percent recombinant.*

the second crossover involves one of the same two chromatids as the first crossover plus one different chromatid. Three-strand double crossovers can occur in twice as many ways as two-strand or four-strand double crossovers (see Fig. 6.10). (3) **Four-strand double crossovers** occur when the second crossover involves the two chromatids not involved in the first crossover. If the second crossover in a tetrad involves two chromatids at random, irrespective of which chromatids were involved in the first crossover, then the ratio of two-strand to three-strand to four-strand double crossovers will be $1:2:1$. Deviations from this $1:2:1$ ratio are said to result from **chromatid interference:** the effect of the involvement of two chromatids in a crossover on which chromatids will be involved in additional crossovers in nearby regions. Ratios of two- to three- to four-strand double crossovers usually approximate $1:2:1$, indicating that little, if any, chromatid interference is occurring.

All of the gametes produced by meiotic events in which two-strand double crossovers have occurred between two loci will contain parental combinations of the genetic markers. Half of the gametes produced by meiotic events in which three-strand double crossovers have occurred between two loci will contain recombinant combinations of the genetic markers; the other half will contain parental combinations. All of the gametes resulting from four-strand double crossovers between the loci will be recombinant. Given the $1:2:1$ ratio of two-strand to three-strand to four-strand double crossovers, the summation of the gamete types produced from meiotic events in which a double crossover always occurs between two loci shows that 50 percent are parental in genotype and 50 percent are recombinant in genotype (Fig. 6.10). Similarly, the frequency of recombination can be shown to equal 50 percent if three crossovers always occur between two loci, or four crossovers, and so on, as long as there is no chromatid interference.

This may be easier to visualize by considering any two nonsister chromatids. As the distance between two loci becomes large, the frequency of multiple crossovers with an **even number of exchanges (yielding parental combinations)** will approximately equal the frequency of multiple crossovers with an **odd number of exchanges (yielding recombinant combinations).** If chromosomes were infinitely long, the two (even number of exchanges and odd number of exchanges) would occur with exactly equal frequency, yielding 50 percent recombinant and 50 percent parental gamete types. Only as a result of very unlikely events, such as a preponderance of four-strand double crossovers, would the observed recombination frequency exceed 50 percent.

THREE-FACTOR CROSSES

How can one detect and take into account the double crossovers that go unrecognized in a two-factor cross? This can be accomplished (and then only in part) only by identifying a third gene with distinguishable alleles that maps between the loci involved in the two-factor cross:

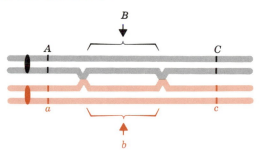

Given a third genetic marker on a chromosome, one can carry out a **three-factor cross,** a cross in which three pairs of alleles are segregating. In addition to allowing the detection of some of the double crossovers not recognizable in two-factor crosses, **three-factor crosses** allow one to **order the markers** involved. The three-factor cross is undoubtedly the most important tool used in chromosome mapping. In diploids, the three-factor cross is usually used in a manner analogous to the use of the two-factor cross in detecting linkage. That is, homozygotes are crossed to produce triple heterozygotes or trihybrids (such as $A\,B\,C/a\,b\,c$), and the trihybrids are then testcrossed (to $a\,b\,c/a\,b\,c$ homozygotes), so that the kinds and frequencies of the F_1 gametes can be determined directly from the phenotypes of the testcross progeny.

In Chapter 2, we showed that trihybrids can produce eight (2^3) different kinds of gametes. If all three genes are assorting independently, the eight gamete types occur with equal frequency. Deviations from this $1:1:1:1:1:1:1:1$ ratio will, of course, occur if two or more of the genes are linked. The exact ratio will be determined by the degree of linkage between the loci involved.

First, and most importantly, how can one determine the order of genes from three-factor crosses? Consider three hypothetical pairs of alleles (X, x; Y, y; and Z, z) of linked genes. Assume that the order of the three genes is known to be X-Y-Z. Now, ask yourself the following questions. (1) What kinds of gametes would a triple heterozygote of genotype $X\ Y\ Z/x\ y\ z$ be expected to produce? (2) In what relative frequencies would these gametes be produced? (3) What different patterns of meiotic crossing over can be detected by examining the testcross progeny of this triple heterozygote?

The eight gamete types produced by the $X\ Y\ Z/x\ y\ z$ triple heterozygote will result from four different patterns with respect to crossing over, as shown in Fig. 6.11. Although crossing over occurs in the tetrad stage of meiosis, for simplicity the diagrams show only the two chromatids involved in each crossover type.

In three-factor crosses of the type diagramed, the **parental gamete types** (as revealed by testcrosses) **will necessarily be present in the highest frequencies.** In the present case, the markers exist in the complete *cis* (coupling)-configuration. Of course, they may be present on the parental chromosomes in any combination. The parental combinations can always be determined, however, by identifying the two classes of testcross progeny that occur at the highest frequencies. These two parental progeny classes are expected to occur in approximately equal frequencies (the deviations representing simply random sampling variations).

The ability to identify the parental progeny classes allows one to determine the linkage relationships of the genetic markers involved, that is, whether the alleles of each pair of genes are present in the *cis*- or *trans*-configuration on the homologous chromosomes of the F_1 progeny. In the above exam-

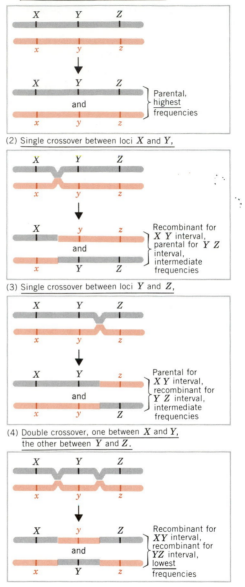

(1) No crossover within the marked region,

Parental, highest frequencies

(2) Single crossover between loci X and Y,

Recombinant for $X\ Y$ interval, parental for $Y\ Z$ interval, intermediate frequencies

(3) Single crossover between loci Y and Z,

Parental for $X\ Y$ interval, recombinant for $Y\ Z$ interval, intermediate frequencies

(4) Double crossover, one between X and Y, the other between Y and Z.

Recombinant for XY interval, recombinant for YZ interval, lowest frequencies

FIGURE 6.11
Gametes produced by the X Y Z/x y z *triple heterozygote.*

ple, the markers are all in the *cis*-configuration ($X\ Y\ Z$ and $x\ y\ z$).

The two reciprocal double crossover progeny classes can also be identified. They will always be

the two classes present in the lowest frequencies. If the two crossovers involved in a double crossover are independent events (we will discuss whether they are later), then the probability of a double crossover occurring should equal the product of the probabilities of the two single crossovers occurring. Stated differently, if crossovers in Interval I occur with a frequency of x and crossovers in Interval II occur with a frequency of y, then double crossovers with one exchange in Interval I and a second exchange in Interval II should occur with a frequency of xy. Since x and y are fractions of less than $\frac{1}{2}$ (usually much less than $\frac{1}{2}$), xy is necessarily smaller than x or y. Thus, the reciprocal double crossover progeny classes will always be less frequent than either of the pairs of reciprocal single crossover classes. In actuality, double crossover types usually occur even less frequently than expected if the two crossovers represent independent events. (This will be discussed further later.)

The ability to **identify the two reciprocal parental progeny classes** and the **two double crossover progeny classes** of a three-factor cross **allows one to determine the order of the three loci.** A double crossover always results in the center marker on each parental chromosome being placed in a recombinant combination with the outside markers from the other parental chromosome. In our hypothetical X-Y-Z example, the double crossover gametes carried X-y-Z and x-Y-z. The middle marker Y of one parent is now present on a chromosome with the outside markers x and z from the other parent, and vice versa. **Only one of the three possible orders of the markers in a three-factor cross will allow the observed double crossover types to be produced from the observed parental types by a double crossover.** Assume, for the moment, that the order of the genes is unknown in our hypothetical X-Y-Z example, but that we know that the parental classes are $X\,Y\,Z$ and $x\,y\,z$ and the double crossover classes are $X\,y\,Z$ and $x\,Y\,z$. The three possible orders of the loci and the double crossover gametes that would be produced from each order are diagrammed in Fig. 6.12. Clearly, the known double crossover gamete types ($X\,y\,Z$ and $x\,Y\,z$) can be produced from the parental chromosomes by a

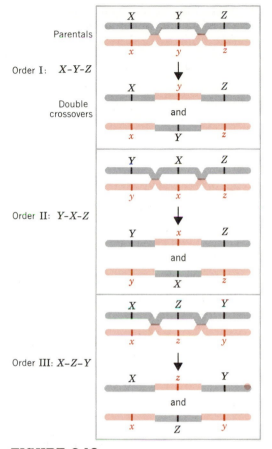

FIGURE 6.12
*Double crossover chromosomes produced for the three **possible** orders of the linked markers* **X**, **Y**, *and* **Z**.

double crossover only in the case of Order I (X-Y-Z).

Once the order of the markers is determined by the procedure detailed above, the linkage distances for each interval can be calculated from the observed frequency of recombination (in percent) for each adjacent pair of markers. This can best be illustrated by an example. Suppose that the data given in Table 6.1 resulted from a testcross of an F_1 triple heterozygote. Note that we do not know before analyzing the data whether the markers are in *cis* (coupling)- or *trans* (repulsion)-configuration, nor do we know the order of the loci. They are arbitrarily written in alphabetical order.

Note that all eight possible genotypes are present, but in very different frequencies, clearly

TABLE 6.1 Three-factor Testcross Data (Note that the Alleles are Listed Arbitrarily in Alphabetical Order.)

GENOTYPE	NUMBER OF PROGENY	
$A\ B\ C/a\ b\ c$	370	Parental
$a\ b\ c/a\ b\ c$	385	
$A\ b\ c/a\ b\ c$	45	Single crossover in Interval I
$a\ B\ C/a\ b\ c$	50	
$A\ B\ c/a\ b\ c$	2	Double crossover: one in Interval I and one in Interval II
$a\ b\ C/a\ b\ c$	3	
$A\ b\ C/a\ b\ c$	75	Single crossover in Interval II
$a\ B\ c/a\ b\ c$	70	
	Total = 1000	

showing that the three genes are linked. The reciprocal types, such as $A\ b\ C$ and $a\ B\ c$, are present in about equal frequency, within the range of random sampling variation. Since the triple recessive ($a\ b\ c$) chromosome was contributed by the gamete from the testcross parent, we need only concern ourselves with the other chromosome (carrying the markers shown on the left). Based on relative frequencies, the **parental chromosomes must have been of genotypes $A\ B\ C$ and $a\ b\ c$** (highest frequencies), and the **double crossover progeny chromosomes are of genotype $A\ B\ c$ and $a\ b\ C$** (lowest frequencies). This tells us that **gene C must be in the center. Gene C was present with A and B on one parental chromosome; it is present with a and b on the double crossover chromosome** (and vice versa for c). If you are uncertain about the order, draw out the predicted double crossover genotypes for all three possible orders as we did with the X-Y-Z example. Only the order A-C-B ($A\ C\ B/a\ c\ b$, F_1) will yield the observed $A\ c\ B$ and $a\ C\ b$ (written $A\ B\ c$ and $a\ b\ C$ above) double crossover classes.

Now that the order of the markers is known, the genetic map can be completed by calculating the linkage distances between loci A and C and between loci C and B. The parental marker combinations for the A-C interval are $A\ C$ and $a\ c$; the recombinants are $A\ c$ and $a\ C$. The frequency of recombination for the A-C interval is thus $\frac{(45 + 50 + 2 + 3)}{1000}$ or .10. Note that the progeny with genotypes resulting from double crossovers ($A\ B\ c$ and $a\ b\ C$) are included in the numerator. One of the two crossovers occurred within the A-C interval. Thus, these progeny carry recombinant ($A\ c$ and $a\ C$) combinations of the alleles of these two loci.

Recall that one map unit equals the distance that yields 1 percent recombination. Thus, A and C are 10 map units apart on the linkage map. Similarly, C and B are 15 map units (calculated from $\frac{[2 + 3 + 75 + 70]}{1000} \times 100$) apart. Again, note that the progeny with genotypes resulting from double crossovers are included in the numerator. One of the crossovers occurred within the C-B interval and produced chromosomes with recombinant ($C\ b$ and $c\ B$) combinations of the alleles of the B and C loci. Thus, the progeny with genotypes produced by double crossovers are included in the numerator (recombinants) in calculating both linkage distances.

The linkage map defined by the data in Table 6.1 is, therefore,

(A, a)   *(C, c)*   *(B, b)*

←— 10 map units —→←— 15 map units —→

In the above example, all of the markers were present on the parental chromosomes in the *cis*-arrangement. One can visualize and remember the products of the various crossover events most easily when the markers are in this arrangement, an important aid in working through a three-factor cross for the first time. However, the markers could just as likely have been present on the chromosomes of the F_1 trihybrid in any of the other three possible arrangements: (1) $A\ C\ b/a\ c\ B$, (2) $A\ c\ b/a\ C\ B$, or (3) $A\ c\ B/a\ C\ b$. The same procedures are used to order the loci and to determine linkage distances in all cases.

Consider, for example, the results (Table 6.2) of a testcross in *Drosophila* between female F_1 flies heterozygous at three loci on the third chromosome, cu/cu^+ (curled versus straight wings), e/e^+ (ebony versus gray bodies), and st/st^+ (scarlet versus red eyes) and completely homozygous recessive ($cu\ e\ st/cu\ e\ st$) male flies.

TABLE 6.2 Results of a Three-factor Testcross in *D. Melanogaster*.

GENOTYPE	NUMBER OF PROGENY	
cu e st⁺/cu e st	366	} Parental
cu⁺ e⁺ st/cu e st	380	
cu e st/cu e st	24	} Single crossover in Interval I
cu⁺ e⁺ st⁺/cu e st	30	
cu⁺ e st/cu e st	89	} Single crossover in Interval II
cu e⁺ st⁺/cu e st	105	
cu e⁺ st/cu e st	2	} Double crossover; one in each interval
cu⁺ e st⁺/cu e st	4	
Total = 1000		

The data in Table 6.2 show that one of the parental chromosomes carried markers *cu*, *e*, and *st⁺*, whereas the other carried *cu⁺*, *e⁺* and *st*. Thus, the alleles at the *curled* and *ebony* loci were in the *cis* (coupling)-configuration, whereas the alleles at the *scarlet* locus were in the *trans* (repulsion)-configuration relative to those of *curled* and *ebony*. Since the parental (most frequent) combinations are *cu e st⁺* and *cu⁺ e⁺ st*, and the double crossover (least frequent) combinations are *cu e⁺ st* and *cu⁺ e st⁺*, the *cu* locus must lie between the *st* and *e* loci. (Note that loci are usually identified using the symbol for their mutant alleles, rather than those for their wild-type alleles.)

Observe that a double crossover moves the middle marker onto a chromosome with the two outside markers from the other parent. By contrast, the two outside markers on each parental chromosome remain together. In our example, *e* and *st⁺* are together on one of the parental chromosomes *and* on one of the double crossover chromosomes. The same is true for *e⁺* and *st*. Thus, these must be the outside markers. The alleles at the *curled* locus, on the other hand, are present on the double crossover chromosomes with the alleles of the *ebony* and *scarlet* loci inherited from the other parent. Thus, the *cu* locus must be in the middle.

With a little practice, you will be able to order

the three loci by simply looking at the parental and double crossover genotypes. Until you reach this level of proficiency, however, diagram the three possible orders of the parental markers and determine which order will give rise to the **observed** double crossover genotypes by means of a double crossover.

Having determined the order of the loci to be *st-cu-e*, we simply have to calculate the distances from *st* to *cu* and from *cu* to *e* to complete the linkage map. For the *st-cu* interval, the recombinant genotypes are *st cu* and *st⁺ cu⁺*. The distance in map units is thus $\frac{(24 + 30 + 2 + 4)}{1000} \times 100 = 6$ map units. For the *cu-e* interval, the recombinant genotypes are *cu e⁺* and *cu⁺ e*. The linkage distance is $\frac{(89 + 105 + 2 + 4)}{1000} \times 100 = 20$ map units. The complete linkage map is, therefore,

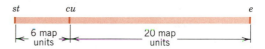

which agrees reasonably well with the known map positions of these three loci (Fig. 6.7).

By combining linkage data from many three-factor crosses that have at least one marker in common, detailed linkage maps, such as those shown in Figs. 6.7 and 6.8, can be derived.

INTERFERENCE

In the three-factor testcross data given in Table 6.1, five progeny resulting from double crossovers in meiosis in the F_1 trihybrid were observed. The **observed frequency of double crossover gametes is thus .005.** If crossovers occur completely at random, such that double crossovers are the result of two independent events as was discussed earlier, then the **expected frequency of double crossover gametes would be .10 × .15 or .015.** Fewer double crossovers are usually observed than would be expected if crossovers were independent events. This phenomenon, which was first observed by H. J. Muller in 1916, is called **chromosome interference or chiasma interference.** It is observed in most three-factor crosses in eukaryotes. Although its molecular basis is unknown, the observed levels of interference indicate that the occurrence of one crossover decreases the likelihood of (or "interferes" with) another crossover occurring nearby. The degree of interference is measured by the **coefficient of coincidence,** which is simply the ratio of the observed to the expected frequency of double crossover gametes. That is,

Coefficient of coincidence

$$= \frac{\text{observed double crossover frequency}}{\text{expected double crossover frequency}}$$

where the expected double crossover frequency is that predicted if the two crossovers are independent events. In the absence of interference, the coefficient of coincidence will equal 1. Sometimes levels of interference are given as the **coefficient of interference,** which equals 1 minus the coefficient of coincidence. The coefficient of interference will thus be 0 in the absence of interference.

Interference values between 0 and 1 indicate that **positive interference** has occurred. For example, in our three-factor testcross example, the coefficient of coincidence was $\frac{.005}{.015}$ or .33. The coefficient of interference for these data is therefore $1 - .33 = .67$. Positive interference indicates that one crossover interferes with the occurrence of a second crossover nearby.

In certain microorganisms, particularly bacteriophages, coefficients of coincidence greater than 1 often occur. Coefficients of coincidence larger than 1 indicate that the occurrence of one crossover increases the likelihood of additional crossovers occurring nearby (so-called "clustering" of crossovers). This second type of interference has been given the rather enigmatic, but accepted, name **negative interference.**

ORDERED TETRAD DATA

As was discussed earlier (see pp. 156–157), in certain fungi all of the products of a single meiotic event can be recovered and analyzed genetically. In some species, the individual haploid products of meiosis are maintained in a linear cylindrical structure (called an **ascus**) in the order in which they are formed during meiosis. In *Neurospora crassa,* a single mitotic division follows meiosis, producing eight

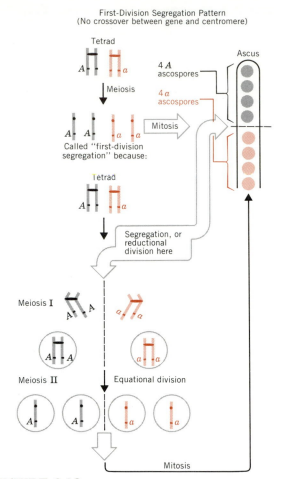

FIGURE 6.13

Diagram showing the first-division segregation ascus pattern that is produced from meiotic events in which no crossover occurs between the segregating locus and the centromere. Note that the alleles (A and a) have segregated into different nuclei at the completion of the first division of meiosis. The second division of meiosis is equational, just like the subsequent mitotic division. The result is an ascus with four ascospores of genotype A at one end and four spores of genotype a at the other end, the so-called first-division segregation pattern.

haploid ascospores (four pairs of identical "ascospore twins") that are maintained in the tubelike ascus (see Fig. 6.5). The individual ascospores can be removed from the ascus and separated to discrete

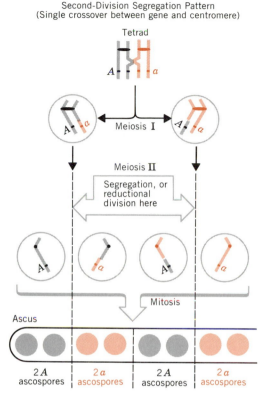

Second-Division Segregation Pattern
(Single crossover between gene and centromere)

Tetrad

A ——— a

Meiosis I

A ——— a A ——— a

Meiosis II

Segregation, or reductional division here

A a A a

Mitosis

Ascus

| $2A$ ascospores | $2a$ ascospores | $2A$ ascospores | $2a$ ascospores |

FIGURE 6.14

Diagram showing the second-division segregation ascus pattern that is produced from meiotic events in which a single crossover occurs between the segregating locus and the centromere. Note that the two alleles (A and a) do not segregate into different nuclei until the second division of meiosis. The result is a 2:2:2:2 ascus pattern with alternating pairs of ascospores of genotype A and pairs of ascospores of genotype a. Compare this ascus pattern with that shown in Fig. 6.13, which resulted when no crossover occurred.

(see pp. 188–191)] and (2) to map the centromere.

Consider the segregation of two alleles (say A and a) of a gene in *N. crassa*. The arrangement of ascospore genotypes in an ascus (called the **ascus pattern**) produced when a crossover occurs in the interval between that gene and the centromere on the chromosome will be very different from the ascus pattern produced when no crossover occurs in that interval (compare Figs. 6.13 and 6.14; see also Fig. 6.15). If no crossover occurs between the A locus and the centromere, the four ascospores at one end of the ascus will all be of one genotype (either A or a), and the four ascospores at the other end will be of the other genotype (Figs. 6.13 and 6.15). This 4:4 pattern is called the **first-division segregation pattern** because the **segregation** of the two alleles **occurs during the first division of meiosis.** If a crossover occurs between the A locus and the centromere, a **second-division segregation pattern** (e.g., $2A:2a:2A:2a$) will result (Figs. 6.14 and 6.15). In this case, the alleles are not segregated into different nuclei until the second division of

FIGURE 6.15

Photomicrograph of the asci from one perithecium of Neurospora, showing both first-division segregation patterns and three of the four second-division segregation patterns expected for the segregation of a single pair of alleles. The dark ascospores carry the wild-type allele ws⁺; the white or colorless ascospores carry the mutant allele ws (white spore). (Reproduced with permission from R. L. Phillips and A. M. Srb, Canadian Journal of Genetics and Cytology *9:766–775, 1967. Original photograph courtesy of R. L. Phillips.)*

cells by micromanipulation, germinated, and then analyzed genetically. Their order within the ascus can be recorded during dissection, providing the geneticist with the so-called **ordered tetrad data.**

Ordered tetrad data are unique in allowing the experimenter (1) to determine whether recombination is reciprocal (as it has been assumed to be in all of the examples discussed so far in this chapter) or nonreciprocal [as in the case of **gene conversion** meiosis. Actually, four different second-division segregation patterns occur, depending on the orientation of the recombinant chromatids during anaphase II of meiosis. Thus, the four second-division segregation patterns shown in Fig. 6.16 occur with approximately equal frequency. Three of these four patterns are represented among the asci shown in the photomicrograph in Fig. 6.15. All four of these patterns provide equivalent information, namely, the occurrence of second-division segregation. They may thus be considered one ascus pattern. In presenting ordered tetrad data, only the genotypes of the four identical pairs of ascospores (rather than all eight ascospores) in each ascus are usually given. The above second-division segregation pattern is often abbreviated as $A:a:A:a$, as shown (top to bottom) in Fig. 6.16.

The recombination frequency for the interval from a gene to the centromere is **one-half the frequency of asci that exhibit second-division segregation patterns for the alleles of the gene.** The **one-half** corrects for the fact that crossing over involves only two of the four chromatids in a tetrad (Fig. 6.6b). Half of the ascospores in each ascus with a second-division segregation pattern will therefore contain chromosomes that were not involved in crossing over. Thus, the map distance from gene to centromere equals

$$\frac{\frac{1}{2} \text{ (number of asci with second-division segregation patterns)}}{\text{total number of asci}} \times 100$$

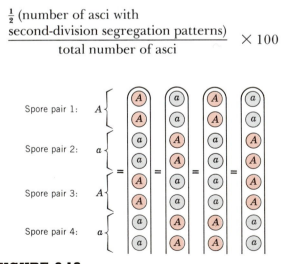

FIGURE 6.16

The four second-division segregation ascus patterns.

The equivalent of a three-factor cross can be done in *N. crassa* by using alleles of just two linked genes, since the centromere can be used as a third marker. That is, ordered tetrad data can be used to order two loci with respect to the centromere. However, before considering how two markers can be ordered relative to the centromere, we should examine the ascus patterns that can occur with two segregating pairs of alleles, for example, *A, a* and *B, b*. Three different ascus patterns occur; these patterns and their formation are illustrated in Figs. 6.17, 6.18, and 6.19.

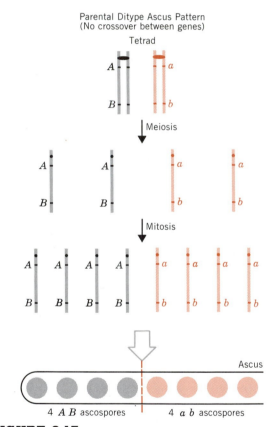

FIGURE 6.17

*Diagram showing the **parental ditype** ascus pattern, which results from meiotic events in which no crossover occurs in the interval between the segregating genes. The four ascospores at one end of the ascus are of one parental genotype; the four at the other end are of the second parental genotype. The same pattern will occur as a result of a two-strand double crossover in the interval between the two genes.*

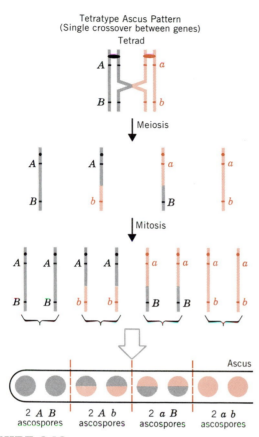

FIGURE 6.18

*Diagram showing the **tetratype** ascus pattern, which results from meiotic events in which a single crossover occurs in the interval between the segregating genes. Four types of ascospores are present in the ascus. Half of the spores are parental in genotype and half are recombinant. The actual sequence of spore pairs within each ascus will depend on the orientation of the homologous chromosomes at metaphase I of meiosis and the orientation of the two chromatids of each chromosome at metaphase II of meiosis. The tetratype pattern will also occur as a result of a three-strand double crossover in the interval between the two genes.*

Meiotic events in which *no* crossover occurs between the two loci will yield asci with four spores of one parental genotype at one end and four spores of the other parental genotype at the other end (Fig. 6.17). A two-strand double crossover between the two loci will yield the same pattern. An ascus with

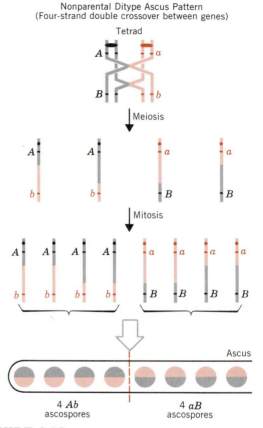

FIGURE 6.19

*Diagram showing the **nonparental ditype** ascus pattern, which results from meiotic events in which a four-strand double crossover occurs in the interval between the segregating genes. Only two types of spores, the two reciprocal recombinant genotypes, are present in the ascus. Note that this type of ascus will be rare in the case of linked genes, but will be just as frequent as the parental ditype class (Fig. 6.17) for independently assorting genes.*

this pattern is called a **parental ditype,** because it contains two types of spores, and both are parental in genotype. When a single crossover or a three-strand double crossover occurs between the two loci, a 2 : 2 : 2 : 2 ascus pattern will result, with two spores being of one parental genotype (e.g., *A B*), two spores being of the second parental genotype (*a b*), and two spores being of each reciprocal recombinant type (two *A b* and two *a B*) (Fig. 6.18). Such an ascus is called a **tetratype** (four different ascospore genotypes). The third ascus pattern results only from a four-strand double crossover between the two loci (Fig. 6.19). This pattern is called the **nonparental ditype** pattern, because such asci contain only two types of spores and both have recombinant genotypes (four *A b* and four *a B*).

The relative frequencies of parental ditype, tetratype, and nonparental ditype asci can be used to calculate the linkage distance between two loci (see the next section, Unordered Tetrad Data). However, with ordered tetrad data, one can also use the centromere as a marker and construct a linkage map showing the order and relative positions of the two genes and the centromere. This is done by determining whether each ascus pattern is the result of first- or second-division segregation of the alleles of each gene, just as was done in calculating the linkage distance between a single segregating gene and the centromere (review Figs. 6.13 and 6.14).

Consider a hypothetical cross in *Neurospora* of the type *AB* × *ab*. If the asci resulting from this cross are dissected so that ordered tetrad data are available, the results of this cross will define two intervals (using the centromere as a third marker). Ordered tetrad data will allow one to determine which of the following three possible orders is correct.

For example, assume that the ordered tetrad data for this cross were as given in Table 6.3. Note that only the genotypes of the four spore pairs in each

ascus are given and that all ascus patterns of the same type are pooled. That is, spore pair 1 may be either the top or the bottom spore pair in an ascus, and all second-division segregation patterns for a locus are pooled under one genotype sequence ($A:a:A:a$ includes $a:A:a:A$, $A:a:a:A$, and $a:A:A:a$).

In crosses involving unlinked genes, the frequency of nonparental ditype asci (Fig. 6.19) is expected to equal the frequency of parental ditype asci (Fig. 6.17). In our example, the frequency of nonparental ditype asci is zero. Thus, the two genes are clearly linked. Of the four types of ascus patterns observed, the parental ditype pattern (ascus type 1) is the most frequent, as expected. Type 4 asci are least frequent ($\frac{2}{200}$); the type 4 pattern clearly must result from meiotic events in which some type of double crossover has occurred. Ascus pattern types 2 and 3 occur at intermediate frequencies. They must result from meiotic events in which a single crossover has occurred in one or the other of the two marked intervals.

If one recalls that first-division segregation occurs when no crossover occurs between a locus and the centromere whereas second-division segregation occurs when a single crossover occurs between a locus and the centromere, it becomes apparent that ascus pattern types 2 and 3 can only be produced by a single crossover for one of the three possible orders. This conveniently allows one to order the two markers relative to the position of the centromere.

The meiotic tetrad that will be involved in our cross can be diagramed as shown in Fig. 6.20 for each of the three possible orders.

Ascus pattern type 2 shows first-division segregation for the alleles at the *A* locus and second-division segregation for the alleles at the *B* locus. This requires that the single crossover occurring in the meiotic events giving rise to these asci be located between the *B* locus and the centromere, but *not* between the *A* locus and the centromere. That, of course, is **not possible for order 2.** Similarly, ascus pattern type 3 exhibits second-division segregation for both loci, *A* and *B*. Thus, the single crossover occurring in the meiotic events giving rise to these asci must be located between both loci and the

TABLE 6.3   Ordered Tetrad Data

SPORE PAIR	TYPE OF ASCUS PATTERN			
	(1)	(2)	(3)	(4)
1	*A B*	*A B*	*A B*	*A B*
2	*A B*	*A b*	*a b*	*a B*
3	*a b*	*a B*	*A B*	*A b*
4	*a b*	*a b*	*a b*	*a b*
Total number of asci:	112	48	38	2

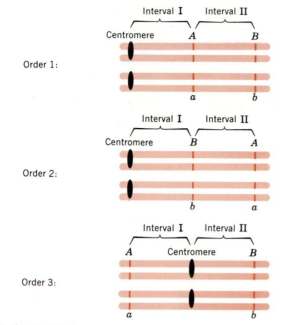

Order 1:

Interval I   Interval II

Centromere   A   B

a   b

Order 2:

Interval I   Interval II

Centromere   B   A

b   a

Order 3:

Interval I   Interval II

A   Centromere   B

a   b

FIGURE 6.20
*The three **possible** orders of locus* A, *locus* B, *and the centromere.*

centromere. That is **not possible if the loci are located on opposite sides of the centromere, order 3.** Ascus pattern types 2 and 3 can both result from single crossovers only in the case of order 1. Given the order centromere-*A-B*, ascus pattern type 2 will be produced when a single crossover occurs in interval II, the interval between loci *A* and *B*. Ascus pattern type 3 will result when a single crossover occurs in interval I, between the *A* locus and the centromere. By the above procedure, one can always eliminate two of the three possible orders and thus deduce the correct order. By a simple diagram, one can easily show that the rare ascus pattern type 4 results from a two-strand double crossover, with one crossover in interval I and the second in interval II.

All that is needed to allow one to draw the linkage map defined by the tetrad data given is the calculation of the linkage distances. We have deduced (1) that ascus pattern type 2 results from a single crossover in interval II, (2) that ascus pattern type 3 results from a single crossover in interval I, and (3) that ascus pattern type 4 results from a double crossover, with one crossover in each inter-

val. The map distance from the *A* locus to the *B* locus (interval II) will thus be [$\frac{1}{2}$ (number of asci of type 2 + number of asci of type 4)/(total number of asci)] × 100, or

$$\frac{\frac{1}{2}(48 + 2)}{200} \times 100 = 12.5 \text{ map units.}$$

These 50 asci all resulted from meiotic events in which there was one crossover in the interval between the *A* and *B* loci. Thus 12.5 percent of the progeny have recombinant combinations of the alleles of the *A* and *B* genes. Note that the 2 (number of 2-strand double crossover asci) is multipled by $\frac{1}{2}$. Even though there were two crossovers in the meiotic events which gave rise to these asci, these crossovers involved **only two of the four chromatids in the tetrads.**

Similarly, the distance from the *A* locus to the centromere (interval I) will be:

$$\frac{\frac{1}{2}(38 + 2)}{200} \times 100 = 10 \text{ map units.}$$

Again, these 40 asci all resulted from meiotic events in which there was one crossover in the interval between the *A* locus and the centromere. Don't forget the factor one-half that corrects for the fact that each crossover involves only two of the four chromatids in a tetrad and thus only one-half of the ascospores in each ascus. Only in the case of nonparental ditype asci (resulting from four-strand double crossovers; see Fig. 6.19) is the factor one-half not used. Thus the linkage map defined by our hypothetical tetrad data is:

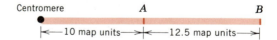

Centromere   A   B

|←—10 map units—→|←—12.5 map units—→|

UNORDERED TETRAD DATA

The dissection of a large number of asci is a tedious task. It is much easier to allow the asci to mature and burst open themselves. By this procedure, all of the ascospores in individual asci can be analyzed, but the order of the ascospores in the asci cannot be determined. Tetrad data obtained in this way are called **unordered tetrad data.** In addition, some Ascomycetes such as the yeasts do not produce ordered tetrads. One can still determine whether recombination events are reciprocal using unordered tetrad data. One can**not,** however, determine whether alleles segregate at the first or the second meiotic division and, thus, one can**not** map the centromere by the procedure described in the preceding section.

Linkage distances between loci can be calculated from unordered tetrad data using the frequencies of parental ditype, tetratype, and nonparental ditype asci (Figs. 6.17, 6.18, and 6.19). The map distance between two loci is:

$$\frac{\frac{1}{2}\ (\text{number of tetratype asci}) + (\text{number of nonparental ditype asci})}{\text{total number of asci}} \times 100$$

This formula gives the frequency of ascospores with recombinant combinations of the alleles of the genes involved. It is equivalent to the formula used to calculate the linkage distance between the *A* and *B* loci in our earlier example of ordered tetrad data. It can**not** be used with ordered tetrad data to calculate linkage distances from genes to centromeres. Unordered tetrad data have been of great value in the study of nonreciprocal recombination or "gene conversion" (see pp. 188–191).

SOMATIC CELL HYBRIDIZATION

Obviously, human genes cannot be mapped by the types of genetic analysis described in earlier sections of this chapter. Prior to about 1965, the only data available for assigning the location of genes to individual linkage groups in humans were pedigree data. During the last two decades, however, much progress has been made in mapping genes in humans. This progress is largely the result of the development of (1) a valuable new technique called somatic cell hybridization, (2) improved cytological techniques for identifying individual human chromosomes and specific segments of individual chromosomes, and (3) recombinant DNA techniques used to isolate and identify the locations of individual genes within DNA molecules and chromosomes.

Somatic cell hybridization, first demonstrated

with mouse cells by G. Barski and colleagues in 1960, involves the fusion of **somatic cells growing *in vitro*** (Fig. 6.21). (Somatic cells include all diploid body cells or nongerm-line cells of an organism.) Cell fusion produces **binucleate hybrid cells** or **heterokaryons.** Cell fusion is followed by nuclear fusion to produce **uninucleate hybrid cells** or **syn-**

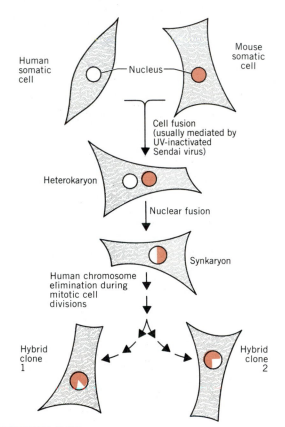

FIGURE 6.21

Schematic illustration of the formation of mouse–human somatic cell hybrids and the **elimination of human chromosomes during subsequent mitotic divisions.** *The established mouse cell culture lines used in these experiments rarely contain the normal diploid number of 40 chromosomes. Usually, they have much higher numbers of chromosomes. The primary human culture cells usually do contain the normal diploid number of 46 chromosomes. After about 30 cell generations, these mouse–human hybrids produce clones that usually contain all of the mouse chromosomes and from 1 to 20 of the original 46 human chromosomes.*

karyons. The frequency of spontaneous cell fusion is very low, about one cell fusion per million cells. Fortunately, the frequency can be greatly increased by the addition of ultraviolet-inactivated **Sendai virus** or the chemical **polyethylene glycol.** These agents stimulate cell fusion by increasing cell contact and altering cell membranes. Sendai viruses are very small relative to the sizes of cells. Thus, when the viruses bind to two cells, they tend to hold them in close juxtaposition to each other. Moreover, the processes involved in normal viral penetration of cells probably facilitate fusion of the membranes of the juxtaposed cells.

Somatic cell hybrids can be produced between two cells of the same species **(monospecific hybrids)** or two cells from different species **(interspecific hybrids). Mouse–human somatic cell hybrids** have been particularly useful for mapping human genes. They are usually made using established mouse cell culture lines and human fibrocytes or leukocytes.

Mouse–human somatic cell hybrids have three important features that make them especially useful for genetic analyses. (1) The **mouse and human chromosomes are easily distinguished,** making the identification of the human chromosomes present in hybrid cells a fairly simple task. (2) All of the mouse chromosomes are usually retained in the hybrid cells, but **only 1–20 of the human chromosomes are retained** in hybrid cell clones. Thus, most of the 46 human chromosomes are eliminated during subsequent mitotic divisions of the hybrid cells, and, most important, they are **eliminated at random or nearly at random.** (3) **Both sets of chromosomes, human and mouse, are expressed in the hybrid cells.** The mouse and human genomes contain many closely related genes that code for gene products with identical, or nearly identical, functions. Nevertheless, these gene products are usually different enough to be distinguished biochemically.

Genes are shown to be located on specific chromosomes by correlating the phenotypes (enzyme activities, etc.) controlled by these genes with the presence of individual human chromosomes in hybrid cell clones. For example, suppose that human cells that make the enzyme thymidine kinase (TK$^+$ cells) were hybridized with mouse cells that can**not** make this enzyme (TK$^-$ cells). Assume that the hybrid cells were cultured for about 30 cell generations, and then five different cell clones were examined to determine which human chromosomes were present. If the results shown in Table 6.4 were obtained, one could deduce that the gene controlling the synthesis of thymidine kinase (gene *TK1*) is located on chromosome 17, because it is the only human chromosome present in all four TK$^+$ hybrid cell clones.

Each of the 24 different human chromosomes can now be accurately identified using improved staining procedures such as G-banding (see Fig. 4.2).

Most somatic cell hybridization experiments are currently done using **selection procedures** that prevent the growth of the parental cells while allowing growth of hybrid cells produced by cell fusion. One widely used method of selection (illustrated in Fig. 6.22) is to fuse mouse cells that cannot make the enzyme thymidine kinase (TK) with human cells that cannot make the enzyme hypoxanthine-guanine phosphoribosyltransferase (HGPRT). [HGPRT

TABLE 6.4   Somatic Cell Hybridization Data

HYBRID CLONE NO.	THYMIDINE KINASE ACTIVITY	HUMAN CHROMOSOMES (+ = PRESENT, − = ABSENT)											
		X	Y	1	2	4	7	9	10	15	17	18	21
1	+	+	+	−	+	−	−	+	+	+	+	−	+
2	+	+	−	+	+	−	+	+	−	+	+	+	−
3	+	−	−	+	+	+	+	−	−	−	+	−	−
4	+	−	−	+	−	−	−	−	+	−	+	−	+
5	−	+	−	−	+	−	−	+	−	+	−	−	−

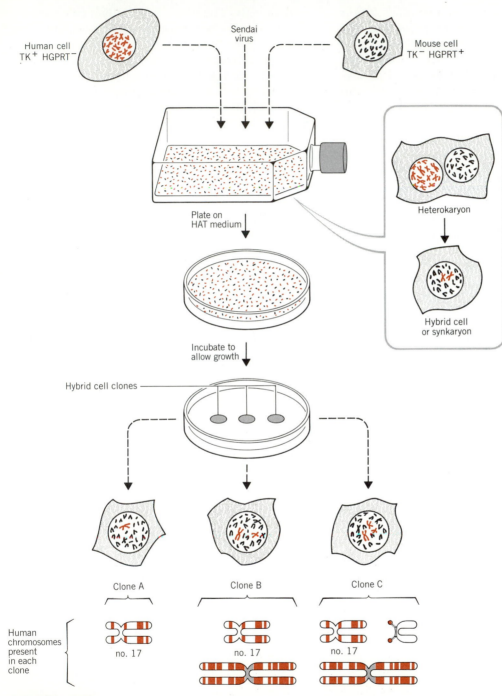

FIGURE 6.22

Identification of the chromosomal location of human genes by mouse–human somatic cell hybridization using the HAT selection procedure. [After F. H. Ruddle and R. S. Kucherlapati, "Hybrid Cells and Human Genes," Sci. Amer. 231(1):36–44, 1974.]

is deficient in cells of individuals with Lesch-Nyhan syndrome (caused by an X-linked recessive allele, see p. 77).]

In the appropriate selective medium, both of these enzymes are required for growth. Thus, neither parent cell can grow, but hybrids produced by cell fusion *can* grow. The TK⁻ mouse cells are HGPRT⁺ (they make enzyme HGPRT) and the HGPRT⁻ human cells are TK⁺ (they make enzyme TK). Since both the mouse and the human chromosomes are expressed in the hybrid cells, they have both enzymes required for growth in the selective medium.

The biochemical basis of this selection procedure involves two important pathways for the biosynthesis of the nucleotides required for DNA synthesis (Fig. 6.23). One is the major pathway by which nucleotides are synthesized *de novo* from simple sugars and amino acids. This pathway is **blocked by the antimetabolite aminopterin.** In the presence of aminopterin, however, cells can still synthesize the nucleotides essential for growth **if certain precursors are added** to the medium and **if** the second pathway, usually called the **salvage pathway, is operative.**

Enzymes TK and HGPRT catalyze key steps in the salvage pathway. As the name suggests, hypoxanthine-guanine phosphoribosyltransferase catalyzes the conversion of hypoxanthine to guanine, and thymidine kinase ("kinases" catalyze phosphate group transfers) catalyzes the phosphorylation of

thymidine. Thus, **TK⁺ HGPRT⁺ cells can grow in the presence of aminopterin if hypoxanthine and thymidine are added to the medium.** Medium containing these three compounds, *h*ypoxanthine, *a*minopterin, and *t*hymidine, is called **HAT medium.** HAT medium selects for TK⁺ HGPRT⁺ hybrid cells, since neither the TK⁻ HGPRT⁺ parental mouse cells nor the TK⁺ HGPRT⁻ parental human cells can grow in it.

If the selection procedure described above is used, human chromosome elimination may occur at random, but those cells that have lost the human chromosome carrying the gene coding for thymidine kinase will **not** be able to grow. Only those hybrid cell clones that have retained this chromosome will grow on HAT medium. The analysis of a set of such clones should reveal that they all have one human chromosome in common. In fact, clones of hybrid cells selected in this way always contain human chromosome 17, indicating that it carries the gene coding for thymidine kinase.

Several selection procedures similar to the HAT procedure described above have been developed for use in somatic cell hybridization experiments. As a result, over 200 human genes have been shown to be located on specific chromosomes. Approximately half of these are X-linked genes, which can be mapped more easily because of their hemizygosity in males. The 24 human "linkage groups" and the locations of a few of the more firmly mapped human genes are shown in Fig. 6.24. Re-

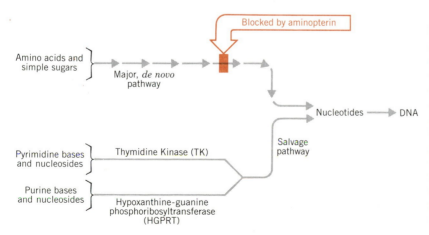

FIGURE 6.23

Pathways of nucleotide biosynthesis, showing only those components important to the HAT (hypoxanthine, aminopterin, thymidine) selection procedure used in somatic cell hybridization experiments. Aminopterin blocks the major, de novo pathway of nucleotide biosynthesis. However, in HAT medium, pyrimidine and purine nucleotides can be synthesized by the salvage pathway from the hypoxanthine and thymidine in this medium if the cells make thymidine kinase and hypoxanthine-guanine phosphoribosyltransferase (TK⁺ HGPRT⁺ cells).

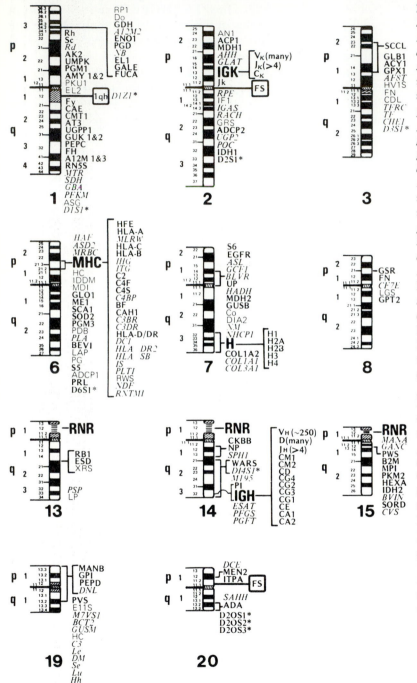

FIGURE 6.24

Status of the human chromosome maps as of March 1982. The short and long arms of each chromosome are designated p *and* q, *respectively. Each arm, in turn, is divided into major and minor segments (numbered intervals) based on chromosome banding patterns. A few of the known genes and gene families (clusters of functionally related genes) are listed below.* Rh, *Rhesus blood group,* 1p32-pter *(chromosome 1, short arm, located within the region from major interval 3, subinterval 2, to the terminus);* IGK, *immunoglobulin kappa light chain gene family,* 2p *(see Chapter 11);* MN, *MN blood group,* 4q; DTS, *diphtheria toxin sensitivity,* 5q15-qter; MHC, *major histocompatibility complex,* 6p21-p23; H, *histones,* 7q32-q36; ABO, *ABO blood group,* 9q34; IFF *and* IFL, *fibroblast and leukocyte forms, respectively, of interferon,* 9; NAG, *non-alpha (β-like) hemoglobin gene family,* 11p12; RNR, *ribosomal RNA gene clusters,* 13p12, 14p12, 15p12, 21p12, 22p12; IGH, *immunoglobulin heavy chain gene family,* 14 *(see Chapter 11);* AG, *alpha hemoglobin gene family,* 16p12-pter; GH, *growth hormone,* 17; TK1, *soluble thymidine kinase,* 17q21-q22; PEPA, *peptidase A,* 18q23-qter; IGL, *immunoglobulin lambda light chain gene family,* 22; HYA, *Y histocompatibility antigen, locus A, Y chromosome;* CBD *and* CBP, *colorblindness, deutan and protan types, respectively,* Xq28; HPRT, *hypoxanthine-guanine phosphoribosyltransferase,* Xq26-qter. *(Based on the map and key compiled by V. A. McKusick, 1982, The human gene map, pp. 327–359, in* Genetic Maps, *Vol. 2, S. J. O'Brien, ed., National Cancer Institute, Frederick, Maryland; for complete descriptions of all the loci listed, see this reference.)*

A confirmed assignment.. ENO1
A provisional assignment .... DHPR

Assignment ''in limbo''
(tentative, inconsistent).. Do

Gene family or cluster .... MHC

* DNA segment, function unknown D14S1*
or restriction fragment length
polymorphism HPA1*

March 1, 1982

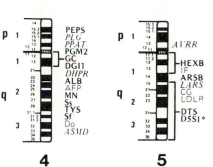

4

5

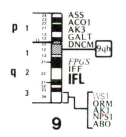

9

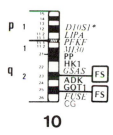

10

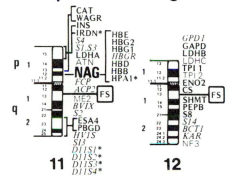

11

12

16

17

18

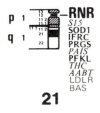

21

22

Y

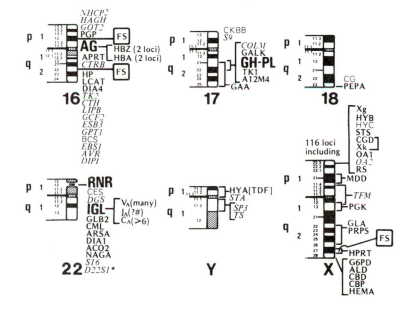

X

member that genes located on the same chromosome are not necessarily linked. Thus, the term **syntenic** has been introduced to refer to genes known to be located on the same chromosome, but for which no information is available regarding their linkage relationship(s).

MOLECULAR MECHANISM OF CROSSING OVER

Despite a wealth of genetic data that describes certain aspects of crossing over and seemingly innumerable models proposed to explain these data, the molecular details of the mechanism of crossing over are still only partly known.

The models that have been proposed to account for crossing over have been of two general types: (1) **breakage and reunion** and (2) **copy choice.** Breakage and reunion involves the breakage of two homologous chromatids and the reunion of the parts in new (recombinant) combinations (as diagrammed in Figs. 6.2 and 6.6). Extensive data now document the occurrence of breakage and reunion during the crossing over process. Copy choice models were based on the assumption that molecules of DNA in the process of being synthesized could switch from using the DNA of one homolog as template in one region to using the DNA of the other homolog as template in another region. Most

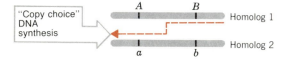

such copy choice models of crossing over were based on the assumption that DNA synthesis was conservative (see Chapter 5, pp. 95–101). Copy choice models rapidly lost support once DNA replication was proven to be semiconservative. [Recalling that a newly synthesized strand of DNA is wound around its template strand in double helix form, draw the structure that would be produced by semiconservative DNA replication and copy choice crossing over, and try to visualize the segregation of the resulting DNA molecules. Segregation will be possible only by (1) completely unwinding the newly synthesized strand from one of the template strands or (2)

breaking the newly synthesized strand at the position of the switch in template (breakage and reunion?).] Pure copy choice (with no associated breakage and reunion) is simply not mechanistically compatible with semiconservative DNA replication. Moreover, pure copy choice requires that crossing over occur during replication, rather than in the tetrad stage of meiosis after replication (see Fig. 6.6). The copy choice mechanism warrants discussion, however, because it now seems quite probable that a small amount of copy choice DNA repair synthesis is associated with crossing over by breakage and reunion. This copy choice repair synthesis may be responsible for the occurrence of **nonreciprocal recombination** or **gene conversion,** which is frequently observed for genetic markers that are located very close to the sites of crossing over (see pp. 188–191). Crossing over, therefore, probably occurs by a complex mechanism that includes some aspects of both breakage and reunion and copy choice models.

The classic experiments of Stern and of Creighton and McClintock (pp. 153–156) were consistent with, but did not prove, the hypothesis that crossing over occurred by breakage and reunion. If the morphology of chromosomes is determined by the genetic information they carry, as seems likely, at least in part, then copy choice models of crossing over would also have predicted Stern's and Creighton and McClintock's results. Their experiments did *not* demonstrate that the recombinant chromosomes contained parts physically derived from (i.e., segments of DNA actually present in) the parental chromosomes.

Direct evidence for breakage and reunion (or possibly breakage and resynthesis) has been obtained in eukaryotes by labeling chromosomes with radioactive (³H) thymidine and following the distribution of radioactive chromatids by autoradiography. Recall that in Taylor's classic experiment, which showed semiconservative chromosome replication in mitosis of *Vicia faba,* one chromatid of each pair was labeled and the other was unlabeled at the second metaphase after labeling (see Fig. 5.57). In subsequent experiments by Taylor involving meiosis in the grasshopper *Romalea,* crossover patterns of labeling were frequently observed. That is,

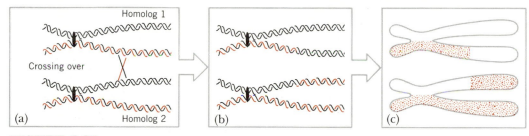

FIGURE 6.25

Schematic diagram of Taylor's autoradiographic evidence for breakage and reunion of homologous chromatids during meiotic crossing over in the grasshopper (Romalea). *Note that for the sake of simplicity each chromatid is shown with a single DNA double helix extending linearly from one end to the other. However,* **remember that the giant DNA molecule in each chromatid actually exists in a highly condensed (coiled and/or folded) configuration** *(see Figs. 5.47, 5.49, and 5.50). Crossing over is shown between a chromatid that is radioactive due to the presence of one strand of DNA containing ³H-thymidine (see Fig. 5.57) and a nonradioactive chromatid. Radioactive strands of DNA are shown in color; nonradioactive strands are shown in black. The centromeres are represented by filled ovals. The tetrad of four chromatids is shown before* (a) *and after* (b) *crossing over. The colored dots on the metaphase chromosomes shown in* (c) *indicate the locations of radioactivity as indicated by autoradiography. (Based on the results of J. H. Taylor,* J. Cell Biol. *25:57, 1965.)*

labeled and unlabeled segments of chromatids of homologous chromosome pairs were often observed to be interchanged (Fig. 6.25). Taylor's experiments thus provided direct evidence for breakage and reunion. They did not exclude, however, the possibility that breakage and **resynthesis** is occurring, that is, that the unlabeled segment is the result of new synthesis rather than the reunion of an unlabeled segment of a parental chromosome.

Definitive evidence for breakage **and reunion,** excluding the possibility of breakage and resynthesis, has been obtained only for crossing over in bacteriophage lambda. M. Meselson crossed $^{13}C^{15}N$-density labeled lambda phage of genotype *h c* (*host range* and *clear* plaque mutant alleles, respectively) with $^{13}C^{15}N$-density labeled lambda phage of genotype $h^+ c^+$. The cross was done in medium containing the common or "light" isotopes (^{12}C, ^{14}N) of carbon and nitrogen. Meselson then analyzed the progeny phage in CsCl density gradients (see Chapter 5, pp. 95–99). He identified $h^+ c$ recombinant phage with the same (or nearly the same) density as the fully "heavy" parental phage. (These phage contained no more than 1–3 percent newly synthesized "light" DNA, probably resulting from a

small amount of repair synthesis.) **The "heavy" recombinants could have been produced only by breakage and reunion,** since breakage and resynthesis would have produced phage of intermediate density (Fig. 6.26).

Although the evidence is convincing that crossing over involves breakage and reunion, probably with a small amount of DNA repair synthesis, the mechanism (or mechanisms) by which the breakage and reunion events occur remains unknown. Clearly the process is complex, especially in eukaryotes. In eukaryotes, crossing over has been associated with the formation of a structure (or set of structures) called the **synaptinemal complex** (Fig. 6.27), which forms during prophase of the first meiotic division. This structure, which is composed primarily of proteins and RNA, has been identified in a large number of eukaryotic species. For unknown reasons, crossing over only rarely occurs, and synaptinemal complexes have not been observed, in *Drosophila* males. (Crossing over does occur in both sexes of most species; the near absence of crossing over in the heterogametic sex is unique to *Drosophila* and a few other species.) A mutation is also known in *Drosophila melanogaster* that sup-

(a) Breakage and reunion:

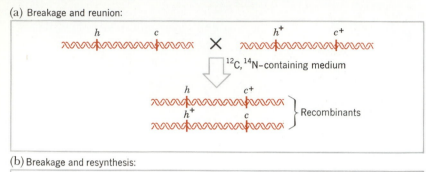

(b) Breakage and resynthesis:

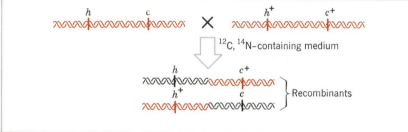

$\text{\footnotesize{}}$ = ^{13}C, ^{15}N–containing DNA; $\text{\footnotesize{}}$ = ^{12}C, ^{14}N–containing DNA

FIGURE 6.26

*Meselson's proof of crossing over by breakage and reunion in phage lambda. Both parental phage (h c and h$^+$ c$^+$) contained "heavy" DNA due to the presence of the heavy isotopes of carbon and nitrogen — ^{13}C and ^{15}N. The cross was carried out in host cells that were growing in medium containing the normal ("light") isotopes ^{12}C and ^{14}N. Progeny phage were separated on the basis of density in a CsCl density gradient and then analyzed for genotype. Meselson demonstrated that rare h$^+$ c recombinant progeny containing fully heavy DNA (DNA in which both strands contain ^{13}C and ^{15}N throughout) were produced. Under these conditions, recombinants containing fully heavy DNA can be produced by breakage and reunion (a), but **not** by breakage and resynthesis (b). Any newly synthesized DNA will be "light," since the cross is being done in medium containing ^{12}C and ^{14}N. (These "heavy" h$^+$ c recombinant progeny are rare since most of the progeny, recombinant and parental, will be "light" or "hybrid" because of several rounds of semiconservative DNA replication in "light" medium. The rare h$^+$ c recombinants were detected by a selection technique not applicable to the reciprocal h c$^+$ recombinant class.)*

presses or prevents crossing over from occurring in females homozygous for this mutation. Such females fail to form synaptinemal complexes during oögenesis. Unfortunately, essentially nothing is known about the functions of the various components of the synaptinemal complex. It is known that a small amount of DNA synthesis (an amount equivalent to less than 1 percent of the total DNA in the genome) occurs at about the time that the synaptinemal complex forms. It seems likely that this DNA synthesis is somehow involved in synapsis and/or

crossing over; however, that remains largely speculation at present.

Despite the paucity of information about the mechanism of recombination, evidence has accumulated, particularly in prokaryotic systems, that indicates that crossing over is an enzymatic process, like most other metabolic processes. In *E. coli*, phage lambda, and phage T4, **recombination-deficient mutants** have been identified and shown to result from mutations in genes coding for nucleases and other enzymes involved in DNA metabolism. How-

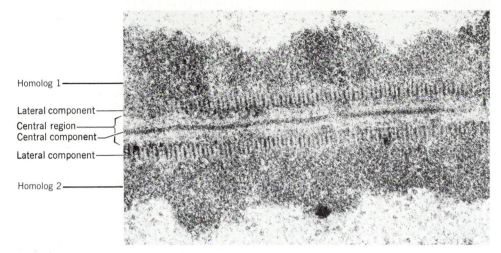

Homolog 1

Lateral component
Central region
Central component
Lateral component

Homolog 2

FIGURE 6.27

Structure of the synaptinemal complex of the ascomycete Neottiella rutilans *(magnified ×61,200). An electron dense "central component" is surrounded by a less electron dense space; together these make up the so-called central region (about 1000 Å in width). The "lateral components" (each about 500 Å in width) are present on each side of the central region. The two homologous chromosomes are juxtaposed to the two lateral elements. (From M. Westergaard and D. von Wettstein,* Ann. Rev. Genet. *6:71–110, 1972. Reproduced with permission, from the* Annual Review of Genetics, *Vol. 6; copyright © 1972 by Annual Reviews Inc. Original photograph courtesy of D. von Wettstein.)*

ever, the complete pathway of enzyme-catalyzed steps has not been worked out for any organism.

Most of the currently popular models of crossing over are derived from the Holliday model and the Whitehouse model, named after the two scientists who proposed them in 1964. These models were the first that took into account all of the various types of genetic data that must be consistent with the mechanism of crossing over in terms of breakage and reunion with associated repair synthesis. Many modifications of these two models have subsequently been proposed. One possible pathway for the occurrence of crossing over, based largely on the Holliday model, is illustrated in Fig. 6.28. This pathway, like most others that have been invoked, begins when an **endonuclease** cleaves single strands of each of the two parental DNA molecules (**breakage**). Segments of the single strands on one side of each cut are displaced from their complementary strands, probably with the aid of **"DNA-binding" proteins, "helix-destabilization" proteins,** and

"DNA-unwinding" proteins (also called **"DNA helicases"**).

The displaced strands then exchange pairing partners, base-pairing with the intact complementary strands of the homologous chromosomes. This process is also aided by certain proteins. In *E. coli*, the product of the *rec A* gene, referred to simply as the ***rec A* protein,** has been shown to promote just such reciprocal exchanges of DNA single strands between two DNA double helices. The *rec A* protein-mediated reaction proceeds in two stages, at least *in vitro*. As a single strand of one double helix "invades" (displaces the identical or homologous strand and base-pairs with the complementary strand) a second, homologous double helix, the displaced single strand of the second double helix, in turn, similarly invades the first double helix. The *rec A* protein mediates such a reaction by binding to the unpaired strand of DNA, somehow "searching" for a homologous DNA sequence, and, once a homologous double helix is found, promoting the displace-

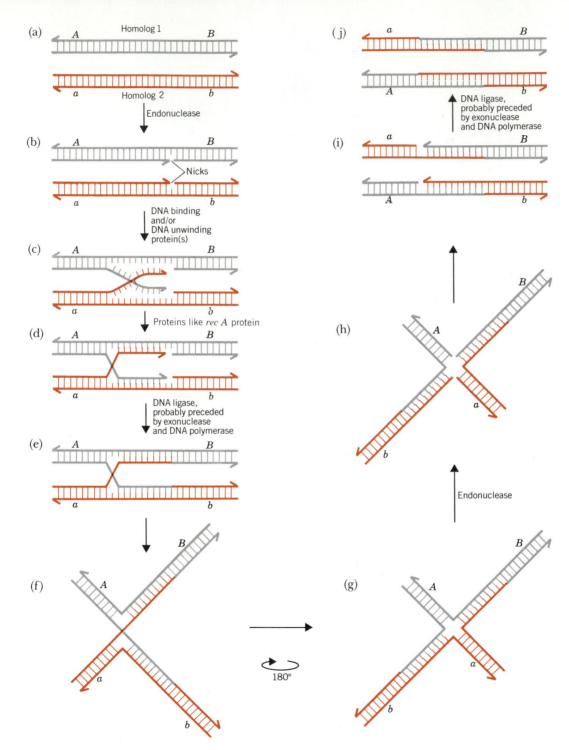

FIGURE 6.28

*A putative pathway for the occurrence of crossing over by breakage and reunion, showing some of the enzymes believed to be involved. This pathway is based on the model of R. Holliday (Genet. Res. 5:282, 1964) with modifications as described by H. Potter and D. Dressler (Proc. Natl. Acad. Sci., U.S. 72:3000, 1976). It involves (a) the synapsis of homolog 1 (shown in gray), carrying markers A and B, with homolog 2 (shown in color), carrying markers a and b. The opposite polarities of the complementary strands of each double helix are indicated by the arrowheads at the ends of each single strand. (Base-pairing must always involve two strands of opposite polarity!) (b) Crossing over is initiated by endonuclease-induced nicks (single-strand cuts) in each DNA molecule. In the Holliday model shown, these cuts occur in strands with the **same polarity**. (c) The free ends on one side of the cuts are displaced from their complementary strands, probably with the aid of "DNA binding," "helix-destabilization," and/or "DNA unwinding" proteins. (d) The displaced strands then base-pair with the complementary intact strands of the other homolog. That is, the cleaved single strand of homolog 1 base-pairs with the complementary intact strand of homolog 2, and vice versa. The protein product of the recA gene of E. coli has been shown to promote this reaction in vitro. (e) The cleaved strands are then covalently rejoined in recombinant combinations by DNA ligase. If the original cuts did not occur between precisely the same two base-pairs, excision by an exonuclease and resynthesis by a DNA polymerase will be required to provide the proper **adjacent** "unlinked-bases" substrate for DNA ligase. At this point, the two homologs are joined by a single-stranded exchange or bridge. The three-dimensional structure of this intermediate in crossing over may be more easily visualized when drawn as in (f) or (g), equivalent structures shown from different planar views. Such X-shaped intermediates [called "Holliday intermediates" or "chi forms" (after the Greek letter chi, written χ)] have been observed in prokaryotes (Fig. 6.29). (h) If the previously intact single strands are then cleaved at the "bridge" or "intersection," the structures shown in (i) will be formed. Covalent closure of the single-stranded interruptions will yield the intact **recombinant chromosomes** shown in (j).*

ment of a segment of one strand of the double helix by the unpaired strand.

If complementary sequences already exist as unpaired strands, as in (*c*) of Fig. 6.28, the presence of the *rec A* **protein will increase the rate of renaturation by over 50-fold.**

The cleaved strands are subsequently covalently joined in recombinant combinations (**reunion**) by DNA ligase. If the original breaks in the two strands do not occur at exactly the same site in the two homologs, some "tailoring" will be required before DNA ligase can catalyze the reunion step; this tailoring involves excision of a limited number of bases by an **exonuclease** and repair synthesis by a **DNA polymerase.** The sequence of events described so far will produce an X-shaped recombination intermediate (called a "chi" form); such inter-

mediates have been observed by electron microscopy in several prokaryotic systems (Fig. 6.29). A similar sequence of enzyme-catalyzed breakage and reunion events, involving the other two single strands, occurs to complete the process of crossing over. While the details of crossing over remain uncertain, enzyme-catalyzed events similar to these are almost certainly involved.

Clearly, proteins such as the *rec A* protein play very important roles in the process of recombination between **homologous chromosomes or homologous DNA molecules.** This type of recombination is called **general recombination.** It should **not** be confused with **site-specific recombination,** recombination that always occurs at specific sites or sequences of DNA molecules, or with **"illegitimate" recombination,** recombination between nonho-

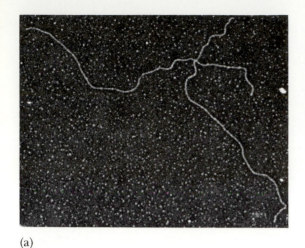

(a)

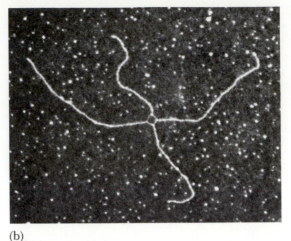

(b)

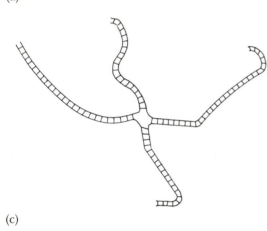

(c)

FIGURE 6.29

Electron micrographs of χ-shaped recombination intermediates ("chi forms" or "Holliday intermediates"). (a) A typical chi form equivalent to that diagramed in Fig. 6.28g. (b) A chi form in which the covalent single-stranded connections at the intersection of the four arms are particularly clear (apparently due to stretching forces during preparation for electron microscopy). (c) Diagram of the structure seen in (b). The DNA molecules are those of the E. coli *"minichromosome" or plasmid Col El. These DNA molecules are circular* in vivo, *but have been cleaved to produce the linear forms shown using a specific endonuclease. (From H. Potter and D. Dressler,* Proc. Natl. Acad. Sci., U.S. *73:3000, 1976.)*

mologous DNA molecules. The latter processes, which will be discussed in subsequent chapters, occur by quite different mechanisms.

GENE CONVERSION

Up to this point, we have only discussed recombination events that can be explained by breakage of homologous chromatids and the **reciprocal exchange** of parts. When crosses are done in Ascomycetes (fungi that produce ascospores) using genetic markers that are very closely linked (such as two mutations at different sites within the same gene; see Chapter 10) and tetrad data are collected and analyzed, recombination is frequently found to be **nonreciprocal.** For example, if crosses are done between two very closely linked mutations in

Neurospora, and asci containing wild-type recombinants are analyzed, these asci frequently do **not contain the reciprocal double mutant recombinant.** Consider a cross involving two very closely linked mutations m_1 and m_2. In a cross of type $m_1 \, m_2^+ \times m_1^+ \, m_2$, asci of the following type are frequently observed:

> Spore pair 1: $m_1^+ \, m_2$
> Spore pair 2: $m_1^+ \, m_2^+$
> Spore pair 3: $m_1 \, m_2^+$
> Spore pair 4: $m_1 \, m_2^+$

The wild-type $m_1^+ \, m_2^+$ spores are present, but the reciprocal recombinant $m_1 \, m_2$ double mutant spores are not present in the ascus. The $m_2^+ : m_2$ ratio is 3 : 1 rather than 2 : 2 as expected. It is as though one of the m_2 alleles is "converted" to the m_2^+ allelic form.

Thus, this type of **nonreciprocal recombination** was called **gene conversion,** and, despite its somewhat misleading connotation, the term has been extensively used for over two decades. One might assume that gene conversion is a result of mutation (see Chapter 9), except that it occurs at a higher frequency than the corresponding mutation events, always produces the allele present on the homologous chromosome, and is correlated about 50 percent of the time with reciprocal recombination of outside markers. The last observation strongly suggests that gene conversion is a result of processes involved in recombination. In fact, it now seems likely that gene conversion is a direct consequence of classical recombination that is observed when very closely linked genetic markers are used, such that the markers are within the region in which the breakage, excision, repair, and rejoining events involved in crossing over are occurring.

With closely linked markers, gene conversion or nonreciprocal recombination is more frequent than reciprocal recombination. In a study of tetrad data from crosses involving mutations at different sites within the histidine-1 gene of yeast, S. Fogel and D. Hurst found that 980 out of 1081 asci containing wild-type his^+ recombinants exhibited gene conversion of one or the other of the segregating mutations, whereas only 101 showed classical reciprocal recombination.

The most striking feature of gene conversion is that the input 1 : 1 allele ratio expected from Mendel's law of segregation (2 : 2 in yeast, 4 : 4 in *Neurospora*) is not maintained. This can most easily be explained if, during the recombination process, short segments of strands of parental DNA are degraded and then resynthesized using a strand of DNA containing the other parental allele as template (a so-called excision-repair reaction). Several DNA repair mechanisms are known in different organisms (see Chapter 9, pp. 293–296). Given mechanisms for the repair of mismatched segments of DNA (segments in which one or more base-pairs cannot properly form hydrogen bonds), the mechanism presented in Fig. 6.28 to explain classical reciprocal recombination would be expected to give rise to gene conversion if genetic markers were present in the immediate vicinity of the crossover. Note that

in Fig. 6.28*d* to 6.28*j* there is a segment of DNA between the *A* and *B* loci where complementary strands of DNA from the two different parents are base-paired. If a third pair of alleles located within this segment were segregating in the cross, mismatches in the two double helices would be present. DNA molecules containing such mismatches, or different alleles in the two complementary strands of a single double helix, are called **heteroduplexes.** Such heteroduplex molecules are now known to occur frequently as intermediates in the process of recombination.

If Fig. 6.28*e* is modified to include a third pair of alleles, and the other two chromatids are added to show the expected tetrad composition, it would appear as follows (again, note that for simplicity, the DNA molecule in each chromatid is shown as an extended linear structure):

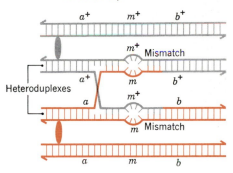

If the mismatches are resolved by an excision-repair pathway, such as that diagrammed in Fig. 9.22, with the *m* strands being excised and resynthesized using the complementary m^+ strands as templates, the following tetrad will result:

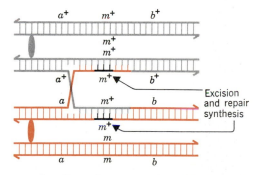

As a result of semiconservative DNA replication during the subsequent mitotic division, this tetrad

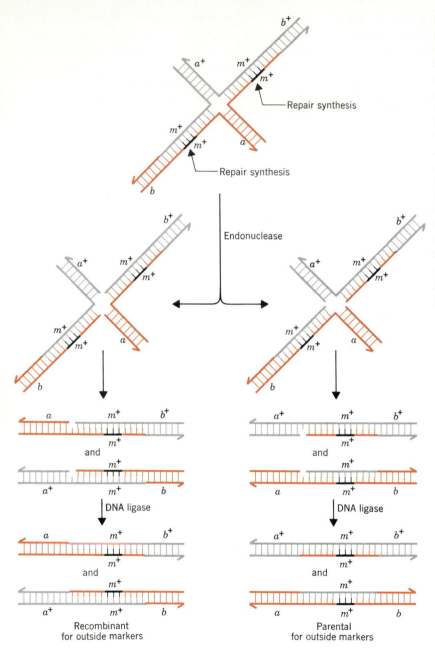

FIGURE 6.30

Diagram showing the formation of either the recombinant combinations (bottom left) or the parental combinations (bottom right) of outside markers in association with gene conversion. The recombination intermediate at the top is the equivalent to Fig. 6.28g for the mismatch repaired chromatids of the tetrad shown in the text (giving rise to the 3 m$^+$ to 1 m gene conversion ascus pattern). Depending on which two of the four strands are cleaved to resolve the bridge and allow segregation, either the parental (a$^+$ b$^+$ and a b) or the recombinant (a$^+$ b and a b$^+$) combinations of outside markers may result. Cleavage in a vertical plane, as shown on the left, yields recombinant combinations of outside markers; cleavage in a horizontal plane, as shown on the right, yields parental combinations of outside markers.

will yield an ascus containing six m$^+$ ascospores and two m ascospores, or the 3 : 1 gene conversion ratio.

Suppose that only one of the two mismatches in the tetrad described above were repaired prior to the mitotic division. In that case, the semiconservative replication of the remaining heteroduplex would yield one m$^+$ homoduplex and one m homoduplex, and the resulting ascus would contain a 5 m$^+$: 3 m ratio of ascospores. Such 5 : 3 gene conversion ratios are in fact observed. They represent postmeiotic or mitotic segregations of unrepaired heteroduplexes.

Another important aspect of gene conversion is the approximately 50 percent correlation with reciprocal recombination of outside markers; this is also nicely explained by the mechanism of recombination presented in Fig. 6.28. If the two recombining chromatids of the above diagramed tetrad (the two center chromatids connected by a single-stranded bridge) are drawn in a planar form equivalent to Fig. 6.28*g*, one can easily see how gene conversion could be associated with either a parental combination of outside markers or a recombinant combination of outside markers. This is illustrated in Fig. 6.30. The analysis of asci produced from tetrads in which gene conversion has occurred has thus provided an important approach by which to study the mechanism of recombination.

SUMMARY

Genes that are located on the same chromosome do not assort independently during meiosis. Instead, they tend to segregate together. Such genes are said to be **linked.** By definition, two genes are linked whenever a dihybrid produces over 50 percent gametes with parental combinations of the segregating pairs of alleles and less than 50 percent gametes with recombinant combinations.

Recombinant combinations of genes located on the same chromosome are produced by **crossing over,** which involves the breakage of individual chromatids and the exchange of parts. This process of **breakage and reunion** is usually associated with a small amount of DNA repair synthesis. Crossing over occurs after chromosome duplication, in the **tetrad** or **four-chromatid** stage of meiosis (early prophase I). A given crossover involves any two of the four chromatids.

The relative locations of genes on chromosomes can be **mapped** by two- and three-factor crosses. **One map unit is defined as the linkage distance that yields 1 percent recombination.** When a large number of genes are mapped in any given species, the genes are observed to occur in **linkage groups, with one linkage group corresponding to each pair of chromosomes** (diploids). All linkage maps exhibit **linearity** (unbranched lines or circles). The maximum frequency of recombination that can occur in any two-factor cross is 50 percent (the value observed for independently assorting genes). In Ascomycetes, such as *Neurospora,* all of the products of a single meiotic event are maintained in a linear, tubelike structure called an **ascus.** By dissecting the haploid products of meiosis, or **ascospores,** out of an ascus, one can determine **whether alleles segregate during the first meiotic division or the second meiotic division.** The alleles will **segregate** during the **second meiotic division only when a crossover occurs between the locus involved and the centromere on the chromosome.** This allows one to map the centromere and to use it as a marker in genetic crosses. Data in which the genotypes of all of the products of a single meiotic event are available are called **tetrad data.** Tetrad data permit one to determine whether recombination is reciprocal or nonreciprocal.

When crosses are performed using **genetic markers that are far apart, recombination is** almost always observed to be **reciprocal.** When **very closely linked markers,** such as two mutations at different sites within the same gene, **are used, recombination is frequently nonreciprocal.** Such nonreciprocal recombination is called **gene conversion.** Gene conversion apparently results from DNA excision and DNA repair synthesis that occurs in heteroduplex regions during the process of breakage and reunion.

Genetic mapping in humans has been facilitated by the development of techniques for the hybridization of somatic cells. **Somatic cell hybridization** involves the fusion of cells (and their nuclei) that are growing *in vitro* to form hybrids. Both monospecific and interspecific cell hybrids can be produced. In rodent–human somatic cell hybrids, most of the human chromosomes are eliminated (for unknown reasons) in the cell divisions following fusion. This elimination of human chromosomes in rodent–human somatic cell hybrids has proven to be extremely advantageous in the determination of linkage relationships in humans.

REFERENCES

Creighton, H. B., and B. McClintock. 1931. "A correlation of cytological and genetical crossing over in *Zea mays.*" *Proc. Natl. Acad. Sci., U.S.* 17:492–497.

Dressler, D., and H. Potter. 1982. "Molecular mechanisms in genetic recombination." *Ann. Rev. Biochem.* 51:727–761.

Hsu, T. C. 1973. "Longitudinal differentiation of chromosomes." *Ann. Rev. Genet.* 7:153–176.

King, R. C. (ed.) 1974. *Handbook of genetics,* Vol. 2, *Plants, plant viruses, and protists.* Plenum, New York.

King, R. C. (ed.) 1975. *Handbook of genetics,* Vol. 3, *Invertebrates of genetic interest.* Plenum, New York.

King, R. C. (ed.) 1975. *Handbook of genetics,* Vol. 4, *Vertebrates of genetic interest.* Plenum, New York.

Lewin, B. 1983. *Genes,* Chapter 35, "Retrieval and recombination of DNA." John Wiley & Sons, New York.

Lindsley, D. L., and E. H. Grell. 1968. *Genetic variations of Drosophila melanogaster.* Carnegie Institution of Washington Publication 627.

McKusick, V. A. 1980. "The anatomy of the human genome." *J. Hered.* 71:370–391.

McKusick, V. A., and F. H. Ruddle. 1977. "The status of the gene map of the human chromosomes." *Science* 196:390–405.

Meselson, M. 1964. "On the mechanism of genetic recombination between DNA molecules." *J. Mol. Biol.* 9:734–745.

Mizuuchi, K., B. Kemper, J. Hays, and R. A. Weisberg. 1982. "T4 endonuclease VII cleaves Holliday structures." *Cell* 29:357–365.

Morgan, T. H., C. B. Bridges, and A. H. Sturtevant. 1925. "The genetics of Drosophila." *Bibliographia Genetica* 2:1–262. (The Hague, Netherlands.)

Morrow, J. 1983. *Eukaryotic cell genetics.* Academic Press, New York.

O'Brien, S. J. (ed.). 1982. *Genetic maps,* Vol. 2. National Cancer Institute, Frederick, Maryland.

Radding, C. M. 1978. "Genetic recombination: strand transfer and mismatch repair." *Ann. Rev. Biochem.* 47:847–880.

Ruddle, F. H., and R. S. Kucherlapati. 1974. "Hybrid cells and human genes." *Sci. Amer.* 231(1):36–44.

Sheridan, W. F. (ed.). 1982. *Maize for biological research.* Plant Molecular Biology Association, Charlottesville, Virginia.

Stahl, F. W. 1979. *Genetic recombination.* W. H. Freeman, San Francisco.

Sturtevant, A. H., and G. W. Beadle. 1939. *An introduction to genetics.* W. B. Saunders Co., Philadelphia. (Republished in 1962 by Dover Publications, New York.)

Vasil, I. K., W. R. Scowcroft, and K. J. Frey (eds.). 1982. *Plant improvement and somatic cell genetics.* Academic Press, New York.

PROBLEMS AND QUESTIONS

6.1. Suggest experiments on some organism to determine genetically (a) whether two genes are located on the same chromosome pair; (b) whether they are in the coupling or repulsion phase. (c) What are the advantages and disadvantages of the testcross method for determining linkage relationships?

6.2. From a cross between individuals with the genotypes $Cc\ Dd\ Ee \times cc\ dd\ ee$, 1000 offspring were produced. The class appearing C-D-ee included 351 individuals. Are the genes c, d, and e on the same or different chromosome pairs? Explain.

6.3. If an animal with the genotype $Rr\ Ss\ Tt$ produced 1020 eggs, of which 127 are $r\ S\ t$, 121 $r\ S\ T$, and 130 $R\ S\ T$, are the three pairs of alleles on the same chromosome or on two different chromosomes or on three different chromosomes? Explain.

6.4. If two loci are 10 map units apart, what proportion of the meiotic events will contain a single crossover in the region between these two loci, assuming that no multiple crossovers occur?

6.5. Genes a and b are linked and are 20 map units apart. An $a^+\ b^+/a^+\ b^+$ individual was mated with an $a\ b/a\ b$ individual. (a) Represent the cross on the chromosomes, illustrate the gametes produced by each parent, and illustrate the F_1. (b) What gametes can the F_1 produce and in what proportions? (c) If the F_1 was crossed with the double recessive, what offspring would be expected and in what proportions? (d) Is this an example of coupling or repulsion?

6.6. If the original cross in the problem above was $a^+\ b/a^+\ b \times a\ b^+/a\ b^+$, diagram (a) the arrangement of the genetic markers on the chromosomes in the F_1; (b) the gametes produced by the F_1 and proportions; and (c) expected testcross results. (d) Is this coupling or repulsion?

6.7. If the recombination frequency in the two problems above were 40 percent instead of 20 percent, what difference would it make in the proportions of gametes and testcross progeny?

6.8. If Problems 6.5 (a) and 6.6 (b) with 20 percent recombination were carried to the F_2 ($F_1 \times F_1$), and a^+ and b^+ were dominant over their alleles, what phenotypic classes would be produced and in what proportions?

6.9. A fully heterozygous F_1 corn plant was red with normal seed. This plant was crossed with a green plant (b) that had tassel seed (ts). The following results were obtained: red, normal 124; red, tassel 126; green, normal 125; and green, tassel 123. (a) Does this indicate linkage? (b) If so, what is the percentage of crossing over? (c) If not, show that the recombination frequency is 50 percent. (d) Diagram the cross showing the arrangement of the genetic markers on the chromosomes.

6.10. A fully heterozygous gray-bodied (b^+), normal-winged (vg^+) female F_1 fruit fly crossed with a black-bodied (b), vestigial-winged (vg) male gave the following results: gray, normal 126; gray, vestigial 24; black, normal 26; and black, vestigial 124. (a) Does this indicate linkage? (b) If so, what is the percentage of crossing over? (c) If not, show that the recombination frequency is 50 percent. (d) Diagram the cross showing the arrangement of the genetic markers on the chromosomes.

6.11. Another fully heterozygous gray-bodied, normal-winged female F_1 fruit fly crossed with a black-bodied, vestigial-winged male gave the following results: gray, normal 23; gray, vestigial 127; black, normal 124; and black, vestigial 26. (a) Does this indicate linkage? (b) If so, what is the percentage of crossing over? (c) Diagram the cross showing the arrangement of the genetic markers on the chromosomes.

6.12. In rabbits, color results from a dominant gene (c^+) and albinism from its recessive allele (c). Black is the result of a dominant gene (b^+), brown of its recessive allele (b). Fully homozygous brown rabbits were crossed with albinos, which carried the gene for black in the homozygous state. The F_1 rabbits were then crossed to double recessive rabbits $\left(\dfrac{c\ b}{c\ b} \text{ or } \dfrac{cb}{cb}\right)$. From many such crosses the results were: black 34; brown 66; and albino 100. (a) Are these genes linked? (b) If so, what is the percentage of crossing over? (c) Diagram the original cross showing the arrangement of the genetic markers on the chromosomes.

6.13. In tomatoes, tall vine (d^+) is dominant over dwarf (d), and spherical fruit shape (p^+) is dominant over pear (p). Vine height and fruit shape are linked with 20 percent crossing over. A certain tall, spherical tomato plant (a) crossed with a dwarf, pear-fruited plant produced 81 tall, spherical; 79 dwarf, pear; 22 tall, pear; and 17 dwarf, spherical. Another tall, spherical plant (b) that was crossed with a dwarf, pear produced 21 tall, pear; 18 dwarf, spherical; 5 tall, spherical; and 4 dwarf, pear. Represent on the chromosomes the arrangements of the genes in these two tall, spherical plants. (c) If these two plants were crossed with each other, what phenotypic classes would be expected and in what proportions?

6.14. Genes a and b are located in chromosome II with a crossover frequency of 20 percent. Genes c and d are located in chromosome III with a crossover frequency of 40 percent. An individual homozygous for $a^+ b^+ c^+ d^+$ was crossed with a fully recessive individual. The F_1 was backcrossed to the full recessive. Diagram (a) the original cross showing the arrangement of genetic markers on the chromosomes, (b) the F_1, and (c) the gametes that the F_1 is expected to produce with their proportions.

6.15. A student has two dominant traits dependent on single genes: cataract (an eye abnormality), which he inherited from his mother, and polydactyly (an extra finger), which he inherited from his father. His wife has neither trait. If the genes for these two traits are closely linked, would the student's child be more apt to have: (a) either cataract or polydactyly, (b) cataract and polydactyly, or (c) neither trait? Explain.

6.16. In *Drosophila*, the recessive genes sr (stripe) and e (ebony body) are located at 62 and 70 map units, respectively, from the left end of the third chromosome. A striped female (homozygous for e^+) was mated with a male having an ebony body (homozygous for sr^+). (a) What kinds of gametes will be produced by the F_1 females and in what proportions? (b) If F_1 females are mated with striped, ebony males, what phenotypes would be expected and in what proportions?

6.17. In *Drosophila*, the gene (vg) for vestigial wing is recessive and is located at 67 units from the left end of the second chromosome. Another gene (cn) for cinnabar eye color is also recessive and is located at 57 units from the left end of the second chromosome. A fully homozygous female with vestigial wings was crossed with a fully homozygous cinnabar male. (a) How many different kinds of gametes could the F_1 females produce and in what proportions. (b) If the females are mated with cinnabar, vestigial males, what phenotypes would be expected and in what proportions?

6.18. In *Drosophila*, the recessive genes st (scarlet eye), ss (spineless bristles), and e (ebony body) are located in the same (third) chromosome in the following positions (map distances), starting from the left end of the chromosomes: st 44, ss 58, e 70. Fully heterozygous females with the genotype $st\ ss\ e^+/st^+\ ss^+\ e$ are mated with fully recessive males $st\ ss\ e/st\ ss\ e$. If many flies are produced and no interference occurs, what phenotypes will be expected and in what percentages?

6.19. In maize, genes Pl for purple (dominant over Pl^+ for green), sm for salmon silk (recessive to sm^+ for yellow silk), and py for pigmy (recessive to py^+ for normal size) are on chromosome 6 (Fig. 6.8). From the following cross:

$$\frac{Pl\ sm\ py}{Pl\ sm\ py} \times \frac{Pl^+\ sm^+\ py^+}{Pl^+\ sm^+\ py^+}$$

and the testcross between the F_1 and fully recessive plants, what phenotypes would be expected and in what proportions? Assume no interference and refer to Fig. 6.8 for map distances.

6.20. In maize, the genes Tu, j_2, and gl_3 are on chromosome 4 (Fig. 6.8). If plants carrying these three genes in the homozygous recessive condition are crossed with plants homozygous for the three dominant alleles, and F_1 plants are testcrossed to fully recessive plants, what genotypes would be expected and in what proportions? Assume no interference and refer to Fig. 6.8 for map distances.

6.21. A cross was made between yellow-bodied (y), echinus (ec), white-eyed (w) female ($y\ ec\ w/y\ ec\ w$) flies and wild males. The F_1 females were mated with $y\ ec\ w$ males. The following genotypes were present in a sample of 1000 progeny flies:

FEMALES	MALES	NUMBER OF PROGENY
$+++/y\ ec\ w$	$+++/Y$	475
$y\ ec\ w/y\ ec\ w$	$y\ ec\ w/Y$	469
$y++/y\ ec\ w$	$y++/Y$	8
$+\ ec\ w/y\ ec\ w$	$+\ ec\ w/Y$	7
$y+w/y\ ec\ w$	$y+w/Y$	18
$+\ ec+/y\ ec\ w$	$+\ ec+/Y$	23
$++w/y\ ec\ w$	$++w/Y$	0
$y\ ec+/y\ ec\ w$	$y\ ec+/Y$	0

Determine the order in which the three loci y, ec, and w occur on the chromosome and prepare a linkage map.

6.22. A cross involving X-linked genes was made between yellow, bar, vermilion female flies and wild males, and the F_1 females were crossed with $y\ B^+\ v$ males. The following phenotypes were obtained when 1000 progeny were examined:

$y\ B\ v$	and $+++$	546
$y++$	and $+B\ v$	244
$y+v$	and $+B+$	160
$y\ B+$	and $++v$	50

Determine the order in which the three loci occur on the chromosome and prepare a linkage map.

6.23. Female *Drosophila* heterozygous for ebony (e^+/e), scarlet (st^+/st), and spineless (ss^+/ss) were testcrossed, and the following progeny were obtained:

PROGENY PHENOTYPES	NUMBER
Wild-type	67
Ebony	8
Ebony, scarlet	68
Ebony, spineless	347
Ebony, scarlet, spineless	78
Scarlet	368
Scarlet, spineless	10
Spineless	54

(a) Are these genes linked? Justify your answer. (b) Write the genes given on a chromosome symbol with the genes in correct order. (c) Write the genotypes of the flies involved in the parental cross and testcross. (d) What is the map distance between the loci for ebony and scarlet? (e) What is the map distance between the loci for ebony and spineless? (f) Calculate the coefficient of coincidence.

6.24. Consider a female *Drosophila* having the following X chromosome constitution:

The genes w and dor are recessive mutations resulting in mutant eye color (white and deep orange, respectively). If these loci (white and deep orange) exhibit 40 percent recombination in two-factor crosses, what proportion of the male progeny of this female will be expected to show a mutant phenotype?

6.25. Assume that in *Drosophila* there are three pairs of alleles, $+/x$, $+/y$, and $+/z$. As shown by the symbols, each mutant gene is recessive to its wild-type allele. A cross between females heterozygous at these three loci and wild-type males yielded the following progeny.

Females:	$+\ +\ +$		1010
Males:	$+\ +\ +$		39
	$+\ +\ z$		430
	$+\ y\ z$		32
	$x\ +\ +$		27
	$x\ y\ +$		441
	$x\ y\ z$		31
		Total =	2010

Draw the appropriate linkage map for the above data showing the order of the three markers and the map distances for each marked interval. Calculate the coefficient of coincidence for these data.

6.26. In the nematode, *Caenorhabditis elegans*, *dpy* (dumpy) and *unc* (uncoordinated) are recessive linked genes with a recombination frequency of *P*. A heterozygote is made with genes in repulsion, for example,

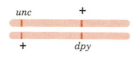

The heterozygote is allowed to self-fertilize. If crossing over in sperm production is independent of crossing over in oöcytes, what fraction of progeny would be expected to be *both* dumpy and uncoordinated?

6.27. In the following testcross, genes *a* and *b* are 20 map units apart and genes *b* and *c* are 10 map units apart:

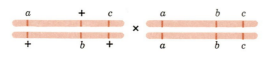

If the coefficient of coincidence is 0.5, how many triple homozygous recessive individuals would you expect to find among 1000 progeny from the cross?

6.28. Consider *a*, *b*, and *c* to be three recessive mutations in *Drosophila*. The data shown below are the results of a testcross in which F_1 females heterozygous at all three loci were crossed to males homozygous for all three recessive mutants.

+++	75	*a* + +	110
+ + *c*	348	*a b* +	306
+ *b c*	96	*a b c*	65
	Total = 1000		

Construct a linkage map showing the correct order of the three genes and the map distances between adjacent genes.

6.29. Singed bristles (*sn*), crossveinless wings (*cv*), and vermilion eye color (*v*) are due to recessive mutant alleles of three sex-linked genes in *Drosophila melanogaster*. When a female heterozygous for each of the three genes was testcrossed with a singed, crossveinless, vermilion male, the following progeny were obtained:

Singed crossveinless, vermilion	3
Crossveinless, vermilion	392
Vermilion	34
Crossveinless	61
Singed, crossveinless	32
Singed, vermilion	65
Singed	410
Wild-type	3
Total =	1000

What is the correct order of the three genes on the chromosome? What are the genetic map distances between *sn* and *cv*, *sn* and *v*, and *cv* and *v*? What is the coefficient of coincidence?

6.30. What will the progeny phenotype ratio be if a plant of the genotype $l_1Al_2^+/l_1^+al_2$ (where *A* and *a* are alleles with *A* dominant to *a*; l_1 and *A* as well as l_1^+ and *a* are completely linked; the *A* and l_2 loci are unlinked; and both l_1 and l_2 are lethal in the homozygous condition but have no detectable effect in the heterozygous condition) is self-fertilized?

6.31. Assume that a cross is made between a plant having the genotype *AAbb* and a second plant having the genotype *aaBB*. The F_1 plants were intercrossed (i.e., *AaBb* × *AaBb*) to produce the following F_2 progeny:

PHENOTYPE	NUMBER OF PLANTS
A-B-	102
A-bb	46
aaB-	50
aabb	2

How can one make a quick rough estimate of the linkage distance between the two loci? Based on the above data, what would this estimate be for the map distance between the *A* locus and the *B* locus?

6.32. On the basis of the following ordered tetrad data from *Neurospora*:

Cross	Asci Patterns				
(A) × (a)	SPORE PAIRS				
	1–2	3–4	5–6	7–8	NUMBER OF ASCI
	(A)	(A)	(a)	(a)	61
	(a)	(a)	(A)	(A)	55
	(a)	(A)	(a)	(A)	40
	(A)	(a)	(A)	(a)	44
				Total =	200

what is the linkage distance in map units between the *A*, *a* locus and the centromere?

6.33. Of what genetic significance is the relatively frequent occurrence of tetratype asci in *Neurospora*?

6.34. The following tetrad data were obtained from the cross $AB \times ab$ in *Neurospora*:

Asci Patterns

SPORE PAIRS				NUMBER OF ASCI
1–2	3–4	5–6	7–8	
(AB)	*(AB)*	*(ab)*	*(ab)*	1766
(AB)	*(aB)*	*(Ab)*	*(ab)*	220
(AB)	*(Ab)*	*(aB)*	*(ab)*	14
			Total =	2000

On the basis of these data, draw a linkage map showing the relative positions of the two genes and the centromere.

6.35. Three mutations (*arg*, *thi*, and *leu*) in *Neurospora* are known to result in blocks in the synthesis of arginine, thiamine, and leucine, respectively. You are given the following data from asci dissections from crosses of *arg* with *thi* and *arg* with *leu*:

Asci Patterns

CROSS	SPORE PAIRS				NUMBER OF ASCI
	1–2 OR 7–8	3–4 OR 5–6	5–6 OR 3–4	7–8 OR 1–2	
(arg +) × *(+ thi)*	*(arg +)*	*(arg +)*	*(+ thi)*	*(+ thi)*	46
	(arg thi)	*(arg thi)*	*(+ +)*	*(+ +)*	54
				Total =	100
(arg +) × *(+ leu)*	*(arg +)*	*(arg +)*	*(+ leu)*	*(+ leu)*	155
	(arg +)	*(arg leu)*	*(+ +)*	*(+ leu)*	44
	(arg +)	*(+ leu)*	*(arg +)*	*(+ leu)*	1
				Total =	200

(Note that the three mutant alleles are represented by *arg*, *thi*, and *leu*; the three wild-type alleles simply by plus signs (+) in each case.) Designate each chromosome involved by a single straight line and each centromere by an open circle, and draw the linkage map(s) dictated by these data; that is, show order and each map distance that can be calculated from the above data.

6.36. Given the following *unordered tetrad* data from *Neurospora*:

Cross (+ + +) × (x y z)

CLASS	UNORDERED SPORE PAIRS				NUMBER OF ASCI
1	x y z	x y z	+ + +	+ + +	24
2	x y z	+ y z	x + +	+ + +	2
3	x y +	x y +	+ + z	+ + z	20
4	x y z	x y +	+ + +	+ + z	8
5	x + +	x + +	+ y z	+ y z	20
6	x + z	x + z	+ y +	+ y +	26

(a) What are the linkage relationships between the genes, that is, which genes are linked and what are the map distances between the linked genes? (b) Which gene is the greatest distance from a centromere? Explain the basis for your answer.

6.37. Two mutant genes (*a* and *t*) in *Neurospora* are known to interfere, respectively, with the synthesis of the amino acid arginine and the vitamin thiamine. After a cross in which these genes were segregating, the following ordered spore arrangements were found in the frequencies noted. (Only one member of each pair of spores is indicated.)

PAIR 1	PAIR 2	PAIR 3	PAIR 4	NUMBER OF ASCI
a t	a t	+ +	+ +	42
+ t	+ t	a +	a +	40
+ +	+ +	a t	a t	39
a +	a +	+ t	+ t	42

How are the *a* and *t* genes located in the chromosomes with respect to their centromere(s) and with respect to each other?

6.38. In *Neurospora* the following is the order of 3 genes and a centromere:

A cross between + + + and *x y z* gave one ascus with the following ordered ascospore composition:

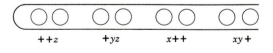

(a) Is this ascus most likely the result of a meiotic event in which 0, 1, 2, or 3 crossovers occurred? (b) In what interval(s) did the crossover(s) most likely occur? (c) If double or triple crossovers were involved, were they 2-strand, 3-strand, or 4-strand multiple crossovers?

6.39. The rosy *(ry)* locus of *Drosophila melanogaster* is located at map position 52.0 on chromosome 3. It controls the structure of the enzyme xanthine dehydrogenase. The rosy locus is particularly susceptible to genetic fine structure analysis because zygotes possessing little or no xanthine dehydrogenase activity are unable to complete development and die before eclosion on standard *Drosophila* culture medium supplemented with purines. The *kar* (karmoisin eyes) locus of *D. melanogaster* is located at map position 53.0 on chromosome 3; it provides an appropriate outside marker in fine structure mapping of the *ry* locus. A. Chovnick and colleagues performed crosses of the following type:

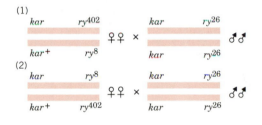

and scored the frequency of the outside marker *kar* among ry^+ progeny (progeny flies surviving on purine-enriched medium; note that ry^{402}, ry^{26}, and ry^8 are all completely recessive alleles of the *rosy* locus).

In cross (1), the majority of the ry^+ progeny were kar^+, whereas in cross (2) the majority of the ry^+ progeny were *kar* (had karmoisin eyes). What is the relative order of *kar*, ry^8, and ry^{402} on chromosome 3 of *D. melanogaster?*

6.40. A patient with Lesch-Nyhan disease was found to be heterozygous for alleles of the *gal K* gene, which codes for the enzyme galactokinase. Fibroblasts of this patient produced two different forms of this enzyme, which could be separated by polyacrylamide gel electrophoresis and identified using an enzyme-specific stain. Somatic cell hybrids were produced between fibroblasts of this patient and thymidine kinase-deficient mouse cells. The hybrid cells were selected on HAT medium (see p. 179) and cultured for about 30 generations to allow human chromosome elimination to occur. Each of 100 independent clones was then examined by electrophoresis to determine its galactokinase phenotype. All of the hybrid clones produced one or the other of the two forms of galactokinase; a few produced both forms. On which human chromosome is the *gal K* gene probably located?

SEVEN

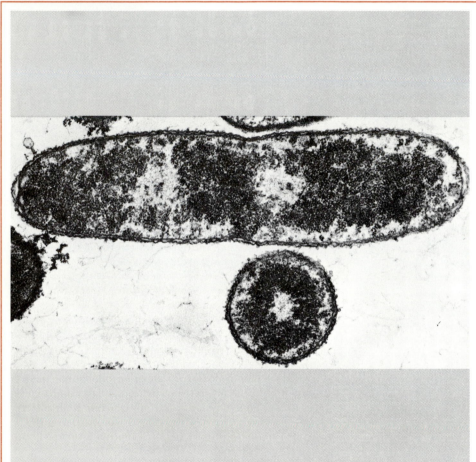

Electron micrograph of a thin section of two phage T4-infected *Escherichia coli* cells showing two separate pools of DNA: (1) the replicating phage DNA in centrally located pools and (2) the disrupted host DNA pools located at the peripheries of the cells. With the staining procedure used, the ribosomes and membranes are dark and the DNA pools are light. Note the phage particle adsorbed to the cell wall at the top left (Reproduced with permission from D. P. Snustad et al., *J. Virology* 18:268–288, 1976.)

RECOMBINATION IN BACTERIA

The genetic information of bacteria is stored in (1) a **single main chromosome,** carrying a few thousand genes, plus (2) from zero to several "minichromosomes" called **episomes** and **plasmids.** Plasmids are highly variable in size, ranging from those carrying no more than three genes to those large enough to carry several hundred genes. Certain bacterial cells are known to harbor as many as 11 different plasmids in addition to the main chromosome.

Each bacterial chromosome characterized to date contains a single, circular molecule of DNA that exists in a highly condensed ("folded," negatively supercoiled; see Chapter 5, pp. 116–118) conformation *in vivo.* The chromosomes of bacteria, like those of other prokaryotes, are not enclosed in nuclear membranes. The nuclear regions of bacteria are thus called **nucleoids,** rather than nuclei, to distinguish them from their eukaryotic counterparts.

Bacteria divide by simple fission, with an equational distribution of their genetic material to the two progeny cells. They are usually haploid (or partially diploid, during recombination processes), but multinucleate. That is, the nucleoid of a bacterium usually contains two or more identical copies of the chromosome (plus plasmids, when present). The chromosomes of bacteria do not go through the mitotic and meiotic condensation cycles associated with cell division and gametogenesis in eukaryotes. The genetic recombination events associated with sexual reproduction in higher organisms (segregation, independent assortment, and meiotic crossing over) thus do not constitute integral parts of the life cycles of bacteria.

Recombination, however, is undoubtedly important in the evolution of bacteria just as it is in the evolution of eukaryotes. Three different processes have evolved that mediate transfer of genetic material from one bacterium to another, making possible the subsequent recombination events. The most obvious difference between these three processes is the mode of transfer of DNA from one cell to another. (1) **Transformation** involves the uptake of naked DNA molecules from one bacterium (the **donor** cell) by another bacterium (the **recipient** cell)

(see Chapter 5). (2) **Transduction** occurs when bacterial genes are carried from a donor cell to a recipient cell by a bacteriophage. (3) **Conjugation** is the process during which DNA from a donor or male cell is transferred to a recipient or female cell through a specialized sex pilus or "conjugation tube."

The three modes of recombination in bacteria can be distinguished by two simple criteria: (1) sensitivity to the presence of deoxyribonuclease (DNase), and (2) dependence on cell contact (Table 7.1). These two criteria can be easily tested experimentally. The first criterion can be tested by simply adding DNase to the medium containing the bacteria involved in recombination. If recombination occurs in the absence of DNase, but not in its presence, then the DNase-sensitive recombination process, transformation, is occurring. The protein coat of the bacteriophage vector and the cell wall and membrane enclosing the conjugation tube protect the donor DNA from degradation by DNase in the medium during transfer to the recipient cell during transduction and conjugation, respectively. Whether cell contact is required is tested by carrying out a so-called U-tube experiment, in which bacteria of two different genotypes are placed in opposite arms of a U-shaped culture tube (Fig. 7.1). The two cultures of bacteria are separated by a semipermeable filter, containing pores large enough to allow DNA molecules and viruses, but **not cells,** to pass through it. Conjugation cannot occur between cells separated in this way. If recombination occurs in the presence of DNase and in the absence of cell contact, it must be due to transduction or an unknown transduction-like process.

TABLE 7.1 Criteria for Determining the Mode of Recombination in Bacteria

RECOMBINATION PROCESS	CRITERION	
	CELL CONTACT REQUIRED?	SENSITIVE TO DNASE?
Transformation	no	yes
Transduction	no	no
Conjugation	yes	no

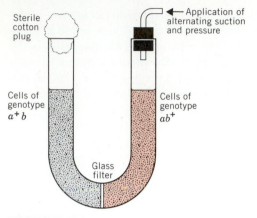

Sterile cotton plug

← Application of alternating suction and pressure

Cells of genotype a⁺ b

Cells of genotype ab⁺

Glass filter

FIGURE 7.1

Diagram of the U-tube experiment used to determine whether cell contact is required for recombination in bacteria. The two arms of the Davis U-tube are separated by a glass filter containing pores of a size that permit the passage of DNA molecules and viruses, but not cells. Bacteria of one genotype (e.g., a⁺ b) are placed in one arm of the Davis U-tube; those of another genotype (e.g., a b⁺) are placed in the other arm. Alternating suction and pressure are applied to one arm of the tube during incubation to mix the medium and any small particles (such as phage) suspended in it. However, the glass filter prevents cells from passing from one arm of the tube to the other. If recombinants (e.g., a⁺ b⁺) are formed during the U-tube experiment, cell contact is not required for the recombination process to occur; thus this excludes conjugation as a possible mechanism.

TRANSFORMATION

As was discussed in Chapter 5 (pp. 81–83), transformation was first discovered in pathogenic strains of *Diplococcus pneumoniae* by Griffith in 1928. The details of Avery, MacLeod, and McCarty's (1944) proof that the "transforming principle" (the cellular component mediating transformation) is DNA were also described in Chapter 5 (pp. 83–84). Some of the details of the mechanism by which transformation occurs are still unknown, 50 years after Griffith's discovery. However, a reasonably complete picture of the overall process of transformation has been established.

The uptake of DNA molecules by recipient bacteria is an active, energy-requiring process. It does not involve passive entry of DNA molecules through permeable cell walls and membranes (although this type of uptake of DNA molecules may be induced by experimental manipulations of bacteria in the laboratory, such as in the *Escherichia coli* recombinant DNA-cloning experiments; see Chapter 10, pp. 353–359). Thus, transformation does not occur "naturally" in all species of bacteria, only in those species possessing the enzymatic machinery involved in the active uptake and recombination processes. Most of the studies on transformation have been done with three species, *D. pneumoniae, Bacillus subtilis,* and *Hemophilus influenzae.* Even in these species, all cells in a given population are not capable of active uptake of DNA. Only **competent** cells, which possess a so-called **"competence factor"** (probably a cell surface protein or enzyme involved in binding or uptake of DNA), are capable of serving as recipients in transformation. The proportion of the bacteria in a culture that are in the physiologically competent state varies with the growth conditions and the stage of the growth curve (becoming maximal in late log-phase).

The process of transformation can be divided into several stages: (1) **reversible binding of double-stranded DNA** molecules to **receptor sites** on the cell surface; (2) **irreversible uptake** of the donor DNA (at which time the donor DNA becomes resistant to DNase in the medium); (3) **conversion** of the double-stranded donor DNA molecules **to single-stranded molecules** by nucleolytic degradation of one strand; (4) **integration (covalent insertion)** of all or part of the single strand of donor DNA into the chromosome of the recipient; and (5) the **segregation** and **phenotypic expression** of the integrated donor gene or genes in the recombinant ("transformed") cell. Steps (2) and (3) may well be coincident effects of a single process. One attractive model, for which there is supporting evidence in the case of pneumococcus, proposes that a specific exonuclease (or **DNA "translocase"**) pulls one strand of donor DNA into the cell using energy derived from the degradation of the complementary strand (Fig. 7.2). Whether degradation of the complementary strand of DNA actually occurs during uptake or immediately after uptake is uncertain. Moreover,

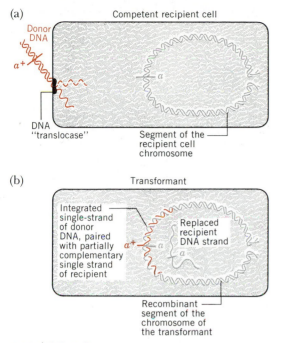

(a) Competent recipient cell

Donor DNA

a^+

DNA "translocase"

Segment of the recipient cell chromosome

a

(b) Transformant

Integrated single-strand of donor DNA, paired with partially complementary single strand of recipient

a^+

a

Replaced recipient DNA strand

a

Recombinant segment of the chromosome of the transformant

FIGURE 7.2

*Diagrammatic illustration of the two key steps, uptake and integration, in transformation of pneumococcus. (a) This model for the uptake of donor DNA was proposed by S. Lacks in 1962. A membrane-bound exonuclease or DNA "translocase" pulls one strand of donor DNA into the cell, using energy derived from degradation of the complementary strand of DNA. The donor DNA fragment (shown in color) carries the genetic marker a^+; the recipient chromosome (shaded, only a segment of which is shown) carries the allelic marker a. (b) The single strand of donor DNA (color) is shown **covalently inserted** (integrated) into the chromosome of the recipient (now called a **transformant**). The donor strand base-pairs over most of its length with the complementary (except at the a site) single strand of the recipient chromosome. A "mismatch" occurs at the a^+/a site where the donor strand carries the a^+ allele, whereas the recipient strand carries the a allele. Such DNA molecules, carrying different alleles in the complementary (partially) strands, are called **DNA heteroduplexes**. DNA heteroduplexes are important intermediates in mutation, recombination, and DNA repair processes. They segregate during subsequent semiconservative replication.*

considerable evidence suggests that these processes may vary in different species.

The first three steps in transformation—bind-ing, uptake, and degradation of one strand of the double-stranded DNA—are not specific for homologous DNA. In fact, competent bacteria will carry out these three processes equally well given calf thymus DNA or other foreign DNAs. However, the **integration,** or **DNA recombination** step, **is specific** for **homologous DNA.** That is not to say that the integration of segments of heterologous (foreign) DNA never occurs. If it does occur, however, it does so at frequencies very much lower than the frequencies observed using homologous DNA. Although very small fragments of DNA are taken up by competent cells, a minimum length of about 500 nucleotide pairs appears to be required for integration to occur. During integration, a single strand (either strand, the previous degradation of one strand is at random) of donor DNA is physically inserted into the recipient chromosome, replacing a segment of one strand of the recipient chromosome (Fig. 7.2 and pp. 220–221). In most transformation experiments, donor DNA fragments are about 20,000 nucleotide pairs (or about $\frac{1}{200}$ of the total chromosome) in length. This means that mapping experiments can be done using transformation only if the genetic markers employed are located close together on the host chromosome.

If two genes are far apart on the chromosome, they will never be carried on the same molecule of transforming DNA. Thus, double transformants for the two genes (say a to a^+ and b to b^+, using an $a^+ b^+$ donor and an a b recipient) will require two independent transformation events (uptake and integration of one DNA molecule carrying a^+ and another molecule carrying b^+). The probability of two such independent events occurring together will equal the product of the probabilities of each occurring alone. Since transformation of any single marker occurs with a low frequency, double independent transformation events of this type will be extremely rare. If, on the other hand, two genes are closely linked, they may be carried on a single molecule of transforming DNA. In this case, double transformants can be formed by the uptake and integration of one molecule of donor DNA carrying both genes. Thus, if two genes or genetic markers are very closely linked, double transformants may be formed at a frequency approaching the frequency of single transformants in comparable single-marker experiments. The frequency with which two genetic markers are cotransformed can thus be used as a crude estimate of the linkage distance between them. Genetic markers can also be ordered by means of three-factor transformation experiments using the same rationale as in three-factor transduction, conjugation, or sexduction experiments (see pp. 215–219).

TRANSDUCTION

Transduction, discovered by N. Zinder and J. Lederberg in 1952, occurs when a **bacteriophage particle carries a segment of the chromosome** from one bacterium (the **donor**) to another bacterium (the **recipient**), facilitating subsequent recombination of the genetic markers of the two cells. Two very different types of transduction are known. (1) In **generalized transduction,** a random or nearly random segment of bacterial DNA is "wrapped up" during phage maturation in place of, or along with, the phage chromosome in a few "progeny" particles, called **transducing particles. Generalized transducing phages** can therefore transport any gene of the donor cell to the recipient cell. Since all of the genes of the donor are represented in a population of these transducing particles (although any one transducing particle contains only one segment of host DNA, representing $\frac{1}{100}$ to $\frac{1}{50}$ of the total donor chromosome), this type of transduction was named "generalized" transduction. In some cases, generalized transducing particles contain only bacterial DNA. In other cases, they contain both phage and bacterial DNA. (2) In **specialized transduction** (also called **restricted transduction**), a recombination event involving the host chromosome and the phage chromosome occurs, producing a phage chromosome containing a segment of bacterial DNA. **Specialized transducing particles** thus always **contain both phage and bacterial DNA.** Specialized transduction is so named because a given virus transduces only genetic markers of the host that are located in one small region of the bacterial chromosome. (The reason for this is discussed on pp. 204–208.) Bacteriophage lambda, the best-known specialized transducing phage, for example, usually mediates transduction of only the *gal* and *bio* genes of *E. coli.*

GENERALIZED TRANSDUCTION

Bacteriophages have been classified into two types on the basis of their interactions with the bacterial cell. **Virulent phages** always multiply and lyse the host cell after infection. **Temperate phages** have a choice between two lifestyles after infection (Fig. 7.3). They can either (1) enter the **lytic cycle,** during which they reproduce and lyse their host cells just like virulent phages, or, alternatively, they can (2) enter the **lysogenic pathway,** during which their chromosomes are integrated into the chromosomes of the host and replicate like any other segments of the host chromosomes. **Generalized transduction** is mediated by some virulent bacteriophages and by certain temperate bacteriophages whose chromosomes are not integrated at specified attachment sites on the host chromosome. Generalized transducing particles are produced during the lytic cycles of these phages.

Not all virulent phages mediate transduction. The T-even bacteriophages (T2, T4, and T6), for example, degrade the host DNA and reutilize the mononucleotides produced for the synthesis of phage DNA. The host DNA is thus not available for packaging in progeny particles. Other viruses may not degrade the host DNA at all, and, since the host chromosome is too large to be packaged intact, they would not be able to form transducing particles. In still other viruses, the maturation process may be highly specific for phage DNA, excluding the packaging of fragments of host DNA. In any case, only a limited number of the virulent phages known mediate transduction. Of the generalized transducing phages, *E. coli* phage P1, *Salmonella* phage P22, and *Bacillus subtilis* phages PBS1 and SP10 have been extensively used for genetic fine structure mapping (mapping mutant sites within individual genes or short segments of the chromosome; see Chapter 10).

After a transducing phage injects a fragment of host DNA into a recipient cell, that DNA may either

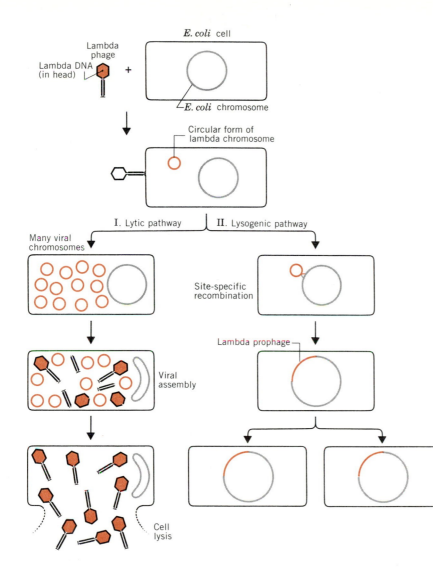

Lambda phage

Lambda DNA (in head)

E. coli cell

E. coli chromosome

Circular form of lambda chromosome

I. Lytic pathway

II. Lysogenic pathway

Many viral chromosomes

Site-specific recombination

Viral assembly

Lambda prophage

Cell lysis

FIGURE 7.3

Schematic diagram illustrating the two alternative life styles of temperate bacterio-phages such as the E. coli *phage lambda (λ). After the injection of the linear lambda chromosome into an* E. coli *cell, the chromosome is converted to its circular form. It then either (I) commences replicating autonomously and lyses the host cell, releasing 100 or more progeny viruses (the **lytic pathway**), or (II) integrates via a site-specific recombination event into the* E. coli *chromosome (the **lysogenic pathway**). In the integrated **prophage** state, the lambda chromosome replicates as part of the host chromosome and is transmitted to both progeny cells at each cell division. With a low probability, the prophage may be spontaneously excised from the host chromosome and enter the lytic pathway. It can be induced to enter the lytic pathway with a high frequency by irradiation with ultraviolet light.*

(1) be integrated into the host chromosome in a manner similar to the integration of transforming DNA, except that the integrated segment is double-stranded, or (2) remain free in the cytoplasm. If it is not integrated, it will not replicate and will be transmitted to only one progeny cell during each cell division. The genes located on the transduced chromosome fragments may be expressed, even if they are not integrated. Cells carrying nonintegrated transducing fragments are called **abortive transductants.** They are partially diploid and can be used to carry out complementation tests. The comple-

mentation test provides the operational definition of the gene. It is used to determine whether different mutations are in the same gene or in different genes (see Chapter 10).

The conclusion that the generalized transducing particles produced by certain phages carry only bacterial DNA (no phage DNA) is based on the results of density transfer experiments (see Chapter 5 pp. 95–99). If bacteria are grown for several generations in medium containing a precursor of DNA that is heavier than normal (such as ^{15}N in place of ^{14}N or 5-bromouracil in place of thymine), their DNA will be heavier than normal and will band at a higher density position in a CsCl density gradient. If such bacteria (containing "heavy" DNA) are transferred to "light" medium (medium containing normal DNA precursors such as ^{14}N) containing a radioactive DNA precursor (such as ^{32}P or ^3H-thymidine), all of the phage DNA synthesized will be "light" and radioactive. If the "progeny" particles produced from such an infected cell are analyzed in a CsCl density gradient, any transducing particles (particles containing bacterial DNA) present should band at a heavier than normal position. This was indeed found to be the case when this experiment was done using coliphage P1. Moreover, no radioactive "phage DNA" (DNA synthesized after infection) was found associated with the band of transducing particles. All of the radioactivity in the gradient was present at the "light" density position, in the band containing the infectious progeny phage. Thus, generalized transducing particles formed during phage P1 infections do not contain any phage DNA. (If this experiment is done using a specialized transducing phage, however, the results are very different.)

Transducing particles are produced at a low frequency. Only one out of $10^5 - 10^7$ of the "progeny" particles present in a lysate contains bacterial DNA. Thus, the probability of a cell being doubly transduced for genetic markers carried in two different transducing particles is negligible. (If cells are simultaneously infected with 100 or more phage particles, they are rapidly killed by a process called "lysis-from-without," which apparently results from simply punching too many holes in the cell membrane.) The cotransduction of two or more genetic markers therefore indicates that the markers are relatively closely linked, and the frequency of cotransduction of any two markers is indicative of the degree of linkage between them. Occasionally, genetic markers can be ordered by cotransduction patterns. If (1) markers a^+ and b^+ are cotransduced, (2) markers b^+ and c^+ are cotransduced, but (3) markers a^+ and c^+ are not cotransduced, then the order of the three markers must be a^+-b^+-c^+. More frequently, however, three-factor transduction experiments must be used to unambiguously order

genetic markers (see pp. 215–219 for the rationale involved).

SPECIALIZED TRANSDUCTION

Specialized transduction is mediated by temperate bacteriophages whose chromosomes are able to **integrate at one** or a **few specified attachment sites** on the host chromosome. The chromosomes of temperate phages of this type are thus capable of both (1) **autonomous replication** (replication independent of the replication of the host chromosome) and (2) **integrated replication** (replication as a segment of the host chromosome). As such, they are examples of genetic elements called **episomes** (see pp. 221–226).

Integration of the chromosome of a specialized transducing phage, such as the coliphage lambda, involves a recombination event between the circular intracellular form of the phage chromosome and the circular bacterial chromosome at **specific attachment sites** on the two chromosomes. This **site-specific recombination** event results in the covalent linear insertion of the phage chromosome into the chromosome of the bacterium (Fig. 7.4). In its integrated state, the phage chromosome is called a **prophage.** The lytic genes of the virus, those involved in viral reproduction and lysis of the host, are **repressed** (turned off) when the chromosome is in the prophage state. (The mechanism by which the prophage genes are repressed is discussed in Chapter 11, pp. 391–392.) A bacterium harboring a prophage is said to be **lysogenic;** the prophage–host relationship is called **lysogeny.** A lysogenic cell is **immune** to secondary infections by the same virus (homologous to the prophage), because the lytic genes of the infecting virus will be repressed just as those of the prophage are repressed.

Temperate phages undergo rare (about one in 10^5 cell divisions) spontaneous transitions from the lysogenic or prophage state to the lytic state. Such transitions can also be **induced,** for example, by irradiation with ultraviolet light. During the switch from the lysogenic state to lytic growth, the prophage is **excised** from the host chromosome (Fig. 7.4) and commences replicating autonomously. The **excision** process is **site-specific,** like the integration

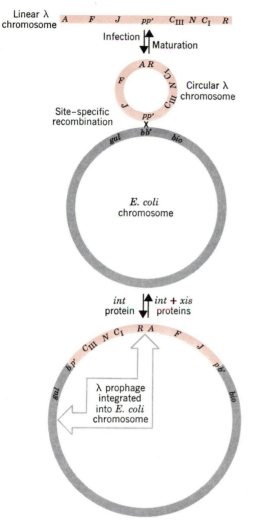

Linear λ chromosome

Infection ⇅ Maturation

Circular λ chromosome

Site–specific recombination

E. coli chromosome

int protein ⇅ $int + xis$ proteins

λ prophage integrated into E. coli chromosome

FIGURE 7.4

Integration and excision of the phage lambda chromosome. When the lambda chromosome is packaged inside the phage head, it exists as a linear molecule with gene sequence A through R, as shown at the top (color). After infection, it is converted to its circular intracellular form by the mechanism described in Fig. 5.17. When lambda enters the lysogenic pathway (Fig. 7.3), it undergoes a site-specific recombination event between the pp' site on the phage chromosome and the lambda attachment site, bb', on the E. coli chromosome (gray). This integrative recombination step is catalyzed by the lambda int gene product. In the prophage state, the lambda gene sequence is p'-C$_{III}$-R-A-J-p. This sequence is a circular permutation of the gene sequence in the linear form packaged in the lambda head (top), which results from a specific endonuclease cut between genes A and R during phage maturation.

During spontaneous or UV-induced excision, essentially the reverse of integration occurs. We can visualize this as the lambda prophage "looping out" to form a "figure-8" structure; this step is followed by a site-specific recombination event between the bp' sequence and the pb' sequence. The site-specific excision step requires the product of int gene plus the product of another lambda gene called xis. Normal excision produces an intact circular lambda chromosome plus an intact circular E. coli chromosome. Landy and Ross have sequenced both the lambda chromosome attachment site (pp') and the E. coli chromosome lambda attachment site (bb'), as well as the lambda phophage, E. coli DNA junctions (bp' and pb'), and found that all four contain the identical 15 nucleotide-pair sequence:

$$— \underset{\overline{}}{GCTTTTTTATACTAA} \rightarrow$$
$$\leftarrow \underset{}{CGAAAAAATATGATT} —.$$

For this to be true, the site-specific recombination events involved in integration and excision must occur within or at one end of these 15 nucleotide-pair sequences.

process. The site-specific integration and excision processes are catalyzed by enzymes coded for by phage genes.

The excision process is usually very precise, cutting out the phage chromosome in exactly the form in which it existed prior to its integration. Occasionally, however, the excision event occurs at a site other than the original attachment site. When this happens, a portion of the phage chromosome is left in the host chromosome and a portion of the bacterial chromosome is excised with the phage DNA (Fig. 7.5). Such "mistakes" during prophage excision are responsible for the formation of special-

ized transducing particles. Only host genes located close to the site of prophage insertion can be excised with the phage DNA and packaged in "phage" particles. Thus specialized transduction is restricted to the transfer of genes located within a short distance on each side of the prophage attachment site. Phage lambda integrates between the *gal* genes (required for the utilization of galactose as an energy source) and the *bio* genes (essential for the synthesis of biotin) on the *E. coli* chromosome (Figs. 7.4 and 7.5); lambda thus usually only transduces *gal* and *bio* markers. Specialized transducing phage Φ80, on the other hand, integrates near the *E. coli trp* genes

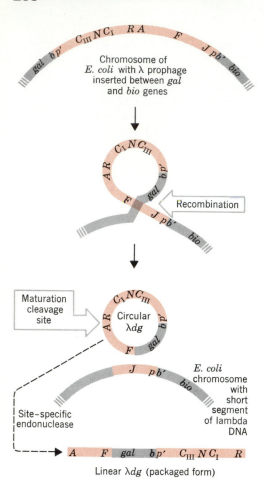

FIGURE 7.5

*Diagram illustrating the formation of a λdg specialized transducing particle. The sequence of events is essentially the same as in normal prophage excision (see Fig. 7.4), except that the recombination event does **not** occur between the bp' and pb' homology sites. Instead, the recombination event occurs at a position within the gal region of the host chromosome and a position in the gene F to gene J region of the lambda chromosome. As a result, the excised lambda chromosome includes a segment of host DNA containing part of the gal region, while a portion of the lambda prophage, containing gene J, is left in the host chromosome. Transducing particles carrying bio markers are similarly produced when a segment of host DNA containing part of the bio region is excised. (Keep the circle the same size and roll it to the right before excision.) The circular λdg chromosome is converted to the linear form (bottom) that is packaged within the head of the phage particle by a site-specific endonuclease that cleaves between genes A and R. This endonuclease makes staggered cuts in the complementary strands of DNA, producing the 12-nucleotide-long cohesive (complementary single-stranded) ends of the mature lambda chromosome.*

(required for the synthesis of the amino acid tryptophan) and transduces *trp* markers.

If specialized transducing particles are formed during prophage excision, as shown in Fig. 7.5, then only phage lysates produced by **induction** of lysogenic cells should have transducing activity. This is indeed the case. If bacteria are infected by specialized transducing phages under conditions where only lytic infections occur, no transducing particles are present in the phage lysates. The frequency of transducing particles in lysates produced by induction of lysogenic cells is about one in 10^6 progeny particles.

The chromosome composition of transductants produced by specialized transduction is quite different from the chromosome composition of transduc-

tants resulting from generalized transduction or transformation. In the latter cases, recombination **replaces** a segment of the recipient's chromosome with a segment of the donor's chromosome. In specialized transduction, the segment of donor DNA and the phage chromosome in which it is covalently inserted are **added** to the recipient's chromosome, producing a **partially diploid transductant** (Fig. 7.6a). This partial diploidy has several important consequences, which are most easily illustrated by considering a specific transducing phage. Since coliphage lambda is by far the best known of the specialized transducing phages, it is the logical choice for discussion. Specifically, consider a lambda chromosome carrying *gal* genes, like the one illustrated in Fig. 7.5. The transducing phage produced is called

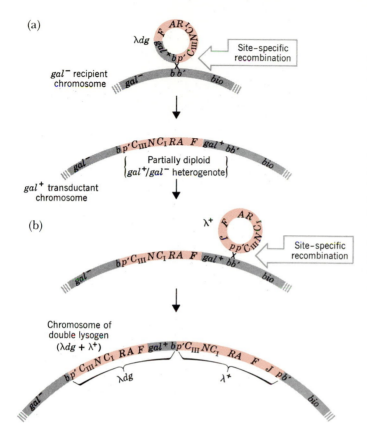

(a)

λdg

gal$^-$ recipient chromosome

Site-specific recombination

gal$^-$

bio

gal$^+$ transductant chromosome

gal$^-$

Partially diploid {gal$^+$/gal$^-$ heterogenote}

bio

(b)

λ^+

Site-specific recombination

gal$^-$

bio

Chromosome of double lysogen (λdg + λ^+)

λdg λ^+

gal$^-$

bio

FIGURE 7.6

(a) *The integration of a λdg chromosome, which carries gal$^+$, into a gal$^-$ recipient chromosome to form a partially diploid gal$^+$/gal$^-$ transductant (heterogenote). (b) The subsequent integration of a λ^+ chromosome ("helper" phage) to form a λ^+/λdg double lysogen. Note that gal$^+$ and gal$^-$ refer to wild-type and mutant alleles, respectively, at the gal locus. Cells that are gal$^+$ can utilize galactose as a carbon source; cells that are gal$^-$ cannot. Induction of a double lysogen of this type with UV will produce an Hft (high frequency transduction) lysate containing 50 percent λdg particles and 50 percent λ^+ phage. If a recipient cell is infected with a single λdg particle (low ratio of phage/bacteria), then the gal$^+$ transductant formed will have the genotype shown in (a). If a recipient is simultaneously infected with both a λdg particle and one or more λ^+ phage (high ratio of phage/bacteria), then the transductant will usually carry both λdg and λ^+ as shown in (b). Integration of the λdg and the λ^+ chromosomes can occur in either order.*

a λdg (for lambda defective *gal*). It is **defective** because genes required for growth and maturation under lytic conditions have been replaced by bacterial DNA. Thus λdg transducing particles can reproduce only in the presence of a wild-type lambda **"helper"** phage. The *g* indicates that the bacterial genes present are *gal* genes, rather than *bio* genes (as in λdb or λbio transducing particles).

When *gal$^+$* λ lysogens (cells lysogenic for λ) are induced by ultraviolet light, rare λdg particles are formed, carrying the donor *gal$^+$* gene or genes (depending on the size of the segment of bacterial DNA that is carried along). When these λdg particles infect *gal$^-$* recipient cells, the λdg integrates by a crossover event within the *gal* region or at the λ attachment site (Fig. 7.6*a*). (Note that microbiologists use a minus superscript to designate a mutant allele of a gene. Thus *gal$^-$* cells are unable to utilize galactose as a carbon source.) If the ratio of phage/

bacteria is high enough, such that the recipient cells are also infected with wild-type λ phage, a wild-type genome usually integrates by recombination within the normal λ attachment site, producing transductants that are double lysogens (carrying one λ^+ prophage and one λdg; see Fig. 7.6*b*). The transductants are thus *gal$^+$*/*gal$^-$* partial diploids. They are called **gal$^+$/gal$^-$ heterogenotes** and contain a **gal$^+$ exogenote** (donor DNA fragment) and a **gal$^-$ endogenote** (recipient chromosome).

The primary transductants, or *gal$^+$*/*gal$^-$* heterogenotes, are unstable; they segregate *gal$^-$* cells at a frequency of about one in 1000 cell divisions. These *gal$^-$* segregants can be explained by the excision of the λdg chromosomes. Since λdg cannot reproduce in the absence of a helper phage, it may simply be lost or diluted out during cell division. Recombination may also occur between the *gal$^+$* exogenote and the *gal$^-$* endogenote, transferring the *gal$^+$* marker

FIGURE 7.7
*J. Lederberg, 1958 Nobel Laureate, who with colleagues discov-
ered transduction and conjugation in bacteria. (Courtesy of The
Rockefeller University.)*

to the endogenote and producing stable *gal*⁺ transductants.

Because of the presence of the λ genes controlling immunity in λ*dg* chromosomes, the partially
diploid primary transductants are immune to subsequent lambda infections. If the transductants are
λ*dg*-λ⁺ double lysogens (Fig. 7.6*b*) and are induced
with ultraviolet light, they will produce lysates containing 50 percent λ*dg* particles and 50 percent λ⁺
particles. Both prophages will be excised and will
replicate with equal efficiency using the gene products coded for by the λ⁺ genome. Such lysates are
called *Hft* (high-frequency transduction) lysates. *Hft*
lysates greatly facilitate genetic analyses using specialized transduction by dramatically increasing the
frequency of transduction events. *Hft* lysates can
also be obtained from *gal*⁺/*gal*⁻ heterogenotes that
are single λ*dg* lysogens by simultaneously infecting
them with wild-type λ and inducing with ultraviolet
light.

CONJUGATION

Conjugation was discovered in 1946 by J. Lederberg (Fig. 7.7) and E. L. Tatum (1958 Nobel Prize
co-recipients). During conjugation, DNA is transferred from a **donor cell to a recipient cell** through
a specialized intercellular connection, or **conjugation tube,** that forms between them (Fig. 7.8). (The
donor and recipient cells are sometimes referred to
as male and female cells, respectively). The transfer
of genetic information is thus a **one-way transfer**
during conjugation, just as in transformation and
transduction, rather than a reciprocal exchange of
genetic material. Cells that have the capacity to
serve as donors during conjugation are differentiated by the presence of specialized cell-surface
appendages called **F pili** (Fig. 7.8). The synthesis of
these F pili is controlled by several (nine, based on
current data) genes that are carried by a small circular molecule of DNA or "minichromosome" (about
94,500 nucleotide pairs long) called an **F factor** (for
fertility factor; also called "sexfactor" and "F plasmid"). Cells carrying an F factor (donor cells) form
conjugation tubes and initiate DNA transfer after
making contact with **cells not carrying an F factor,**
called **F⁻ cells** (recipient cells).

The F factor can exist in two different states: (1)
the **autonomous state,** in which it replicates independently of the host chromosome, and (2) the
integrated state, in which it is covalently inserted
into the host chromosome and replicates along with
the host chromosome like any other set of chromosomal genes (Fig. 7.9). The F factor is thus, like the
chromosomes of specialized transducing phages, an
example of a class of genetic elements called **episomes** (see pp. 221–226).

A donor cell containing the F factor in the
autonomous state is called an **F⁺ cell. When an F⁺
donor cell conjugates with an F⁻ recipient cell,
only the autonomous F factor is transferred.** Both
exconjugants (cells that have been involved in conjugation) become F⁺ because the F factor replicates
during transfer. Thus, mixing a population of F⁺
cells with a population of F⁻ cells results in virtually
all of the cells in the new population becoming F⁺.

The F factor can integrate into the host chromosome at any one of many sites by a mechanism

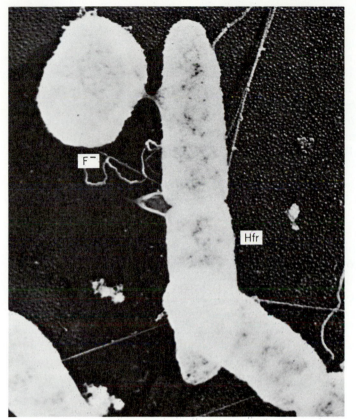

FIGURE 7.8

An electron micrograph of conjugating E. coli *cells. The elongated donor (Hfr) cell (on the right) is undergoing cell division. It is joined to the short, plump recipient (F⁻) cell (top left) by a conjugation tube, through which the Hfr chromosome passes. These morphological differences are **not** characteristic differences between Hfr cells and F⁻ cells. Cells with these different morphologies were used in this experiment to aid in distinguishing donor and recipient cells. The only characteristic morphological difference between donor (Hfr and F⁺) cells and recipient (F⁻) cells is the presence of F pili on donor cells. The ropelike fibers on the Hfr cell are F pili; their synthesis is controlled by genes located on the F factor. Magnification is ×32,700. (Photograph by T. F. Anderson, E. Wollman, and F. Jacob, from* Annales Institute Pasteur *93:450–455, 1957.)*

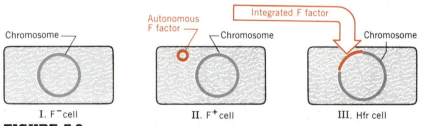

FIGURE 7.9

The three states of an E. coli *cell with respect to the F factor: (I) F⁻ cell, no F factor present; (II) F⁺ cell, containing an autonomously replicating F factor (color); (III) Hfr cell, containing an F factor (color) integrated into the* E. coli *chromosome (shaded). Conjugation occurs when an Hfr or an F⁺ cell contacts an F⁻ cell and forms a conjugation tube. In F⁺ by F⁻ matings, only the F factor is transferred. In Hfr by F⁻ matings, the Hfr chromosome is nicked within the integrated F factor, and a sequential transfer of chromosomal genes ensues.*

that appears analogous to the integration of the chromosome of a specialized transducing phage, namely, a site-specific recombination event (Fig. 7.10). The integration of the F factor is believed to be mediated by short DNA sequences called **IS elements** (see pp. 221–226). A cell carrying an integrated F factor is called an **Hfr** (for **high frequency recombination**). **In the integrated state, the F factor mediates the transfer of a chromosome of the Hfr cell to a recipient (F⁻) cell.** Usually, only a portion of the Hfr chromosome is transferred before the cells separate, breaking the chromosome. Only rarely will an entire Hfr chromosome be transferred.

The mechanism of transfer of DNA from a donor cell to a recipient cell during conjugation appears to be the same whether just the F factor is being transferred, as in F⁺ by F⁻ matings, or the chromosome is being transferred, as in Hfr by F⁻ matings. Transfer is believed to be initiated by an endonucleolytic nick in one strand at a specific site (the "origin" of transfer) on the F factor. The 5′-end of the nicked strand is then transferred through the conjugation tube into the recipient cell. Transfer is believed to be coupled to **rolling circle replication** (see Chapter 5, pp. 112–116), with the intact circular strand being replicated in the donor cell and the displaced strand being replicated in the recipient cell as it is transferred (Fig. 7.11). Because the origin of transfer is within the integrated F factor, one portion of the F factor is transferred from an Hfr cell to an F⁻ cell prior to the sequential transfer of chromosomal genes. The remaining part of the F factor, however, is the last segment of DNA to be transferred. Thus, in Hfr by F⁻ matings, the recipient F⁻ cell acquires a complete F factor (thus becoming an Hfr donor) only in those rare cases when an entire Hfr chromosome, with its integrated F factor, is transferred.

Several of the details of conjugation were worked out using one particular Hfr strain celled **HfrH** or **Hfr Hayes** (for W. Hayes, who isolated it). In this strain, the F factor is integrated near the *thr* (threonine) and *leu* (leucine) loci, as shown in Fig. 7.10. The most famous and most informative conjugation experiment was the **interrupted mating ex-**

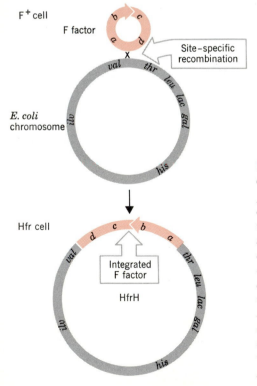

FIGURE 7.10

Conversion of an F⁺ cell to an Hfr cell by the integration of the autonomous F factor into the host chromosome. The F factor is covalently inserted into the host chromosome by a site-specific recombination event like that occurring during the integration of the phage lambda chromosome. Integration of the F factor appears to be mediated by IS elements. The markers a, b, c, and d shown on the F factor are hypothetical; they were arbitrarily positioned to illustrate the linear insertion occurring during integration. The arrowhead indicates the site of origin and direction of the sequential transfer that occurs during conjugation. The site of integration illustrated is that occurring during the formation of the classical Hfr strain, HfrH (or Hfr Hayes). A few E. coli genes are shown for the sake of orientation. Genes thr, leu, his, ilv, and val are involved in the biosynthesis of the amino acids threonine, leucine, histidine, isoleucine plus valine, and valine, respectively. Genes lac and gal are required for the utilization of lactose and galactose, respectively, as energy sources.

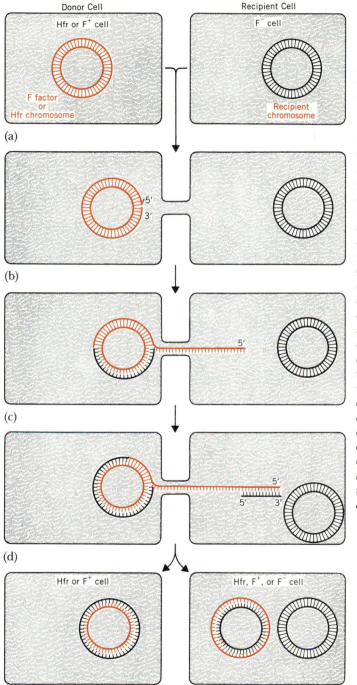

(a)

Donor Cell
Hfr or F⁺ cell
F factor
or
Hfr chromosome

Recipient Cell
F⁻ cell
Recipient
chromosome

(b)

5′
3′

(c)

5′

(d)

5′
5′ 3′

(e)

Hfr or F⁺ cell

Hfr, F⁺, or F⁻ cell

FIGURE 7.11

Mechanism of DNA transfer during conjugation. (a) The recipient chromosome is shown in gray. The chromosome shown (color) in the donor is either carrying the integrated F factor (Hfr cell) or is the F factor (F⁺ cell). (b) After cell contact occurs, a conjugation tube forms between the two cells. An endonuclease cleaves one strand of DNA at a unique site (the "origin" of transfer) on the F factor (either integrated in the case of an Hfr or autonomous in the case of an F⁺ cell). (c) The 5′-end of the cleaved strand is then displaced, as in normal rolling circle replication, except that during conjugation the 5′-end moves through the conjugation tube into the recipient cell. (d) Transfer occurs concurrently with, possibly driven by, rolling circle replication, with the intact circular strand serving as a template for the synthesis of a complementary strand in the donor cell and the transferred linear strand being replicated discontinuously (by synthesis of short 5′→3′ Okazaki fragments then joined by DNA ligase in an overall 3′→5′ reaction) in the recipient cell immediately after transfer. (DNA strands synthesized during the conjugation process are shown in black; parental DNA strands are shown in color.) (e) In F⁺ by F⁻ matings, both exconjugants will be F⁺, since both will have a complete copy of the F factor. In Hfr by F⁻ matings, the donor cell will remain an Hfr and the recipient cell will usually remain F⁻, since a portion of the integrated F factor is the last segment of DNA to be transferred. Although the diagram shows complete chromosome transfer to be comparable to F factor transfer in F⁺ by F⁻ matings, the conjugation tube and chromosome usually break spontaneously in Hfr by F⁻ matings before the entire chromosome is transferred. In rare cases where the entire Hfr chromosome is transferred, the recipient cell becomes an Hfr after conjugation.

periment of E. Wollman and F. Jacob. They crossed HfrH cells of genotype *thr⁺ leu⁺ azi-s T1-s lac⁺ gal⁺ str-s* with F⁻ cells of genotype *thr⁻ leu⁻ azi-r T1-r lac⁻ gal⁻ str-r.* [The *thr* gene and the *leu* gene are responsible for the syntheses of the amino acids threonine and leucine, respectively. Allele pairs *azi-s/azi-r, T1-s/T1-r,* and *str-s/str-r* control sensitivity (*s*) or resistance (*r*) to sodium azide, bacteriophage T1, and streptomycin, respectively. Alleles *lac⁺* and *lac⁻* and alleles *gal⁺* and *gal⁻* govern the ability (+) or inability (−) to utilize lactose and galactose, respectively, as energy sources.] At varying times after the HfrH and F⁻ cells were mixed to initiate matings, samples were removed and agitated vigorously in a blender to break the conjugation tubes and separate the conjugating cells. The cells were then plated on medium containing the antibiotic streptomycin, but lacking the amino acids threonine and leucine. On this **selective medium, only recombinant cells** carrying the *thr⁺* and *leu⁺* genes of the HfrH parent and the *str-r* gene of the F⁻ parent **can grow.** The HfrH donor cells are killed by the streptomycin; the F⁻ recipient cells cannot grow in the absence of threonine and leucine. Colonies produced by *thr⁺ leu⁺ str-r* recombinants were then **replica-plated** (see Chapter 9, Fig. 9.1) to a series of plates containing different selective media to determine which of the other donor markers were present. The series included medium containing (1) sodium azide, to score cells as *azi-r* or *azi-s,* (2) bacteriophage T1, to score cells as *T1-r* or *T1-s,* (3) lactose as the sole carbon source, to score recombinants as *lac⁺* or *lac⁻,* and (4) galactose as the sole carbon source, to score recombinant cells as *gal⁺* or *gal⁻.*

When the conjugating cells were interrupted (separated by agitating in a blender) at any time prior to 8 minutes after mixing the HfrH cells and the F⁻ cells, **no** *thr⁺ leu⁺ str-r* recombinants were detected. Recombinants (*thr⁺ leu⁺ str-r*) first appeared about 8½ minutes after mixing the HfrH and F⁻ cells and accumulated with linear kinetics up to a maximum frequency within a few minutes. Most important, when the presence of the other donor (HfrH) markers was scored for the *thr⁺ leu⁺ str-r* recombinants from subcultures interrupted at varying times after the initiation of mating, the donor markers were observed to be transferred in a spe-

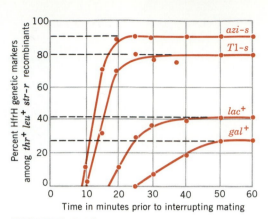

FIGURE 7.12

Kinetics of appearance of the unselected HfrH markers azi-s, T1-s, lac⁺, *and* gal⁺ *in* thr⁺ leu⁺ str-r *recombinants produced during conjugation between an HfrH strain of genotype* thr⁺ leu⁺ azi-s T1-s lac⁺ gal⁺ str-s *and an F⁻ strain of genotype* thr⁻ leu⁻ azi-r T1-r lac⁻ gal⁻ str-r. *Matings were interrupted by agitating in a blender at the times shown on the abscissa. The frequencies of the unselected donor markers (in percentages), as determined by replica plating to various indicator media, are shown on the ordinate. The dashed lines indicate the plateau frequencies observed for the various donor markers; they correspond to the frequencies observed in comparable uninterrupted mating experiments. (Based on the data of F. Jacob and E. L. Wollman, Sexuality and the Genetics of Bacteria, Academic Press, New York, 1961.)*

cific temporal sequence (Fig. 7.12). The HfrH *azi-s* gene first appeared among recombinants formed by conjugating cells that were separated by blending at about 9 minutes after the matings were initiated. The *T1-s, lac⁺,* and *gal⁺* markers first appeared after 11, 18, and 25 minutes of mating, respectively. These results indicated that the HfrH genes were being transferred to the F⁻ cells in a specific linear sequence (Fig. 7.13).

Subsequent studies with different Hfr strains revealed similar fixed transfer sequences, although different Hfr's initiated transfer from different sites on the chromosome. It is now clear that the F factor can integrate at many different sites in the circular *E. coli* chromosome (Fig. 7.14), and the site of integration determines the origin of transfer characteristic of a given Hfr. Moreover, the orientation of F integration (either *d c b a* reading clockwise, as shown for HfrH in Fig. 7.10, or *a b c d* reading

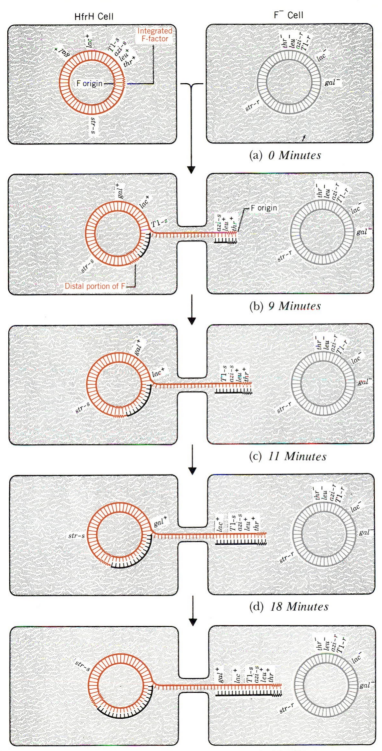

HfrH Cell

F⁻ Cell

(a) *0 Minutes*

(b) *9 Minutes*

(c) *11 Minutes*

(d) *18 Minutes*

(e) *25 Minutes*

FIGURE 7.13

Interpretation of the interrupted mating data of Jacob and Wollman (Fig. 7.12) in terms of the linear transfer of the HfrH chromosomes to the F⁻ cells. The genetic markers carried by the HfrH strain are all dominant or codominant (both alleles expressed) to the alleles carried by the F⁻ strain. The times given indicate the number of minutes since conjugation was initiated. The diagram is not drawn to scale; distances between markers are only approximate.

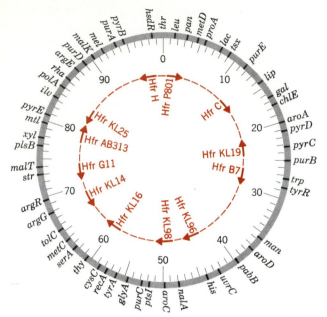

FIGURE 7.14
The circular linkage map of E. coli *strain K12. The map is
divided into 100 1-minute intervals based on conjugation
chromosome transfer times. Only 52 of the over 650 loci that have
been mapped in* E. coli *K12 are shown. The loci shown were
selected by B. J. Bachmann, K. B. Low, and A. L. Taylor on the
basis of greatest accuracy of map position, utility as reference
points in further mapping experiments, and/or familiarity as
well-known landmarks on the K12 linkage map. The arrows on
the dashed inner circle indicate the origin and direction of
transfer for a few well-known Hfr strains. The origin and direction of transfer (clockwise or counterclockwise) are determined by
the site and the orientation, respectively, of integration of the F
factor. Genetic symbols, descriptions of the loci, and references to
the mapping studies on which the map positions are based are
given in Bachmann, Low, and Taylor, 1976. (After B. J.
Bachmann, K. B. Low, and A. L. Taylor,* Bact. Rev. 40:116,
1976.)

clockwise) determines whether the sequence of
chromosome markers transferred is clockwise relative to the *E. coli* linkage map (Fig. 7.14) or counterclockwise. The inner circle in Fig. 7.14 shows the
sites of F integration and the direction of chromosome transfer for a few Hfr strains.

The transfer of a complete chromosome from
an Hfr cell to an F⁻ cell takes from 90 to 100
minutes, depending on the strain. Chromosome
transfer appears to proceed at a fairly constant rate.
Thus, the time interval between the transfer of
any two markers (easily determined by interrupted
mating experiments) is a good estimate of the physical distance separating the markers on the chromosome. It has therefore proven convenient to use the
minute, representing the time interval between the
transfer of markers in interrupted mating experiments, as the standard unit for measuring linkage in
E. coli. A map distance of 1 minute corresponds to
the length of the segment of the chromosome transferred in 1 minute during conjugation. The standard *E. coli* linkage map (Fig. 7.14) is thus divided
into minute intervals from 0 (arbitrarily set at the
thrA gene) to 100 minutes on the basis of interrupted mating experiments.

Linkage relationships can also be determined
from uninterrupted mating experiments. Consider,

for example, the HfrH by F⁻ cross discussed earlier,
except now with the matings being allowed to proceed uninterrupted for 1–2 hours. If *thr⁺ leu⁺ str-r*
recombinants are selected and scored for the presence of the other segregating markers by replica-
plating as before, what will the frequencies of the
donor *azi-s, T1-s, lac⁺,* and *gal⁺* markers be among
the recombinants? Will these donor markers all be
present with the same frequency? The frequencies
of these donor markers are observed to vary, with
the frequency of a marker decreasing as a function
of its distance from the selected (*thr⁺* and *leu⁺*)
donor markers. The frequencies will, in fact, be
identical to the plateau frequencies observed in the
interrupted mating experiment (Fig. 7.12). Donor
markers *azi-s, T1-s, lac⁺,* and *gal⁺* will occur among
thr⁺ leu⁺ str-r recombinants with percent frequencies
of 90, 80, 40, and 25, respectively. The farther a
marker is from the selected donor markers (in the
HfrH experiment, *thr⁺* and *leu⁺*), the lower its frequency among the recombinants. The marker frequency gradient is caused by two major factors: (1)
the approximately constant probability per unit
time of spontaneous rupture of the conjugation
tube and the chromosome, and (2) the decreasing
probability that any two donor markers will be incorporated into the recipient chromosome by a sin-

gle pair of recombination events (incorporation of a donor fragment into a recipient chromosome always requires **two** recombination events; see pp. 215–219) as the distance separating the two markers increases. Although uninterrupted conjugation experiments of this type can be used to determine linkage relationships, interrupted mating experiments are simpler and more direct. Thus, when a new mutation is identified, its approximate location is usually first determined by interrupted conjugation mapping. Its exact location is then usually determined by transduction mapping.

F-MEDIATED SEXDUCTION

Specialized transducing particles are occasionally formed during the excision of prophages (Fig. 7.5). The Hfr strains are formed when the F factor integrates into the chromosome (Fig. 7.10) by a mechanism similar to prophage formation. Moreover, rare F$^+$ cells are present in Hfr cultures, indicating that excision of the F factor also occurs, presumably by a mechanism analogous to the excision of a prophage (Fig. 7.4). Therefore, one might expect that an occasional anomalous excision event would produce an F factor carrying bacterial genes (Fig. 7.15). E. A. Adelberg and S. N. Burns first identified such modified F factors, called F$'$ ("F-prime") factors or F$'$ elements in 1959. Since F$'$ factors do not require packaging in a phage head like specialized transducing elements (e.g., λdg's), the size of the segment of the bacteria chromosome carried by the F$'$ is not restricted. The F$'$ factors range in size from those carrying a single bacterial marker to those carrying up to half of the bacterial chromosome (Fig. 7.16).

Transfer of F$'$ factors to recipient (F$^-$) cells apparently occurs by the same mechanism as F factor transfer in F$^+$ by F$^-$ matings and chromosome transfer in Hfr by F$^-$ matings (Fig. 7.11). The F$'$ factors have some interesting and useful properties, however. Consider an F$'$ thr^+ leu^+ factor generated by anomalous excision of the F factor from HfrH, as shown in Fig. 7.15. Matings between F$'$ thr^+ leu^+ donor cells and thr^- leu^- recipient (F$^-$) cells result in essentially all of the recipient cells being converted to thr^- leu^-/F$'$ thr^+ leu^+ partial diploids (or heterogenotes). These partial diploids are unstable like the partial diploids formed by specialized transduction. The F$'$ factor may be lost, producing thr^- leu^- haploids, or recombination may occur between the chromosome and the F$'$, producing stable thr^+ leu^+ recombinants. Recombination of this type, mediated by F$'$ factors, is called **sexduction** or **F-duction.** Because of the partial diploidy resulting from sexduction, it provides an important method for determining dominance relationships between alleles and defining genes by complementation tests in bacteria (see Chapter 10). Since F$'$ factors are available carrying almost any segment of the *E. coli* chromosome (Fig. 7.16), sexduction analysis can be carried out with almost any mutation of interest.

RATIONALE OF FINE STRUCTURE MAPPING IN MEROZYGOTES

Definitive fine structure mapping in bacteria, particularly the ordering of closely linked markers, is usually done using three-factor crosses, just as it is in eukaryotes (see Chapter 6). The rationale behind such mapping experiments with bacteria is essentially the same as for eukaryotes. The key differences between mapping in eukaryotes and in prokaryotes will be briefly discussed in this section. Those mapping techniques unique to bacteria, such as interrupted mating, cotransformation, cotransduction, and sexduction, have been described in the preceding sections of this chapter and will not be discussed further here.

The recombination events occurring in bacteria usually take place (except for rare occurrences of complete chromosome transfer during conjugation) in **partial zygotes** called **merozygotes** rather than in true zygotes or diploid cells as in eukaryotes. With the above-mentioned exception, the recipient cells in transformation, transduction, conjugation, and sexduction are converted to merozygotes, containing a fragment of the donor chromosome, the **exogenote,** and a complete recipient chromosome, the **endogenote.** Recombination in merozygotes does not take place between two complete chromosomes. **In merozygotes, the recombination process must incorporate a segment of the exogenote into the endogenote** (the complete chromosome of the recipient cell) **to yield an intact recombinant chro-**

(a) *E. coli* HfrH
chromosome:

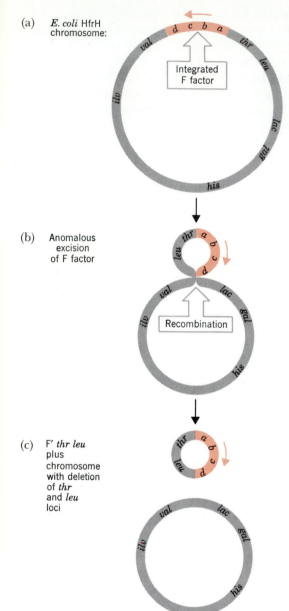

(b) Anomalous
excision
of F factor

(c) F′ *thr leu*
plus
chromosome
with deletion
of *thr*
and *leu*
loci

FIGURE 7.15

Anomalous excision of the F factor from an HfrH chromosome producing an F′ factor carrying the thr *and* leu *loci. (a) The HfrH chromosome is shown with the F factor integrated between the* val *and* thr *loci. The arrowhead indicates the origin and direction of chromosome transfer during conjugation;* a, b, c, *and* d *are hypothetical genetic markers on the F factor. (b) The F factor "loops out" in a "figure-8" structure as in normal F excision, but in this case the "loop" containing the F factor includes the segment of the chromosome carrying* thr *and* leu *loci. Recombination produces (c) an F′ factor carrying* thr *and* leu *plus a chromosome with a deletion of the* thr-leu *segment. An E.* coli *cell carrying the F′* thr leu *and the deletion chromosome remains haploid for all genetic loci, like the Hfr cell from which it was derived; the genes are just present in a new arrangement. When such an F′ cell conjugates with a normal (nondeletion) F⁻ cell, however, the recipient cell becomes partially diploid, with two copies of the* thr-leu *region.*

mosome. **Such an incorporation requires two crossovers or recombination events** (Fig. 7.17). Single crossovers between exogenotes and endogenotes will **not** yield structurally intact chromosomes. Since the incorporation of each donor DNA segment requires two crossovers, only recipient chromosomes involved in an **even number of crossovers with exogenotes will remain structurally intact.**

This restriction to paired recombination events in the case of merozygotes has little effect on the overall rationale of fine structure mapping. Consider, for example, a standard three-factor cross. In diploids, genetic markers are ordered by using three-factor cross data and by comparing the frequencies of the various recombinant progeny genotypes (Chapter 6). The rationale for diploids (or

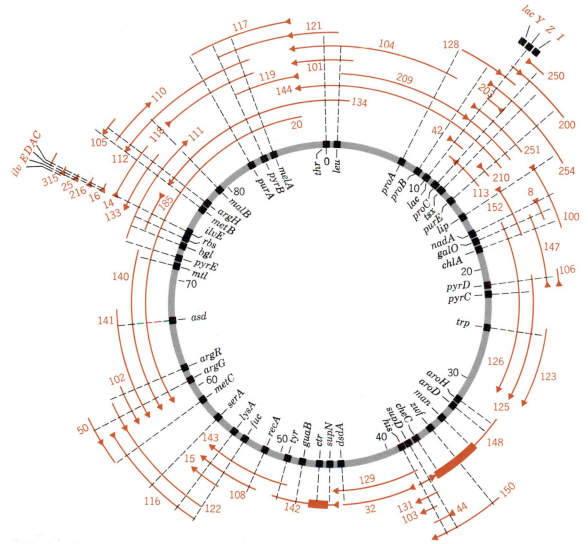

FIGURE 7.16

Summary of some of the F' factors of E. coli K12. The chromosomal genes carried by each F' factor are indicated by the lines paralleling the circular genetic map (center), which shows the locations of a few selected loci. Note, however, that all of the character-ized F's exist as circular DNA molecules; the true structures of these F' factors thus require joining of the ends of the line representing each numbered F' factor. The arrowheads on each F' factor indicate the origin and direction of transfer during conjugation. F' 142 and F' 148 carry deletions, as indicated by the boxes. This map was prepared before the revision of the E. coli linkage map (Fig. 7.14); it is thus divided into 90 1-minute intervals instead of the currently accepted 100 1-minute intervals. [From B. Low and J. O. Falkinham, Handbook of Microbiology, Vol. IV, p. 593 (A. I. Laskin and H. Lechevalier, eds.). The Chemical Rubber Co., Cleveland, Ohio, 1974. Copyright © 1974, The Chemical Rubber Co., CRC Press, Inc.]

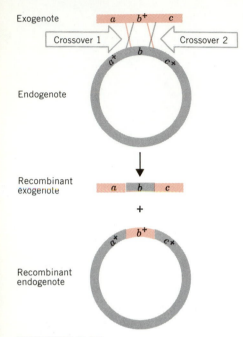

Exogenote

Crossover 1   Crossover 2

Endogenote

Recombinant exogenote

+

Recombinant endogenote

FIGURE 7.17

Incorporation of a genetic marker (b⁺) *on an exogenote (donor chromosome fragment, shown in color) into an endogenote (recipient chromosome, shaded) in a merozygote. The exogenote is only a fragment of the donor chromosome and is not capable of autonomous replication. Two crossovers are therefore required to incorporate a segment of the exogenote into the recipient chromosome, if the latter is to remain structurally intact and capable of replication.*

other complete zygotes) is that a genotype that requires two crossover events to be produced should be less frequent than a genotype that requires just one crossover. Specifically, if each single crossover type occurs with a frequency of x, then each double crossover type should occur with a frequency of about x^2. (Recall that the maximum frequency of each of the reciprocal recombinants produced by a crossover and thus the maximum value **of x** is $\frac{1}{4}$; therefore, x^2 is always much smaller than x.) Exactly the same rationale holds for merozygotes, except that the frequencies of genotypes formed by two crossovers (say, frequencies of y) are compared with the frequencies of genotypes formed by four crossovers (frequencies of about y^2; see Fig. 7.18).

In merozygotes, each three-factor cross can

be performed in two ways: (1) double mutant donor × single mutant recipient, and (2) single mutant donor × double mutant recipient. The progeny genotype frequencies of these reciprocal crosses are not necessarily the same. In fact, the results of such reciprocal crosses can be used to order the markers involved. Suppose that one wishes to order two markers (b_1 and b_2) in a particular gene (gene b) relative to a marker (a) in a nearby gene. The two reciprocal crosses are:

(1) $a^+ b_1^+ b_2$ donor × $a\ b_1\ b_2^+$ recipient
(2) $a\ b_1\ b_2^+$ donor × $a^+\ b_1^+\ b_2$ recipient

[One may also consider crosses (3) $a^+\ b_1\ b_2^+$ donor × $a\ b_1^+\ b_2$ recipient, and (4) $a\ b_1^+\ b_2$ donor × $a^+\ b_1\ b_2^+$ recipient; but only crosses (1) and (2) need be considered to illustrate the rationale involved. Note that the order $a\ b_1\ b_2$ is arbitrary; the crosses are being done to determine the real order of the b_1 and b_2 markers relative to the known outside marker a.] Usually, such crosses are done with conditional lethal mutants so that one can use selective media or conditions under which only the $a^+\ b_1^+\ b_2^+$ recombinants can grow. Such mutants carry mutations that are lethal to the organism when grown in one set of environmental conditions. In another set of environmental conditions, the mutations are not lethal, allowing the mutant organisms to grow and reproduce. Such mutations are thus called **conditional lethal mutations;** they are extremely useful in many kinds of genetic analysis (see Chapters 9 and 10).

Two orders are possible for the two gene b mutations relative to the outside marker in gene a: (1) a-b_1-b_2 and (2) a-b_2-b_1. If the correct order is a-b_1-b_2, $a^+\ b_1^+\ b_2^+$ recombinants will occur with about the same frequency in both crosses. They will be formed by two crossovers in both crosses. If, on the other hand, the order is a-b_2-b_1, $a^+\ b_1^+\ b_2^+$ recombinants will occur much more frequently in cross (2) than in cross (1). If the order is a-b_2-b_1, four crossovers will be required to produce $a^+\ b_1^+\ b_2^+$ recombinants in cross (1) whereas only two crossovers will be required in cross (2). This is diagramed in Fig. 7.18.

The same rationale can be applied to three-factor merozygote crosses in which the markers are not conditional lethal mutations. Such mapping experiments are much more laborious, however, because

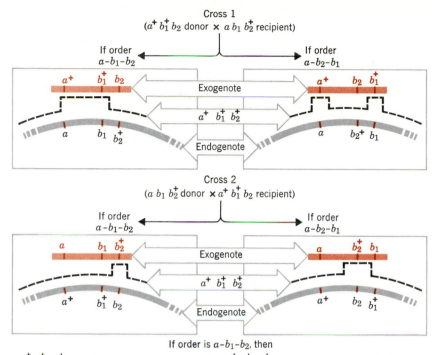

Cross 1
($a^+ b_1^+ b_2$ donor $\times$ $a\, b_1\, b_2^+$ recipient)

If order
$a\text{–}b_1\text{–}b_2$

If order
$a\text{–}b_2\text{–}b_1$

Cross 2
($a\, b_1\, b_2^+$ donor $\times$ $a^+ b_1^+ b_2$ recipient)

If order
$a\text{–}b_1\text{–}b_2$

If order
$a\text{–}b_2\text{–}b_1$

If order is $a\text{–}b_1\text{–}b_2$, then

$a^+\, b_1^+\ b_2^+$ recombinant frequency in Cross 1 $\cong a^+\, b_1^+\ b_2^+$ recombinant frequency in Cross 2.

If order is $a\text{–}b_2\text{–}b_1$, then

$a^+\, b_1^+\ b_2^+$ recombinant frequency in Cross 1 $\lll a^+\, b_1^+\ b_2^+$ recombinant frequency in Cross 2.

FIGURE 7.18

Diagram illustrating the rationale used in ordering markers by three-factor crosses involving merozygotes. The rationale is essentially the same as that for true zygotes, except that recombination events must always occur in pairs so as to incorporate markers present on the exogenote (donor fragment, shown in pale color) into the endogenote (recipient chromosome, shown in gray). In the example illustrated, conditional lethal mutations are used as genetic markers, and the mapping is based on the frequencies of selected wild-type recombinants ($a^+\, b_1^+\, b_2^+$ genotypes) formed in the two reciprocal crosses. Mutant organisms carrying conditional lethal mutations are viable when grown in one environment, but are nonviable in another environment. Mutations b_1 and b_2 are ordered relative to the known outside marker a. **The dashed lines indicate the segments of the exogenote and the endogenote that must be present in a recombinant of the selected $a^+\, b_1^+\, b_2^+$ genotype.** *If the order of the markers is a-b_1-b_2, as is shown on the left, then $a^+\, b_1^+\, b_2^+$ recombinants will be formed by two crossovers in both crosses. If the order of the markers is a-b_2-b_1, as shown on the right, then $a^+\, b_2^+\, b_1^+$ recombinants will be formed by two crossovers in Cross 2, but will require four crossovers in Cross 1. Thus, if the frequencies of wild-type recombinants are about the same in the two reciprocal crosses, the order is a-b_1-b_2. If, on the other hand, the frequency of wild-type recombinants is much higher in Cross 2 than in Cross 1, the correct order is a-b_2-b_1.*

the genotypes of all of the progeny must be individually determined by direct visualization of colony phenotypes or other screening techniques. Most of

the detailed fine structure maps in bacteria are thus based on data obtained using conditional lethal mutations.

BREAKAGE AND REUNION AS THE MECHANISM OF RECOMBINATION IN BACTERIA

During recombination in bacteria, be it during transformation, transduction, conjugation, or sexduction, a gene or a set of closely linked genes is transferred from a donor cell to a recipient cell. If the appropriate recombination events occur (see the preceding section), the donor gene or set of genes becomes a stable part of the genetic material of the recipient cell and is transmitted to its descendents, generation after generation, just like any of its other genes. This indicates that the incorporated genetic material of the exogenote becomes associated with the endogenote in a stable way. In fact, there is extensive evidence indicating that the donor DNA is physically inserted, by covalent linkages, into the chromosome of the recipient.

In Chapter 6, evidence was presented which demonstrated that crossing over occurred by the actual breakage of parental chromosomes (or chromatids) and the reunion of their parts, usually, if not always, with the occurrence of small amounts of repair synthesis. Recombinant chromosomes are composed, therefore, of parts derived from two different parental chromosomes. Recombination in

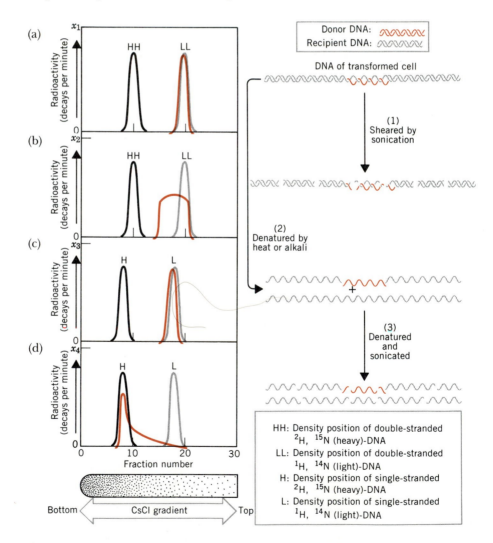

HH: Density position of double-stranded ^{2}H, ^{15}N (heavy)-DNA

LL: Density position of double-stranded ^{1}H, ^{14}N (light)-DNA

H: Density position of single-stranded ^{2}H, ^{15}N (heavy)-DNA

L: Density position of single-stranded ^{1}H, ^{14}N (light)-DNA bacteria also occurs by breakage and reunion, with limited amounts of DNA repair synthesis. This has been experimentally demonstrated in several systems by density transfer experiments (see Chapter 5, pp. 95–99). These experiments are done using donor cells containing DNA that is both radioactive and "heavy," that is, has a higher than normal density, and recipient cells containing nonradioactive DNA of normal density. Donor cells with heavy, radioactive DNA are prepared by growing them on medium containing one or more heavy isotopes such as ^{2}H and ^{15}N and a radioactive isotope such as ^{3}H or ^{32}P. After recombination (transformation, transduction, or conjugation) has occurred, the state of the donor DNA, which can be followed because of its radioactivity, is examined in the recombinants by CsCl density-gradient analysis combined with various biochemical techniques. The

details of one such study, the pneumococcus transformation experiment of M. S. Fox and M. K. Allen, are described in Fig. 7.19. Similar conjugation and transduction studies have shown that the donor DNA is covalently inserted into the chromosome of the recipient by a breakage and reunion mechanism, probably involving a small amount of DNA repair synthesis, during these processes.

PLASMIDS, EPISOMES, IS ELEMENTS, AND Tn ELEMENTS

In the introductory section of this chapter, we stated that the genetic material of bacteria is carried in one main chromosome plus, in many cases, from one to several extrachromosomal DNA molecules or "minichromosomes" called **plasmids.** By definition, **a plasmid is a replicon** (unit of genetic mate-

FIGURE 7.19

Diagram illustrating the results (left) and interpretation (right) of the experiment by M. S. Fox and M. K. Allen, demonstrating the covalent insertion of a single strand of donor DNA into the recipient chromosome during transformation in Diplococcus pneumoniae. *Donor cells were grown for several generations in medium containing the radioactive isotope of phosphorus, ^{32}P, and the heavy isotopes of hydrogen and nitrogen, ^{2}H and ^{15}N, respectively. The DNA of donor cells was thus both heavy and radioactive (shown in color). Recipient cells were grown in "normal" (^{1}H, ^{14}N, nonradioactive) medium; their DNA was therefore light and nonradioactive (shown shaded). Recipient cells were transformed with heavy, radioactive donor DNA, and the state of the donor DNA fragments in the transformed cells was analyzed using cesium chloride density gradients. (a) When the DNA of transformants was simply extracted and analyzed in CsCl density gradients, the donor DNA (followed by its radioactivity) banded (peak shown in color) at a density position only slightly heavier than that of light recipient DNA (shown in gray). This indicates that short segments of donor DNA were associated with long segments of light recipient DNA (top right). (b) If the DNA of transformants was first sheared into short double-stranded fragments by sonication [exposure to sound waves (1)], the donor DNA exhibited densities ranging from "hybrid" density (one strand heavy and one strand light) to near-light density (both strands light). No fully heavy donor DNA was observed after shearing. (c) Denaturation (2) of the DNA from transformed cells had no significant effect on the density position of the donor DNA, except the denatured (single-stranded) DNA has a slightly higher density than native (double-stranded) DNA. The donor DNA still banded at the near-light density. This result demonstrates that the donor DNA is **covalently inserted** into the recipient chromosome. (d) Only after the DNA of transformants had been both sonicated and denatured (3) was donor DNA recovered in the gradient at the heavy density position. These results can be explained only if a single strand of donor DNA is covalently inserted into the chromosome of the recipient cell during transformation in pneumococcus. (Based on the results of M. S. Fox and M. K. Allen,* Proc. Natl. Acad. Sci., U.S. *52:412, 1964.)*

rial capable of independent replication) that is **stably inherited** (maintained without specific selection) **in an extrachromosomal state. Most,** but not all, **plasmids are dispensable;** that is, they are not required for survival of the cell in which they reside. In many cases, however, they are essential under certain environmental conditions, such as in the presence of an antibiotic.

The importance of plasmids has become increasingly recognized during the last two decades. Plasmids have been identified in almost all strains of bacteria tested. They are known to have major practical significance in two areas: (1) the spread of multiple antibiotic and drug resistance in pathogenic bacteria, and (2) the instability of industrially important microorganisms. Multiple antibiotic and drug resistance will be discussed in some detail later in this chapter. In *Streptococcus lactis* and related bacteria used in cheese processing, multiple plasmids have been identified and shown to carry genes coding for enzymes important in the fermentation processes involved in making cheeses. It now seems likely that these observations explain why the cheese "starter cultures" of these bacteria are unstable and frequently must be discarded, at considerable expense to the cheese-making industry.

Three major types of bacterial plasmids have been extensively studied: (1) **F and F′ plasmids,** the conjugation fertility factors previously discussed; (2) **R plasmids** (previously called RTF, or resistance transfer factors), plasmids carrying genes for resistance to antibiotics or other antibacterial drugs; and (3) **Col plasmids** (previously called colicinogenic factors), plasmids coding for **colicins,** proteins that kill sensitive *E. coli* cells. Plasmids are also known in bacteria that code for bacteriocins other than colicins. For example, plasmids are known that code for **vibriocins;** these are proteins that kill sensitive *Vibrio cholerae* cells. They appear analogous to Col plasmids.

Strictly speaking, the chromosomes of mitochondria and chloroplasts in eukaryotes also fit the definition of plasmids. They will be discussed in Chapter 14, along with other examples of extranuclear inheritance in eukaryotes.

Plasmids may be divided into two groups on the basis of whether or not they mediate conjugative self-transfer. **Conjugative** or **transmissible plasmids mediate transfer of DNA by conjugation** (as in F$^+$ by F$^-$ matings; see pp. 208–215). All F and F′ plasmids, many R plasmids, and some Col plasmids are conjugative. The conjugative nature of many R plasmids has major significance in the rapid spread of antibiotic and drug resistance genes through populations of pathogenic bacteria. **Nonconjugative** or **nontransmissible plasmids are those that do not mediate transfer of DNA by conjugation.** Many R and Col plasmids are nonconjugative.

Some plasmids, such as F factors, also fit the definition of genetic elements called episomes. **Episomes are genetic elements that can replicate in either of two alternative states: (1) as an integrated (covalently inserted) part of the main host chromosome, or (2) as an autonomous genetic element, independent of the main host chromosomes.** The terms plasmid and episome are **not** synonyms. Many plasmids do not exist in integrated states and are thus not episomes. Similarly, many temperate phage chromosomes, such as the phage λ genome, are episomes but are not plasmids.

Spectacular progress has been made in our understanding of the structures and properties of plasmids and episomes during the last few years. Many of their properties are now known to depend on the presence of short DNA sequences called **IS elements** or insertion sequences. The IS elements are also present in the main host chromosomes (see Chapter 9, pp. 310–312). These short sequences (from about 800 to about 1400 nucleotide pairs in length) are **transposable;** that is, they can move from one position to another within a chromosome, or move from one chromosome to a different chromosome. In addition, IS elements mediate recombination between otherwise nonhomologous genetic elements within which they reside. Considerable evidence indicates that IS elements mediate the integration of episomes into host chromosomes. This is particularly clear in the case of the integration of the *E. coli* K12 F plasmid (F factor) during the formation of Hfr's (Fig. 7.20).

The first four IS elements to be extensively

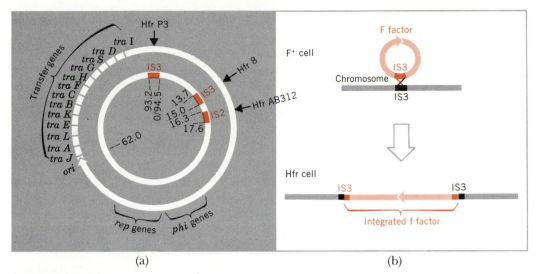

(a)   (b)

FIGURE 7.20

Abbreviated genetic map of the E. coli *K12 F factor (F plasmid) (a), and the postulated IS element-mediated integration of the F factor during Hfr formation (b). (a) The inner circle shows the location of one IS2 element and two IS3 elements. The distances given within the inner circle are in kilobases (1000 nucleotide-pair units); the* E. coli *K12 F factor is 94,500 nucleotide pairs in length. The approximate locations of 13 genes involved in chromosome transfer (*tra *genes), the origin of transfer (*ori*), the genes required for replication (*rep *genes), and the genes involved in the inhibition of growth of F⁻-specific bacteriophages such as phage T7 (*phi*, for phage inhibition, genes) are shown on the outer circle. The sites of recombination with the host chromosome during integration are shown by the arrows for three Hfr strains. Note that these sites coincide with the known locations of the three IS elements. (b) Proposed mechanism of integration of the F factor mediated by the homology of an IS element in the chromosome and in the F factor. Such integration results in the F factor being flanked by identical IS elements when covalently inserted into the chromosome of an Hfr cell. (Based on the data summarized by J. A. Shapiro, p. 671. In* DNA Insertion Elements, Plasmids, and Episomes, *A. I. Bukhari, J. A. Shapiro, and S. L. Adhya, eds., Cold Spring Harbor Laboratory Press, Cold Spring Harbor, New York, 1977.)*

characterized and sequenced—IS1, IS2, IS3, and IS4—are 768, 1327, 1300, and 1426 nucleotide pairs in length, respectively. Several other IS elements have subsequently been identified, characterized, and, in some cases, sequenced.

The *E. coli* K12 chromosome apparently contains eight copies of IS1 and five copies of IS2, plus one or more copies of IS3 and IS4. The *E. coli* K12 F factor contains one copy of IS2 and two copies of IS3 (Fig. 7.20). The positions of the IS elements in the various F factors and in the chromosomes of various *E. coli* strains are believed to determine the

sites of integration of the F factor during the formation of Hfr strains (Fig. 7.20).

The **Tn elements** or **transposons** are a group of more complex transposable elements that behave genetically much like the IS elements. Unlike IS elements, however, Tn elements are usually over 2000 nucleotide pairs in length and carry one or more genes unrelated to their transposability. Usually, the additional gene(s) provides resistance to one or more antibiotics. (Some geneticists consider IS elements to be a class of transposons.)

The most striking structural feature of trans-

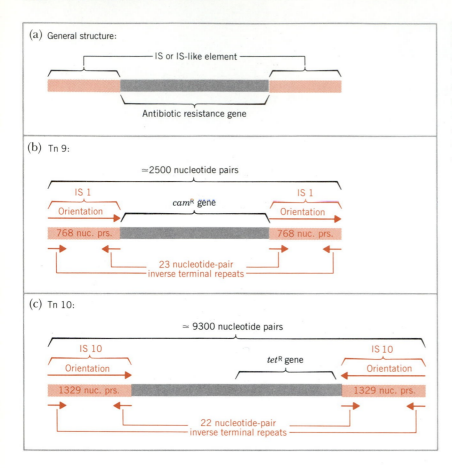

(a) General structure:

IS or IS-like element

Antibiotic resistance gene

(b) Tn 9:

≈2500 nucleotide pairs

IS 1

cam^R gene

IS 1

Orientation

Orientation

768 nuc. prs.

768 nuc. prs.

23 nucleotide-pair inverse terminal repeats

(c) Tn 10:

≈ 9300 nucleotide pairs

IS 10

tet^R gene

IS 10

Orientation

Orientation

1329 nuc. prs.

1329 nuc. prs.

22 nucleotide-pair inverse terminal repeats

FIGURE 7.21

Structure of Tn elements. (a) Common structural features of many Tn elements. (b) Structure of Tn 9. The short inverse terminal repeats at the ends of the two IS1 elements are imperfect repeats — 18 of the 23 nucleotide pairs are the same, 5 are different. (c) Structure of Tn 10. Again, the short inverse repeats are imperfect — 17 of 22 nucleotide pairs being identical.

posable elements is that they all have the same or very similar sequences repeated at both ends (so-called **terminal repeats**). This is true for the transposable elements of eukaryotes also. The IS elements have from 9 to 40 nucleotide-pair **inverse repeats** (the same sequence reading in opposite directions) at their ends. The more complex Tn elements have IS elements or IS-like elements repeated at their ends (Fig. 7.21). These may be either in the same orientation, in which case they are called **direct terminal repeats,** or in the opposite orientation, in which case they are called **inverse terminal repeats.** The terminal repeats undoubtedly play a key role in the transposability of these genetic elements.

When the transposable elements insert themselves into a chromosome, they somehow cause the duplication of a short segment, usually five or nine nucleotide pairs, of DNA at the insertion site. The duplicate sequences are found directly adjacent to the ends of the inserted transposable element.

The IS and Tn elements are clearly responsible for the transposition of genes controlling resistance to antibiotics and other drugs from one genetic element to another. They are believed to play a major role in the observed rapid evolution of R plasmids. All conjugative R plasmids have at least two components, one segment carrying a set of genes involved in conjugative DNA transfer (probably analogous to the *tra* genes of F plasmids; see Fig. 7.20) and a second segment carrying the antibiotic- and/or drug-resistance gene or genes (Fig. 7.22). The segment carrying the transfer genes is called the **RTF** (resistance transfer factor) **component;** the segment carrying the resistance gene or genes is called the **R-determinant.** The RTF components of

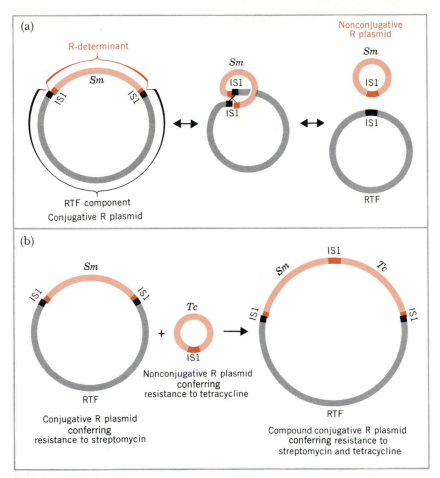

FIGURE 7.22

Structure of conjugative and nonconjugative R plasmids (a) and a proposed mechanism for the evolution of compound R plasmids, which provide multiple-antibiotic resistance and drug resistance to bacterial cells (b). (a) Simple conjugative R plasmids contain two major components: (1) the RTF component (shown shaded), which contains the tra *genes responsible for conjugative transfer of the plasmid, and (2) the R-determinant component (shown in color), which carries the gene or genes responsible for antibiotic or drug resistance. The R-determinant in several characterized conjugative R plasmids is flanked by identical IS elements. These IS elements are believed to mediate the transposition of R-determinants from one R plasmid to* *another (or to other genetic elements). Nonconjugative R plasmids (top right) do not carry the RTF (*tra *genes) component. (b) IS element-mediated formation of a compound R plasmid, carrying genes that provide resistance to both streptomycin and tetracycline. The mechanism is as shown in* a, *from right to left. This process may continue with R-determinants carrying genes for resistance to other antibiotics and drugs until conjugative R plasmids have evolved, which provide the host cell with resistance to a whole battery of antibiotics and drugs. Sm and Tc symbolize the plasmid-carried genes providing streptomycin- and tetracycline-resistance, respectively.*

several different conjugative R plasmids appear to have a large amount of homology, based on DNA-DNA cross-hybridization experiments. The R-determinant components exhibit more divergence.

In several R plasmids, the R-determinant is flanked by homologous IS elements. In some cases, they are present in the same orientation (both clockwise or both counterclockwise); in other cases, they are inserted with opposite orientations (one clockwise, the other counterclockwise). In either case, they can mediate the transposition of an R-determinant from one R plasmid to another (Fig. 7.22). Several compound R plasmids have been characterized as carrying two or more R-determinants, each flanked by IS elements. These IS elements are almost certainly responsible for the rapid evolution of bacterial plasmids that carry multiple antibiotic and drug resistance.

Transposable elements have now been isolated from yeast and *Drosophila* and shown to have structures very similar to those of the transposable elements of prokaryotes. In addition, the integrated DNA genomes or proviruses of certain cancer-causing viruses (so-called retroviruses — see Chapter 11, pp. 422–430) are known to have structures essentially identical to those of the transposons shown in Fig. 7.21. This suggests that these proviruses might themselves be transposable elements.

The transmissibility of R plasmids, the transposability of the R-determinants, and the rapid evolution of compound R plasmids, which carry genes for resistance to a whole battery of our most effective antibiotics and drugs, is of great concern to medical practitioners. Not only are these plasmids rapidly dispersed within a bacterial species, but they are also transmitted across species and even generic lines. For example, *E. coli* R plasmids are known to be transferred to several genera, including *Proteus, Salmonella, Hemophilus, Pasteurella,* and *Shigella* (which causes dysentery). The increased frequency of bacteria carrying plasmids with R-determinants, which result in resistance to antibiotics such as penicillin, tetracycline, streptomycin, and kanamycin, in hospital populations (which are continuously exposed to these antibiotics), has been extensively documented. Of even greater concern are the results of studies in Japan that show that, in less than 10 years, natural populations of bacteria (in sewers and in polluted lakes and streams) have evolved from very low frequencies (less than 1 percent) of R plasmid-mediated antibiotic resistance to relatively high frequencies (50–80 percent).

The results of these studies make it eminently clear that we should restrict our use of antibiotics to serious bacterial infections and not use them for every minor infection that comes along. If we do not restrict use, the antibiotics and drugs that are so effective today may have little, if any, utility in the future.

SUMMARY

In bacteria, there are three different mechanisms for the transfer of genetic material from one cell, the **donor** cell, to a second cell, the **recipient** cell. During (1) **transformation,** naked donor DNA molecules are taken up enzymatically by **competent** recipient cells. (2) **Transduction** occurs when a fragment of the donor chromosome is carried to and injected into the recipient cell by a bacterial virus (bacteriophage). (3) **Conjugation** requires direct cell contact and involves the transfer of donor DNA to recipient cells through a **conjugation tube** that forms between the two cells.

Transduction is of two types: (1) **generalized transduction,** in which all genetic markers of the donor cell are represented in a population of transducing phage, and (2) **restricted transduction,** in which only genetic markers near the **prophage** (integrated phage chromosome) site are transduced.

A **plasmid** is an extrachromosomal DNA molecule or "minichromosome" that can replicate independently of the main cellular chromosome. The three main types of plasmids are: (1) **F plasmids,** the F factors responsible for DNA transfer during conjugation; (2) **R plasmids,** DNA molecules carrying genes for resistance to various antibiotics and antibacterial drugs; and (3) **Col plasmids,** plasmids that code for proteins called **colicins,** which kill sensitive *E. coli* cells. All known plasmids are circular molecules of DNA.

Episomes are genetic elements that can replicate (1) in an **integrated state,** covalently inserted in the host chromosome, and (2) in an **autonomous** or **extrachromosomal state.** The *E. coli* K12 F factor and the phage λ chromosome are the best-known episomes. A cell carrying the F factor in the autonomous state is called an **F⁺ donor.** During conjugation between an F⁺ donor and an **F⁻ recipient,** only the F factor is transferred. A cell carrying the F factor in the integrated state is called an **Hfr** (for high frequency recombination). During conjugation between an Hfr cell and an F⁻ cell, the Hfr chromosome undergoes linear transfer to the F⁻ cell. Usually, only part of the Hfr chromosome is transferred before the cells separate. The origin and direction of transfer are determined by the site and orientation of the F factor in the chromosome.

Occasionally, anomalous excisions of the F factor from Hfr chromosomes occur, producing recombinant F factors, called **F′ factors,** that carry chromosomal genes. The conjugative transfer of donor chromosomal genes carried by F′ factors to recipient cells is called **sexduction.**

Transformation, transduction, and conjugation almost always produce cells that are partial zygotes or partial diploids called **merozygotes.** Merozygotes contain only a part of the donor chromosome, the **exogenote,** plus the intact recipient chromosome, the **endogenote.** Crossovers in merozygotes must therefore always occur **in pairs,** to yield intact chromosomes. Recombination in bacteria occurs by **breakage and reunion** of parental chromosomes.

The integration of episomes and the evolution of plasmids, particularly R plasmids, are mediated by short (about 800–1400 nucleotide-pairs long) DNA sequences called **insertion sequences** or **IS elements** and slightly longer (>2000 nucleotide-pairs long) DNA sequences called **transposons** or **Tn elements.** These IS and Tn elements are **transposable;** that is, they can move from one position to another in the genome of a cell. The IS elements can also mediate recombination between genetic elements in which they are inserted.

REFERENCES

Adelberg, E. A. (ed.). 1966. *Papers on bacterial genetics,* 2nd ed. Little, Brown, Boston.

Bachmann, B. J., 1983. "Linkage map of *Escherichia coli* K-12, edition 7." *Microbiological Revs.* 47:180–230.

Bukhari, A. I., J. A. Shapiro, and S. L. Adhya (eds.). 1977. *DNA insertion elements, plasmids, and episomes.* Cold Spring Harbor Laboratory Press, Cold Spring Harbor, New York.

Davis, R. W., D. Botstein, and J. R. Roth. 1980. *Advanced bacterial genetics.* Cold Spring Harbor Laboratory Press, Cold Spring Harbor, New York.

Fox, M. S., and M. K. Allen. 1964. "On the mechanism of deoxyribonucleate integration in pneumococcal transformation." *Proc. Natl. Acad. Sci., U.S.* 52:412–419.

Freifelder, D. 1983. *Molecular biology, a comprehensive introduction to prokaryotes and eukaryotes.* Science Books International, Boston.

Goodgal, S. H. 1982. "DNA uptake in *Haemophilus* transformation." *Ann. Rev. Genet.* 16:169–192.

Hayes, W. 1968. *The genetics of bacteria and their viruses,* 2nd ed. John Wiley & Sons, New York.

Hershey, A. D. (ed.). 1971. *The bacteriophage lambda.* Cold Spring Harbor Laboratory Press, Cold Spring Harbor, New York.

Kleckner, N. 1981. "Transposable elements in prokaryotes." *Ann. Rev. Genet.* 15:341–404.

Lewin, B. 1974. *Gene expression-1, bacterial genomes.* John Wiley & Sons, New York.

Lewin, B. 1977. *Gene expression-3, plasmids and phages.* John Wiley & Sons, New York.

Lewin, B. 1983. *Genes.* John Wiley & Sons, New York.

Matthews, C., E. M. Kutter, G. Mosig, and P. Berget (eds.). 1983. *The bacteriophage T4.* American Society for Microbiology, Washington, D.C.

Meynell, G. G. 1973. *Bacterial plasmids.* The MIT Press, Massachusetts Institute of Technology, Cambridge, Massachusetts.

Miller, J. H. 1972. *Experiments in molecular genetics.* Cold Spring Harbor Laboratory Press, Cold Spring Harbor, New York.

Nash, H. A. 1981. "Integration and excision of bacteriophage λ: the mechanism of conservative site-specific recombination." *Ann. Rev. Genet.* 15:143–167.

Novick, R. P. 1980. "Plasmids." *Sci. Amer.* 243(6):102–127.

Reznikoff, W. S. 1982. "Tn5 transposition and its regulation." *Cell* 31:307–308.

Starlinger, P., and H. Saedler. 1976. "IS-elements in microorganisms." *Current Topics Microbiol. Immunol.* 75:111–152.

Watson, J. D. 1976. *Molecular biology of the gene,* 3rd ed. W. A. Benjamin, Menlo Park, California.

QUESTIONS AND PROBLEMS

7.1. A nutritionally defective *E. coli* strain grows only on a medium containing thymine, whereas another nutritionally defective strain grows only on medium containing leucine. When these two strains were grown together, a few progeny were able to grow on a minimal medium with neither thymine or leucine. How can this result be explained?

7.2.   Assume that you have just demonstrated genetic recombination (e.g., when a strain of genotype $a\ b^+$ is present with a strain of genotype $a^+\ b$, some recombinant genotypes, $a^+\ b^+$ and $a\ b$, are formed) in a previously undescribed species of bacteria. Describe how you would go about (operationally) determining whether the observed recombination was the result of a process similar to transformation, a process similar to transduction, or a process similar to conjugation.

7.3.   Compare, in table form, similarities and differences of the mechanisms through which (a) transduction, (b) sexduction, and (c) transformation may occur.

7.4.   (a) What are the genotypic differences between F⁻ cells, F⁺ cells, and Hfr cells? (b) What are the phenotypic differences? (c) By what mechanism are F⁻ cells converted to F⁺ cells; F⁺ cells to Hfr cells?

7.5.   (a) Of what use are F′ factors in genetic analysis? (b) How are F′ factors formed? (c) By what mechanism does sexduction occur?

7.6.   What are the basic differences between generalized transduction and specialized transduction?

7.7.   (a) What is the difference between a phage and a prophage? (b) Are the linkage maps of the mature phage lambda chromosome and the lambda prophage identical? (c) If not, in what way(s) do they differ?

7.8.   (a) What are IS elements? (b) What role(s) do IS elements play in recombination? (c) What role(s) do they play in the evolution of genetic elements?

7.9.   How can genes be mapped by interrupted mating conjugation experiments?

7.10.   What does the term cotransduction mean? How can cotransduction frequencies be used to map genetic markers?

7.11.   An F⁺ strain, marked at 10 loci, gives rise spontaneously to Hfr progeny whenever the F factor becomes incorporated into the chromosome of the F⁺ strain. The F factor can integrate into the circular chromosome at many points, so that the various Hfr segregants transfer the genetic markers in different sequences. For any Hfr strain, the order of markers entering early can be determined by interrupted mating experiments using a Waring blender. From the following data for several Hfr strains derived from the same F⁺, determine the order of markers in the F⁺ strain.

HFR STRAIN	MARKERS DONATED IN ORDER
1	—Z-H-E-R →
2	—O-K-S-R →
3	—K-O-W-I →
4	—Z-T-I-W →
5	—H-Z-T-I →

7.12.   In *E. coli*, the ability to utilize lactose as a carbon source requires the presence of the enzymes β-galactosidase and β-galactoside permease. These enzymes are coded for by two closely linked genes, *lacZ* and *lacY*, respectively. Another gene, *proC*, controls, in part, the ability of *E. coli* cells to synthesize the amino acid proline. The alleles str^r and str^s control resistance and sensitivity, respectively, to streptomycin. HfrH is known to transfer the two *lac* genes, *proC*, and *str*, in that order, during conjugation.

A cross is made between HfrH of genotype *lacZ⁻ lacY⁺ proC⁺ str^s* and an F⁻ of genotype *lacZ⁺ lacY⁻ proC⁻ str^r*. After about two hours, the mixture is diluted and plated out on medium containing streptomycin, but no proline. When the resulting *proC⁺ str^r* recombinant colonies were checked for their ability to grow on medium containing lactose as the sole carbon source, very few of them were capable of fermenting lactose. When the reciprocal cross (HfrH *lacZ⁺ lacY⁻ pro C⁺ str^s* × F⁻ *lacZ⁻ lacY⁺ proC⁻ str^r*) was done, many of the *proC⁺ str^r* recombinants were able to grow on medium containing lactose as the sole carbon source. What is the order of the *lacZ* and *lacY* genes relative to *proC*?

7.13.   The data in the table below were obtained from three-point transduction tests made to determine the order of mutant sites in the *A* gene for tryptophan synthetase in *E. coli*. *Anth* is a linked, unselected marker. In each cross, *trp⁺* recombinants were selected and then scored for the *anth* marker (*anth⁺* or *anth⁻*). What is the linear order of *anth* and the three mutant alleles of the *A* gene indicated by the data in the table below?

CROSS	DONOR MARKERS	RECIPIENT MARKERS	*Anth* ALLELE IN RECOMBINANTS	PERCENT *Anth⁺*
1	*anth⁺*-A34	*anth⁻*-A223	72 *anth⁺* : 332 *anth⁻*	18
2	*anth⁺*-A46	*anth⁻*-A223	196 *anth⁺* : 180 *anth⁻*	52
3	*anth⁺*-A223	*anth⁻*-A34	380 *anth⁺* : 379 *anth⁻*	50
4	*anth⁺*-A223	*anth⁻*-A46	60 *anth⁺* : 280 *anth⁻*	20

7.14. Two additional mutations in the *trp A* gene of *E. coli, trp A58* and *trp A487*, were also ordered relative to *trp A223* and the outside marker *anth* by three-factor transduction crosses as described in Problem 7.13. The results of these crosses are summarized in the table below. What is the linear order of *anth* and the three mutant sites in the *trp A* gene?

CROSS	DONOR MARKERS	RECIPIENT MARKERS	*Anth* ALLELE IN *trp*+ RECOMBINANTS	% *anth*−
1	*anth*+*-A487*	*anth*−*-A223*	72 *anth*+ : 332 *anth*−	82
2	*anth*+*-A58*	*anth*−*-A223*	196 *anth*+ : 180 *anth*−	48
3	*anth*+*-A223*	*anth*−*-A487*	380 *anth*+ : 379 *anth*−	50
4	*anth*+*-A223*	*anth*−*-A58*	60 *anth*+ : 280 *anth*−	80

7.15. Mutations *nrd* 11 (gene *nrd B*, coding for the B subunit of the enzyme ribonucleotide reductase), *am* M69 (gene 63, coding for a protein that aids tail-fiber attachment), and *nd* 28 (gene *denA*, coding for the enzyme endonuclease II) are known to be located between gene 31 and gene 32 on the bacteriophage T4 chromosome. Mutations *am* N54 and *am* A453 are located in genes 31 and 32, respectively. Given the three-factor cross data in the table below, what is the linear order of the five mutant sites?

THREE-FACTOR CROSS DATA	
CROSS	% RECOMBINATION[a]
1. *am* A453-*am* M69 × *nrd* 11	2.6
2. *am* A453-*nrd* 11 × *am* M69	4.2
3. *am* A453-*am* M69 × *nd* 28	2.5
4. *am* A453-*nd* 28 × *am* M69	3.5
5. *am* A453-*nrd* 11 × *nd* 28	2.9
6. *am* A453-*nd* 28 × *nrd* 11	2.1
7. *am* N54-*am* M69 × *nrd* 11	3.5
8. *am* N54-*nrd* 11 × *am* M69	1.9
9. *am* N54-*nd* 28 × *am* M69	1.7
10. *am* N54-*am* M69 × *nd* 28	2.7
11. *am* N54-*nd* 28 × *nrd* 11	2.9
12. *am* N54-*nrd* 11 × *nd* 28	1.9

[a] All recombination frequencies are given as

$$\frac{2 \text{ (wild-type progeny)}}{\text{total progeny}} \times 100.$$

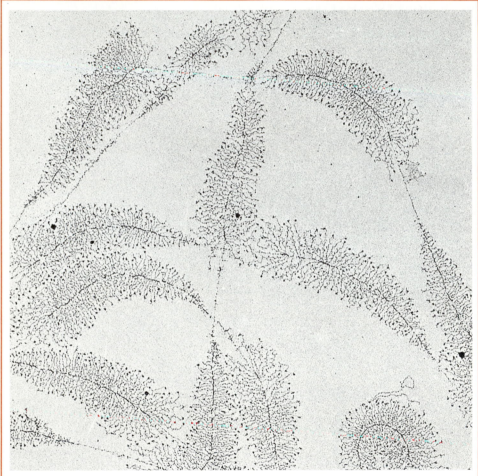

Electron microscopic "visualization" of the transcription of tandemly repeated rRNA genes in the extrachromosomal nucleoli of oöcytes in the newt *Triturus viridescens*. A gradient of rRNA fibril lengths is seen for each rRNA gene. The small spherical structures at the bases of the RNA fibrils are molecules of RNA polymerase. The tandemly repeated genes are separated by nontranscribed segments of DNA. (From O. L. Miller, Jr., and B. R. Beatty, "Visualization of Nucleolar Genes," *Science* 164:955–957, May 23, 1969. Copyright © 1969 by the American Association for the Advancement of Science. Original photograph courtesy of O. L. Miller, Jr.)

GENE EXPRESSION

In the preceding chapters, we have considered the patterns of transmission of independently assorting and linked genes, the chromosomal location of genes, the chemical structure of genes and chromosomes, and the mechanism of replication of the genetic material. In this chapter, we will discuss how genes perform their phenotypic functions. That is, how do genes exert their effects on the phenotype of a virus, a cell, or an organism? This question does not have a simple answer. Different genes clearly exert their effects in different ways. Moreover, all of the genes of an organism are located in the same cells and in the same nuclei. They do not function independently. The ultimate phenotype of an organism is the product of the action of all of the genes (including many gene-product interactions) and their interaction with the environment. The pathway of transfer or flow of information — from the gene to the final effect of the gene on the phenotype — is thus frequently very complex (Fig. 8.1).

Each gene also has an effect on the population of which the organism carrying the gene is a part; these effects will be discussed in Chapter 16. Ultimately, each gene has a potential effect, small though it may be, on the cumulative phenotype of our biosphere, as each gene may affect the ability of the organism, or the population, or the species, to compete for an ecological niche in our biosphere (see Chapters 15 and 16).

While the pathway of information flow by which a gene exerts its effect on phenotype is complex (Fig. 8.1), rather spectacular progress has been made in the last two decades in elucidating this pathway. The molecular details of the first part of this pathway, DNA → RNA → protein, are now quite well understood, and we are currently extending our knowledge of this pathway to the cellular level. However, this extension to the cellular level involves an increase in organizational complexity (protein → cell) and promises to challenge several generations of biologists before all the details are clearly understood.

We know that the information stored in the form of the nucleotide-pair sequence in a gene is transferred by a process called **transcription** to an intermediary, a single-stranded **messenger RNA (mRNA)** molecule, which carries that information from the genes in the chromosomes to the **ribosomes,** the cytoplasmic sites of protein synthesis. We also understand the most important features of the mechanism by which the information stored in the sequences of nucleotides in the mRNA molecules is converted, in a process called **translation,** into the sequences of amino acids in the protein gene products. The **genetic code** that governs this translation process has been worked out in full. Finally, these protein gene products (enzymes and structural proteins) are known to control the metabolic processes that occur in cells. Enzymes catalyze the numerous metabolic reactions that occur in living systems (see the following section, Genetic Control of Metabolism). Other proteins are important structural constituents of various cellular components, such as membranes, ribosomes, and chromosomes. In the following sections of this chapter, we will focus on these two key processes, transcription and translation, on the properties of the genetic code, and on the overall genetic control of metabolism.

EVOLUTION OF THE ONE GENE– ONE POLYPEPTIDE CONCEPT

At the time of the rediscovery of Mendel's work in 1900, the English physician–biochemist Sir Archibald E. Garrod was studying several congenital metabolic diseases in humans. One of these was the inherited disease alcaptonuria, which is easily detected because of the blackening of the urine upon exposure to air. The substance responsible for this blackening is alcapton or homogentisic acid, an intermediate in the degradation of the aromatic amino acids tyrosine and phenylalanine (see Fig. 8.4). Garrod believed that the presence of alcapton or homogentisic acid in the urine was due to a block in the normal pathway of metabolism of this compound. Moreover, on the basis of family pedigree studies, Garrod proposed that alcaptonuria was inherited as a single recessive gene. The results of Garrod's studies of alcaptonuria and a few other congenital diseases in humans, such as albinism, were presented in detail in the first edition of Gar-

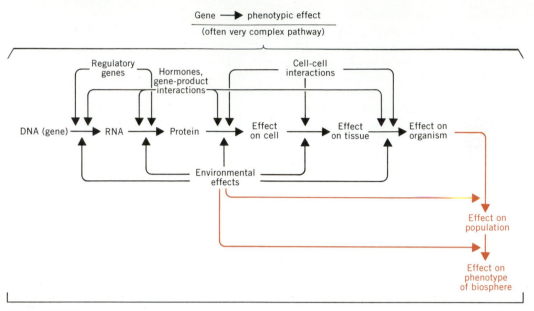

FIGURE 8.1

The complex pathway from the gene to its effect on the phenotype of a higher plant or animal and, ultimately, on the phenotype of the biosphere. Some of the known interactions involved in the frequently very complex pathway of gene expression in eukaryotes are indicated. The population and biosphere levels of gene effects are included for completeness.

rod's book, *Inborn Errors of Metabolism*, published in 1909. The title of his book is clear evidence of Garrod's insight into the genetic control of metabolism. Although the details of the biochemical pathway affected by the recessive mutation responsible for alcaptonuria were not worked out until many years later, Garrod clearly understood the gene–metabolic reaction relationship. His concept might be best stated as **one mutant gene–one metabolic block.** Garrod's concept was the forerunner to the one gene–one enzyme concept and our present one gene–one polypeptide concept. Like Mendel's work, Garrod's results were largely unknown or ignored prior to their rediscovery many years later (in Garrod's case, by Beadle and Tatum in 1941).

As was discussed in Chapter 6, the fungus *Neurospora crassa* can grow on medium containing only (1) inorganic salts, (2) a simple sugar, and (3) one vitamin, biotin. Medium containing only these components is called "minimal medium." G. W. Beadle (Fig. 8.2) and E. L. Tatum reasoned that *Neurospora*

must be capable of synthesizing all of the other essential metabolites, such as the purines, pyrimidines, amino acids, and other vitamins, *de novo*. Furthermore, they reasoned that the biosynthesis of these growth factors must be under genetic control. If so, mutations in genes whose products are involved in the biosynthesis of essential metabolites would be expected to produce mutant strains with additional growth-factor requirements.

Beadle and Tatum tested this prediction by irradiating asexual spores (conidia) of wild-type *Neurospora*, with X rays or ultraviolet light, and screening the clones produced by the mutagenized spores for new growth-factor requirements (Fig. 8.3). Only mutant strains that yielded a 1:1 mutant to wild-type ratio when crossed to wild-type were studied further, so as to select strains with a mutation in only one gene. Mutant clones that could grow on medium supplemented with all of the amino acids, purines, pyrimidines, and vitamins (called "complete medium"), but could not grow on minimal

FIGURE 8.2

George Beadle, distinguished researcher, teacher of genetics, and university administrator. Dr. Beadle, along with E. L. Tatum and J. Lederberg, was Nobel Laureate in medicine and physiology in 1958. (Photo courtesy G. W. Beadle.)

Beadle and Tatum received a Nobel Prize in 1958, was soon verified by similar studies of many other organisms in many laboratories. The **one gene–one enzyme** concept thus became a central tenet of molecular genetics.

Appropriately, in his Nobel Prize acceptance speech, Beadle stated:

"In this long, roundabout way, we had discovered what Garrod had seen so clearly so many years before. By now we knew of his work and were aware that we had added little if anything new in principle. . . . Thus we were able to demonstrate that what Garrod had shown for a few genes and a few chemical reactions in man was true for many genes and many reactions in *Neurospora*" [Beadle, 1958].

Subsequently, many enzymes, as well as the hemoglobins (see Chapter 9, pp. 280–282), were shown to consist of two or more different polypeptide chains, and each polypeptide was found to be the product of a separate gene. Tryptophan synthetase of *E. coli*, for example (see Chapter 10, pp. 340–343), contains an α-polypeptide, the product of the *trpA* gene, and a β-polypeptide, the product of the *trpB* gene. It was necessary, therefore, to change the concept of one gene–one enzyme to **one gene–one polypeptide.**

medium, were further analyzed for their ability to grow on medium supplemented with just amino acids, or just vitamins, and so on (Fig. 8.3*a*). Strains that grew when just vitamins were added, for example, were in turn analyzed for their ability to grow on media supplemented with each of the vitamins separately (Fig. 8.3*b*). In this way, Beadle and Tatum demonstrated that each mutation resulted in a requirement for **one** growth factor. By correlating their genetic analyses with biochemical studies of the mutant strains, Beadle and Tatum demonstrated in several cases that one mutation resulted in the loss of one enzyme activity. This work, for which

GENETIC CONTROL OF METABOLISM

Even if the one gene–one polypeptide concept requires modification or refinement in the future, it is clear that most genes exert their effects on phenotype via the polypeptides for which they code. Metabolism occurs by sequences of chemical reactions, each step of which is catalyzed by a specific enzyme. Each enzyme is, in turn, specified by one or more genes. The genetic control of a metabolic pathway may thus be diagrammed as shown at the bottom of this page. The number of steps in a pathway may vary from two to 10 or more.

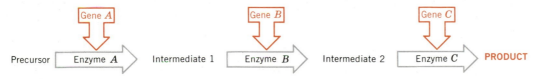

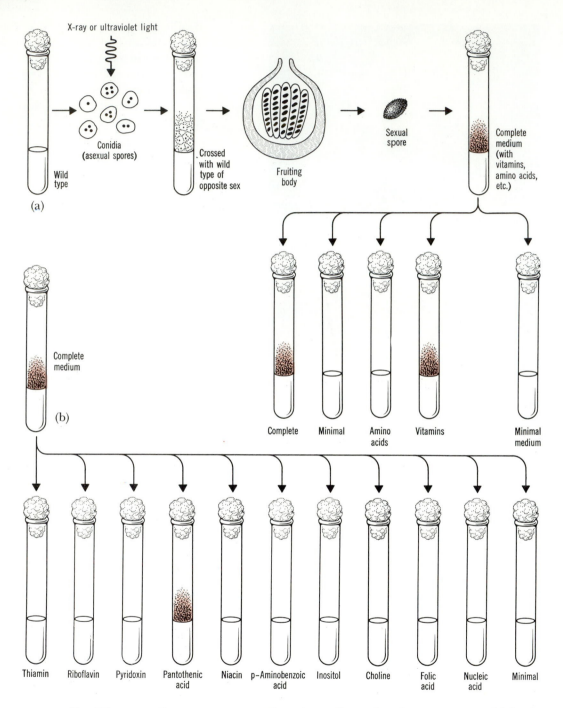

(a)

(b)

Enzymes are very specific. They usually catalyze only one or a few very similar chemical reactions. This specificity depends on their unique, complex three-dimensional structures, which, in turn, are determined by their primary structures (amino acid sequences). Most importantly, their

FIGURE 8.3

Diagram of the classic experiment of Beadle and Tatum, showing that one mutation results in a new requirement for one growth factor. **Neurospora** *conidia were mutagenized with X rays or ultraviolet light and then screened for their ability to grow on "minimal medium" (containing only inorganic salts, a simple sugar, and biotin) and "complete medium" (supplemented with all of the vitamins, purines, pyrimidines, and amino acids). Wild-type* **Neurospora** *will grow on either minimal or complete medium. Mutant strains that could grow on complete medium, but not minimal medium, were analyzed further to determine their exact growth-factor requirements (*a *and* b*). A backcross to wild-type (shown at top) was done with each mutant clone to select mutant strains carrying only one mutation. Single mutants should yield a 1 : 1 mutant to wild-type segregation pattern. Correlated biochemical analyses demonstrated that each mutation resulted in the loss or inactivation of one enzyme. (From G. W. Beadle, "Genes and the Chemistry of the Organism,"* American Scientist *34:31 – 53, 1946. Reprinted by permission of* American Scientist, Journal of Sigma Xi, *"The Scientific Research Society.")*

amino acid sequences are controlled by the base-pair sequences in the structural genes (the genes coding for the particular polypeptides; see the section in this chapter entitled Protein Synthesis).

When mutations occur in genes that result in the synthesis of inactive or otherwise defective enzymes (see Chapter 9), they produce what Garrod so aptly called "inborn errors of metabolism." Hundreds of "inborn errors" or congenital diseases are now known in humans (some of these are described in Chapter 4). We can illustrate the genetic control of metabolism and the consequences of gene defects in organisms by considering the metabolism of phenylalanine and tyrosine in humans (Fig. 8.4).

Alcaptonuria (Fig. 8.4a) is caused by a recessive mutation that results in the loss of activity of the enzyme homogentisic acid oxidase. In the absence of this enzyme, which catalyzes cleavage of the benzene ring of homogentisic acid, homogentisic acid accumulates and is excreted in the urine.

Another defect in phenylalanine-tyrosine metabolism occurs in individuals with the inherited disorder phenylketonuria (PKU; Fig. 8.4b). PKU results from a recessive mutation that causes a loss of phenylalanine hydroxylase activity. Phenylalanine hydroxylase converts phenylalanine to tyrosine, the first step in the catabolism of phenylalanine. As protein is consumed, phenylalanine accumulates in the blood of individuals with PKU, sometimes up to 100 times the normal level. As a result, metabolic derivatives of phenylalanine, such as phenylpyruvic acid, are formed. Some of the derivatives are toxic to the central nervous system and produce irreversible brain damage. If PKU is diagnosed in infancy, however, subsequent mental retardation can be avoided by placing the child on a carefully controlled diet. Phenylalanine is an essential amino acid in humans; it cannot be synthesized *de novo* as it can in *Neurospora*. Thus, individuals with PKU must be provided with enough phenylalanine in their diet to provide for the synthesis of body proteins, but not enough to build up toxic levels of derivatives of phenylalanine. Successful treatment of infants with PKU, avoiding mental retardation, is one of the more dramatic achievements of the application of genetics and biochemistry in modern medicine. It should be emphasized, however, that successful treatment of PKU is not a "cure" of the disease. The defective gene is still present and is transmitted to the progeny.

Albinism (Fig. 8.4c) is another recessive condition caused by a defect in phenylalanine-tyrosine metabolism. In certain types of albinism, the enzyme tyrosinase is inactive or lacking, resulting in a block in the pathway of conversion of tyrosine to the dark-colored pigment melanin. Last, tyrosinosis (Fig. 8.4d) appears to result from a mutation in another gene coding for an enzyme involved in the degradation of phenylalanine and tyrosine.

While we have focused here on the genetic

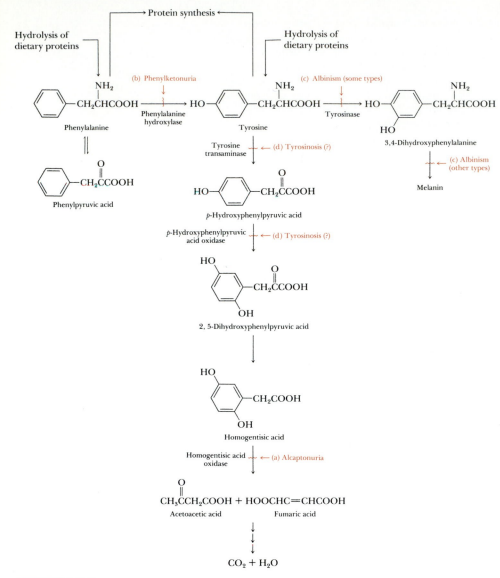

FIGURE 8.4

Inherited human diseases ("inborn errors") with defects in phenylalanine-tyrosine metabolism. (a) *alcaptonuria, extensively studied by Garrod at the turn of the century, which results in blackening and hardening of cartilage and blackening of urine upon exposure to air;* (b) *phenylketonuria, which results in severe mental retardation if untreated;* (c) *albinism, the inability to synthesize melanin pigments, and* (d) *tyrosinosis, the inability to convert tyrosine to 2,5-dihydroxyphenylpyruvic acid, an early stage in the tyrosine degradative pathway. At least two different types of albinism occur in humans. One type is caused by a mutation resulting in the loss of tyrosinase activity. A second type of albinism apparently results from a mutational block in a subsequent step in melanin synthesis or in the ability to deposit melanin pigment in melanocytes.*

control of the metabolism of phenylalanine and tyrosine, and the "inborn errors" resulting in blocks in this metabolic pathway, this picture of the genetic control of metabolism and the consequences of mutations is valid for all metabolic pathways in humans and in all other organisms. Moreover, given this picture, it should be obvious that epistatic interactions will be important components of the genetic control of most phenotypic traits.

VARIABILITY OF GENE EFFECTS AND PHENOCOPIES

Given (1) the complexity of the pathways by which some genes exert their effects on phenotype (Fig. 8.1), (2) the effects of regulator genes on the expression of many genes (see Chapter 11), and (3) the occasionally complex interactions among environmental factors, regulator genes, and pathways of gene expression, it is not surprising that the phenotypic effects of some genes are variable. Two terms, **expressivity** and **penetrance,** are frequently used in discussing variability of the effects of specific genes on phenotype. When the degree of phenotypic expression of a gene varies from individual to individual, the gene is said to have **variable expressivity.** When the variability in expressivity of a given gene is such that the presence of the gene does not always result in a detectable phenotypic effect, the gene is said to have **incomplete penetrance.** For example, if a dominant gene is expressed (has a detectable phenotypic effect) in only 70 percent of the individuals known to carry it, its penetrance is said to be 70 percent. Clearly, incomplete penetrance is merely an extension of variable expressivity of genes (where the effects on phenotype are so small that they **cannot** be detected). A common type of variable expressivity is the variable "age of onset" observed with some inherited diseases. It should be apparent that genetic analyses involving genes with incomplete penetrance and/or variable expressivity are much more difficult to interpret than those involving genes with fairly invariant phenotypic effects.

Another potentially confounding factor in genetic analyses, particularly in diagnosing inherited abnormalities in humans, is that environmental factors sometimes induce **nonhereditary phenotypic changes,** called **phenocopies,** that are indistinguishable or closely resemble conditions resulting from mutant genes. Epilepsy, for example, may result from either genetic or environmental causes. As a result, the definitive diagnosis of an inherited abnormality requires family pedigree data or other information indicating that the condition is inherited, in addition to the appropriate phenotypic manifestations in an individual.

PROTEIN SYNTHESIS

Most genes exert their effects on phenotype through the proteins (enzymes and structural proteins) whose structures they specify. Proteins are complex macromolecules that exhibit a high degree of functional specificity. A given enzyme, for example, will usually only catalyze one specific reaction. This explains why a given gene usually has a specific effect on the phenotype of an organism. The one primary effect of each gene may, however, lead to many secondary effects.

Proteins are composed of from one to several polypeptides, with each species of polypeptide being coded for by one gene (or, in a few cases, multiple copies of a redundant gene). Each polypeptide consists of a long sequence of amino acids linked together by **peptide bonds.** Twenty different amino acids are commonly found in natural proteins. Their structures (Fig. 8.5), with one exception, proline, include a free amino group and a free carboxyl group, as shown below.

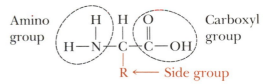

The amino acids differ from each other by the side groups that are present (Fig. 8.5). During protein synthesis, the amino acids become covalently linked by peptide bonds formed by hydrolysis from the amino and carboxyl groups:

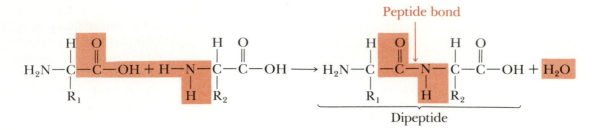

A **peptide** is a compound composed of two or more amino acids. **Polypeptides** are long sequences of amino acids, ranging in length from 51 amino acids in insulin to over 1000 amino acids in the silk protein fibroin. Given the 20 different amino acids commonly found in polypeptides, the number of different polypeptides that are possible is enormous. Calculate, for example, the number of different sequences that can occur in a polypeptide that is 100 amino acids long (20^{100})!

The sequence of amino acids in a polypeptide is called its **primary structure.** Proteins, however, have highly varied and very complex three-dimensional structures, a few of which have been determined by X-ray diffraction studies. The **secondary structure** of a polypeptide refers to the spatial interrelationships of the amino acids in segments of the polypeptide. For example, a segment of a polypeptide may exhibit a helical arrangement referred to as an α-helix. The **tertiary structure** of a protein refers to the folding of the polypeptide in three-dimensional space. Last, **quaternary structure** refers to the association of two or more polypeptides in a multimeric protein.

Hemoglobin provides an excellent example of the complexity of proteins, exhibiting all four levels of structural organization (Fig. 8.6). It consists of four polypeptides, two identical α-chains and two identical β-chains, all of which are highly folded. In most if not all proteins, the secondary, tertiary, and quaternary structures are determined by the primary structure(s) of the polypeptide(s) involved. We will concern ourselves, therefore, for the rest of this chapter with the mechanism by which genes control the primary structure of polypeptides.

The **central dogma** of molecular genetics is that genetic information is transferred (1) from DNA to DNA (replication; see Chapter 5) during its transmission from generation to generation and (2) from DNA to RNA to protein during its phenotypic expression in an organism (Fig. 8.7). (Recently, genetic information has also been shown to be transferred from RNA to DNA in provirus formation by RNA tumor viruses.) The transfer of genetic information from DNA to RNA to protein, or protein synthesis, involves (1) **transcription,** the transfer of the genetic information from DNA to RNA, and (2) **translation,** the transfer of information from RNA to protein.

TRANSCRIPTION

In eukaryotic organisms, the chromosomal genes, consisting of DNA, remain in the nuclei of cells, whereas proteins are synthesized in the cytoplasm. DNA **cannot,** therefore, serve directly as a template for protein synthesis. Instead, one strand of DNA, called the **sense strand,** is used as a template for the synthesis of a complementary strand of RNA, called **messenger RNA** (or **mRNA**) or **pre-messenger RNA** (if it requires processing prior to translation), in a process called **transcription** (Fig. 8.8). The transcribed (sense) strand of two different genes, even adjacent genes, is not always the same strand. However, for a given gene only one strand is usually transcribed. The mRNA then carries the genetic information from its site of synthesis in the nucleus to the sites of protein synthesis, the ribosomes in the cytoplasm. The synthesis of mRNA (or pre-mRNA) in the nucleus and its subsequent transport to the cytoplasm can be documented by pulse-labeling experiments, pulse-chase labeling experiments, and autoradiography (Fig. 8.9). If a cell is exposed to a

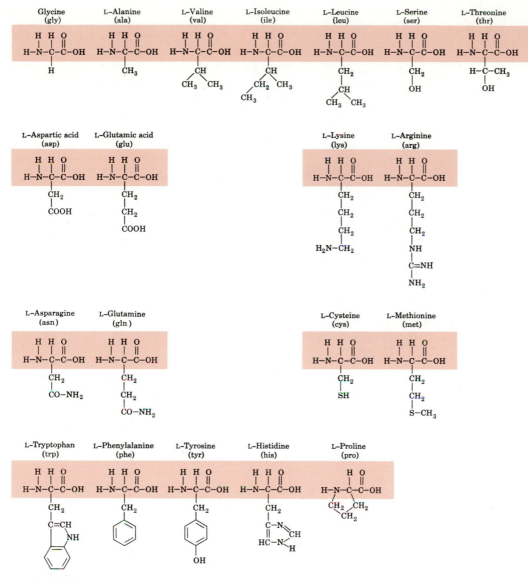

FIGURE 8.5

Structures of the 20 amino acids commonly found in proteins, arranged in groups with similar chemical properties of the side groups (shown below the shaded areas). The amino and carboxyl groups, which participate in peptide bond formation during protein synthesis, are shown in the shaded areas. The commonly used abbreviation is shown below the name of each amino acid.

radioactive RNA precursor (such as ³H-uridine or ³H-cytidine) for a few minutes, and the intracellular location of the incorporated radioactivity is deter-

mined by autoradiography, almost all of the nascent, labeled RNA is found in the nucleus (Fig. 8.9a). If, on the other hand, the short exposure

FIGURE 8.6

The three-dimensional structure of hemoglobin as deduced from X-ray diffraction studies. The hemoglobin molecule consists of two α-polypeptides (shown in light gray), two β-polypeptides (shown in dark gray), and four heme groups (shown as flat disks; only three are visible from this side view). (Reproduced with permission from "The Hemoglobin Molecule," by M. F. Perutz, Sci. Amer. 211(5):64–67. Copyright © by Scientific American, Inc. All rights reserved.)

("pulse") to the labeled RNA precursor is followed by a period of growth in nonradioactive medium (a "pulse-chase experiment") before doing the autoradiography, most of the incorporated radioactivity is found transported to the cytoplasm (Fig. 8.9b). [Pulse-chase experiments are done by either (1) sedimenting the cells from the radioactive medium or (2) adding a vast excess of nonradioactive precursor so as to dilute the radioactive precursor to negligible concentrations.]

Strong evidence for an RNA intermediary in protein synthesis also resulted from studies on T2 phage-infected *E. coli* cells. The phage proteins were shown to be synthesized on ribosomes present in the cell prior to infection. The specificity determining the amino acid sequences of polypeptides was not therefore an integral part of ribosome structure. E. Volkin and L. Astrachan demonstrated that there is a large burst of RNA synthesis shortly after T2 phage infection, and these short-lived (half-lives of only a few minutes) RNA molecules have nucleotide compositions like the T2 phage DNA and not like the host DNA. Shortly thereafter, these unstable RNA molecules were shown to be complementary to segments of strands of DNA in the phage chromosome. Many different phage mRNA molecules have now been isolated and shown to direct the synthesis of specific phage proteins by *in vitro* pro-

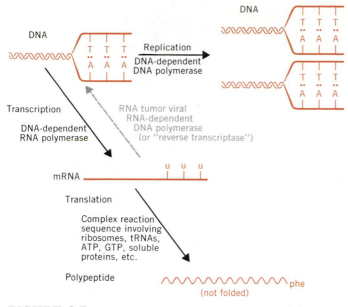

FIGURE 8.7

Routes of transfer of genetic information according to the central dogma of molecular biology: (1) DNA to DNA during transmission from one cell and/or one generation to the next, and (2) DNA → mRNA → protein during gene expression. The segment of three base-pairs on the right end of each DNA molecule is enlarged to show the base-pair sequence. During transcription, the bottom (AAA-containing) strand of DNA is used as a template for the synthesis of a complementary (UUU-containing) molecule of mRNA. Transcription is catalyzed by DNA-dependent RNA polymerase. During translation, the sequence of nucleotides in the mRNA is decoded in a complex sequence of reactions to produce the polypeptide gene product. The amino acid sequence of a polypeptide is determined by the nucleotide sequence of the mRNA according to the specifications of the genetic code.

Within the last decade, genetic information has also been shown to be transferred from RNA to DNA (dashed line) in cells infected with RNA tumor viruses. The genetic information stored in the RNA of these tumor virus genomes must apparently be converted to DNA-form before it can be stably integrated into the chromosomes (DNA) of the host cells as a provirus. These RNA tumor viruses thus direct the synthesis of an RNA-dependent DNA polymerase (also called a reverse transcriptase, since it catalyzes the reaction that is the reverse of transcription) to carry out the RNA to DNA conversion.

tein synthesis (see the following section).

The DNA-dependent RNA polymerases that catalyze transcription are usually complex, multimeric proteins. The *E. coli* RNA polymerase, the most extensively studied of all the RNA polymerases, has a molecular weight of about 490,000 and consists of six polypeptides (Fig. 8.10). Two of these are identical; thus the enzyme contains five distinct polypeptides. One of these subunits, the sigma (σ) factor, is only involved in the initiation of transcription; it does not have a catalytic function. The complete RNA polymerase molecule, called the "holo-

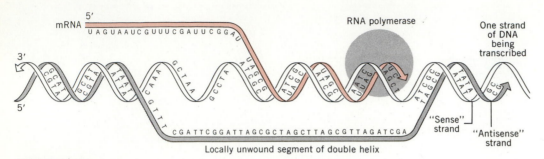

mRNA 5′
U A G U A A U C G U U U C G A U U C G G A U

RNA polymerase

One strand of DNA being transcribed

3′
5′

"Sense" strand
"Antisense" strand

CGATTCGGATTAGCGCTAGCTTAGCGTTAGATCGA

Locally unwound segment of double helix

FIGURE 8.8

Transcription, or messenger RNA synthesis. The "sense" strand of DNA functions as a template for the synthesis of a complementary molecule of RNA (messenger RNA or mRNA). The process is catalyzed by the enzyme DNA-dependent RNA polymerase, usually shortened to just RNA polymerase. RNA synthesis, like DNA synthesis, always occurs in a 5′ to 3′ direction. Base sequences are shown along the DNA and mRNA strands to illustrate their complementary nature. (Recall that uracil is present in RNA at positions where thymine would occur in DNA.) Because of the complementarity of the two strands of DNA, their base sequences cannot evolve independently. It is not surprising, therefore, that only one of the two strands, the sense strand, of any given gene specifies the amino acid sequence of the polypeptide gene product. The other, nontranscribed, strand is usually referred to as the "antisense" strand. The sense and antisense strands of DNA play equally important roles in replication; however, only the sense strand is involved in transcription.

enzyme," contains two α-polypeptides and one polypeptide of each of the following types: β, β′, ω, and σ. After initiation of the synthesis of an RNA chain, the σ-factor is released and chain elongation is catalyzed by the so-called core enzyme, which has the composition 2α, β, β′, ω. The function of sigma is to recognize and bind RNA polymerase to the correct initiation sites, the **promoter** sites (see Chapter 11), on the DNA. RNA core polymerase (minus sigma) will catalyze RNA synthesis from DNA templates *in vitro,* but, in so doing, it initiates at random sites on both strands of DNA. The holo-enzyme (plus sigma), on the other hand, initiates *in vitro* only at sites used *in vivo.*

Over 50 promoters have now been sequenced in *E. coli,* and these sequences have surprisingly little in common. Two short sequences within these promoters are sufficiently conserved to be recognized, but even these are seldom identical in two different promoters. The midpoints of the two conserved sequences occur at about 10 and 35 nucleotide pairs, respectively, before the transcription-initia-

tion site. Thus, they are often called the **"−10 sequence"** and the **"−35 sequence,"** respectively. The most common or **consensus** −10 sequence (also called the **"Pribnow box"**) is **TATAAT.** The consensus −35 sequence (also called the **"recognition sequence"**) is **TTGACA.**

In eukaryotes, there are three different RNA polymerases, I, II, and III. RNA polymerase I is located in the nucleolus and catalyzes the synthesis of rRNA. RNA polymerases II and III are present in the nucleoplasm (outside the nucleolus). RNA polymerase III transcribes the genes for small nuclear RNAs and tRNAs. Its binding sites seem to be located within these genes, rather than on the 5′ or "upstream" sides of the genes.

RNA polymerase II transcribes the majority of the nuclear structural genes; it is responsible for hnRNA or pre-mRNA synthesis. Comparisons of the sequences of the promoters or RNA polymerase II-binding sites of about 60 different eukaryotic genes reveal consensus sequences located at about 25 and 75 nucleotide pairs, respectively, before the

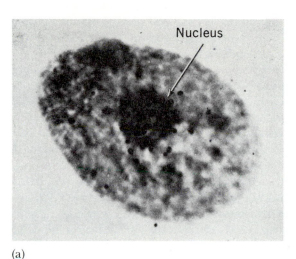

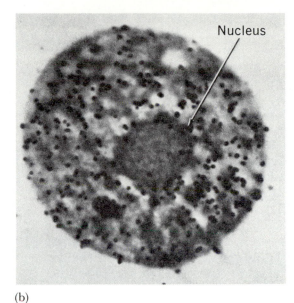

(a)   (b)

FIGURE 8.9

Autoradiographs demonstrating the synthesis of RNA in the nucleus (a) and its subsequent transport to the cytoplasm (b). Each autoradiograph is superimposed on a photograph of a thin section of the cell. Each black dot represents a silver granule in the autoradiographic emulsion that has reacted with an electron emitted by the decay of an ³H-atom. (a) A Tetrahymena *cell pulse-labeled with ³H-cytidine for 15 minutes. (b) A* Tetrahymena *cell that was grown on nonradioactive medium for 88 minutes after exposure to ³H-cytidine for 12 minutes ("pulse-chase" experiment). Note that most of the label is present in the nucleus of the cell shown in (a), whereas the nucleus of the cell shown in (b) is almost free of radioactivity, all of the labeled RNA molecules having been transported to the cytoplasm. (From D. M. Prescott, "Cellular Sites of RNA Synthesis."* Progress in Nucleic Acid Research and Molecular Biology *3:33–57, 1964.)*

transcription-start site. The consensus sequences for the **"−25 sequence"** (also called the **"Hogness box"** and the **"TATA box"**) and the **"−75 se-** quence" (also called the **"CAAT box"**) are **TATAAAA** and **GGCCAATCT,** respectively.

The mechanism of RNA synthesis is analogous

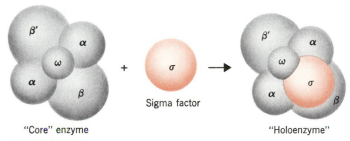

"Core" enzyme   Sigma factor   "Holoenzyme"

FIGURE 8.10

Schematic illustration of the complex structure of the RNA polymerase of E. coli. *The "core" enzyme, which catalyzes covalent chain extension, consists of two α-polypeptides (each of molecular weight about 41,000), one β-polypeptide (molecular weight about 155,000), one β'-polypeptide (molecular weight about 165,000), and one ω-polypeptide (molecular weight about 12,000). The "holoenzyme" contains, in addition, a σ-polypeptide (molecular weight about 95,000). The presence of the sigma factor is required for initiation at the proper transcription initiation (or promoter) sites. After each RNA chain-initiation event, sigma is released, and the core enzyme catalyzes chain elongation.*

to DNA synthesis (see Chapter 5) except that (1) the precursors are ribonucleoside triphosphates, (2) only limited segments of single strands are copied, and (3) the complementary RNA is released from the template as it is synthesized (see this chapter's frontispiece). Covalent extension occurs, as with DNA synthesis (see Chapter 5), by the addition of mononucleotides to the 3'-end of the chain, with the release of pyrophosphate.

The termination of transcription occurs at specific **terminator sequences** in DNA. Most prokaryotic mRNAs terminate with the sequence 5'-UUUUUUA-3', suggesting that the sequence 3'-AAAAAAT-5' in the sense strand of DNA is at least part of the transcription terminator sequence (see Fig. 11.11a). Some transcription-termination signals require the presence of a protein called **rho,** others do not.

Although only one of the two DNA strands is transcribed in any given region, both strands of DNA in a chromosome may participate in transcription, with some mRNAs being transcribed from one strand and other mRNAs (from different genes) being transcribed from the other strand. Because the two strands of a DNA double helix have opposite polarity, transcription events using opposite strands as templates will proceed in opposite directions along the DNA molecule. Even in simple viruses such as λ and T4 (see Fig. 10.10), transcription occurs off both DNA strands (but only rarely in the same region).

TRANSLATION

The process during which the genetic information (which is stored in the sequence of nucleotides in an mRNA molecule) is translated, following the dictations of the genetic code, into the sequence of amino acids in the polypeptide gene product is complex, requiring the functions of a large number of macromolecules. These include (1) approximately 50 polypeptides and from 3 to 5 RNA molecules present in each ribosome (the exact composition varies from species to species), (2) at least 20 amino acid-activating enzymes (aminoacyl-tRNA synthetases), (3) from 40 to 60 different tRNA molecules, and (4) at least 9 soluble proteins involved in poly-

peptide chain initiation, elongation, and termination. Since many of these macromolecules, particularly the components of the ribosome, are present in large quantities in each cell, the translation system makes up a major portion of the metabolic machinery of each cell.

An overview of protein synthesis, illustrating its complexity and the major macromolecules involved, is presented in Fig. 8.11. The translation process occurs on **ribosomes,** complex macromolecular structures located in the cytoplasm. Translation involves three types of RNA, all of which are transcribed from DNA templates (chromosomal genes). In addition to mRNA (see the preceding section), three to five RNA molecules (**rRNA molecules**) are present as part of the structure of each ribosome, and 40–60 small (70–80 nucleotides) RNA molecules (**tRNA molecules**) function as adaptors mediating the incorporation of the proper amino acids in response to specific codons in mRNA. The ribosomes may be thought of as workbenches, complete with machines and tools needed to make a polypeptide. They are nonspecific in the sense that they can synthesize any polypeptide (any amino acid sequence) specified by a particular mRNA molecule.

Given the above superficial overview of protein synthesis, we will next more closely examine some of the more important components and steps involved in the translation process.

Early studies using pulse-labeling (with radioactive amino acids) and autoradiography showed that proteins are synthesized largely in the cytoplasm on small, but complex, macromolecular structures called ribosomes. In prokaryotes, the ribosomes are distributed throughout the cells; in eukaryotes, they are located in the cytoplasm, frequently on an extensive intracellular membrane network called the **endoplasmic reticulum** (Fig. 8.12).

Ribosomes are approximately half protein and half RNA (Fig. 8.13). They are composed of two subunits, one large and one small, which dissociate when the translation of an mRNA molecule is completed; they reassociate during the initiation of translation. Ribosome sizes are most frequently expressed in terms of their rates of sedimentation

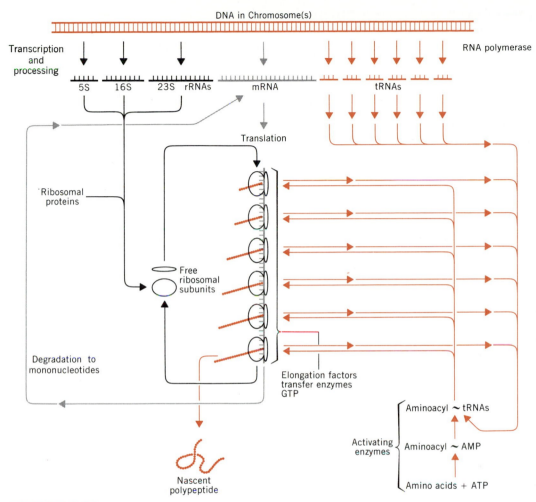

FIGURE 8.11

Diagrammatic overview of protein synthesis showing (1) the origin of the three types of RNA (rRNA, mRNA, and tRNA) by transcription from a DNA template, (2) translation of the mRNA on polyribosomes (several ribosomes simultaneously translating a single mRNA molecule), (3) the formation of ribosomal subunits from ribosomal RNAs and proteins, and the ribosome cycle (shown in black), (4) the synthesis and degradation of mRNA molecules (cycle shown in gray), (5) the formation of aminoacyl-tRNAs (catalyzed by activing enzymes) and the release of the tRNAs after each peptide bond is formed (cycle shown in color), and (6) the covalent extension and eventual release of the nascent polypeptide chain. For simplicity, all of the different species of RNA are shown being transcribed from contiguous segments of a single DNA molecule. In different organisms, the various RNA molecules are actually transcribed from genes located at different positions on from one to many chromosomes.

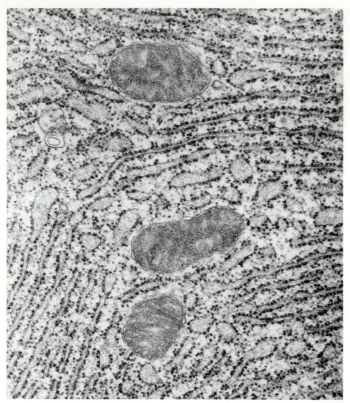

FIGURE 8.12

Electron micrograph of a thin section of a secretory cell from rat pancreas. Ribosomes (the small densely stained structures) are seen associated with membranous sheets of endoplasmic reticulum (sectioned in a crosswise manner). Three mitochondria are also present. Magnification ×28,900. (Courtesy of Ross Johnson, Department of Genetics and Cell Biology, University of Minnesota.)

during centrifugation, in units called **S** or **Svedberg units.** The *E. coli* ribosome, like those of most prokaryotes, has a molecular weight of 2.7×10^6 and a "size" of "70S." The ribosomes of eukaryotes are larger (usually about 80S); however, size varies from species to species. The ribosomes present in organelles (mitochondria and chloroplasts) of eukaryotic cells are smaller (usually about 60S).

In all cases, each ribosome consists of two subunits. In the case of *E. coli,* the small (30S) ribosomal subunit contains a 16S (mol. wt. about 6×10^5) RNA molecule plus 19 different polypeptides, and the large (50S) subunit contains two RNA molecules (5S, mol. wt. about 4×10^4, and 23S, mol. wt. about 1.2×10^6) plus 30 polypeptides. In eukaryotic ribosomes, the small subunit contains an 18S (average size) RNA molecule and the large subunit contains a 5S, a 5.8S, and a 28S (average size) RNA molecule. *Drosophila* ribosomes, but not those of several other

eukaryotes examined, also appear to contain a small 2S RNA molecule. In organelles, the corresponding rRNA sizes are 5S, 13S, and 21S.

In the case of *E. coli,* M. Nomura and his colleagues have been able to completely dissociate the 30S ribosomal subunit into the individual macromolecules and then reconstitute functional 30S subunits from the components. This has allowed them to study the function(s) of individual RNA and protein molecules.

The ribosomal RNA molecules are transcribed from a DNA template, just like mRNA molecules. In eukaryotes, however, rRNA synthesis occurs in the nucleolus and is catalyzed by a special RNA polymerase present only in the nucleolus. Moreover, transcription of the rRNA genes produces rRNA precursors that are larger than the RNA molecules found in ribosomes. These rRNA precursors undergo post-transcriptional processing to

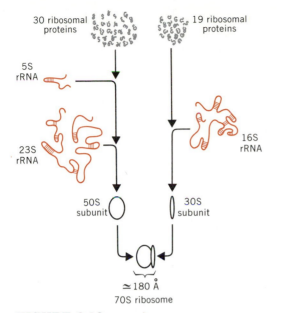

FIGURE 8.13

Macromolecular composition of the E. coli *ribosome. All other ribosomes, prokaryotic and eukaryotic, have a similar structure and composition, although their sizes, and the sizes and number of individual components, vary. The ribosomal RNA molecules exhibit considerable secondary structure because of intrastrand hydrogen bonding. Their actual structures within the ribosome are unknown. The molecular weights of the 5S, 16S, and 23S rRNA molecules are 4×10^4, 6×10^5, and 1.2×10^6, respectively. The complete 70S ribosome has a molecular weight of about 2.7 million.*

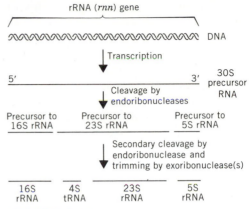

FIGURE 8.14

Transcription of the E. coli *rRNA (rnn) genes to produce the 30S rRNA precursor and its subsequent processing (cleavage and trimming) to form the 5S, 16S, and 23S rRNA molecules, plus one 4S tRNA molecule. The 30S gene transcript is rapidly cleaved (in fact, intact 30S rRNA precursors can only be isolated from ribonuclease III-deficient mutant strains of* E. coli*) to three smaller molecules, one containing the 5S rRNA sequence, another containing the 16S rRNA sequence, and the third containing the 23S rRNA sequence. These three intermediates are then further processed to produce the mature RNA molecules present in the ribosome. During the second stage of rRNA maturation, a 4S tRNA sequence is cleaved from the intermediate precursor to either the 16S or the 23S rRNA.*

produce the rRNA molecules involved in translation. In *E. coli*, the rRNA gene transcript is a 30S precursor, which undergoes cleavage by endoribonucleases to produce the 5S, 16S, and 23S rRNAs plus a 4S transfer RNA molecule (Fig. 8.14). In eukaryotes, the 2S (when present), 5.8S, 18S, and 28S rRNAs are cut from a 40S to 45S (depending on the species) precursor, while the 5S rRNA is produced by post-transcriptional processing of a separate gene transcript.

In addition to the post-transcriptional cleavages of rRNA precursors, post-transcriptional methylation of many of the nucleotides in rRNA occurs. Presumably, the methylation protects the rRNA molecules from intracellular ribonucleases

that are involved in mRNA degradation; its exact function is not yet clear, however.

Multiple copies of the genes for rRNA are present in the genomes of all organisms studied to date. This is not surprising considering the large number of ribosomes present per cell. In *E. coli*, there are estimated to be 5–10 copies of the rRNA (*rrn*) gene, with at least one copy at each of three distinct sites on the chromosome. In eukaryotes, the rRNA genes are present in hundreds to thousands of copies. The 5.8S-18S-28S rRNA genes of eukaryotes are present in tandem duplications in the **nucleolar organizer** regions of the chromosomes. In some eukaryotes, such as maize, there is a single pair of nucleolar organizers (on chromosome pair 6 in maize). In *Drosophila* and the extensively studied South African clawed toad, *Xenopus laevis,* the sex chromosomes carry the nucleolar organizers. Humans, on the other hand, have five pairs of nucleolar organizers, located on the short arms of chromosomes 13, 14, 15, 21, and 22 (see Fig. 6.24). Careful studies indicate that there are about 500 copies of the 18S-28S rRNA gene per nucleolar organizer in *Xenopus laevis*. Similar levels of redundancy have been estimated to occur in several other animals. Plants exhibit a greater variation in rRNA gene redundancy, with several thousand copies present in some genomes. Intraspecies variation in the amount of rRNA gene redundancy has also been documented in several species.

The 5S rRNA genes in eukaryotes are not located in the nucleolar organizer regions. Instead, they are usually distributed over several chromosomes. They are, however, highly redundant, like the 5.8S-18S-28S rRNA genes.

While the ribosomes provide the work-benches and much of the machinery required for protein synthesis, and the specifications for each polypeptide are encoded in an mRNA molecule, the translation of a coded mRNA message into a sequence of amino acids in a polypeptide requires one additional class of RNA molecules, the **transfer RNA (tRNA)** molecules. Chemical considerations suggest that a direct interaction between the amino acids and the nucleotide sequences or **codons** in mRNA is unlikely. (A **codon** is a nucleotide sequence in mRNA that specifies the incorporation of one amino acid.) In 1958, Crick, therefore, proposed that some kind of an "adapter" molecule mediates amino acid-codon recognition during protein synthesis. The "adapter" molecules were soon identified and found to be small (4S, 70–80 nucleotides long) RNA molecules. These molecules, first called "soluble RNA" (sRNA) molecules and, subsequently, transfer RNA (tRNA) molecules, contain a triplet base sequence, called the **anticodon** sequence, which is complementary to and recognizes the **codon** sequence in mRNA during translation. There are from one to four known tRNAs for each amino acid.

The amino acids are attached to the tRNAs by high-energy (very reactive) bonds between the carboxyl groups of the amino acids and the 3'-hydroxyl termini of the tRNAs. These reactive **aminoacyl-tRNAs** are formed in a two-step process, both steps

being catalyzed by a specific "activating enzyme" or **aminoacyl-tRNA synthetase.** There is at least one aminoacyl-tRNA synthetase for each of the 20 amino acids. The first step in aminoacyl-tRNA synthesis involves the **activation** of the amino acid using energy from adenosine triphosphate (ATP):

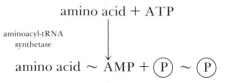

The amino acid~AMP intermediate is not normally released from the enzyme before undergoing the second step in aminoacyl-tRNA synthesis, namely, the reaction with the appropriate tRNA:

$$\text{amino acid} \sim \text{AMP} + \text{tRNA}$$

$$\big\downarrow \text{\scriptsize aminoacyl-tRNA synthetase}$$

$$\text{amino acid} \sim \text{tRNA} + \text{AMP}$$

The aminoacyl~tRNAs (amino acid~tRNAs) are the immediate precursors of polypeptide synthesis on ribosomes, each activated tRNA recognizing the correct mRNA codon and presenting the amino acid in a steric configuration (three-dimensional structure) that facilitates peptide bond formation.

The tRNAs are transcribed from chromosomal genes. As in the case of rRNAs, the tRNAs are transcribed in the form of larger precursor molecules which undergo post-transcriptional processing (cleavage, trimming, methylation, etc.). The mature tRNA molecules contain several nucleosides not present in mRNA or in the primary tRNA gene transcripts. These unusual nucleosides, such as inosine, pseudouridine, dihydrouridine, 1-methylguanosine, and several others, are produced by post-transcriptional, enzyme-catalyzed modifications of the four nucleosides incorporated into RNA during transcription.

Because of their small size (70–80 nucleotides long), tRNAs have been more amenable to structural analysis than the other, larger molecules of RNA involved in protein synthesis. The complete

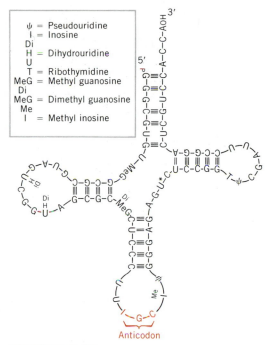

FIGURE 8.15

Nucleotide sequence and cloverleaf configuration of the yeast alanine tRNA. The secondary structure results from hydrogen bonding between bases in different segments of the molecule as shown. The structure contains three loops, within which no base-pairing is possible. One of these loops contains the anticodon sequence; in the case of the yeast alanine tRNA, the sequence is 3'-CGI-5'. The names of the modified nucleosides present in the yeast alanine tRNA are shown in the inset. (Adapted from R. W. Holley et al., Science *147:1462–1465, 1965. Copyright © 1965 by the American Association for the Advancement of Science.)*

nucleotide sequence and proposed "cloverleaf" structure of the alanine tRNA of yeast (Fig. 8.15) was published by R. W. Holley and colleagues in 1965; Holley was awarded a 1968 Nobel Prize in physiology and medicine for this work. Since then, many tRNAs have been sequenced, and the yeast alanine tRNA gene has even been synthesized *in vitro* from mononucleotides by H. G. Khorana (another 1968 Nobel Prize winner; in Khorana's case, for work on the nature of the genetic code) and co-workers. The three-dimensional structure of the phenylalanine tRNA of yeast has been determined

by X-ray diffraction studies (Fig. 8.16). The anticodons of the alanine (Fig. 8.15) and phenylalanine (Fig. 8.16) tRNAs of yeast occur within a loop (non-hydrogen-bonded region) near the center of the molecule. In fact, the anticodons of all of the tRNAs sequenced to date (over 70 from all organisms) have been found within comparably located anticodon loops.

Each ribosome has two tRNA binding sites (Fig. 8.17). The **A** or **aminoacyl site** binds the incoming aminoacyl-tRNA, the tRNA carrying the amino acid that is next to be added to the growing polypeptide chain. The **P** or **peptidyl site** binds the tRNA to which the growing polypeptide is attached. The specificity for aminoacyl-tRNA binding in these sites is provided by the mRNA codons that make up part of the A and P binding sites. As the ribosome moves along an mRNA (or as the mRNA is shuttled across the ribosome), the specificity for the aminoacyl-tRNA binding in the A and P sites changes as different mRNA codons move into register in the binding sites. The ribosomal binding sites by themselves (minus mRNA) are thus capable of binding any aminoacyl-tRNA.

It should be quite apparent that the tRNA molecules contain a great deal of specificity despite their small size. Not only must they (1) have the correct anticodon sequences, so as to respond to the right codons, but they must also (2) be recognized by the correct aminoacyl-tRNA synthetases, so that they are activated with the right amino acids, and (3) bind to the A and P sites on the ribosomes. F. Chapeville and G. von Ehrenstein and colleagues have proven, by means of a simple and direct experiment (Fig. 8.18), that the specificity for codon recognition resides in the tRNA portion of an aminoacyl-tRNA, rather than in the amino acid. They treated cysteyl-tRNA$_{cys}$ (the cysteine tRNA activated with cysteine) with a strongly reducing nickel powder (Raney nickel), which converted (reduced) the cysteine to alanine—still attached, however, to the cysteine tRNA. When this "hybrid" aminoacyl-tRNA, **alanyl-tRNA$_{cys}$**, was placed in *in vitro* protein synthesizing systems, alanine was found to be incorporated into positions in polypeptides normally occupied by cysteine.

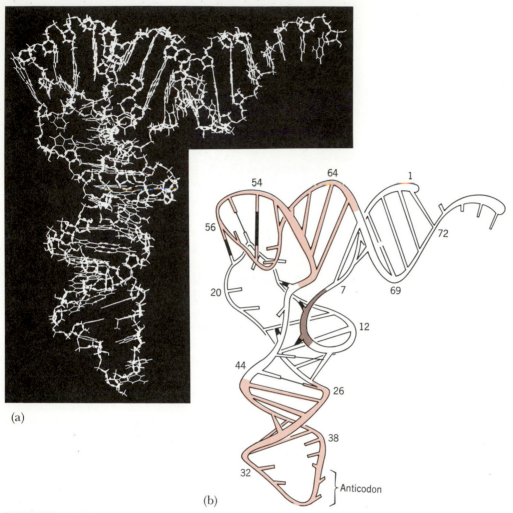

(a)

(b)

FIGURE 8.16

Photograph (a) and interpretative drawing (b) of a molecular model of the yeast phenylalanine tRNA based on X-ray diffraction data. The anticodon loop is at the bottom. (b) The ribose-phosphate backbone of the tRNA is drawn as a continuous cylinder. The crossbars indicate hydrogen-bonded base-pairs. Unpaired bases are indicated by shortened rods. (From S. H. Kim, F. L. Suddath, G. J. Quigley, A. McPherson, J. L. Sussman, A. H. J. Wang, N. C. Seeman, and A. Rich, Science *185:435–440, 1974. Copyright © 1974 by the American Association for the Advancement of Science.)*

Protein synthesis is initiated by a special **initiator tRNA,** designated **tRNA$_f^{Met}$.** This means that all polypeptides begin with methionine during synthesis. The amino terminal methionine is subsequently cleaved from many polypeptides. Thus, functional proteins need not have an amino terminal methionine. In prokaryotes and in eukaryotic organelles, the **methionine** on the initiator tRNA$_f^{Met}$ has the

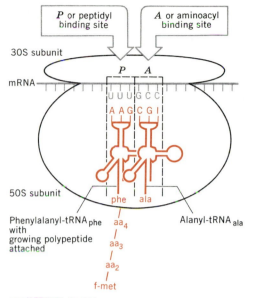

FIGURE 8.17

The aminoacyl-tRNA binding sites formed by each ribosome-mRNA complex. The A or aminoacyl-tRNA site is shown occupied by an alanyl-tRNA$_{ala}$ complex. The P or peptidyl site is shown occupied by a phenylalanyl-tRNA$_{phe}$ complex, with the growing polypeptide chain covalently linked to the phenylalanine tRNA. The next step in protein synthesis will involve the formation of a peptide bond between phenylalanine and alanine, followed by the translocation of the alanine tRNA (with the growing polypeptide now attached) from the A site to the P site.

amino group blocked with a formyl $\left(\begin{array}{c} O \\ \parallel \\ -C-H \end{array}\right)$

group. In eukaryotic cytoplasmic systems, a special initiator tRNA$_i^{Met}$ also exists, but the amino group is not formylated. A distinct methionine tRNA, tRNA^{Met}, which responds to internal methionine codons, exists in both prokaryotic and eukaryotic systems. Both methionine tRNAs have the same anticodon and both respond to the same codon (AUG) for methionine. In prokaryotes, the formylated amino group on methionyl-tRNA$_f^{Met}$ prevents the formation of a peptide bond between the amino group and the carboxyl group of the amino acid at the end of the growing polypeptide chain. In eukaryotes, however, the amino group of methionyl-

tRNA$_i^{Met}$ is not blocked. What, then, prevents methionyl-tRNA$_i^{Met}$ from responding to internal AUG codons and methionyl-tRNA^{Met} from responding to initiator AUG codons in eukaryotic mRNAs? Apparently, only methionyl-tRNA$_i^{Met}$ will react with the protein initiation factors, IF-1, IF-2, and IF-3, and only methionyl-tRNA^{Met} will interact with the protein elongation factors, Ef-T$_s$ and Ef-T$_u$. In any case, only methionyl-tRNA$_i^{Met}$ responds to AUG initiation codons and only methionyl-tRNA^{Met} responds to AUG internal codons. Methionyl-tRNA$_f^{Met}$ also responds to an alternate initiator codon, GUG (a valine codon when present at internal positions), known to be present in certain natural mRNAs.

Polypeptide chain initiation occurs with the formation of a complex between mRNA, methionyl-tRNA$_f^{Met}$, and the 30S ribosomal subunit (Fig. 8.19, *a* to *b*). The formation of this initiation complex requires the activity of three protein initiation factors, designated **IF-1, IF-2,** and **IF-3,** plus guanosine triphosphate (GTP). It may be facilitated by a base-pairing interaction between a base sequence near the 3′-end of the 16S rRNA and a base sequence in the **"leader sequence"** of the mRNA. (**Leader sequences** of mRNA molecules are the nontranslated sequences from the 5′-end to the first AUG or GUG initiation codon. These nontranslated leader sequences vary in length from a few nucleotides to several hundred nucleotides. Little is known about their biological significance.)

The above-described initiation complex then combines with the 50S ribosomal subunit, and the methionyl-tRNA$_f^{Met}$ becomes bound to the peptidyl site (Fig. 8.19, *b* to *c*). This requires the hydrolysis of one molecule of GTP. The alignment of the AUG initiation codon with the anticodon of tRNA$_f^{Met}$ (in the P site) fixes the codon present at the A site, thus establishing the specificity for aminoacyl-tRNA binding at the A site (for alanyl-tRNA$_{ala}$ in the example diagramed in Fig. 8.19*d*). The binding of alanyl-tRNA$_{ala}$ in the A site (and all subsequent aminoacyl-tRNA binding) requires the hydrolysis of one molecule of GTP and the protein elongation factors designated **Ef-T$_s$** and **Ef-T$_u$** (Fig. 8.19, *c* to *d*). Peptide bond formation between the carboxyl

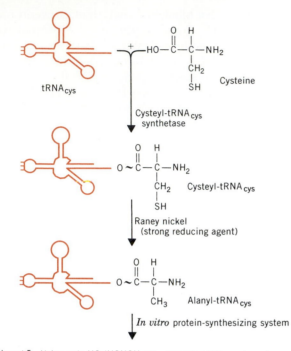

tRNA$_{cys}$

$\overset{+}{HO-}\overset{\overset{O}{\|}}{C}-\overset{\overset{H}{\|}}{C}-NH_2$ Cysteine

CH$_2$

SH

Cysteyl-tRNA$_{cys}$
synthetase

$O\sim\overset{\overset{O}{\|}}{C}-\overset{\overset{H}{\|}}{C}-NH_2$ Cysteyl-tRNA$_{cys}$

CH$_2$

SH

Raney nickel
(strong reducing agent)

$O\sim\overset{\overset{O}{\|}}{C}-\overset{\overset{H}{\|}}{C}-NH_2$ Alanyl-tRNA$_{cys}$

CH$_3$

In vitro protein-synthesizing system

Experiment I: Using poly-UG (UGUGU etc., repeating UG copolymer) as an artificial messenger RNA. Alanine attached to tRNA$_{cys}$ was incorporated, despite the fact that the alanine codons are GCU, GCC, GCA, and GCG. UGU = cysteine codon!

Experiment II: Using the hemoglobin-synthesizing rabbit reticulocyte system containing native hemoglobin mRNAs. Demonstrated that alanine from alanyl-tRNA$_{cys}$ was incorporated into positions in the rabbit hemoglobin chains normally occupied by cysteine.

FIGURE 8.18

Proof that the codon-recognizing specificity of an aminoacyl-tRNA complex resides in the tRNA rather than in the amino acid. The tRNA specific for cysteine was activated with cysteine using cysteyl-tRNA$_{cys}$ synthetase. The cysteine, attached to tRNA$_{cys}$, was then reduced to alanine with Raney nickel (fine powdered nickel). The "hybrid" alanyl-tRNA$_{cys}$ was then placed in an in vitro *protein-synthesizing system. In one experiment, protein synthesis was directed by a synthetic poly-UG mRNA. Poly-UG usually stimulates the incorporation of cysteine (codon UGU), but not alanine (codon GCX, where X equals any one of the four common bases in RNA). When the "hybrid" alanyl-tRNA$_{cys}$ was present, poly-UG stimulated the incorporation of the alanine into polypeptides. In a second experiment, protein synthesis was directed by native rabbit hemoglobin mRNAs. When the alanyl-tRNA$_{cys}$ "hybrid" was added, alanine was found incorporated into hemoglobin chains at positions normally occupied by cysteine. (Based on the results of Chapeville et al.,* Proc. Natl. Acad. Sci., U.S. *48:1086–1092, 1962, and of von Ehrenstein et al.,* Proc. Natl. Acad. Sci., U.S. *49:669–675, 1963.)*

group of f-methionine bound to the tRNA$_f^{Met}$ in the P site and the amino group of the alanine molecule bound to tRNA$_{ala}$ in the A site is then catalyzed by **peptidyl transferase,** an enzyme bound to the 50S ribosomal subunit (Fig. 8.19, *d* to *e*). This reaction leaves the f-met-ala dipeptide attached to tRNA$_{ala}$ bound to the A site of the ribosome (Fig. 8.19*e*)

The next step in translation, called **translocation,** involves (1) movement of f-met-ala-tRNA$_{ala}$ from the A site to the P site and (2) movement of the mRNA molecule exactly three nucleotides, relative to the position of the ribosome, so that the codon previously in register with the A site moves into register with the P site (Fig. 8.19, *e* to *f*). In Fig. 8.19, *e* to *f*, the alanine codon GCC moves from its position at the A site into register with the P site, and the subsequent codon, the serine codon UCC,

moves into register with the A site. Translocation requires the activity of the elongation factor designated **Ef-G** and the hydrolysis of one molecule of GTP.

The next aminoacyl-tRNA specified by the mRNA codon at the A site (seryl-tRNA$_{ser}$ in Fig. 8.19, *f* to *g*) then binds at the A site, and the peptide bond formation and translocation steps are repeated. The above-described sequence is repeated for each codon of the mRNA (about 300 codons on the average) until a **chain-termination codon** is reached (Fig. 8.19, *g* to *h*). The formyl group on the amino-terminal methionine is usually removed by a deformylase (Fig. 8.19, *g* to *h*) before synthesis of the polypeptide is completed. When one of the three polypeptide chain-termination codons (UAG, UAA, or UGA; see pp. 265–266) comes into register with the *A* site (Fig. 8.19*h*), the nascent polypeptide, the tRNA in the *P* site, and the mRNA are released, and the ribosomal subunits dissociate (Fig. 8.19, *h* to *i*). Termination requires the activity of one of two protein **release factors,** designated RF-1 and RF-2. The dissociated ribosomal subunits are then free to initiate the translation of another mRNA molecule (Fig. 8.11).

Rather than each mRNA molecule being translated by a single ribosome, most mRNAs are simultaneously translated by several ribosomes, spaced about 90 nucleotides apart along the mRNA molecule. The size of these translation complexes, called **polyribosomes** or **polysomes,** is highly variable but is often correlated with the size of the polypeptide being synthesized. Hemoglobin chains (about 150 amino acids), for example, are synthesized on pentaribosome (average size) complexes.

COUPLED TRANSCRIPTION AND TRANSLATION IN PROKARYOTES

In prokaryotes, the translation of an mRNA molecule frequently begins before its synthesis (transcription) is complete. This is possible because mRNA molecules are both synthesized and translated in the 5′ to 3′ direction, and because there is no nuclear membrane separating transcription from translation as in eukaryotes. This coupling between transcription and translation facilitates the very rapid and efficient "turn-on" and "turn-off" of gene expression that is observed in prokaryotes (see Chapter 11).

O. L. Miller, B. A. Hamkalo, and colleagues have developed techniques by which transcription and translation can be visualized directly using electron microscopy. Figure 8.20, for example, shows the coupled transcription and translation of a gene in *E. coli.*

TRANSCRIPTION, RNA PROCESSING AND TRANSPORT, AND TRANSLATION IN EUKARYOTES

In eukaryotes, transcription and translation cannot be coupled, since transcription occurs in the nucleus and translation occurs in the cytoplasm. This poses the questions of how gene transcripts are trans-

ported from the nucleus to the cytoplasm and what determines the time and place of mRNA translation. Unfortunately, we do not yet have answers to these questions. We do know that the transcription and translation processes in eukaryotes are more complex than those in prokaryotes, involving several intermediate mRNA processing steps, as well as transport from nucleus to cytoplasm.

The mRNAs of eukaryotes are derived from the primary gene transcripts by several types of processing (Fig. 8.21). These include: (1) cleavages of **large mRNA precursors (pre-mRNAs)** to smaller mRNA molecules, (2) the addition of **7-methyl guanosine groups (mRNA "caps")** to the 5′-ends of the molecules, (3) the addition of approximately **200-nucleotide long sequences of adenylate nucleotides ("poly-A tails")** to the 3′-ends of the molecules, and (4) formation of complexes with specific proteins. The cleavages involved in the conversion of pre-mRNAs to mRNAs frequently involve the removal of **leader sequences,** sequences from the 5′-end to the translation initiation codon, and **noncoding sequences** (called **"intervening sequences"** or **"introns"**) that are located between coding sequences (see the following section and Chapter 10, pp. 361–365). Individual gene transcripts may undergo some or all of these four types of processing. Not all mRNAs contain the 5′-cap, nor do all of them have poly-A tails. This has made it difficult to determine the functions of these posttranscriptional modifications.

Most of the nonribosomal RNAs synthesized in the nuclei of eukaryotic cells consist of very large molecules that are highly variable in size (10S–200S, or about 1000–50,000 nucleotides in length). This RNA has been called **heterogeneous nuclear RNA,** abbreviated **hnRNA.** It now seems clear that much, if not most, of this hnRNA is really pre-mRNA. Rapid processing of these giant hnRNA or pre-mRNA molecules in the nucleus soon after transcription apparently results in (1) the bulk of the nonribosomal RNA synthesized in the nucleus (probably large segments of each primary transcript) being rapidly degraded (average half-life of about 30 minutes) and (2) the formation of the smaller mRNA molecules that are transported to

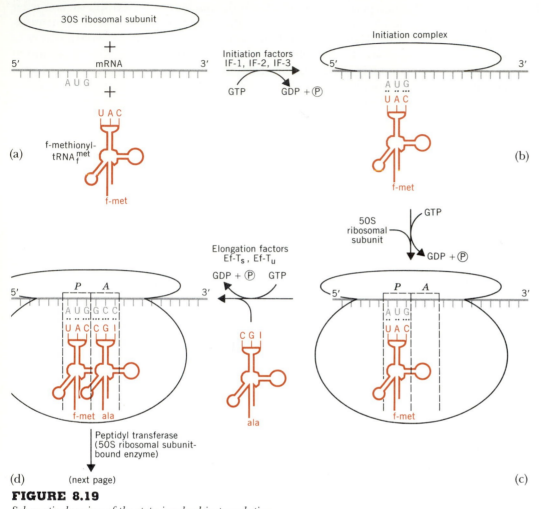

FIGURE 8.19

Schematic drawing of the steps involved in translation.

the cytoplasm. However, it is not yet clear whether all, most, or only part of the hnRNA molecules synthesized in the nuclei of eukaryotic cells are, in fact, pre-mRNA molecules.

Definitive evidence for pre-mRNA processing in the formation of eukaryotic mRNAs has been obtained in the case of the β-globin gene transcript of the mouse. In this case, a 15S hnRNA (or pre-mRNA; 1200–1500 nucleotides long) is processed to a 9S (about 600 nucleotides long) β-globin mRNA.

Similar evidence for hnRNA or pre-mRNA processing in the formation of mature mRNA molecules is now available for many other eukaryotic gene transcripts. This processing often involves the excision of noncoding intervening sequences or introns located between coding sequences (called **exons** for **ex**pression).

In addition, the mRNAs of some eukaryotic viruses are known to contain leader sequences (the sequence from the 5′-ends to the translation-initiation codons of mRNAs) that are transcribed from

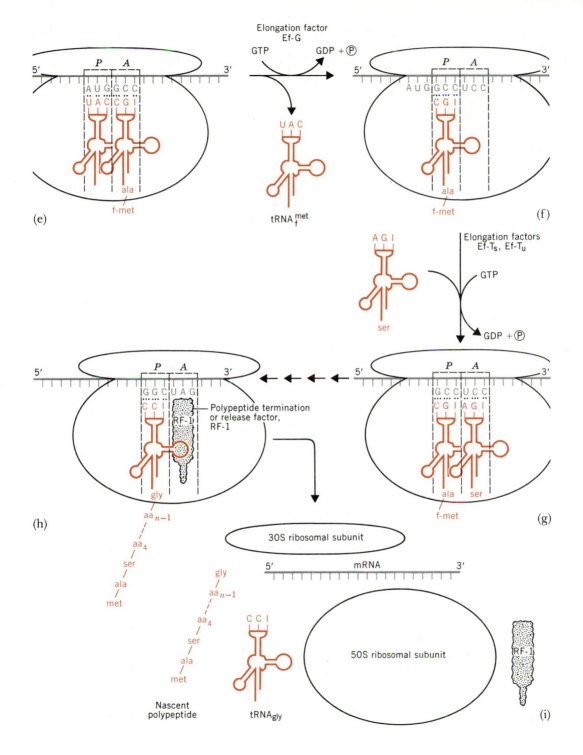

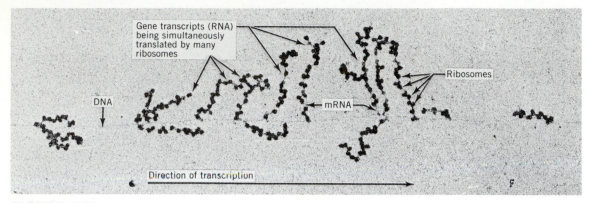

FIGURE 8.20

Electron micrograph showing the coupled transcription and translation of a gene in E. coli. *The nascent polypeptide chains being synthesized on the ribosomes are not visible as they fold into their three-dimensional configuration during synthesis. However, the silk protein fibroin, synthesized in the silk gland of the silkworm, remains rodlike on the ribosome and can be seen during synthesis by electron microscopy (see Fig. 8.22). (From O. L. Miller, Jr., B. A. Hamkalo, and C. A. Thomas, Jr.,* Science *169:392–395, 1970. Copyright © 1970 by the American Association for the Advancement of Science. Original micrograph courtesy of O. L. Miller, Jr.)*

DNA sequences that are *not* contiguous with the structural genes. Several different mRNAs may, in fact, contain identical leader sequences. These leader sequences are apparently spliced onto the 5′-ends of gene transcripts during processing. It is believed that these leader sequences must be involved in the regulation of translation. However, their function(s) is not yet known.

Translation in eukaryotes appears to be analogous to translation in prokaryotes, except that (1) the amino group of methionyl-tRNA$_i^{Met}$ (the initiation tRNA) is not formylated, and (2) most eukaryotic mRNAs studied to date appear to be monogenic, such that only one polypeptide species is translated from each mRNA. In prokaryotes, many mRNAs are **multigenic;** that is, two or more different polypeptides are synthesized from nonoverlapping segments of a single mRNA (see Chapter 11, pp. 382–387).

The synthesis of one eukaryotic protein, the silk protein fibroin, can be visualized by electron microscopy using the techniques developed by O. L. Miller, B. A. Hamkalo, and colleagues. Fibroin does not fold up on the surface of the ribosome as other

polypeptides do under the conditions used. As a result, nascent polypeptide chains of increasing length can be seen attached to the ribosomes as one scans from one end (the mRNA 5′-end) of the giant polysomes (containing 50–80 ribosomes; fibroin has a molecular weight of over 200,000) to the other end (Fig. 8.22).

"SPLICING" OUT INTRONS DURING RNA PROCESSING IN EUKARYOTES

Many eukaryotic genes contain **noncoding intervening sequences** or **introns separating the coding sequences or exons.** (See Chapter 10 for a discussion of the fine structure of eukaryotic genes.) In the case of these "split" or "mosaic" genes (coding sequences separated by noncoding sequences), the primary transcript contains the entire sequence of the gene and the **noncoding sequences are "spliced" out during processing** (Fig. 8.23).

There is a great deal of interest in the mechanism by which this splicing occurs, in part because it must be accurate to the single nucleotide to assure that codons in exons distal to introns are read in the correct reading frame. Accuracy to this degree

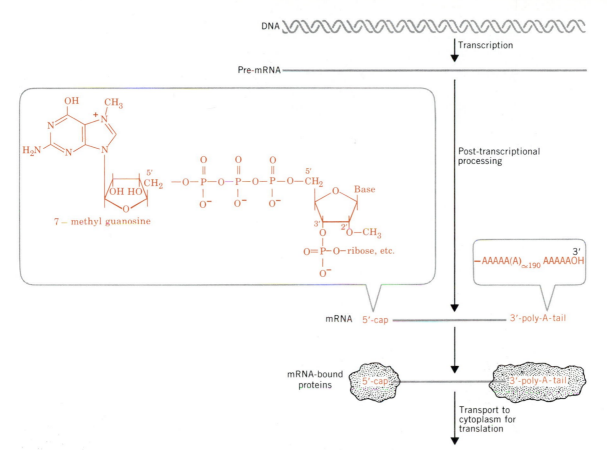

FIGURE 8.21

Processing of primary gene transcripts (pre-mRNA) in eukaryotes. Processing involves: (1) the cleavage of the pre-mRNA, with the removal of ends and/or the excision of noncoding introns, (2) the enzymatic addition of a 5'-cap, a 7-methyl guanosine group linked to a 5' subterminal, 2'-O-methyl nucleotide by an unusual 5'-5' triphosphate linkage, (3) the addition of a 200-nucleotide long poly-A tail (average size) to the 3'-end, and (4) the association with two specific proteins. The addition of the 3'-poly-A tail is catalyzed by an enzyme called poly-A polymerase.

would seem to require very precise splicing signals, presumably nucleotide sequences within introns and at the exon–intron junctions. However, in primary transcripts of nuclear genes of higher animals, the only highly conserved sequences of different introns appear to be the dinucleotide sequences at the ends of introns, namely,

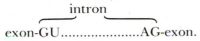

The introns of mitochondrial genes of yeast have been found to contain additional conserved sequences not observed in introns of nuclear genes of higher animals.

It now seems probable that there is no single splicing mechanism. The introns of yeast nuclear tRNA gene transcripts are removed by a process that can be separated into two steps *in vitro*. The tRNA precursor is first cleaved into two halves and

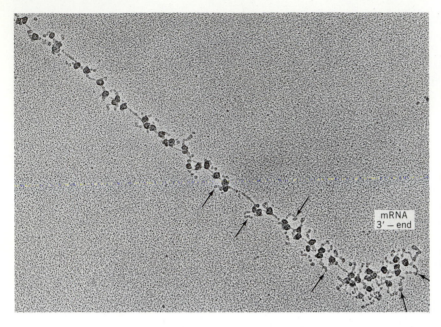

mRNA
3' — end

FIGURE 8.22

Electron micrograph showing the translation of the fibroin mRNA in cells of the posterior silk gland of the silkworm Bombyx mori. *The arrows point to putative nascent fibroin polypeptides. Note their increasing length as one approaches the 3'-end of the mRNA molecule. Each fibroin polysome consists of 50–80 ribosomes that simultaneously translate each giant fibroin mRNA; only a portion of one polyribosome is shown. (See* S. L. McKnight, N. L. Sullivan, and O. L. Miller, Jr., Progress in Nucleic Acid Research and Molecular Biology *19:313–318, 1976. Micrograph courtesy of S. L. McKnight and O. L. Miller, Jr., University of Virginia.)*

the intron cut out by a nuclease reaction. Then, the two halves are joined in an RNA ligation reaction (analogous to the DNA ligations described in Chapter 5). At present, it is not clear whether the nuclease and ligase activities reside in a single enzyme or in two separate enzymes. However, the evidence indicates that one or more enzymes are required to catalyze the splicing reaction.

By contrast, the reaction by which the intron of the rRNA gene of the ciliate *Tetrahymena thermophila* is spliced out of the rRNA precursor appears **not** to require the intervention of an enzyme. In-

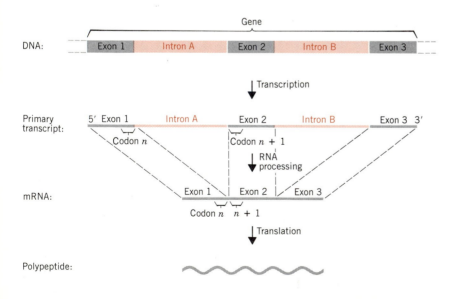

FIGURE 8.23

Splicing intron sequences out of primary transcripts during RNA processing. The splicing process must be accurate to the single nucleotide so that codons in exons distal to introns are in the proper reading frame. (See Chapter 10, p. 361, for evidence that intron sequences are removed from primary transcripts during processing.)

stead, the transcript appears to be **self-splicing.** If an enzyme is involved, it is a most unusual one, being resistant to removal or inactivation by agents that are known to remove or inactivate other enzymes. T. R. Cech and colleagues have introduced the term **ribozyme** to name such RNAs with self-splicing or autocatalytic properties.

Very little is known about the mechanism by which introns are spliced out of primary transcripts of nuclear genes in higher animals. There is evidence that a small nuclear RNA (snRNA) called U-1 may be involved. U-1 is highly conserved in many different animal species and has a sequence near the 5'-end that could assist in aligning intron–exon junctions, as shown in Fig. 8.24. Moreover, antibodies specific to ribonucleoprotein particles containing snRNA U-1 have been shown to block splicing reactions. Whether or not snRNA U-1 is actually involved in splicing reactions remains to be determined.

Interestingly, the introns of genes containing multiple introns are not spliced out in a specific sequence, nor are they spliced out at random. Instead, there are preferred, but variable sequences for splicing out the different introns. Finally, some introns are spliced out in two steps, part of the intron being excised prior to the removal of the rest of the intron. Clearly, we still have a great deal to learn about the mechanisms by which introns are spliced out of primary transcripts.

THE GENETIC CODE

As it became evident that genes controlled the structure of polypeptides, attention focused on how the sequence of the four base-pairs in DNA could control the sequence of the 20 amino acids found in proteins. With the discovery of the mRNA intermediary, the question became one of how the sequence of the four bases present in mRNA molecules could specify the amino acid sequence of a polypeptide. What is the nature of the **genetic code** relating mRNA base sequences (or DNA base-pair sequences) to amino acid sequences? Clearly, the symbols or "letters" used in the code must be the bases; but what comprises a **codon,** the unit or "word" specifying one amino acid (or, actually, one aminoacyl-tRNA complex)?

THREE NUCLEOTIDES PER CODON

Twenty different amino acids are incorporated during translation. Thus, at least 20 different codons must be formed using the four symbols (bases) available in the "message" (mRNA). Two bases per codon would yield only 4^2 or 16 possible codons — clearly not enough. Three bases per codon yields 4^3 or 64 possible codons — an apparent excess.

The first strong evidence that the genetic code was in fact a **triplet code** (three nucleotides per codon) resulted from a genetic analysis of proflavin-induced mutations in the rII locus of phage T4, carried out by F. H. C. Crick and colleagues in 1961.

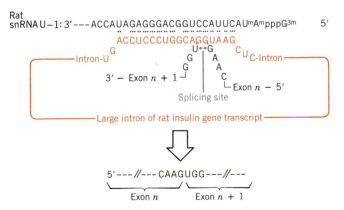

FIGURE 8.24

Possible role of the snRNA U-1 in aligning exon–intron junctions during RNA splicing. The example shown is for the snRNA U-1 of the rat and the large intron (shown in color) of the rat insulin primary transcript. (After J. Rogers and R. Wall, Proc. Natl. Acad. Sci., U.S. 77:1877–1879, 1980.)

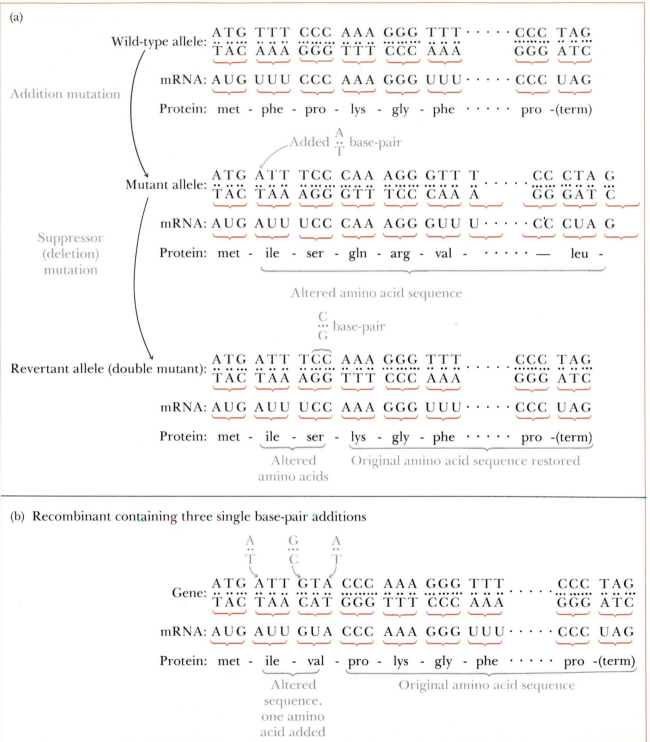

(a)

Wild-type allele:
ATG TTT CCC AAA GGG TTT · · · · CCC TAG
TAC AAA GGG TTT CCC AAA   GGG ATC

Addition mutation

mRNA: AUG UUU CCC AAA GGG UUU · · · · CCC UAG

Protein: met - phe - pro - lys - gly - phe · · · · · pro -(term)

Added $\overset{A}{\underset{T}{\cdot\cdot}}$ base-pair

Mutant allele:
ATG ATT TCC CAA AGG GTT T · · · · CC CTA G
TAC TAA AGG GTT TCC CAA A   GG GAT C

Suppressor (deletion) mutation

mRNA: AUG AUU UCC CAA AGG GUU U · · · · CC CUA G

Protein: met - ile - ser - gln - arg - val - · · · · · — leu -

Altered amino acid sequence

$\overset{C}{\underset{G}{\cdot\cdot\cdot}}$ base-pair

Revertant allele (double mutant):
ATG ATT TCC AAA GGG TTT · · · CCC TAG
TAC TAA AGG TTT CCC AAA   GGG ATC

mRNA: AUG AUU UCC AAA GGG UUU · · · · CCC UAG

Protein: met - ile - ser - lys - gly - phe · · · · · pro -(term)

Altered amino acids

Original amino acid sequence restored

(b) Recombinant containing three single base-pair additions

$\overset{A}{\underset{T}{\cdot\cdot}}$ $\overset{G}{\underset{C}{\cdot\cdot\cdot}}$ $\overset{A}{\underset{T}{\cdot\cdot}}$

Gene:
ATG ATT GTA CCC AAA GGG TTT · · · · CCC TAG
TAC TAA CAT GGG TTT CCC AAA   GGG ATC

mRNA: AUG AUU GUA CCC AAA GGG UUU · · · · CCC UAG

Protein: met - ile - val - pro - lys - gly - phe · · · · · pro -(term)

Altered sequence, one amino acid added

Original amino acid sequence

FIGURE 8.25

*Schematic illustration of Crick and co-workers' proof that the genetic code is a triplet code (three bases per codon). Crick and colleagues studied a series of suppressor mutations of a mutation at the rII locus of phage T4. The original rII mutation had been induced with the acridine dye proflavin; it was most likely, therefore, the result of a single base-pair addition or deletion. In (a), the original mutation is shown arbitrarily as a single base-pair addition, specifically, an AT base-pair insertion (wild-type allele → mutant allele). The nucleotide-pair sequence shown for the wild-type allele (and, thus, the mRNA base sequence and amino acid sequence of the polypeptide) is hypothetical. Crick and co-workers selected phenotypic revertants of this mutant and demonstrated by backcrosses that these revertants resulted from suppressor mutations, not back-mutations at the original mutant site. If the original mutation is an addition, gene-product activity might be restored by a deletion (single base-pair) mutation in a nearby region, for example, the deletion of a CG base-pair as shown (mutant allele → revertant allele). The original addition mutation will change the **reading frame** (determining codons in phase for translation) for all codons distal (relative to the direction of translation) to the site of the mutation. The subsequent deletion (suppressor mutation) will restore the reading frame for the distal portion of the gene. If the altered amino acid sequence is not critical to function, the protein produced by the doubly mutant gene will be active. When the suppressor mutations were isolated in single mutant strains by backcrosses to wild-type, all were found to yield mutant phenotypes, just like the original mutations that they suppressed. Crick and colleagues then isolated proflavin-induced suppressor mutations of the previously isolated suppressor mutations (present in single mutants recovered from backcrosses). After repeating this process for several cycles, all of the mutations were classified as plus (for single base-pair addition) or minus (for single base-pair deletion) on the basis that a plus mutation would suppress a minus mutation, but not another plus mutation, and vice versa. Actually, Crick and colleagues had no idea whether the plus group of mutations represented additions or not; they could just as likely have been deletions. The only important point was that all deletions ended up in one group (be it the plus group or the minus group), and all additions ended up in the other group.*

*Having so classified the mutations, Crick and co-workers next did the critical experiment. They isolated recombinants carrying various combinations of plus and minus mutations. When two plus mutations were present in a recombinant, its phenotype was always mutant. The same was true in the case of recombinants with two minus mutations. However, recombinants containing **three plus mutations** (b) or **three minus mutations** frequently had wild-type phenotypes. Thus, the wild-type reading frame for the distal portion of the gene was restored by three single base-pair additions or three single base-pair deletions, but was altered by either one or two base-pair additions or deletions. These results are most easily explained if each codon contains three bases.*

Crick and colleagues isolated proflavin-induced revertants of a proflavin-induced mutation. (Proflavin, an acridine dye, induces single base-pair additions and deletions; see Chapter 9, pp. 300–301) These revertants were shown (by backcrosses to wild-type; see Fig. 9.9) to result from the occurrence of suppressor mutations rather than from back-mutation at the original site of mutation. Crick and colleagues reasoned that if the original mutation was a single base-pair addition or deletion, then the suppressor mutations must be single base-pair deletions or additions, respectively, occurring at a site or sites near the original mutation. A single base-pair addition or deletion will alter the **reading frame** of the gene and mRNA (the codons in phase during translation) for that portion of the gene distal to the mutation (relative to the direction of translation). This is illustrated in Fig. 8.25a. When the suppressor mutations were isolated as single mutants by screening progeny of backcrosses to wild-type, they were found to produce mutant phenotypes, just like the original mutation. Crick and colleagues next isolated proflavin-induced suppressor mutations of the original suppressor mutations, and so on.

All of the isolated mutations were then classified into two groups, plus (+) and minus (−) (for additions and deletions, although Crick et al. had no idea which group was which) using the reasoning that a (+) mutation would suppress a (−) mutation, but not another (+) mutation, and vice versa (see Fig. 8.25 and legend for additional details). Next, Crick et al. constructed recombinants that carried various combinations of the (+) and the (−) mutations. Recombinants with two (+) mutations or two (−) mutations always had mutant phenotypes, just like the single mutants. Recombinants carrying **three (+) mutations** (Fig. 8.25*b*) or **three (−) mutations,** on the other hand, often had wild-type phenotypes. This indicated that the addition of three base-pairs or the deletion of three base-pairs left the distal portion of the gene with the correct (wild-type) reading frame, a result that would be expected only if each codon contained three nucleotides.

Confirmation that the coding ratio (nucleotides to amino acids) is indeed three has come from many sources. Considerable evidence favoring a triplet code evolved from studies using *in vitro* translation systems. The following observations were of major importance. (1) Trinucleotides were found sufficient to stimulate specific binding of aminoacyl-tRNAs to ribosomes. For example, 5′-UUC-3′ stimulates ribosomal binding of phenylalanyl-tRNA$_{\text{phe}}$. (2) Chemically synthesized RNA molecules, containing repeating dinucleotide sequences, directed the synthesis of copolymers with alternating amino acid sequences. Poly (UG)n$_1$, for example, when used as an artificial mRNA in an *in vitro* system, directed the synthesis of the repeating copolymer (cys-val)n$_2$. (3) Molecules with repeating trinucleotide sequences, on the other hand, directed the synthesis of a mixture of three homopolymers (initiation being random on such an mRNA in an *in vitro* system). Poly (UCG)n, for example, directed the synthesis of a mixture of polyserine, polyarginine, and polyvaline. Again, these results are only consistent with a triplet code.

Ultimately, the triplet nature of the code was definitively established by the results of correlated nucleic acid and protein sequencing (e.g., see Figs. 8.26 and 10.27).

DECIPHERING THE CODE

The deciphering of the genetic code — that is, determining (1) which codons specify which amino acids, (2) how many of the 64 possible codons are used, (3) how the code is punctuated, and (4) whether different species use the same or different codons — took place during the early 1960s and was one of the most exciting periods in the history of science.

The first major breakthrough came in 1961 when M. W. Nirenberg (1968 Nobel Prize recipient) and J. H. Matthaei and then S. Ochoa (1959 Nobel Prize recipient) and co-workers demonstrated that synthetic RNA molecules could be used as artificial mRNAs to direct *in vitro* protein synthesis. That is, when ribosomes, aminoacyl-tRNAs, and the soluble protein factors required for translation are purified free of natural mRNAs, these components can be combined *in vitro* and stimulated to synthesize polypeptides by the addition of chemically synthesized RNA molecules. If these synthetic mRNA molecules are of known composition, the composition of the polypeptides synthesized can be used to deduce which codons specify which amino acids.

The first codon assignment (UUU for phenylalanine) was made when Nirenberg and Matthaei demonstrated that **polyuridylic acid [poly U = (U)$_n$]** directed the synthesis of polyphenylalanine [(phenylalanine)$_n$]. Ochoa and others continued this approach using synthetic RNAs with random sequences of known nucleotide composition, such as 50 percent U and 50 percent G. The frequencies of the different triplets in such a random copolymer can be easily calculated. For example, the 50 percent U/50 percent G copolymer will contain 12.5 percent ($\frac{1}{2} \cdot \frac{1}{2} \cdot \frac{1}{2} = \frac{1}{8}$) of each of the eight possible codons: UUU, UUG, UGU, GUU, UGG, GUG, GGU, and GGG. These can then be compared with the amino acids incorporated (phenylalanine, leucine, cysteine, valine, tryptophan, and glycine) when this random copolymer is used in an *in vitro* protein-synthesizing system. By varying the composition, for example, to 75 percent U and 25 percent G, one can vary the relative frequencies of the eight codons and correlate them with the relative frequencies of the amino acids in the polypeptides synthesized. Such experiments provided a great deal of information about the nature of the code.

More definitive data were later obtained by H. G. Khorana using *in vitro* systems that were activated by synthetic mRNAs of known nucleotide sequence. Khorana's experiments permitted direct comparisons between nucleotide sequences and the amino acids incorporated in response to these se-

FIGURE 8.26

5′ leader ending in the coat protein initiation codon (AUG):

```
                                          Coat Protein initiation codon
                                                                 ⌢
· · · (G)· AUA· GAG· CCC· UCA· GGC· ACU· GUC· GAC· GGC· GCC· GGU· UGA· AGC· AUG·
```

Coat protein gene (codon · amino acid; residue numbers every fifth position):

```
GCU· UCU· AAC· UUU· ACU· CAG· UUC· GUU· CUC· GUC· GAC· AAU· GGC· GGA· ACU· GGC· GAC· GUC· ACU· GUC· GCC· CCA· AGC· AAC· UUC·
Ala  Ser  Asn  Phe  Thr  Gln  Phe  Val  Leu  Val  Asp  Asn  Gly  Gly  Thr  Gly  Asp  Val  Thr  Val  Ala  Pro  Ser  Asn  Phe
 1             5                   10                  15                  20                  25

GCU· AAC· GGG· GUC· GCU· GAA· UGG· AUC· AGC· UCU· AAC· UCG· CGU· UCA· CAG· GCU· UAC· AAA· GUA· ACC· UGU· AGC· GUU· CGU· CAG·
Ala  Asn  Gly  Val  Ala  Glu  Trp  Ile  Ser  Ser  Asn  Ser  Arg  Ser  Gln  Ala  Tyr  Lys  Val  Thr  Cys  Ser  Val  Arg  Gln
           30                  35                  40                  45                  50

AGC· UCU· GCG· CAG· AAU· CGC· AAA· UAC· ACC· AUC· AAA· GUC· GAG· GUG· CCU· AAA· GUG· GCA· ACC· CAG· ACU· GUU· GGU· GGU· GUA·
Ser  Ser  Ala  Gln  Asn  Arg  Lys  Tyr  Thr  Ile  Lys  Val  Glu  Val  Pro  Lys  Val  Ala  Thr  Gln  Thr  Val  Gly  Gly  Val
      55                  60                  65                  70                  75

GAG· CUU· CCU· GUA· GCC· GCA· UGG· CGU· UCG· UAU· UUA· AAU· AUG· GAA· CUA· ACC· AUU· CCA· AUU· UUC· GCU· ACG· AAU· UCC· GAC·
Glu  Leu  Pro  Val  Ala  Ala  Trp  Arg  Ser  Tyr  Leu  Asn  Met  Glu  Leu  Thr  Ile  Pro  Ile  Phe  Ala  Thr  Asn  Ser  Asp
      80                  85                  90                  95                  100

UGC· GAG· CUU· AUU· GUU· AAG· GCA· AUG· CAA· GGU· CUC· CUA· AAA· GAU· GGA· AAC· CCG· AUU· CCC· UCA· GCA· AUC· GCA· GCA· AAC·
Cys  Glu  Leu  Ile  Val  Lys  Ala  Met  Gln  Gly  Leu  Leu  Lys  Asp  Gly  Asn  Pro  Ile  Pro  Ser  Ala  Ile  Ala  Ala  Asn
      105                 110                 115                 120                 125

UCC· GGC· AUC· UAC· UAA· UAG·
Ser  Gly  Ile  Tyr
      129          Tandem ochre and amber termination codons
```

Intergenic region and replicase gene start:

```
                        Termination            Replicase
                        codon (UGA)            initiation codon (AUG)
                                                        ⌐
... UGA ...                        AUG· UCG· AAG· ACA· ACA· AAG· AAG· (U)
                                        Ser  Lys  Thr  Thr  Lys  Lys
                                         1              5
```

Correlated nucleotide sequence of the coat protein gene of the RNA bacteriophage MS2 and the amino acid sequence of the coat polypeptide (coat protein) that it specifies. The initial sequence of the MS2 replicase (RNA polymerase) gene and the correlated six amino-terminal amino acids are also shown. Untranslated intergenic sequences separate the genes. Translation proceeds from left to right as drawn, and from the coat gene to the replicase gene. Both polypeptides are initiated by ƒ-methionine at the AUG codons indicated. The methionine is cleaved off following (or during) translation, yielding the alanine terminus on the coat protein and the serine terminus on the replicase. The coat protein gene has two tandem "periods" (two termination codons) at the end, as though to make absolutely certain that translation terminates at this point. In addition, a third termination codon is located, in proper reading frame, seven base triplets from the second tandem termination. Each of the three termination codons is present once between the translated sequence of the coat gene and the translated sequence of replicase gene. Note that the amino acid sequence of this protein, synthesized in vivo, is precisely that predicted from the nucleotide sequence using the codon assignments presented in Table 8.1. (Data from W. Min Jou, G. Haegeman, Y. Ysebaert, and W. Fiers, *Nature* 237:82–88, 1972.)

quences. The ultimate "cracking" of the code occurred when trinucleotides were found to function as "mini-mRNAs," directing the specific binding of aminoacyl-tRNAs to ribosomes. By using all of the 64 possible trinucleotide sequences in such aminoacyl-tRNA binding experiments, it was possible to verify the codon assignments made from data of earlier experiments.

On the basis of extensive data accumulated over several years, the codon assignments shown in Table 8.1 became firmly established. Two important questions remained to be answered. (1) Are the assignments based on *in vitro* experiments valid *in vivo?* (2) Is the code **universal;** that is, do the codons specify the same amino acids in all organisms? Several lines of evidence now indicate that these codon assignments are correct for protein synthesis *in vivo* for most, if not all, species. When the amino acid substitutions that result from mutations induced with chemical mutagens with specific mutagenic effects (see Chapter 9) are determined by amino acid sequencing, the substitutions are almost always consistent with the codon assignments given in Table 8.1 and the known effect of the mutagen. More convincingly, when the nucleotide sequences of genes or of mRNAs are determined and compared with the amino acid sequences of the polypeptides coded for by those genes or mRNAs, the observed correlations are always found to be those predicted from the accepted codon assignments (Table 8.1). This can be illustrated by comparing the nucleotide sequence of the gene coding for the protein coat or capsid of bacteriophage MS2 with the amino acid sequence of the capsid polypeptide (Fig. 8.26). Phage MS2 stores its genetic information in RNA (like TMV virus; see Chapter 5, pp. 85–87). Its chromosome is the equivalent to an mRNA molecule in organisms with DNA genomes. (Also, see Chapter 10, Fig. 10.27).

DEGENERACY AND WOBBLE

All of the amino acids except methionine and tryptophan are specified by more than one codon (Table 8.1). Three amino acids, leucine, serine, and arginine, are each specified by six different codons. Isoleucine has three codons. The other amino acids

each have either two or four codons. The occurrence of more than one codon per amino acid is called **degeneracy** (though the usual connotations of the term are hardly appropriate). The degeneracy in the genetic code is not at random; instead, it is highly ordered. Usually, the multiple codons specifying an amino acid differ by only one base, the third or 3′ base of the codon. The degeneracy is primarily of two types. (1) Partial degeneracy occurs when the third base may be either one of the two pyrimidines (U and C) or, alternatively, either one of the two purines (A and G). With partial degeneracy, changing the third base from a purine to a pyrimidine, or vice versa, will change the amino acid specified by the codon. (2) In the case of complete degeneracy, any of the four bases may be present at the third position in the codon, and the codon will still specify the same amino acid. For example, valine is specified by GUU, GUC, GUA, and GUG (Table 8.1).

It has been speculated that the order in the genetic code has evolved as a way of minimizing mutational lethality. Many base substitutions at the third position of codons do not change the amino acid specified by the codon. Moreover, amino acids with similar chemical properties (such as leucine, isoleucine, and valine) have codons that differ from each other by only one base. Thus, many single base-pair substitutions will result in the substitution of one amino acid for another amino acid with very similar chemical properties (e.g., valine for isoleucine). In most cases, such substitutions will not result in inactive gene products; again, this minimizes the effects of mutations.

Because of the degeneracy of the genetic code, there must either be several different tRNAs that recognize the different codons specifying a given amino acid, or the anticodon of a given tRNA must be able to base-pair with several different codons. Actually, both of the above occur. Several tRNAs exist for certain amino acids, and some tRNAs recognize more than one codon. The hydrogen bonding between the bases in the anticodon of tRNA and the codon of mRNA appears to follow strict base-pairing rules (i.e., be "tight") only for the first two bases of the codon. The base-pairing involving the third base of the codon is apparently less stringent,

TABLE 8.1 The Genetic Code[a]

Second letter

		U	C	A	G	
First (5') letter	**U**	UUU⎫ UUC⎭ Phe UUA⎫ UUG⎭ Leu	UCU⎫ UCC⎪ UCA⎬ Ser UCG⎭	UAU⎫ UAC⎭ Tyr UAA *Ochre* (terminator) UAG *Amber* (terminator)	UGU⎫ UGC⎭ Cys UGA *Opal* (terminator) UGG Tryp	U C A G
	C	CUU⎫ CUC⎪ CUA⎬ Leu CUG⎭	CCU⎫ CCC⎪ CCA⎬ Pro CCG⎭	CAU⎫ CAC⎭ His CAA⎫ CAG⎭ GluN	CGU⎫ CGC⎪ CGA⎬ Arg CGG⎭	U C A G
	A	AUU⎫ AUC⎬ Ileu AUA⎭ AUG Met (initiator)	ACU⎫ ACC⎪ ACA⎬ Thr ACG⎭	AAU⎫ AAC⎭ AspN AAA⎫ AAG⎭ Lys	AGU⎫ AGC⎭ Ser AGA⎫ AGG⎭ Arg	U C A G
	G	GUU⎫ GUC⎪ GUA⎬ Val GUG⎭ (initiator)	GCU⎫ GCC⎪ GCA⎬ Ala GCG⎭	GAU⎫ GAC⎭ Asp GAA⎫ GAG⎭ Glu	GGU⎫ GGC⎪ GGA⎬ Gly GGG⎭	U C A G

Third (3') letter

[a] Each triplet nucleotide sequence or codon refers to the nucleotide sequence in **mRNA** (not DNA) that specifies the incorporation of the indicated amino acid or polypeptide chain termination.

allowing what Crick has called **wobble** at this site.

On the basis of molecular distances and steric (three-dimensional structure) considerations, Crick proposed that wobble would allow several types, but not all types, of base-pairing at the third codon base in the codon–anticodon interaction. His proposal has since been strongly supported by experimental data. Table 8.2 shows the base-pairing predicted by the wobble hypothesis. It necessitates that there be at least two tRNAs for each amino acid whose

codons exhibit complete degeneracy at the third position. This has indeed been found to be true. The wobble hypothesis predicted the occurrence of three tRNAs for the six serine codons. Three serine tRNAs have now been characterized: (1) tRNAser$_1$ (anticodon AGG) binds to codons UCU and UCC, (2) tRNAser$_2$ (anticodon AGU) binds to codons UCA and UCG, and (3) tRNAser$_3$ (anticodon UCG) binds to codons AGU and AGC. These specificities were verified by the trinucleotide-stimulated bind-

TABLE 8.2 Base-pairing Between the 5′ Base of the Anticodon of tRNAs and the 3′ Base of Codons of mRNAs According to the Wobble Hypothesis

BASE IN ANTICODON	BASE IN CODON
G	U or C
C	G
A	U
U	A or G
I	A, U, or C

ing of purified aminoacyl-tRNAs to ribosomes *in vitro.*

Finally, several tRNAs contain the base inosine (produced by post-transcriptional enzymatic modification). Crick's wobble hypothesis predicted that inosine could pair (at the wobble position) with adenine, uracil, or cytosine (in the codon). In fact, purified alanyl-tRNA containing inosine (I) at the 5′ position of the anticodon (Fig. 8.15) binds to ribosomes activated with GCU, GCC, or GCA trinucleotides. The same result has been obtained with other purified tRNAs with inosine at the 5′ position of the anticodon. The wobble hypothesis thus fits several observations; whether it is entirely accurate remains unknown.

INITIATION AND TERMINATION CODONS

The genetic code also provides for punctuation of genetic information at the level of translation. Three codons, UAA, UAG, and UGA, specify polypeptide chain termination. These codons are recognized by protein release factors, rather than by tRNAs. One of these proteins, designated RF-1, is apparently specific for UAA and UAG. The other, RF-2, causes termination at UAA and UGA codons. Two codons, AUG and GUG, are recognized by the initiator tRNA, $tRNA_f^{Met}$, but apparently only when they follow an appropriate nucleotide sequence in the leader segment of an mRNA molecule. At internal positions, AUG is recognized by $tRNA^{Met}$, and GUG is recognized by a valine tRNA. In the case of the initiation codons AUG and GUG and $tRNA_f^{Met}$, the wobble base appears to be the first or 5′ base of the codon. Since wobble at the first base is unique to initiation, it may be related to base-pairing at the *P* site rather than at the *A* site on the ribosome.

UNIVERSALITY OF THE CODE

A vast amount of data is now available from *in vitro* studies, from amino acid replacements due to mutations and from correlated nucleic acid and polypeptide sequencing—all suggesting that the genetic code is the same or very nearly the same in all organisms. These data (e.g., see the human hemoglobin substitutions, Fig. 9.8; the correlated nucleotide and amino acid sequences in the overlapping genes of the DNA bacteriophage ΦX174, Fig. 10.27; and of the RNA bacteriophage MS2, Fig. 8.26) all indicate that the genetic code is largely **universal.**

The major exception to the universality of the code occurs in mitochondria of humans, yeast, and several other species, where UGA is a tryptophan codon. UGA is a termination codon in nonmitochondrial systems. Also, in yeast mitochondria, CUA specifies threonine instead of the usual leucine, and, in mammalian mitochondria, AUA specifies methionine instead of the usual isoleucine. Excluding these exceptions, the code appears to be universal.

SUPPRESSOR MUTATIONS PRODUCING tRNAs WITH ALTERED CODON RECOGNITION

The nonmitochondrial genetic code is not absolutely universal, at least not in the most restrictive sense of the word, because minor variations in codon recognition and translation are well documented in a few cases. In *E. coli* and yeast, for example, mutations occur in tRNA genes, resulting in altered codon recognition by the anticodons of tRNAs. These mutations were initially detected as suppressor mutations, mutations that suppressed the effects of other mutations. They were subsequently shown to be located in tRNA genes and to change the specificity of codon recognition by the tRNAs.

The best-known examples of suppressor mutations that alter tRNA specificity are the suppressors of mutations that produce UAG chain-termination triplets within the coding sequences of structural genes. Such mutations, called *amber* mutations, result in the synthesis of amino-terminal fragments of the polypeptides specified by the affected genes.

Mutations that produce chain-termination triplets within genes have come to be known as **nonsense** mutations, in contrast to **missense** mutations, which change a triplet to another triplet specifying a different amino acid. Missense mutations thus result in complete polypeptides, but with an amino acid substitution (such as the mutant hemoglobins described in Fig. 9.8). Nonsense mutations result in fragments of the polypeptides, whose lengths depend on the position of the mutation within the gene (see Fig. 8.28).

Nonsense mutations frequently result from single base-pair substitutions (as shown in Fig. 8.27). The polypeptide fragments produced from genes containing nonsense mutations are almost always totally nonfunctional.

Suppression of nonsense mutations has been shown to result from mutations in tRNA genes that cause the mutant tRNAs to recognize the nonsense (UAG, UAA, or UGA) codons, albeit with varying efficiencies. These mutant tRNAs are referred to as suppressor tRNAs. When one of the **amber** (UAG) suppressor tRNAs, the one resulting from the **amber su3 mutation,** was sequenced, it was found to have an altered anticodon. This particular *amber* suppressor mutation occurs in the $tRNA_{tyr2}$ gene (specifying one of the two tyrosine tRNAs in *E. coli*). The anticodon of the wild-type (nonsuppressor) $tRNA_{tyr2}$ was shown to be 5'-G'UA-3' (where G' is a derivative of guanine). The anticodon of the mutant (suppressor) $tRNA_{tyr2}$ is 5'-CUA-3'. This mutant anticodon is thus able to base-pair with the 5'-UAG-3' *amber* codon (pairing with opposite polarity, of course); that is,

3'-AUC-5' (anticodon)
5'-UAG-3' (codon)

The suppressor tRNAs thus permit complete polypeptides to be synthesized from mRNAs containing nonsense codons. These polypeptides will be functional as long as the amino acid inserted by the suppressor tRNA is acceptable at that position, yielding a functional gene product (Fig. 8.27c).

E. coli strains containing *amber*-suppressor genes (and tRNAs) exhibit growth rates comparable to strains without suppressor genes. This might not be expected if the UAG codon is frequently used for the normal termination of translation at the ends of coding sequences of mRNAs. In suppressor strains, one would expect translation occasionally (how frequently depends on the efficiency of the particular suppressor) to continue through into normally untranslated intergenic spacer sequences and even into other coding sequences located translationally distal on the mRNAs. This would generate long "poly-proteins," which one would not expect to be very functional. Why suppressor strains do not show deleterious effects of such "read-through" is not understood. Perhaps most genes are like the phage MS2 coat gene (Fig. 8.26) and terminate with two or more different translation-termination codons. The nucleotide sequences of several genes, however, reveal the presence of single termination codons. The apparent lack of deleterious effects of suppressor mutations thus remains an enigma.

COLINEARITY OF GENE AND POLYPEPTIDE

The genetic information is stored in linear sequences of nucleotide-pairs in DNA (or nucleotides in RNA, in some cases). Transcription and translation convert this genetic information into polypeptides (linear sequences of amino acids), which function as the key intermediaries in the genetic control of phenotype.

We now know that the amino acid sequences of the polypeptides and the nucleotide-pair sequences of the genes coding for these polypeptides are **colinear.** That is, the first three base-pairs of a gene specify the first amino acid of the polypeptide, the next three base-pairs (4 to 6) specify the second amino acid, and so on, in a colinear fashion. The first strong evidence for colinearity of genes and polypeptides resulted from studies on one of the two polypeptides in tryptophan synthetase of *E. coli*. C. Yanofsky and colleagues demonstrated that there was a perfect correlation between the map positions of mutations in the tryptophan synthetase *A* gene and the positions of the resultant amino acid substitutions in the tryptophan synthetase α polypeptide. Yanofsky's results are summarized in Fig. 10.18.

At about the same time, A. Sarabhai and asso-

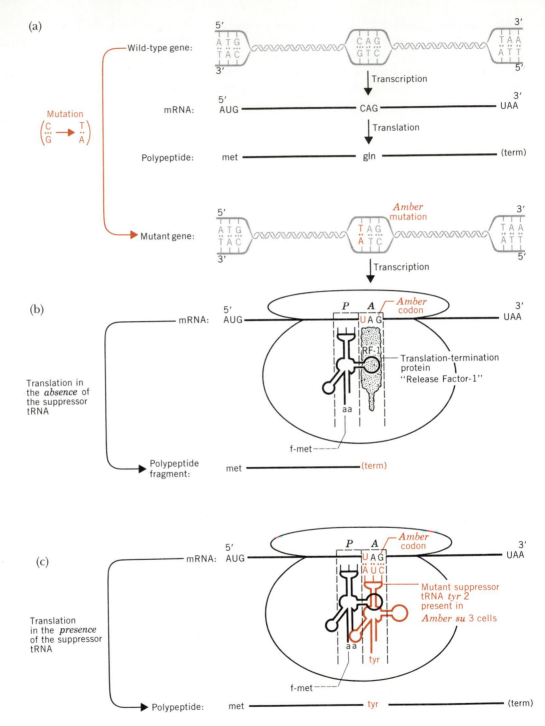

FIGURE 8.27

Schematic illustration showing the formation of an amber *(UAG) chain-termination mutation (a to b) and its effect in (b) the absence of a suppressor tRNA and (c) the presence of a suppressor tRNA. The* amber *mutation shown changes a CAG mRNA codon that specifies glutamine to a UAG chain-termination codon. In the absence of a suppressor mutation resulting in a suppressor tRNA in the cell, the* amber *mutation causes the synthesis of only a fragment of the polypeptide gene product. When the presence of the suppressor mutation yields a suppressor tRNA (c), a modified tyrosine tRNA mediates the insertion of tyrosine into the polypeptide at the site specified by the UAG codon. Other* amber *suppressors insert serine, lysine, leucine, or glutamine in response to UAG codons.* Amber *suppressors will yield a functional product of a mutant gene containing a UAG codon only if the amino acid inserted by the suppressor is an acceptable substitute for the original ("wild-type") amino acid present at that position. For example, the tyrosine inserted in (c) replaces a glutamine at that position in the wild-type polypeptide (a). The polypeptide containing tyrosine may or may not be functional. Suppression will, of course, be observed only if the tyrosine-containing polypeptide is active.*

ciates demonstrated a similar colinearity between the positions of mutations in the gene of bacteriophage T4 which codes for the major structural protein of the phage head and the positions in the polypeptide affected by the mutations. Sarabhai and colleagues studied *amber* (UAG chain-termination) mutations and demonstrated a direct correlation between the length of the polypeptide fragment produced and the position of the mutation within the gene (Fig. 8.28).

Definitive evidence for colinearity has been provided by correlated nucleic acid sequencing and polypeptide sequencing studies. These studies (see Figs. 8.26 and 10.27) show a direct colinearity between the nucleotide sequences of the genes and the amino acid sequences of the polypeptide gene products.

In eukaryotes, the data available to date also support colinearity, **but the linear sequence of nucleotide-pairs in a gene that specifies a colinear polypeptide may not always consist of contiguous nucleotide-pairs.** Instead, there frequently are **noncoding intron sequences intervening between coding sequences** (see Chapter 10, pp. 361–365). This does not violate the concept of colinearity. It only demonstrates that the sequences of base-

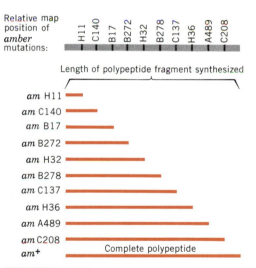

FIGURE 8.28

Colinearity between the map positions of amber *(UAG chain-termination) mutations in gene 23 (the gene coding for the major structural protein of the head) of bacteriophage T4 and the positions of the resulting polypeptide chain-terminations in the mutant gene products. The wild-type (am⁺) polypeptide is represented at the bottom for comparison. (Based on the data of A. S. Sarabhai, A. O. W. Stretton, and S. Brenner,* Nature *201:13–17, 1964.)*

pair triplets in the gene specifying mRNA codons and amino acids in the colinear polypeptide are **not** always **uninterrupted.** In one case, F. Sherman and colleagues have demonstrated an uninterrupted co-linear relationship between the map positions of mutations in the *CYC1* gene of yeast and the amino acid substitutions in the mutant forms of iso-1-cy-tochrome c, the polypeptide specified by the *CYC1* gene. Whether colinearity is a universal feature of gene-polypeptide relationships is still very much an open question.

SUMMARY

The pathway of information flow by which a gene exerts its effect on the phenotype of an organism is often very complex. The main features of the first two steps in this pathway, **transcription** and **trans-lation,** are now quite well understood. Transcrip-tion involves the synthesis of a **messenger RNA** or **pre-messenger RNA** molecule, which functions as an intermediary in protein synthesis, using one of the two strands of DNA in the gene as a template. Transcription is catalyzed by a complex, multimeric enzyme, **DNA-dependent RNA polymerase.** In eu-karyotes, the primary transcripts usually undergo several types of processing. Processing frequently includes excision of specific segments of the primary transcript, addition of a 5′-cap and a 3′-poly-A tail, and association with specific proteins. The messen-ger RNA produced is then **translated,** according to the specifications of the **genetic code,** into the se-quence of amino acids in the **polypeptide** gene product. Translation takes place on complex, mac-romolecular structures called **ribosomes** that are located in the cytoplasm of cells. Each ribosome consists of two subunits, which together contain 3–5 different RNA molecules and about 50 differ-ent proteins. The exact composition varies from species to species. Translation requires the partici-pation of 40–60 small RNA molecules called **transfer RNAs.** Each transfer RNA is activated for protein synthesis by the covalent attachment of a specific amino acid. This attachment is catalyzed by an enzyme called **aminoacyl-tRNA synthetase.** There is at least one specific tRNA and one specific aminoacyl-tRNA synthetase for each of the 20 amino acids commonly found in natural proteins. Translation also requires several soluble proteins called **initiation, elongation, and termination fac-tors.** The result of the transcription of a gene and the subsequent translation of the mRNA produced during transcription is the production of a polypep-tide whose amino acid sequence is **colinear** with the sequence of nucleotide pairs in the gene. In eukary-otes, this sequence of nucleotide pairs in the gene may be interrupted by **noncoding spacer sequences** that intervene between coding sequences.

The **genetic code** is a **triplet code,** a sequence of three nucleotides making up each **codon.** The code is **degenerate** in the sense that all 64 possible triplet nucleotide sequences are used and several different codons frequently code for the same amino acid. Each of the 61 amino acid-specifying codons in mRNA is capable of base-pairing with a three-nucleotide sequence present at the **anticodon** site of one or more tRNAs. The base-pairing be-tween the third (3′) base of the codon and the 5′ base of the anticodon does not follow the normal strict base-pairing rules; instead, there is **"wobble"** at this site, permitting base-pairs to form other than the usual four base-pairs. Thus, the anticodon of a single tRNA may recognize one, two, or three dif-ferent codons. Two codons are used for polypeptide chain initiation, and three codons specify polypep-tide chain termination. The genetic code is nearly **universal;** the codons appear to have almost the same meaning in all species.

The polypeptides produced by transcription and translation make up the structural proteins and enzymes that control metabolic processes in cells.

Blocks in metabolic processes resulting from genetic defects frequently have severe conse-quences. When the molecular bases of such congeni-tal diseases are known, the symptoms of the disease can sometimes be relieved by appropriate treat-ment.

REFERENCES

Breathnach, R., and P. Chambon. 1981. "Organization and expression of eucaryotic split genes coding for proteins." *Ann. Rev. Biochem.* 50:349–383.

Brown, D. D. 1981. "Gene expression in eukaryotes." *Science* 211:667–674.

Busch, H., R. Reddy, L. Rothblum, and Y. C. Choi. 1982. "SnRNAs, SnRNPs, and RNA processing." *Ann. Rev. Biochem.* 51:617–654.

Crick, F. H. C. 1966. "Codon-anticodon pairing; the wobble hypothesis." *J. Mol. Biol.* 19:548–555.

———. 1966. "The genetic code: III." *Sci. Amer.* 215:(4)55–62.

Davies, R. W., R. B. Waring, J. A. Ray, T. A. Brown, and C. Scazzocchio. 1982. "Making ends meet: a model for RNA splicing in fungal mitochondria." *Nature* 300:719–724.

Fiers, W., R. Contreras, F. Duerinck, G. Haegeman, D. Iserentant, J. Merregaert, W. Min Jou, F. Molemans, A. Raeymekers, A. Van den Berghe, G. Volckaert, and M. Ysebaert. 1976. "Complete nucleotide sequence of bacteriophage MS2 RNA: primary and secondary structure of the replicase gene." *Nature* 260:500–507.

Freifelder, D. 1983. *Molecular biology, a comprehensive introduction to prokaryotes and eukaryotes.* Science Books International, Boston.

Garen, A. 1968. "Sense and nonsense in the genetic code." *Science* 160:149–159.

Gold, L., D. Pribnow, T. Schneider, S. Shinedling, B. S. Singer, and G. Stormo. 1981. "Translational initiation in prokaryotes." *Ann. Rev. Microbiol.* 35:365–403.

Goodman, H. M., J. Abelson, A. Dandry, S. Brenner, and J. D. Smith. 1968. "*Amber* suppression: a nucleotide change in the anticodon of a tyrosine transfer RNA." *Nature* 217:1019–1024.

Hall, B. D., S. G. Clarkson, and G. Tocchini-Valentini. 1982. "Transcription initiation of eukaryotic transfer RNA genes." *Cell* 29:3–5.

Holley, R. W. 1966. "The nucleotide sequence of a nucleic acid." *Sci. Amer.* 214(2):30–39.

Kruger, K., P. J. Grabowski, A. J. Zaug, J. Sands, D. E. Gottschling, and T. R. Cech. 1982. "Self-splicing RNA: autoexcision and autocyclization of the ribosomal RNA intervening sequence of *Tetrahymena.*" *Cell* 31:147–157.

Lewin, B. 1974. *Gene expression-1, bacterial genomes.* John Wiley & Sons, New York.

———. 1977. *Gene expression-3, plasmids and phages.* John Wiley & Sons, New York.

———. 1980. *Gene expression-2, eucaryotic genomes,* 2nd ed. John Wiley & Sons, New York.

———. 1983. *Genes.* John Wiley & Sons, New York.

Maitra, U., E. A. Stringer, and A. Chaudhuri. 1982. "Initiation factors in protein synthesis." *Ann. Rev. Biochem.* 51:869–900.

Nirenberg, M. W. 1963. "The genetic code: II." *Sci. Amer.* 208(3):80–94.

Nirenberg, M. W., and J. H. Matthaei. 1961. "The dependence of cell-free protein synthesis in *E. coli* upon naturally occurring or synthetic polyribonucleotides." *Proc. Natl. Acad. Sci., U.S.* 47:1588–1602.

Nomura, M. (ed). 1974. *Ribosomes.* Cold Spring Harbor Laboratory Press, Cold Spring Harbor, New York.

Pederson, T. 1981. "Messenger RNA biosynthesis and nuclear structure." *Amer. Scientist* 69:76–84.

Sarabhai, A. S., A. O. W. Stretton, and S. Brenner. 1964. "Co-linearity of the gene with the polypeptide chain." *Nature* 201:13–17.

Sharp, P. A. 1981. "Speculations on RNA splicing." *Cell* 23:643–646.

"The genetic code." 1966. *Cold Spring Harbor Symp. Quant. Biol.,* Vol. 31. Cold Spring Harbor Laboratory Press, Cold Spring Harbor, New York.

"The mechanism of protein synthesis." 1970. *Cold Spring Harbor Symp. Quant. Biol.,* Vol. 34. Cold Spring Harbor Laboratory Press, Cold Spring Harbor, New York.

Watson, J. D. 1976. *Molecular biology of the gene,* 3rd ed. Benjamin, Menlo Park, California.

Weissbach, H., and S. Ochoa. 1976. "Soluble factors required for eukaryotic protein synthesis." *Ann. Rev. Biochem.* 45:191–216.

Woese, C. R. 1967. *The genetic code, the molecular basis for genetic expression.* Harper & Row, New York.

Yanofsky, C. 1967. "Gene structure and protein structure." *Sci. Amer.* 216(5):80–94.

PROBLEMS AND QUESTIONS

8.1. Distinguish between DNA and RNA (a) chemically, (b) functionally, and (c) by location in the cell.

8.2. What bases on the mRNA transcript would represent the following DNA sequence: 5'-TGCAGACA-3'?

8.3. What bases in the transcribed strand of DNA would give rise to the following mRNA base sequence: 5'-CUGAU-3'?

8.4. (a) How can phenocopies be used to study gene action? (b) What values and limitations do they have for investigations of this kind? (c) How can an investigator determine whether an altered phenotype is due to a mutation or a phenocopy?

8.5. What significance may phenocopies have in medical genetics?

8.6. A series of crosses in a particular controlled environment, from which all progeny were known to carry a dominant autosomal gene *A* (causing an eye abnormality), resulted in 1400 abnormal and 600 normal flies. What is the penetrance of gene *A* in this experiment?

8.7. From what evidence was the messenger RNA hypothesis established?

8.8. In a general way, describe the molecular organization of proteins and distinguish proteins from DNA, chemically and functionally. Why is the synthesis of proteins of particular interest to geneticists?

8.9. At what different locations in the cell may protein synthesis occur?

8.10. Characterize ribosomes in general as to size, location, function, and macromolecular composition.

8.11. (a) Where in the cells of higher organisms do ribosomes originate? (b) Where in the cells are ribosomes most active in protein synthesis?

8.12. Identify three different major kinds of RNA and give the principal locations, characteristics, and functions of each in the living cell.

8.13. (a) How is messenger RNA related to polysome formation? (b) How does rRNA differ from mRNA and tRNA in specificity? (c) How does the tRNA molecule differ from that of DNA and mRNA in size and helical arrangement?

8.14. Outline the process of activation of amino acids.

8.15. (a) How was the genetic code first decoded? (b) What refinements have since been incorporated in the technique?

8.16. In what sense and to what extent is the genetic code (a) degenerate and (b) universal?

8.17. What evidence supports the hypothesis of colinearity between the nucleotide sequence in a gene and the amino acid sequence in a polypeptide?

8.18. Why is colinearity between codons and polypeptides significant?

8.19. Draw an analogy between the processes of transcription and translation and the process of building a house.

8.20. The thymine analog 5-bromouracil is a chemical mutagen that induces single base-pair substitutions in DNA called transitions (substitutions of one purine for another purine and one pyrimidine for another pyrimidine; see Chapter 9). Using the known nature of the genetic code (Table 8.1), which of the following amino acid substitutions should you expect to be induced by 5-bromouracil with the highest frequency: (a) Met → Val; (b) Met → Leu; (c) Lys → Thr; (d) Lys → Gln; (e) Pro → Arg; or (f) Pro → Gln? Why?

8.21. Using the information referred to in Problem 8.20 above, would you expect 5-bromouracil to induce a higher frequency of His → Arg or His → Pro substitutions? Why?

8.22. How is transcription (a) initiated and (b) terminated? How is translation (c) initiated and (d) terminated?

8.23. If the average molecular weight of an amino acid is assumed to be 100 daltons, about how many nucleotides will be present in an mRNA coding sequence specifying a single polypeptide with a molecular weight of 27,000 daltons?

8.24. How can a mutation in a tRNA gene lead to the incorrect translation of a specific codon? Could mutations in any genes other than tRNA genes lead to incorrect or ambiguous translation of a particular codon? If so, in genes coding for what?

8.25. What base-pair sequence in a segment of a gene in *Drosophila melanogaster* would code for the dipeptide sequence (NH$_2$) Met-Trp (COOH)?

8.26. If you were to (1) purify cysteine transfer RNA and charge it with labeled cysteine (i.e., activate it by attaching ^{3}H-labeled cysteine), (2) use Raney nickel (a highly reducing nickel powder) to convert the cysteine to alanine still attached to the cysteyl-specific transfer RNA, and (3) place the alanine-charged cysteyl transfer RNA into an *in vitro* protein-synthesizing system activated with poly UG templates that normally stimulate the incorporation of cysteine **but not alanine** into polypeptide chains (i.e., when you use the normal alanine-charged alanyl and cysteine-charged cysteyl transfer RNAs), what result would you expect?

8.27. A. Garen extensively studied a particular nonsense (chain termination) mutation in the alkaline phosphatase gene of *E. coli*. This mutation resulted in the termination of the alkaline phosphatase polypeptide chain at a position where the amino acid tryptophan occurred in the wild-type polypeptide. Garen induced revertants (in this case, mutations altering the same codon) of this mutant with chemical mutagens that induced single base-pair substitutions (see Chapter 9) and sequenced the mutant polypeptides. Seven different types of revertants were found, each with a different amino acid at the tryptophan position of the wild-type polypeptide (termination position of the mutant polypeptide fragment). The amino acids present at this position in the various revertants included tryptophan, serine, tyrosine, leucine, glutamic acid, glutamine, and lysine. Was the nonsense mutation studied by Garen an *amber* (UAG), an *ochre* (UAA), or an *opal* (UGA) nonsense mutation? Explain the basis of your deduction.

8.28. The goal of many genetic engineering experiments is to place eukaryotic genes into bacteria and to have the bacteria synthesize the eukaryotic gene-products, preferably in large amounts. Based on what is known about gene expression in bacteria and eukaryotes, what obstacles might have to be overcome to achieve this goal?

(Also see Chapter 9, Problems 9.28–9.35.)

NINE

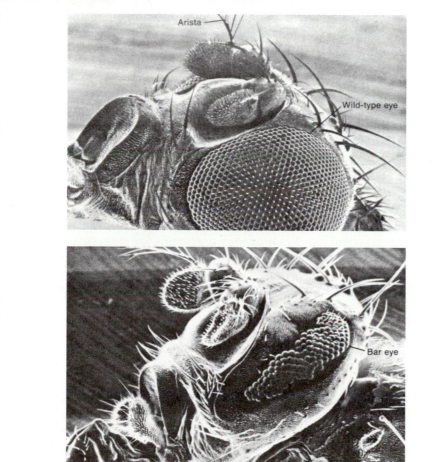

Scanning electron micrographs of *Drosophila melanogaster* showing the wild-type eye (top) and the bar eye (below) resulting from a dominant mutation on the X chromosome. (From D. T. Suzuki, A. J. F. Griffiths, R. C. Lewontin, *An Introduction to Genetic Analysis*, second edition, W. H. Freeman and Company, 1981.)

MUTATION

Inheritance is based on genes that are faithfully transmitted from parents to offspring during reproduction (see Chapter 2). In higher organisms, these genes are located in chromosomes, which are duplicated and passed on to the progeny through the gametes during sexual reproduction (see Chapter 3). Structurally, these genes are almost always DNA, encoding their information in sequences of base-pairs that are accurately duplicated during semiconservative replication. The DNA polymerases that catalyze the replication of DNA contain built-in $3' \rightarrow 5'$ exonuclease activities to proofread the progeny DNA molecules and correct mistakes made during the initial polymerization reaction (see Chapter 5). Mechanisms have thus evolved to facilitate the faithful transmission of genetic information from generation to generation. Nevertheless, "mistakes" or changes in the genetic material do occur. **Such sudden, heritable changes in the genetic material are called mutations.**

The term **mutation** refers both to the **change in the genetic material** and to the **process by which the change occurs.** An organism exhibiting a novel phenotype as a result of the presence of a mutation is referred to as a **mutant.** Used in its broad historical sense, mutation refers to any sudden, heritable change in the genotype of an organism not explainable by recombination of preexisting genetic variability (see Chapters 6 and 7). Such genotypic changes include changes in chromosome number (**euploidy** and **aneuploidy;** see Chapter 13), gross changes in the structure of chromosomes (**chromosome aberrations;** see Chapter 12), and changes in individual genes. The term mutation is now frequently used in a narrow sense to refer only to changes in individual genes (excluding changes in chromosome number or structure). This chapter will deal with mutation in this narrower sense.

Many mutations involve changes in single base-pairs, the substitution of one base-pair for another or the duplication or deletion of single base-pairs. Such mutations are referred to as **point mutations.**

Mutation is the ultimate source of all genetic variation; it provides the raw material for evolution. Recombination (independent assortment plus recombination of genetic variability present in individual chromosomes; see Chapters 6, 7, and 10) merely rearranges this genetic variability into new combinations, and natural (or artificial) selection simply preserves the combinations best adapted to the existing environmental conditions (or desired). **Without mutation, all genes would exist in only one form. Alleles would not exist, and, thus, genetic analysis would not be possible.** Most important, **organisms would not be able to evolve and adapt to environmental changes.** Mutation, then, is an important phenomenon. Some level of mutation is essential to provide new genetic variability to allow organisms to adapt to new environments. At the same time, if mutations occurred too frequently, they would totally disrupt the transmission of genetic information from generation to generation. What, then, is an appropriate level of mutation? This complex question will be returned to later in this chapter.

SPONTANEOUS VERSUS INDUCED MUTATION

Spontaneous mutations are those that occur without a known cause. They may be truly spontaneous, resulting from an inherent low level of metabolic errors, that is, mistakes during DNA replication, or they may actually be caused by mutagenic agents present in the environment. **Induced mutations** are those resulting from exposure of organisms to mutagenic agents such as ionizing irradiation, ultraviolet light, or various chemicals that react with DNA (or RNA, in RNA viruses). Operationally, it is impossible to prove that a particular mutation occurred spontaneously or that a specific mutation was induced by a mutagenic agent. The potential mutagenicity of both human-made and naturally occurring chemicals is only beginning to be assessed. Solar radiations are one cause of mutation, one that life can hardly do without. Fortunately, the ultraviolet rays in sunlight are of very low energy and do not penetrate tissue far enough to expose the gonads or germ cells of higher organisms to possible mutagenic effects. Ultraviolet light is mutagenic only to the skin cells of humans. This is probably of only little comfort, however, to individuals with the inherited disease xeroderma pigmentosum, who often develop severe skin cancers following exposure to sunlight (see pp. 296–297). Thus, many, if not most, of the apparently "spontaneous" mutations may actually be induced by physical and chemical agents.

On the basis of chemical considerations, the intrinsic mistake frequency in nucleotide selection during DNA polymerization has been estimated to be about 10^{-5}. One cycle of proofreading by the $3' \rightarrow 5'$ exonuclease activity of DNA polymerases, assuming the same degree of accuracy in replacing the mismatched bases, would reduce the intrinsic mistake frequency to about $10^{-10}(10^{-5} \times 10^{-5})$ errors per incorporated nucleotide. Additional repair mechanisms may reduce the intrinsic error frequency even further. Nevertheless, it seems clear that some minimal intrinsic error frequency does exist, and that this inherent frequency is subject to both genetic and environmental influences.

One cannot, therefore, distinguish between "spontaneous" and "induced" mutations when considering individual mutations. Such distinctions must be restricted to the population level. If the mutation rate is increased a hundredfold by treatment of a population with a mutagen, then an average of 99 out of every 100 mutations present in the population will have been induced by the mutagen. Valid comparisons between spontaneous and induced mutations can thus be made statistically by comparing populations exposed to a mutagenic agent with control populations that have not been exposed to the mutagenic agent.

Spontaneous mutations occur infrequently, although the observed frequencies vary from gene to gene and from organism to organism. Measurements of spontaneous forward mutation frequencies for various genes of phage and bacteria range from about 10^{-8} to 10^{-10} detectable mutations per nucleotide pair per generation. For eukaryotes, estimates of forward mutation rates range from about 10^{-7} to 10^{-9} detectable mutations per nucleotide pair per generation (considering only those genes for which extensive data are available). (In comparing mutation rates **per nucleotide** with mutation rates **per gene,** the average gene is usually assumed to be 1000 nucleotide pairs in length.)

Treatment with mutagenic agents can increase mutation frequencies by orders of magnitude. The mutation frequency **per gene** in bacteria and viruses, for example, can easily be increased to over 1 percent by treatment with potent chemical mutagens. That is, over 1 percent of the genes of the treated organisms will contain a mutation, or, stated differently, over 1 percent of the individual phage or bacteria in the population will have a mutation in a given gene.

MUTATION: RANDOM RATHER THAN DIRECTED BY THE ENVIRONMENT

Mouse populations in many cities are no longer affected by the anticoagulants that have traditionally been used as rodent poisons. Many cockroach populations are insensitive to Chlordane, the poison used to control them in the 1950s. Housefly populations often exhibit high levels of resistance to insecticides like DDT. More and more pathogenic microorganisms are becoming resistant to antibiotics such as penicillin and streptomycin developed by modern medicine to control them. The introduction of these pesticides and antibiotics by humans produced a new environment for the organisms involved. These organisms responded to the imposed environmental changes by evolving to forms resistant to these chemicals. Mutations producing resistance to these pesticides and antibiotics occurred, and the mutant organisms were at a large selective advantage in environments where these agents were present. The sensitive organisms were killed, and the mutants multiplied to produce new resistant populations. Many such cases of evolution via mutation and natural selection are now well documented.

The examples mentioned above raise a basic question about the nature of mutation. Is mutation a purely random event with the environmental stress merely preserving preexisting mutations? Or is mutation directed by the environmental stress? As a specific example, consider a population of bacteria such as *E. coli* that is isolated from an environment where streptomycin has not been present. Upon exposure of the population of bacteria to streptomycin, the vast majority of them will be killed. If the population is large enough, however, it will soon be repopulated with bacteria, all of which are resistant

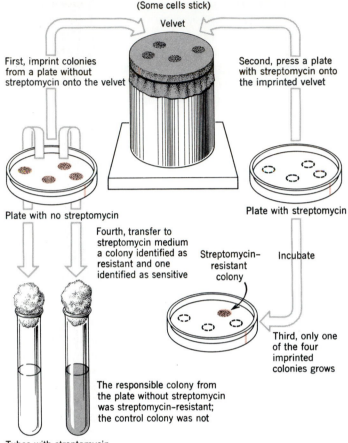

(Some cells stick)

Velvet

First, imprint colonies from a plate without streptomycin onto the velvet

Second, press a plate with streptomycin onto the imprinted velvet

Plate with no streptomycin

Plate with streptomycin

Fourth, transfer to streptomycin medium a colony identified as resistant and one identified as sensitive

Streptomycin-resistant colony

Incubate

Third, only one of the four imprinted colonies grows

The responsible colony from the plate without streptomycin was streptomycin-resistant; the control colony was not

Tubes with streptomycin

FIGURE 9.1

Diagram illustrating the use of the replica-plating technique to demonstrate the random or nondirected nature of mutation. For simplicity, only four colonies are shown on each plate, and only two are tested for streptomycin resistance in step 4. Actually, each plate will contain 50–100 colonies, and a large number of plates may be used to find one mutant colony. Also, many colonies are tested in step 4. The definitive result is that all of the colonies from the nonselective plates that give rise to colonies on the selective plates are found to contain streptomycin-resistant mutants, and all of the colonies on the nonselective plates that do not grow on the selective plates contain only streptomycin-sensitive cells. (After R. Sager and F. J. Ryan, Cell Heredity, *John Wiley & Sons, 1961.)*

to streptomycin. Does streptomycin simply select rare, randomly occurring mutants preexisting in the large population, or do all of the cells have some low probability of developing resistance in response to the environmental stress—in this case streptomycin? Distinguishing between these two possibilities is not a trivial task. The evidence of a resistance to streptomycin is usually ascertained only by treatment of the culture with the antibiotic. How can one determine whether resistant mutants are present **prior** to exposure to streptomycin?

This was first accomplished by J. and E. M. Lederberg by a technique called **replica plating** (Fig. 9.1). Replica-plating experiments are carried out as follows. Bacterial cultures are diluted, and the cells are spread on the surface of semisolid nutrient

agar medium in petri dishes. After a sufficient period of growth, each bacterium will produce a visible colony on the surface of the agar. Each plate is then inverted and pressed onto sterile velvet, which is covering a wood block. Some of the cells from each colony stick to the velvet. A sterile plate of nutrient agar medium containing streptomycin (or any other antibiotic) is then pressed onto the velvet. This replica-plating procedure is repeated with a large number of plates, each containing a large number of bacterial colonies. After incubation of the selective plates (containing streptomycin), rare streptomycin-resistant colonies will have formed. When the colonies on the nonselective plates (*not* containing streptomycin) are subsequently tested for streptomycin resistance, those that gave rise to colonies on the selective replica plates are always found to contain resistant cells, while those that did not grow on the selective medium seldom contain any streptomycin-resistant cells (Fig. 9.1). The existence of streptomycin-resistant mutants in a population of bacteria prior to their exposure to the antibiotic can thus be demonstrated by replica-plating experiments. The environmental stress does not direct or cause genetic changes; it simply selects rare preexisting mutations that result in phenotypes that are better adapted to the new environment.

PHENOTYPIC EFFECTS OF MUTATIONS

Mutations must normally cause some detectable **phenotypic change** for their presence to be recognized. The effects of mutations on phenotype range from alterations so minor that they can be detected only by special genetic or biochemical techniques to gross modifications of morphology to lethals. A gene is a specific sequence of nucleotide pairs coding for a particular polypeptide. Any mutation occurring within a given gene will thus produce a new form or **new allele** of that gene. Because of the degeneracy of the genetic code (see Chapter 8), some base-pair changes do not change the protein products coded for by the genes in any way. Genes containing mutations with small effects that can be recognized only by special techniques are called "**isoalleles.**" Other mutations result in total loss of gene-product activity. If mutations of the latter type occur in essential genes (genes required for viability), they will, of course, be lethal.

Mutations may be either recessive or dominant. In haploid (or, more accurately, monoploid) organisms like viruses and bacteria, both recessive and dominant mutations can be recognized by their effects on the phenotype of the organism in which they originated. The dominance or recessiveness of mutations in bacteria can be determined only by studying partial diploids (see Chapter 7). In diploid (or polyploid) organisms, recessive mutations will be recognized only when present in the homozygous condition. Most recessive mutations in diploids will not be recognized at the time of their occurrence, since they will be present in the heterozygous state. Sex-linked recessive mutations are an exception, since they will be expressed in the hemizygous state in the heterogametic sex (males in humans and fruit flies; females in birds). Sex-linked recessive lethal mutations will alter the sex ratio, since hemizygous individuals carrying the lethal will not survive (Fig. 9.2).

The most useful mutations for the genetic analyses of many biological processes are **conditional lethal mutations.** These are mutations that are (1) **lethal** in one environment, the so-called **restrictive conditions,** but are (2) **viable** in a second environment, the **permissive conditions.** Such mutations allow geneticists to identify and study mutations in essential genes that result in complete loss of gene-product activity even in haploid organisms. Mutants carrying conditional lethals can be propagated under permissive conditions, and information about the functions of the gene products can be deduced by studying the consequences of their absence under the restrictive conditions. The use of conditional lethal mutations to dissect biological pro-

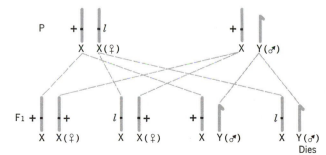

FIGURE 9.2

Altered sex ratio that results from a sex-linked recessive lethal. Heterozygous individuals of the homogametic sex will produce progeny with a 2:1 (homogametic sex to heterogametic sex) sex ratio.

cesses is illustrated on pp. 313–315. Conditional lethal mutations also provide valuable selective sieves for genetic fine structure analysis (see Chapter 10).

The three major classes of mutants with conditional lethal mutations are (1) **auxotrophic mutants,** (2) **temperature-sensitive mutants,** and (3) **suppressor-sensitive mutants.** Auxotrophic mutants (as opposed to prototrophic "wild-types") are mutants that are unable to synthesize an essential metabolite (amino acid, purine, pyrimidine, vitamin, etc.) that is synthesized *de novo* by wild-type individuals of the species. Such auxotrophic mutants will grow and reproduce when the metabolite is supplied in the medium (the permissive condition); they will not grow when the essential metabolite is absent (the restrictive condition). Temperature-sensitive mutants will grow at one temperature but not at another temperature. Most temperature-sensitive mutants are heat-sensitive; some, however, are cold-sensitive. The temperature sensitivity usually results from increased heat or cold lability of the mutant gene product, for example, an enzyme which is active at low temperature but partially or totally inactive at higher temperatures. Occasionally, only the synthesis of the gene product is sensitive to temperature, and, once synthesized, the mutant gene product may be as stable as the wild-type gene product. Suppressor-sensitive mutants are viable when a second genetic factor, a **suppressor,** is present, but are nonviable in the absence of the suppressor. The suppressor gene may correct or compensate for the defect in phenotype that is caused by the suppressor-sensitive mutation, or it may render the gene product, altered by the mutation, nonessential. Special types of suppressor-sensitive mutations are discussed in Chapters 8 and 10.

Most of the thousands of mutations that have been identified and studied by geneticists have been found to be deleterious and recessive. This is to be expected, considering what we know about the genetic control of metabolism and the techniques available for identifying mutations. Metabolism occurs by sequences of chemical reactions, each step of which is catalyzed by a specific enzyme coded for by one or more genes (see Chapter 8). Mutations in these genes frequently produce blocks in metabolic pathways (Fig. 9.3). These blocks occur because changes in the base-pair sequences of genes often (but not always) cause changes in the amino acid sequences of polypeptides (Fig. 9.4), which may, in turn, result in loss of function (Fig. 9.3). This, in fact, has been the most commonly observed effect of easily detected mutations. Given a wild-type allele coding for an active enzyme and mutant alleles coding for less active or totally inactive enzymes, it is apparent why most of the observed mutations might be recessive, as is observed. If the enzyme is metabolically important, such mutations will also be deleterious. If the enzyme catalyzes an essential reaction, the mutations causing total loss of activity will be recessive lethals.

But why should most mutations with phenotypically recognizable effects result in decreased gene-product activity or no gene-product activity? This result can be predicted if one accepts the effectiveness of natural selection and if one assumes the existence of a semiconstant environment during the recent (on the evolutionary scale) evolution of life forms on earth. A "wild-type" allele of a gene coding for a "wild-type" enzyme or structural protein will have been selected for optimal activity for many generations. Mutations resulting in amino acid changes that increased the efficiencies of enzymes in carrying out particular functions will have been preserved by natural selection, and, as the most "fit," they will have become the new "wild-types." Given a sufficient period of time in a semiconstant environment, most, if not all, sequences of amino acids (at least all sequences of amino acids that differ from the "wild-type" by a single mutation) will have been tried, and natural selection will have preserved the most efficient one. It will now be the wild-type. Mutations, which produce changes in these very specific sequences of amino acids, will usually result in less activity or no activity at all. As such, they will most frequently be recessive and deleterious.

An analogy can be made with any complex, carefully engineered machine. If you randomly modify any one essential component (e.g., of a watch or an automobile), it seldom performs as well as it did prior to the random change.

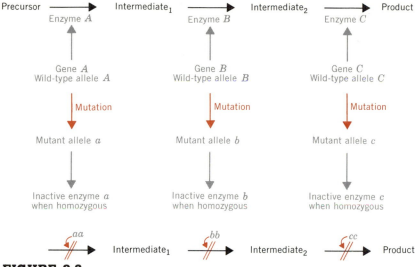

FIGURE 9.3

Blocks in metabolic pathways caused by recessive mutant alleles. Metabolic pathways may be only a few steps long, as diagramed here, or many steps in length. The wild-type form of each gene (or wild-type allele) codes for an active enzyme that efficiently catalyzes the appropriate reaction. Most mutations occurring in a wild-type gene result in the synthesis of an altered form of the enzyme with reduced activity or no activity. Such mutant alleles are usually recessive to their wild-type alleles. In the homozygous state, such mutant alleles often result in metabolic blocks (─//→) because of the inability of the mutant form of the enzyme to catalyze the requisite reaction. This picture of the genetic control of metabolism is valid for both anabolic (biosynthetic) and catabolic (degradative) pathways. Consider, for example, that the product is an anthocyanin pigment, yielding colored flowers in plants. Homozygosity for the recessive mutant allele of any one of the three genes (aa, bb, or cc) will result in the inability to synthesize the pigment.

SOMATIC AND GERMINAL MUTATIONS

Mutations may occur in any cell and at any state in the cell cycle. The immediate effect of the mutation and its ability to produce a phenotypic change are determined by its dominance, the type of cell in which it occurs, and when it happens relative to the life cycle of the organism.

If the mutation occurs in a somatic cell (all cells other than reproductive cells), which can produce cells like itself but not the whole organism, the mutant change will be perpetuated only in **somatic cells** that descend from the original cell in which the mutation occurred. The "Delicious" apple and the navel orange, for example, originally were mosaics

in somatic tissues. Changes that give these two fruits their desirable qualities apparently followed spontaneous mutation in single cells. In each case, the cell carrying the mutant gene reproduced, eventually producing an entire branch that had the characteristics of the mutant type. Fortunately, **vegetative propagation** was feasible for both the Delicious apple and the navel orange, and today numerous progeny from grafts and buds have perpetuated the original mutation. Descendants of the mutant types are now widespread in apple orchards and orange groves.

If dominant mutations occur in **germ** cells, their effects may be expressed immediately in pro-

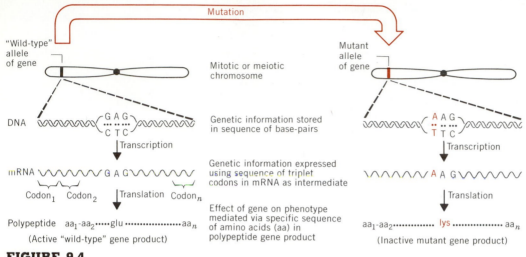

FIGURE 9.4

Mutation and expression of wild-type and mutant alleles. Mutations occurring in genes alter the base sequences and thus usually change the amino acid sequences of the gene products. In the example shown, a GC base-pair (top, left) has mutated to an AT base-pair (top, right), changing one codon in the mRNA from GAG to AAG and one amino acid in the polypeptide gene product from glutamic acid (glu) to lysine (lys) (see Table 8.1). Such changes frequently cause the gene products to be nonfunctional.

geny. If the mutations are **recessive,** their effects are often obscured in diploids. Germinal mutations, like somatic mutations, may occur at any stage in the reproductive cycle of the organism, but they are more common during some stages than others. If the mutation arises in a **gamete,** only a single member of the progeny is likely to have the mutant gene. If, on the other hand, a mutation occurs in gonial cells, several gametes may receive the mutant gene and thus enhance its potential for perpetuation. In any case, the dominance of the mutant allele and the stage in the reproductive cycle when the mutation occurs are major factors in determining the likelihood of the mutant allele being manifest in an organism and a population.

The earliest recorded dominant germinal mutation in domestic animals was that observed by Seth Wright in 1791 on his farm by the Charles River in Dover, Massachusetts. Wright noticed a peculiar male lamb with unusually short legs in his flock of sheep. It occurred to him that it would be an advantage to have a whole flock of these short-legged sheep (Fig. 9.5), which could not jump over the low stone fences in his New England neighborhood. Wright used the new short-legged ram for breeding his 15 ewes in the next season. Two of the 15 lambs produced had short legs. Short-legged sheep were then bred together, and a line was developed in which the new trait was expressed in all individuals. The mutation that gave rise to the short-legged sheep was obviously of the germinal type, because the cell carrying the mutation had the capacity to reproduce the entire organism. Germinal mutations have since been described in a variety of animals and plants.

MUTANT HEMOGLOBINS

Hemoglobin is the oxygen-transporting macromolecule present in the red corpuscles of chordate animals. It serves the essential function of transporting oxygen from the lungs to all of the various tissues of the body. Hemoglobin is a complex protein consist-

(a)   (b)

FIGURE 9.5

Effect of a dominant mutation. (a) *Short-legged sheep of the Ancon breed;* (b) *sheep with normal length legs. (Courtesy of Australian News and Information Bureau.)*

ing of four protein globin chains combined with the iron pigment heme. Two major types of hemoglobin occur in humans at different stages of development: fetal hemoglobin (hemoglobin F) and adult hemoglobin (hemoglobin A). Each molecule is composed of four polypeptide chains, two of each of two different polypeptides, plus a heme group. Hemoglobin A contains **two identical alpha chains** and **two identical beta chains.** Hemoglobin F contains two alpha chains and two **identical gamma chains.** Each polypeptide chain — alpha, beta, or gamma — is coded for by a specific gene. Hemoglobin F is present in the developing fetus, but is normally replaced by hemoglobin A within the first six months after birth. Of course, this raises the interesting question of how the expression of genes is regulated. How are genes turned on and off? This intriguing question is discussed in Chapter 11.

Each alpha polypeptide consists of a specific sequence of 141 amino acids. The beta and gamma chains are 146 amino acids long. Because of similarities in their amino acid sequences, all of the hemoglobin chains (and, thus, their structural genes) are believed to have evolved from a common progenitor (see Chapter 16, pp. 552–553).

Many different variants of the hemoglobins have been identified in human populations, most being variants of hemoglobin A. Some of the human hemoglobin variants (or mutant forms) have severe phenotypic effects; others have very minor phenotypic effects. Many of the variants were initially detected by their altered electrophoretic behavior (movement in an electric field due to charge differences). The hemoglobin variants provide an excellent illustration of the effects of mutation on the structures and functions of gene products and, ultimately, on the phenotypes of the affected individuals.

Sickle cell hemoglobin (hemoglobin S) is one such variant. Individuals homozygous for the hemoglobin S allele ($Hb_{\beta}^{S}/Hb_{\beta}^{S}$; the β subscript is used because hemoglobin S is a variant of the beta chain) develop severe hemolytic anemia. Hemoglobin S molecules precipitate when deoxygenated, forming crystalloid aggregates that distort the morphology of red blood corpuscles. The erythrocytes (red blood corpuscles) elongate and form sickle-shaped cells (Fig. 9.6). These sickle-shaped cells clog small blood vessels and cut off oxygen transport to various tissues. In many but not all cases, sickle-cell anemia results in death during childhood.

When the amino acid sequences of the beta chains of hemoglobin A and hemoglobin S were determined and compared, the hemoglobin S beta chain was found to differ from that of hemoglobin A by only one amino acid (Fig. 9.7). The sixth amino acid from the amino (NH_2-) terminal end of the beta chain of hemoglobin A is glutamic acid (a negatively

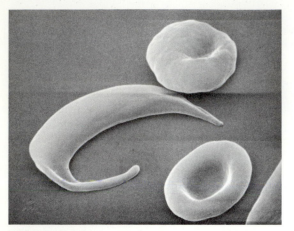

FIGURE 9.6

Effects of hemoglobin S on erythrocyte morphology. A sickle-shaped erythrocyte is shown between two normal erythrocytes. The cells are magnified several thousand-fold in this scanning electron micrograph. (Courtesy of John A. Long, University of California at San Francisco.)

charged amino acid). The beta chain of hemoglobin S contains valine (no charge at neutral pH) at that position. The alpha chains of hemoglobin A and hemoglobin S are identical. Thus, changing a single amino acid in one polypeptide can have severe effects on phenotype. A large number of similar effects of mutation on protein structure, and ultimately on phenotype, are now well documented.

In the case of hemoglobin S, it is hypothesized that the substitution of valine for glutamic acid at the sixth position in the beta chain allows a new bond to form which changes its conformation and leads to molecular stacking or aggregation of hemoglobin molecules. In any case, the change results in the gross abnormality of the morphology of the red blood cells. Although the nucleotide-pair sequences of the Hb_β^A and Hb_β^S genes are not known, the mutational change in the Hb_β^A gene that gave rise to Hb_β^S can be deduced from our knowledge about the genetic code (the triplet base sequences or codons in mRNA that specify the amino acids in the polypeptide gene product; see Chapter 8). The glutamic acid to valine change is nicely explained by a substitution of an AT base-pair for a TA base-pair (where

the T is in the transcribed strand in the first case and the A is in the transcribed strand in the second case; see Fig. 9.7). Over 75 hemoglobin variants with amino acid changes in the beta chain are now known. All of them except one differ from the normal beta chain (of hemoglobin A) by a single amino acid substitution. A few examples are illustrated in Fig. 9.8. Numerous variants of the alpha and gamma polypeptides have also been identified. Mutation, then, is a process in which changes in gene structure, often changes in one or a few base-pairs, cause changes in the amino acid sequences of the polypeptide gene products. These, in turn, cause the changes in phenotype that are recognized as mutant phenotypes. The pathways by which the polypeptides, the primary gene products, exert their effects on the phenotype are often very complex, particularly in higher plants and animals (see Chapter 8).

PLEIOTROPY

The term **pleiotropy** refers to the situation in which a gene influences more than one trait. Many such instances have been discovered. In fact, all genes (whether mutant or wild-type allelic forms) may be pleiotropic, with their various effects simply not yet recognized. Even though a structural gene may have many end effects, it has only one primary function, that of producing one polypeptide (in some cases, one RNA molecule). This polypeptide may give rise to different expressions at the phenotypic level.

The Hb_β^S allele provides a classic example of pleiotropy. It not only causes hemolytic anemia (in the homozygous state) but also results in increased resistance to one type of malaria, that caused by the parasite *Plasmodium falciparum*. Because the increased resistance to falciparum malaria occurs in $Hb_\beta^A Hb_\beta^S$ heterozygotes, such heterozygous individuals have a selective advantage in geographical regions where this type of malaria is prevalent (see Chapter 16, pp. 554–557). The sickle-cell allele also has pleiotropic effects on the development of many tissues and organs such as bones, the lungs, the kidneys, the spleen, and the heart.

Cystic fibrosis is a hereditary, metabolic dis-

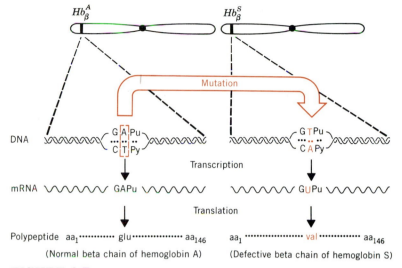

DNA

mRNA

Transcription

Translation

Polypeptide

FIGURE 9.7

Mutational origin of sickle-cell hemoglobin (hemoglobin S). The amino acid sequences of the "normal" (hemoglobin A) beta polypeptide and the sickle-cell beta polypeptide have been determined by direct chemical analysis (amino acid sequencing). From our knowledge of the nature of the genetic code (which triplet codons specify which amino acids; see Table 8.1) and the observed substitution of valine at amino acid position 6 in the beta chain of hemoglobin S, one can deduce that the mutation giving rise to hemoglobin S (or, more accurately, to the Hb_β^S gene) is the substitution of an adenine for a thymine in the transcribed strand of DNA. Or stated in base-pair terms, the substitution of a $\leftarrow T$ — base-pair for an $\leftarrow A$ — base-pair, where the bottom strand is the one $— \ddot{A} \rightarrow$ $— \ddot{T} \rightarrow$ transcribed (complementary to mRNA).

order in children that is controlled by a single auto-somal recessive gene. The gene apparently specifies an enzyme that produces a unique glycoprotein. This glycoprotein results in the production of mucus with abnormally high viscosity. Overly viscous mucus interferes with the normal functioning of several exocrine glands, including those in the skin (sweat), lungs (mucus), liver, and pancreas. The **syndrome (group of symptoms that characterizes the disease)** is related directly or indirectly to the abnormal mucus. Abnormally high levels of sodium chloride occur in the sweat. Mucus stagnates in tubules of the lungs, which frequently become infected, giving rise to bronchitis. Secreting cells in the liver and the pancreas are impaired, curtailing

production of fat-emulsifying agents and digestive enzymes and thus interfering with digestion and absorption of food. Several different phenotypic effects may thus result from the action of a single pleiotropic gene.

BACK-MUTATIONS AND SUPPRESSOR MUTATIONS

The mutation of a "wild-type" gene to a form that results in a mutant phenotype is usually referred to as "forward mutation." Sometimes, however, the designation of the "wild-type" and the "mutant" is quite arbitrary. They may simply represent two **different** phenotypes. For example, we consider the

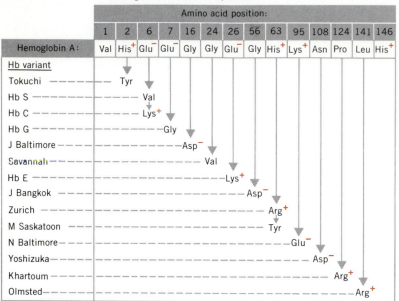

FIGURE 9.8

A few of the large number of hemoglobin beta chain variants that have been determined by amino acid sequencing. All of the examples shown differ from the sequence of the beta chain of hemoglobin A by a single amino acid substitution as indicated. They all can be explained by a single base-pair substitution as the mutational event in the Hb_β^A gene giving rise to the variant (as illustrated for hemoglobin S in Fig. 9.7). In fact, almost all of the over 100 hemoglobin variants known in the alpha, beta, and gamma polypeptides can be explained by single base-pair substitutions. This not only provides insight into the mechanism of mutation in humans, but also indicates that the codon assignments (see Table 8.1) determined experimentally with lower organisms are in large part valid for humans. Note that many of the amino acid substitutions involve changes in charge, even though only 5 of the 20 common amino acids have a net charge at neutral pH. This may be explained in part because of the ease of detecting hemoglobins with altered charges by electrophoresis. In addition, however, amino acid substitutions involving charge changes are probably more likely to alter protein structure and function than substitutions not involving charge changes.

two alleles controlling brown and blue eye color in humans both to be "wild-type." However, in a population composed almost entirely of brown-eyed individuals, the allele for blue eyes might be thought of as a mutant allele. In any case, the mutational events are often reversible. That is, a subsequent mutation may occur that restores the original "wild-type" phenotype. This is referred to as **back-mutation, reverse mutation,** or **reversion.** Reversion may occur in two different ways. Restoration of the original phenotype may occur (1) by a true back-mutation at the same site in the gene as the original mutation, restoring the wild-type nucleotide sequence, or (2) by the occurrence of a second mutation at a different location in the genome, which somehow compensates for the first mutation (Fig. 9.9). Mutations of the latter kind are called **suppressor mutations:** they "suppress" the effects of

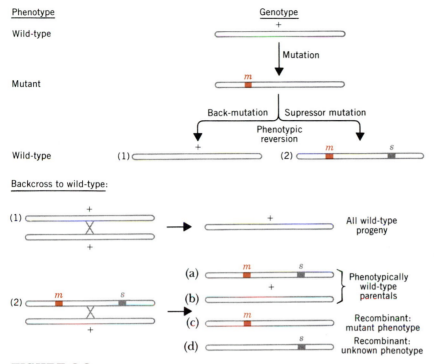

FIGURE 9.9

Restoration of the original "wild-type" phenotype by (1) true back-mutation and (2) suppressor mutation. Many mutants can revert to the wild-type phenotype by both mechanisms. Revertants of the two types can be distinguished by backcrosses to the original wild-type. All progeny with parental genotypes (a and b) will have wild-type phenotypes. Revertants containing suppressor mutations will produce some recombinant progeny with mutant phenotypes (c). Their frequency will depend on whether the initial mutation and the suppressor mutation are linked (as shown) and, if linked, how far apart they are on the chromosome. In higher eukaryotes, it may be difficult to analyze enough progeny to assure detection of the recombinants with mutant phenotype if the suppressor mutation and the original mutation are closely linked. The recombinant progeny carrying the suppressor mutation but not the initial mutation (d) may have either the mutant or the wild-type phenotype, depending on the nature of the mutation. Note that only single chromosomes carrying the various genetic markers being discussed are shown. For diploids, homologous chromosomes (not shown) will also be present. In diploids, the backcross will involve crossing a homozygous phenotypic revertant with a homozygous wild-type organism and examining testcross progeny of the F_1 heterozygotes. At the chromosome level, however, the scheme is applicable to both haploid and diploid organisms.

the original mutations. Suppressor mutations may occur at distinct sites in the same gene as the original mutation or in different genes, even in different chromosomes. Suppressor mutations can be distinguished from true back-mutations by backcrossing the revertants to the original wild-type. If the wild-type phenotype was restored by a suppressor mutation, the original mutation will still be present and can be separated from the suppressor mutation by recombination (Fig. 9.9).

THE MOLECULAR BASIS OF MUTATION

When Watson and Crick described the double helix structure of DNA and proposed its semiconservative replication based on specific base-pairing to explain the faithful transmission of genetic information from generation to generation, they also proposed a mechanism to explain spontaneous mutations. Watson and Crick pointed out that the structures of the bases in DNA are not static. Hydrogen atoms can move from one position in a purine or pyrimidine to another position, for exam-

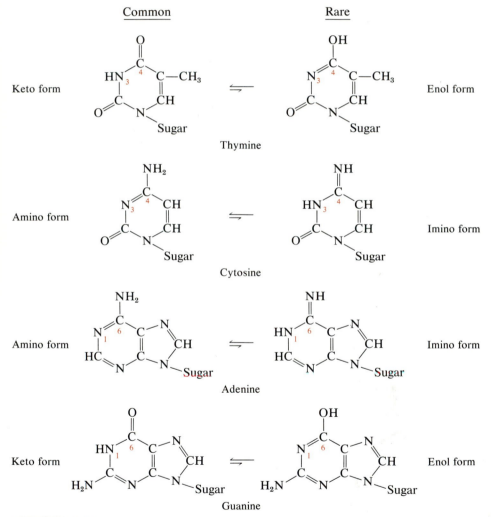

Common   Rare

Keto form   Thymine   Enol form

Amino form   Cytosine   Imino form

Amino form   Adenine   Imino form

Keto form   Guanine   Enol form

FIGURE 9.10

Tautomeric forms of the four common bases in DNA. The shifts of hydrogen atoms between the number 3 and number 4 positions of the pyrimidines and between the number 1 and number 6 positions of the purines change the base-pairing potential of the bases.

ple, from an amino group to a ring nitrogen. Such chemical fluctuations are called **tautomeric shifts.** Although tautomeric shifts are rare, they may be of considerable importance in DNA metabolism since they alter the base-pairing potential of the bases. The structures of the bases shown in Fig. 5.5 are the common, more stable forms, in which adenine always pairs with thymine and guanine always pairs with cytosine (Fig. 5.11). The **more stable keto forms** of thymine and guanine and **amino forms** of adenine and cytosine may **infrequently** undergo tautomeric shifts to **less stable enol and imino forms,** respectively (Fig. 9.10). The bases would be expected to exist in their less stable tautomeric forms for only very short periods of time. However, if a base existed in the rare form at the moment that it was being replicated or being incorporated into a nascent DNA chain, a mutation might result. When the bases are present in their rare imino or enol states, they can form adenine-cytosine and guanine-thymine base-pairs (Fig. 9.11). The net effect of such an event, and the subsequent replication required to segregate the "mismatched" base-pair, is an AT to GC or a GC to AT base-pair substitution (Fig. 9.12).

Mutations resulting from tautomeric shifts in the bases of DNA involve the replacement of a purine in one strand of DNA with the other purine and the replacement of a pyrimidine in the complementary strand with the other pyrimidine. Such base-pair substitutions are called **transitions.** Base-pair substitutions involving the substitution of a purine for a pyrimidine and vice versa are called **transversions.** Four different transitions and eight different transversions are possible (Fig. 9.13). A third type of point mutation involves the addition or deletion of one or a few base-pairs. Base-pair additions and deletions are collectively referred to as **frameshift mutations,** because they alter the reading frame of all base-pair triplets (specifying codons in mRNA and amino acids in the polypeptide gene product) in the gene distal to the mutation (Fig. 9.14).

All three types of point mutations—transitions, transversions, and frameshift mutations—are present among spontaneously occurring muta-

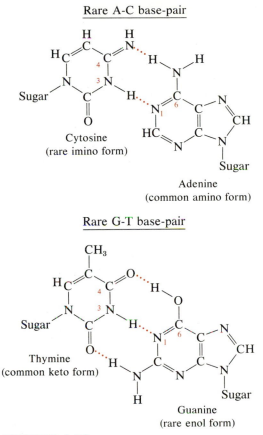

Rare A-C base-pair

Cytosine
(rare imino form)

Adenine
(common amino form)

Rare G-T base-pair

Thymine
(common keto form)

Guanine
(rare enol form)

FIGURE 9.11

Examples of mismatched AC and GT base-pairs that can form when one of the purines or pyrimidines exists in a rare tautomeric form. Similar mismatched base-pairs can form when thymine exists in its rare enol form or adenine exists in its rare imino form.

tions. A surprisingly large proportion of the spontaneous mutations studied in prokaryotes are found to be single base-pair additions and deletions rather than base-pair substitutions.

While there is still a great deal to be learned about the causes, molecular mechanisms, and frequencies of various types of spontaneous mutations, the three key factors appear to be (1) the **accuracy of the DNA replication machinery,** (2) the **efficiency** of the numerous mechanisms that have evolved for the **repair of damaged DNA** (see pp.

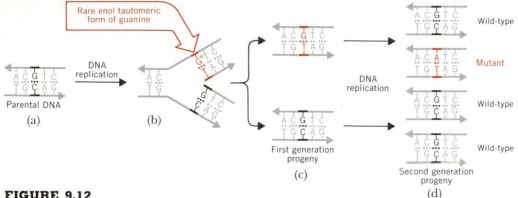

FIGURE 9.12

Mutation via tautomeric shifts in the bases of DNA. In the example diagramed, a guanine (a) undergoes a tautomeric shift to its rare enol form (G*) at the time of replication (b). In its enol form, it pairs with thymine (b). During the next replication (c to d), the guanine shifts back to its more stable keto form. The thymine incorporated opposite the enol form of guanine (b) directs the incorporation of adenine during the subsequent replication (c to d). The net result is a GC to AT mutation. If a guanine undergoes a tautomeric shift from the common keto form to the rare enol form at the time of incorporation (as a nucleoside triphosphate, rather than in a template strand as diagramed above), it will be incorporated opposite thymine in the template strand and cause an AT to GC mutation.

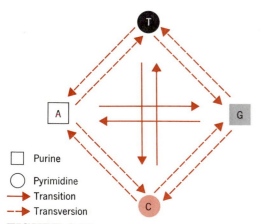

☐ Purine
◯ Pyrimidine
→ Transition
--→ Transversion

FIGURE 9.13

Diagram illustrating the base substitutions that are possible in DNA. These include four transitions (purine for purine or pyrimidine for pyrimidine; solid arrows) and eight transversions (purine for pyrimidine and pyrimidine for purine; dashed arrows). T = thymine, A = adenine, C = cytosine, and G = guanine.

293–296), and (3) the degree of **exposure to mutagenic agents present in the environment.** Any perturbations of the DNA replication apparatus or DNA repair systems clearly increase mutation rates.

RADIATION-INDUCED MUTATION

That portion of the electromagnetic spectrum containing wavelengths that are shorter and of higher energy than visible light (wavelengths below about 0.1 μm) can be subdivided into ionizing radiation (X rays, gamma rays, and cosmic rays) and nonionizing radiation (ultraviolet light). Ionizing radiations such as X rays (about 0.1 to 1 nm) are of high energy and thus are useful for medical diagnosis because they can penetrate living tissues. In the process of penetrating matter, these high-energy rays collide with atoms and cause the release of electrons, leaving positively charged free radicals or ions. These ions, in turn, collide with other mole-

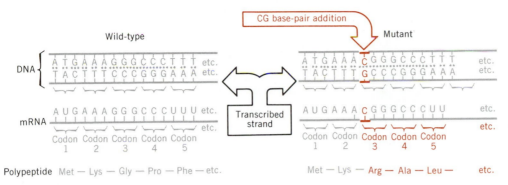

FIGURE 9.14

Diagram of a frameshift mutation that results from the addition of a single base-pair to a structural gene. The mutant gene (top, right) was produced by the insertion of a CG base-pair between the sixth and seventh base-pairs of the wild-type gene (top, left). This alters the "reading frame" of that portion of the gene distal, relative to the direction of transcription and translation (left to right, as diagramed), to the mutation. As a result, all of the codons of the mRNA and all of the amino acids of the polypeptide that correspond to base-pair triplets distal to the mutation are altered.

cules, causing the release of further electrons. The net result is that a "core" of ions is formed along the track of each high-energy ray as it passes through matter or living tissues. This process of ionization (thus the name ionizing radiation) is induced by machine-produced X rays, protons, and neutrons, as well as by the alpha, beta, and gamma rays released by radioactive isotopes of the elements (e.g., ^{32}P, ^{35}S, radium, cobalt-90, etc.). Ultraviolet rays, having lower energy, penetrate only the surface layer of cells in higher plants and animals and do not induce ionizations. Ultraviolet rays dissipate their energy to atoms that they encounter, raising the electrons in the outer orbitals to higher energy levels, a state referred to as **excitation.** Molecules containing atoms in either **ionic** forms or **excited** states are chemically more reactive than those containing atoms in their normal stable states. The increased reactivity of atoms present in DNA molecules is the basis of the mutagenic effects of ultraviolet light and ionizing radiation.

IONIZING RADIATION

In 1927, H. J. Muller (Fig. 9.15) first demonstrated that mutation could be induced by an external fac-

tor. Muller demonstrated that X-ray treatment markedly increased the frequency of sex-linked recessive lethal mutations in *D. melanogaster*. Muller's unambiguous demonstration of the mutagenicity of X rays became possible by his development of a technique facilitating the simple and accurate identification of lethal mutations in the X chromosome of *Drosophila*. This technique, called the **ClB method,** involves the use of females heterozygous for a normal X chromosome and an X chromosome (the *ClB* chromosome) specifically constructed for Muller's experiment.

The *ClB* chromosome has three essential components. (1) The **C** (for **crossover** "suppressor") refers to the presence of a long inversion (see Chapter 12) that prevents recombination of genetic markers on the *ClB* chromosome and alleles on the normal X chromosome. The inversion does not actually prevent crossing over, but causes gametes containing X chromosomes produced by crossing over between the *ClB* chromosome and the normal X chromosome to be inviable. Chromosomes resulting from crossing over between a chromosome containing an inversion (an inverted segment of the chromosome) and a normal chromosome will con-

FIGURE 9.15

H. J. Muller, who first demonstrated the mutagenic effects of X rays in 1927. Professor Muller made many contributions to our understanding of basic genetics through his extensive research. He was awarded the Nobel Prize in 1949. (National Archives.)

tain duplications and deficiencies (repeated and missing sets of genes; see Chapter 12). The inversion is required in Muller's experiment to assure that the markers on the *ClB* chromosome stay together through meiosis. (2) The *l* refers to a recessive **lethal** in the *ClB* chromosome. (3) The **B** refers to the presence of the **partially dominant** mutation that causes the **bar** eye phenotype, which is a narrow, slit-shaped eye. Because it is partially dominant, it allows females heterozygous for the *ClB* chromosome to be readily identified. Both the recessive lethal (*l*) and the bar eye mutation (*B*) are located within the inverted segment of the *ClB* chromosome.

Given females heterozygous for the *ClB* chromosome, Muller's experiment was operationally quite simple (Fig. 9.16). Male flies were irradiated and mated with *ClB* females. The bar-eyed daughters of this mating will all carry the *ClB* chromosome of the female parent and the irradiated X chromosome of the male parent. Since the entire population of reproductive cells of the males was irradiated, each bar-eyed daughter carries a potentially mutated X chromosome. That is, each male is likely to produce some sperm that contain X chromosomes carrying new lethal mutations and some sperm without X-linked lethal mutations. The bar-eyed daughters were then mated individually (in separate bottles) with wild-type males. If the irradiated X chromosome carried by a bar-eyed daughter contains a sex-linked lethal, all of the progeny of the mating will be female. Since males are hemizygous for the X chromosome, those receiving the *ClB* chromosome will die due to the recessive lethal (*l*) that it carries. Those receiving the irradiated X chromosome will also die if a recessive lethal has been induced in it (Fig. 9.16). Matings of bar-eyed daughters carrying an irradiated X chromosome, in which no lethal mutation has been induced, with wild-type males will produce female and male progeny in a ratio of 2:1 (only the males with the *ClB* chromosome will die). Scoring for the presence of recessive sex-linked lethals is thus unambiguous and error free using the *ClB* technique — simply scoring for the presence or absence of male progeny. By using this technique, Muller was able to demonstrate an increase in mutation rate of up to 150-fold after X-ray treatment.

Another technique that facilitates the detection of mutations in *Drosophila*, in this case sex-linked mutations with visible effects on phenotype (often called "visible mutations"), involves the use of **attached-X chromosomes.** Attached-X chromosomes undergo compulsory nondisjunction (failure of homologous chromosomes or chromatids to disjoin or separate during anaphase) since the two X chromosomes are joined to a single centromere. If females with two attached-X chromosomes plus a Y chromosome ($\widehat{XX}Y$) are mated to normal males, any mutation that occurs on the X chromosome of the male will be expressed in the surviving male pro-

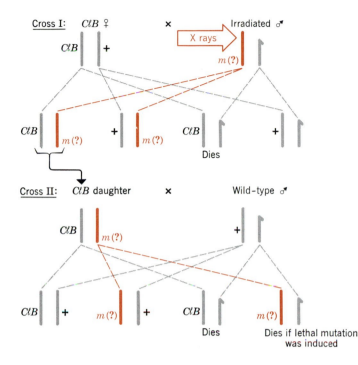

FIGURE 9.16

Muller's ClB *method for detecting sex-linked recessive lethal mutations in* Drosophila. *Females heterozygous for the ClB X chromosome are mated with irradiated males (Cross I). ClB daughters, which will always carry the ClB chromosome of the female parent and the irradiated X chromosome of the male parent, are then mated with wild-type males (Cross II). This second mating will produce only female progeny if an X-linked recessive lethal has occurred on that particular X chromosome. It will produce both female and male progeny (in a ratio of 2:1) if no X-linked lethal was induced on the irradiated X chromosome that the ClB daughter received. Scoring for the presence or absence of sex-linked recessive lethal mutations thus involves simply scoring the presence or absence of male progeny from matings of Cross II.*

geny (Fig. 9.17). In such attached-X matings, the male progeny receive their X chromosome from their male parent, rather than from their female parent as in a normal mating. If the male parent is treated with a mutagenic agent such as X rays, the increased frequency of recessive visible mutations can be easily assessed by screening the male progeny of the attached-X mating.

X rays and most other forms of ionizing radia-

tion are quantitated in **roentgen units** (**r units,** pronounced "rŭntgen"), which are measured in terms of the number of ionizations per unit volume under a standard set of conditions. More specifically, one roentgen unit is the quantity of ionizing radiation that produces one electrostatic unit of charge in a one cm^3 volume. Note that the dosage of irradiation in roentgen units does **not** involve a time scale. The **same dosage** may be obtained by a **low**

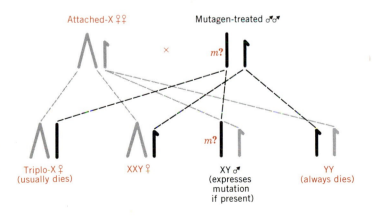

FIGURE 9.17

Attached-X method for detecting sex-linked visible mutations. The mutagen-treated males are crossed with $\widehat{XX}Y$ females (the two X chromosomes are attached to a single centromere). In such a mating, the viable male progeny receive their X chromosomes from their male parents. If a mutation with a visible effect on phenotype has been induced in the X chromosome of a mutagenized male, it should be expressed in all viable male progeny.

intensity of irradiation over a long period of time or **a high intensity of irradiation for a short period of time.** This is very important because in most studies the **frequency of induced point mutations is directly proportional to the dosage of irradiation** (Fig. 9.18). In *Drosophila* sperm, for example, there is an increase of approximately 3 percent in the mutation rate for each 1000 r increase in irradiation dosage. This linear relationship between mutation frequency and dosage is indicative of so-called "single-hit kinetics." That is, only one event (ionization?) or one "hit" is required to cause a mutation. Or, stated differently, every ionization has some fixed (under a specific set of conditions) probability of inducing a mutation. If the cumulative effects of many ionizations (or other radiation-induced events) were required to induce a mutation, the relationship shown in Fig. 9.18 would plot as a curve that was concave upward.

The linear relationship between mutation rate and radiation dosage is important because it speaks directly to the frequently asked question of **What is a safe level of irradiation?** Even very low levels of irradiation have certain low, but very real, probabilities of inducing mutations. The question is thus meaningless. There is no such thing as a safe level. In *Drosophila* sperm, for example, very low levels of irradiation over long periods of time (chronic irradiation) are as effective in inducing mutations as the same total dosage of irradiation administered at

high intensity for short periods of time (acute irradiation). This clearly has major practical significance in evaluating the effects of the increased exposure of living organisms to radiation that results from the testing and use of nuclear weapons and nuclear reactors in generators, spaceships, and so on.

In mice, chronic irradiation has been found to induce somewhat fewer mutations than the same dosage of acute irradiation. Moreover, when mice were treated with intermittent doses of irradiation, the mutation frequency was slightly lower than when they were treated with the same total amount of irradiation in a continuous dose. It should be emphasized that **all** of these irradiation treatments were mutagenic, albeit to different degrees, to both *Drosophila* and mice. The different responses of fruit flies and mice to chronic irradiation may result from differences in their ability to repair damaged DNA. Repair mechanisms (see pp. 293–296) may exist in the spermatogonia and oöcytes of mice that do not exist in *Drosophila* sperm.

The single-hit theory implies that one ionization can produce one mutation, but it does not imply anything about the **efficiency** with which this happens. Several factors have been shown to affect the efficiency with which irradiation induces mutation. The receptivity of a cell at different stages of its metabolic cycle is an important factor in determining rates for induced mutations. A. H. Sparrow has shown marked variations in numbers of chromosome fragments, assumed to be directly related to mutational changes, at different stages in meiosis and cleavage (Fig. 9.19) in the plant genus *Trillium*. Chromosome aberrations were induced about 60 times more frequently at metaphase than at interphase. Nondividing cells in *Trillium* showed little radiation damage, whereas rapidly dividing cells were very sensitive.

Oxygen tension and temperature change, when associated with irradiation, also may significantly alter the frequency of mutations. Low-oxygen tension decreases mutations. Oxygen can magnify the effect of radiation, but only if it is present during the irradiation. Oxygen has less effect with intense conditions than with moderate conditions of

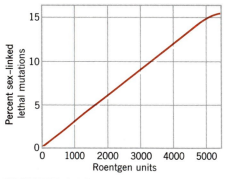

FIGURE 9.18

Relationship between the frequency of sex-linked lethal mutations induced in Drosophila *sperm and ionizing irradiation dosage.*

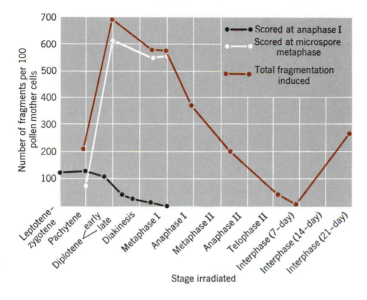

FIGURE 9.19

Relationship between the number of chromosomal fragments induced in meiotic cells of Trillium erectum *at a specific dosage of radiation (50 r) and the stage at which the irradiation occurred. (From A. H. Sparrow,* New York Academy of Sciences Annals *51:1508–1540, 1951.)*

ionization. Environmental agents that protect germ cells from radiation damage often do so by lowering the oxygen concentration of tissue, and those that enhance the effectiveness of radiation add oxygen.

Ionizing radiation also induces various kinds of gross changes in chromosome structure (chromosome aberrations) such as deletions, duplications, inversions, and translocations (see Chapter 12). These changes in chromosome structure result from breaks in chromosomes caused by ionizing radiation. Since they require two breaks, the kinetics of induction are two "hit" as expected, rather than single "hit."

ULTRAVIOLET RADIATION

Ultraviolet rays do not possess sufficient energy to induce ionizations. They are, however, readily **absorbed** by certain substances such as **purines** and **pyrimidines,** which then enter a more reactive or **excited** state. Because of their lower energy, they penetrate tissues only slightly, usually only the surface layer of cells in multicellular organisms. Nevertheless, ultraviolet light (UV) is a potent mutagen for unicellular organisms. The maximum absorption of UV by DNA is at a wavelength of 254 nm. Maximum mutagenicity also occurs at 254 nm, suggesting that the UV-induced mutation process is mediated directly by the absorption of UV by purines and pyrimidines. *In vitro* studies show that the pyrimidines (especially thymine) absorb strongly at 254 nm and, as a result, become very reactive. The two major products of UV absorption by pyrimidines appear to be pyrimidine hydrates and pyrimidine dimers (Fig. 9.20). Several lines of evidence indicate that thymine dimerization is probably the major mutagenic effect of UV. Thymine dimers appear to cause mutations indirectly in two ways. (1) Dimers apparently perturb the DNA double helix and interfere with accurate DNA replication. (2) Occasional errors are made during the processes that cells possess for the repair of "damaged" DNA, such as DNA containing thymine dimers (see the following section).

The relationship between mutation rate and UV dosage is highly variable, depending on the type of mutation, the organism, and the conditions employed. "Single-hit kinetics" are only occasionally observed, in contrast to ionizing radiation.

DNA REPAIR MECHANISMS

An indication of the importance of keeping mutation, both somatic and germ line, at a tolerable level is the multiplicity of repair mechanisms in organisms ranging from the simple bacterial viruses to

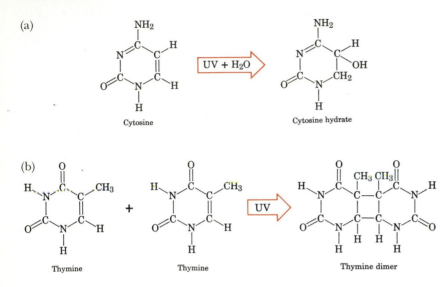

(a)

Cytosine

UV + H₂O

Cytosine hydrate

(b)

Thymine + Thymine

UV

Thymine dimer

FIGURE 9.20

Pyrimidine photoproducts of UV irradiation. (a) Hydrolysis of cytosine to a hydrate form that may cause mispairing of bases during replication. (b) Crosslinking of adjacent thymine molecules to form thymine dimers, which block replication.

humans. Mechanisms for the repair of damaged DNA are probably universal. *E. coli,* for example, possesses at least three different mechanisms for the repair of DNA containing thymine dimers: (1) **photoreactivation,** (2) **excision repair,** and (3) **postreplication recombination repair.**

Photoreactivation involves an enzyme that splits thymine dimers directly without the removal of any nucleotides (Fig. 9.21). This enzyme will bind to thymine dimers in DNA in the dark, but it cannot catalyze cleavage of the bonds joining the thymine molecules without energy derived from visible light, specifically light within the blue region of the spectrum. The enzyme is also active on cytosine dimers and cytosine-thymine dimers. Thus, when ultraviolet light is used as an experimental mutagen, the treatment is usually carried out in the dark in order to maximize the mutation frequency.

Excision repair involves a sequence of enzyme-catalyzed steps in which the thymine dimers are removed from the DNA molecule and a new segment of DNA is synthesized (Fig. 9.22). Excision repair occurs as efficiently in the dark as in the presence of blue light. The first step in excision repair is catalyzed by an endonuclease that recognizes thymine dimers, or the distortion in the double helix that they cause, and cleaves the phosphodiester backbone of the DNA strand containing the

dimer at a site near the damage. An exonuclease, probably the 5′ → 3′ exonuclease activity of DNA polymerase I (at least it catalyzes the removal of thymine dimers *in vitro*), then removes a segment of the strand adjacent to the endonuclease cut, including the dimer. A DNA polymerase, probably DNA polymerase I, then fills in the gap using the complementary strand as template. DNA ligase then catalyzes covalent closure or formation of the final phosphodiester linkage between adjacent nucleotides. Although the exact mechanism of excision repair is not yet known, the general features of the pathway described above are well documented. DNA polymerase I appears to be involved because mutants lacking this enzyme are defective in excision repair. In addition, the products of three genes of *E. coli* — *uvrA, uvrB,* and *uvrC* — are required for this process.

The second type of dark repair of UV damage that occurs in *E. coli* involves both replication and recombination and has thus been called postreplication recombination repair. The exact details of this repair process are still uncertain. However, the process clearly involves replication of the damaged chromosomes followed by recombination. A simplified model of postreplication recombination repair is shown in Fig. 9.23. When DNA molecules containing thymine dimers are replicated, gaps are

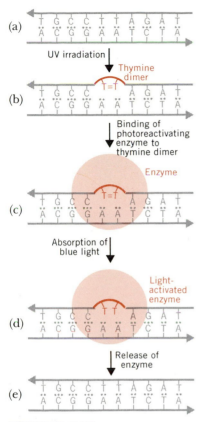

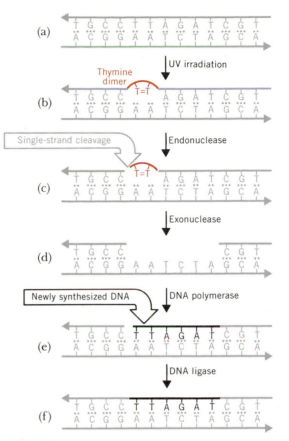

FIGURE 9.21

Cleavage of thymine dimer crosslinks by photoreactivation. (a) Segment of DNA containing two adjacent thymine bases. (b) Thymine dimer formation as a result of UV irradiation. (c) Binding of the photoreactivating enzyme to the thymine dimer-containing segment of the DNA molecule. (d) Cleavage of the dimer crosslinks by the photoreactivating enzyme using energy from the absorption of blue light. (e) Release of the photoreactivating enzyme, leaving a repaired, structurally normal DNA molecule. The arrows indicate the opposite polarity of the complementary strands.

FIGURE 9.22

Diagram of the excision repair pathway for the removal of thymine dimers from DNA. (a) Segment of DNA containing two adjacent thymine bases. (b) Thymine dimer formation as a result of UV radiation. (c) An endonuclease-induced single-strand cut adjacent to the thymine dimer. (d) An exonuclease-catalyzed removal of six nucleotides, including the thymine dimer. (e) Resynthesis of the excised segment by a DNA polymerase. (f) Formation of the final phosphodiester bond by DNA ligase. The arrows indicate the opposite polarity of the two complementary strands.

formed in the nascent complementary strands opposite the dimers because DNA polymerase cannot use the distorted strands as templates. After a lag due to the thymine dimer blockage, replication is reinitiated at secondary initiation sites beyond the dimer. This results in progeny double helices with thymine dimers in one strand and gaps in the complementary strand. If these two "sister" chromo-

somes recombine such that the dimers and gaps end up in one chromosome and the intact, undamaged segments end up in the other chromosome, the latter will be functional and produce a viable cell.

Recently, a new DNA repair enzyme, **uracil-DNA glycosylase,** has been discovered in *E. coli.* This enzyme efficiently removes uracil from DNA (thus its name). It is called a glycosylase rather than a

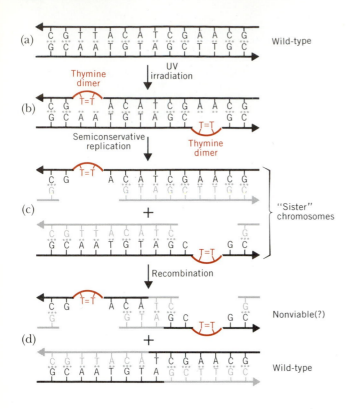

(a) ... Wild-type

UV irradiation

Thymine dimer

(b) ...

Semiconservative replication

Thymine dimer

(c) ... "Sister" chromosomes

Recombination

Nonviable(?)

(d) ... Wild-type

FIGURE 9.23

Simplified version of postreplication recombination repair of UV-induced damage in DNA. (a) Segment of a DNA molecule containing adjacent thymine bases. (b) Formation of thymine dimers during UV irradiation. (c) Replication of dimer-containing DNA. Gaps are present in the nascent strands (gray) because of the inability of the distorted dimer-containing regions of the parental strands to function as templates for DNA polymerase. After a lag period, synthesis is reinitiated past the dimers, leaving the indicated gaps. (d) Recombination between the "sister chromosomes," producing one undamaged DNA molecule and one doubly damaged DNA molecule. The net effect is the production of one undamaged chromosome from two damaged ones or, quite possibly, one viable product from two inviable ones. Details of the recombination process have been purposely omitted since they are not yet clear. Nucleases, a DNA polymerase, and DNA ligase are undoubtedly involved.

nuclease because it removes uracil by cleaving the glycosidic bond between the base and the sugar rather than phosphodiester bonds in the DNA chain. The enzyme thus produces apyrimidinic sites in the DNA. Such apyrimidinic sites have been shown to undergo an excision repair mechanism much like that described in Fig. 9.22. Uracil is produced in DNA by deamination of cytosine (e.g., by nitrous acid; see Fig. 9.27). *E. coli* mutants deficient in uracil-DNA glycosylase exhibit increased sensitivity to deaminating agents such as nitrous acid, indicating that the enzyme is important in the repair of this type of damage to DNA.

Mechanisms for the efficient repair of DNA damaged by mutagenic agents are clearly important even in bacteria and viruses. One can argue that they should be increasingly important with the increasing biological complexity and the correlated increase in genome size of higher organisms. The **larger the number of essential genes,** the more sites at which a lethal mutation can occur, and, given equivalent repair processes, the **greater the proba-**

bility that a given dose of mutagen such as irradiation **will cause a lethal mutation.**

XERODERMA PIGMENTOSUM

DNA repair mechanisms similar to those described above for *E. coli* have also been demonstrated in humans in studies done with cell cultures. In addition, their existence is indicated by genetic studies on individuals with inherited sensitivity to mutagenic agents such as ultraviolet light and/or ionizing radiation. The best known example of an inherited deficiency in the repair of radiation damage in humans is the autosomal recessive disease called **xeroderma pigmentosum.** The skin of individuals with xeroderma pigmentosum exhibits extreme sensitivity to sunlight, resulting in a high frequency of tumors in the epidermal cells of exposed areas such as the face (Fig. 9.24). The cells of many individuals with xeroderma pigmentosum have been shown to be deficient in the repair of UV-induced damage to DNA, such as thymine dimers. It seems

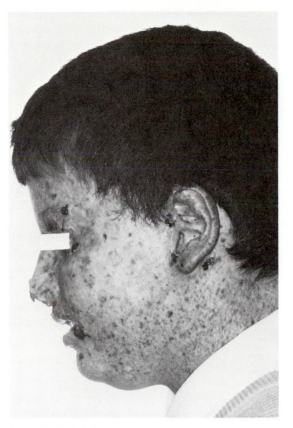

FIGURE 9.24

Phenotypic effects of the inherited disease xeroderma pigmentosum. Individuals with this malignant disease develop extensive skin tumors after exposure to sunlight. Homozygotes for the autosomal recessive mutation responsible for xeroderma pigmentosum are less efficient in the repair of DNA damaged by exposure to ultraviolet light. (Photography by P. E. Polani, Guy's Hospital, London. Reprinted with permission of Macmillan Publishing Co., Inc., from Human Genetics *by E. Novitski. Copyright © 1977 by Edward Novitski.)*

very likely, therefore, that the high incidence of skin cancer results directly from the inability of individuals with xeroderma pigmentosum to repair damage in DNA induced by UV in sunlight. In some cases, the defective enzyme appears to be the endonuclease that recognizes the thymine dimer and catalyzes the first step in excision repair (see Fig. 9.22). Genetic analyses of cells from individuals with xeroderma pigmentosum suggest that mutations in as many as six different genes can cause the disease.

This is not unexpected since (1) many enzymes are known to be composed of two or more different polypeptides (gene products) and (2) a mutation in any of the genes coding for polypeptides involved in the multiple-step repair processes might be expected to cause a block (see Fig. 9.3) in a repair pathway.

Some individuals with xeroderma pigmentosum develop neurological abnormalities that appear to result from premature death of nerve cells. This effect on the very long-lived nerve cells has potentially very interesting implications for the mechanism of aging. One theory of aging is that it results from the accumulation of somatic mutations. If so, a defective repair system would be expected to speed up the aging process; this appears to be the case with the nerve cells of some xeroderma pigmentosum patients. F. M. Burnet has emphasized that a low but significant level of spontaneous mutation must be maintained in living cells to provide sufficient flexibility for evolution to occur, and that aging may be an unavoidable consequence of this low level of mutation. Certainly, somatic mutations would be expected to contribute to the aging process. At present, however, essentially nothing is known about the relative importance of somatic mutation in senescence.

Fanconi's anemia, ataxia-telangiectasia, and Bloom's syndrome are three other inherited (autosomal recessive) diseases in humans where the primary defect may be in a DNA repair pathway. All three conditions result in a high risk of malignancy, particularly leukemia in the case of Fanconi's anemia and ataxia-telangiectasia. Cells of patients with ataxia-telangiectasia are abnormally sensitive to ionizing radiation, suggesting a defect in the repair of radiation-induced DNA damage. Cells of individuals with Fanconi's anemia are impaired in the removal of DNA interstrand cross-links, such as those formed by the antibiotic mitomycin C. Individuals with Bloom's syndrome exhibit a high frequency of chromosome breaks that result in aberrations (see Chapter 12) and sister chromatid exchanges. While all three of these malignant-prone, inherited diseases are probably caused by defects in DNA metabolism, possibly the repair processes, the primary lesions are still unknown.

CHEMICALLY INDUCED MUTATION

Muller's discovery of the mutagenic effects of irradiation provided a method for the induction of large numbers of mutations with a minimal amount of effort. However, because of the large number of effects of radiation on living tissues, both primary and secondary, studies on radiation-induced mutation provided little information about the molecular details of mutation processes. Subsequent discoveries of the mutagenic effects of a large variety of chemicals, many with very specific effects on DNA, provided the tools needed to work out many of the details of mutation processes at the molecular level. Hundreds of chemicals are now known to have from slight to very large mutagenic effects. Only a few of those with very strong mutagenic effects will be discussed here. Some of the more potent chemical mutagens are listed in Table 9.1.

The first chemical mutagen discovered was mustard gas (sulfur mustard). When C. Auerbach and her associates first discovered the mutagenic effects of mustard gas and related compounds during World War II, their data were classified. These compounds are examples of a large class of chemical mutagens that transfer alkyl ($CH_3—$, $CH_3CH_2—$, etc.) groups to the bases in DNA; thus they are called **alkylating agents.**

Chemical mutagens can be divided into two classes: (1) those that are mutagenic to both replicating and nonreplicating DNA, such as the alkylating agents and **nitrous acid,** and (2) those that are mutagenic only to replicating DNA. The latter class includes the **acridine dyes,** which bind to DNA and increase the probability of mistakes during replication, and **base analogs,** purines and pyrimidines with structures similar to the normal bases of DNA. The base analogs must be incorporated into DNA chains in the place of normal bases during replication to exert their mutagenic effects.

BASE ANALOGS

The base analogs that are mutagenic have structures sufficiently similar to the normal bases so that they are metabolized and incorporated into DNA during replication, but sufficiently different such that they increase the frequency of mispairing and thus mutation. The two most commonly used base analogs are **5-bromouracil** and **2-aminopurine.** The pyrimidine 5-bromouracil is a thymine analog, the bromine at the 5 position being similar in several respects to the methyl ($—CH_3$) group at the 5 position in thymine. The presence of the bromine, however, changes the charge distribution and increases the probability of tautomeric shifts (see Fig. 9.10). In its more stable keto form, 5-bromouracil pairs with adenine. After a tautomeric shift to its enol form, 5-bromouracil pairs with guanine (Fig. 9.25). The mutagenic effect of 5-bromouracil should be the same as that predicted for tautomeric shifts in normal bases (see Fig. 9.12), namely, transitions. If 5-bromouracil is present in its less frequent enol form as a nucleoside triphosphate at the time of its incorporation into a nascent strand of DNA, it will be incorporated opposite guanine in the template

(a) 5-Bromouracil: adenine base-pair

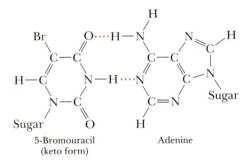

5-Bromouracil
(keto form)   Adenine

(b) 5-Bromouracil: guanine base-pair

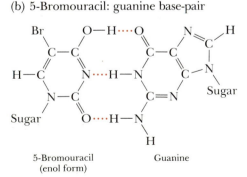

5-Bromouracil
(enol form)   Guanine

FIGURE 9.25

Base-pairing of 5-bromouracil with adenine (a) and guanine (b) in its two tautomeric forms.

TABLE 9.1   Some of the More Potent Chemical Mutagens

CHEMICAL NAME	COMMON NAME OR ABBREVIATION	STRUCTURE
(I) Alkylating agents		
di-(2-chloroethyl) sulfide	Mustard gas or sulfur mustard	$Cl-CH_2-CH_2-S-CH_2-CH_2-Cl$
di-(2-chloroethyl) methylamine	Nitrogen mustard	$Cl-CH_2-CH_2-\overset{\overset{\displaystyle CH_3}{\displaystyle \vert}}{N}-CH_2-CH_2-Cl$
ethylmethane sulfonate	EMS	$CH_3-CH_2-O-SO_2-CH_3$
ethylethane sulfonate	EES	$CH_3-CH_2-O-SO_2-CH_2-CH_3$
N-methyl-N'-nitro-N-nitrosoguanidine	NTG	$HN=C-NH-NO_2$ $O=N-N-CH_3$
(II) Base analogs		
5-bromouracil	5-BU	
2-aminopurine	2-AP	
(III) Acridines		
2,8-diamino acridine	Proflavin	
(IV) Deaminating agents		
nitrous acid	—	HNO_2
(V) Miscellaneous		
hydroxylamine	HA	NH_2OH

strand and cause a GC $\rightarrow$ AT transition (Fig. 9.26a). If, on the other hand, 5-bromouracil is incorporated in its more frequent keto form opposite adenine (in place of thymine) and undergoes a tautomeric shift to its enol form during a subsequent replication, it will cause an AT $\rightarrow$ GC transition (Fig. 9.26b). Thus, 5-bromouracil induces transi-

tions in both directions, AT $\rightleftharpoons$ GC. An important consequence of the bidirectionality of 5-bromouracil-induced transitions is that mutations originally induced with this thymine analog can also be induced to revert with it. The purine base analog 2-aminopurine is believed to induce mutations in a similar manner.

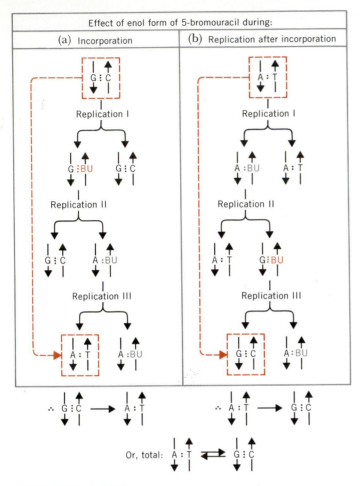

FIGURE 9.26

Summary of the mutagenic effects of 5-bromouracil. When 5-bromouracil (abbreviated BU in the diagram) is in its less frequent enol form (abbreviated in color) at the time of incorporation into DNA, it induces GC → AT transitions (a). When it is incorporated into DNA in its more common keto form (shown in gray) and shifts to its enol form during a subsequent replication, it induces AT → GC transitions (b). The summation of these two effects is that 5-bromouracil can induce transitions in both directions, AT ⇌ GC.

NITROUS ACID

Nitrous acid (HNO$_2$) is a very potent mutagen that acts directly on either replicating or nonreplicating DNA, by oxidative deamination of the bases, adenine, guanine, and cytosine, which contain amino groups. Conversion of the amino groups to keto groups changes the hydrogen-bonding potential of the bases (Fig. 9.27). Adenine is deaminated to hypoxanthine, which base-pairs with cytosine rather than thymine. Cytosine is converted to uracil, which base-pairs with adenine instead of guanine. Deamination of guanine produces xanthine, but xanthine base-pairs with cytosine just like guanine. Thus, the deamination of guanine is not directly mutagenic like that of adenine and cytosine. Since the deamination of adenine leads to AT → GC transitions, and the deamination of cytosine results in GC → AT transitions, nitrous acid induces transitions in both directions, AT ⇌ GC. Nitrous acid-induced mutations are thus also induced to revert with nitrous acid.

ACRIDINES

The acridine dyes such as proflavin (see Table 9.1), acridine orange, and a whole series of compounds called ICR-170, ICR-191, and so on, are very powerful mutagens that induce frameshift mutations (see Fig. 9.14). The ICR compounds have acridine moieties with various side chains, often alkylating agents. The positively charged acridines intercalate or "sandwich themselves" between the stacked base-pairs in DNA (Fig. 9.28). In so doing, they increase the rigidity and alter the conformation of the double helix, possibly causing slight "kinks" in

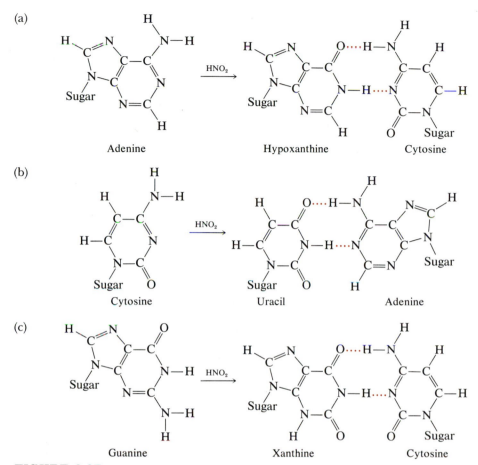

FIGURE 9.27

Mutagenic action of nitrous acid. Oxidative deamination converts (a) *adenine to hypoxanthine, causing AT → GC transitions;* (b) *cytosine to uracil, causing GC → AT transitions; and* (c) *guanine to xanthine, which is not directly mutagenic. The combined effects of nitrous acid on adenine and cytosine explain its ability to induce transitions in both directions, AT ⇌ GC.*

the molecule. In any case, when DNA molecules containing intercalated acridines replicate, additions and deletions of from one to a few base-pairs occur. As would be expected, these small additions and deletions, usually of a single base-pair, result in reading frameshifts for the portion of the gene distal to the mutation (see Fig. 9.14).

ALKYLATING AND HYDROXYLATING AGENTS

Alkylating agents such as nitrogen and sulfur mustards, methyl and ethyl methanesulfonate (MMS

and EMS), nitrosoguanidine (NTG; see Table 9.1), and many others have several effects on DNA. One major mechanism of mutagenesis by alkylating agents involves the transfer of methyl or ethyl groups to the bases such that their base-pairing potentials are altered and transitions result. For example, ethylation at the 7-*N* position and at the 6-*O* position are believed to be two effects of EMS. 7-Ethylguanine is then believed to base-pair with thymine (Fig. 9.29a). Other base alkylation products are believed to somehow "activate" repair processes in much the same way as thymine dimers do

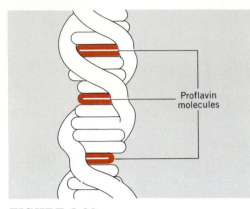

FIGURE 9.28

Intercalation of the acridine proflavin into the DNA double helix. X-ray diffraction studies have shown that the positively charged acridines become "sandwiched" between the stacked base-pairs. Genetic studies have shown that acridines such as proflavin induce frameshift mutations by the addition or deletion of one or a few nucleotide pairs during the replication of DNA containing such intercalated molecules. [After L. S. Lerman, J. Cellular and Comparative Physiology, 64 (Suppl. 1): 1, 1964.]

(see Figs. 9.21, 9.22, and 9.23). The occasional errors occurring in these repair processes may lead to transversions and frameshift mutations in addition to transitions. Finally, some alkylating agents, particularly difunctional alkylating agents (those with two reactive alkyl groups), cross-link DNA strands and/or molecules and induce chromosome breaks and the various kinds of chromosomal aberrations (see Chapter 12) found correlated with breaks. Alkylating agents as a class therefore exhibit less specificity in their mutagenicity than base analogs, nitrous acid, or acridines. They induce all types of mutations, including transitions, transversions, frameshifts, and even chromosome aberrations, with various relative frequencies, depending on the specific alkylating agent employed.

Nitrosoguanidine (NTG), one of the most potent chemical mutagens known, has been found to induce clusters of closely linked mutations in the segment of the chromosome that is replicating during the mutagenic treatment. Mutants isolated after NTG treatment often carry multiple, closely linked mutations, making them less useful for most genetic studies.

The hydroxylating agent **hydroxylamine, NH_2OH**, in contrast to many of the alkylating agents, has a very specific mutagenic effect. It induces **only GC → AT transitions.** Although the exact mechanism of hydroxylamine mutagenesis is still somewhat uncertain, it apparently acts by hydroxylating the amino group of cytosine (Fig. 9.29*b*). The resulting hydroxylaminocytosine can base-pair with adenine to produce the observed GC → AT transitions. Because of its specificity, hydroxylamine has been very useful in classifying transition mutations. Mutants that are revertible by nitrous acid or base analogs, and therefore resulted from transitions, can be divided into two classes on the basis of their revertibility with hydroxylamine. (1) Those with an AT base-pair at the mutated site will not be induced to revert by hydroxylamine. (2) Those with a GC base-pair at the mutated site will be induced to revert by hydroxylamine. Thus, hydroxylamine can be used to determine whether a particular mutation was an AT → GC or a GC → AT transition.

CORRELATION BETWEEN MUTAGENICITY AND CARCINOGENICITY

For many years it has been recognized that most of the strongly mutagenic agents, such as ionizing radiations, ultraviolet light, and chemicals like those discussed in the preceding sections, are also **carcinogenic,** that is, **induce cancers.** In recent years, sensitive techniques have been developed to test chemicals and other agents for mutagenicity and carcinogenicity. Carcinogenicity tests are usually done with rodents, most frequently newborn mice. These studies involve feeding or injecting the substance being tested and subsequent examination of the animals for tumor formation. The mutagenicity tests have often been done in similar fashion. However, because mutation is a low frequency event and because maintaining large populations of animals such as mice is an expensive undertaking, such animal tests have usually been quite insensitive. That is, low levels of mutagenicity would seldom have been detected.

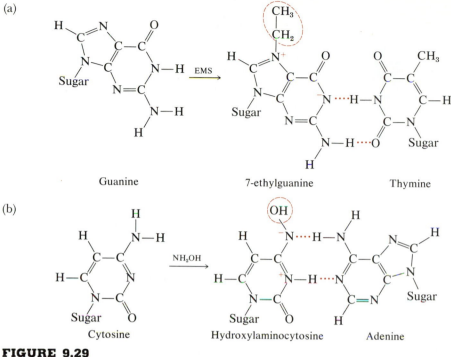

(a)

Guanine → EMS → 7-ethylguanine   Thymine

(b)

Cytosine → NH₂OH → Hydroxylaminocytosine   Adenine

FIGURE 9.29

One proposed mutagenic effect of (a) *the alkylating agent ethyl methanesulfonate (EMS) and* (b) *the hydroxylating agent hydroxylamine (NH₂OH). Alkylating agents such as EMS clearly induce mutations by other mechanisms as well (see text). The mutagenic action of hydroxylamine is known to involve cytosine, but its exact effect on cytosine is still somewhat disputed.*

B. Ames and his associates have recently developed very sensitive techniques by which large numbers of chemicals and other agents can be screened for mutagenicity. Their procedures are rapid and inexpensive. Ames' procedures involve the use of specially constructed strains of *Salmonella typhimurium* (a bacterium related to *E. coli*) that carry auxotrophic mutations of various types: transitions, transversions, and frameshifts. The reversion of these auxotrophic mutants to prototrophy, either spontaneous or induced with various agents, can easily be monitored by placing a known number of mutant cells in a petri plate containing medium supplemented with a trace of the growth factor required by the auxotroph (enough to support a few cell divisions, but not enough to allow the formation of visible colonies) and counting the number of colonies produced by prototrophic revertants after

an appropriate period of incubation. (Some growth is required in order to detect the effects of mutagens that are only mutagenic to replicating DNA.) The frequency of reversion induced by a particular substance can be compared directly with the spontaneous reversion frequency to obtain an estimate of its mutagenicity. Its ability to induce different types of mutations can be assessed by using a set of tester strains that carry different types of mutations—one strain with a transition, one with a frameshift mutation, and so forth. The procedures developed by Ames and his colleagues are very sensitive (Fig. 9.30), facilitating the detection of even weak mutagens.

Over a period of several years during which hundreds of different chemicals were tested, Ames and his colleagues have observed a greater than 90 percent correlation between the mutagenicity and

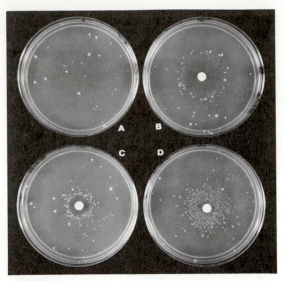

FIGURE 9.30

Salmonella *mutagenicity tests. Each petri dish contains a layer of agar medium containing only a trace of histidine and a known number of* Salmonella *"tester strain" cells, which are carrying, in this case, a frameshift mutation that results in histidine auxotrophy. Plates* C *and* D *also contain the rat-liver microsomal activation system. The potential mutagens are applied to 6-mm diameter filter paper disks, which are placed on the surface of the agar in the centers of plates* B, C, *and* D. (A) *Control plate with no addition, showing the background level of spontaneous reversion to prototrophy.* (B) *Plate to which the carcinogen furylfuramide (a food additive) has been added.* (C) *Plate to which the carcinogen aflatoxin B (a fungal toxin) has been added.* (D) *Plate to which the carcinogen 2-aminofluorene has been added. (From B. N. Ames, J. McCann, and E. Yamasaki,* Mutation Research *31:347, 1975. Photograph courtesy of B. N. Ames.)*

the carcinogenicity of the substances tested. Initially, several potent carcinogens were found to be nonmutagenic to the tester strains. Subsequently, many of these carcinogens, although nonmutagenic per se, were found to be metabolized to strongly mutagenic derivatives in eukaryotic cells. Ames and his associates therefore added a rat liver microsomal fraction to their assay systems in an attempt to detect the mutagenicity of metabolic derivatives of substances being tested. Coupling of the rat liver microsomal activation system to the microbial mutagenicity tests expanded the utility of the system

considerably. Nitrates, for example, are not mutagenic or carcinogenic themselves. *In vivo*, however, in eukaryotic cells, nitrates are converted via a series of enzyme-catalyzed reactions to nitrosamines, which are highly mutagenic and carcinogenic.

Ames' *Salmonella* mutagenicity tests demonstrated the presence of frameshift mutagens in several different components of chemically fractionated cigarette smoke condensates. In some cases, activation by the liver microsomal preparation was required for mutagenicity; in other cases, activation was not required.

The observed correlation between mutagenicity and carcinogenicity is consistent with the theory that cancer is caused by somatic mutations. Most geneticists would agree that somatic mutations **can** cause cancer. This has received strong support from the recent discovery of cellular oncogenes (cancer-causing genes) and the demonstration that the oncogene responsible for human bladder carcinoma resulted from a single base-pair change in its normal cellular counterpart (see Chapter 11, pp. 422–430). The common feature of all of the many different types of cancer is that the malignant cells, whatever their type, continue to divide when their normal counterparts would no longer be undergoing cell division. That is, all cancers exhibit a loss of the normal control of cell division, with the resultant formation of tumors. Cell division is undoubtedly, at least in part, under genetic control. Thus, a mutation in a gene involved in the control of cell division, like a mutation in any other gene, can cause a loss of function and thus a loss of the normal control of cell division. One can also easily envision epigenetic causes of cancer, that is, changes in states of differentiation, rather than permanent changes in the genetic material per se. The key question would appear to be, therefore, not whether somatic mutations can cause cancer, but what proportion of human cancers are, in fact, caused by somatic mutations.

MUTATION FREQUENCY

Mutation is necessary to provide the genetic variability required for the evolutionary adaptation of species to environmental changes. On the other hand, most mutations are deleterious. Thus, if mutation were to become too frequent in a species, it would create a sizable "genetic load" of deleterious effects. Clearly, if this "genetic load" became too large, the species would face extinction. Human technology has already contaminated the earth with increased levels of radiation and chemicals that are known to be mutagenic. While the consequences of the present increased level of mutagens in the environment cannot yet be accurately assessed, most scientists agree that further increases in the levels of mutagens in the environment should be avoided. Yet significant quantities of hundreds of new chemicals are introduced into the environment each year, most of them with insufficient, if any, mutagenicity tests. Whether they will cause harmful increases in mutation rate (and/or cancer incidence, see pp. 302–304) is a question of utmost concern to everyone.

Each gene probably has its own characteristic mutational behavior (e.g., see Fig. 10.16). Some genes undergo mutations more frequently than others in the same organism. Those with unusually high mutation rates are called unstable or mutable, but a wide range of mutation rates exists among genes that are considered stable. The mutation rate per gene in bacteria is of the order of 1 in 100,000 to 1 in 10 million (10^{-5} to 10^{-7}) per cell generation. For fruit flies, the average for mutation in a particular gene is in the order of 1 in 100,000, with a range from 1 in 20,000 to 1 in 200,000 gametes. Since most of the data on mutation rates in fruit flies have been obtained from experiments with males, questions have arisen concerning a possible sex difference in overall mutation rates. B. Wallace has shown through extensive experiments that mutation rates are not significantly different in the two sexes of *D. melanogaster*. However, mutation rates do differ in different strains.

Estimates of mutation rates for humans indicate a somewhat greater frequency than those cited for stable genes in most other organisms. Samples collected thus far have been small, and the methods used were indirect and subject to large errors. Genes associated with such human traits as intestinal polyposis and muscular dystrophy have been estimated to mutate once in 10^4 to 10^5 people. A human

generation is equal to about 50 cell generations. By expressing any mutation rate as a probability of mutation per cell per generation, a mutation rate is defined independently of exact physiological conditions and stage in life cycle. This definition is based on a time unit proportional to a cell's division time. When expressed in terms of cell generations, rates for fruit flies and humans are generally comparable with those for bacteria.

ESTIMATING MUTATION RATES

Mutation rates are usually defined in terms of mutational events per generation or, less frequently, mutations per unit time. The direct experimental determination of mutation rates is complex. Because mutation is a low-frequency event, large numbers of individuals must be sampled to get accurate estimates of mutation rates. In addition, recurrent sampling is required unless the initial population is sufficiently small that its probability of containing a mutant organism is negligible. Simply enumerating the number of mutant individuals in a population does not indicate how many mutational events have occurred since the mutants will usually undergo exponential growth like the nonmutants, although often at a slower rate than the latter (Fig. 9.31).

Consider a "wild-type" allele $a+$ mutating to a mutant allele a with a mutation rate μ (defined as a constant probability of a mutational event per cell duplication or per organism duplication), that is,

$$a^+ \xrightarrow{\mu} a$$

Clearly, the number of mutation events will depend on the number of a^+ alleles in the population. For simplicity, consider a haploid organism. **If** one starts with a population in which the number of mutant organisms equals M and the total population size equals N, then

$$dM = \left(\mu + \frac{M}{N}\right) dN$$

This **assumes** that (1) M is very small relative to N, so that the difference between N and $N - M$ is negligible; (2) the mutant and nonmutant organisms duplicate at the same rate; and (3) reversion of $a \rightarrow a^+$ does not occur or is negligible. If one starts with a population in which M is zero, and makes all the

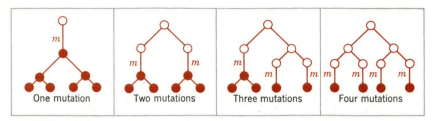

FIGURE 9.31

A given number of mutant organisms in a population can result from various numbers of mutational events, depending on when the mutation occurs during the exponential growth of the population. For example, the presence of four mutant organisms in a population may be the result of one, two, three, or four mutational events, as shown.

same assumptions, then the relationship simplifies to

$$dM = \mu dN$$

These equations can be integrated and used to obtain estimates of mutation rates in experimental populations. More frequently, however, populations are analyzed in terms of **allele frequencies,** or **changes in allele frequencies** due to mutation, selection, or migration (see Chapter 16). Moreover, mutation is not a unidirectional process; reverse mutations occur. A more realistic picture is thus

$$a^+ \underset{v}{\overset{\mu}{\rightleftharpoons}} a$$

in which the "wild-type" allele a^+ mutates to a at some frequency μ and the mutant allele a mutates back to a^+ at some frequency v. Given a sufficient period of time and the absence of selection for or against either allele, a mutational equilibrium will be reached at which the frequencies of the two alleles will be determined by μ and v (see Chapter 16, pp. 545–547).

When the mutation rate of a "wild-type" gene to a mutant form (usually a nonfunctional or only slightly functional form) is investigated, what is actually measured is the summation of many different mutation events at many different sites in the gene (see Figs. 9.4, 9.8, and Chapter 10). The "average gene" is usually considered to be a DNA sequence of about 1000 nucleotide pairs coding for a polypeptide that is about 333 amino acids long. Amino

acid changes at many different positions in the polypeptide (resulting from base-pair substitutions at many different positions in the structural gene) may be expected to result in an inactive gene product. Thus, when one measures the mutation rate of a "wild-type" gene to the nonfunctional mutant form, one will actually be measuring the sum of many different mutations, each occurring at its own specific rate (e.g., see Fig. 10.16).

MUTATION "LOAD" VERSUS GENOME SIZE

In the preceding section, an average mutation rate of 1 per 100,000 (or 10^{-5}) per gene per generation was cited for *Drosophila*. Estimates of the total mutation rate — the summation of mutation rates for the entire genome — in *Drosophila* are of the order of 5 percent (or 5×10^{-2}) per generation (per haploid complement, or per gamete). These values can be used to estimate the number of genes in the *Drosophila* genome (haploid complement). If the mutation rate per gene is m and the mutation rate per total genome is M, then the number of genes in the genome (N) will equal M/m. Using the above estimates, $N = 5 \times 10^{-2} / 10^{-5} = 5000$. This estimate agrees well with estimates of the number of genes in *Drosophila* obtained by other genetic criteria. On the other hand, one biochemical estimate suggests that *Drosophila* has about three times that many genes (Chapter 10, pp. 350–351).

If the *Drosophila* genome (haploid) contains

only 5000–15,000 genes, there is a major difference in the organization of the *Drosophila* genome and the genomes of the extensively studied bacteria and viruses. This difference, in fact, appears to be a basic difference between prokaryotes and eukaryotes. In prokaryotes, essentially all of the DNA of the genome appears to represent structural genes. Estimates of the number of genes from mutation studies (in some cases, identification of all or most of the genes of the organism) agree well with estimates obtained by dividing the total DNA content (in nucleotide pairs) of the genome by 1000 (10^3 nucleotide pairs per "average" gene). The phage lambda chromosome consists of 45×10^3 nucleotide pairs and about 40 known genes. The phage T4 chromosome contains about 200×10^3 nucleotide pairs. Over 100 genes have been mapped in phage T4, and a significant number remain to be identified. The *E. coli* genome contains about 4000×10^3 nucleotide pairs, while mutation analyses suggest that it contains about 3000 genes. In these prokaryotic organisms, then, the genomes are composed almost entirely of structural genes. The *Drosophila melanogaster* haploid chromosome complement contains about $120,000 \times 10^3$ nucleotide pairs. If all of the DNA represented structural genes of average size (10^3 nucleotide pairs), then *Drosophila* should have about 120,000 genes, rather than 5000 to 15,000 genes as indicated by the calculations above (also, see Chapter 10, pp. 350–351). It now seems clear that most of the DNA of *Drosophila* and other eukaryotes, including humans, does not represent structural genes. What is the function of this "excess" or noncoding DNA? Many geneticists believe that this noncoding DNA is regulatory, playing important roles in the regulation of gene expression (see Chapter 11). Others believe that it has an important role in some aspect of chromosome structure. Still others believe that much of the noncoding DNA is just "junk" DNA with no important function, possibly a reservoir of nucleotide-pair sequences available for the evolution of new genes. In any case, this DNA does not code for proteins. This "noncoding" DNA is of two types: (1) DNA that is transcribed, but for which the transcripts never leave the nuclei, and (2) nontranscribed

DNA. These two types of DNA may have very different functions. The first type of noncoding DNA clearly includes many of the noncoding intervening sequences or introns of eukaryotic genes (see Chapters 8, pp. 256–259, and 10, pp. 361–365).

An estimate of the maximum number of essential genes per genome can also be made from considerations of average mutation rates and the maximum mutation "loads" that a species could be expected to survive. For example, consider an average mutation rate of 4×10^{-6} recessive lethal mutations per essential gene per generation (per gamete). J. L. King, and others, have emphasized that it is unlikely that a species could tolerate more than 0.8 new lethal mutations per zygote (or 0.4 per gamete). Dividing 0.4 new recessive lethal mutations per gamete by 4×10^{-6} new recessive lethal mutations per essential gene per gamete gives a **maximum** number of 10^5 essential genes per gamete. At 10^3 nucleotide pairs per gene, the 10^5 essential genes would represent a total of 10^8 nucleotide pairs. The haploid chromosome complement of mammals, including humans, contains about 3×10^9 nucleotide pairs. Thus, according to these estimates, essential genes can represent a **maximum** of only about 3 percent of the total DNA in the genome of mammals. The actual number of essential genes in mammals is probably closer to $1–2 \times 10^4$, or less than 1 percent of the total DNA.

MUTABLE GENES

Most genes are relatively stable and mutate infrequently, but in some organisms a few genes mutate spontaneously so often that individuals carrying them are mosaics of mutated and unmutated genes. The *R* locus, which is involved with anthocyanin pigment synthesis in maize, for example, was found by R. A. Emerson to undergo alteration much more frequently than other loci. The change from R^r to r^r, for example, was found to occur at the rate of 1 per 20,000 gametes. These "mutable" genes are either more unstable than others or they are influenced by other factors in the genetic environment. McClintock's conclusion from extensive studies on maize was that a mutable gene is not an autonomous entity, but is derived from an agent that is integrated at the site of the mutable gene to cause instability.

Examples of highly mutable genes have been found in both plants and animals, but they appear to be more common in plants. Mutations in highly mutable genes occur frequently in somatic tissue and occasionally in germ cells. Somatic mutations may show their effects as color variegations (mosaics) in such plant parts as endosperm, leaves, and petals. Many common plants—including the larkspur, snapdragon, sweet pea, four-o'clock, and morning glory—have color variegations suggesting unstable or mutable genes.

The classical investigations of M. Demerec (1941) on a mutable gene called miniature-alpha in *Drosophila virilis* provided the first substantial data on the genetic properties of mutable genes in animals. These properties are: (1) mutation occurs primarily before meiosis; (2) mutation occurs in both females and males, implying that meiotic crossing over is not involved (crossing over occurs only rarely in *Drosophila* males; see Chapter 6); and (3) mutation is strongly influenced by neighboring genes. More recently, M. M. Green and others have described several mutable gene systems involving the white locus (eye color) and other genes in *D. melanogaster*. White-crimson (w^c), for example, mutates to wild-type and phenotypes other than white-crimson at a frequency of 10^{-3}. This mutable system and several others at the w locus are associated with chromosome alterations, particularly deficiencies. This led to the hypothesis that a "*controlling element*" from the cytoplasm is integrated into the chromosome at the site of mutability, and that this agent is responsible for chromosome aberrations.

Many of these "controlling elements" have now been identified by G. Rubin, P. M. Bingham, and colleagues, and shown to be **transposable elements** with structures similar to the Tn elements of bacteria (see Fig. 7.21). The unstable allele w^c mentioned above has been shown to result from the insertion of an approximately 10,000 nucleotide-pair-long transposable element belonging to the class called FB (for "foldback") elements.

Another unstable allele, w^{a1}, has been shown to have a transposable element called *copia* (because its sequence is present in RNA in *copious* amounts) inserted at the site of mutation. *Copia* has been isolated and sequenced. It is about 5000 nucleotide pairs long with 276 nucleotide-pair perfect direct repeats at its ends. These 276 nucleotide-pair repeats, in turn, have 17 nucleotide-pair inverse repeats at each end. Note the similarity of *copia* to *E. coli* elements Tn 9 and Tn 10 (Fig. 7.21).

MUTATOR AND ANTIMUTATOR GENES

Genes in maize were shown by McClintock to influence the stability of other genes. These have been called **mutator genes.** A striking example of the action of a mutator gene with a specific effect on a basic color gene in maize was described by M. M. Rhoades. The color of maize leaves and other plant parts is dependent on a complex of three complementary genes, symbolized *A*, *C*, and *R*. A color other than green is produced only when the dominant alleles (*A*, *C*, and *R*) of all three genes are present. The color may be purple if gene *P* is also present, or red (*pp*), or some other color depending on what other genes are included along with *A-C-R-*. Plants of genotype *aa*, *cc*, or *rr* have green leaves regardless of the alleles present at the other loci. Plants with the genotype *aaC-R-* would be expected to be green, but in the presence of a mutator gene called *Dt*, they are variegated. Light-colored corn kernels have purple spots at locations where somatic mutations have occurred (Fig. 9.32). The *Dt* gene produces its effect by influencing one or more of the *a* alleles of the *aa* genotype to mutate to *A*. Patches of cells scattered throughout the plant carry *A* alleles resulting from these mutations. On the leaves and kernels, these patches give a speckled appearance. The size of the spots depends on the stage of development at which the mutations occurred. Green cells contain unmutated genes; that is, they are of genotype *aa*. In this case, the allele (*Dt* or *dt*) present at one locus thus influences the mutation rate of an allele (*a*) present at another locus.

J. F. Speyer and others have found broad-spectrum mutator genes in bacteriophage T4. Temperature sensitive (*ts*) mutants of gene 43, for example, alter the mutation rates of point mutations in other genes. Some mutator genes exert their mutagenic

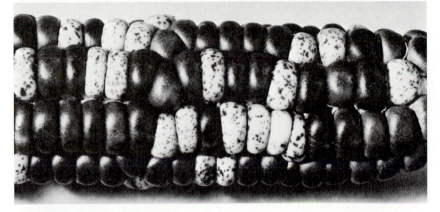

FIGURE 9.32

Effect of a mutator gene on phenotype in maize. Light-colored kernels with purple dots carry the mutator gene **Dt**. *This gene increases the rate of mutation of allele* **a** *(light-colored) to allele* **A** *(purple-colored). The deep purple-colored kernels carry allele* **A** *as well as other genes necessary for purple color. (From Crop Science, 1968,* The Mutants of Maize, *by M. G. Neuffer, L. Jones, and M. S. Zuber. Courtesy of M. G. Neuffer and the Crop Science Society of America.)*

effects during DNA replication by altering polymerase activity, producing mutagenic base analogs, or modifying DNA bases, thus influencing the mutation rates of other genes. One mutator in phage T4 gene 43 (*ts* L88) produces a DNA polymerase that utilizes incorrect nucleotides at a higher frequency than does the wild-type enzyme.

Several mutations in gene 43 exhibit powerful negative or antimutator activities, particularly during the formation of $A:T \rightarrow C:G$ substitutions. Experiments have shown that DNA polymerases (gene 43 products) that are isolated from *E. coli* that were infected with mutator, antimutator, and wild-type strains of T4 bacteriophage discriminate between adenine and 2-aminopurine to different degrees during DNA synthesis *in vitro*. Significantly larger amounts of 2-aminopurine are incorporated into DNA by wild-type and mutator than by antimutator enzymes. This indicates that organisms have the potential to evolve mechanisms that result in mutation rates that are lower than those presently existing in wild-type strains of some organisms.

When the mutator and antimutator DNA polymerases of phage T4 were studied *in vitro*, they were found to have altered ratios of polymerase activity to $3' \rightarrow 5'$ exonuclease activity (see Chapter 5, pp. 105–110). Mutators have increased polymerase/exonuclease activity. Antimutators exhibit decreased polymerase/exonuclease activity. This suggests that the mutation rate is, at least in part, controlled by the relative rates of polymerization and proofreading. Increased $3' \rightarrow 5'$ exonuclease activity and, thus, increased proofreading rates decrease mutation rates (antimutator). Decreased proofreading efficiencies (or increased polymerization rates) increase mutation frequencies (mutator).

Several mutator genes have also been identified and characterized in *E. coli*. Some increase the frequency of only certain types of base-pair substitutions. Others cause an increase in all types of point mutations. R. C. von Borstell and others measured spontaneous mutation rates in yeast and isolated many mutator genes with different kinds of mutator and antimutator activity.

Decreases as well as increases in mutation rates, as compared with wild-type, can thus result from new mutations. With both mutators and antimutators operating in the same system, particular spontaneous mutation rates may be optimized through natural selection. Mutators and antimutators thus become relative terms, because no standards are available for optimum rates. Results of studies on yeast have borne out the conclusions of those on viruses and bacteria; **that spontaneous mutation rates are under the genetic control of the cell itself.**

INSERTION SEQUENCES ("IS ELEMENTS") AS MUTATORS

A significant proportion of the mutations that occur in bacteria and some bacteriophages are now known to be effects of small sequences of DNA called **IS elements** or **insertion sequences** (see Chapter 7). These DNA sequences, from about 800 to about 1400 nucleotide pairs in length (Fig. 9.33), are present in *E. coli* chromosomes, for example, in several copies. The first five IS elements to be characterized were: IS1 (768 nucleotide pairs long) and IS2, IS3, IS4, and IS5 (all between 1200 and 1400 nucleotide pairs in length). **The IS elements are transposable;** that is, they have the ability to move from one position in the chromosome to other positions, either in the same chromosome or to other chromosomes in the same cell. They can move, for example, from the *E. coli* chromosome to bacteriophage chromosomes or to **plasmids** (small circular molecules of DNA, or "minichromosomes," often present in bacteria; see Chapter 7).

The chromosomes of the strains of *E. coli* that have been studied contain about eight copies of IS1 and five copies of IS2. The IS elements are physically and covalently inserted as linear sequences into the chromosomes (Fig. 9.34). When an IS element is inserted (or inserts itself?) into a gene, it destroys the function of, or **mutates,** that gene. While insertion of IS elements occurs at many sites in the *E. coli* chromosomes, it is not completely at random. Preferred sites of insertion have been demonstrated in some cases.

Detectable IS-induced mutations occur with a frequency of about 10^{-6} to 10^{-7} per cell per generation in *E. coli*. IS-induced mutations are revertible. They can be distinguished from all mutations with

similar effects in that their reversion frequency (usually in the range 10^{-6} to 10^{-8}) **cannot** be enhanced with mutagens. The fact that IS-induced mutations are revertible means that IS elements can excise themselves from the chromosome with precision to the single nucleotide pair. Reversion to the functional state requires restoration of the original nucleotide-pair sequence of the gene; no base-pairs can be added or deleted during the insertion-excision process. The failure of mutagens to enhance reversion is to be expected. Reversion in this case occurs by a recombination mechanism (Fig. 9.34), not a mutation process (see pp. 283–285).

The "controlling elements" and mutable loci (or unstable genes) in higher plants and animals (see pp. 307–308) are probably caused by sequences of DNA analogous to the IS elements in prokaryotes.

VIRAL CHROMOSOMES AS MUTATORS

Bacterial viruses are of two general types. **Virulent** viruses, such as phage T2, always multiply in the host cells and, in so doing, kill them. **Temperate** viruses, like phage lambda, have a choice of two life styles. (1) They may multiply and kill the host like virulent phages, the **lytic cycle,** or (2) they may insert their chromosome into the host chromosomes (much like the insertion of IS elements diagramed in Fig. 9.34). (For details of the mechanism of insertion or integration of the phage lambda chromosome into the *E. coli* chromosome, see Chapter 7, pp. 204–205). Once covalently inserted into the host chromosome, the viral chromosome, now called a **prophage** or **provirus,** replicates once each cell generation and behaves much like any other segment of DNA of the chromosome. Cells carrying integrated viral chromosomes, called **lysogenic cells,** grow, divide, and behave in most respects like uninfected cells.

The chromosomes of most temperate phages integrate into one or a few specific sites of homology on the host chromosome. One particular bacteriophage, **phage mu,** is most remarkable in this respect. Phage mu has the unique ability to insert its chromosome into the host chromosome at ran-

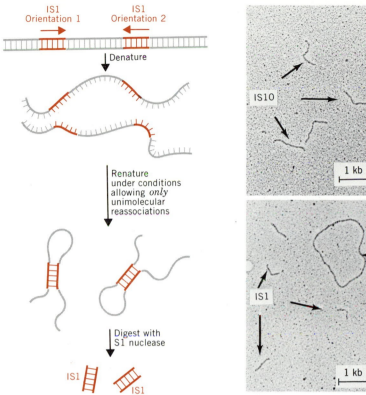

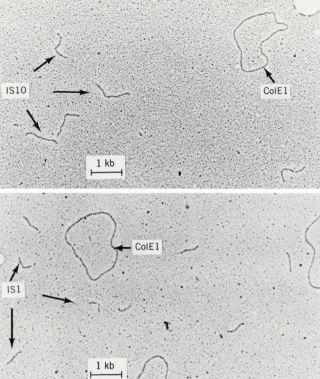

FIGURE 9.33

Electron micrographs (right) of IS1 (bottom) and IS10 (top) DNA sequences, and diagram (left) illustrating how they were prepared for electron microscopy. Plasmids (extrachromosomal DNA molecules or "minichromosomes" present in many bacterial cells) containing two homologous IS elements present in opposite orientations (inverted with respect to each other) were denatured and then renatured under conditions permitting unimolecular (one molecule) reassociations, but not bimolecular (two molecule) reassociations. Under these conditions only complementary sequences contained in a single strand of DNA will reform a double helix structure. After renaturation, the DNA molecules were treated with S1 nuclease, which degrades all regions remaining single-stranded. Because the two IS1 elements are present in opposite orientations, the top strand of the IS1 on
the left will have the same sequence as the bottom strand of the IS1 on the right, and vice versa. Thus, the two sequences in each single strand are complementary and renature, as shown. Col E1 is a plasmid DNA reference. (The Col E1 plasmid is a "colicinogenic factor" that codes for a protein—a "colicin"— that kills other E. coli *cells not containing this plasmid.) One kilobase = 1 kb = 1000 base-pairs. Note the relative sizes of the IS elements: IS1 is 768 base-pairs long and IS10 is about 1400 base-pairs long. [Electron micrographs from H. Ohtsubo and E. Ohtsubo, pp. 49–63, in* DNA Insertion Elements, Plasmids, and Episomes *(A. I. Bukhari, J. A. Shapiro, and S. L. Adhya, eds.). Cold Spring Harbor Laboratory Press, Cold Spring Harbor, N.Y. Copyright © 1977 by Cold Spring Harbor Laboratory.]*

domly distributed sites, if not all sites (between any two adjacent nucleotide-pairs?). The consequence of this random, or nearly random, integration of the

mu chromosome is that **phage mu acts as a very powerful mutagen.** Its insertion within a gene inactivates that gene just like the insertion of an IS

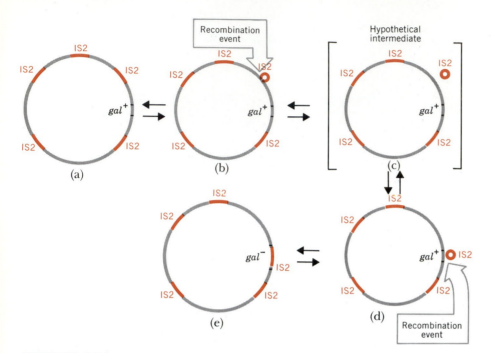

FIGURE 9.34

Proposed mechanism of mutation induced by IS element transposition. (a) An E. coli *chromosome is shown containing five IS2 elements (each 1400 nucleotide pairs long). For simplicity, the eight IS1 elements plus IS3 and IS4 elements are not shown. The IS elements have been arbitrarily positioned in the chromosome. (b) One IS2 element is shown "looped out" in a figure-8 configuration. A recombination event results in the excision of this IS2 element (c), which may then move to another position such as the gal locus (d). Recombination between the chromosome and the* IS2 element (d) inserts the IS2 element into one of the gal genes (e), resulting in the loss of gal function. The net effect of the IS2 element transposition into the gal gene is the formation of a gal⁻ auxotroph (or gal⁻ mutant). While the general features of the above model are probably accurate, no autonomous circular IS forms have yet been demonstrated. If IS elements exist as circular intermediates during their transposition, they must be rapidly integrated into their new locations in the chromosomes.

element (Fig. 9.34). From **1 to 3 percent** of the lysogenic cells produced by infection with phage mu acquire a **detectable, new auxotrophic mutation** somewhere in the genome. If mutations in every gene in the host chromosome could be detected, 100 percent of the lysogenic cells would probably be found to harbor a new mutation. In addition to the mutations caused by mu chromosome insertions, phage mu mediates, by mechanisms not yet fully understood, deletion formation, the integration of other genetic elements such as plasmids into the *E. coli* chromosome, and complex transpositions of segments of DNA. In other words, **phage mu promotes** a high frequency of **recombination events between nonhomologous chromosomes** (often called "illegitimate" recombination to contrast it with the usual or "legitimate" recombination involving **homologous** chromosomes).

The insertion of viral chromosomes into eukaryotic chromosomes has also been demonstrated in several animal systems. Whether such viral chromosome insertions account for a significant number of mutations in higher plants and animals remains uncertain.

PRACTICAL APPLICATIONS OF MUTATIONS

As was discussed earlier, mutations are invaluable to the process of evolution since they provide the raw material required for its occurrence. Moreover, mutations provide the alleles required for the various types of genetic analysis, from Mendel's two-factor crosses (Chapter 2) to chromosome mapping (Chapters 6, 7, and 10) to studies of genetic structures of populations (Chapter 16). Without mutation, genetic analyses of the types described in this textbook would be impossible. Mutations, and the ability to induce them with mutagenic agents, have thus proven useful in many ways.

BENEFICIAL MUTATIONS

Even though most mutations make the organism less efficient and are thus disadvantageous, the possibility of developing new desirable traits through induced mutations has intrigued many plant breeders. Plant breeders have reported induced mutants in barley, wheat, oats, soybeans, tomatoes, and fruit trees that may improve presently cultivated strains. Barley mutants, for example, have been obtained that provide increased yield, resistance to smut (Fig. 9.35), stiff straw, increased protein content, and hull-less seeds.

One application of induced mutations came from concentrated efforts to improve the yield of penicillin by the mold *Penicillium*. When penicillin was first discovered, the yield was low and production was seriously limited. Then millions of spores were irradiated and a few of the surviving colonies produced considerably more penicillin than the average. Such mutant penicillin-overproducers have proven invaluable in the commercial production of this important antibiotic.

HIGH-RESOLUTION DISSECTION OF BIOLOGICAL PROCESSES VIA MUTATION ANALYSIS

Mutations have been used extensively to elucidate the pathways by which biological processes occur. Metabolism occurs via sequences of enzyme-catalyzed reactions. By isolating and studying mutations in the genes coding for the enzymes involved, the sequence of steps in a pathway can often be determined. Morphogenesis frequently involves the sequential addition of proteins in the formation of specific three-dimensional structures. Again, the sequence of protein additions can often be deter-

FIGURE 9.35

Heads of barley demonstrating the effects of resistance to loose smut. (Left) A head from a strain that is smut resistant. (Center) Heads from a strain that carries some resistance to loose smut. (Right) Head from a susceptible strain destroyed by loose smut. (Courtesy of W. Dewey.)

mined by isolating and studying mutant organisms with mutations in the genes coding for the proteins involved. Because an appropriate mutation will eliminate the activity of a single polypeptide, mutations provide an extremely powerful probe with which to dissect biological processes.

Consider, for example, the following simple pathway:

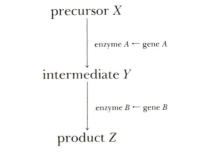

Intermediate *Y* is produced from precursor *X* by the action of enzyme *A*, the product of gene *A*. Intermediate *Y*, however, may be rapidly converted to product *Z* by enzyme *B*, the product of gene *B*. If so, intermediate *Y* may be present in very low quantities and be very difficult to identify biochemically. However, in a mutant organism that has a mutation in gene *B*, resulting in the synthesis of either an inactive form of enzyme *B* or no enzyme *B*, intermediate *Y* will often accumulate to much higher concentrations, facilitating its isolation and identification. Similarly, a mutation in gene *A* may aid in the identification of precursor *X*. Furthermore, the functions and mechanisms of action of individual gene products can often be deduced by comparative biochemical and biophysical analyses of mutant and wild-type organisms.

The resolving power of mutational dissection in biological processes has been emphatically demonstrated by the elucidation of the pathway of morphogenesis in bacteriophage T4. This complex process involves the products of about 50 genes. Each gene codes for a structural protein of the virus or for an enzyme that catalyzes one or more steps in the morphogenetic pathway. By (1) isolating mutant strains of phage T4 with temperature-sensitive and suppressor-sensitive conditional lethal mutations (see p. 278) in each of the approximately 50 genes, and (2) analyzing the structures that accumulate when these mutant strains are grown under the restrictive conditions by electron microscopy and biochemical techniques, R. S. Edgar, W. B. Wood, J. King, and colleagues have been able to work out almost the entire pathway of phage T4 morphogenesis (Figs. 9.36 and 9.37).

Other biological processes that have been successfully dissected by mutational analysis include the photosynthetic electron transport chains in *Chlamydomonas reinhardi* and maize and nitrogen fixation in bacteria. This approach is currently being used to dissect differentiation and development in higher plants and animals. S. Benzer and colleagues are using mutations to dissect behavior and learning in *Drosophila*. In theory, mutational dissection should be applicable to any process that is under genetic control. The existence of a gene that does not mu-

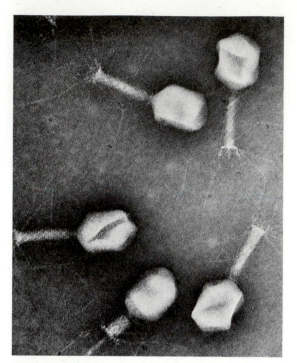

FIGURE 9.36
Electron micrograph of negatively stained T4 bacteriophages, showing their intricate morphology. The DNA-filled heads, the tails, and the six-pointed baseplates with their attached tail fibers are easily recognized. On some of the phage (particularly the one at the lower right), the collar between the head and the tail is visible. The tail sheath of the phage at the lower left is contracted, revealing the distal portion of the inner core or tube of the tail through which the DNA is ejected during infection. The tail sheath functions like a little muscle. After the tail fibers and spikes on the baseplate have become firmly attached to receptor sites on the bacterial cell, the tail sheath contracts, pushing the tail core through the cell envelope, much like an injection needle. Magnification is ×37,000. (Courtesy E. Boy de la Tour and E. Kellenberger.)

tate to a nonfunctional state seems extremely unlikely. Thus, mutational dissection of biological processes is limited only by the ingenuity of scientists in identifying mutations of the desired types.

MUTATIONS AND HUMANS

Purposeful artificial selection is not practiced in humans (except possibly in the broadest sense, in-

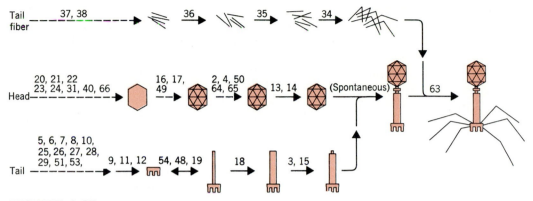

FIGURE 9.37

Summary of the pathway of morphogenesis in bacteriophage T4, as revealed by mutational dissection and electron microscopic and biochemical analyses. The head, the tail, and the tail fibers are synthesized by means of separate branches of the pathway, which are then joined in terminal steps of the pathway. The numbers beside the arrows indicate the T4 genes whose products are required at that particular step in the pathway. The pathway is obligatorily sequential in all but a few steps; that is, the morphogenesis must proceed in the sequence indicated. The tail fibers will not attach to headless tails, for example, even though the structure of the baseplate appears complete prior to the joining of that tail to the head. (After W. B. Wood, R. S. Edgar, J. King, I. Lielausis, and M. Henninger, Federation Proceedings *27:1160–1166, 1968.)*

volving selective mating and personal decisions to limit reproduction) and, therefore, the possible advantages cited for domestic animals and plants do not apply to humankind. Variations do exist in populations, however, and presumably they originated through past mutations. Since mutations, on the average, are detrimental, it would seem advantageous, from the standpoint of short-term effects, for humans to avoid excessive exposure to mutagenic agents.

In the case of acute irradiation, two types of danger should be considered: (1) the immediate damage to the exposed person, and (2) the more insidious damage to the DNA in his or her reproductive cells, which would affect future generations. The immediate damage is indicated by burns and other direct or secondary effects on body tissues. When doses are on the order of 50 mr (milliroentgens) or lower, no immediate damage can be detected, although some unseen harmful effects such as induction of leukemia and general shortening of

the life span may occur. Doses that exceed safety standards may be prescribed by physicians as therapeutic measures, such as for cancer treatment. In these cases, possible benefits must be carefully weighed against possible damage.

Effects of the second type of damage will be observed only in future generations. There is reason to believe, however, that exposure to high-energy irradiations of any kind, at any dosage level, is potentially harmful. In the few clear experimental results to date, mutations have generally been proportional to the dosage and the effects have been cumulative.

The relation between dosage and effects cannot be accurately measured in humans at present because of the complexity of the subject and the special difficulties of dealing with the genetics of humans. Problems have been recognized in investigations concerning the survivors of the Hiroshima and Nagasaki bombings. The Atomic Bomb Casualty Commission is investigating the bomb's effect on the people exposed to atomic bomb irradiation and their descendants. Preliminary reports, including data on children born to parents who survived the bombing, have revealed a significant increase in the incidence of leukemia. The normal sex ratio has been altered, and the changes have been interpreted as resulting from induced sex-linked lethals. Most of the data available on mutation rates and the nature of mutations have come from other organisms and only by inference are they applied to humans. However, the genetic material in humans is DNA, just as it is in most other organisms. It should not be surprising, therefore, that the effects of irradiation seem fairly comparable for most organisms. In the absence of specific data, the facts learned from other organisms should be considered relevant to humans. All such facts indicate that increased exposure to irradiation will be detrimental to future generations.

SUMMARY

Sudden, heritable changes in the genetic material are called **mutations.** Mutation also refers to the process by which such changes are produced. Mutations may occur "spontaneously" (for unknown reasons) or may be induced by agents that interact with DNA and RNA. Various kinds of irradiation and many chemicals that react with DNA and RNA are very potent **mutagenic agents.**

New mutations provide the genetic variability used for evolution. Some level of mutation is thus required to provide the raw material for evolution. However, most mutations are detrimental. High frequencies of mutation would thus be disadvantageous to a species, except possibly in a rapidly changing environment.

The potential benefits of the use of irradiation (solar irradiation, X rays, nuclear reactors) must be carefully weighed against the known and estimated potential risks. Similar precautions must be taken to prevent the continued pollution of our environment with mutagenic (and/or carcinogenic) chemicals. These risk estimates and precautions must take into account the potential harm to future generations of living organisms, including humans, keeping in mind the increased frequencies of deleterious recessive mutations that may result.

REFERENCES

Ames, B. N., J. McCann, and E. Yamasaki. 1975. "Methods for detecting carcinogens and mutagens with the *Salmonella*/mammalian-microsome mutagenicity test." *Mutation Research* 31:347–364.

Auerbach, C. 1962. *Mutation, an introduction to research on mutagenesis.* Oliver and Boyd, Edinburgh.

Bukhari, A. I., J. A. Shapiro, and S. L. Adhya (eds.). 1977. *DNA insertion elements, plasmids, and episomes.* Cold Spring Harbor Laboratory Press, Cold Spring Harbor, New York.

Burnet, F. M. 1974. *Intrinsic mutagenesis: a genetic approach to aging.* Medical and Technical Publishing Co., Great Britain.

Cleaver, J. E. 1967. "Defective repair replication of DNA in xeroderma pigmentosum." *Nature* 218:652–656.

Couturier, M. 1976. "The integration and excision of the bacteriophage mu-1." *Cell* 7:155–163.

Cox, E. C. 1976. "Bacterial mutator genes and the control of spontaneous mutation." *Ann. Rev. Genet.* 10:135–156.

Denniston, C. 1982. "Low level radiation and genetic risk estimation in man." *Ann. Rev. Genet.* 16:329–355.

Drake, J. W. 1970. *The molecular basis of mutation.* Holden-Day, San Francisco.

———(ed.). 1973. "The genetic control of mutation." *Genetics* 73, supplement: 1–205.

Drake, J. W., and R. H. Baltz. 1976. "The biochemistry of mutagenesis." *Ann. Rev. Biochem.* 45:11–37.

Howard-Flanders, P. 1981. "Inducible repair of DNA." *Sci. Amer.* 245(5):72–80.

Kikuchi, Y., and J. King. 1975. "Genetic control of bacteriophage T4 baseplate morphogenesis. I, II, and III." *J. Molecular Biology* 99:645–672, 673–694, 695–716.

King, J. L. 1976. "Progress in the neutral mutation—random drift controversy." *Federation Proceedings* 35:2087–2091.

Krieg, D. R. 1963. "Specificity of chemical mutagenesis." *Progress Nucleic Acid Res.* 3:125–168.

Lederberg, J., and E. M. Lederberg. 1952. "Replica plating and indirect selection of bacterial mutants." *J. Bacteriology* 63:399–406. (Reprinted in E. A. Adelberg, ed., 1966. *Papers on bacterial genetics,* 2nd ed., Little, Brown, Boston.)

Lindahl, T. 1982. "DNA repair enzymes." *Ann. Rev. Biochem.* 51:61–87.

Little, J. W., and D. W. Mount. 1982. "The SOS-regulatory system of *Escherichia coli.*" *Cell* 29:11–22.

Muller, H. J. 1927. "Artificial transmutation of the gene." *Science* 66:84–87. (Reprinted in J. A. Peters, ed., 1959. *Classical papers in genetics.* Prentice-Hall, Englewood Cliffs, N. J.)

Schull, W. J., M. Otake, and J. V. Neel. 1981. "Genetic effects of the atomic bombs: a reappraisal." *Science* 213:1220–1227.

Shortle, D., D. DiMaio, and D. Nathans. 1981. "Directed mutagenesis." *Ann. Rev. Genet.* 15:265–294.

Singer, B., and J. T. Kuśmierek. 1982. "Chemical mutagenesis." *Ann. Rev. Biochem.* 51:655–693.

Starlinger, P., and H. Saedler. 1976. "IS-elements in microorganisms." *Current Topics in Microbiology and Immunology* 75:111–152.

Vogel, F. 1970. *Chemical mutagenesis in mammals and man.* Springer-Verlag, New York.

Witkin, E. M. 1969. "Ultraviolet-induced mutation and DNA repair." *Ann. Rev. Genet.* 3:525–552.

Wood, W. B., R. S. Edgar, J. King, I. Lielausis, and M. Henninger. 1968. "Bacteriophage assembly." *Federation Proceedings* 27:1160–1166.

PROBLEMS AND QUESTIONS

9.1. Identify the following point mutations represented in DNA and in RNA as (1) transitions, (2) transversions, or (3) reading frameshifts. (a) A to G; (b) C to T; (c) C to G; (d) T to A; (e) UAU ACC UAU to UAU AAC CUA; (f) UUG CUA AUA to UUG CUG AUA.

9.2. Both lethal and visible mutations are expected to occur in fruit flies that are subjected to irradiation. Outline a method for detecting (a) sex-linked lethals and (b) visible mutations in irradiated *Drosophila*.

9.3. How can mutations in bacteria causing resistance to a particular drug be detected? How can it be determined whether a particular drug causes mutations or merely identifies mutations already present in the organisms under investigation?

9.4. Published spontaneous mutation rates for humans are generally higher than those for bacteria. Does this indicate that individual genes of humans mutate more frequently than those of bacteria? Explain.

9.5. A precancerous condition (intestinal polyposis) in a particular human family group is determined by a single dominant gene. Among the descendants of one woman who died with cancer of the colon, 10 people have died with the same type of cancer and 6 now have intestinal polyposis. All other branches of the large kindred have been carefully examined and no cases have been found. Suggest an explanation for the origin of the defective gene.

9.6. Juvenile muscular dystrophy in humans is dependent on a sex-linked recessive gene. In an intensive study, 33 cases were found in a population of some 800,000 people. The investigators were confident that they had found all cases that were well enough advanced to be detected at the time the study was made. The symptoms of the disease were expressed only in males. Most of those who had it died at an early age and none lived beyond 21 years of age. Usually only one case was detected in a family, but sometimes two or three cases occurred in the same family. Suggest an explanation for the sporadic occurrence of the disease and the tendency for the gene to persist in the population.

9.7. Products of somatic mutation, such as the navel orange and the Delicious apple, have become widespread in citrus groves and apple orchards, but they are uncommon in animals. Why?

9.8. If a single short-legged sheep should occur in a flock, suggest experiments to determine whether it is the result of mutation or an environmental modification, and if it is a mutation, is it dominant or recessive?

9.9. How might enzymes such as DNA polymerase be involved in the mechanism of both mutator and antimutator genes?

9.10. How could spontaneous mutation rates be optimized by natural selection?

9.11. A mutator gene *Dt* in maize increases the rate at which the gene for colorless aleurone (*a*) mutates to the dominant allele (*A*) for colored aleurone. When reciprocal crosses were made (i.e., seed parent *dt/dt*, *a/a* × *Dt/Dt*, *a/a* and seed parent *Dt/Dt*, *a/a* × *dt/dt*, *a/a*), the cross with *Dt/Dt* seed parent produced three times as many dots per kernel as the reciprocal cross. Explain these results.

9.12. A single gene change blocks the normal conversion of phenylalanine to tyrosine. (a) Is the mutant gene expected to be pleiotropic? (b) Explain.

9.13. How can normal hemoglobin (hemoglobin A) and hemoglobin S be distinguished?

9.14. If CTT is a DNA triplet (transcribed strand of DNA) specifying glutamic acid, what DNA and mRNA base triplet alterations could account for valine and lysine in position 6 of the beta hemoglobin chain?

9.15. Why is sickle-cell anemia called a molecular disease?

9.16. Assuming that the beta hemoglobin chain originated in evolution from the alpha chain, what mechanisms might explain the differences that now exist in these two chains (see Fig. 16.13)? What changes in DNA and mRNA codons would account for the differences that have resulted in unlike amino acids in corresponding positions?

9.17. In a strain of bacteria, all organisms are usually killed when a given amount of streptomycin is introduced into the medium. Mutations sometimes occur that make the bacteria resistant to streptomycin. Resistant mutants are of two types: some can live with or without streptomycin; others cannot live unless this drug is present in the medium. Given a nonresistant strain, outline an experimental procedure by which resistant strains of the two types might be established.

9.18.   One sample of fruit flies was treated with X rays at 1000 roentgens (r). The mutation rate of a particular gene was found to be increased by 2 percent. What percentage increases would be expected at 1500 r, 2000 r, and 3000 r?

9.19.   Why does the frequency of chromosome breaks induced by X rays vary with the total dosage and not with the rate at which it is delivered?

9.20.   One person was in an accident and received 50 roentgens (r) of X rays at one time. Another person received 5 r in each of 20 treatments. Assuming no intensity effect, what proportionate number of mutations would be expected in each person?

9.21.   How does ultraviolet light produce revertible mutations?

9.22.   How does nitrous acid induce mutations? What specific end results might be expected on DNA and mRNA from treatment of viruses with nitrous acid?

9.23.   Are mutational changes induced by nitrous acid more likely to be transitions or transversions?

9.24.   How does the action and mutagenic effect of 5-bromouracil differ from that of nitrous acid?

9.25.   How has induced mutation been put to practical use in improving crops of economic value?

9.26.   Evaluate the effects, immediate and potential, that might come from intense, mass irradiation of people.

9.27.   How do acridine-induced changes in DNA result in inactive proteins?

Use the known codon-amino acid assignments given in Chapter 8 on p. 265 to work the following problems.

9.28.   Bacteriophage MS2 carries its genetic information in RNA. Its chromosome is analogous to a polycistronic molecule of mRNA in organisms that store their genetic information in DNA. The MS2 "minichromosome" codes for 4 polypeptides (i.e., it has 4 genes). One of these 4 cistrons codes for its coat protein, a polypeptide 129 amino acids long. The entire nucleotide sequence in the RNA of MS2 is known. Codon 112 of the coat protein gene is CUA, which specifies the amino acid leucine. If you were to treat a replicating population of bacteriophage MS2 with the mutagen 5-bromouracil, what amino acid substitutions would

you expect to be induced at position 112 of the MS2 coat protein (i.e., Leu → other amino acid)? (*Note:* Bacteriophage MS2 RNA replicates using a complementary strand of RNA and base-pairing like DNA.)

9.29.   Would the different amino acid substitutions induced by 5-bromouracil at position 112 of the coat polypeptide that you indicated in Problem 9.28 above be expected to occur with equal frequency? If so, why? If not, why not? Which one(s), if any, would occur more frequently?

9.30.   Would such mutations occur if a nonreplicating suspension of MS2 phage was treated with 5-bromouracil?

9.31.   Recalling that nitrous acid deaminates adenine, cytosine, and guanine (adenine → hypoxanthine, which base-pairs with adenine; and guanine → xanthine, which base-pairs with cytosine), would you expect nitrous acid to induce any mutations that result in the substitution of another amino acid for a glycine residue in a wild-type polypeptide (i.e., glycine → another amino acid) if the mutagenesis was carried out on a suspension of mature (nonreplicating) T4 bacteriophages. (*Note:* After the mutagenic treatment of the phage suspension, the nitrous acid is removed. The treated phage are then allowed to infect *E. coli* cells to express any induced mutations.) If so, by what mechanism? If not, why not?

9.32.   Keeping in mind the known nature of the genetic code, the information given about phage MS2 in Problem 9.28, and the information we have learned about nitrous acid in Problem 9.31, would you expect nitrous acid to induce any mutations that would result in amino acid substitutions of the type *glycine → another amino acid* if the mutagenesis were carried out on a suspension of mature (nonreplicating) MS2 bacteriophage? If so, by what mechanism? If not, why not?

9.33.   Would you expect nitrous acid to induce a higher frequency of Tyr → Ser or Tyr → Cys substitutions? Why?

9.34.   Which of the following amino acid substitutions should you expect to be induced by 5-bromouracil with the highest frequency? (a) Met → Leu; (b) Met → Thr; (c) Lys → Thr; (d) Lys → Gln; (e) Pro → Arg; or (f) Pro → Gln. Why?

9.35.   Acridine dyes such as proflavin are known to induce primarily single base-pair additions and deletions. Suppose that the wild-type nucleotide sequence in the mRNA of a gene is:

(5′-end) AUGCCCUUUGGGAAAGGG-

UUUCCCUAA (3′-end)

Also, assume that a mutation is induced within this gene by proflavin and, subsequently, a revertant of this mutation is likewise induced with proflavin and shown to be a second-site suppressor mutation within the same gene. If the amino acid sequence of the polypeptide coded for by this gene in the revertant (double mutant) strain is:

(amino end)

Met-Pro-Phe-Gly-Glu-Arg-Phe-Pro

(carboxyl end)

what would be the most likely nucleotide sequence in the **mRNA** of this gene in the revertant (double mutant)?

TEN

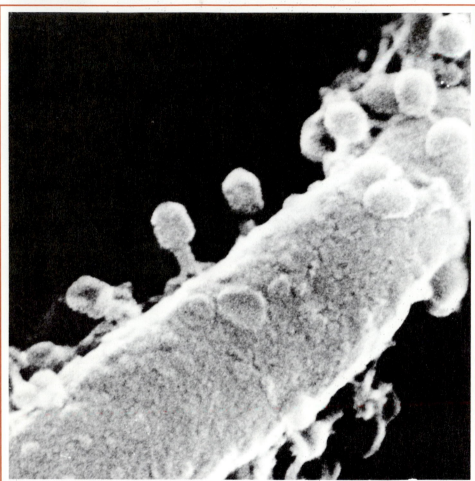

GENETIC FINE STRUCTURE

Pre-1940 studies of the gene were carried out primarily with higher plants and *Drosophila*. These studies focused on the gene as the unit of inheritance controlling one phenotypic trait. The genes were found to undergo rare changes or mutations to new forms or new alleles. New combinations of the wild-type and mutant forms of genes located at different positions on a chromosome were shown to result from crossing over or recombination between homologous chromosomes. These studies provided no information about the structure of the gene itself, however.

Prior to 1940, recombination was not believed to occur within genes. The genes in a chromosome were considered analogous to beads on a string. Recombination was believed to occur only between the beads or genes. The gene was not considered subdivisible.

Since recombination frequencies are correlated with the physical distances separating genetic markers on chromosomes, the failure to detect recombination between markers located at different positions in the same gene is not surprising. Such recombinants occur very infrequently. Their detection involves screening thousands to hundreds of thousands of progeny. With higher plants, and even with *Drosophila*, examining enough progeny to detect such rare recombinants is a formidable task.

More recent studies with microorganisms using selection techniques that permit millions and even billions of progeny to be screened within one to a few days have provided a detailed picture of the fine structure of the gene. In particular, the use of conditional lethal mutations (see pp. 277–278) has greatly facilitated the detection of rare recombination events occurring within genes — **intragenic** recombination.

CLASSICAL VERSUS MOLECULAR CONCEPT OF THE GENE

According to the pre-1940 "beads-on-a-string" concept, the gene was the basic unit of inheritance defined by three criteria: (1) function, (2) recombination, and (3) mutation. More specifically, the gene was:

1. **The unit of function:** the unit of genetic material that controlled the inheritance of one "unit character" or one attribute of phenotype.

2. **The unit of structure:** this could be operationally defined in two ways:

    a. **By recombination:** as the unit of inheritance not subdivisible by recombination.

    b. **By mutation:** as the smallest unit of genetic material capable of independent mutation.

The classical view was that all three criteria defined the same basic unit of inheritance, namely, the gene.

We now know that these criteria define two very different units of inheritance. According to our current molecular concept, the gene is:

The unit of function: the unit of inheritance (1) coding for one polypeptide chain, and (2) operationally defined by the *cis-trans* or complementation test (see pp. 322–325). The other two criteria, recombination and mutation, define the **unit of structure,** which is **equivalent to the single nucleotide pair** (see pp. 340–343). Since it clearly does not make sense to call each nucleotide pair a gene, emphasis has been shifted to the original definition of the gene as the unit of function. Defining the gene as the unit of function, while discarding the recombination and mutation criteria, clearly seems to be the most appropriate solution to the dilemma posed by the demonstration that the original three criteria define two very different units. The emphasis in Mendel's work was on the "anlage" (factor, or gene, as we now call it) controlling one phenotypic characteristic. Thus, Mendel's unit of inheritance correlates more directly with the gene as a unit of function than with the unit of structure or single nucleotide pair.

EARLY EVIDENCE THAT THE GENE IS SUBDIVISIBLE

The first evidence that the gene is subdivisible by mutation and recombination came from studies of the X-linked lozenge locus of *Drosophila melanogaster* by C. P. Oliver in 1940. The two mutations lz^s ("spectacle" eye) and lz^g ("glassy" eye) were considered to be alleles, that is, different forms of the same gene. The data available prior to 1940 indicated that they mapped at the same position on the X chromosome. They had similar effects on the phenotype of the eye, and, most important, lz^s/lz^g heterozygous females had lozenge rather than wild-type eyes. When lz^s/lz^g females were crossed with either lz^s or lz^g males and large numbers of progeny were examined, however, wild-type progeny occurred with a frequency of about 0.2 percent.

These rare wild-type progeny could be explained by reversion of either the lz^s or the lz^g mutation. But there were two arguments against the reversion explanation. (1) The frequency of reversion of lz^s or lz^g to wild-type in hemizygous lozenge males is much lower than 0.2 percent. (2) When the lz^s/lz^g heterozygotes carried outside genetic markers bracketing the lozenge locus, the rare progeny with wild-type eyes always carried an X chromosome with lz^+ that was flanked by a **recombinant combination of outside markers.** Moreover, the same combination of outside markers always occurred, as though the sites of lz^s and lz^g were fixed relative to each other and reciprocal recombination was occurring. Different sets of outside markers were used and yielded the same result. If the lz^s/lz^g heterozygous female carried X chromosomes of the type

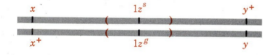

the rare progeny with wild-type eyes all (with one exception) contained an X chromosome with the following composition:

Among progeny of these matings, the reciprocal combination of outside markers (x^+-y^+) was never found in combination with lz^+. This strongly suggested that recombination was involved. Definitive evidence of the involvement of recombination required the recovery and identification of the $lz^s lz^g$ double mutant with the reciprocal combination of outside markers, that is,

Oliver was not able to identify this double mutant, reciprocal recombinant because of the inability to

distinguish it from the parental single mutant phenotypes. The identification of both products, the wild-type and the double mutant, of recombination between two functionally allelic mutations was first accomplished by two of Oliver's students, E. B. Lewis and M. M. Green, in studies of the star-asteroid (small rough eye) locus and the white (eye color) locus, respectively, of *D. melanogaster*.

These pioneering studies first indicated that the gene was in fact more complex than a "bead-on-a-string." They were the first steps toward our present concept of the gene as a long sequence of nucleotide pairs, capable of mutating and recombining at many different sites along its length.

THE *CIS-TRANS* OR COMPLEMENTATION TEST FOR FUNCTIONAL ALLELISM

The functional allelism of any two recessive mutations is determined experimentally by the *cis-trans* or **complementation** test. (To be completely accurate, we should state that the complementation test is only part, the *trans*-component, of the *cis-trans* test.) The *cis-trans* **test provides an operational definition of the gene as the unit of function — the unit controlling the synthesis of one polypeptide.**

The **cis-test** involves putting the two mutations being examined **in a common protoplasm in the *cis*- or coupling configuration** (Fig. 10.1*a*). The

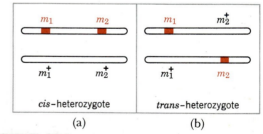

FIGURE 10.1

Composition of cis- *and* trans-*heterozygotes for two mutations* m_1 *and* m_2. *The corresponding wild-type sequences indicated by* m_1^+ *and* m_2^+ *will represent one sequence in* (a) *and will not exist in* (b) *if* m_1 *and* m_2 *overlap a common site, that is, have mutational defects that overlap by one nucleotide pair or more.*

organism or cell containing the two mutations in the *cis*- or coupling configuration (on the same chromosome) is called a **cis-heterozygote.** The *cis*-heterozygote must have the **wild-type phenotype** (Fig. 10.2, right) for the results of the *trans*-test with the same pair of mutations to be valid. If the *cis*-heterozygote has the mutant phenotype, the *trans*-test **cannot** be used to determine whether the two mutations are in the same gene. Thus the *trans*-test can never be used with dominant mutations. The *cis*-test is an important control, which establishes the validity of the correlated *trans*- or complementation test.

The **trans-test,** more commonly called the complementation test, involves putting the **two mutations** to be tested **in a common protoplasm in the trans- or repulsion configuration,** that is, on separate chromosomes (Fig. 10.1*b*). This is accomplished in different ways in different organisms. With diploids like fruit flies and corn plants, it simply involves crossing two homozygotes: one homozygous for one mutation, the other homozygous for the second mutation. In the case of bacteriophage, it involves simultaneously infecting host cells with two different mutants. The organism or cell containing the two mutations in the *trans*- or repulsion configuration (on different chromosomes) is called a **trans-heterozygote.**

1. **If the *trans*-heterozygote has the mutant phenotype** (the phenotype of organisms or cells homozygous for either one of the two mutations), **then the two mutations are in the same unit of function, the same gene** (Fig. 10.2, top left).
2. **If the *trans*-heterozygote has the wild-type phenotype, then the two mutations are in two different units of function, two different genes** (Fig. 10.2, bottom left). In this case the two mutations are said to **complement** each other.

Because construction of the *cis*-heterozygotes requires the often difficult identification of chromosomes carrying both mutations (the "double mutants"), the *cis*-controls are often omitted in routine complementation analyses. Omission of the *cis*-tests may occasionally result in erroneous conclusions. In most cases, however, the *trans*-tests can be interpreted correctly without the *cis*-controls. Because of the time and expense that would usually be required

to construct all of the *cis*-heterozygotes, their omission is usually justified.

The rationale behind the *cis-trans* or complementation test is illustrated in Fig. 10.2. The production of a wild-type phenotype depends on whether functional copies of all of the relevant gene products (usually polypeptides) are produced in the common protoplasm (cell or organism). If no functional product is produced from one or more of the relevant genes, then the cell or organism will exhibit a mutant phenotype. The *cis*-heterozygotes (Fig. 10.2, right) carry one chromosome with wild-type copies of all of the relevant genes. These should produce functional gene products and thus a wild-type phenotype, unless their activity is somehow interfered with or inhibited by the mutant gene products produced from the double-mutant chromosome. In the latter case, the results of the correlated *trans*-tests will not be informative. The *trans*-heterozygotes (Fig. 10.2, left), on the other hand, will produce active forms of all of the gene products only when the two mutations are in two different genes (Fig. 10.2, bottom left). If the two mutations are in the same gene, the *trans*-heterozygote will contain only mutant (nonfunctional) products of that particular gene (Fig. 10.2, top left). It will therefore have the mutant phenotype, the same phenotype as cells or organisms homozygous for one or the other of the two mutations.

S. Benzer introduced the term **cistron** to refer to the unit of function operationally defined by the *cis-trans* test. The concept of the gene has now evolved to where most geneticists use the terms **gene** and **cistron as exact synonyms.**

The results of *cis-trans* or complementation tests are usually unambiguous when mutations that result in the synthesis of no gene product, partial gene products, or totally defective gene products are used, for example, deletions of segments of genes, frameshift mutations, or polypeptide chain-terminating mutations. The mutations must, of course, be recessive. When mutations causing single amino acid substitutions are used, and when the active form of the gene product is a multimer composed of two or more polypeptides, the results of *cis-trans* tests are frequently ambiguous due to the occurrence of a distinct phenomenon called **intra-**

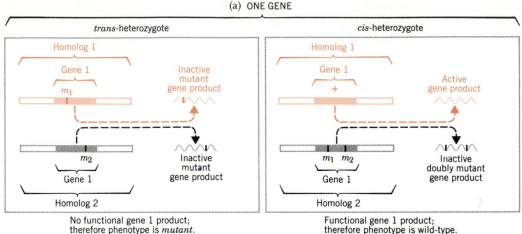

(a) ONE GENE

trans-heterozygote

No functional gene 1 product; therefore phenotype is *mutant*.

cis-heterozygote

Functional gene 1 product; therefore phenotype is wild-type.

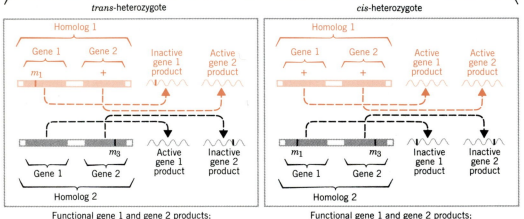

(b) TWO GENES

trans-heterozygote

Functional gene 1 and gene 2 products; therefore phenotype is wild-type.

cis-heterozygote

Functional gene 1 and gene 2 products; therefore phenotype is wild-type.

FIGURE 10.2

Rationale of the cis-trans *or complementation test for functional allelism. The two mutations* m$_1$ *and* m$_2$ *are both in the same gene (a); mutation* m$_3$ *is in a second gene (b). The phenotype of the* cis-*heterozygote (right) must be wild-type or the* trans-*test (left) will not be valid. If the two mutations are located in the same functional unit (the same gene), the* trans-*heterozygote will have the mutant phenotype, because no functional gene product will be synthesized (a, left). If the two mutations are in two different genes, the phenotype of the* trans-*heterozygote will be wild-type, since both gene products will be synthesized in the common protoplasm (b, left). In the case shown, mutations* m$_1$ *and* m$_2$ *are both in gene 1; they do not complement each other. Mutations* m$_1$ *and* m$_3$, *on the other hand, are in two different genes and do complement each other. Homologous chromosome 1 and its gene products are shown in color; homologous chromosome 2 and its gene products are shown in gray.*

genic complementation (see pp. 325–327).

The *cis-trans* operational definition of the gene is one of the most important concepts of molecular genetics. **The information provided by *cis-trans* or complementation tests is totally distinct from that obtained from recombination analyses. Complementation and recombination should not be confused.** Figures 10.3 and 10.4 illustrate complementation (Fig. 10.3) and recombination (Fig. 10.4) tests with the same mutant strains of phage T4.

Complementation is the result of the interaction of the **gene products** produced by chromosomes carrying two different mutations when they are present in a **common protoplasm.** It does not depend on recombination of the two chromosomes or involve any direct interaction between the chromosomes. **Complementation, or the lack of it, is determined by the phenotype (wild-type or mutant) of each *trans*-heterozygote.**

Recombination, on the other hand, involves **direct interaction between the chromosomes** carrying the mutations, the actual **breakage of chromosomes and reunion of parts** in such a way that wild-type and double-mutant chromosomes are produced (see Chapter 6). **Complementation should occur in every *trans*-heterozygote** containing mutations in two different genes. **Recombination is normally detected by examining the progeny of heterozygotes,** and a *trans*-heterozygote will never produce more than 25 percent gametes (or progeny for haploids) with wild-type chromosomes. If the mutations are closely linked, as, for example, in Fig. 10.4*b*, this frequency will be much lower than 25 percent.

LIMITATIONS OF THE CIS-TRANS TEST

The *cis-trans* or complementation test has proven to be extremely useful in delimiting genes. In many cases, two or more mutations producing the same phenotype can be unambiguously assigned to one or more genes on the basis of the results of complementation tests. There are four cases in which the results of complementation tests **cannot** be used to unambiguously delimit genes. These are (1) dominant or codominant mutations, (2) genes in which mutations occur that exhibit **intra**genic complementation (see below), (3) "polar mutations" (mutations that affect the expression of adjacent genes) within "operons" (coordinately regulated multigenic units of expression; see Chapter 11), and (4) *cis*-acting genes (most commonly genes that do not code for diffusible products).

Cis-heterozygotes will **not** have wild-type phenotypes (as required for valid *cis-trans* tests) unless both mutations present are completely recessive. Thus, dominant or codominant mutations **cannot** be analyzed by *cis-trans* tests.

The functional forms of certain enzymes are dimers or higher multimers consisting of two or more polypeptides. These polypeptides may be either homologous, the products of a single gene, or nonhomologous, the products of two or more distinct genes. When the active form of the enzyme contains two or more homologous polypeptides (it may or may not also contain nonhomologous polypeptides), **intragenic complementation** may occur. **Intra**genic complementation is a phenomenon totally distinct from **inter**genic complementation, on which the *cis-trans* test is based. In organisms that are homozygous for the wild-type allele of a given gene, all of the protein dimers or higher multimers will contain identical wild-type polypeptides. Similarly, organisms that are homozygous for any mutation in that gene will contain dimers or higher multimers, all of which contain identical copies of the mutant polypeptide (Fig. 10.5). An organism that is **heterozygous for two different mutations in the gene** will usually produce **some dimers or higher multimers that contain one or more of each of the two different mutant polypeptides** (Fig. 10.5). These "heteromultimers" (protein multimers composed of the polypeptide products of two different alleles of a gene) **may** have partial or complete (wild-type) activity. In the case of **noncomplementing** mutations in a gene coding for a multimeric enzyme, the "heteromultimers" are **inactive** just like the mutant "homomultimers" (protein multimers composed of two or more identical mutant polypeptides).

In several known cases of **intra**genic comple-

(a) Complementation between mutations *am*B17 and *am*E18.

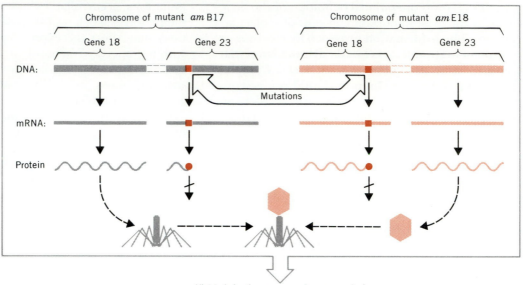

Yields infective progeny phage upon lysis.

(b) Lack of complementation between mutations *am*B17 and *am*H32.

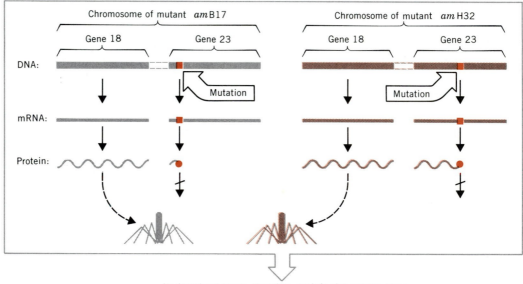

No functional heads; therefore, no infective progeny phage.

mentation, the active form of the enzyme in the heterozygote has been purified and shown to be a heterodimer or heterotetramer containing the two distinct mutant polypeptides. Why such heteromultimers should be active whereas the two corresponding homomultimers are inactive is not clear.

FIGURE 10.3

Diagram illustrating complementation (a) *or lack of complementation* (b) *in* trans-*heterozygotes.* (a) *Complementation between a mutation* (am *B17*) *in the gene (gene 23) coding for the major structural protein of the phage T4 head and a mutation* (am *E18*) *in the gene (gene 18) coding for a major structural protein of the phage tail. Both heads and tails are synthesized in the common protoplasm with the result that infective progeny phage are produced.* (b) *When both mutations* (am *B17 and* am *H32*) *are in the gene (gene 23) coding for the major head protein, no heads are synthesized. Thus, no infective progeny phage can be assembled. In both cases, the* trans-*heterozygotes are produced by simultaneously infecting* E. coli *strain B with T4 phage strains that carry the mutations being analyzed. In this host,* amber *mutations cause only a fragment of the polypeptide gene product to be synthesized, since these mutations yield mRNA codons that are chain-terminating codons. (After D. P. Snustad and D. S. Dean,* Genetics Experiments with Bacterial Viruses, *W. H. Freeman, San Francisco, 1971.)*

Apparently, the sequence of amino acids in the "nonmutant" segment of one mutant polypeptide stabilizes or somehow compensates for the mutant segment of the polypeptide coded for by the second mutant allele, and vice versa (Fig. 10.5). Most proteins, however, have very complex three-dimensional structures (see Chapter 8, pp. 237–240), and until the exact structures of a wild-type homomultimer, two mutant homomultimers, and an active heteromultimer composed of the two mutant polypeptides have been determined, the molecular basis of **intra**genic complementation will continue to be a subject of speculation.

A **"polar mutation"** is a mutation that not only results in a defective product of the gene in which it is located, but also interferes with the expression of one or more adjacent genes. The latter are always located on one side of the gene carrying the mutation (thus the term "polar mutation"). Such polar mutations are frequently observed in coordinately regulated sets of genes called **operons** in prokaryotes (see Chapter 11). Most commonly, they are mutations resulting in polypeptide chain-termination signals (nucleotide-pair triplets yielding UAA, UAG, and UGA codons in mRNA; see Table 8.1) within structural genes (genes coding for polypeptide products). One would not expect a polar mutation to complement a mutation in another gene if the latter cannot be expressed due to the epistatic effect of the polar mutation. *Cis-trans* analyses of polar mutations thus frequently yield ambiguous results.

In addition, "***cis*-acting" genes** or segments of genetic material such as "operator regions" (binding sites for regulator proteins) or "promoter regions" (binding sites for RNA polymerase) cannot be analyzed by the *cis-trans* test. Clearly, the *cis-trans* test depends on the gene products being diffusible. Regulatory binding sites on chromosomes like operators and promoters do not code for diffusible products. They only affect the expression of genes located *cis* to them, that is, on the same chromosome (see Chapter 11). Such "*cis*-acting" genes or sequences are not tractable by *cis-trans* analysis.

FINE STRUCTURE OF THE PHAGE T4 rII LOCUS

The most extensive fine structure map constructed to date is that of the phage T4 *r*II locus. S. Benzer isolated over 3000 independent mutant strains carrying mutations that map within the *r*II locus. His detailed fine structure analysis of these *r*II mutants showed that the mutations mapped at over 300 distinct sites, all located with the two genes that make up the *r*II locus.

Phage T4, like phage T2 (see Chapter 5, pp. 84–86), is an obligate parasite that grows in the common colon bacillus *Escherichia coli*. It consists of a single linear molecule of DNA about 200,000

(a) Recombination between chromosomes carrying mutations *am*B17 and *am*E18.

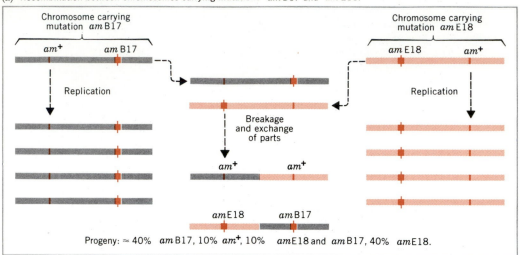

(b) Recombination between chromosomes carrying mutations *am*B17 and *am*H32.

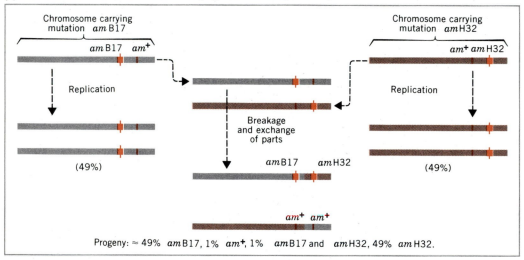

FIGURE 10.4

Diagram illustrating recombination between (a) *the complementing mutations* am *B17 (gene 23) and* am *E18 (gene 18) and* (b) *the noncomplementing mutations* am *B17 and* am *H32 (both in gene 23). In both cases, recombination is analyzed by simultaneously infecting* E. coli *strain CR63* (amber *permissive) with the two mutant T4 phages. In this host, the* amber *mutations are suppressed, allowing the mutants to grow. The genotypes of the progeny are determined by their ability or inability to grow on restrictive (suppressor negative) hosts, such as* E. coli *strain B. (After D. P. Snustad and D. S. Dean,* Genetics Experiments with Bacterial Viruses, *W. H. Freeman, San Francisco, 1971.)*

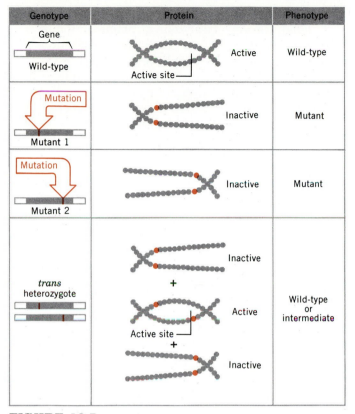

Genotype	Protein	Phenotype
Gene / Wild-type	Active — Active site	Wild-type
Mutation / Mutant 1	Inactive	Mutant
Mutation / Mutant 2	Inactive	Mutant
trans heterozygote	Inactive + Active (Active site) + Inactive	Wild-type or intermediate

FIGURE 10.5

Schematic illustration of the mechanism of intragenic complementation. Intragenic complementation may occur whenever the active form of the enzyme (or protein) is a multimer that **contains at least two copies of any one gene product (polypeptide).** *In the case illustrated, the functional form of the enzyme is a dimer composed of two polypeptide products of a single gene. In mutant organisms carrying either mutation 1 or mutation 2 (either homozygous diploids or haploids), only nonfunctional dimers are produced (thus the mutant phenotype). In a* **trans**-*heterozygote carrying both mutant alleles (1 and 2), part of the dimers may contain one copy of each of the mutant polypeptides. Such "heterodimers" (one mutant 1 polypeptide plus one mutant 2 polypeptide) (bottom, center) are sometimes functional. These heterodimers may be totally active (wild-type level of activity), partially active (some activity, but less than the wild-type level), or inactive. As a result, such* **trans**-*heterozygotes may have a wild-type phenotype or a phenotype intermediate between the mutant and wild-type. The positions of the two mutations within the gene are indicated by boxes (in color). The amino acids altered by the mutations are shown as filled circles (in color) in the polypeptide chains.*

nucleotide pairs long, which is packed within the head of the protein phage coat (Figs. 10.6 and 10.7; also see Chapter 9, pp. 314–315). When a wild-type

T4 phage particle infects an *E. coli* (Fig. 10.8) under optimal conditions, it will produce 200–400 progeny viruses and lyse the host cell in about 20–25

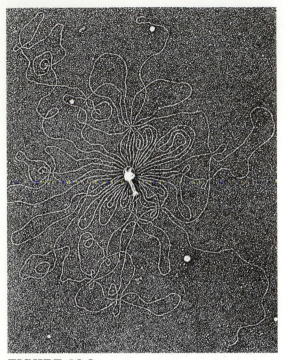

FIGURE 10.6

Electron micrograph of a bacteriophage T2 particle (center) from which the DNA (a single linear molecule 20 Å in diameter and about 500,000 Å long) has been released by osmotic shock. The DNA molecule has been shadowed with platinum to make it visual. The diameter of the visible DNA-platinum thread is thus much greater than 20 Å, making it appear impossible to pack all of the DNA inside the phage head. Note the two ends of the linear molecule—one at the top center and the other at the bottom right. [Courtesy of Dr. A. K. Kleinschmidt. Reproduced with permission from Biochem. Biophys. Acta *61, p. 861 (Fig. 1), 1962.]*

minutes (Fig. 10.9). If a T4 phage particle is placed on the surface of semisolid agar nutrient medium seeded with a confluent growth ("lawn") of *E. coli* cells in a petri dish, the phage will infect one cell, lyse it, and release about 300 progeny phage particles. Each of these progeny phage will, in turn, infect another *E. coli* cell, lyse it, and release about 300 progeny viruses. After a few hours and several cycles of infection, all of the cells in the immediate vicinity of the original phage particle will have been lysed, leaving a clear zone or **plaque** in the lawn of bacteria (Fig. 10.10). (Each plaque will ultimately contain up to 5×10^8 phage particles.)

Plaques produced by wild-type T4 phage have fuzzy or turbid margins or halos (Fig. 10.10). When a wild-type T4-infected *E. coli* cell is superinfected with a second T4 phage prior to lysis, synthesis of new cell wall material is triggered, delaying lysis for up to two hours. This phenomenon, called "lysis inhibition," is responsible for the turbid halos or margins around plaques produced by wild-type T4 phage. Progeny phage, released by cells lysing early, superinfect surrounding cells and delay their lysis. The margins of wild-type T4 plaques thus contain a mixture of lysed and lysis-inhibited cells, producing a partially clear zone (Fig. 10.10).

Certain mutants of phage T4, designated *r* mutants for **rapid lysis,** do not exhibit lysis inhibition. As a result, *r* mutants produce easily recognized plaques with clear margins (Fig. 10.10). The *r* mutant plaques are also larger under certain conditions. The mutations in these *r* mutants map primarily at three different locations on the T4 chromosome (Fig. 10.11).

Benzer discovered that mutants carrying mutations located at one of these three loci, the *r*II locus, were unable to grow on strains of *E. coli* K12 carrying an integrated (covalently inserted, see pp. 204–205) chromosome of another bacteriophage, phage lambda [designated *E. coli* K12(λ)]. They grew rapidly, however, on other strains of *E. coli*, like strain B or K12 strains not carrying the integrated lambda chromosome. While neither Benzer nor anyone else to date has been able to explain the basis of the lethality of *r*II mutants in *E. coli* K12(λ), Benzer recognized that the *r*II mutants provided him with a conditional lethal system (see pp. 277–278) that could be exploited to study genetic fine structure.

Phage carrying *r*II mutations can be easily isolated by serially transferring inocula with sterile toothpicks from *r*-type plaques on *E. coli* B to lawns of *E. coli* K12(λ) and *E. coli* B. All *r*II mutants will produce *r*-type plaques on *E. coli* B, but will not grow on *E. coli* K12(λ) (Fig. 10.12). The *r*I and *r*III mu-

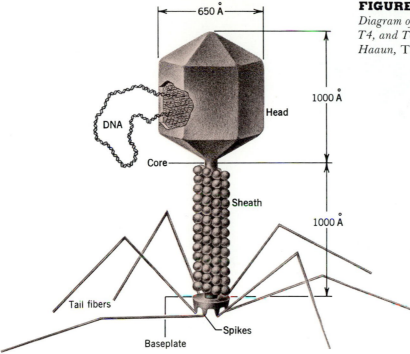

FIGURE 10.7

Diagram of the intricate morphology of the T-even (T2, T4, and T6) bacteriophages. (From Nason and De-Haaun, The Biological World, *1973.)*

650 Å

1000 Å

Head

DNA

Core

Sheath

1000 Å

Tail fibers

Spikes

Baseplate

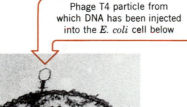

Phage T4 particle from which DNA has been injected into the *E. coli* cell below

FIGURE 10.8

Electron micrograph showing a cross section of an E. coli *cell that has been infected by a T4 bacteriophage. The tail sheath has condensed like a small muscle, and the tail core protrudes into the host cell. The DNA that was packaged inside the phage head has been injected into the cell through an opening in the tail core. Magnification approximately* ×*30,000. (Micrograph by D. P. Snustad.)*

tants will grow on both B and K12(λ). As mentioned earlier, Benzer isolated over 3000 independent *r*II mutants. The first question that one might ask, as Benzer did, is how many genes do these mutations identify—one, two, three, . . . , or 3000? When analyzed by *cis-trans* or complementation tests (see Fig. 10.2), all of the *r*II mutations were found to be located in one of two genes or cistrons. Benzer designated these two genes *r*IIA and *r*IIB. Fortunately, not all pairwise combinations of the over

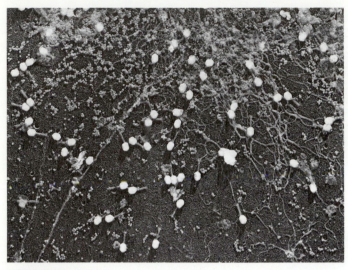

FIGURE 10.9

A lysate of a phage T4-infected E. coli *cell, showing many T4 progeny particles (large white heads with gray tails), clusters of ribosomes (small gray spherical structures), DNA molecules (gray threads), and cell debris. Magnification ×30,000. (Courtesy of E. Kellenberger, Biozentrum der Universität, Basel.)*

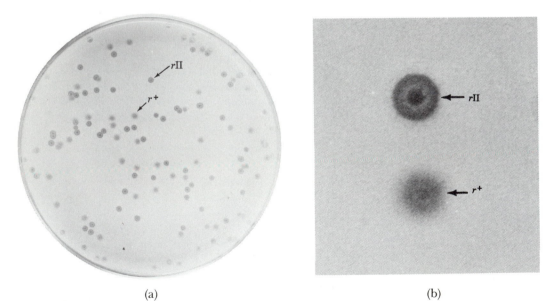

(a)   (b)

FIGURE 10.10

(a) *Photograph of a mixture of phage T4* rII *mutant and* r+ *plaques on a confluent lawn of* E. coli *strain B cells.* (b) *Higher magnification of individual* rII *and* r+ *plaques, showing the sharp margin on the* rII *plaque and the fuzzy margin on the* r+ *plaque. (Photographs by R. Wagner and D. P. Snustad.)*

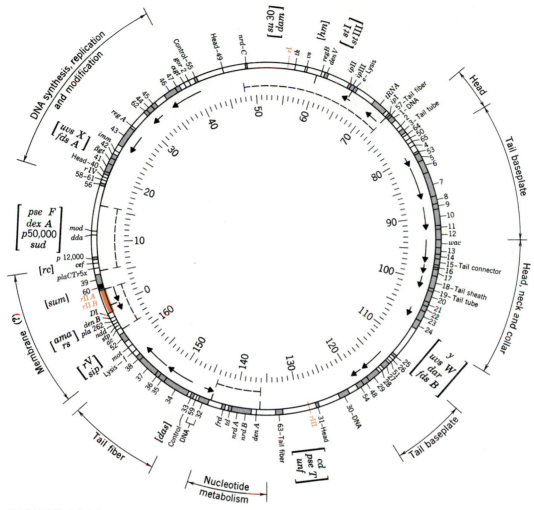

FIGURE 10.11

The circular linkage map of phage T4, showing the location of the rII locus (genes A and B, color) studied by Benzer as well as the rI and rIII loci (color). Most r mutations map at these three loci. The interior numerical scale gives physical distances in 1000 base-pair ("kilobase") units, starting arbitrarily at the junction between the rIIA and rIIB genes. The map includes most of the known genes of phage T4. Genes whose exact order is not known are enclosed in parentheses in arbitrary order. The arrows indicate the direction of transcription where known. Functional descriptions of clusters of genes are included around the periphery of the map. (From W. B. Wood and H. R. Revel, Bacteriological Revs. *40:847, 1976.)*

3000 mutants needed to be tested. Once a mutation in each gene had been identified, it was used as a reference for that gene. Each new mutation that was isolated only had to be tested for complementation with the *A* gene and *B* gene reference mutations. Additional complementation tests were unneces-

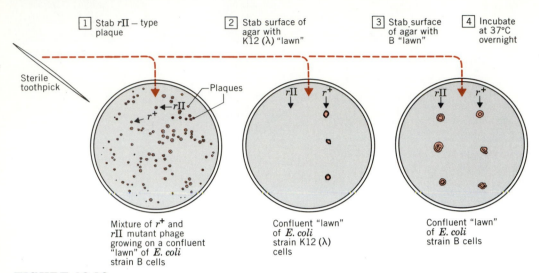

1 Stab rII – type plaque

2 Stab surface of agar with K12 (λ) "lawn"

3 Stab surface of agar with B "lawn"

4 Incubate at 37°C overnight

Sterile toothpick

Plaques

rII

r⁺

rII   r⁺

rII   r⁺

Mixture of r⁺ and rII mutant phage growing on a confluent "lawn" of *E. coli* strain B cells

Confluent "lawn" of *E. coli* strain K12 (λ) cells

Confluent "lawn" of *E. coli* strain B cells

FIGURE 10.12

Identification of rII *mutants by sterile toothpick transfers of phage from individual plaques growing on* E. coli *strain B (rII-permissive) "lawns" to lawns of* E. coli *strain K12(λ) (rII-restrictive) and lawns of* E. coli *strain B. Each plaque to be tested (left) is stabbed with a sterile toothpick, which is subsequently touched to a marked area in a petri dish with a K12(λ) lawn (center) and then to an identically marked area in a dish with an* E. coli *B lawn (right). Mutants that fail to grow (are lethal) on K12(λ) (center plate, left side) can be recovered from the plaques on the* E. coli *B plates (right). Mutants at the* rI *and* rIII *loci as well as* r⁺ *phage (center plate, right side) will grow on both K12(λ) and B. (After D. P. Snustad and D. S. Dean,* Genetics Experiments with Bacterial Viruses, *W. H. Freeman, San Francisco, 1971.)*

sary since all of the *r*II mutations identified failed to complement with one of the two reference mutations. That is, one group of mutations complemented all gene *A* mutations and failed to complement all other gene *B* mutations. The second group of mutations complemented all gene *B* mutations and failed to complement all other gene *A* mutations. (A few mutations, "deletion" mutations, failed to complement all other *r*II mutations; see pp. 335–340).

The *r*II and *r*⁺ phenotypes are the most distinct in *E. coli* K12(λ), *r*II mutants being lethal and *r*⁺ phage exhibiting normal growth. Thus, complementation tests between *r*II mutants are done by simultaneous pairwise infections of *E. coli* K12(λ) cells. After allowing a period of time for phage growth and lysis of the host cells (usually 90–120

minutes), the infected K12(λ) cells, or lysates, are spread on lawns of *E. coli* B cells to see if any progeny phage were produced. If complementation has occurred, 200–400 progeny phage will have been produced per infected cell. In the absence of complementation, no progeny phage will have been produced.

Benzer also developed a simplified procedure, the so-called **complementation spot test,** for examining complementation qualitatively. This procedure involves placing a small drop of a mixed suspension of each pair of mutants being tested on a lawn of *E. coli* K12(λ) cells. Alternatively, small drops, each containing a different mutant, may be placed on a lawn containing the appropriate ratio of *E. coli* K12(λ) cells and a reference mutant. If complementation occurs, all of the cells in the area covered by the drop of suspended phage will be lysed, yielding a large plaque (Fig. 10.13). If complementation does not occur, the area will be overgrown with a lawn of *E. coli* cells.

While complementation tests showed that all of the *r*II mutations were located in two genes, recombination tests demonstrated that these mutations were located at many distinct sites within these two genes. Recombination between *r*II mutations is an-

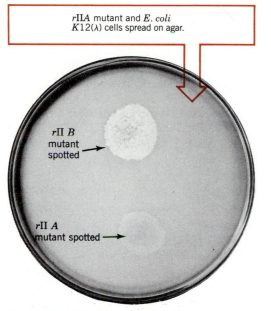

*r*IIA mutant and *E. coli* K12(λ) cells spread on agar.

*r*II B mutant spotted

*r*II A mutant spotted

FIGURE 10.13

Complementation spot tests with phage T4 rII mutants. T4 phage particles with a mutation in the A gene of the rII locus and enough E. coli K12(λ) cells to yield a confluent growth of cells after incubation were spread on the surface of nutrient agar medium. A drop of solution that contained a suspension of phage particles (about $10^6 - 10^7$ phage) with a mutation in the B gene was then applied to a spot at the top of the plate. Similarly, a drop containing phage particles with a mutation in the A gene was applied to a spot at the bottom of the plate. The large number of plaques (clear zones) formed in the top "spot" after incubation is evidence of complementation. The absence of plaques in the bottom "spot" indicates that complementation did not occur between the two rIIA mutants. (Photograph by R. Wagner and D. P. Snustad.)

alyzed by simultaneously infecting *E. coli* B cells (**permissive** host cells) with the two *r*II mutants in question and "plating" (spreading on the surface of agar medium in a petri dish) the progeny phage on a lawn of *E. coli* K12(λ) to see if any wild-type recombinants have been produced. The total number of progeny phage produced is determined by plating the lysate on an *E. coli* B lawn. As many as 10^8 progeny phage may be examined in a single petri dish. If only one r^+ recombinant is present, only one plaque will be formed on K12(λ). Thus, recombination events as rare as 1 in 100 million can easily be detected. Such a conditional lethal system thus provides an extremely powerful sieve for selecting rare recombinants.

Only the r^+ recombinants form plaques on K12(λ). The equally frequent double-mutant recombinants do not produce plaques on K12(λ). Thus, the

Recombination frequency

$$= \frac{2[\text{number of plaques on K12}(\lambda)]}{\text{number of plaques on B}}$$

However, to map over 3000 independent mutations by standard two- and three-factor crosses (see Chapter 6), even with a phage system, is a formidable task. All possible two-factor crosses of 3000 mutants is about $4\frac{1}{2}$ million [(3000)(2999)/2] crosses. Benzer was able to avoid such a laborious undertaking by developing a shortcut method of mapping that used overlapping deletion mutations.

DELETION MAPPING

Many of the spontaneous *r*II mutants isolated by Benzer failed to recombine with two or more mutants that recombined with each other. They behaved as though they contained "multisite" mutations or defects that extended over a segment of the *r*II locus. Some of them appeared to involve only short segments within either the *r*IIA gene or the *r*IIB gene. Others were much longer; a few appeared to span the entire *r*II locus. Benzer proposed that these multisite mutations resulted from the deletion or loss of segments of DNA. His proposal that these multisite mutations were deletions of segments of DNA was supported by the observation that they did not undergo reverse mutation back to wild-type. Point mutations (base-pair substitutions and single base-pair additions and deletions) are capable of reverse mutation, albeit at highly variable rates. Benzer's proposal has subsequently been verified directly by analysis of several of these putative deletion mutations by electron microscopy (see the following section).

The extents of the deleted segments can be determined by crossing the deletion mutants to a set of reference point mutants previously mapped by two- and three-factor crosses (Fig. 10.14). Once a set of overlapping deletions has been mapped, their end points will divide the region spanned by the longest deletion into a set of intervals (such as A, B, C, and D in Fig. 10.14). When a new mutant carrying a point mutation is isolated, the mutation can be quickly mapped to a defined interval by crossing the mutant strain with each of the overlapping deletion mutants. For example, suppose that five point mutants, r_1, r_2, r_3, r_4, and r_5, are crossed with the four deletion mutants, I, II, III, and IV, shown in Fig. 10.14. Furthermore, suppose that the results of these two-factor crosses are:

POINT MUTANTS	DELETIONS			
	I	II	III	IV
r_1	0	+	+	+
r_2	0	0	+	+
r_3	0	0	0	+
r_4	0	0	0	0
r_5	0	0	0	0

where + indicates that recombination has occurred and 0 indicates that no recombinants have been formed. These results would immediately allow one to localize the mutation present in mutant r_1 to interval A. Similarly, the mutations in r_2 and r_3 could be assigned to intervals B and C, respectively. Finally, the defects in both r_4 and r_5 must be in interval D. The r_4 and r_5 mutations cannot be ordered with respect to each other from these data. If a set of smaller overlapping deletions in interval D were available, positions of the mutations present in mutants r_4 and r_5 might be further localized by crossing each of these point mutants with each mutant of a set of smaller overlapping deletion mutants. Benzer characterized a large number of deletion mutants that **divided the rII locus into 47 small segments** (Fig. 10.15). By using these to localize the rII point mutants in newly isolated mutant strains, Benzer was able to greatly reduce the number of crosses needed to establish the detailed fine structure map of the rII locus. After the shortcut deletion mapping was completed, the mutations located within each of the 47 intervals had to be mapped relative to each other by standard two- and three-factor crosses. Since most of the intervals contained a relatively small number of mutations, this was a manageable task, in contrast to the $4\frac{1}{2}$ million

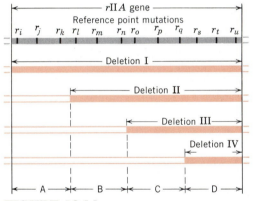

FIGURE 10.14

Illustration of the use of overlapping deletions in genetic fine structure mapping. Four hypothetical deletion mutants are shown; the shaded areas indicate the extents of the deleted segments. Their end points have been determined by crossing them to a set of rII mutant strains that carry point mutations (top) previously mapped by standard two- and three-factor crosses. These deletions divide the rIIA gene into four intervals. The interval A, B, C, or D in which an unknown point mutation is located can be determined by crossing a mutant that carries the unknown mutation with each of the four deletion mutants. A mutation in interval D will not produce any wild-type recombinant progeny in any of the four crosses. A mutation in interval C will recombine with deletion IV, but not with the other three deletions, and so on.

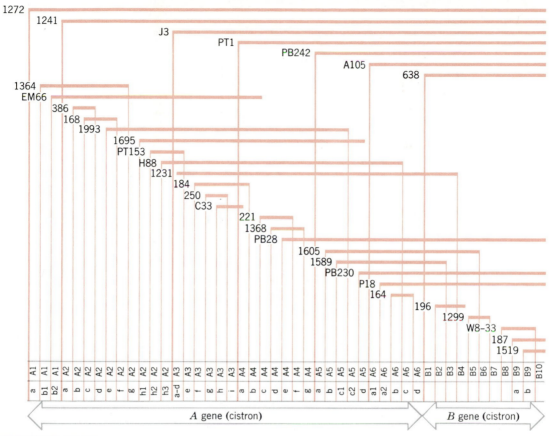

FIGURE 10.15

The overlapping deletions used by Benzer and co-workers to divide the rII locus into 47 smaller intervals. Benzer's designations for each of the 47 segments are given in the boxes near the bottom; they correspond to the segments shown on the map in Fig. 10.16. The limits of the A and B genes, as defined by complementation tests, are shown by the arrows at the bottom. (After S. Benzer, Proc. Natl. Acad. Sci., U.S. *47:410, 1961.)*

two-factor crosses required without the deletion mapping step.

Benzer and colleagues identified **over 300 sites of mutation that were separable by recombination.** Figure 10.16 shows the 250 sites at which spontaneous mutations occurred. Note that the **frequency of mutation at different sites is highly variable.** In particular, two sites, one in segment A6c and one in segment B4, are highly mutable (so-called "hot spots" for mutation). It would be

very interesting to know why these two sites mutate so frequently.

ELECTRON MICROSCOPE HETERODUPLEX MAPPING

The existence of genetically well-defined deletion mutations at the *r*II locus has permitted E. K. F. Bautz and colleagues to determine the physical size of the *r*II locus. This was done using a technique called **heteroduplex mapping. A DNA heterodu-**

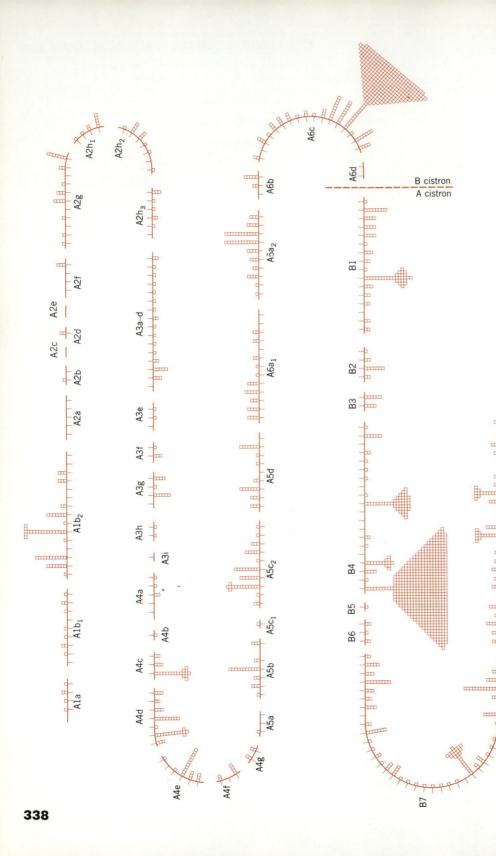

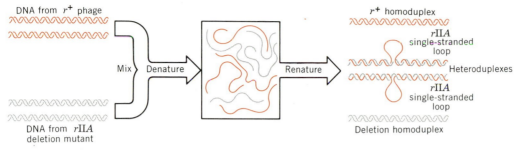

FIGURE 10.17

Diagram illustrating the use of deletion heteroduplex mapping of genes or other segments of DNA. The example shown illustrates the formation of heteroduplex molecules in which one strand of DNA is r$^+$ and the other strand contains a deletion of the entire rIIA gene. Such DNA heteroduplexes will contain single-stranded loops, the length of which corresponds to the length of the rIIA gene. If equal numbers of the two kinds of DNA molecules are mixed, the renatured double helices should contain, on an average, 50 percent heteroduplexes.

plex is a DNA molecule in which the two strands
are not entirely complementary. One strand of a
DNA double helix may contain one allele of a gene
and the other strand (largely, but not totally, complementary) may carry a different allele of the gene.
The noncomplementary regions of DNA heteroduplexes vary in size from one mismatched base-pair to
large segments of the molecule. DNA heteroduplexes occur *in vivo* as intermediates in certain processes of mutation (see Chapter 9) and recombination (see Chapter 6).

Heteroduplex mapping involves the *in vitro*
preparation of DNA heteroduplexes and their analysis by electron microscopy. If, for example, a preparation of DNA from wild-type T4 phage and a
preparation of DNA from an *r*II mutant with a

deletion of the *r*IIA gene are denatured (as by heat-
ing to 100°C or by alkali at pH 11.4), are mixed
together, and are then allowed to renature (e.g.,
60°C for 8–12 hours in the appropriate solution),
some of the renatured DNA double helices (half of
them if equal amounts of the two DNA types are
mixed) will be heteroduplexes. These heteroduplexes (Fig. 10.17) will contain one strand of DNA
from the wild-type T4 and one strand from the *r*IIA
deletion mutant. The wild-type strands of these
heteroduplexes will not have complementary base
sequences with which to pair in the segment of the
molecule corresponding to the *r*IIA gene. The
wild-type strands will thus form single-stranded
loops, the lengths of which can be measured on
electron micrographs.

FIGURE 10.16

A genetic map of the spontaneous mutations in the rII locus studied by Benzer and colleagues. Each small square symbolizes one mutation that occurred at a particular mutable site. Mutable sites that are separable by recombination are indicated by short vertical lines. Mutable sites at which no squares occur were identified by mutations induced with mutagenic agents; no spontaneous mutations were detected at these sites. Each short segment of the map (defined by the discontinuities) represents one of the 47 intervals defined by

the overlapping deletions shown in Fig. 10.15. No spontaneous mutations were detected that mapped in intervals A2a, A2c, A2e, or A6d; in fact, no mutations, spontaneous or induced, were identified that mapped in intervals A2c and A2e. Spontaneous mutations occurred very frequently at two sites ("hot spots"), one in interval A6c and one in interval B4. The boundary between the A and B genes is indicated by the dashed line. (After S. Benzer, Proc. Natl. Acad. Sci., U.S. 47:410, 1961.)

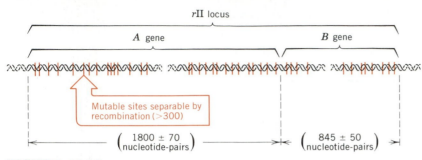

FIGURE 10.18

Summary of the fine structure of the phage T4 rII locus, based largely on the results of Benzer and co-workers and Bautz and co-workers. The total length of the rII locus is, according to Bautz's data, 2645 ± 100 nucleotide pairs of the DNA double helix. For simplicity, only the distal portions of each gene are shown. Benzer identified over 300 distinct (by recombination) sites of mutation within these two genes.

Bautz and co-workers prepared DNA heteroduplexes between DNA from T4 r^+ phage and DNA from each of several genetically well-characterized rII deletion mutants; they then analyzed them by electron microscopy. Their results yielded estimates of **1800 ± 70 nucleotide pairs and 845 ± 50 nucleotide pairs for the sizes of the rIIA and rIIB genes, respectively.** These results, combined with the extensive genetic data of Benzer and colleagues, provide a fairly clear picture of the fine structure of the rII locus (Fig. 10.18).

Since the mutable site is clearly equivalent to the single nucleotide pair (see Chapter 9 and the following section), the over 300 mutable sites identified by Benzer represent only a small fraction of the total number of "sites" or nucleotide pairs (about 2645) in the rII locus. A mutable site can be identified, of course, only when a base-pair substitution at that site results in a mutant phenotype, usually as a result of the production of an inactive or partially inactive gene product. Because of the "degeneracy" in the genetic code (more than one triplet codon per amino acid; see Chapter 8), many base-pair substitutions will not change the amino acid sequences of the polypeptide gene products at all. Many other base-pair substitutions will result in the substitution of one amino acid for another, where both amino acids have very similar chemical properties. Such substitutions usually do not significantly alter the activity of the protein. Thus, while it is clear that many mutable sites in the rII locus were not detected in Benzer's study, it is also clear that **many nucleotide pairs are not potential sites for changes that lead to a mutant phenotype.**

THE UNIT OF STRUCTURE IS THE SINGLE NUCLEOTIDE PAIR

Benzer's extensive genetic analysis of the rII locus of phage T4 demonstrated that many mutable sites that are separable by recombination can exist within a single gene (defined by the *cis-trans* test as the unit of function coding for one polypeptide). This picture of the fine structure of the gene was soon verified by the results of studies of many different genes in many different organisms. Based on the known structure of DNA and the nature of base-pair substitution in mutagenesis (Chapter 9), it seemed likely that the unit of structure not divisible by recombination was the single nucleotide pair. This was first demonstrated experimentally by C. Yanofsky and co-workers in studies of mutations in the *trpA* gene of *E. coli*.

The enzyme tryptophan synthetase catalyzes the final step in the biosynthesis of the amino acid tryptophan. In *E. coli*, this enzyme is composed of two different polypeptides, designated A and B (the products of genes *trpA* and *trpB*, respectively). Yan-

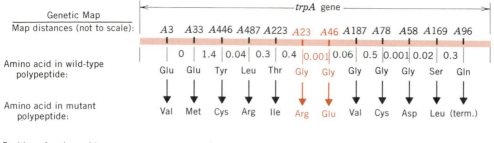

FIGURE 10.19

Genetic map of the trpA *gene of* E. coli, *and the amino acid replacements that have been shown to occur in the tryptophan synthetase A polypeptide as a result of the indicated mutations (A3, A33, etc.). C. Yanofsky and colleagues have worked out the entire sequence of this 268 amino acid-long polypeptide. The positions (numbered from the amino terminus) of the amino acid changes that occurred due to the mutations shown are given at the bottom. Note that mutations A23 and A46 (shown in color) do recombine despite the fact that they both alter the same amino acid, namely, the glycine at position number 211 of the wild-type polypeptide. The tryptophan synthetase A polypeptide of mutant A23 has arginine at position 211; that of mutant A46 has glutamic acid at position 211. Thus both mutations must have occurred within the sequence of three base-pairs of the* trpA *gene that specifies (via the mRNA triplet codon) the amino acid present at position 211 of the tryptophan synthetase A polypeptide. (After C. Yanofsky and V. Horn,* J. Biol. Chem. *247:4494–4498, 1972.)*

ofsky and colleagues isolated and characterized a large number of tryptophan auxotrophic mutants with mutations in the *trpA* gene. The *trpA* gene codes for a polypeptide (the tryptophan synthetase A polypeptide) that is 268 amino acids long, and Yanofsky and associates used the techniques of protein sequencing to work out the complete amino acid sequence of this polypeptide. They also determined the amino acid sequences of several mutant forms of the tryptophan synthetase A polypeptide. When the positions of the amino acid changes in the polypeptide were compared with the locations of the mutations on the genetic map, a precise **colinear** relationship was observed (Fig. 10.19).

Convincing evidence for **colinearity** between the sequence of nucleotide pairs in the gene and the sequence of amino acids in the polypeptide gene product has also been obtained in several other studies, in both prokaryotes and eukaryotes (see Chapter 8, pp. 267–270). The recent demonstrations of "noncoding segments" or "introns" within

certain eukaryotic genes (see pp. 361–365), however, indicate that the concept of an **uninterrupted** sequence of triplet base-pairs in the gene specifying a colinear sequence of amino acids in the polypeptide gene product may have to be modified in many cases.

Yanofsky's correlated genetic and biochemical data for the *trpA* gene and the tryptophan synthetase A polypeptide (Fig. 10.19) showed that independent mutations that alter the same amino acid may either (1) affect the same mutable site and not recombine (such as *A3* and *A33*) or (2) occur at different mutable sites (such as *A23* and *A46* or *A78* and *A58*). Moreover, the data show that **independent mutations at the same mutable site** (as shown by their failure to give rise to *trp*⁺ recombinants in crosses) **are not always identical.** Mutation *trpA3* changes glutamic acid at position 49 to valine. Mutation *trpA33* changes the glutamic acid at position 49 to methionine. No recombinants are produced in crosses between mutant strains carrying mutations

$A3$ and $A33$. Thus, **$trpA3$ and $trpA33$ are structurally allelic (as determined by recombination tests) and functionally allelic (as determined by *cis-trans* tests). Both mutations have occurred at the same unit of structure, the same nucleotide pair.** H. Roman has proposed that such mutations be called **"homoalleles"** ("homo" = same, or **alleles at the same site**), to distinguish them from **"heteroalleles"** ("hetero" = different, or **alleles at different sites**). **Heteroalleles are functionally allelic (in the same gene or unit of function, as operationally defined by the *cis-trans* test), but are structurally nonallelic (in different units of structure or at different sites, as operationally defined by the recombination test).**

Mutations $trpA23$ and $trpA46$ both result in the substitution of another amino acid (arginine in the case of $A23$, glutamic acid in the case of $A46$) for the glycine present at position 211 of the wild-type tryptophan synthetase A polypeptide (Fig. 10.19). However, these two mutations occur at different mutable sites. That is, the $A23$ and $A46$ sites are separable by recombination. By determining the amino acids present at position 211 of the tryptophan synthetase A polypeptide in other mutants as well as in revertants and partial revertants of the $trpA23$ and $trpA46$ mutants, Yanofsky and co-workers were able to determine which of the glycine, arginine, and glutamic acid codons (see p. 265) were present in the messenger RNA (mRNA) at the position corresponding to amino acid number 211 of the polypeptide gene product in trp^+, $trpA23$, and $trpA46$ cells, respectively (Fig. 10.20). (A partial revertant is the result of a mutation that occurs in a mutant strain and that restores some, but not wild-type, levels of gene-product activity. The partial revertants referred to here are capable of slow growth on medium lacking tryptophan. These partial revertants can further mutate to "full revertants" that grow at wild-type rates in the absence of tryptophan.) Once the specific codons (triplet base sequences in mRNA) are known, the corresponding

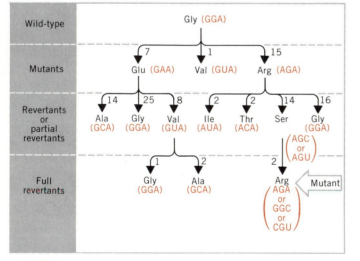

FIGURE 10.20

Pedigree of the tryptophan synthetase A polypeptide position number 211 (211th amino acid from the NH_2 terminus), the position of the trpA23 and trpA46 mutation-induced amino acid substitutions (see Fig. 10.19). The triplet codons shown in parentheses (in color) are the only codons that are specific for these amino acids and that will also permit all of the observed amino acid replacements at position 211 to occur as a result of single base-pair substitutions. The number beside each arrow indicates the number of times that particular substitution has been observed to occur. These data strongly indicate that the
arginine and glutamic acid codons used to specify amino acid number 211 of the tryptophan synthetase A polypeptide in trpA23 and trpA46 are AGA and GAA, respectively. The corresponding base-pair sequences in the trpA gene would be $- AGA \rightarrow$ and $- GAA \rightarrow$ for A23 and A46, respectively, $\leftarrow TCT -$ $\leftarrow CTT -$ indicating that these two mutations alter adjacent base-pairs in this gene. (After C. Yanofsky, J. Ito, and V. Horn, Cold Spring Harbor Symp. Quant. Biol. 31:151, 1966.)

base-pair sequences in the structural gene from which the mRNA is transcribed are also known. One strand of DNA will be complementary to the mRNA, and the second strand of DNA will, of course, be complementary to the first strand. Thus, Yanofsky's data demonstrated that the mutation events producing the *trpA*23 and *trpA*46 mutants were GC to AT transitions occurring at adjacent nucleotide pairs. The *trp*⁺ cells produced by recombination between chromosomes carrying mutations *A*23 and *A*46 (Fig. 10.21) demonstrate that recombination can occur between adjacent nucleotide pairs. This strongly suggests that the unit of genetic material not divisible by recombination is the single nucleotide pair. Clearly, then, the unit of structure of genetic material, as defined by the operational criteria of mutation and recombination, is the nucleotide pair (or single nucleotide in genomes consisting of single-stranded DNA or RNA).

FINE STRUCTURE OF GENES AND "COMPLEX LOCI" IN EUKARYOTES

Examining enough progeny of a cross to detect rare **intra**genic recombination in eukaryotes is a laborious project. Nevertheless, intragenic recombination was first detected in studies with *Drosophila* (p. 321). Genetic fine structure maps have now been constructed for many genes of *Drosophila*. Fine structure maps have also been worked out for several other higher animals and higher plants. In almost every case where an extensive analysis of a gene has been carried out, mutable sites that are separable by recombination have been detected. Clearly, then, the genes of eukaryotes are sequences of nucleotide pairs specifying the amino acid sequences of polypeptides. In eukaryotes, however, some genes have interesting structural features not observed in any prokaryotic genes to date (see pp. 361–365).

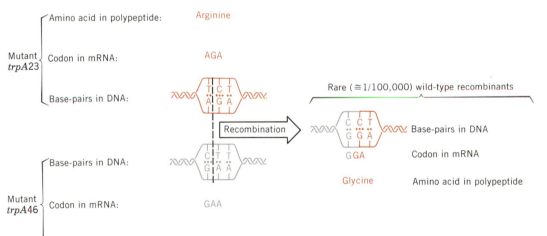

FIGURE 10.21

Recombination between mutations at adjacent base-pairs. Mutations A23 and A46 both result in an amino acid substitution at position number 211 of the tryptophan synthetase A polypeptide. A23 causes a glycine to arginine substitution; A46 causes a glycine to glutamic acid substitution. Analysis of the pedigree of amino acid substitutions that have been observed at position 211 as a result of mutations in the trpA *gene have allowed Yanofsky and colleagues to deduce the exact sequence of base-pairs at the position in the* trpA *gene corresponding to amino acid number 211 in the A polypeptide of both mutants (A23 and A46) as well as wild-type. (Based on the data of C. Yanofsky, J. Ito, and V. Horn,* Cold Spring Harbor Symp. Quant. Biol. *31:151, 1966.)*

In many cases, determining how many genes are present at a locus has proven difficult in eukaryotes. Attempts to delimit genes by complementation tests have often yielded ambiguous results. In some cases, the ambiguities resulted from **intragenic** complementation (see pp. 325–329), as at the *rosy* locus (discussed later in this section) in *Drosophila*. In other cases, it is still not clear whether a particular locus (a location on the chromosome at which mutations causing similar mutant phenotypes map) contains one or more than one gene. For these cases, where two or more mutable sites that are separable by recombination are known, but where the number of genes involved is not known (because of complex or ambiguous complementation data), the term **complex locus** has been introduced. This term is appropriate because **locus** just implies a region on the chromosome and does not have a precise operational definition such as that for **gene** (coding for one polypeptide and defined by the *cis-trans* test).

Sometimes the ambiguities that have arisen in attempts to operationally define functional units may be the result of differential effects of different mutations in a gene on pleiotropic interactions of the gene product. It seems reasonable to expect more pleiotropic effects in eukaryotic organisms with their greater developmental complexity than in prokaryotic organisms. Finally, some of the ambiguities may result from the structural organization of genes in eukaryotes. What effect(s) the noncoding "introns" within the coding sequences of structural genes in eukaryotes might have on *cis-trans* tests is still not clear.

As an example of a complex locus, consider the *white* (white eyes) locus in *Drosophila*. A large number of mutations that have similar, but phenotypically distinguishable, effects on eye color are known to map at this locus (Fig. 10.22). Recombination analyses have demonstrated that these mutations map at seven or more distinct sites within this locus. Complementation tests suggest that two units of function exist. Some of the mutations at the four sites on the left (identified by w^{Bwx}, w^{bf}, w^a, and w^{ch}) complement mutations at the site on the right (the *sp-w* site). This could mean that two genes are present within the *white* locus. A different functional subdivision of the locus is suggested by the ability of the mutations to act as dominant suppressors of mutations at another locus, *zeste* (a nearby, but distinct, eye color locus on the X chromosome). Mutations at the two sites on the right of the *white* locus

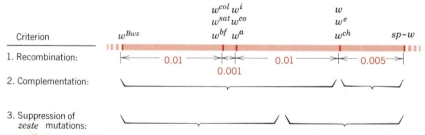

FIGURE 10.22

An early map of the white *locus of* D. melanogaster, *showing a few of the phenotypically distinguishable mutations. Those shown are* w^{Bwx} *(white Brownex),* w^{bf} *(white buff),* w^{sat} *(white satsuma),* w^{col} *(white colored),* w^a *(white apricot),* w^{co} *(white coral),* w^i *(white ivory),* w^{ch} *(white cherry),* w^e *(white eosin),* w *(white), and* sp-w *(spotted-white). Others mapping at the* white *locus and not shown include* w^t *(white tinged),* w^{bl} *(white blood),* w^w *(white wine),* w^{cf} *(white coffee), and* w^{crr} *(white carrot). Different mutations occurring at the* white *locus thus produce a large number of similar, but distinguishable, effects on phenotype. The subdivision of the locus based on recombination analysis (1) as well as the functional subdivisions suggested by complementation analysis (2) and the ability of the various mutations to suppress mutations at a different (*zeste*) locus (3) are also indicated.*

(identified by wch and *sp-w*) are dominant suppressors of *zeste* mutations. Mutations at the three sites on the left (w^{Bwx}, w^{bf}, and w^a) do not suppress *zeste* mutations.

Does the *white* locus contain one, two, three, or seven genes? The definitive evidence needed to answer this question—the nucleotide sequence of the *white* locus—is about to become available. Much of the sequence is known and many of the classical mutations at the locus, like w^a, are known to be due to the insertion of transposable elements. Our best guess at present, based partly on the one gene–one band concept (see pp. 350–351), is that it contains a single gene. The observed complementation may well be **intra**genic complementation. The ability of the mutations to suppress *zeste* mutations may involve protein–protein interactions and depend on the region in the polypeptide that is altered by the mutation.

Early support for the above interpretation of the structure of the *white* locus was provided by W. J. Welshons' study of the *Notch* locus in *Drosophila*. The *Notch (N)* locus corresponds to a single salivary chromosome band (see pp. 350–351) located near the centromere-distal end of the X chromosome. *Notch* mutants have "notched" wings, thickened wing veins, and minor bristle abnormalities. *N* mutations are classified as **dominants,** since they exert these phenotypic effects in $N/+$ heterozygous females. These *N* mutations are **recessive lethals** in that N/N females and N/Y males are inviable. Other mutations at the *Notch* locus exhibit recessive visible effects on phenotype and are nonlethal. The recessive visible mutations, such as *fa* (facet), *spl* (split), fa^{no} (facet-notchoid), and *nd* (notchoid), map

interspersed among the various recessive lethal and dominant visible *N* mutations (Fig. 10.23).

Welshons has shown that when the recessive lethal *N* mutations are analyzed by *cis-trans* tests, they define a single gene; that is, they do not complement each other. He refers to these as **amorphic mutations—mutations that result in a complete loss of gene-product function.** Welshons presents evidence that indicates that the nonlethal, recessive visible mutations (*fa, spl*, etc.) are **hypomorphic mutations—mutations that result in reduced levels, but still some gene-product activity.** (The terms amorphic and hypomorphic were initially introduced by H. J. Muller.) These hypomorphic mutations do complement each other in certain combinations in *trans*-heterozygotes. Taken at face value, such complementation might suggest that the *Notch* locus contains more than one gene. Based on his study of the *Notch* locus, Welshons makes a very important point, namely, that **complementation data obtained using hypomorphic mutations often are not useful in delimiting genes.** This has become eminently clear in studies with prokaryotes. Attempts to delimit genes using temperature-sensitive mutations, which are often hypomorphic, have frequently yielded ambiguous results due to **intra**genic complementation. When *amber* mutations (UAG polypeptide chain-termination mutations resulting in the absence of the carboxyl terminus), which are usually amorphic, are used, clear-cut results are almost always obtained.

If the *white* locus of *Drosophila* is considered, keeping in mind Welshons' data for the *Notch* mutants, it is apparent that most of the *white* locus mutations, such as w^{ch}, w^a, and so on, are hypo-

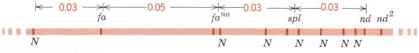

FIGURE 10.23
Map of the Notch *locus of* Drosophila melanogaster. *The map positions of several dominant visible and recessive lethal* N *mutations are given below the line. The locations of recessive visible mutations are shown above the line. Distances are in map units; total length of the defined interval is about 0.14 map units. (Data are from W. J. Welshons,* Science *150:1122, 1965.)*

morphs, since they have phenotypes intermediate between the phenotype of the most extreme mutation, *w* (white), and wild type. As such, they might be expected to exhibit **intra**genic complementation in certain combinations. However, whether this explanation of the complementation observed with mutations at the *white* locus is correct is not yet known.

The most complete picture available for a locus in eukaryotes is of the *rosy (ry)* locus in *Drosophila*. Rosy mutants have brownish-colored eyes; they are deficient in the normal red pigment drosopterin. This results from the absence of a single enzyme, xanthine dehydrogenase. Studies in the laboratories of A. Chovnick and E. Glassman have shown that xanthine dehydrogenase is a dimer composed of two identical polypeptides. As might be expected, some of the *ry* mutant alleles (Fig. 10.24) exhibit **intra**genic complementation, just as was observed for certain mutations at the *white* and *Notch* loci. In the case of *rosy*, however, definitive data are available, indicating that the locus contains a single structural gene.

Drosophila embryos deficient in xanthine dehy-

drogenase activity (e.g., *ry/ry* homozygotes) are inviable on culture medium supplemented with purine. This has provided Chovnick and colleagues with a very powerful selective sieve with which to carry out fine structure analysis. Only those progeny of *trans*-heterozygotes (e.g., ry^1/ry^2) that carry a rare wild-type recombinant chromosome will be viable on purine-supplemented medium. *Rosy* mutants thus provide a specialized (applicable only to this one locus) conditional lethal system, which is ideal for genetic fine structure analysis.

Similar fine structure maps have been obtained in higher plants. **Intra**genic recombination in plants was first detected at the *waxy (wx)* locus in *Zea mays* (corn) by O. Nelson. Wild-type (*Wx*, "starchy") corn plants contain two types of starch, called amylose and amylopectin, deposited in the endosperm of the kernels and in the pollen. *Waxy (wx)* mutants contain amylopectin, but no amylose; they lack a single enzyme, ADPG transferase, required for the synthesis of amylose. Fine structure mapping of *wx* mutations was facilitated by the fact that this phenotype can be scored in the male gametes or pollen

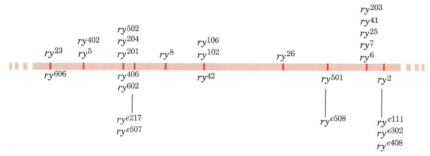

FIGURE 10.24

Map of the rosy *locus of* Drosophila melanogaster, *showing the map positions of mutations in the structural gene for the enzyme xanthine dehydrogenase. Noncomplementing xanthine dehydrogenase-deficient* ry *mutations are shown above the line. Mutations that exhibit intragenic complementation are shown directly below the line. Alterations in the* ry *gene that result in active xanthine dehydrogenase molecules, but with altered electrophoretic mobilities (changes in the charge of the polypeptide gene product), are shown at the bottom (superscripts with* e *prefixes). Chovnick and colleagues have also identified and mapped mutations that alter the amount of xanthine dehydrogenase synthesized. These mutations are probably in control or regulatory regions of the* rosy *locus; they map outside the structural gene in the region just to the left of* ry²³ *as shown here. (Based on the data of A. Chovnick, W. Gelbart, and M. McCarron,* Cell *11:1, 1977.)*

grains. *Wx* pollen grains turn blue whereas *wx* pollen grains turn light red after staining with iodine. Since each corn plant produces millions of pollen grains, rare wild-type recombinants can be detected by screening pollen produced by a wx^1/wx^2 *trans*-heterozygote. Thus, Nelson was able to demonstrate that the gene in higher plants, like the gene in prokaryotes and in *Drosophila,* contains mutable sites that are separable by recombination.

MULTIPLE ALLELES

An allele is a specific form or sequence of nucleotide pairs of a given gene. An allele may be a mutant allele, resulting in an altered phenotype, or a wild-type allele, producing an active gene product and a "normal" phenotype. Different wild-type alleles often produce active protein gene products with different amino acid sequences. These sometimes have altered electrophoretic mobility (see Fig. 10.24). When more than two different forms of a given gene exist in a species, they are referred to as **multiple alleles.** It should be clear from the preceding sections of this chapter that multiple alleles probably exist for most, if not all, genes.

Members of a series of alleles are conventionally represented by the same letter or symbol, with appropriate superscripts to identify particular alleles, for example, w, w^a, w^{ch}, and so on (see Fig. 10.22) or ry^5, ry^7, ry^{e111} (see Fig. 10.24) for alleles at the *white* and *rosy* loci of *Drosophila.* Different alleles may produce gene products with different levels of activity. For example, if the quantity of eye pigment present in mutants homozygous for different alleles of the white locus is measured spectrophotometrically, it is found to vary from a very low level in w/w homozygotes to about 25 percent of the "wild-type" level in w^{col}/w^{col} homozygotes (Table 10.1). Actually, flies homozygous for different wild-type alleles vary almost two-fold in the amount of eye pigment present (Table 10.1). Thus, the proportion of "wild-type level" of pigment present in various eye mutants depends on which wild-type one selects for comparison. For many genes, alleles probably exist with gene-product activity levels ranging from the complete **amorph or "null" mutation** (**no** activity, as will always be the case for a deletion of the gene)

TABLE 10.1 **Relative Amounts of Eye Pigment Present in Flies Homozygous for Different Alleles of the *white* Locus of *Drosophila melanogaster***

GENOTYPE	RELATIVE CONCENTRATION OF TOTAL PIGMENT[a]
w/w (white)	0.0044
w^a/w^a (white apricot)	0.0197
w^e/w^e (white eosin)	0.0324
w^{ch}/w^{ch} (white cherry)	0.0410
w^{co}/w^{co} (white coral)	0.0798
w^{sat}/w^{sat} (white satsuma)	0.1404
w^{col}/w^{col} (white-colored)	0.1636
w^{+s}/w^{+s} (wild-type, Stellenbusch strain)	0.6854
w^{+c}/w^{+c} (wild-type, Canton S strain)	0.9895
w^{+G}/w^{+G} (wild-type, Graff-Reinet strain)	1.2548

[a] Data from D. J. Nolte. *Heredity* 13:219, 1959. For map positions of mutant alleles, see Fig. 10.22.

to the wild-type allele producing the maximum observed level of gene-product activity.

The number of different genotypes possible in diploid organisms is, of course, a function of the number of alleles that exist for any given gene. If n is the number of alleles of a gene, the number of different genotypes possible is $n(n + 1)/2$. Thus, with 2, 3, 4, or 5 alleles, there are 3, 6, 10, and 15 possible genotypes, respectively.

A classic example of multiple alleles involves coat color in rabbits. Four alleles of the rabbit coat color (c) gene have been studied: c^+ (wild-type or full color), c^h ("himalayan," white with black tips on the extremities), c^{ch} ("chinchilla," mixed colored and white hairs), and c (albino). These alleles show a gradation in dominance of $c^+ > c^{ch} > c^h > c$. That is, c^+ is dominant to each of the three mutant alleles, while c^h is recessive to c^{ch}, but dominant to c, and so on.

ABO BLOOD TYPE ALLELES IN HUMANS

The most firmly established series of multiple alleles in humans occurs for the locus controlling the blood types: A, B, AB, and O. The ABO locus has three common alleles: I^A, I^B, and I^O (Table 10.2). I^A and I^B

TABLE 10.2 Genotypes and the Corresponding Phenotypes (Blood Group Types) for the ABO Locus in Humans

GENOTYPE	PHENOTYPE
$I^A I^A$ or $I^A I^O$	A
$I^B I^B$ or $I^B I^O$	B
$I^A I^B$	AB
$I^O I^O$	O

are codominant ($I^A I^B$ heterozygotes have both A and B antigens on their red blood cells); I^O is recessive ($I^O I^O$ homozygotes have no ABO antigens on their red blood cells; $I^A I^O$ and $I^B I^O$ heterozygotes have A and B antigens, respectively, on their red blood cells).

The ABO locus controls the type of glycolipids found on the surface of erythrocytes, apparently by specifying the type of glycosyl-transferases (enzymes catalyzing the synthesis of polysaccharides) synthesized in the red blood cells. The specific types of glycolipids on the red cell surface in turn provide the antigenic determinants that react with specific antibodies present in the blood serum. Humans, like all other mammals, produce antibodies and circulate them in the blood serum as a defense mechanism against foreign substances. Fortunately, no antibodies are synthesized (in normal individuals) that react with antigens present on the individual's own cells. However, when type A blood and type B blood are mixed, the anti-A antibodies in the type B blood serum react with the antigens on the type A blood cells, and vice versa, causing agglutination or clumping of cells (Fig. 10.25). Cross-matching blood types to determine compatibility is thus essential in blood transfusions.

Table 10.3 summarizes the cell surface antigenic determinants and the serum antibodies present in the four major ABO blood types. Individuals with blood type AB have both A and B antigens on their erythrocytes, but no anti-A or anti-B antibodies in their blood serum. Type O individuals lack both antigens, but carry both anti-A and anti-B antibodies in their blood serum. Type O individuals are referred to as universal donors; type O blood can be used in transfusions for individuals of any blood type if the blood is introduced slowly enough

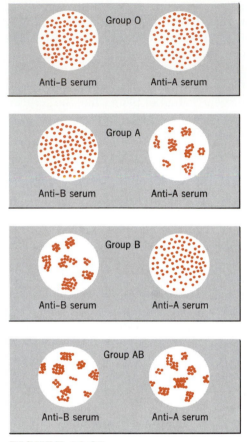

FIGURE 10.25

Agglutination reactions controlled by the ABO blood-type locus in humans. Red blood cells (erythrocytes) of the type indicated at the top of each slide are mixed with blood serum of the type indicated below each reaction mixture (circle). A clumped pattern of cells within a circle indicates that agglutination occurs.

to permit sufficient dilution of the anti-A and anti-B antibodies present in the serum of the donor.

RH FACTOR ALLELES IN HUMANS

The Rh factor was discovered in 1940 by K. Landsteiner and A. S. Wiener from rabbits immunized with the blood of a monkey *Macaca rhesus*. The resulting antibodies were found to agglutinate not only the red corpuscles of the monkey but those of a high percentage of the Caucasian people of New

TABLE 10.3   Blood Transfusion Compatibilities for the ABO Blood Groups

BLOOD GROUP	ANTIGENS PRESENT	ANTIBODIES PRESENT	RED CELL TYPES AGGLUTINATED	TRANSFUSIONS ACCEPTED FROM
A	A (galactosamine)	Anti-B	B, AB	A or O
B	B (galactose)	Anti-A	A, AB	B or O
AB	A (galactosamine) plus B (galactose)	None	None	A, B, AB, or O
O	None	Anti-A plus anti-B	A, B, and AB	O

York. Individuals whose blood cells react with the Rh-antibody are termed Rh-positive; those who do not react are Rh-negative. The symbol "Rh" came from the first two letters of the species name of the monkey. A test for Rh incompatibility is accomplished by placing a drop of blood from the subject on a slide and introducing anti-Rh serum. Agglutination of erythrocytes indicates incompatibility, whereas an even distribution of erythrocytes indicates no reaction.

The original antigen, now symbolized Rh_0, is highly antigenic to humans. Thus, cross-matching of the Rh factor, as well as of the ABO types, of donor and recipient blood is now used to avoid incompatibility agglutination reactions following transfusions. Blood is frequently exchanged between the mother and the fetus during childbirth. Thus, Rh-negative mothers are often immunized by blood from Rh-positive fetuses (which may result when the fathers are Rh-positive) to which they give birth. (Rh-positive is dominant; Rh-negative is recessive.) Usually no ill effects are associated with exposure of the mother to the Rh-positive antigen during the first childbirth (unless the mother has been previously exposed to Rh-antigen by transfusion). Subsequent Rh-positive children carried by the same mother, however, may be exposed to antibodies against the Rh-antigen, which are carried across the placenta in blood serum. Such children may develop symptoms of hemolytic jaundice and anemia, a condition referred to as erythroblastosis fetalis. The symptoms may be mild or severe, even resulting in the death of the fetus or newborn infant if appropriate steps are not taken by the physician.

Passive immunity for Rh-hemolytic anemia can now be accomplished by use of an incomplete antibody against the Rh_0 antigen. This antibody does not agglutinate Rh-positive red blood corpuscles. Instead, the antibodies attach to antigen receptors on red cell surfaces and coat the cells. These incomplete antibodies may be injected into an Rh-negative mother immediately after she has given birth to an Rh-positive child. The coating of any Rh-antigens from the Rh-positive fetus with incomplete antibodies inhibits the mother's capacity to form Rh-antibodies. The injection of incomplete antibodies thus prevents the fetal antigens from eliciting the normal immune response of the mother. Injected antibodies dissipate within a few months and present no danger to the mother or fetuses in subsequent pregnancies.

At first, the genetic control of the Rh-system seemed simple. A single pair of alleles, R and r, was postulated to account for the difference between Rh-positive and Rh-negative individuals. New antibodies were soon discovered, however, and additional genes were postulated to explain the more complicated situation. Wiener developed a hypothesis based on a series of multiple alleles (Table 10.4). Eight alleles were initially included in the series and more have since been added. Considerable evidence has been presented in support of this hypothesis. On the other hand, R. R. Race, R. A. Fisher, and other investigators explained the same data on the basis of three genes (C, D, and E) that are closely linked. Such genes would be expected to act like alleles in most situations that can be analyzed in human populations.

TABLE 10.4 Hypotheses for the Genetic Control of the Rh Factors in Human Blood

RH BLOOD TYPE	WIENER HYPOTHESIS (ONE GENE) ALLELES PRESENT	FISHER AND RACE HYPOTHESIS (THREE GENES) ALLELES PRESENT	APPROXIMATE FREQUENCY OF TYPE AMONG CAUCASIANS
Rh_1	R^1	CDe	41%
Rh_2	R^2	cDE	14%
Rh_0	R^0	cDe	3%
Rh_z	R^z	CDE	Rare
rh	r	cde	39%
rh′	$r′$	Cde	1%
rh″	$r″$	cdE	1%
rh_y	r^y	CdE	Very rare

ONE GENE–ONE BAND IN *DROSOPHILA* SALIVARY GLAND CHROMOSOMES?

The somatic chromosomes of the salivary glands of *Dipteran* flies such as *Drosophila* are giant structures with a banded appearance resulting from the alternation of regions of highly condensed DNA (densely staining regions) with regions of less condensed DNA (lightly staining regions; see Figs. 12.2 and 12.3). Each giant salivary gland chromosome contains over 1000 times as much DNA as the homologous meiotic germ line chromosome. These giant chromosomes are produced by replication of the single molecule of DNA in each germ line chromosome some 10 times without separation of the replication products into different chromosomes. Each chromosome thus contains about 2^{10} or 1024 DNA molecules, which are associated in a lateral array. Such "multistranded" (containing many DNA molecules running longitudinally throughout the chromosome) chromosomes are called **polytene** chromosomes. (Details of the structure of the polytene salivary gland chromosomes of *Drosophila melanogaster* are discussed in Chapter 12.)

Extensive genetic and cytological analyses of specific regions of the *D. melanogaster* salivary gland chromosomes have provided strong evidence that there is a 1:1 correlation between genes and bands (or **chromomeres,** as they are often called). In a very detailed study, B. H. Judd and colleagues identified and characterized a large number of mutations in one region of the X chromosome. They identified 16 genes ("complementation groups") in a segment of the chromosome containing 15 distinct bands (Fig. 10.26). Similarly, the small chromosome 4 of *Drosophila* has been shown to contain 43 genes (units of function) to date. Chromosome 4 has somewhere between 33 and 50 bands, again indicating a good fit to the one gene–one band hypothesis.

In total, the salivary gland chromosomes contain about 5000–6000 bands. If the one gene–one band hypothesis is correct, *Drosophila* has 5000–6000 genes. Note that this estimate of the number of genes is consistent with the **independent** estimate of the number of genes in *Drosophila* based on mutation frequencies (see Chapter 9, pp. 306–307). However, estimates of the number of different mRNA molecules indicate that there are about 17,000 structural genes—or about three times as many genes as polytene chromosome bands—in *Drosophila*.

The haploid chromosome complement of *D. melanogaster* contains about 10^8 nucleotide pairs. If all of this DNA represented structural genes of an average size of 1000 nucleotide pairs, the *Drosophila* genome would contain 100,000 genes, and each band (containing about 30,000 nucleotide pairs) would contain about 30 genes. What is (are) the function(s) of this apparent excess of DNA in *Drosophila* and all other eukaryotes? We know that part of this DNA resides in noncoding introns within eukaryotic genes (pp. 361–365), but what is (are) the role(s) of the rest of this DNA? This is one of the most intriguing questions challenging geneticists today (see Chapter 5, pp. 131–136, and Chapter 9, pp. 306–307, for further discussion).

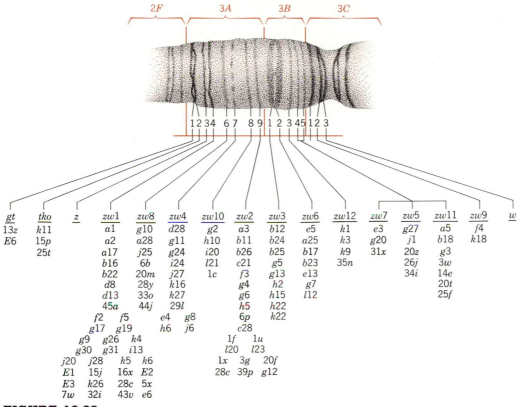

FIGURE 10.26

Correlation between the number of genes (complementation groups) and the number of bands in one segment (region 3A1 to 3C2) of the polytene X chromosome in the salivary glands of Drosophila melanogaster. *The mutations identified and localized in the 16 complementation groups are shown below the schematic drawing of the 3A1 to 3C2 segment of the X chromosome. (Data are from B. H. Judd, M. W. Shen, and T. C. Kaufman,* Genetics *71:139, 1972.)*

GENES-WITHIN-GENES IN PHAGE ΦX174

Bacteriophage ΦX174 is a small virus that stores its genetic information in a single-stranded, circular molecule of DNA. Phage ΦX174 replicates its DNA using a double-stranded replicative intermediate (see Chapter 5, pp. 112–116). Until recently, the ΦX174 genome was believed to contain nine genes coding for nine different polypeptides. All nine of these gene products had been characterized and their molecular weights determined. By using the known molecular weights of the nine polypeptides, and knowing that the genetic code is a triplet code (three nucleotides per amino acid), it was possible to calculate the length of DNA required to code for all nine polypeptides. Assuming that the nine structural genes represented nonoverlapping nucleotide sequences, the minimum length of DNA required to code for the nine polypeptides was shown to be 6078 nucleotides. However, the ΦX174 chromosome is only 5387 nucleotides long. It is not long enough to carry the coding sequences for all nine

Polypeptide D amino acid sequence → [(F-Met)]

DNA nucleotide sequence → [A-T-G]   ↑ *D start*

D Ser - Gln - Val - Thr - Glu - Gln - Ser - Val - Arg - Phe - Gln - Thr - Ala - Ser - Leu - Ile -
A-G-T-C-A-A-G-T-T-A-C-T-G-A-A-C-A-A-C-C-G-T-T-C-C-T-A-T-T-A-A-G-C-T-C-A-T-T-

D Gln - Ala - Ser - Ala - Val - Leu - Asp - Leu - Thr - Glu - Asp - Phe - Leu - Thr - Ser - Lys -
C-A-G-G-C-T-T-C-T-G-C-C-G-T-T-C-T-G-G-A-T-C-T-G-A-C-A-G-A-A-G-A-T-T-T-C-C-T-G-A-C-A-A-G-T-A-A-A-A-C-A-A-

Polypeptide E amino acid sequence → (F-Met)-Val-

D Val - Trp - Ile - Ala - Thr - Ala - Asp - Arg - Ser - Arg - Cys - Val - Tyr - Gly -
G-T-T-T-G-G-A-T-T-G-C-T-A-C-T-G-C-T-G-A-C-C-G-C-T-C-G-C-C-T-G-C-G-T-T-A-T-G-G-T-
↑ *E start*

E Arg - Trp - Thr - Leu - Trp - Asp - Thr - Leu - Ala - Phe - Leu - Leu - Pro - Ser -

D Thr - Leu - Asp - Phe - Val - Gly - Tyr - Pro - Arg - Phe - Pro - Ala - Ile - Ala - Ala - Val -
A-C-G-C-T-G-G-A-C-T-T-T-G-T-G-G-G-A-T-A-C-C-C-G-C-G-T-T-T-C-C-C-T-G-C-T-A-T-T-G-C-T-G-C-C-G-T-C-

E Leu - Leu - Ile - Met - Phe - Ile - Pro - Ser - Thr - Phe - Lys - Arg - Ser - Trp - Lys - Ala - Leu -

D Ile - Ala - Tyr - Tyr - Val - His - Pro - Val - Asn - Ile - Gln - Thr - Cys - Leu - Met - Glu - Gly - Ala -
A-T-T-G-C-T-T-A-T-T-A-T-G-T-T-C-A-T-C-C-C-G-T-C-A-A-C-A-T-C-C-A-A-A-C-G-T-G-C-C-T-G-A-T-G-G-A-G-G-G-C-G-C-T-

E Asn - Leu - Arg - Lys - Thr - Leu - Leu - Met - Ala - Ser - Ser - Val - Arg - Leu - Pro - Leu - Asn - Cys - Ser -

D Glu - Phe - Thr - Glu - Asn - Ile - Ile - Asn - Gly - Val - Glu - Arg - Pro - Val - Ala - Glu - Leu - Phe -
G-A-A-T-T-T-A-C-G-G-A-A-A-A-C-A-T-T-A-T-T-A-A-T-G-G-C-G-T-C-G-A-A-C-G-C-C-C-G-G-T-A-G-C-T-G-A-A-C-T-T-T-T-C-

E Arg - Leu - Pro - Cys - Val - Tyr - Ala - Tyr - Arg - Ala - Val - Thr - Gln - Glu - Leu - Thr - Gln - Lys - Thr -

D Ala - Phe - Thr - Leu - Arg - Val - Leu - Thr - Asp - Asn - Thr - Asp - Val - Leu - Thr - Asp - Val - Glu - Glu - Asn -
G-C-G-T-T-T-A-C-C-C-T-T-C-G-G-G-T-A-C-T-G-A-C-G-G-A-T-A-A-T-A-C-C-G-A-C-G-T-A-C-T-G-A-C-C-G-A-T-G-T-A-G-A-A-A-A-C-

Polypeptide J amino acid sequence
└→(F-Met)- Ser - Lys - Gly - Lys - Lys - Arg - Ser -

E Cys - Val - Lys - Asn - Tyr - Val - Arg - Lys - Glu -

D Val - Arg - Gln - Lys - Leu - Arg - Ala - Glu - Gly - Val - Met
G-T-G-C-G-C-A-A-A-A-A-T-A-C-G-G-T-G-C-C-G-G-A-A-G-G-A-A-G-A-A-G-A-G-A-A-G-A-A-A-A-C-G-T-T-C-T-G

↑ *E stop*   ↑ [T-A-A]-T-G   *D stop* ↑   ↑ *J start*

polypeptides in nonoverlapping genes. For several years (until 1976), this presented a puzzling dilemma. This dilemma was not resolved until nucleotide sequence data became available.

The nucleotide sequence data for ΦX174 resolved the dilemma of not enough DNA by showing that the coding sequences of two genes are located within the coding sequences of two other genes with the reading frames (the triplet sequences read during translation; see Chapter 8) offset by one nucleotide pair in each case (Figs. 10.27 and 10.28). This surprising result has important genetic implications. Clearly, the amino acid sequences of genes with overlapping coding sequences cannot evolve totally independently of one another. (They have some independence due to the degeneracy of the genetic code; see Chapter 8.) Moreover, a single mutation (e.g., a single base addition or deletion) can result in the loss of two gene-product activities (pleiotropy of the most direct type).

How common overlapping genes or genes-within-genes are in various organisms remains unknown. Recently, a fourth gene has been discovered in phage MS2 that overlaps two of the three previously characterized genes of MS2. Moreover, partial-coding-sequence overlaps do occur in the Simian virus 40 (SV40) chromosome. Whether such overlaps exist in the chromosomes of organisms other than viruses remains a question to challenge present and future generations of geneticists.

Of course, the ultimate fine structure map is the nucleotide sequence of a gene complete with the nucleotide changes occurring during mutation events that result in the loss of specific functions. Such information allows one to correlate particular functions with specific signals encoded in nucleotide sequences. We now know the complete nucleotide sequence of (1) the single-stranded RNA chromosome of phage MS2 (3569 nucleotides long), (2) the double-stranded DNA chromosome of SV40 (5226 nucleotide pairs long), (3) the single-stranded DNA chromosome of phage ΦX174 (5387 nucleotides long), and (4) the double-stranded DNA chromosome of phage λ (48,502 nucleotide pairs long).

"RECOMBINANT DNA" AND "GENE CLONING" AS A TOOL FOR FINE STRUCTURE ANALYSIS

Nucleic acid sequencing is a very powerful technique for studying the structure and function of genes and chromosomes (e.g., see the preceding section on phage ΦX174). However, sequencing is presently practicable only with relatively short nucleic acid sequences, such as those of the chromosomes of small viruses. Sequencing the entire chromosome of even a bacterium such as *E. coli*, about 4 million nucleotide pairs in length, let alone the giant DNA molecules in chromosomes of eukaryotes, would be an almost inconceivable task. One might suggest isolating a particular segment that contained only one to a few thousand nucleotide pairs

FIGURE 10.27

The nucleotide sequence for the segment of the phage ΦX174 chromosome containing the coding sequence of gene E, which is located within the coding sequence of gene D. The reading frame of gene E is offset to the right by one nucleotide from the reading frame of gene D. The reading frame of the D gene and the correlated amino acid sequence of the D polypeptide are shown (in color) immediately above the nucleotide sequence. The reading frame of the E gene and the amino acid sequence of the E polypeptide predicted from the nucleotide sequence are shown (in gray) above the reading frame of the D gene and the amino acid sequence of the D polypeptide, respectively. The amino-terminal segment of the J polypeptide and the corresponding nucleotide sequence of the J gene are also shown. Note that the initiation triplet (ATG) of the J gene also overlaps the termination triplet (TAA) of the D gene by one nucleotide. (Based on the data of B. G. Barrell, G. M. Air, and C. A. Hutchison, III, Nature 264:34, 1976.)

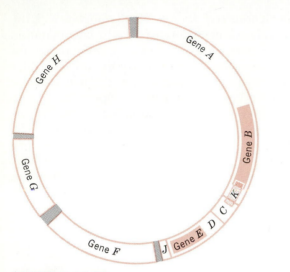

FIGURE 10.28

Organization of the 10 genes present in the circular chromosome of phage ΦX174, showing the two cases of a gene-within-a-gene. The coding sequences of genes B and E (shown in color) are located entirely within the coding sequences of genes A and D, respectively. The coding sequence of gene K (shown in color, stippled) overlaps the coding sequences of genes A, B and C. It only overlaps with the termination codon of gene B. In all cases, the reading frames of the two overlapping genes are offset from each other (see Fig. 10.27). The four short noncoding segments of the chromosome are shown in gray. Relative distances are only approximate. (Based on the data of F. Sanger, G. M. Air, B. G. Barrell, N. L. Brown, A. R. Coulson, J. C. Fiddes, C. A. Hutchison, III, P. M. Slocombe, and M. Smith, Nature 265:687, 1977, and E. S. Tessman, I. Tessman, and T. J. Pollock, J. Virology 33:557, 1980.)

and just sequencing that segment. The haploid genome of mammals, for example, contains about 3×10^9 nucleotide pairs. Any one gene or sequence of one to a few thousand nucleotide pairs represents only about one unit out of a total of approximately a million such units in the genome. Thus isolating any one gene is a bit like searching for a needle in a haystack. Nucleic acid sequencing techniques require that the sequence be available in significant quantities in pure or essentially pure form. How, then, can one hope to extend the nucleic acid sequencing techniques to the more complex chromosomes of eukaryotes?

"Recombinant DNA" and "gene cloning" technology now provide molecular biologists with a method by which genes or other segments of large chromosomes can be isolated, replicated, and studied by nucleic acid sequencing techniques, electron microscopy, and other analytical techniques. Various procedures for gene cloning are now in use. In all cases, however, the overall procedure involves (1) **the *in vitro* incorporation of the gene or segment of DNA of interest into a small self-replicating chromosome** such as that of phage lambda or a plasmid (see Chapter 7, pp. 202–205 and 221–226), and (2) **the introduction of the recombinant "minichromosome" into a host cell where it will replicate.** Step one involves the *in vitro* synthesis of **recombinant DNA,** for example, an *E. coli* plasmid containing one gene of *Drosophila.* Step two is the **gene cloning** step, in which the recombinant DNA (or recombinant minichromosome) is replicated or "cloned" to produce many identical copies for subsequent biochemical analysis. Thus, while the entire procedure is often referred to as the "recombinant DNA" or "gene cloning" technique (particularly by the popular press), these terms actually refer to two separate steps in the overall process, both of which are essential to its utility.

THE *IN VITRO* SYNTHESIS OF RECOMBINANT DNA AND GENE CLONING TECHNIQUES

The original cloning experiments by A. Chang and S. Cohen in 1973, and many subsequent cloning experiments, have been done using a special class of endonucleases (nucleases that make internal cuts in nucleic acids) called **restriction endonucleases.** Many endonucleases make random cuts in DNA. The **restriction endonucleases are site-specific,** however. They cleave DNA molecules only at **specific nucleotide sequences called restriction sites.** Different restriction endonucleases, present in different organisms, recognize different nucleotide sequences (Table 10.5).

A function served by the restriction nucleases is to protect the genetic material of the organism from "invasion" by foreign DNA. All restriction sites in the host chromosome of an organism must be protected from its own restriction nucleases to prevent

TABLE 10.5   Recognition Sequences and Cleavage Sites of Representative Restriction Endonucleases

ENZYME	SOURCE	RECOGNITION SEQUENCE[a] AND CLEAVAGE SITES[b]	NUMBER OF RECOGNITION SEQUENCES PER CHROMOSOME OF:	
			PHAGE λ	SV40 VIRUS
*Eco*R1	*Escherichia coli*	GAA TTC CTT AAG	5	1
*Hind*II	*Hemophilus influenzae*	GTPy PuAC[c] CAPu PyTG	34	7
*Hind*III	*Hemophilus influenzae*	AAG CTT TTC GAA	6	6
*Hpa*I	*Hemophilus parainfluenzae*	GTT AAC CAA TTG	11	4
*Hpa*II	*Hemophilus parainfluenzae*	CC GG GG CC	750	1

[a] The axis of symmetry in each palindromic recognition sequence is indicated by the colored dot.
[b] The position of each bond cleaved is indicated by a colored arrow. Note that the cuts are staggered (at different positions in the two complementary strands) with some restriction nucleases.
[c] Pu indicates that either purine (adenine or guanine) may be present at this position: Py indicates that either pyrimidine (thymine or cytosine) may be present.

suicidal self-degradation. This is accomplished by methylation of one or more nucleotides in each nucleotide sequence that is recognized by the organism's own restriction nuclease(s). Methylation occurs rapidly after replication, catalyzed by site-specific methylases coded for by the host chromosome. Each restriction nuclease will cleave a foreign DNA molecule (a DNA molecule from another species) into a fixed number of fragments, the number depending on the number of restriction sites in the particular DNA molecule.

An interesting feature of restriction endonucleases is that they commonly recognize DNA se-

quences that are **palindromes,** that is, nucleotide-pair sequences that are the same reading forward or backward from a central axis of symmetry, like the nonsense phrase:

$$\longleftarrow \quad \longrightarrow$$

AND MADAM DNA

In addition, a very useful feature of many restriction nucleases is that they make "staggered" cuts; that is, they cleave the two strands of a double helix at different points. Most important, because of the palindromic nature of the restriction sites, the staggered cuts produce segments of DNA with **comple-**

mentary single-stranded ends. For example, cleaving a DNA molecule of the following type:

with the restriction endonuclease *Eco*RI (an *E. coli* enzyme) will yield:

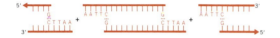

Because all of the fragments produced will have complementary single-stranded 5′-ends, they can be rejoined under appropriate renaturation conditions using the enzyme DNA ligase (see Chapter 5, pp. 101–116). The single-stranded ends of "*Eco*RI fragments" (as fragments produced by digestion of DNA with *Eco*RI restriction enzyme are called) will be complementary regardless of the source of DNA. Thus, *Eco*RI fragments cut from mouse chromosomes, chicken chromosomes, and so on, can be incorporated into phage lambda chromosomes or plasmids that have been similarly cleaved with *Eco*RI restriction enzyme (Fig. 10.29). Once a gene or restriction enzyme fragment is inserted into a plasmid, the recombinant plasmid is introduced into a host cell, such as *E. coli*, by transformation (uptake of naked DNA by cells; see Chapter 7, pp. 200–202). When *E. coli* is used, the cells must first be made permeable to DNA by treatment with a calcium salt. An important advantage of this procedure over some other cloning procedures (see Fig. 10.30) is that after cloning, to produce a large amount of recombinant DNA, the cloned sequence can be precisely cut out of the recombinant chromosome by using the restriction enzyme that was used to incorporate it (the restriction sites will remain at the same positions).

A second procedure for the *in vitro* construction of recombinant DNA molecules involves the synthesis of complementary single-stranded tails on the 3′-ends of strands of DNA using the enzyme **terminal transferase** (Fig. 10.30). This procedure

has an advantage over the restriction enzyme procedure (see Fig. 10.29) in that it can be used to join DNA molecules that terminate with any DNA sequence as well as fragments produced by cleavage with restriction enzymes that cut both strands of DNA at the same point. Once recombinant DNA molecules are produced using the terminal transferase procedure, they are cloned in the same way that those produced using complementary ends generated by restriction enzymes are cloned.

The two procedures described above are two of the original techniques used to construct recombinant DNA molecules. Several modifications of the above procedures as well as new techniques for synthesizing recombinant DNA molecules are currently in use.

Cloning experiments can be divided into two types: (1) those where the cloned DNA segment is of known composition and function, and (2) those where the cloned DNA segment is of unknown composition and function, such as uncharacterized restriction fragments.

When pure mRNA or pre-mRNA transcripts of a gene can be obtained (such as for hemoglobin genes using reticulocytes and the ovalbumin gene using chicken oviduct cells), they can be used to carry out cloning experiments of the first type. The purified mRNA is used as a template for the *in vitro* synthesis of a complementary strand of DNA (called *cDNA*) by the enzyme **reverse transcriptase** (because it catalyzes the reaction that is the reverse of transcription, that is, RNA → DNA rather than DNA → RNA as in transcription). The complementary DNA strand in turn serves as a template for DNA polymerase-catalyzed synthesis of a second strand of DNA (complementary to the first). Thus, **reverse transcriptase mediates the conversion of the genetic information present in a single-stranded molecule of RNA into a double-stranded molecule of DNA.** It is therefore more accurately called an **RNA-dependent DNA polymerase.** The double-stranded DNA molecule synthesized with reverse transcriptase and DNA polymerase can then be inserted into a phage chromosome or plasmid by the terminal transferase procedure (Fig. 10.30) and cloned.

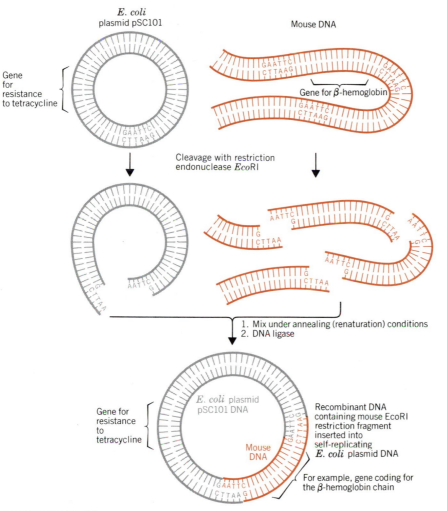

FIGURE 10.29

Schematic illustration of the EcoRI *restriction nuclease technique for the* in vitro *construction of recombinant DNA molecules for subsequent cloning. In the example illustrated, the autonomously replicating* E. coli *plasmid pSC101 is used as the cloning vector. It has two important features: (1) it carries a gene that provides tetracycline resistance to the host cell; (2) it has a single* EcoRI *restriction site (see Table 10.5), facilitating the insertion of a fragment of foreign DNA after cleavage by* EcoRI *endonuclease.* EcoRI *restriction fragments from any DNA (mouse DNA is used for illustrative purposes above) may be inserted into the cleaved pSC101 plasmid by mixing the two*

DNA preparations under annealing conditions and sealing with the enzyme DNA ligase. The recombinant DNA molecules are then used to transform E. coli *cells. If tetracycline-sensitive* E. coli *cells are used and the cells are subsequently grown in medium containing tetracycline, only cells that were "transformed" and thus contain a pSC101 plasmid will be able to grow. Once an* E. coli *cell that contains a recombinant plasmid carrying the desired gene or nucleotide sequence has been identified (the most difficult step in the gene cloning procedure), that cell can be used to grow a population of identical cells (or "clone"), all containing the desired gene or sequence.*

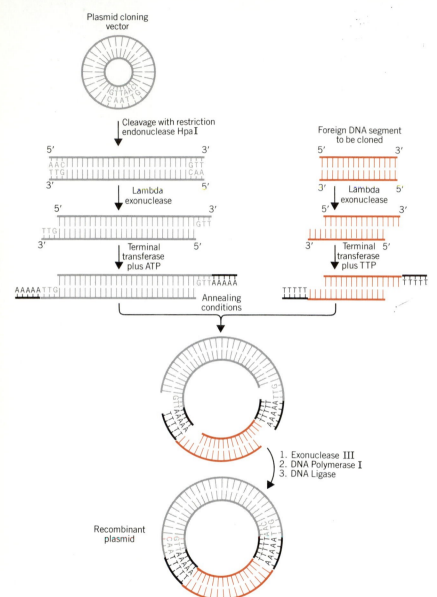

Plasmid cloning
vector

Cleavage with restriction
endonuclease Hpa I

Lambda
exonuclease

Terminal
transferase
plus ATP

Foreign DNA segment
to be cloned

Lambda
exonuclease

Terminal
transferase
plus TTP

Annealing
conditions

1. Exonuclease III
2. DNA Polymerase I
3. DNA Ligase

Recombinant
plasmid

FIGURE 10.30

Diagram illustrating the terminal transferase procedure for the construction of recombinant DNA molecules. A segment of foreign DNA (color) is cloned in a plasmid (gray) containing a single endonuclease HpaI restriction site (see Table 10.5). After cleavage of the plasmid with HpaI, the linear plasmid DNA molecules and the segments of foreign DNA are treated briefly with phage lambda exonuclease to prepare appropriate substrates for the terminal transferase reaction. (Lambda exonuclease cleaves off nucleotides from the 5'-ends of each chain, leaving double helices with single-stranded 3' tails.) Poly-A tails are then added to the 3'-ends of each plasmid DNA molecule, using terminal transferase in the presence of ATP (with no other nucleoside triphosphates present). Poly-T tails are similarly added to the 3'-ends of the foreign DNA molecules, using TTP as substrate. The two DNA preparations are then mixed under conditions facilitating the annealing of the complementary poly-A and poly-T tails. Phosphates are removed from the 3'-ends with the enzyme exonuclease III, and the gaps are filled by using the four nucleoside triphosphates and DNA polymerase I. Covalent closure is catalyzed by DNA ligase. The recombinant, self-replicating plasmids are then cloned following introduction into E. coli *cells by transformation.*

Reverse transcriptases are coded for by many RNA tumor viruses. Their normal function is to convert the genetic information of the virus from the RNA form in which it is stored in the mature virion into a double-stranded DNA form after infection of a host cell. The DNA form is called a **provirus,** and it is this DNA provirus form that integrates by covalent linear insertion into the chromosome of the host (see Chapter 11, pp. 422–430). In its integrated provirus form, the viral chromosome replicates along with the host chromosome and segregates in a Mendelian fashion during sexual reproduction (just like any other gene or set of genes of the host). The *in vitro* synthesis of cDNA from mRNA and other RNA molecules using reverse transcriptases has become a very important tool in molecular genetics.

The second approach to cloning has been to prepare clonable-sized fragments of DNA by digestion of the entire genome of an organism with one or more restriction enzymes or by shearing forces, and to use these to clone random fragments of the genome. Such experiments are commonly referred to as "shotgun" cloning experiments. The difficulty with the shotgun approach is in identifying a cell or clone containing the gene or sequence of interest (which can be expected to occur at maximum frequencies of about 10^{-6} in the case of higher eukaryotes, see pp. 353–354). Where selection can be applied, this is no problem. For example, cloning the *E. coli* gene that confers resistance to penicillin could easily be accomplished by "shotgun" cloning the entire genome of a pen^r strain using a tetracycline-resistant plasmid (like pSC101; see Fig. 10.29), and selecting for penicillin-resistant, tetracycline-resistant transformants (where the transformed cells were pen^s). In other cases, complementation can be used as a basis of selection. *His* genes of yeast, for example, were cloned in *E. coli* histidine auxotrophs (containing nonreverting deletion mutations), and the desired clones were identified by their ability to grow on medium containing no histidine. Of course, this complementation selection procedure depends on the correct transcription and translation of the cloned gene or genes in the new host, which may not always be the case. The signals that regulate gene expression are different in *E. coli* and eukaryotes.

Another approach to identifying the desired clones after "shotgun" cloning, which does not depend on the correct expression of the cloned genes, involves replica plating the colonies formed by transformed cells to nitrocellulose filters, *in situ* hybridization with a radioactive probe (labeled DNA or RNA molecule) and autoradiography (Fig. 10.31). If purified mRNA containing the gene or sequence of interest can be obtained, it can be used as a template for the *in vitro* synthesis of radioactive cDNA (using labeled nucleoside triphosphate precursors) by reverse transcriptase. The labeled cDNA is then used as a probe for *in situ* hybridization to denatured DNA of colonies that are lysed right on nitrocellulose filters (Fig. 10.31). After allowing a sufficient period of time for hybridization to occur, the filters are washed to remove nonhybridized cDNA (the DNA from the lysed cells is tightly bound to the nitrocellulose filter before hybridization so that it won't wash off) and then autoradiographed (see Chapter 5, pp. 97–101). Only colonies containing DNA sequences complementary to the radioactive cDNA will give rise to radioactive spots (Fig. 10.31). These are used in turn to identify colonies that contain the desired sequence on the original replicated plates.

APPLICATIONS AND POTENTIAL HAZARDS OF RECOMBINANT DNA AND GENE CLONING TECHNOLOGY

As the technology for constructing recombinant DNA molecules and cloning was developing, many molecular biologists recognized that this technology would provide a powerful tool with which to study the structure and function of genetic material. They also wondered whether it might introduce potential hazards. For example, if the gene or genes responsible for transformation to malignancy carried by a tumor virus were cloned in an *E. coli* plasmid, might the *E. coli* strain carrying the viral genes serve as a vector for the transmission of cancer-causing genes? Thus, several of the leading molecular biologists called for a voluntary moratorium on all research of this type. Extensive conferences and discussions on the pros and cons of doing research involving recombinant DNA and cloning were held. A complete discussion of the pros and cons of recombinant DNA and cloning research is beyond the scope of this book. The reader is referred to the published proceedings of a forum entitled *Research with Recombinant DNA*, which was organized by the National Academy of Sciences (U.S.) for a discussion of the arguments presented by proponents and opponents of recombinant DNA research.

Clearly, a major concern of opponents of recombinant DNA work is that a pathogenic or otherwise harmful new organism might be produced and introduced into the ecosystem. Proponents argue

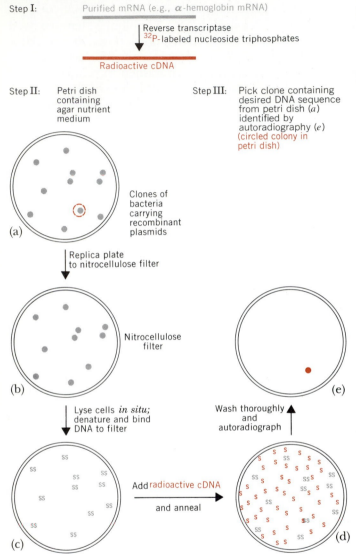

Step I: Purified mRNA (e.g., α-hemoglobin mRNA)

Reverse transcriptase
^{32}P-labeled nucleoside triphosphates

Radioactive cDNA

Step II: Petri dish containing agar nutrient medium

Step III: Pick clone containing desired DNA sequence from petri dish (*a*) identified by autoradiography (*e*) (circled colony in petri dish)

Clones of bacteria carrying recombinant plasmids

(a)

Replica plate to nitrocellulose filter

Nitrocellulose filter

(b)

Lyse cells *in situ;* denature and bind DNA to filter

Wash thoroughly and autoradiograph

(e)

Add radioactive cDNA and anneal

(c)   (d)

FIGURE 10.31

Diagrammatic illustration of the autoradiographic technique for identifying an E. coli *clone carrying a recombinant plasmid with a DNA sequence homologous to a purified mRNA. Step I: A labeled DNA sequence complementary to the mRNA is synthesized using reverse transcriptase. Step II: (a) Cells transformed with recombinant DNA molecules (such as recombinant plasmids) are plated on agar medium and incubated until visible colonies form. (b) The colonies are replica plated onto nitrocellulose filters. (c) The cells on the nitrocellulose replica are lysed in situ, and their DNA is denatured and bound to the filter (e.g., 0.1 N NaOH for a few seconds, followed by heat and vacuum). (d) Then the filter is placed in a solution containing the radioactive cDNA under annealing condition for several hours. (e) The filter is washed thoroughly to remove nonannealed cDNA and is autoradiographed. The autoradiograph will show the exposure to radioactive decays only at the locations of colonies formed by cells that contain DNA sequences complementary to the cDNA probe (only one such colony is shown). Step III: The colony at the corresponding position in the petri dish is picked for subsequent cloning and analysis.*

against the likelihood of this occurring, and emphasize the potential benefits of recombinant DNA research. Proponents emphasize the information that can be obtained about the structure, mode of expression, and regulation of expression of the genetic material of eukaryotes. The importance of potential practical applications, particularly in medicine and agriculture, is also emphasized.

After extensive hearings and discussions, the National Institutes of Health (NIH) of the U.S. Department of Health, Education, and Welfare (now Health and Human Services) established specific guidelines under which recombinant DNA research of various types could be continued with **prior** approval from NIH. These guidelines emphasized both **physical** and **biological containment** of the recombinant molecules constructed. Physical containment includes the use of sterile techniques, containment hoods, and specially designed laboratories to prevent vectors containing recombinant DNA molecules from being transferred or "escaping" from laboratories to natural ecosystems. Biological containment involves the use of organisms with specially constructed, "weakened" genotypes as vectors in cloning experiments. Ideally, these organisms should be unable to survive under conditions existing in any natural ecosystem.

As more work with recombinant DNA was done, it became evident that bacteria and viruses that carry foreign genes are simply not very "healthy." They have been found not to survive in competition with wild-type organisms under natural ecosystem conditions. Thus, the NIH guidelines have been gradually relaxed for the more routine types of gene cloning experiments. Other kinds of experiments still require approval by an NIH panel and by a local institutional biosafety committee **prior** to their initiation.

The potential of recombinant DNA and gene cloning methodology has been rapidly realized. New knowledge about the structures of eukaryotic genes and chromosomes has accumulated at an unprecedented rate since the development of these techniques. (See, for example, the following section on eukaryote gene structure and Chapter 11, pp. 413–422, on antibody genes, and pp. 422–430, on oncogenes.) As for practical applications, human insulin and human growth hormone produced in bacteria engineered by recombinant-DNA and gene-cloning techniques are both now commercially available. For diabetics who are allergic to animal insulins or for individuals with defects in the synthesis of human growth hormone, these achievements of recombinant technology are very important, indeed.

NONCODING INTERVENING SEQUENCES OR INTRONS WITHIN EUKARYOTIC GENES

The application of the recombinant DNA and cloning technology to the organization of the genetic material in eukaryotes has already yielded important and surprising results. In prokaryotes, we know that all of the well-characterized genes consist of continuous sequences of nucleotide pairs, which specify colinear sequences of amino acids in the polypeptide gene products (see Chapter 8 and pp. 340–343). However, analyses of a large number of cloned eukaryotic genes—including genes of animals, plants, and eukaryotic microorganisms—have shown that the **coding sequences** (called **"exons,"** for **ex**pressed sequences) **of these genes** (sequences specifying the functional gene products, RNA or protein) **are often, but not always, interrupted by noncoding "intervening sequences" or "introns."** Several kinds of evidence, the most important being via electron microscopy, the location of restriction enzyme cleavage sites, and correlated nucleotide and amino acid sequences, document the existence of these intervening sequences within the structural genes.

When purified RNA is annealed with double-stranded DNA containing the gene from which the RNA was transcribed using conditions that favor DNA-RNA duplex formation (high concentrations of formamide, etc.), the RNA strands will displace the homologous DNA strands. As a result, single-stranded loops of DNA, called **R-loops,** will form spanning that segment of the DNA molecule (Fig. 10.32a). R-loops can be characterized directly by electron microscopy. When P. Leder and co-workers hybridized purified mouse β-globin mRNA to a recombinant DNA molecule that contained the mouse β-globin gene, cloned in a phage lambda chromosome, they observed two R-loops separated by a loop of double-stranded DNA (Fig. 10.32b). Their results demonstrated the presence of a sequence of nucleotide pairs in the middle of the β-globin gene that is not present in β-globin mRNA and does not, therefore, code for amino acids in the β-globin polypeptide (Fig. 10.33). Leder and colleagues verified this interpretation of their results by sequencing a segment of the gene spanning one junction between a coding sequence and an intron and correlating the nucleotide sequence with the known amino acid sequence of mouse β-hemoglobin.

When Leder and co-workers repeated the R-loop experiments using purified β-globin hnRNA

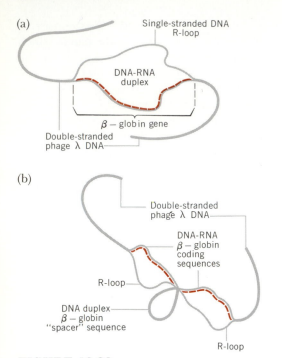

(a)

Single-stranded DNA
R-loop

DNA-RNA
duplex

β — globin gene

Double-stranded
phage λ DNA

(b)

Double-stranded
phage λ DNA

DNA-RNA
β — globin
coding
sequences

R-loop

DNA duplex
β — globin
"spacer" sequence

R-loop

FIGURE 10.32

Diagram illustrating the R-loop DNA-RNA hybridization technique used by P. Leder and colleagues to demonstrate the 653-nucleotide long "intron" (noncoding sequence) in the mouse β-globin gene. (a) When the hnRNA (heterogeneous nuclear RNA) precursor to β-globin mRNA was hybridized to a phage lambda chromosome carrying a segment of mouse DNA encompassing the β-globin gene under conditions favoring DNA-RNA hybridization, a single R-loop was observed. (The RNA strand in DNA-RNA hybrids is shown by the dashed colored lines.) This R-loop corresponds to the nontranscribed strand of the segment of double-stranded DNA from which the mouse β-globin hnRNA was transcribed. (b) When β-globin mRNA was used in the experiment, two R-loops were observed that were separated by a double-stranded DNA loop. This result indicates that the mouse β-globin gene contains a noncoding sequence or intron for which no complementary nucleotide sequence exists in the β-globin mRNA. The intron nucleotide sequence is excised during the conversion of hnRNA to mRNA. This excision must occur by a very precise mechanism to produce the correct codon reading frame in the distal portion of the β-globin mRNA.

(heterogeneous nuclear RNA), the large nuclear precursor to mRNA (believed to be the primary gene transcript; see Chapter 8), in place of β-globin mRNA, only one R-loop was observed (Fig. 10.32a). This indicates that during processing of hnRNA to form mRNA, the intron sequence is cut out and the coding sequences are spliced together.

The large intron sequence in the mouse β-globin gene is 653 nucleotide pairs long. Shortly after the discovery of the large intron, Leder and colleagues found that the mouse β-globin gene contains a second, smaller intron, 116 nucleotide pairs in length (Fig. 10.33). While Leder and co-workers were demonstrating the presence of two introns in the mouse β-globin gene, work in two laboratories, O'Malley's and Chambon's, showed that the chicken ovalbumin gene contains seven intervening sequences or introns separating eight exons (Fig. 10.34).

Subsequent studies have demonstrated the presence of noncoding introns in numerous other eukaryotic genes. The rat serum albumin gene contains 13 introns, and the *Xenopus laevis* vitellogenin (which ends up as egg yolk protein) *A2* gene contains 33 introns. The record for introns per gene is presently held by the chicken α2 collagen gene, which has over 50 introns. This gene spans 38,000 nucleotide pairs, but gives rise to an mRNA molecule only about 5000 nucleotides long. The first evidence for introns in plant genes was the demonstration that the phaseolin (major storage protein) gene of the French bean contains three introns.

Studies by Maniatis, Flavell, and many others have shown that the structure of the human β-globin gene is very similar to that of the mouse β-globin gene, with two introns separating three exons. The two human α-globin genes also each contain two introns. In fact, the human genes coding for the embryonic α-like (ζ) and β-like (ε) hemoglobin chains, the fetal β-like (γ) hemoglobin chain, and the adult minor β-like (δ) hemoglobin chain, all contain two introns and three exons. Moreover, all of the human genes coding for α-like globins have the introns at the same positions (separating codon 31 from 32 and codon 99 from 100), as do all the human genes coding for β-like globins (separating

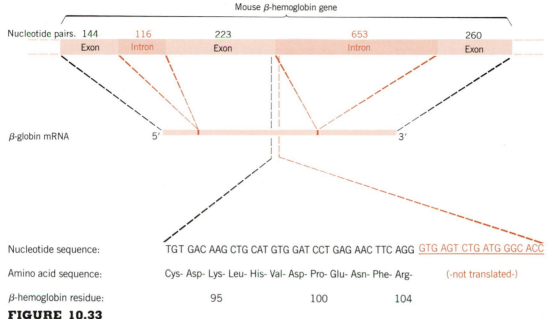

FIGURE 10.33

Organization of the mouse β-globin gene, showing the locations of noncoding intron sequences (in color) and coding ("exon") sequences (gray). The correspondence between segments of the gene and segments of the mRNA is indicated by dashed diagonal lines. The nucleotide sequence of the "antisense" (nontranscribed) strand of DNA for the indicated juncture (dashed lines) between the second coding sequence and the large intron sequence is shown at the bottom, along with the amino acid sequence of the segment of the β-globin chain for which it codes. Amino acid number 104 (numbering from the amino terminus) is arginine (Arg). Amino acids number 105 through 110 are Leu-Leu-Gly-Asn-Met-Ile. If the contiguous (intron) nucleotide sequence (underlined in color on the right) was transcribed and translated, amino acids 105 through 110 should be Val-Ser-Leu-Met-Gly-Thr. (Based on the data of S. M. Tilghman, D. C. Tiemeier, J. G. Seidman, B. M. Peterlin, M. Sullivan, J. V. Maizel, and P. Leder, Proc. Natl. Acad. Sci., U.S. 75:725, 1978, and the data summarized in T. Pederson, Amer. Scientist 69:76, 1981.)

codon 30 from 31 and codon 104 from 105).

The human genes coding for the α-like globins and the β-like globins are located in clusters on chromosomes 16 and 11, respectively. The globin genes in each cluster are lined up on their respective chromosomes in the order in which they are expressed during development. We don't yet know how this relates to the regulation of their expression, if at all, but it is a most interesting correlation.

The only structural features that seem to be shared by different introns are the dinucleotide sequences at their ends. The primary transcripts of genes almost always begin (5′) introns with the sequence GT and end (3′) them with the sequence AG; these consensus sequences at intron–exon junctions must be important in the mechanism by which introns are spliced out of primary transcripts (Chapter 8, pp. 256–259).

We do not yet understand much about the biological significance of the exon–intron structure of eukaryotic genes. Introns are highly variable in size, ranging from a few nucleotide pairs to thou-

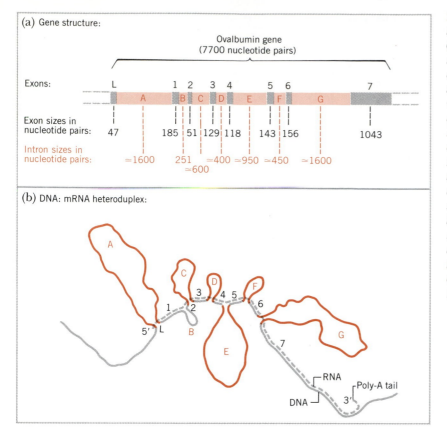

(a) Gene structure:

Ovalbumin gene
(7700 nucleotide pairs)

Exons: L   1 2   3 4   5 6   7

A   B C D   E   F   G

Exon sizes in
nucleotide pairs: 47   185 51 129 118   143 156   1043

Intron sizes in
nucleotide pairs: ≈1600   251   ≈400 ≈950 ≈450 ≈1600
                              ≈600

(b) DNA: mRNA heteroduplex:

A

C

D

3   4 5   F

1   2   6

5' L   B   7   G

E

┌ RNA

┌ Poly-A tail

DNA ┘   3'

FIGURE 10.34

*Structure of the chicken ovalbumin gene.
(a) Diagram showing the 7 introns (A–G,
shown in color) and 8 exons (L and 1–7)
of the ovalbumin gene. Exon L specifies
the 5′ leader sequence of the ovalbumin
mRNA; exons 1–7 code for amino acid sequences of ovalbumin. (b) Tracing of a
DNA:mRNA heteroduplex containing the
ovalbumin mRNA and a single strand of
DNA carrying the transcribed strand of
the chicken ovalbumin gene, as observed by
electron microscopy. The mRNA and DNA
strands are shown as dashed and solid
lines, respectively. The unpaired intron
segments of the DNA strand are shown in
color; they correspond to the segments of
the gene shown in color in (a). [After P.
Chambon, Sci. Amer. 244(5):60–71,
1981.]*

sands of nucleotide pairs in length. In several genes
the introns have been shown to accumulate new
mutations much more rapidly than the exons. This
suggests that their specific nucleotide-pair sequences, excluding the ends, are not very important.

In some cases, the different exons of genes code
for different functional domains of the protein gene
products. This is most apparent in the case of the
genes coding for antibodies (Chapter 11, pp.
413–422). In the case of the globin genes, the
middle exon codes for the domain of the protein
containing the heme-binding site. This has led to
speculation that the exon–intron structure of eukaryotic genes results from the evolution of new
genes by bringing together exons that were ances-

tral genes. If this is correct, introns may be merely
relics of the evolutionary process.

Alternatively, introns may provide a selective
advantage by increasing the rate at which coding
sequences in different exons of a gene can reassort
by recombination, thus speeding up the process of
evolution. In the case of the mitochondrial gene of
yeast coding for cytochrome b, the introns appear to
be part of exons for genes coding for enzymes ("maturases") involved in processing of the primary
transcript of the gene. (For further details, see
Lewin, *Genes,* Chapter 20). Thus, different introns
may indeed play different roles, and some introns
may have no biologically important function. Finally, we should emphasize that not all eukaryotic
genes contain introns. For example, the sea urchin

histone genes and four *Drosophila* heat shock genes were among the first to be shown **not** to contain introns.

MAPPING RESTRICTION ENZYME CLEAVAGE SITES OF CHROMOSOMES

Many of the **restriction endonucleases cleave DNA molecules in a site-specific manner** (Table 10.5). As a result, they can be used to generate **physical maps of chromosomes** that are of tremendous value in assisting researchers in isolating small segments of chromosomes carrying genes or other DNA sequences of interest. These maps are invaluable in DNA sequencing experiments because the sequencing techniques can be used only with small DNA molecules, usually sequences of one to a few hundred nucleotide pairs in length.

During their early work on the characterization of restriction endonucleases, H. Smith and D. Nathans used the DNA chromosome of **simian virus 40 (SV40)** for their studies. (They shared the 1978 Nobel Prize in Physiology and Medicine with

W. Arber for the discovery of restriction enzymes.) SV40 is an animal virus that can transform cells to the cancerous state. Its circular chromosome (now completely sequenced) is only 5226 nucleotide pairs in length, making it well suited for restriction enzyme studies. As a result of Smith and Nathans' work, **restriction enzyme cleavage site maps,** now commonly called simply **restriction maps,** of the SV40 chromosome were among the first such maps constructed.

Restriction enzyme *Eco*RI cleaves the SV40 chromosome at only one site (Table 10.5 and Fig. 10.35*a*). This site has been arbitrarily set as position O on the SV40 chromosome and restriction map. Restriction enzymes *Hpa*I and *Hind*III cleave the SV40 chromosome at four and six sites, respectively (Fig. 10.35, *b* and *c*). When the SV40 chromosome is digested with all three of the above enzymes, 11 distinct restriction fragments are produced (Fig. 10.35*d*).

The sizes of the restriction fragments can be determined by **polyacrylamide or agarose gel electrophoresis** (Fig. 10.36) as described in the legend to Fig. 5.41. Because of the nucleotide subunit

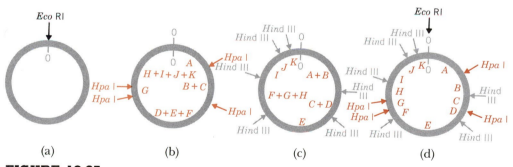

(a)   (b)   (c)   (d)

FIGURE 10.35

Restriction enzyme cleavage site maps of the simian virus 40 (SV40) chromosome. The restriction enzymes EcoRI *(a),* HpaI *(b), and* HindIII *(c) cleave the SV40 chromosome at 1, 4, and 6 sites, respectively. The 11 fragments produced when the SV40 chromosome is cut with all three restriction enzymes are shown in (d); they are arbitrarily labeled alphabetically starting at the unique* EcoRI *cleavage site and progressing in the clockwise direction. The sizes of the various fragments can be determined by polyacrylamide or agarose gel electrophoresis (see Fig. 5.41 and legend) as shown in Fig. 10.36. (Based on the data summarized by D. Nathans and H. O. Smith,* Ann. Rev. Biochem. *44:273–293, 1975.)*

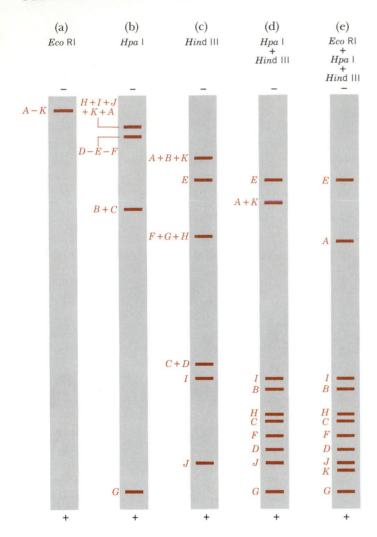

(a) *Eco* RI (b) *Hpa* I (c) *Hind* III (d) *Hpa* I + *Hind* III (e) *Eco* RI + *Hpa* I + *Hind* III

FIGURE 10.36

Separation of SV40 restriction fragments by polyacrylamide gel electrophoresis. See the legend to Fig. 5.41 for a description of the procedure. The fragments are separated on the basis of size—the smaller the fragment, the greater the distance that it migrates in the gel. SV40 chromosomes were cleaved with (a) EcoRI, (b) HpaI, (c) HindIII, (d) HpaI + HindIII, *or* (e) EcoRI + HpaI + HindIII. *The fragments produced by digestion with all three restriction enzymes have been arbitrarily labeled* A *through* K, *in clockwise order, starting at the* EcoRI *cleavage site. The restriction fragment bands are visualized under ultraviolet light after staining with ethidium bromide. Ethidium bromide binds to DNA and in the bound state exhibits an increased fluorescence under UV-light, making DNA bands easily detected. The positions of the various restriction fragment bands shown above are only approximate. (The relative sizes of the restriction fragments are based on the data summarized by D. Nathans and H. O. Smith,* Ann. Rev. Biochem. 44:273–293, 1975.)

structure of DNA, with one phosphate group per nucleotide, DNA has an essentially constant charge per unit of mass. Thus, the rates of migration of DNA fragments during electrophoresis provide accurate estimates of their lengths, with the rate of migration being inversely proportional to length. The polyacrylamide or agarose gel simply acts as a molecular sieve, with small fragments able to move through the sieve faster than large molecules. Agarose works somewhat better than polyacrylamide for large fragments as it has a slightly larger pore size; polyacrylamide gels yield better resolution of short fragments.

There are three major approaches to ordering restriction fragments—all of which depend on the use of polyacrylamide or agarose gel electrophoresis to obtain accurate estimates of fragment size. These are: (1) **sequential digestion of chromosomes** with two or more different restriction enzymes, (2) **partial digestion of chromosomes** after labeling the ends with a radioactive isotope, and (3) determining whether the strands of **different restriction fragments can hybridize with each other** after denaturation.

If the SV40 chromosome is cleaved with *Hin*-dIII (Fig. 10.35*c*), a large fragment *E* is produced

(Fig. 10.36*c*). This same fragment is still present after subsequent digestion with *Hpa*I (Fig. 10.36*d*), indicating that there are no *Hpa*I cleavage sites within the *Hin*dIII *E* fragment. By contrast, when *Hpa*I restriction fragment *B + C* (Figs. 10.35*b* and 10.36*b*) is subsequently digested with *Hin*dIII, it is cleaved into two fragments *B* and *C* (Figs. 10.35*d* and 10.36*d*). Similarly, subsequent digestion of *Hin*dIII fragment *C + D* (Figs. 10.35*c* and 10.36*c*) with *Hpa*I yields fragments *C* and *D* (Figs. 10.35*d* and 10.36*d*). These data, taken together, show that the common sequence (or overlap sequence) in the *Hpa*I fragment *B + C* and *Hin*dIII fragment *C + D* is fragment *C* and establish the order *B-C-D* (or *D-C-B*). By continuing to analyze the fragments produced by sequential digestions with these and other restriction enzymes, the order of the fragments can usually be unambiguously established.

Partial restriction digests of radioactively end-labeled DNA molecules provide an even simpler mechanism for ordering restriction sites. The radioactive isotope ^{32}P is added to the ends of a linear DNA molecule (such as an *Eco*RI-cleaved SV40 chromosome, Fig. 10.35*a*) using ^{32}P-ATP and the enzyme polynucleotide kinase. The molecule is then usually cut into two unequal halves by using a restriction enzyme that cleaves the molecule only once, in an asymmetrical manner. Separation of the two halves by electrophoresis yields molecules with only one end labeled; the two halves are then analyzed separately.

The fragments produced by **partial digestion** of the isolated half-chromosomes are separated by gel electrophoresis, and the radioactive fragments are identified by autoradiography (see Chapter 5, p. 97). Only those fragments that contain an original radioactive end will be detected by this procedure. The shortest radioactive fragment will indicate how far the first restriction enzyme cleavage site is from the labeled end of the chromosome. The second shortest fragment will give the distance from the labeled end to the second restriction enzyme cleavage site, and so on.

Finally, overlapping sequences of different restriction fragments can be detected by their ability to hybridize with each other ("cross-hybridization"). Two fragments that cross-hybridize must overlap a common sequence, whereas two fragments that do not cross-hybridize cannot overlap by more than a few nucleotide pairs.

By combining these approaches, detailed restriction maps (Fig. 10.37) have been worked out for many different chromosomes. In the future, restriction maps of this kind will be available for entire genomes, even those of the most complex eukaryotes.

NUCLEOTIDE SEQUENCES: THE ULTIMATE FINE STRUCTURE MAPS

The **ultimate fine structure map of a gene or a chromosome is its nucleotide-pair sequence,** complete with a chart of all nucleotide-pair changes that alter the function of that gene or chromosome. Prior to 1975, the thought of trying to sequence entire chromosomes was barely conceivable—at best, a laborious task requiring years of work. By late 1976, the entire 5387 nucleotide-long chromosome of phage ΦX174 had been sequenced. Today, the entire chromosomes of several viruses, including the complete 48,502 nucleotide-pair sequence of the circular form of the phage λ chromosome, and segments of eukaryotic chromosomes tens of thousands of nucleotide pairs in length have been sequenced. Within the next few years, the nucleotide sequences of hundreds of genes and even entire eukaryotic chromosomes will be determined and stored in computer data-banks for subsequent reference.

Our present ability to sequence essentially any DNA molecule is the result of four major developments. (1) The most important breakthrough was the discovery of restriction enzymes and their use in preparing homogeneous samples of specific segments of chromosomes. (2) A second major advance was the improvement of gel electrophoresis procedures to the point where DNA fragments that differed in length by a single nucleotide could be resolved. (3) The development of gene cloning techniques greatly facilitated the preparation of large quantities of a particular gene or DNA sequence of interest. (4) Finally, two different approaches to the rapid sequencing of DNA molecules were developed in the period from 1974 to 1977.

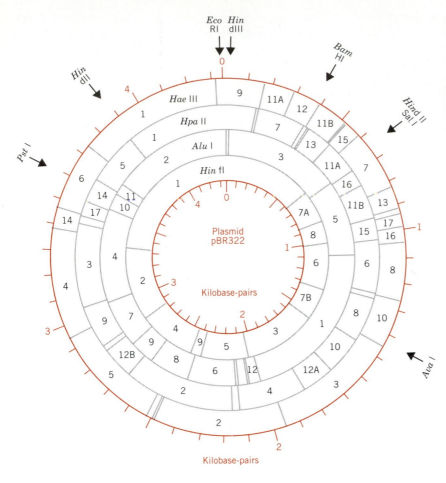

Eco Hin
RI dIII

Bam
HI

Hind II
Sal II

Hin
dII

Pst I

Ava I

Plasmid
pBR322

Kilobase-pairs

Kilobase-pairs

FIGURE 10.37

Restriction enzyme cleavage map of plasmid pBR322, one of the more popular cloning vehicles for recombinant DNA research. The cleavage sites for enzymes HinfI, AluI, HpaII, and HaeIII are shown in the concentric rings going from the inside to the outside, respectively. The restriction fragments produced by each enzyme are numbered 1 through n in order of decreasing size. The cleavage sites for enzymes that cut the pBR322 DNA molecule only once or twice are shown around the perimeter. The scale given on the inner and outer rings shows distances in kilobase-pairs (1000 nucleotide-pair units). The complete 4362 nucleotide-pair sequence of plasmid pBR322 has been worked out by J. G. Sutcliffe. (After J. G. Sutcliffe, Nucleic Acid Res. *5:2721–2728, 1978.)*

Both approaches depend on the generation of a **population of DNA fragments** that all have **one end in common** (all end at exactly the same nucleotide) and **terminate at all possible positions** (every consecutive nucleotide) **at the other end.** These fragments are then separated on the basis of chain length by polyacrylamide gel electrophoresis. In both cases, four separate parallel reactions are carried out, each of which generates a set of fragments terminating at one of the four bases (A, G, C, or T) in DNA.

The **Maxam and Gilbert procedure,** named after A. Maxam and W. Gilbert, who developed it, uses four different **chemical reactions** to cleave DNA chains specifically at As, Gs, Cs, or Cs + Ts. The second approach, developed by F. Sanger and

colleagues, uses an **enzymatic procedure** and **specific chain-terminators** to generate four populations of fragments that terminate at As, Gs, Cs, and Ts, respectively.

2',3'-Dideoxyribonucleoside triphosphates (Fig. 10.38) are the chain-terminators most commonly used in DNA sequencing studies. Recall that DNA polymerases have an absolute requirement for a free 3'-OH on the DNA primer strand (see Fig. 5.26). If a 2',3'-dideoxynucleotide is added to the end of a chain, it will block subsequent extension of that chain since the **2',3'-dideoxynucleotides have no 3'-OH.** By using (1) 2',3'-dideoxythymidine triphosphate (ddTTP), (2) 2',3'-dideoxycytidine triphosphate (ddCTP), (3) 2',3'-dideoxyadenosine triphosphate (ddATP), and (4) 2',3'-dideoxyguan-

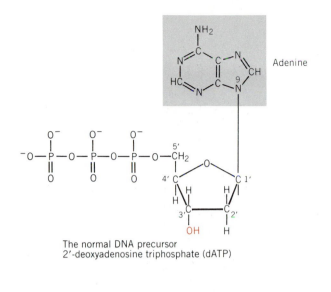

The normal DNA precursor
2'-deoxyadenosine triphosphate (dATP)

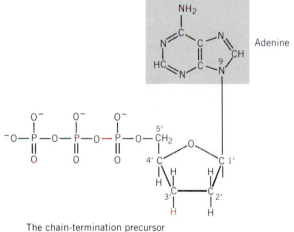

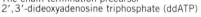

The chain-termination precursor
2',3'-dideoxyadenosine triphosphate (ddATP)

FIGURE 10.38

*Comparison of the structures of the normal DNA precursor
2'-deoxyadenosine triphosphate (dATP) and the chain-terminator 2',3'-dideoxyadenosine triphosphate (ddATP) used in DNA
sequencing reactions (see Fig. 10.39). Note the absence of the
3'-OH on ddATP. Once a 2',3'-dideoxynucleotide is added to
the 3'-terminus of a nascent DNA chain, no further covalent
extension of the chain is possible.*

osine triphosphate (ddGTP) as chain-terminators in
four separate DNA biosynthetic reactions, one can
generate four populations of fragments, each popu-

lation containing chains that all terminate with the
same base (T, C, A, or G) (Fig. 10.39).

In a given reaction, the ratio of dXTP : ddXTP
(where X can be any one of the four bases) is kept at
approximately 100 : 1 so that the probability of terminating at a given "X" in the nascent chain is about
$\frac{1}{100}$. This yields a population of fragments terminating at all potential ("X") termination sites within a
distance of more than 100 nucleotides from the
original primer terminus (Figs. 10.39 and 10.40).

After the nascent DNA fragments generated in
the four parallel reactions are released from the
template strands by denaturation, they can be separated by polyacrylamide gel electrophoresis and
their positions in the gel detected by autoradiography. The bands on the autoradiograms will correspond to radioactive chains of different lengths;
they will produce a "ladder" defining the nucleotide sequence of the longest nascent chain synthesized (Figs. 10.39 and 10.40).

The shortest fragment will migrate the greatest
distance, giving rise to the band nearest the anode
(the positive electrode). Each successive band will
contain chains that are one nucleotide longer than
the chains in the preceding band of the "ladder."
The 3'-terminal nucleotide of the chain in each
band will be the dideoxynucleotide chain-terminator present in the reaction mixture (1, 2, 3, or 4) in
which that specific chain was produced (see Fig.
10.39). By "reading" the "ladder" produced by
autoradiography of the polyacrylamide gels used to
separate the fragments generated in each of the four
parallel reactions, the complete nucleotide sequence of a DNA chain can be determined. This is
illustrated for a hypothetical nucleotide sequence in
Fig. 10.39. A photograph of an autoradiogram of
an actual dideoxynucleotide chain-terminator sequencing gel is shown in Fig. 10.40. It shows that
long sequences of nucleotide chains can be "read"
from a single sequencing gel.

SUMMARY

Traditionally, the **gene** has been defined as the unit
of genetic material controlling the inheritance of
one phenotypic characteristic or **one trait**. Today,
the **gene** is more precisely defined as the unit of

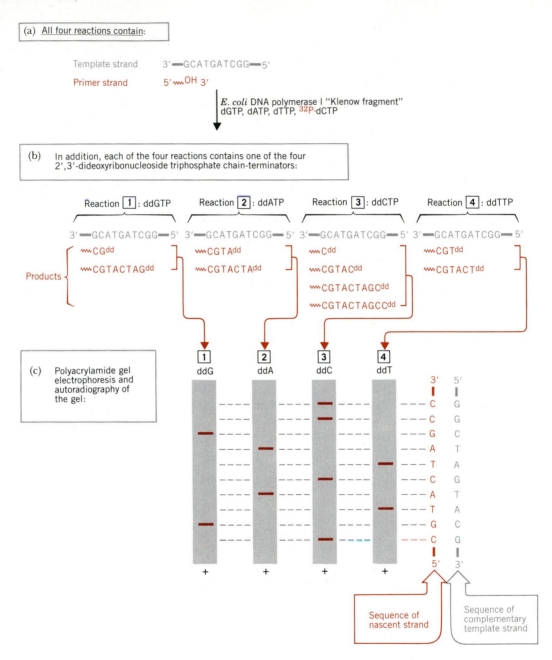

(a) <u>All four reactions contain:</u>

Template strand   3′ ▬GCATGATCGG▬5′

Primer strand   5′ ⌇⌇OH 3′

E. coli DNA polymerase I "Klenow fragment"
dGTP, dATP, dTTP, ³²P-dCTP

(b)   In addition, each of the four reactions contains one of the four
2′,3′-dideoxyribonucleoside triphosphate chain-terminators:

Reaction ☐1 : ddGTP   Reaction ☐2 : ddATP   Reaction ☐3 : ddCTP   Reaction ☐4 : ddTTP

3′ ▬GCATGATCGG▬5′   3′ ▬GCATGATCGG▬5′   3′ ▬GCATGATCGG▬5′   3′ ▬GCATGATCGG▬5′

Products
⌇⌇CGdd   ⌇⌇CGTAdd   ⌇⌇Cdd   ⌇⌇CGTdd
⌇⌇CGTACTAGdd   ⌇⌇CGTACTAdd   ⌇⌇CGTACdd   ⌇⌇CGTACTdd
                                                 ⌇⌇CGTACTAGCdd
                                                 ⌇⌇CGTACTAGCCdd

☐1   ☐2   ☐3   ☐4
ddG   ddA   ddC   ddT

(c)   Polyacrylamide gel
electrophoresis and
autoradiography of
the gel:

				3′	5′
		▬		C	G
		▬		C	G
▬				G	C
	▬			A	T
			▬	T	A
		▬		C	G
	▬			A	T
			▬	T	A
▬				G	C
		▬		C	G
			▬	5′	3′

+   +   +   +

Sequence of nascent strand

Sequence of complementary template strand

genetic material **coding for one polypeptide** (or one RNA molecule in the case of genes not coding for proteins). Prior to 1941, the **gene** was also believed **not to be subdivisible by mutation or recombination.** The unit of genetic material **not subdivisible** by mutation or recombination is now known to be the **single nucleotide pair** (or single nucleotide in the case of single-stranded nucleic acid genomes).

FIGURE 10.39

*Diagram illustrating the technique of DNA sequencing using 2′,3′-dideoxynucleoside triphosphate chain-terminators (see Fig. 10.38). Four reactions are carried out in parallel, each of which contains one of the four 2′,3′-dideoxy chain-terminators: ddGTP, ddATP, ddCTP, and ddTTP. (a) All of the reaction mixtures contain (1) a template strand, the nucleotide sequence of which is to be determined, (2) a primer strand with a free 3′-hydroxyl (usually obtained from a restriction fragment or by chemical synthesis), (3) the four DNA precursors: dGTP, dATP, dTTP, and dCTP, at least one of which is radioactive (^{32}P-dCTP in the scheme shown), and the "Klenow fragment" of E. coli DNA polymerase I. The "Klenow fragment" is the "large half" of E. coli DNA polymerase I produced by cleavage with the proteolytic enzyme trypsin. The "Klenow fragment" retains the 5′→3′ polymerase and 3′→5′ exonuclease activities of DNA polymerase I, but has **no 5′→3′ exonuclease activity** (see Chapter 5, pp. 105–110). (The 5′→3′ exonuclease activity is on the "small half" of the molecule cleaved off with trypsin.) The absence of 5′→3′ exonuclease activity is critical to the technique; if this activity is present, it will cut back the primer strand from the 5′-end. **The 5′ ends must remain fixed at the same nucleotide position on all nascent chains so that their lengths will be determined by the position of the dideoxy chain-termination events at the 3′-ends.** (b) The key to the sequencing technique is to use the correct ratio of the normal deoxyribonucleoside triphosphate (e.g., dGTP in reaction 1) and the terminator dideoxynucleoside triphosphate (e.g., ddGTP in reaction 1), so as to obtain a population of nascent chains terminating at all possible nucleotide positions (e.g., all template-strand Cs in reaction 1). The ratio of 100 deoxyribonucleoside triphosphates to 1 dideoxyribonucleoside triphosphate usually gives the desired frequency of termination at each potential termination site. (c) The products of the four reactions are separated in parallel by polyacrylamide gel electrophoresis and the positions of the radioactive reaction products (nascent DNA chains) are determined by autoradiography. Since the shortest chain migrates the greatest distance in the gel, the nucleotide sequence defined by the nascent chains is given for the 5′→3′ direction by "reading" from the bottom (anode) to the top (cathode) of the gel. This sequence is shown in color to the immediate right of the autoradiogram. The sequence of the template strand (shown in black) will be complementary to that of the longest nascent chain. (For additional details, see F. Sanger, S. Nicklen, and A. R. Coulson,* Proc. Natl. Acad. Sci., U.S. *74:5463– 5467, 1977.)*

Many **mutable sites separable by recombination thus exist within each gene.**

The gene is operationally defined by the *cis- trans* or complementation test. This test is to determine whether organisms heterozygous for two mutations have mutant or wild-type phenotypes. *cis*-Heterozygotes (organisms where the two mutations are on one chromosome and the corresponding wild-type alleles are on another chromosome) must exhibit the wild-type phenotype for the *cis- trans* test to be informative. If **both the *cis*-heterozygote and** the *trans*-heterozygote (a cell or organism where the two mutations are on different chromosomes) containing a given pair of mutations have **wild-type phenotypes,** then the **two mutations are in two different genes.** If the *cis*-heterozygote has the **wild-type phenotype** and the *trans*-heterozygote has the **mutant phenotype,** then the **two mutations are in the same gene.**

The utility of the *cis-trans* test is limited in some

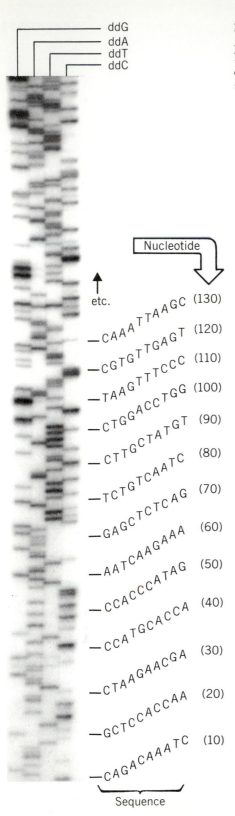

FIGURE 10.40

Photograph of a section of a 2′,3′-dideoxynucleotide chain-terminator sequencing gel. The sequence shown is for a segment of one strand of an 18S ribosomal RNA gene of corn. (Original photograph courtesy of John Carlson and Joachim Messing, Department of Biochemistry, University of Minnesota.)

cases by (1) **intragenic complementation**, (2) **polar effects** of chain-termination mutations in **operons**, and (3) ***cis*-acting genes or gene products**. **Intragenic complementation** frequently occurs when the active form of the protein gene product is a **multimer** consisting of **two or more polypeptide products of a given gene.** Delimiting genes by complementation tests should be done whenever possible using **amorphic** or **null mutations** (mutations resulting in no gene product, a partial gene product, or an otherwise totally nonfunctional gene product) to minimize the possibility of confounding effects of **intra**genic complementation.

Whenever more than two forms of a particular gene are present in a species, they represent **multiple alleles** of that gene. The **ABO blood types in humans** represent a classical case of multiple allelism.

Detailed genetic and cytological analyses of segments of the giant salivary chromosomes in *D. melanogaster* support the idea that each band of the chromosome contains one gene. If so, *Drosophila* would have only 5000–6000 genes, rather than 100,000 as estimated from the total amount of DNA in the genome, and an average of 1000 base-pairs per gene. However, estimates of the total number of mRNA molecules suggest that *Drosophila* has about 17,500 genes.

The complete nucleotide sequence of the bacteriophage ΦX174 chromosome is known. Surprisingly, of the 10 genes of ΦX174, **two are located entirely within the coding sequences of two different genes.** A third overlaps the sequences of three different genes.

Technology is now available for constructing **recombinant DNA molecules that contain sequences from totally unrelated species and cloning these molecules in appropriate host cells.** The application of recombinant DNA technology to the question of the structure of genes has shown that **eukaryotic genes** frequently **contain noncoding "introns" or "intervening sequences" located between coding sequences** or **exons.** The **different exons** of eukaryotic genes often code for **different functional domains** of the protein gene products. In some cases, the introns of one gene may contain exons of a different gene.

Restriction maps of chromosomes are now routinely prepared using various restriction enzymes. **Nucleotide sequences** of chromosomes provide the ultimate fine structure maps.

REFERENCES

Anderson, W. F., and E. G. Diacumakos. 1981. "Genetic engineering in mammalian cells." *Sci. Amer.* 245(1): 60–93.

Benzer, S. 1959. "On the topology of the genetic fine structure." *Proc. Natl. Acad. Sci., U.S.* 45:1607–1620.

———. 1961. "On the topography of the genetic fine structure. *Proc. Natl. Acad. Sci., U.S.* 47:403–415.

Breathnach, R., and P. Chambon. 1981. "Organization and expression of eucaryotic split genes coding for proteins." *Ann. Rev. Biochem.* 50:349–383.

Brink, R. A. (ed.). 1967. *Heritage from Mendel.* University Wisconsin Press, Madison.

Carlson, E. A. 1966. *The gene: a critical history.* Saunders, Philadelphia.

Chambon, P. 1981. "Split genes." *Sci. Amer.* 244(5):60–71.

Chovnick, A., W. Gelbart, and M. McCarron. 1977. "Organization of the *rosy* locus in *Drosophila melanogaster.*" *Cell* 11:1–10.

Cohen, S. N. 1975. "The manipulation of genes." *Sci. Amer.* 233(1):24–33.

Fiddes, J. C. 1977. "The nucleotide sequence of a viral DNA." *Sci. Amer.* 237(6):54–67.

Fincham, J. R. S. 1966. *Genetic complementation.* Benjamin, Menlo Park, Calif.

Judd, B. H., M. W. Shen, and T. C. Kaufman. 1972. "The anatomy and function of a segment of the X chromosome of *Drosophila melanogaster.*" *Genetics* 71:139–156.

Lewin, B. 1983. *Genes.* John Wiley & Sons, New York.

Maniatis, T., E. F. Fritsch, and J. Sambrook. 1982. *Molecular cloning, a laboratory manual.* Cold Spring Harbor Laboratory Press, Cold Spring Harbor, New York.

Maxam, A. M., and W. Gilbert. 1977. "A new method for sequencing DNA." *Proc. Natl. Acad. Sci., U.S.* 74:560–564.

Sanger, F., G. M. Air, B. G. Barrell, N. L. Brown, A. R. Coulson, J. C. Fiddes, C. A. Hutchison, III, P. M. Slocombe, and M. Smith. 1977. "Nucleotide sequence of bacteriophage ΦX174 DNA." *Nature* 265:687–695.

Sanger, F., A. R. Coulson, G. F. Hong, D. F. Hill, and G. B. Petersen. 1982. "Nucleotide sequence of bacteriophage λ DNA." *J. Mol. Biol.* 162:729–773.

Sanger, F., S. Nicklen, and A. R. Coulson. 1977. "DNA sequencing with chain-terminating inhibitors." *Proc. Natl. Acad. Sci., U.S.* 74:5463–5467.

Smith, H. O. 1979. "Nucleotide sequence specificity of restriction endonucleases." *Science* 205:455–462.

Stent, G. S., and R. Calendar. 1978. *Molecular genetics, an introductory narrative,* 2nd ed. W. H. Freeman, San Francisco.

Tilghman, S. M., D. C. Tiemeier, J. G. Seidman, B. M. Peterlin, M. Sullivan, J. V. Maizel, and P. Leder. 1978. "Intervening sequence of DNA identified in the structural portion of a mouse β-globin gene." *Proc. Natl. Acad. Sci., U.S.* 75:725–729.

Watson, J. D., and J. Tooze, 1981. *The DNA story.* W. H. Freeman, San Francisco.

Welshons, W. J. 1965. "Analysis of a gene in *Drosophila.*" *Science* 150:1122–1129.

Yanofsky, C., and V. Horn. 1972. "Tryptophan synthetase α chain positions affected by mutations near the ends of the genetic map of *trpA* of *Escherichia coli.*" *J. Biol. Chem.* 247:4494–4498.

Zachar, Z., and P. M. Bingham. 1982. "Regulation of *white* locus expression: the structure of mutant alleles at the *white* locus of *Drosophila melanogaster.*" *Cell* 30:529–541.

PROBLEMS AND QUESTIONS

10.1. In what ways does our present concept of the gene differ from the pre-1940 or classical concept of the gene?

10.2. What was the first evidence that indicated that the unit of function and the unit of structure of genetic material were not the same?

10.3. What is the currently accepted operational definition of the gene?

10.4. Of what value are conditional lethal mutations for genetic fine structure analysis?

10.5. You are given five female *E. coli* strains, each carrying a different deletion in the *z* gene (for β-galactosidase). The **boxes illustrated here represent the rela-**

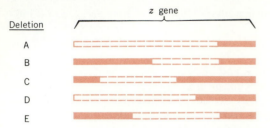

Deletion

A
B
C
D
E

tive positions and extents of the five deletions. Seven independent z^- point mutations were induced by 5-bromouracil in an Hfr (male) strain. When each of the seven mutants is mated with each of the five deletion strains, the following results are obtained ($+$ indicates the occurrence of recombination with the formation of z^+ recombinants, and 0 indicates that no recombination is detected).

DELETION	POINT MUTANTS						
	1	2	3	4	5	6	7
A	0	0	+	0	0	0	0
B	0	0	+	0	+	+	+
C	+	+	+	0	0	0	+
D	0	+	+	0	0	0	0
E	0	0	+	0	+	0	+

What is the linear order of the seven point mutations?

10.6. Eight independently isolated mutants of *E. coli*, all of which are unable to grow in the absence of histidine (his$^-$), were examined in all possible *cis-* and *trans-*heterozygotes (partial diploids). All of the *cis-*heterozygotes were able to grow in the absence of histidine. The *trans-*heterozygotes yielded two different responses: some of them grew in the absence of histidine; some did not. The experimental results, using $+$ to indicate growth and 0 to indicate no growth, are given in the following table. How many genes are defined by these eight mutations? Which mutant strains carry mutations in the same gene(s)?

	GROWTH OF *TRANS*-HETEROZYGOTES (−HISTIDINE)							
MUTANT	1	2	3	4	5	6	7	8
8	0	0	0	0	0	0	+	0
7	+	+	+	+	+	+	0	
6	0	0	0	0	0	0		
5	0	0	0	0	0			
4	0	0	0	0				
3	0	0	0					
2	0	0						
1	0							

10.7. Assume that the mutants described in Problem 10.6 yielded the following results. How many genes would they have defined? Which mutations would have been in the same gene(s)?

	GROWTH OF *TRANS*-HETEROZYGOTES (−HISTIDINE)							
MUTANT	1	2	3	4	5	6	7	8
8	+	+	+	+	+	+	0	0
7	+	+	+	+	+	+	0	
6	+	+	+	+	0	0		
5	+	+	+	+	0			
4	+	+	0	0				
3	+	+	0					
2	0	0						
1	0							

10.8. What determines the maximum number of different alleles that can exist for a given gene?

10.9. What is the difference between a pair of homoalleles and a pair of heteroalleles?

10.10. Two different inbred varieties of a particular plant species have white flowers. All other varieties of this species have red flowers. What experiments might be done to obtain evidence to determine whether the difference in flower color in these varieties is the result of different alleles of a single gene or the result of genetic variation in two or more genes?

10.11. Why are mutant (alternative) alleles essential for identifying wild-type alleles and locating the positions of gene loci on chromosomes?

10.12. (a) How do multiple alleles originate? (b) How should a series of multiple alleles be symbolized? (c) To what extent do they represent alterations of the same basic phenotype?

10.13. The following, listed in order of dominance, are four alleles in rabbits: c^+, colored; c^{ch}, chinchilla; c^h, himalayan; and c, albino. What phenotypes and ratios would be expected from the following crosses: (a) $c^+c^+ \times cc$; (b) $c^+c \times c^+c$; (c) $c^+c^h \times c^+c^{ch}$; (d) $c^{ch}c \times cc$; (e) $c^+c^h \times c^+c$; and (f) $c^hc \times cc$?

10.14. In mice, a series of five alleles has been associated with fur pattern. These alleles are, in order of dominance, A^Y (homozygous lethal) for yellow fur; A^L, agouti with light belly; A^+; agouti; a^t, black and tan; and a, black. For each of the following crosses, give the coat color of the parents and the phenotypic ratios expected among the progeny: (a) $A^YA^L \times A^YA^+$; (b) $A^Ya \times A^La^t$; (c) $a^ta \times a^Ya$; (d) $A^La^t \times A^LA^L$; (e) $A^LA^L \times A^YA^+$; (f) $A^+a^t \times a^ta$; (g) $a^ta \times aa$; (h) $A^YA^L \times A^+a^t$; and (i) $A^Ya^L \times A^YA^+$.

10.15. If a series of four alleles is known to exist in a given diploid ($2n$) species, how many would be present in: (a) a chromosome; (b) a chromosome pair; and (c) an individual member of the species? (d) How many different combinations might be expected to occur in the entire population?

10.16. Assume that in a certain animal species four alleles (c^+, c^1, c^2, and c) have their locus in chromosome I and another series of two alleles (d^+ and d) have their locus in chromosome II. How many different genotypes, with respect to these two series of alleles, are theoretically possible in the population?

10.17. A series of multiple alleles in a certain species of fish that breeds readily in the laboratory was listed by Myron Gordon as follows: P^o, one spot; P^m, moon complete; P^c, crescent; P^{cc}, crescent complete; P^{co}, comet; P^t, twin spot; and P, plain. (a) How many combinations of these alleles might be expected to occur in the population? (b) How could the allelic nature of these genes be indicated by genetic methods?

10.18. In several plants, such as tobacco, primrose, and red clover, combinations of alleles in eggs and pollen have been found to influence the reproductive compatibility of the plants. Homozygous combinations such as S_1S_1 do not develop because S_1 pollen is not effective on S_1 stigmas. S_1 pollen, however, is effective on an S_2S_3 stigma. What progeny might be expected from the following crosses (seed parent always written first): (a) $S_1S_2 \times S_2S_3$; (b) $S_1S_2 \times S_3S_4$; (c) $S_4S_5 \times S_4S_5$; and (d) $S_3S_4 \times S_5S_6$?

10.19. In humans, a series of alleles has been associated with the blood typing groups as follows: I^A, A type; I^B, B type; I^O, O type. I^A and I^B are codominant; I^AI^B heterozygotes have AB type blood; I^O is recessive to both I^A and I^B. What phenotypes and ratios might be expected from the following matings: (a) $I^AI^A \times I^BI^B$; (b) $I^AI^B \times I^OI^O$; (c) $I^AI^O \times I^BI^O$; and (d) $I^AI^O \times I^OI^O$?

10.20. A case was brought before a certain judge in which a woman of blood group O presented a baby of blood group O, which she claimed as her child, and brought suit against a man of group AB whom she claimed was the father of the child. What bearing might the blood-type information have on the case?

10.21. In another case, a woman of blood group AB presented a baby of group O, which she claimed as her baby. What bearing might the blood-type information have on the case?

10.22. A homozygous Rh-positive man (RR) married an Rh-negative (rr) woman. Their first child was normal and their second child had the hemolytic disease of the newborn. (a) What genetic explanation might be offered? (b) What prediction might be made concerning future children by this couple?

10.23. A heterozygous Rh-positive man (Rr) married an Rh-negative (rr) woman. Their first child was normal and their second child showed the effects of Rh incompatibility. (b) What prediction might be made concerning future children by this couple?

10.24. (a) What are some of the genetic implications of overlapping genes or "genes-within-genes"? (b) What is the maximum number of different amino acid sequences that can be produced from the same segment of one strand of DNA? (c) From both strands of a DNA double helix?

10.25. (a) In what ways is the introduction of recombinant DNA molecules into host cells similar to mutation? (b) In what ways is it different?

10.26. What is the function of the "introns" or "intervening sequences" (noncoding) located between coding sequences within genes in eukaryotic organisms?

10.27. Ten micrograms of a decanucleotide-pair *Hpa*I restriction fragment were isolated from the double-stranded DNA chromosome of a small virus. Octanucleotide poly-A tails were then added to the 3'-ends of both strands using terminal transferase and dATP, i.e.,

5'-X X X X X X X X X X -3'
3'-X'X'X'X'X'X'X'X'X'X'-5'
| terminal transferase
↓ dATP
5'-X X X X X X X X X X A A A A A A A A-3'
3'-A A A A A A A A X'X'X'X'X'X'X'X'X'X'-5'

type; I^O, O type. I^A and I^B are codominant; I^AI^B heterozygotes have AB type blood; I^O is recessive to both I^A and I^B. What phenotypes and ratios might be expected from the following matings: (a) $I^AI^A \times I^BI^B$; (b) $I^AI^B \times I^OI^O$; (c) $I^AI^O \times I^BI^O$; and (d) $I^AI^O \times I^OI^O$?

where X and X' can be any of the four standard nucleotides, but X' is always complementary to X.

The two complementary strands (Watson strand and Crick strand) were then separated and sequenced by the 2',3'-dideoxyribonucleoside triphosphate chain-termination method. The reactions were all primed using a radioactive (^{32}P-labeled) synthetic poly-T octamer, i.e.,

Watson strand

3′-A A A A A A A X′X′X′X′X′X′X′X′X′X′-5′
5′-^{32}P-T T T T T T T TOH

Crick strand

5′-X X X X X X X X X A A A A A A A A-3′
HOT T T T T T T T-^{32}P-5′

The usual four parallel reactions: (1) ddTTP, (2) ddCTP, (3)ddATP, and (4) ddGTP (plus the Klenow fragment of DNA polymerase I and all other substrates and required components) were carried out for both strands. Each reaction mixture was applied to a lane in a polyacrylamide slab gel, electrophoresed, and autoradiographed. The autoradiogram of the sequencing gel for one of the strands is shown on the left below.

Draw the banding pattern that would be expected on the autoradiogram of the gel of reaction mixtures for the complementary strand on the diagram shown on the right below.

10.28. **Assume** that the gene *mut A* and its polypeptide product, the enzyme trinucleotide mutagenase, have been intensively studied during December 1982 in the previously unknown organism *Escherichia mutaphilium.* This organism has been shown to use the established,

nearly universal genetic code, and to behave in all other respects relevant to molecular genetics like *Escherichia coli.*

The sixth amino acid from the amino terminus of the wild-type trinucleotide mutagenase is histidine, and the wild-type *mut A* gene has the triplet nucleotide-pair sequence 5′-GTA-3′ at the position corresponding to
3′-CAT-5′
the sixth amino acid of the gene product. Seven independently isolated mutants with single nucleotide-pair substitutions within the triplet nucleotide sequence of the *mut A* gene given above have also been characterized. Moreover, the mutant mutagenases have all been purified and sequenced. All seven are different; they contain glutamine, tyrosine, asparagine, aspartic acid, arginine, proline, and leucine as the sixth amino acid from the amino terminus.

Mutants *mut A*1, *mut A*2, and *mut A*3 will not recombine with each other, but each will recombine with each of the other four mutants (*mut A*4, *mut A*5, *mut A*6, and *mut A*7) to yield true wild-type recombinants. Similarly, mutants *A*4, *A*5, and *A*6 will not recombine with each other, but will each yield true wild-type recombinants in crosses with each of the other four mutants. Finally, crosses between *mut A*1 and *mut A*7 yield about twice as many true wild-type recombinants as do crosses between *mut A*6 and *mut A*7.

Mutants *A*1 and *A*6 can be induced to revert to true wild-type by mutagenesis with 5-bromouracil; mutants

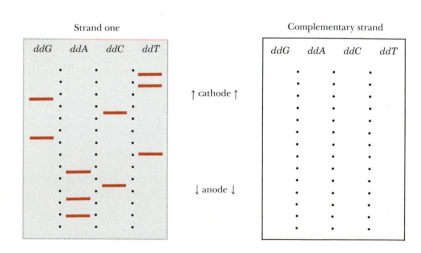

$A2$, $A3$, $A4$, $A5$, and $A7$ can**not** be induced to revert with 5-bromouracil. Mutants $A2$ and $A4$ are capable of slow growth on minimal medium, whereas mutants $A3$ and $A5$ carry null mutations (producing completely inactive gene products) and are incapable of growth on minimal medium. This has been used to select for mutation events from genotypes *mut A*3 and *mut A*5 to genotypes *mut A*2 and *mut A*4. Mutants $A3$ and $A5$ can be induced to mutate to $A2$ and $A4$, respectively, by treatment with 5-bromouracil or hydroxylamine. How-

ever, mutant $A3$ can**not** be induced to mutate to $A4$, nor $A5$ to $A2$, by treatment with either mutagen.

Given that (a) 5-bromouracil induces both GC $\rightarrow$ AT and AT $\rightarrow$ GC transitions and (b) hydroxylamine induces only GC $\rightarrow$ AT transitions, and using the nature of the genetic code (Chapter 8, p. 265), deduce which mutant allele specifies the mutant polypeptide with each of the seven different amino acid substitutions at position 6 of trinucleotide mutagenase, and describe the rationale behind each of your deductions.

ELEVEN

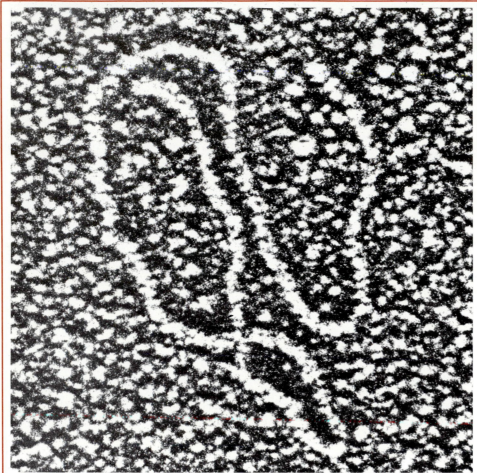

Electron micrograph of the purified *lac* operon (DNA) of *Escherichia coli*. (Original micrograph courtesy of L. A. MacHattie. Reprinted with permission from J. Shapiro, L. MacHattie, L. Eron, G. Ihler, K. Ippen, and J. Beckwith, *Nature* 224:768–774, 1969.)

REGULATION OF GENE EXPRESSION AND DEVELOPMENT

An amoeba is not likely to be confused with an *E. coli* cell. The phenotype of a corn plant is very different from the phenotype of an oak tree. Different human beings, excluding identical twins, can usually be readily distinguished. These different phenotypes, we know, result from different genes and different forms of genes (different alleles) in the genomes of various organisms and individuals.

Similarly, if we microscopically examine the nerve cells, kidney cells, liver cells, bone cells, blood cells, hair cells, skin cells, and so forth, of any one organism or any one individual, we find vast differences in the phenotypes of these cells. Some are short and fat; some are long and thin; some have appendages; and others are roughly spherical. Yet all of these cells (with a few notable exceptions) contain the same set of genes. All of them are produced from a single cell, the zygote in sexually reproducing species, by purely equational mitotic divisions. How, then, can their highly diverse phenotypes be explained?

At present, this question can be answered only very superficially. A detailed answer to this question is one of the major goals of many present-day geneticists. The superficial answer is that not all of the genes present in a nucleus are ever expressed at the same time. Furthermore, in differentiated cells of higher eukaryotes, only a small proportion (10 percent or less) of the genes are ever expressed at the same time.

Certain genes—for example, the genes specifying ribosomal RNAs, ribosomal proteins, and transfer RNAs—are undoubtedly expressed at some time in virtually all cells. However, many other genes are apparently expressed for only a short period of time in one or a few cell types at a specific stage in development. In other words, the **expression of genes is regulated.** Genes are continually being "turned on" and "turned off" in various cells at various times.

Although gene expression can be (and is) regulated at several different levels—for example, transcription, mRNA processing, mRNA transport, translation, and enzyme function—extensive data now indicate that gene expression is regulated in both prokaryotes and eukaryotes primarily at the levels of transcription and RNA gene transcript processing. That is not to say that regulation does not occur at other levels. Regulatory fine tuning at translational levels is clearly important in the overall control of metabolic processes in living organisms. The regulatory mechanisms with the largest effects on phenotype, however, have been shown to act at the level of transcription and RNA processing. (For a detailed discussion of the evidence supporting this statement, the reader is referred to E. H. Davidson's *Gene Activity in Early Development,* 2nd ed.)

Unfortunately, we know very little about the regulation of gene expression in eukaryotes. In higher eukaryotes, we still do not understand, at the molecular level, the mechanism responsible for the regulation of transcription of a single gene (excluding the yeasts, fungi, and other eukaryotic microorganisms). In prokaryotes, on the other hand, great progress has been made in elucidating the mechanisms involved in the regulation of gene expression.

Based on what is presently known about the regulation of transcription in both prokaryotes and eukaryotes, the various regulatory mechanisms seem to fit into two general categories. The first, and best understood, category includes mechanisms involved in the **rapid turn-on and turn-off of gene expression in response to environmental changes.** Regulatory mechanisms of this type are very important in microorganisms because of the frequent exposure of these organisms to sudden changes in environment. They provide microorganisms with a great deal of "plasticity," an ability to rapidly adjust their metabolic processes in order to achieve maximal growth and reproduction under highly variable environmental conditions. These quick responding on-off switches seem to be less important in higher eukaryotes. This might be expected since the circulatory systems of higher eukaryotes buffer their cells against many sudden environmental changes.

The second major category of regulatory mechanisms includes what might be called **preprogramed circuits of gene expression.** In these cases, some event (such as infection by a virus, release of a hormone in the blood stream, or fertilization of an egg) triggers the expression of one set of genes. The product (or products) of one (or more) of these genes functions by turning off the transcription of the first set of genes and/or turning on the transcription of a second set of genes. In turn, one or more of the products of the second set acts by turning on a third set, and so on. In these cases, the sequential expression of sets of genes is genetically preprogramed, and the genes *cannot* usually be turned on out of sequence. Such preprogramed sequences of gene expression in viral infections are well documented. There is also evidence, far from definitive, that preprogramed sequences of gene expression are involved in the control of differentiation in higher eukaryotes. In most of these preprogramed sequences it seems the circuitry is cyclical. For example, in viral infections some event associated with the packaging of the viral DNA or RNA inside the protein coat somehow seems to reset the program so that the first set of genes will again be expressed when a progeny virus subsequently infects another host cell.

In eukaryotes, we know that hormones can trigger the sequential expression of sets of genes. In addition, we know that regulatory genes are involved in the control of patterns of differentiation. How these hormones and regulatory genes act to control gene expression remains a question to challenge present and future geneticists.

INDUCTION AND REPRESSION IN PROKARYOTES

Certain gene products, such as tRNA molecules, rRNA molecules, ribosomal proteins, RNA polymerase components (polypeptides), and other enzymes catalyzing metabolic processes that are frequently referred to as cellular "housekeeping" functions, are essential components of almost all living cells. Genes that specify products of this type are **continually being expressed** in most cells. Such genes are said to be expressed **constitutively** and are frequently referred to as **constitutive genes.**

Other gene products are needed for cell growth only under certain environmental conditions. Constitutive synthesis of such gene products would clearly be wasteful, using energy that could

otherwise be utilized for more rapid growth and reproduction under the existing environmental conditions. The evolution of regulatory mechanisms that would provide for the synthesis of such gene products only when and where they were needed would clearly provide organisms possessing these regulatory mechanisms with a selective advantage over organisms lacking these mechanisms. This undoubtedly explains why presently existing organisms, including the "primitive" bacteria and viruses, exhibit highly developed and very efficient mechanisms for the control of gene expression.

E. coli and most other bacteria are capable of growth using any one of several carbohydrates (e.g., glucose, sucrose, galactose, arabinose, lactose) as an energy source. If glucose is present in the environment, it will be preferentially metabolized by *E. coli* cells. In the absence of glucose, however, *E. coli* cells can grow very well on other carbohydrates. Cells growing in medium containing the sugar lactose, for example, as the sole carbon source synthesize two enzymes, β-galactosidase and β-galactoside permease, that are uniquely required for the catabolism of lactose. (A third enzyme, β-galactoside transacetylase, is also synthesized. It has no **known** metabolic function, however.) β-Galactosidase cleaves lactose into glucose and galactose, and β-galactoside permease pumps β-galactosides into the cell. Neither of these enzymes is of any use to *E. coli* cells when present in environments not containing lactose. The synthesis of these two enzymes, of course, requires the utilization of considerable energy (in the form of ATP and GTP; see Chapter 8). Thus, *E. coli* cells have evolved a regulatory mechanism by which the synthesis of these lactose-catabolizing enzymes is turned on in the presence of lactose and turned off in its absence.

In natural environments (intestinal tracts and sewers), *E. coli* cells probably encounter an absence of glucose and the presence of lactose relatively infrequently. Most of the time, therefore, the *E. coli* genes coding for the enzymes involved in lactose utilization are not being expressed. If cells growing on a carbohydrate other than lactose are transferred to medium containing lactose as the only carbon source, they rapidly begin synthesizing the

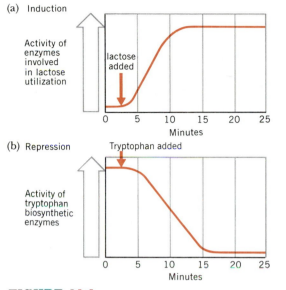

(a) Induction

Activity of enzymes involved in lactose utilization

lactose added

Minutes

(b) Repression

Tryptophan added

Activity of tryptophan biosynthetic enzymes

Minutes

FIGURE 11.1

Induction (a) *and* *repression* (b) *of enzyme synthesis in bacteria. Induction is characteristic of catabolic (degradative) pathways; repression is characteristic of anabolic (biosynthetic) pathways.* (a) *The induction of the synthesis of the enzymes required for the utilization of the sugar lactose as an energy source in* E. coli *is illustrated. In the absence of lactose in the environment,* E. coli *cells synthesize only very minute amounts of the lactose-utilizing enzymes. When such cells are transferred to an environment containing lactose as the sole carbon source (occurring at the time indicated by the arrow labeled* **lactose added**), *the synthesis of the enzymes required for lactose catabolism is rapidly* **induced** *(turned on).* (b) *The repression of the synthesis of the enzymes required for the biosynthesis of tryptophan in* E. coli *is illustrated. When tryptophan is not present in the environment, the* E. coli *cells synthesize the enzymes required for tryptophan biosynthesis. If tryptophan is added to the environment of such cells (e.g., at the time indicated by the arrow labeled* **tryptophan added**), *the synthesis of the tryptophan biosynthetic enzymes is rapidly* **repressed** *(turned off). The kinetics shown are only approximate.*

enzymes required for lactose utilization (Fig. 11.1*a*). This process, by which the expression of genes is turned on in response to a substance in the environment, is called **induction.** Genes whose expression are so regulated are called **inducible genes;** their products, if enzymes, are called **induci-**

ble **enzymes.** The substances or molecules responsible for induction are known as **inducers.**

Enzymes that are involved in **catabolic (degradative) pathways,** such as in lactose, galactose, or arabinose utilization, are characteristically inducible. As will become apparent in the following sections of this chapter, **induction occurs at the level of transcription.** It alters the rate of synthesis of enzymes, not the activity of existing enzyme molecules. Induction should not be confused with **enzyme activation,** in which the binding of a small molecule to an enzyme increases the **activity** of the enzyme (but does not affect its rate of synthesis).

Bacteria possess the metabolic capacity to synthesize most of the organic molecules (such as amino acids, purines, vitamins, etc.) required for their growth. For example, *E. coli* has five genes coding for enzymes that are required in the synthesis of tryptophan. These five genes must be expressed in *E. coli* cells growing in an environment devoid of tryptophan in order to provide adequate amounts of this amino acid for ongoing protein synthesis.

When *E. coli* cells are present in an environment containing concentrations of tryptophan sufficient to support optimal growth, the continued synthesis of the typtophan biosynthetic enzymes would be a waste of energy, because these bacteria have the capacity to take in external tryptophan. Thus, a regulatory mechanism has evolved in *E. coli* by which the synthesis of the tryptophan biosynthetic enzymes is turned off when tryptophan is present in the external milieu (Fig. 11.1*b*). This process of "turning off" the expression of sets of genes is called **repression.** A gene whose expression has been turned off in this way is said to be **repressed;** when its expression is turned on, a gene of this type is said to be **derepressed.**

Enzymes that are components of **anabolic (biosynthetic) pathways** are frequently subject to **repression** (are **repressible**). **Repression,** like induction, **occurs at the level of transcription.** Repression should not be confused with **feedback inhibition,** in which the binding of an end product to the first enzyme in a biosynthetic pathway **inhibits** the **activity** of the enzyme (but does not affect its synthesis).

THE OPERON MODEL

Induction and repression of gene expression can be accomplished by essentially the same mechanism. This mechanism was first accurately described in 1961 when F. Jacob and J. Monod, both 1965 Nobel Prize recipients, proposed the **operon model** to explain the regulation of genes coding for the enzymes required for lactose utilization in *E. coli*. Jacob and Monod proposed that the transcription of one or a set of contiguous structural genes (genes coding for polypeptides) is regulated by two controlling elements (Fig. 11.2*a*). One of these elements, called the **regulator gene** (or **repressor gene**), codes for a protein called the **repressor;** under the appropriate conditions, the repressor binds to the second element, the **operator** (or **operator sequence**). The operator is always located contiguous to the structural gene or genes whose expression it regulates. **When the repressor is bound to the operator, transcription of the structural genes cannot occur.** We now know that this results because the binding of the repressor to the operator sterically prevents RNA polymerase from binding at the **promoter site** (the RNA polymerase binding site; see Chapter 8), which is always located contiguous with (or even overlapping) the operator sequence. The operator is usually located between the promoter and the structural genes (Fig. 11.2*a*). (The promoter was not recognized at the time of Jacob and Monod's proposal, but has since been shown to be an essential component of an operon.) The complete contiguous unit, including the structural gene or genes, the operator, and the promoter, is called an **operon.**

Whether the repressor will bind to the operator and turn off the transcription of the structural genes in an operon is determined by the presence or absence of **effector molecules** (small molecules such as amino acids, sugars, etc.) in the environment. In the case of inducible operons, these effector molecules are called **inducers.** Those active on repressible operons are called **co-repressors.** These effector molecules (inducers and co-repressors) act by binding to (or forming a complex with) the repressors.

The only essential difference between inducible operons and repressible operons is whether the naked repressor or the repressor-effector molecule complex is active in binding to the operator. (1) In the case of an inducible operon, the **free repressor binds to the operator,** turning off transcription (Fig. 11.2*b*). When the effector molecule (the inducer) is present, it binds to the repressor, releasing the repressor from the operator; that is, the **repressor-inducer complex cannot bind to the operator.** Thus, the addition of inducer turns on or induces the transcription of the structural genes in the operon (Fig. 11.2*b*). (2) In the case of a repressible operon, the situation is just reversed. **The free repressor cannot bind to the operator. Only the repressor-effector molecule (co-repressor) complex is active in binding to the operator** (Fig. 11.2*c*). Thus, transcription of the structural genes in a repressible operon is turned on in the absence of and turned off in the presence of the effector molecule (co-repressor). Except for this difference in the operator-binding behavior of the repressor, inducible and repressible operons are comparable.

A single mRNA transcript carries the coding information of an entire operon. Thus, the mRNAs of operons consisting of more than one structural gene are multigenic. For example, the tryptophan operon mRNA of *E. coli* is a huge macromolecule carrying the coding sequences that specify five different polypeptides (see Fig. 11.5). Because of their co-transcription, all of the structural genes in an operon are **coordinately expressed.** Whereas the molar quantities of the different gene products need not be the same (due to different rates of initiation of translation), the relative amounts of the different polypeptides specified by genes in an operon remain the same, regardless of the state of induction or repression.

Because the product of the regulator gene, the repressor, acts by shutting off the transcription of structural genes, the operon model, as originally proposed by Jacob and Monod, is referred to as a **negative control** system. In **positive control** systems, the products of regulator genes are required to turn on transcription. We will discuss examples of positive control mechanisms later in this chapter.

LAC, AN INDUCIBLE OPERON

Jacob and Monod proposed the operon model largely as a result of their studies of the *lac* operon of

E. coli. More is known about the *lac* operon than any other operon. The *lac* operon contains a promoter, an operator, and three structural genes, *z*, *y*, and *a*, coding for the enzymes β-galactosidase, β-galactoside permease, and β-galactoside transacetylase, respectively (Fig. 11.3). β-Galactoside permease "pumps" lactose into the cell, where β-galactosidase cleaves it into glucose and galactose (Fig. 11.4). The function of the transacetylase is still not clear.

The *lac* regulator gene, designated the *i* gene, codes for a repressor that is 360 amino acids in length. The active form of the *lac* repressor, however, is a tetramer that contains four copies of the *i* gene product. In the absence of inducer, the repressor binds to the *lac* operator sequence, preventing RNA polymerase from binding to the promoter and transcribing the structural genes (see Fig. 11.2*b*). A few molecules of the *z*, *y*, and *a* gene products are synthesized in the uninduced state, providing a low background level of enzyme activity. However, this background activity is essential for induction of the *lac* operon, because the **inducer** of the operon, **allolactose,** is derived from lactose in a reaction that is catalyzed by β-galactosidase (Fig. 11.4). Once formed, allolactose binds to the repressor, causing it to be released from the operator; in so doing, it **induces** transcription of the *z*, *y*, and *a* structural genes (see Fig. 11.2*b*).

The *lac i* gene, operator, and promoter were all initially identified genetically by the isolation of mutations within these genetic units which rendered them nonfunctional. Mutations within the *i* gene and the operator frequently result in the constitutive synthesis of the lactose-utilizing enzymes. These mutations are designated i^- and o^c, respectively. The i^- and o^c constitutive mutations can be distinguished not only by map position, but by their behavior in F′ merozygotes (see Chapter 7, p. 215) in which they are located in *cis*- and *trans*-configurations relative to mutations in *lac* structural genes.

Merozygotes (partial diploids) of genotype $F′ i^+ o^+ z^+ y^+ a^+/i^+ o^+ z^- y^- a^-$ or of genotype $F′ i^+ o^+ z^- y^- a^-/i^+ o^+ z^+ y^+ a^+$ are inducible for the utilization of lactose as a carbon source, just like haploid wild-type ($i^+ o^+ z^+ y^+ a^+$) cells. The wild-type alleles (z^+, y^+, and a^+) of the three structural

genes are dominant to their mutant alleles (z^-, y^-, and a^-). This is to be expected because the wild-type alleles produce functional enzymes, whereas the mutant alleles produce no enzymes or defective (inactive) enzymes. F′ merozygotes that have the genotype $i^+ o^+ z^+ y^+ a^+/i^- o^+ z^+ y^+ a^+$ are also inducible for the synthesis of the three enzymes specified by the *lac* operon. Thus i^+ is dominant to i^- as expected, since i^+ specifies an active protein (repressor molecule) and i^- yields an inactive protein.

Merozygotes that have the genotype $F′ i^+ o^+ z^+ y^+ a^+/i^- o^+ z^- y^- a^-$ or the genotype $F′ i^+ o^+ z^- y^- a^-/i^- o^+ z^+ y^+ a^+$ are **inducible** for β-galactosidase, β-galactoside permease, and β-galactoside transacetylase, like wild-type cells. This indicates that the *i* gene codes for a diffusible product, since it affects the expression of structural genes located either in *cis*- or *trans*-configuration in relation to itself. On the other hand, the operator constitutive mutations, o^c mutations, **act only in *cis*.** That is, o^c mutations cause only the constitutive expression of structural genes located on the same chromosome. This is, of course, to be expected if the operator is the repressor binding site; as such, it does not code for any product, diffusible or otherwise. A merozygote of genotype $F′ i^+ o^c z^- y^- a^-/i^+ o^+ z^+ y^+ a^+$ is thus inducible for the three enzymes specified by the structural genes of the *lac* operon, whereas a merozygote of genotype $F′ i^+ o^c z^+ y^+ a^+/i^+ o^+ z^- y^- a^-$ synthesizes these enzymes constitutively.

Some of the *i* gene mutations, those designated i^{-d}, are dominant to the wild-type allele (i^+). This dominance apparently results from the inability of heteromultimers (recall that the *lac* repressor functions as a tetramer), which contain both wild-type and mutant polypeptides, to bind to the operator sequence. Other *i* gene mutations, those designated i^{-s} (for "superrepressed"), cause the *lac* operon to be **uninducible.** (Strains carrying these i^{-s} mutations can usually be induced to some degree by using extremely high concentrations of inducer. They are not induced at normal inducer concentrations.) When studied *in vitro*, the mutant i^{-s} polypeptides form tetramers that bind to *lac* operator DNA. They either do not bind inducer, however, or exhibit a very low affinity for inducer. The i^{-s} muta-

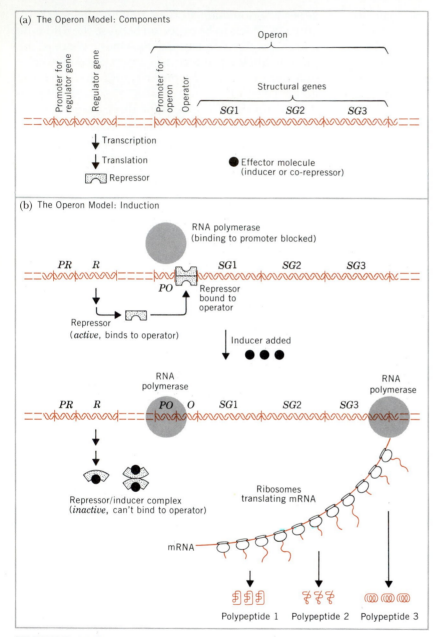

(a) The Operon Model: Components

Operon

Promoter for regulator gene

Regulator gene

Promoter for operon

Operator

Structural genes

SG1   SG2   SG3

↓ Transcription

↓ Translation

Repressor

● Effector molecule (inducer or co-repressor)

(b) The Operon Model: Induction

RNA polymerase (binding to promoter blocked)

PR   R   SG1   SG2   SG3

PO   Repressor bound to operator

Repressor (*active*, binds to operator)

↓ Inducer added
● ● ●

RNA polymerase

PR   R   PO   O   SG1   SG2   SG3   RNA polymerase

Repressor/inducer complex (*inactive*, can't bind to operator)

Ribosomes translating mRNA

mRNA

Polypeptide 1   Polypeptide 2   Polypeptide 3

FIGURE 11.2

*The operon model for regulation of gene expression. (a) Diagram showing the essential components of regulation specified by the operon model. The **operon** consists of one or more **structural genes** (three—SG1, SG2, and SG3—are arbitrarily shown) and the adjoining **operator** and **promoter** sequences. The promoter for the operon (PO) is the site at which RNA polymer-*

*ase must bind to initiate transcription of the structural genes. The operator (O) is the site at which the protein **repressor**—the product of the **regulator gene** (or **repressor gene**)—binds. The regulator gene need not be closely linked to the operon; in fact, it can be located at any position in the genome. The transcription of the regulator gene is initiated by RNA polymer-*

(continued on next page)

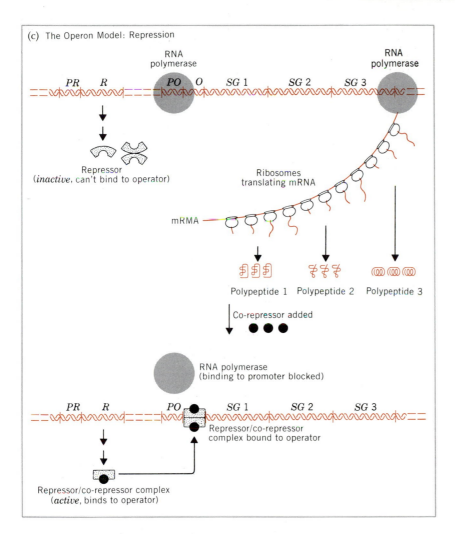

(c) The Operon Model: Repression

RNA polymerase

PR   *R*   *PO*   *O*   *SG* 1   *SG* 2   *SG* 3

RNA polymerase

Repressor
(*inactive*, can't bind to operator)

Ribosomes
translating mRNA

mRMA

Polypeptide 1   Polypeptide 2   Polypeptide 3

Co-repressor added

RNA polymerase
(binding to promoter blocked)

PR   *R*   *PO*   *SG* 1   *SG* 2   *SG* 3

Repressor/co-repressor
complex bound to operator

Repressor/co-repressor complex
(*active*, binds to operator)

ase, which binds to its promoter (labeled **PR**, *for promoter for regulator gene). When the repressor is bound to the operator, it sterically prevents RNA polymerase from binding to the adjoining promoter* (**PO**) *and from initiating transcription of the structural genes. Whether the repressor binds to the operator or not depends on the presence or absence of a metabolite called an* **effector molecule.** (b) *Mode of regulation of gene expression for an* **inducible operon.** *The product of the regulator gene* (**R**), *the repressor, in the* **absence** *of the effector molecule (called an* **inducer,** *for inducible operons), binds to the operator, preventing RNA polymerase from binding to the promoter for the operon* (**PO**). *Thus, transcription of the structural genes* **cannot** *occur. When inducer is added, it binds to the repressor, causing it to be released from the operator* (**O**). *This, in turn, allows RNA*

polymerase to bind to the promoter (**PO**) *and initiate transcription of the structural genes. The resulting multigenic mRNA is rapidly translated by ribosomes, producing the three polypeptide products of the structural genes.* (c) *Mode of regulation of gene expression for a* **repressible operon.** *In this case, the repressor can* **only bind to the operator** *in the* **presence** *of the* **effector molecule** *(called a* **co-repressor,** *for repressible operons). In its absence, the operator is free, permitting RNA polymerase to bind at the adjoining promoter* (**PO**) *and to initiate transcription of the structural genes. When co-repressor is added, it forms a complex with the repressor. This repressor/co-repressor complex then binds to the operator* (**O**). *This, in turn, prevents RNA polymerase from binding at* **PO** *and transcribing the three structural genes.*

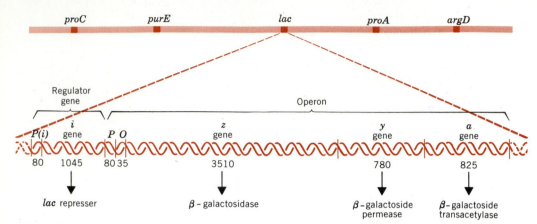

FIGURE 11.3

The lac operon of E. coli, an inducible operon. The lac operon consists of three structural genes, z, y, and a, plus the promoter (P) and operator (O) regions adjoining the z gene. The regulator gene (i) is contiguous with the operon in the case of lac. Regulator genes of other operons are frequently located at a considerable distance from the operon. (Note that the regulator gene is normally not considered to be part of the operon proper, which consists of the structural genes plus their promoter and operator.) The regulator gene has its own promoter [P(i)]. The

numbers below the various genes indicate their approximate lengths in nucleotide pairs. The nucleotide sequences for P(i), P, and O are known (see Fig. 11.7). The lac operon functions as shown in Fig. 11.2b, with allolactose, a derivative formed from lactose in the presence of β-galactosidase, acting as inducer. Thus, the lac inducer is produced by the small amounts of β-galactosidase and β-galactoside permease that are present in uninduced cells.

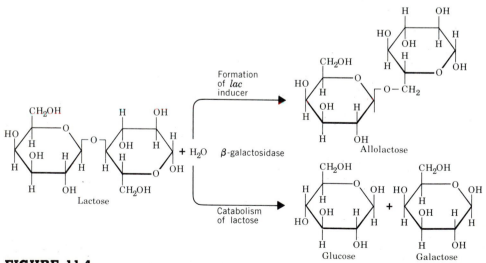

FIGURE 11.4

The two physiologically important reactions that are catalyzed by β-galactosidase: (1) conversion of lactose to the lac operon inducer allolactose, and (2) cleavage of lactose to produce the

monosaccharides glucose and galactose, the first step in the degradation of lactose.

tions therefore modify the inducer binding site of the *lac* repressor.

Promoter mutations do not alter the inducibility of the *lac* operon. Instead, they modify the levels of gene expression in the induced and uninduced state by changing the frequency of initiation of *lac* operon transcription (i.e., the efficiency of RNA polymerase binding).

The *lac* promoter actually contains two functionally distinct components: (1) the RNA polymerase binding site and (2) a binding site for another protein, called **Catabolite Activator Protein** (abbreviated **CAP**), that functions such that the *lac* operon is not transcribed in the presence of glucose at concentrations sufficient to support optimal growth. This second control circuit assures the preferential utilization of glucose as an energy source when it is available.

TRP, A REPRESSIBLE OPERON

The *trp* (tryptophan) operon of *E. coli* is probably the best known repressible operon. The organization of the five structural genes and the adjacent regulatory sequences of the *trp* operon (Fig. 11.5) has been analyzed in detail by Charles Yanofsky and colleagues. The regulation of transcription of the *trp* operon occurs as diagramed in Fig. 11.2c. The *trp* operon repressor is the product of gene *trpR*, which is not closely linked to the *trp* operon (Fig. 11.5).

In the absence of tryptophan (the co-repressor), RNA polymerase binds to the promoter region and transcribes the structural genes of the operon. In the presence of tryptophan, the co-repressor/repressor complex binds to the operator region and prevents the binding of RNA polymerase to the promoter. The operator sequence of the *trp* operon lies entirely within the promoter region (Fig. 11.5).

The rate of transcription of the *trp* operon in the derepressed state (absence of tryptophan) is 70 times the rate that occurs in the repressed state (presence of tryptophan). In *trpR* mutants that cannot make repressor, the rate of synthesis of the tryptophan biosynthetic enzymes (the products of the structural genes of the *trp* operon) is still reduced about 10-fold by the addition of tryptophan

to the medium. This reduction is due to a second level of regulation of *trp* operon expression called attenuation. Attenuation occurs by tryptophan-mediated termination of transcription in the *trpL* (mRNA leader) region of the operon (see pp. 391–396).

POSITIVE CONTROL OF THE *LAC* OPERON BY CAP AND CYCLIC AMP

The operon model was proposed by Jacob and Monod to explain the induction of the biosynthesis of the enzymes involved in lactose utilization when this sugar is added to the medium in which *E. coli* cells are growing. The presence of glucose, however, has long been known to prevent the induction of the *lac* operon, as well as other operons controlling enzymes involved in carbohydrate catabolism (for example the arabinose and galactose operons). This phenomenon, called **catabolite repression** (or the **glucose effect**), has apparently evolved to assure that glucose is metabolized when present, in preference to other, less efficient, energy sources.

Catabolite repression of the *lac* operon is now known to be mediated via positive control of transcription by a regulatory protein called **CAP** (for Catabolite Activator Protein) and a small effector molecule called **cyclic AMP** (adenosine-3′,5′-phosphate; Fig. 11.6). (CAP is also sometimes called **cyclic AMP receptor protein**.) As mentioned earlier, the *lac* promoter contains two separate binding sites: (1) one for RNA polymerase and (2) one for the CAP-cAMP complex (Fig. 11.7; cyclic AMP is abbreviated cAMP).

The CAP-cAMP complex must be bound to its binding site in the *lac* promoter in order for the operon to be induced. The CAP-cAMP complex thus exerts positive control over the transcription of the *lac* operon. It has an effect exactly opposite to that of repressor binding to an operator. Although the precise mechanism by which CAP-cAMP stimulates RNA polymerase binding to the promoter is still uncertain, its positive control of *lac* operon transcription is firmly established by both *in vivo* and *in vitro* experiments. CAP is known to function as a dimer; thus, like the *lac* repressor, it is multimeric in its functional state.

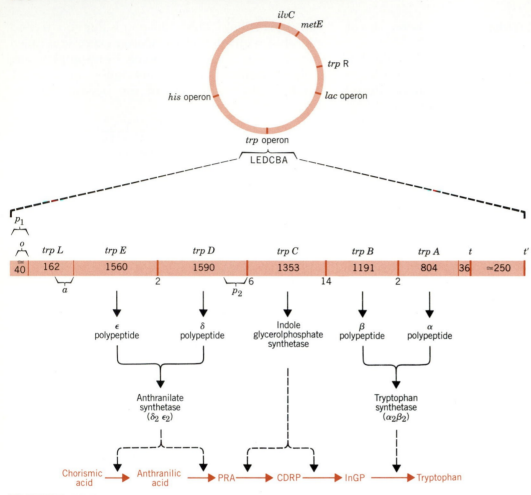

FIGURE 11.5

Organization of the tryptophan (trp) *operon of* E. coli. *(The Salmonella typhimurium trp operon is essentially identical.) The operon contains five structural genes that code for the enzymes involved in the biosynthesis of tryptophan as shown at the bottom. The trpR gene, which codes for the trp repressor, is not closely linked to the trp operon (top). The operator (o) region of the trp operon lies entirely within the primary promoter (p₁) region. There is also a weak promoter (p₂), at the operator-distal end of the trpD gene, that results in a somewhat increased basal level of constitutive transcription of the trpC, B, and A genes. There are two transcription termination sequences (t and t') downstream from trpA. The trpL region specifies a 162 nucleo-* tide-long mRNA leader sequence; it contains the attenuator (a) region that provides a second level of control of trp operon expression (see pp. 391–396). The p₁ promoter actually extends about 18 nucleotide pairs into trpL. The length of each gene or region is given in nucleotide pairs; the intergenic distances are given in nucleotide pairs below the gene sequence. The abbreviations used are: PRA = phosphoribosyl anthranilate; CDRP = carboxyphenylamino-deoxyribulose phosphate; InGP = indole-glycerol phosphate. (Based on the data summarized by Yanofsky, *Nature* 289:751–758, 1981, and by Platt, *Cell* 24:10–23, 1981.)*

Only the CAP-cAMP complex binds to the *lac* promoter; in the absence of cAMP, CAP does not bind. Thus, cAMP acts as the effector molecule, determining the effect of CAP on *lac* operon transcription. The intracellular cAMP concentration is sensitive to the presence or absence of glucose. High

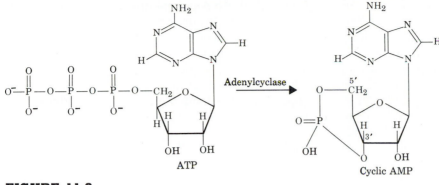

FIGURE 11.6

Formation of the regulatory molecule cyclic AMP (adenosine-3′,5′-phosphate) from ATP by the action of the enzyme adenylcyclase. Cyclic AMP binds to the CAP protein; the CAP-cAMP complex then binds to the CAP-site of the lac *promoter, stimulating the binding of RNA polymerase to its binding site in the* lac *promoter. In the absence of sufficient quantities of cAMP (which results for unknown reasons in the presence of high concentrations of glucose), CAP* **cannot** *bind to the* lac *promoter and the* lac *operon* **cannot** *be induced.*

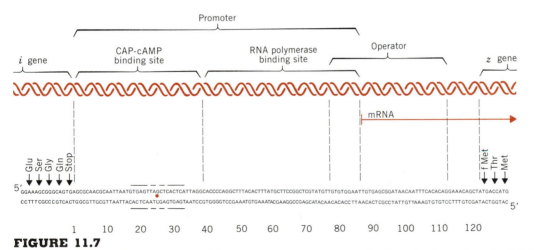

FIGURE 11.7

Organization and nucleotide sequence of the promoter-operator region of the lac *operon. The promoter consists of two components: (1) the site that binds the catabolite activator protein (abbreviated CAP)-cyclic AMP (or cAMP) complex and (2) the RNA polymerase binding site. Note that the promoter and the operator (repressor binding site) overlap slightly. The adjacent segments of the* i *(repressor) and* z *(β-galactosidase) structural genes are also shown. The horizontal line labeled mRNA indicates the position at which transcription of the operon begins (5′-end of mRNA). The numbers at the bottom give distances in* nucleotide pairs from the end of the i gene (beginning of the promoter). The dot between the nucleotide chain sequences (in the CAP-cAMP binding site) indicates the center of symmetry of an imperfect palindrome [a sequence where the nucleotide-pair sequences read almost the same in both directions (indicated by the parallel bars); see Chapter 10, p. 355]. This is a potential site of primary interaction with the CAP-cAMP complex. (Based on the data of R. C. Dickson, J. Abelson, W. M. Barnes, and W. S. Reznikoff, Science 187:27–35, 1975).

concentrations of glucose cause sharp decreases in the intracellular concentration of cAMP. How glucose controls the cAMP concentration is not clear. Perhaps glucose, or some metabolite that forms in the presence of sufficient concentrations of glucose, inhibits the activity of **adenylcyclase,** the enzyme that catalyzes the formation of cAMP from ATP. Whatever the mechanism, the presence of glucose results in a decrease in the intracellular concentration of cAMP. In the absence of (or in the presence of a low concentration of) cAMP, CAP **cannot** bind to the *lac* operon promoter. In turn, RNA polymerase cannot bind efficiently to the *lac* promoter in the absence of bound CAP. The overall result of the positive control of transcription of the *lac* operon by the CAP-cAMP complex is that in the presence of glucose *lac* operon transcription never exceeds 2 percent of the induced rates observed in the absence of glucose.

The complete nucleotide-pair sequence of the *lac* operon regulatory region (the promoter and operator sequences) is now known (Fig. 11.7). Comparative nucleotide-sequence studies of mutant and wild-type promoters and operators (plus *in vitro*

CAP-cAMP, RNA polymerase, and repressor binding studies) are beginning to provide detailed information about the nature of these important **sequence-specific protein – nucleic acid interactions.** These studies provide an excellent example of the advantages of integrating genetic and biochemical approaches in attempting to understand biological phenomena.

POSITIVE REGULATION OF THE *ARA* OPERON

In the *lac* and *his* operons discussed above, the product of the regulator gene, the repressor, functions in a negative manner, turning off transcription of the operon. On the other hand, the catabolite activator protein (CAP) exerts a positive control over the *lac* operon, stimulating transcription of the operon. The **arabinose (*ara*) operon** of *E. coli*, containing three structural genes coding for enzymes involved in the catabolism of arabinose (Fig. 11.8), differs from the *lac* and *his* operons in that the product of the regulator gene, *araC*, exerts a positive (stimulatory) effect on transcription of the *ara* operon in the presence of arabinose. Since the *ara* operon is also

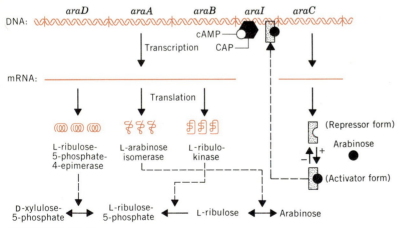

FIGURE 11.8

The arabinose (ara) *operon of* E. coli, *illustrating positive control of transcription by the regulator* (araC) *gene product and by the CAP-cAMP complex. The* araC *gene product (protein) exists in a form that represses transcription of the* **ara** *operon when arabinose is absent. When arabinose is present, it binds to the* araC *protein, converting it to a form that stimulates transcription of the operon. The* **ara** *operon is subject to cataboli*te *repression like the* lac *operon (see text); thus binding of the CAP-cAMP complex is required for induction. The binding sites for the* araC *gene product, CAP-cAMP, and RNA polymerase are probably all located in the region historically designated* araI. *(The structures shown for the various protein molecules are all totally hypothetical.)*

subject to catabolite repression, and thus to positive control by CAP and cAMP, induction of the *ara* operon depends on the positive regulatory effects of two proteins, the *araC* protein and CAP. The binding sites for these two proteins and for RNA polymerase all appear to be located in a region of the *ara* operon historically called *araI*, located between the three structural genes of the operon (*araD*, *araA*, and *araB*) and the regulator gene (*araC*) (Fig. 11.8). In the absence of arabinose, the *araC* protein has a negative effect on transcription of the *ara* operon. Thus, depending on the presence or absence of the effector molecule arabinose, the *araC* gene product may exert either a positive or a negative effect on transcription.

LAMBDA PROPHAGE REPRESSION DURING LYSOGENY

When a temperate bacteriophage such as lambda exists in the prophage state in a lysogenic cell (see Chapter 7, pp. 203–205), the genes coding for products involved in the lytic pathway—namely, the genes controlling phage DNA replication, phage morphogenesis, and lysis of the host cell—must **not** be expressed. This is accomplished by a repressor-operator-promoter circuit, much like that involved in bacterial operons. Specifically, the C_1 gene of phage lambda codes for a repressor, a well-characterized protein with a molecular weight of 27,000, which in the dimer or tetramer state binds to two operator regions that control transcription of the lambda genes involved in lytic growth (Fig. 11.9). These two operator regions, termed O_L (for transcription in a leftward direction) and O_R (for transcription in the rightward direction), overlap with promoter sequences at which RNA polymerase binds and initiates transcription of the genes controlling lytic development. Thus, with the repressor bound to the two operators, RNA polymerase **cannot** bind to the two promoters and **cannot,** therefore initiate transcription. In this way, the phage genes are kept repressed, allowing the "dormant" prophage to be transmitted from parental host cells to progeny cells generation after generation.

In experiments in which the operator and promoter regions of phage lambda were sequenced, each operator was unexpectedly found to contain three repressor binding sites with similar, but not identical, sequences of 17 nucleotide pairs. Each repressor binding site has partial twofold symmetry around the central base-pair (i.e., is partially palindromic; see Chapter 10, pp. 355–356). It has been speculated that this partial symmetry may facilitate interaction with repressor dimers, which might also have twofold symmetry. While this is an attractive possibility, it is nothing more than that at present.

The interaction of lambda repressor with DNA sequences $O_L P_L$ and $O_R P_R$ nicely explains how the lambda prophage genes are maintained in the repressed state. The mechanism responsible for the decision between lytic development and lysogenic development after infection of an *E. coli* cell with a lambda phage is considerably more complex, involving interactions among several other lambda regulatory genes. The reader is referred to one of the papers by Ptashne and co-workers (see the References for this chapter) for a discussion of the mechanism by which the choice between the lytic pathway and the lysogenic pathway is made.

CONTROL OF THE *TRP* OPERON BY ATTENUATION

Repression and derepression can change the level of expression of the structural genes of the *trp* operon by about 70-fold. There is a second level of regulation of *trp* operon expression, however. In *trpR* mutants that cannot make repressor, the addition of tryptophan to a culture of cells growing in the absence of tryptophan will cause an 8- to 10-fold decrease in the rates of synthesis of the tryptophan biosynthetic enzymes. Moreover, deletions that remove part of the *trpL* region (Fig. 11.5) result in increased rates of expression of the *trp* operon. The effects of these deletions are independent of repression; the increase occurs in both the repressed and the derepressed state.

This second level of regulation of the *trp* operon is called **attenuation,** and the sequence within *trpL* that controls this phenomenon is called the **attenuator** (Fig. 11.10). Attenuation occurs by

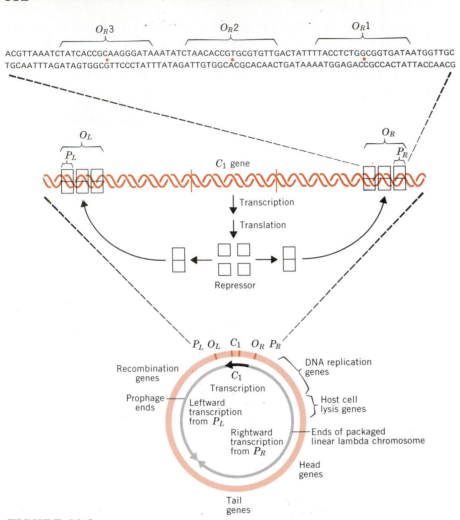

O_R3 O_R2 O_R1

ACGTTAAATCTATCACCGCAAGGGATAAATATCTAACACCGTGCGTGTTGACTATTTTACCTCTGGCGGTGATAATGGTTGC
TGCAATTTAGATAGTGGCGTTCCCTATTTATAGATTGTGGCACGCACAACTGATAAAATGGAGACCGCCACTATTACCAACG

O_L

P_L

C_1 gene

O_R

P_R

Transcription

Translation

Repressor

P_L O_L C_1 O_R P_R

DNA replication genes

Recombination genes

C_1
Transcription

Prophage ends

Leftward transcription from P_L

Host cell lysis genes

Rightward transcription from P_R

Ends of packaged linear lambda chromosome

Head genes

Tail genes

FIGURE 11.9

Repression of the phage lambda genes that control lytic development by the binding of the C_1 gene product (repressor) to the two operator sequences (O_L and O_R) that control leftward and rightward transcription of the lambda chromosome. Transcription of the C_1 gene itself is in a leftward direction (solid black arrow), beginning at the C_1 promoter (not shown), which is located between the C_1 gene and O_R. The circular intracellular form of the lambda chromosome (see Chapter 5, pp. 101–103) is shown at the bottom, with the approximate location of the C_1 gene, $P_L O_L$ (promoter leftward, operator leftward), $O_R P_R$ (operator rightward, promoter rightward), the ends of the linear forms of the lambda chromosome (prophage and mature forms), and clusters of some of the genes controlling lytic development.

The shaded arrows indicate the transcription of genes from P_L and P_R, respectively. The repression of the genes controlling lytic development by the binding of repressor molecules to O_L and O_R is illustrated in the center enlargement. Note that each operator sequence has three repressor binding sites and that the operator and promoter (RNA polymerase binding site) overlap. The nucleotide-pair sequence of the O_R region is shown at the top, with the 17 nucleotide-pair sequence of each repressor binding site in brackets. The colored dot between the two strands of DNA within each repressor binding site indicates the axis of partial symmetry. (The nucleotide-pair sequence data are from M. Ptashne, K. Backman, M. Z. Humayun, A. Jeffrey, R. Maurer, B. Meyer, and R. T. Sauer, Science 194:156, 1976.)

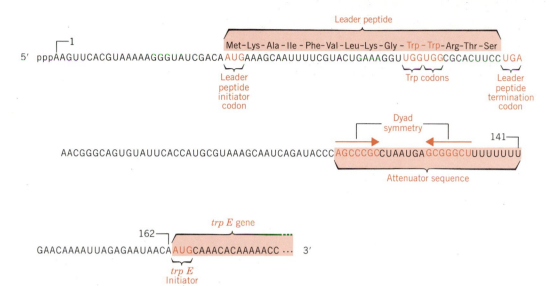

FIGURE 11.10

Nucleotide sequence of the leader of the trp *operon mRNA. The region of dyad symmetry in the attenuator forms the transcription-termination "hairpin" shown in Fig. 11.11c. Note the two tandem tryptophan codons in the sequence coding for the putative leader peptide. These Trp codons are believed to be responsible for the control of attenuation by tryptophan (see Fig. 11.11,* b *and* c*). (Based on the data summarized by Yanofsky, Nature 289:751–758, 1981.)*

control of the termination of transcription at a site near the end of the mRNA leader sequence. This "premature" termination of *trp* operon transcription occurs only in the presence of tryptophan-charged tRNA$_{trp}$. It yields a 140 nucleotide-long leader-sequence transcript (Fig. 11.10).

The attenuator region has a nucleotide-pair sequence essentially identical to the transcription-termination signals found at the ends of most bacterial operons (including the *trp* operon, see Fig. 11.11*a*). These termination signals contain a GC-rich palindrome followed by several AT base-pairs. Transcription of these termination signals yields a nascent RNA with the potential to form a **hydro-gen-bonded "hairpin" structure followed by several U's** (Fig. 11.11*a*). When a nascent transcript forms this hairpin structure, it is believed to cause a conformational change in the associated RNA polymerase resulting in termination of transcription within the following, more weakly hydro-

gen-bonded [(A : U)$_n$] region of DNA-RNA base-pairing.

The nucleotide sequence of the attenuator therefore explains its ability to prematurely terminate *trp* operon transcription. But how can this be regulated by the presence or absence of tryptophan?

First, recall that **transcription and translation are coupled in prokaryotes.** That is, ribosomes begin translating mRNAs while they are still being produced by transcription. Thus, events occurring during translation may also affect transcription.

Second, note that the 162 nucleotide-long leader sequence of the *trp* operon mRNA (Fig. 11.10) contains sequences that can base-pair to form alternate secondary structures. Two of these sequences form the previously mentioned transcription-termination hairpin (Fig. 11.11*c*). This hairpin is formed by base-pairing between nucleotide sequences 114–121 and 126–134 (nucleotide 1 is at the 5′ terminus). An alternate secondary structure

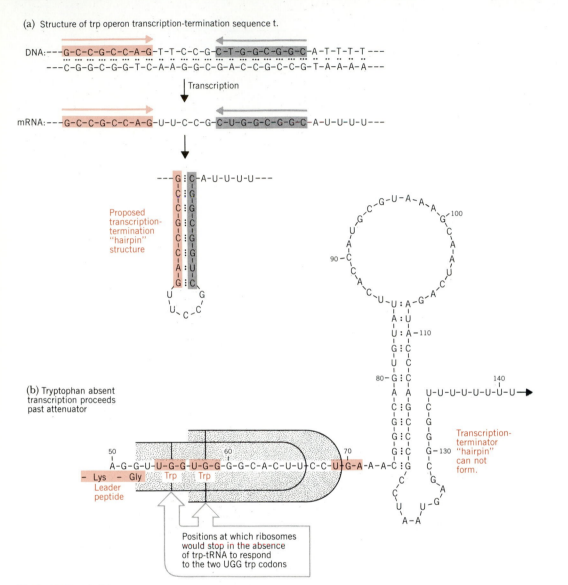

(a) Structure of trp operon transcription-termination sequence t.

Proposed transcription-termination "hairpin" structure

(b) Tryptophan absent transcription proceeds past attenuator

Transcription-terminator "hairpin" can not form.

Lys – Gly
Leader peptide

Positions at which ribosomes would stop in the absence of trp-tRNA to respond to the two UGG trp codons

FIGURE 11.11

Control of the trp *operon by attenuation. Attenuation occurs by tryptophan-regulated termination of transcription within* trpL *(controlling the* trp *operon mRNA leader sequence; see Fig. 11.5). The normal transcription-termination signal in bacteria is a region of dyad symmetry (two sequences that read the same in opposite directions) followed by 4–8 AT base-pairs. The region of dyad symmetry results in mRNA sequences that can base-pair to form "hairpin" structures as shown in (a) and (c). The sequence shown in (a) is, in fact, the base-pair sequence for the first transcription-termination signal (t in Fig. 11.5) at the end*

of the trp *operon. Note the similarity of the mRNA hairpin structures formed by the normal* t *terminator (a) and by the attenuator (c). (b) In the absence of tryptophan, the ribosome translating the mRNA to form the leader peptide becomes stalled at one or the other of the two Trp codons. This allows the base-pairing formed between leader sequences 74–85 and 108–119 to remain intact, which, in turn, prevents the transcription-termination hairpin from forming. Thus, transcription continues through the rest of the* trp *operon. (c) If tryptophan is present, the ribosome can translate on past the Trp codons to the*

(continued on next page)

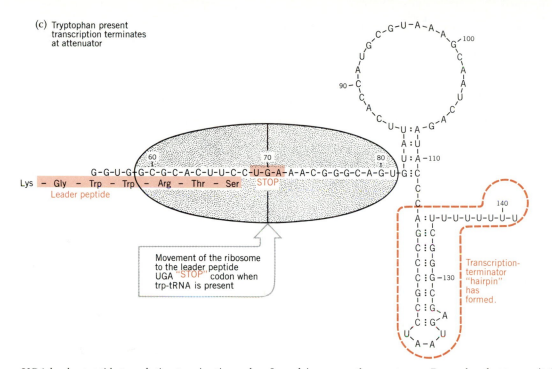

(c) Tryptophan present transcription terminates at attenuator

Lys – Gly – Trp – Trp – Arg – Thr – Ser
Leader peptide

Movement of the ribosome to the leader peptide UGA "STOP" codon when trp-tRNA is present

Transcription-terminator "hairpin" has formed.

UGA leader-peptide translation-termination codon. In so doing, it disrupts the base-pairing between leader sequences 74–85 and 108–119. This, in turn, leaves the latter sequence free to base-pair with sequence 126–134, forming the transcription-termination hairpin. Thus, transcription is terminated at the attenuator sequence and does not continue through the rest of the trp operon. Remember that transcription starts at the 5′-end of the mRNA. For this reason, the 74–85 and 108–119 sequences will be produced and be able to base-pair before the 126–134 sequence has even been synthesized. (After Yanofsky, Nature 289:751–758, 1981, and Platt, Cell 24:10–23, 1981.)

results from base-pairing between leader sequences 74–85 and 108–119 (Fig. 11.11b). Obviously, only one of these structures can exist at one time, since nucleotides 114–119 are part of both. Thus, **if sequences 74–85 and 108–119 are base-paired, the attenuator transcription-termination hairpin cannot form.**

Third, note that the leader sequence contains an AUG translation-initiation codon, followed by 13 codons for amino acids, followed, in turn, by a UGA translation-termination codon (Fig. 11.10). Moreover, the *trp* leader sequence has been shown to contain an efficient ribosome-binding site located in the appropriate position for the initiation of translation at the leader AUG initiation codon. It seems very likely that a 14 amino acid–long "leader peptide" is synthesized as diagramed in Fig. 11.10. This putative leader peptide has not yet been detected *in vivo;* but short peptides of this type are very rapidly degraded in *E. coli,* so failure to detect it is not unexpected.

Note that the leader peptide contains two contiguous tryptophan residues. The two Trp codons are positioned such that in the absence of tryptophan (and thus the absence of Trp-tRNA$_{trp}$), the ribosome will become stalled before it encounters the base-paired structure formed by leader sequences 74–85 and 108–119 (Fig. 11.11b). This base-pairing precludes the formation of the transcription-termination hairpin. Thus, in the absence of tryptophan, transcription will continue past the attenuator into the *trpE* gene.

In the presence of tryptophan, the ribosome can translate past the Trp codons to the leader-peptide termination codon. In the process, it will have to disrupt the base-pairing between leader sequences 74–85 and 108–119. This, in turn, frees the 114–121 sequence, allowing it to base-pair with the 126–134 sequence and form the transcription-termination hairpin (Fig. 11.11c). Thus, in the presence of tryptophan, transcription frequently terminates at the attenuator, reducing the amount of mRNA for the *trp* structural genes.

The transcription of the *trp* operon can be regulated over a range of almost 700-fold by the combined effects of repression (up to 70-fold) and attenuation (up to 10-fold).

Regulation of transcription by attenuation is not unique to the *trp* operon. Six operons (*trp, thr, ilv, leu, phe,* and *his*) are known to be regulated by attenuation. Of these, *trp* and possibly *phe* are also regulated by repression. The *his* operon, which has long been thought to be repressible, is now believed to be regulated entirely by attenuation. While minor details vary from operon to operon, the main features of attenuation are the same for all six operons.

FEEDBACK INHIBITION AND ALLOSTERIC ENZYMES

Earlier in this chapter, we described the mechanism by which the transcription of bacterial genes coding for enzymes in a biosynthetic pathway is repressed when the end product of the pathway is present in the medium in which the cells are growing. A second, and more rapid, regulatory fine tuning of metabolism often occurs at the level of enzyme activity. The presence of sufficient concentrations of an end product (such as histidine, tryptophan, etc.) of a biosynthetic pathway will frequently result in the inhibition of the first enzyme in the pathway. This phenomenon is called **feedback inhibition** or **end-product inhibition; it should not be confused with repression (inhibition of enzyme synthesis).** Feedback inhibition results in an almost instantaneous arrest of the synthesis of an end product when it is added to the medium.

Feedback inhibition-sensitive enzymes have been shown to have an end-product binding site (or sites) in addition to the substrate binding site (or sites). In the case of some multimeric enzymes, the **end-product or regulatory binding site** is on a different subunit (polypeptide) than the substrate site. Upon binding the end product, such enzymes are believed to undergo changes in conformation, called **allosteric transitions,** that reduce their affinity for their substrates. Proteins that undergo such conformational changes are usually referred to as **allosteric proteins.** Many examples are known, including numerous feedback inhibition-sensitive enzymes and the repressor molecules discussed in the preceding sections.

TEMPORAL SEQUENCES OF GENE EXPRESSION DURING PHAGE INFECTION

Regulation of gene expression during the lytic life cycles of bacteriophages is quite different from the reversible on-off switches characteristic of bacterial operons. Instead, viral genes are expressed in **genetically preprogramed sequences,** possibly analogous to the preprogramed sequences of gene expression putatively involved in differentiation in higher organisms. Although different bacterial viruses exhibit variations of the specific mechanisms involved, a common picture emerges. One set of phage genes, usually called "early" genes, is expressed immediately after infection. The product(s) of one or more of the "early" genes is responsible for turning off the expression of the "early" genes and turning on the expression of the next set of genes, and so on. Two to four sets of genes, depending on the virus, are characteristically involved. In all cases studied so far, the regulation of **sequential gene expression** during phage infection **occurs primarily at the level of transcription.**

In three of the most extensively studied bacterial viruses—*E. coli* phages T4 and T7 and *Bacillus subtilis* phage SP01—the sequential gene expression is controlled by modifying the promoter specificity of RNA polymerase, either by the synthesis of a new RNA polymerase (T7) or by phage-induced alterations of the host cell's RNA polymerase (T4 and SP01).

In phage T7-infected cells, the "early" genes are transcribed by the *E. coli* RNA polymerase. One of the "early" genes codes for a T7 RNA polymerase, which then transcribes all of the "late" genes (coding for T7 structural proteins, lysozyme, etc.). *B. subtilis* phage SP01 exhibits a slightly more complex pathway of sequential gene expression, including three sets of genes. These three sets of genes are called "early," "middle," and "late" genes, in reference to their time of expression during the phage reproductive cycle. The SP01 "early" genes are transcribed by the *B. subtilis* RNA polymerase. One of the "early" gene products is a polypeptide that binds to the host cell's RNA polymerase, changing its specificity such that it transcribes the "middle" genes of SP01. Two of the products of "middle" genes are, in turn, polypeptides that associate with the *B. subtilis* RNA polymerase, further changing its specificity so that it then transcribes the "late" genes of SP01.

Phage T4 exhibits an even more complex pattern of sequential gene expression, involving several different modifications of the host cell's RNA polymerase. Thus, in the case of these bacterial viruses, the control of the observed sequential gene expression occurs primarily at the level of transcription and is mediated by specific RNA polymerase-promoter sequence interactions.

CONTROL OF GENE EXPRESSION IN EUKARYOTES

At present, we know very little about the mechanisms by which gene expression is regulated in eukaryotes. We do know that different sets of genes are transcribed in different cell types in higher eukaryotes. This indicates that regulatory mechanisms acting at the level of transcription are important in cell differentiation. What the genetic bases of these regulatory mechanisms are, and how they function, are two of the foremost questions challenging biologists today.

In higher eukaryotes, it does seem very clear that operons are **not** important, if they exist at all.

While there is evidence for operons or operonlike units in the lower eukaryotes (e.g., fungi), operons appear to be rare or nonexistent in higher eukaryotes. All of the mRNAs of higher eukaryotes characterized to date are **monogenic** (contain the coding sequence of one structural gene). In a few cases, the primary transcripts are multigenic and are cleaved to produce monogenic mRNAs.

It also is quite evident that cell differentiation involves, in part, the preprogramed expression of sets of genes, with various kinds of signals (various cytoplasmic molecules, hormones, environmental stimuli, etc.) triggering the reading of different programs at the appropriate times and places during development. The strongest evidence for this is the occurrence of mutations that lead to dramatic alterations in the normal sequences of differentiation — for example, the development of a wing on the head of a fruit fly in place of an eye. Geneticists hope that by studying these mutations they will be able to determine how gene expression is regulated during normal development.

REGULATION OF GENE EXPRESSION DURING DIFFERENTIATION

During the development of a higher eukaryote, a single cell — the zygote — gives rise by mitotic cell divisions to a vast array of cell types (in animals, skin cells, nerve cells, bone cells, blood cells, etc.) with highly divergent morphologies and macromolecular compositions. These different cell types are often highly specialized, carrying out only a few specific metabolic functions. For example, red blood cells are highly specialized for the synthesis and storage of hemoglobin. Over 90 percent of the protein molecules synthesized in red blood cells during their period of maximal biosynthetic activity are hemoglobin chains. Nerve cells are apparently the only cells capable of synthesizing neurotransmitters. What, then, is the mechanism by which these various cells differentiate from one another during the growth and development of an organism? By what mechanism is the expression of the hemoglobin genes in red blood cells or the neurotransmitter genes in neurons, and the lack of expression of these genes in other cell types, brought about?

The only definitive answer that can be given to these questions at present is that the expression of these genes is controlled, at least in part, at the levels of transcription and transcript processing. Hemoglobin mRNA molecules are present in red blood cells, but are absent in other types of cells not synthesizing hemoglobin.

Differentiation occurs by the regulation of gene expression, rather than by changes in genome composition. This has been demonstrated by various techniques in many different organisms. In amphibians, for example, nuclei from differentiated cells can be transplanted into enucleated eggs (eggs from which the original nuclei have been removed) and shown to direct the development of normal embryos. Thus none of the genetic information required for the normal development of entire embryos is lost during the differentiation of the amphibian nuclei-donor cells. In addition, biochemical analyses of the DNA from the nuclei of various differentiated cells have in almost all cases shown that the genomes contain the same set of nucleotide-pair sequences. Rare exceptions are known. Mammalian red blood cells, for example, extrude their nuclei during the last stages of differentiation. By this time, large intracellular pools of stable hemoglobin mRNA have already been synthesized, so that the nuclei are no longer needed.

In higher eukaryotes, only a small proportion of the genome is represented among mRNA molecules in any given cell type. This has been demonstrated by **RNA-DNA saturation hybridization experiments.** RNA is extracted from cells of a particular type and allowed to hybridize with total nuclear DNA (denatured). The RNA is added to the hybridization reaction in large excess (relative to the DNA concentration) so that all DNA sequences complementary to sequences represented in the RNA population will form DNA-RNA hybrids. The proportion of the total genomic DNA present as DNA-RNA hybrids is then determined, providing an estimate of the proportion of the genome represented by sequences in the mRNA population in that particular cell type. (Quantification is usually accomplished by using radioactively labeled DNA, although several other procedures have also been used.)

RNA-DNA saturation hybridization experiments have been done on a number of eukaryotic species using RNA from several different cell types. The results of these experiments show that **less than 10 percent of the DNA in the genome is represented by mRNA molecules in the cytoplasm of any one cell type.** In mice, for example, from 2 to 5 percent of the DNA sequences are represented in the mRNA molecules present in liver cells. Brain cells appear to contain the maximum variety of RNA transcripts. In the toad *Xenopus,* 8 percent of the DNA sequences are represented in mRNA from brain cells. The mRNA from *Xenopus* oöcytes, on the other hand, contains sequences homologous to less than 1 percent of the genomic DNA sequences. It is clear, therefore, that the **majority of the DNA sequences in the genome of a higher eukaryote are not represented among the mRNA populations of any one tissue or cell type.**

Different sets of genes are transcribed and different primary transcripts are processed into mRNA in different types of differentiated cells. Usually, some of the same genes and some different genes are transcribed in different tissues. This can be demonstrated by **RNA-DNA competitive hybridization experiments.** In these experiments, one measures the amounts of radioactively labeled RNA from one cell type that hybridize to total genomic DNA (1) in the presence and (2) in the absence of an excess of competing nonradioactive RNA from a second cell type. If the two cell types contain totally different ("nonoverlapping") RNA populations, in the presence or absence of the competing RNA, the same amount of labeled RNA will hybridize with DNA. If the two RNA populations overlap (share common sequences), the amount of hybridized labeled RNA will be decreased in proportion to the degree of overlap (decreased by a proportion equal to the proportion of shared RNA sequences).

RNA-DNA competitive hybridization experiments of this type indicate that the RNA sequences present in RNA populations taken from different tissues or different cell types differ by from 10 to 100 percent. E. Davidson and colleagues detected no mRNA sequences present in both oöcytes and blastulae of *Xenopus laevis.* Mouse liver, spleen, and kidney cells contain mRNA populations estimated to differ in sequence composition by from 15 to 70 percent. While these are relatively crude estimates, they do indicate that different sets of genes are transcribed and that different transcripts are processed into mRNA in different types of cells.

Since over 90 percent of the DNA sequences in the genome are **not** represented among mRNA populations in any given cell, it has been hypothesized that eukaryotic genes are packaged in chromatin (see Chapter 5, pp. 119–124) in a nonspecifically repressed state, and that the regulation of transcription and/or transcript processing occurs by a **positive** mechanism involving specific gene **activators.** These activators are proposed to function by somehow turning on or "activating" the transcription of specific genes, or sets of genes, at the proper time in the appropriate cells. What these activators are and how such activation might occur (if indeed this hypothesis is correct) are unknown at present. Some evidence indicates that certain nonhistone chromosomal proteins function as specific activators of transcription; however, additional evidence is needed. Moreover, recent evidence suggests that the regulation of RNA transcript processing may be very important in eukaryotes.

The histones are thought to be responsible, at least in part, for the nonspecific repression of eukaryotic genes. The histones, excluding histone H1, have been highly conserved throughout the evolution of eukaryotic organisms, and these histones are known to be tightly complexed with the DNA in nucleosomes. Moreover, histone-complexed DNA is transcribed much less efficiently in *in vitro* transcription systems than is the same DNA after removal of the histones.

The same histones are usually present in the chromatin of various types of differentiated cells. Thus many scientists believe that the histones do not function as specific repressors or activators of transcription. Yet other scientists believe that certain modifications of histones, such as phosphorylations and acylations of key amino acids, are involved in the regulation of transcription. Clearly, conclusions regarding the possible roles of histones in the regulation of transcription must await further evidence.

TRANSCRIPTION ON LAMPBRUSH CHROMOSOMES IN AMPHIBIAN OÖCYTES

In all higher organisms studied so far, the fertilization of a mature egg by a sperm triggers a dramatic increase in protein synthesis, followed by the rapid nuclear and cell divisions of early cleavage stages of embryogenesis. In most eukaryotes, this protein synthesis is not accompanied by RNA synthesis. Instead, all of the components required for protein synthesis during early embryogenesis are present in the egg prior to fertilization. Gene transcripts, in the form of mRNA or pre-mRNA molecules, must therefore be stored in the egg in a dormant state. Translation of these preformed mRNA molecules must somehow be triggered by events associated with fertilization. (A discussion of the evidence supporting these hypotheses is beyond the scope of this text. The reader is referred to E. H. Davidson's *Gene Activity in Early Development* for an excellent review of the pertinent experiments.)

Therefore, the informational molecules that direct protein synthesis during the early cleavage stages following fertilization must be synthesized during oögenesis. Studies of oögenesis in vertebrates, particularly amphibians, reveal that extensive transcription occurs during prophase I (specifically diplotene) of meiosis. During this stage, the chromosomes exist as large **lampbrush** structures. (The structure of these oöcyte **lampbrush chromosomes** is described in Chapter 5, pp. 125–126.)

Most of the DNA in lampbrush chromosomes exists in a highly condensed, transcriptionally inactive state in the so-called axial regions of the chromosomes. Certain segments of the DNA in each lampbrush chromosome, however, exist in highly extended lateral loops (see Figs. 5.44 and 5.45). Each loop consists of a central molecule of DNA that is surrounded by a matrix of newly synthesized RNA and protein. By means of pulse-labeling with ^{3}H-uridine and autoradiography, the loops of lampbrush chromosomes have been shown to be regions of active transcription. The lampbrush chromosomes of oöcytes thus appear to be an excellent example of the correlation between structure and function—the lampbrush morphology being the structural correlate of the transcription of a specific set of chromosomal genes. The transcribed genes of the lampbrush chromosomes apparently are those whose products are required during the early stages of embryogenesis. The gene transcripts synthesized during oögenesis must be stored in an **inactive but stable form** (possibly in RNA-protein complexes) until fertilization occurs. Clearly, regulatory mechanisms are involved which act at a post-transcriptional (mRNA processing?) or translational level.

In addition, particular gene transcripts and/or other gene products must become **localized** in specific areas of the egg cytoplasm during oögenesis. This is evident from experiments that show that the destiny of a particular cell depends on the section of the egg cytoplasm that the cell receives in the early cleavage divisions.

In amphibians, then, and probably in most vertebrates, the genetic programs controlling early development (up to about the blastula stage) are established during oögenesis. Later stages of development, when cell differentiation begins (from about the gastrula stage on), require new programs of gene expression.

rRNA GENE AMPLIFICATION IN AMPHIBIAN OÖCYTES

Despite the rapid initiation of protein synthesis following fertilization, no rRNA is synthesized in amphibian embryos until the gastrula stage. This means that large amounts of rRNA must also be synthesized during oögenesis. In fact, the large eggs of amphibians contain vast quantities of ribosomes, on the order of 10^{12} per mature egg. The requirement for the synthesis of such enormous amounts of a particular gene transcript (the 40S amphibian rRNA precursor) in a single cell has resulted in the evolution of a novel mechanism of **specific gene amplification.** In amphibian oöcytes, the rRNA genes are selectively amplified about a thousandfold in order to facilitate the synthesis of the huge quantities of rRNA stored in mature eggs.

As discussed in Chapter 8, the rRNA genes are normally present as tandemly repeated copies located within the nucleolar organizer regions of the chromosomes. D. Brown, J. Gurdon, and colleagues have shown that there are about 500 copies of the rRNA gene in each of the two nucleolar organizer regions of diploid nuclei of *Xenopus laevis.* The rRNA precursors are synthesized and processed in the nucleoli.

Given about 1000 rRNA genes per diploid nucleus, it has been estimated that over 450 years would be required to synthesize the large number of rRNA molecules present in mature *Xenopus* eggs, hardly a plausible situation given the average life expectancy of a toad. This potential dilemma has been resolved by the evolution of a mechanism by which the rRNA genes are selectively replicated in oöcytes.

D. Brown and I. Dawid have shown that the nuclei of oöcytes of *Xenopus laevis* contain hundreds of nucleoli (Fig. 11.12), each containing circular DNA molecules carrying tandemly repeated copies of the rRNA gene. Once formed, these extrachromosomal DNA molecules appear to replicate by the rolling circle mechanism (Chapter 5, pp. 112–116). However, how the nucleolar regions of the chromosomes are selectively replicated to produce the first extrachromosomal DNA molecules is unknown.

The selective replication of the rRNA genes in oöcytes is the best known example of this type of specific gene amplification. When large quantities of a specific protein are required, as in the case of hemoglobin in red blood cells, extensive amplification can be accomplished at the translation level; each mRNA molecule can be translated many times. Of course, this is not possible when the required gene product is an RNA molecule.

HORMONAL CONTROL OF TRANSCRIPTION

Intercellular communication is a very important phenomenon in higher plants and animals. Signals originating in various glands and/or secretory cells somehow stimulate **target tissues** or **target cells** to undergo dramatic changes in their metabolic patterns. These changes frequently include altered

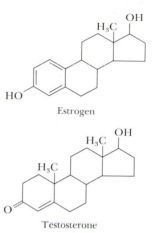

Estrogen

Testosterone

FIGURE 11.13

Chemical structures of the steroid sex hormones estrogen (female) and testosterone (male). The steroid hormones are relatively small molecules (molecular weights around 300) with a conjugated four-ring structure. The various steroid hormones have different side chains and different bonding patterns within the rings. These differences permit them to be recognized by different receptor proteins that are present in the cytoplasm of various target cells.

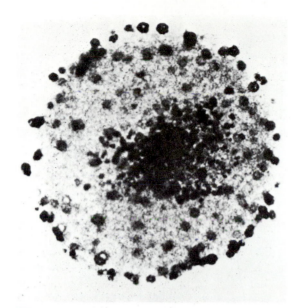

FIGURE 11.12

Photomicrograph of a nucleus from an oöcyte of the South African clawed toad Xenopus laevis, *showing a large number of supernumerary nucleoli that contain circular extrachromosomal molecules of DNA carrying tandemly repeated rRNA genes. (From D. W. Brown and I. B. Dawid,* Science *160:272–280, 1968. Copyright © 1968 by the American Association for the Advancement of Science.)*

patterns of differentiation that are dependent, at least in some cases, on altered patterns of gene expression. What kinds of molecules carry these signals from one cell to another? How do they trigger the altered patterns of gene expression?

Peptide hormones such as insulin and **steroid hormones** such as estrogen and testosterone (Fig. 11.13) represent two types of signal systems utilized in intercellular communication. In higher animals, hormones are synthesized in various specialized secretory cells and are released into the bloodstream. The peptide hormones do not normally enter cells because of their relatively large size. Their effects appear to be mediated by receptor proteins located in target cell membranes and by the intracellular levels of cyclic AMP. The steroid hormones, on the other hand, are small molecules (Fig. 11.13) that readily enter cells through the plasma membranes. Once inside the appropriate target cells, the steroid

hormones become tightly bound to **specific receptor proteins.** These receptor proteins are present only in the cytoplasm of target cells (an example of cell differentiation at the molecular level).

Autoradiographic studies using radioactively labeled steroid hormones have shown that the **hormone-receptor protein complexes rapidly accumulate in the nuclei of target cells.** What primary effects these complexes have on target cell nuclei is still a subject of considerable controversy. Studies by G. Tomkins and colleagues on mice and by B. W. O'Malley and associates on chickens have provided evidence that these hormone-receptor protein complexes somehow activate the transcription of specific genes or sets of genes (Fig. 11.14). How they do this is by no means clear. Some evidence supports the hypothesis that these complexes interact with specific nonhistone chromosomal proteins (specific nonhistone proteins present only in the chromatin of target cells?). This interaction then supposedly stimulates the transcription of the correct genes. If this hypothesis is correct, these complexes may serve as positive regulators (or "activators") of tran-

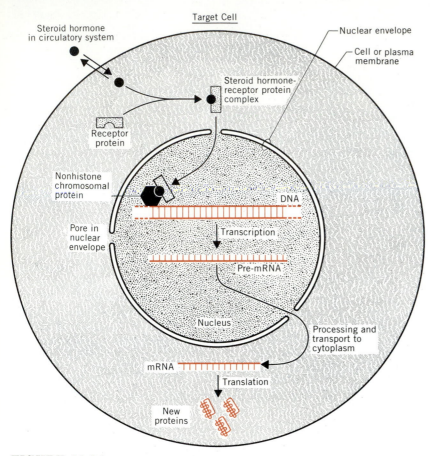

FIGURE 11.14

*Diagram illustrating the putative effects of steroid hormones. Hormones are synthesized in specialized secretory cells and are distributed to the various tissues of the organism through the circulatory system. Their small size (see Fig. 11.13) permits them to readily pass into cells through plasma membranes. The **target cells** (the cells responding to the presence of the specific steroid hormone) contain **receptor proteins** that specifically bind the hormone molecules. The steroid hormone-receptor protein complexes then pass through the pores in the nuclear envelope and accumulate in target cell nuclei. The sequence of events described to this point appears to be well established. How the hormone-receptor protein complexes stimulate transcription of specific* genes *once they have accumulated in the target cell nuclei is less clear. One attractive hypothesis, for which there is some rather convincing support, is that the hormone-receptor protein complexes interact specifically with certain nonhistone chromosomal proteins. Supposedly, these nonhistone chromosomal proteins would be associated with the promoter regions of specific genes, so that the hormone-receptor protein binding could somehow stimulate their transcription by RNA polymerase. Further evidence is needed, however, before this mechanism can be considered an accurate description of the effect of steroid hormones on gene expression.*

scription, much like the CAP-cAMP complexes in prokaryotes.

Evidence that nonhistone chromosomal proteins can control the transcriptional state of particu-

lar genes has, in fact, been obtained by J. Stein, G. Stein, and L. Kleinsmith. Histones are synthesized, like DNA, during the *S* phase of the cell cycle. When chromatin from *S*-phase (DNA synthesis phase) cells is transcribed *in vitro*, **histone mRNA** is synthesized. When chromatin from G_1 phase (the period after the completion of mitosis, but prior to S) is used, no histone mRNA is synthesized. When the nonhistones are removed from G_1-phase chromatin and replaced with nonhistone chromosomal proteins from S-phase chromatin, and this reconstituted chromatin is transcribed *in vitro*, histone mRNA is synthesized. On the other hand, when the nonhistones in reconstituted chromatin are from G_1-phase cells and the DNA and histones are from S-phase cells, no histone mRNA is synthesized. These results indicate that the nonhistone proteins in chromatin determine whether the genes coding for histones are transcribed. It seems likely, therefore, but is by no means proven, that the nonhistone chromosomal proteins play important roles in the regulation of gene expression in eukaryotes. Evidence of this type certainly does not exclude the involvement of histones in the regulation of transcription. The regulation of transcription in eukaryotes may well involve specific interactions among DNA, histones, and nonhistone chromosomal proteins.

In the giant salivary gland chromosomes of certain *Dipteran* flies, such as *Drosophila* species and *Chironomus tentans*, individual chromosome bands undergo striking morphological changes at particular times during development. The individual bands expand into diffuse, less densely staining structures called **"puffs"** (Fig. 11.15); the phenomenon is frequently referred to as "puffing." Each puff almost certainly represents a segment of the chromosome that is in a highly extended state to facilitate the transcription of the resident gene or genes. By means of *in situ* hybridization and autoradiography (see Chapter 5, pp. 131 and 97), the puffs have been shown to contain DNA sequences that are complementary to the RNA sequences present in newly synthesized cytoplasmic mRNA (see Fig. 11.15).

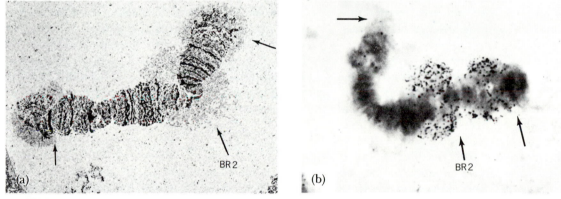

FIGURE 11.15

Photomicrograph (a) *and autoradiograph* (b) *of the polytene salivary chromosome IV of the fly* Chironomus tentans, *showing the three giant puffs (arrows) characteristic of this chromosome. These three puffs are called Balbiani rings (BR1, BR2, and BR3, respectively) after the cytologist who first described them. Balbiani ring 2 (labeled BR2) is a particularly large puff, which has been shown to direct the synthesis of a huge 75S RNA transcript. This giant mRNA molecule is believed to specify the synthesis of one of the large salivary polypeptides. The template for the giant 75S mRNA has been shown to reside in BR2; this was demonstrated by purifying radioactively labeled 75S mRNA from cytoplasmic polyribosomes and using this labeled mRNA to carry out* in situ *hybridization and autoradiography (see Chapter 5, p. 131 and pp. 97–101). Note the heavy labeling of BR2 in* (b). *(From B. Daneholt, S. T. Case, J. Hyde, L. Nelson, and L. Wieslander,* Progress in Nucleic Acid Research and Molecular Biology *19:319–334, 1976).*

During development of *Dipteran* flies, the steroid hormone **ecdysone** is released and triggers molting. Very specific patterns of salivary chromosome puffing occur during these moltings. If larvae of *D. melanogaster* and *C. tentans* are treated with ecdysone at stages of development prior to or between moltings, patterns of chromosome puffing occur that are identical to those occurring during natural moltings. These **ecdysone-induced patterns of sequential puffing** provide convincing demonstrations of the **effect of a steroid hormone on gene expression.** The patterns of puffing observed are very specific and are completely repeatable from experiment to experiment.

During early larval stages of development in *D. melanogaster,* the puffs that existed prior to the ecdysone treatment regress, and a small number of new puffs form within 5 minutes after treatment. These initial puffs regress within a few hours and some 100–125 new puffs appear. By using inhibitors of protein synthesis, such as cycloheximide, the formation of the "late" puffs has been shown to require protein synthesis after ecdysone treatment. However, "early" puffs form in the absence of posttreatment protein synthesis. This suggests that the "late" ecdysone-induced puffing pattern is triggered by one or more of the proteins coded for by gene transcripts synthesized in "early" puffs (Fig. 11.16).

In addition to illustrating the effect of steroid hormones on gene expression, the ecdysone-induced puffing patterns provide evidence for the existence of preprogramed patterns of gene expression in eukaryotes (possibly analogous to the preprogramed patterns of gene expression observed with bacterial viruses; see pp. 396–397).

DETERMINATION AND HOMEOTIC MUTANTS IN *DROSOPHILA*

In some ways, development in insects is more complex than development in many other higher eukaryotes. Genetic programs for the development of the immature, wormlike larvae and pupae, as well as genetic programs for the adult insects, are required. Metamorphoses from larva to pupa and from pupa to adult require sudden changes in programs of gene expression. These changes in gene expression are triggered, at least in part, by the steroid hormone ecdysone. Interestingly, ecdysone has been shown to stimulate the transcription of different sets of genes in different tissues. The genes activated by ecdysone in cells destined to form adult wings or adult legs, for example, are different from those activated by ecdysone in early larval salivary glands.

Preparations for insect metamorphosis are made during early embryogenesis. *Drosophila* blastoderm cells differentiate into two kinds of cells: (1) those that will give rise to larval tissues, and (2) those that will develop into the various tissues and organs of the adult fly. Specifically, certain groups of undifferentiated cells form structures called **imaginal discs.** In all, 17 imaginal discs form, each of which is destined to give rise to a specific organ of the adult fly (Fig. 11.17). These include pairs of labial discs, antenna discs, eye discs, wing discs, haltere discs, three different pairs of leg discs, and a genital disc. In preparation for the bilateral symmetry of the adult fly, each disc, except the genital disc, is present in duplicate.

Although the cells of imaginal discs are (for the most part) undifferentiated, their destinies have already been decided. This process by which cells (actually groups of cells; usually groups of from 5 to 50 cells in *Drosophila*) make irreversible commitments to specific patterns of differentiation is called **determination. Determination** thus occurs at the time that the **destiny of a cell lineage is determined.** Imaginal disc cells are already determined; in fact, imaginal disc determination occurs in the cellular blastoderm stage. The time at which cell determination occurs can be ascertained by experiments in which genetic and embryological approaches are combined. Such experiments allow one to work out **fate maps**—diagrams relating the fates (or destinies) of various cell lineages. (The reader is referred to E. Davidson's *Gene Activity in Early Development* for excellent descriptions of the techniques used in fate mapping.)

The determined state of imaginal disc cells can be demonstrated by dissecting them from larvae and transplanting them into abnormal positions in other larvae. Regardless of their new locations, transplanted imaginal discs develop into the pre-

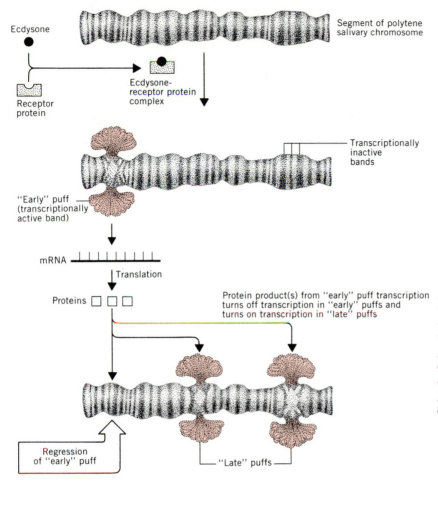

FIGURE 11.16

Schematic illustration of the ecdysone-induced sequences of chromosome puffing in Drosophila *larvae. Ecdysone is the steroid hormone responsible for triggering the events associated with molting in many insects. Its putative mode of action is as diagramed in Fig. 11.14. Evidence that the puffs are sites of active transcription is presented in Fig. 11.15b. Within about 5 minutes after injection of* Drosophila *larvae with ecdysone, a specific set of bands of the polytene salivary gland chromosomes begin to puff; these are always the same bands. Puffing at these sites terminates within about 4 hours, and new bands begin to develop this puffed morphology. By about 10 hours after the ecdysone injection, some 100–125 bands have formed puffs. The "late" puffs require one or more of the protein products that are synthesized from gene transcripts produced in "early" puffs; these "late" puffs do not form if protein synthesis is inhibited during the ecdysone treatment. Thus, ecdysone is a signal that triggers specific programs of sequential gene expression in target cells.*

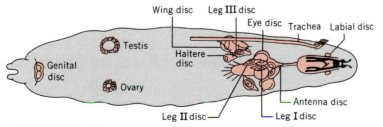

FIGURE 11.17

The location of the imaginal discs of Drosophila *larvae. Each disc gives rise to the indicated organ in the adult fly. (After D. Bodenstein,* Biology of Drosophila, *M. Demerec, ed. John Wiley & Sons, New York, 1950.)*

viously **determined** organ. If an eye disc, for example, is transplanted into another larvae, the fly that emerges following metamorphosis will have three eyes, one of which may occur in a totally bizarre location.

Clearly, then, a key step in understanding differentiation is understanding the mechanism(s) by which cell determination occurs. Determination is clearly under genetic control; this is demonstrated by the existence of **homeotic mutations**—that is, mutations that cause altered states of determination in specific imaginal discs. The mutation *antennapedia*, for example, transforms the determination of the antenna disc to that of a leg disc, which results in a fly with a leglike appendage extending from its head. *Ophthalmoptera* results in the development of wing structures from eye discs, and so on.

Only those homeotic mutations that cause a certain subset of the possible transformations in imaginal disc determination have been observed. Interestingly, the transformations observed in homeotic mutants are the same as the changes occasionally observed following repeated transplantations of larval imaginal discs in adult flies. E. Hadorn, for example, observed that after many serial transplantations of genital discs to adult abdomens, some of the genital discs would develop into leg structures when subsequently implanted in metamorphosing larvae. These events were called **transdeterminations.** They did not occur at random; only specific transdeterminations were observed. A genital disc occasionally underwent transdetermination to a leg structure or an antenna structure, but never to an eye or a wing structure.

The same transformations in determination are observed in homeotic mutations and in transdetermination events. This suggests that certain preprogramed circuits of gene expression have more closely related controlling elements. For example, mutations in a single **"program selector gene"** may divert the cells of a leg imaginal disc from the expression of the genes in the "leg program" to the expression of the genes in the "wing program," but **not** to the expression of the genes in the "eye program." The transformation from the "leg program" to the "eye program" would supposedly re-

quire changes in two or more "program selector genes."

Regardless of the molecular details, the existence of homeotic mutations provides clear evidence of the role of preprogramed patterns of gene expression in determination and differentiation.

ARE DNA METHYLATION PATTERNS REGULATORY?

In most higher plants and animals, the DNA is often modified after synthesis by the enzymatic conversion of many cytosine bases to **5-methylcytosine** bases (Fig. 11.18). The extent of methylation varies from species to species; in mammals, from 2 to 7 percent of the cytosine residues are methylated.

Methyl groups on the 5-carbons of pyrimidines occupy exposed positions within the major grooves of DNA molecules. Thus, they have the potential to play influential roles in the interactions of DNA with specific proteins. In fact, studies on the binding of the repressor for the *lac* operon of *E. coli* to *lac* operator DNA have shown that the addition or removal of a single methyl group can sharply change the affinity of the repressor for the DNA. Thus, the potential regulatory role of 5-methyl groups on pyrimidine bases is well established.

To date, there has been no definitive proof of the role of methylation in the regulation of the expression of any eukaryotic gene. Instead, the arguments for the involvement of methylation in the control of gene expression in eukaryotes are based primarily on three kinds of indirect evidence. (1)

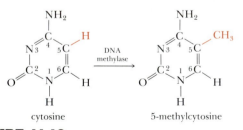

FIGURE 11.18

Structures of cytosine and 5-methylcytosine. The DNA methylases that catalyze the addition of the methyl groups act after DNA replication has incorporated cytosine into nascent DNA chains.

Numerous studies have demonstrated a correlation between the level of gene expression and the degree of methylation — low methylation → high gene expression, high methylation → low gene expression. (2) Methylation patterns are tissue-specific, at least in some cases. (3) The drug (base analog) 5-azacytidine, which can**not** be methylated after it is incorporated into DNA, has been shown to result in the expression of genes in tissues where they normally are not expressed.

An important aspect of methylation patterns is that they appear to be clonally heritable. Over 90 percent of the methylation in the DNA of most eukaryotes occurs in CG dinucleotide sequences, and these sequences are symmetrically methylated, i.e.,

Semiconservative replication of such a symmetrically methylated sequence will yield two half-methylated sequences, i.e.,

5-methyl
|
—CG—   —CG—
—GC—   and   —GC—
                        |
                     5-methyl

In the presence of a DNA methylase that is specific for half-methylated sequences, such methylation patterns will be clonally heritable (Fig. 11.19). That is, once formed, the pattern will be passed on to all of the cells of a given lineage. Some bacterial methylases are indeed known to act primarily on half-methylated sites, and considerable evidence points to the existence of eukaryotic methylases with similar properties.

The above features of DNA methylation patterns nicely explain their maintenance, once formed. However, the key step in any model for the regulation of gene expression or differentiation through DNA methylation involves the **formation of the tissue-specific methylation patterns.** The most popular hypothesis is that the patterns are formed during development by **tissue-specific demethylases,** which remove methyl groups from critical sites in genes that are scheduled to be expressed in a particular cell type (Fig. 11.19). While this is an attractive model, we should emphasize that it is just that: a model. No such tissue-specific demethylase has yet been identified.

Recently, the methylation-blocking drug 5-azacytidine has been shown to result in the expression of the fetal (γ-hemoglobin) and, to a lesser extent, the embryonic (ε-hemoglobin) β-like hemoglobin genes (see Chapter 10, p. 362) of anemic adult baboons and adult humans with severe β-thalassemia (an inherited disease characterized by the inability to synthesize the β-hemoglobin chain of adult hemoglobin) and with sickle cell anemia (see Chapter 9, pp. 280–282). These embryonic and fetal genes are normally not expressed in red blood cells of adults. In one of these studies, the DNA in the region of the γ-hemoglobin and ε-hemoglobin genes was shown to contain fewer methyl groups (to be "undermethylated" or "hypomethylated") in the red blood cells of the individuals after treatment. These results not only support the hypothesis that methylation is important in the regulation of gene expression, but also suggest a possible approach to the treatment of these inherited diseases.

Clearly, additional data are needed before any conclusions can be drawn about the role of DNA methylation as a regulatory mechanism in eukaryotes. Nevertheless, the above-mentioned and other correlations between methylation patterns and levels of gene expression suggest many interesting questions. For example, of what importance is the stabilizing effect of methylation on the Z-form of DNA, and is the correlation between undermethylation and nuclease sensitivity of DNA in chromatin biologically significant (see the following two sections)?

DOES Z-DNA PLAY A REGULATORY ROLE?
One of the more interesting recent discoveries is that segments of DNA that have sequences in which purines and pyrimidines alternate along each strand can form **left-handed** double helices. The Watson-

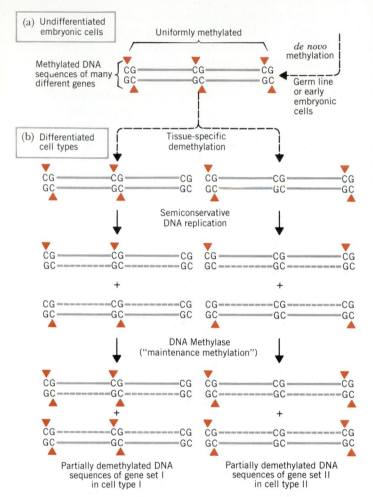

(a) Undifferentiated embryonic cells

Uniformly methylated

de novo methylation

Methylated DNA sequences of many different genes

CG / GC   CG / GC   CG / GC

Germ line or early embryonic cells

(b) Differentiated cell types

Tissue-specific demethylation

CG GC   CG GC   CG GC   CG GC   CG GC   CG GC

Semiconservative DNA replication

DNA Methylase ("maintenance methylation")

Partially demethylated DNA sequences of gene set I in cell type I

Partially demethylated DNA sequences of gene set II in cell type II

FIGURE 11.19

Model for the regulation of gene expression by specific patterns of DNA methylation in different cell types. Each colored triangle represents a 5-methyl group on cytosine; dashed lines represent nascent DNA chains. (a) The DNA in germ line and early embryonic cells is uniformly methylated. (b) During differentiation, tissue-specific patterns of methylation are established by the action of specific demethylases. These patterns are clonally inherited as a result of semiconservative DNA replication and the subsequent maintenance methylation. Note that the DNA methylase involved acts only on half-methylated sequences. (After Razin and Riggs, Science *210:604– 609, 1980.)*

Crick B-form of DNA is a **right-handed** double helix. The novel left-handed double-helical form of DNA has been named **Z-DNA** for the zigzagged paths of the sugar–phosphate backbones of the molecule (Fig. 11.20).

Normally, the Z-form of alternating purine and pyrimidine DNA sequences occurs only at high salt concentrations. When some of the bases in the potential Z-form sequences are methylated (see the preceding section), however, the Z-conformation is stable at lower salt concentrations. Thus, Z-DNA composed of alternating purine-pyrimidine sequences containing methylated bases may be stable *in vivo.* Moreover, its stability is enhanced by cations

including polyamines such as spermine, by negative supercoiling (see Chapter 5, pp. 116–118), and by DNA binding proteins specific for Z-DNA.

In fact, there is evidence that Z-DNA exists in the interband regions of the giant salivary gland chromosomes of *Drosophila melanogaster* and in the transcriptionally active macronucleus of the ciliated protozoan *Stylonychia mytilus.* A. Rich and colleagues have prepared antibodies specific for Z-DNA; these antibodies do not react with B-form DNA. Rich and co-workers have shown that the Z-DNA-specific antibodies bind to the interband regions of the polytene chromosomes of *D. melanogaster* (Fig. 11.21). It will be of great interest to determine the sequences

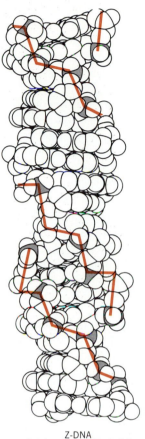

Z-DNA
(left-handed double helix)

B-DNA
(right-handed double helix)

FIGURE 11.20

Comparison of the structures of DNA in the well-known B-form (right) and the recently discovered Z-form (left). The heavy lines show the paths of the sugar–phosphate backbones of the molecules. Note the zigzagged nature of the backbones in Z-DNA. (This drawing of the structure of Z-DNA is based on that of A. H.-J. Wang et al., Science *211:171–176, 1981.)*

and methylation patterns of the DNA in the interband regions of these polytene chromosomes. Do these interband sequences control the expression ("puffing") of the genes located in the adjacent bands? Or are these sequences merely structural or involved in synapsis?

Another hint of the possible involvement of Z-DNA in regulating gene expression is that the structures of certain regulatory proteins suggest that they may bind in the major groove of left-handed double-helices, but not right-handed helices. In fact, McKay and Steitz have proposed that the catabolite activator protein (CAP; see pp. 387–390) stabilizes its CAP-binding sequence in a left-handed conformation. They further propose

that this right-handed to left-handed transition in the double helix unwinds the adjacent promoter or RNA polymerase-binding site and thus activates transcription of the adjacent structural genes. Repressor proteins might act in the opposite direction, stabilizing regulatory sequences in the right-handed B-form and preventing transcription. Although their functions are still unknown, Z-DNA-specific binding proteins have been isolated from *Drosophila.*

When alternating purine-pyrimidine sequences are complexed with histones, these sequences do not display B-DNA to Z-DNA transitions. This and other observations suggest that Z-DNA sequences must be stabilized by Z-DNA-specific binding pro-

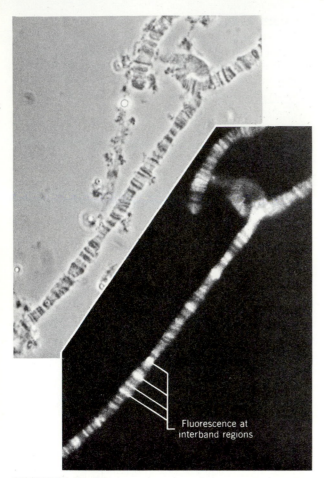

Fluorescence at
interband regions

FIGURE 11.21

Binding of antibodies specific for Z-DNA to the interband regions of the polytene chromosomes of D. melanogaster. *Phase (top) and fluorescence (bottom) micrographs are shown for a segment of a giant salivary gland chromosome. The specific binding of the antibodies to the interband regions of the chromosomes was detected by ultraviolet light fluorescence microscopy. A fluorescent compound was attached to a second antibody (prepared in a goat) which, in turn, is specific for the original anti-Z-DNA immunoglobulin (prepared in a rabbit). The secondary fluorescein-labeled antibody allows the location of the bound Z-DNA antibody to be visualized as shown in the lower photograph. (For additional details, see A. Nordheim et al. Nature 294:417–422, 1981; original photographs courtesy of M. L. Pardue and A. Nordheim.)*

teins, and that Z-DNA sequences may not be found in nucleosomes. Indications of nuclease sensitivity of B-DNA to Z-DNA junctions has led to speculation that such junctions may be related to the nuclease-sensitive sites near the promoter regions of transcriptionally active genes (see the following section).

Clearly, much more information is needed before the potential validity of these proposals can be evaluated. While transitions from sequences of DNA in the B-conformation to sequences in the Z-conformation have been shown to occur within individual plasmids, the biological significance of these transitions remains unknown. Experiments designed to test the possibilities (1) that B-form to Z-form transitions in DNA are involved in the regulation of gene expression and (2) that regulatory proteins may act by binding to and stabilizing one or the other of these conformations may lead to some exciting developments during the next few years.

CHROMATIN STRUCTURE: NUCLEASE SENSITIVITY OF ACTIVE GENES

The demonstration that much, if not all, of the chromosomal DNA of eukaryotes is packaged into nucleosomes and that the 146-nucleotide-pair length of DNA in the nucleosome core is protected to a considerable degree from nuclease digestion (Chapter 5, pp. 121–124) immediately raised the question of whether transcriptionally active DNA is similarly packaged. Since RNA polymerases are very large, complex enzymes (eukaryotic RNA polymerases are larger than nucleosomes) and since DNA is locally unwound during transcription (Chapter 8, p. 242), it seemed likely that the nucleosomes would have to disassemble or at least undergo conformational changes during transcription of the resident DNA sequences.

Does the DNA of a gene remain packaged in nucleosomes during transcription, and, if so, what structural changes, if any, occur in these nucleosomes? Electron microscope and nuclease digestion studies of transcriptionally active genes and chromatin have shown that genes that are being transcribed are indeed packaged into nucleosomes displaying the same frequency and spacing as nucleosomes containing the DNA of genes that are not being transcribed. However, the structures of nucleosomes containing active genes are not identical to those of nucleosomes containing inactive genes. This is shown by the **increased nuclease sensitivity of transcriptionally active genes.**

In 1976, M. Groudine and H. Weintraub showed that the hemoglobin genes present in chromatin from red blood cells of 18-day-old chickens were more sensitive to degradation by pancreatic deoxyribonuclease I (DNase I) than were the ovalbumin genes (not expressed in red blood cells) in chromatin from these same cells or the hemoglobin genes in chromatin isolated from fibroblasts or brain cells of the same chickens. These experiments were done using globin and ovalbumin cDNAs (DNA sequences synthesized *in vitro* by reverse transcriptase using purified globin and ovalbumin mRNAs as templates; see Chapter 10, p. 360) as hybridization probes to measure the quantities of globin and ovalbumin gene sequences in isolated chromatin before and after partial digestion with DNase I. In these experiments, over 50 percent of the DNA sequences of transcriptionally active genes had already been degraded at a time when only 10 percent of the total DNA had been hydrolyzed by DNase I.

Subsequent studies have demonstrated the nuclease sensitivity of transcriptionally active genes in several other organisms. In addition, the nuclease sensitivity of active genes has been found to depend on the presence of two nonhistone chromosomal proteins called HMG14 and HMG17 (HMG for *high mobility group*; small proteins with high mobility during polyacrylamide gel electrophoresis). When these proteins are removed from active chromatin, nuclease sensitivity is lost. When they are added back, sensitivity is restored.

When isolated chromatin containing transcriptionally active genes is treated with very low concentrations of DNase I, the DNA molecules are cleaved at a few **specific sites.** Some of these **hypersensitive sites** have been shown to lie "upstream" (adjacent to the end of the gene homologous to the 5'-end of the mRNA) to transcriptionally active genes. In a few cases, these hypersensitive sites have been shown to be located right at the upstream ends of the promoters of the transcribed genes.

The nature of these hypersensitive sites adjacent to promoters of transcriptionally active genes is still unknown. In the case of the active chicken globin genes, however, the hypersensitive sites are cleaved by S1 nuclease, an endonuclease (isolated from *Aspergillus*) that is specific for single-stranded DNA. This suggests that the DNA is not precisely base-paired or has some other structural modification (B-DNA to Z-DNA junction?) at these hypersensitive sites, very possibly in preparation for RNA polymerase binding at the promoter.

MODELS OF REGULATION OF GENE EXPRESSION IN EUKARYOTES

In the absence of precise information about the mechanisms that regulate gene expression in eukaryotes, many models and much speculation have ensued. One of the more popular models is that proposed by R. J. Britten and E. H. Davidson. Britten and Davidson's model (Fig. 11.22) proposes an integrated regulation of sets of structural genes by means of moderately repetitive regulator genes. It therefore takes into account the observed interspersion of single-copy DNA sequences and repetitive DNA sequences (see Chapter 5, pp. 131–136).

According to the Britten–Davidson model, specific **sensor genes** represent sequence-specific binding sites (analogous to the CAP-cAMP binding site in the *E. coli lac* promoter?; see pp. 387–390) that respond to specific signals (such as hormone-receptor protein complexes?). When sensor genes receive the appropriate signals, they activate the transcription of the adjacent **integrator genes.** The integrator gene products then interact in a sequence-specific manner with **receptor genes.** Britten and Davidson proposed that the integrator gene products were **activator RNAs** that interacted directly with the receptor genes to trigger the transcription of the contiguous **producer genes** (analogous to the structural genes in prokaryotic operons). They pointed out, however, that, formally, it would make no difference whether the active integrator gene products were RNA molecules or proteins.

By making either the receptor genes (Fig.

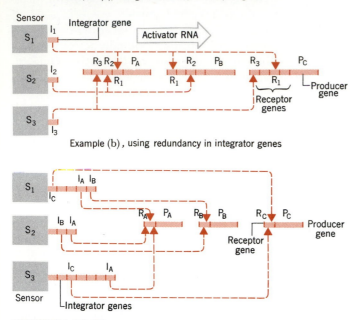

FIGURE 11.22

The Britten and Davidson model of regulation of gene expression in eukaryotes. Two variations of integrated regulation are shown: (a) *a system based on redundancy of "receptor" genes, and* (b) *a system based on redundancy of "integrator" genes. The three "sensor" genes (S_1, S_2, and S_3) respond to three different signals (perhaps hormone-receptor complexes). The diagrams schematize the events proposed to occur after the sensor genes have triggered the transcription of their respective integrator genes (I_1, I_2, I_3 or I_A, I_B, I_C). The integrator gene products, "activator RNAs," diffuse (symbolized by dotted lines) from their sites of synthesis (integrator genes) to their sites of action (receptor genes). The binding of the various activator RNAs to the respective receptor genes somehow triggers the transcription of the contiguous producer genes (P_A, P_B, P_C). Depending on which integrator gene (or genes) is activated by its sensor gene (or genes), one, two, or all three of the producer genes (structural genes) may be turned on. (From R. J. Britten and E. H. Davidson, "Gene Regulation for Higher Cells: A Theory." Science 165:349–357. Copyright © 1969 by the American Association for the Advancement of Science.)*

11.22*a*) or the integrator genes (Fig. 11.22*b*) redundant, various combinations of the producer genes can be turned on in response to different signals. If both the integrator genes and the receptor genes are redundant, complex integrated circuits of gene expression can easily be devised. Unfortunately, testing the validity of such models is far more difficult than devising them. The most attractive feature of the Britten and Davidson model is that it provides a plausible reason for the observed pattern of inter-

spersion of moderately repetitive DNA sequences and single-copy DNA sequences. Direct evidence indicates that most structural genes (producer genes) are indeed single-copy DNA sequences. The adjacent moderately repetitive ("middle-repetitive") DNA sequences would contain the various kinds of regulator genes (sensor, integrator, and receptor genes), according to Britten and Davidson's model.

As more studies have been done comparing

the complexity of heterogeneous nuclear RNA (hnRNA) populations and mRNA populations in different types of cells, it has become evident that hnRNA populations are almost always more complex (contain more distinct sequences) than mRNA populations. These results suggested that considerable **regulation is occurring post-transcriptionally during RNA processing,** that is, in the **hnRNA → mRNA stage.**

These observations led Davidson and Britten to propose a second model, the **"Davidson–Britten model,"** in which gene expression is regulated at the RNA processing level (Fig. 11.23). According to this model, most of the structural genes are located in **"constitutive transcription units,"** which are transcribed at basal levels in all cells. These constitutive transcripts are processed, however, only in cells that contain the appropriate **"integrating regulatory transcripts."** The latter are transcribed in a **cell-specific** manner and must be present before the **"constitutive transcripts" of the structural genes can be processed into mRNAs.**

These "integrating regulatory transcripts" contain repetitive sequences that interact with different structural gene transcripts much like the repetitive "integrator" genes interacted with different "receptor" genes in the original Britten–Davidson model. The key difference is that the regulation occurs post-transcriptionally during RNA processing in the new Davidson–Britten model, rather than transcriptionally as in the original model.

GENOME REARRANGEMENTS DURING DIFFERENTIATION: THE GENETIC CONTROL OF ANTIBODY STRUCTURE

When foreign substances called **antigens** (for example, the coat proteins of viruses) enter the bloodstream of a mammal, they trigger a defense mechanism, the **immune response,** which results in the synthesis of an extremely important group of proteins called **antibodies.** These **antibodies bind to the antigens with exceptional specificity,** facilitating their removal from the circulatory system.

The most remarkable aspect of the immune response, at least from a genetics standpoint, is the **seemingly infinite variety of antibodies that can be synthesized in response to antigens that the animal has not previously encountered.** How can an organism have prepared to synthesize an antibody designed to bind very specifically to a particular antigen without ever having made contact with the antigen? Moreover, how can an organism store enough genetic information to code for the amino acid sequences of a virtually unlimited variety of antibodies? These and related questions about the immune response have puzzled geneticists for several decades. Within the last few years, however, the main features of the answers to these questions have become clear.

We do not know how many different antibodies a mouse or a human can produce; but we do know that the number is very large, almost certainly in the millions. This presents a paradox. The complete mammalian genome (one of each of the 23 pairs of human chromosomes, for example) contains about 3×10^9 nucleotide pairs. If all of this DNA were in the form of uninterrupted coding sequences of genes each 1000 nucleotide-pairs long (of course, we know that much of it is not), the genome would contain a maximum of about 3 million genes. Since we know that many of these genes code for various RNA molecules, enzymes, and structural proteins, and we know that many of these genes contain long noncoding introns, how can we account for the genetic information needed to code for the plethora of different antibodies?

Past attempts to explain the genetic basis of antibody diversity can be roughly grouped into three different hypotheses.

1. The **"germ line" hypothesis** stated that there is a **separate germ line gene for each antibody.** (This fit with our early knowledge about protein synthesis, but presented the paradox of not enough DNA.)
2. The **"somatic mutation" hypothesis** stated that there is only one or a few germ line genes specifying each major class of antibodies and that the **diversity is generated by a high frequency of somatic mutation**—mutation occurring in the antibody-producing somatic cells or in cell lineages leading to antibody-producing cells. (There was no precedent for

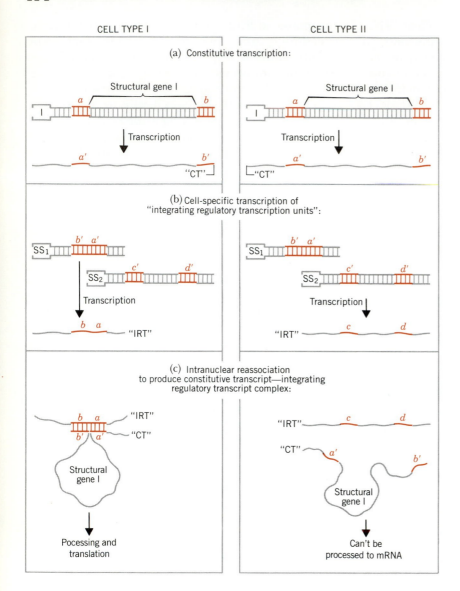

a high frequency of mutation occurring in only certain genes and in only certain types of cells. By what mechanism could this occur, and how could it be regulated?)

3. The **"minigene" hypothesis** stated that the diversity is generated by the **shuffling of many small segments of a few genes into a multitude of possible combinations.** The shuffling would occur by recombination processes in somatic cells. (This re-

quired totally novel mechanisms for rearranging segments of DNA.)

We now know that the minigene hypothesis explains a great deal of the observed diversity. However, we also know that somatic mutation contributes additional diversity. Finally, we know that one segment (the "constant" region; see below) of each antibody chain is specified by a "gene" or "gene

FIGURE 11.23

The Davidson and Britten model for the regulation of gene expression at the RNA processing level in eukaryotes. (a) The majority of the structural genes are assumed to be located in **"constitutive transcription units,"** *which are transcribed continually in all cells. I denotes the transcription initiation site. The sequences labeled a, a', b, etc., and shown in color are middle-repetitive sequences that are proposed to be involved in regulating gene expression. Sequences a', b', c', and d' are complementary to sequences a, b, c, and d, respectively. (b) Structural gene expression is regulated by repetitive RNA sequences transcribed in a tissue- or cell-specific manner as components of* **"integrating regulatory transcription units"** *(IRTUs). Transcription of the IRTUs is controlled by nucleoprotein* **"sensors"** *(SS) that respond to specific external signals. Different IRTUs are transcribed in different tissues or cell types. (c)* **A given structural gene constitutive transcript (CT) can be processed into mRNA** *and thus expressed* **only if it forms an appropriate complex with an "integrating regulatory transcript" (IRT).** *Thus, the expression of a particular structural gene in any given cell requires the presence of the appropriate nuclear IRT. Different cell types will have IRTs carrying overlapping populations of repetitive sequences, such that they process some of the same structural gene transcripts and some different structural gene transcripts. No attempt has been made to illustrate the complex nature of the IRT populations of different cells in the above diagram. (Modified from Davidson and Britten,* Science *204:1052–1059, 1979.)*

segment" that is present in the genome in only a few copies. Thus, all three hypotheses were correct in certain respects.

Antibodies belong to the class of proteins called **immunoglobulins.** Each antibody is a tetramer composed of four polypeptides, **two identical light chains and two identical heavy chains,** joined by disulfide bonds (Fig. 11.24). The light chains are about 220 amino acids long, and the heavy chains are about 440–450 amino acids long. Every chain, heavy and light, has an amino-terminal **variable region,** within which the amino acid sequence varies among antibodies specific for different antigens, and a carboxyl-terminal **constant region,** within which the amino acid sequence is the same for all antibodies of a given immunoglobulin (Ig) class,

regardless of antigen-binding specificity. The variable regions of all antibody chains are about 110 amino acids long.

Regions of proteins that carry out particular functions are called **domains.** Each antibody has **two antigen-binding sites or domains,** each of which is formed by the variable regions of one light chain and one heavy chain (Fig. 11.25). In addition, the constant regions of the two heavy chains interact to form a third domain, called the **effector function domain,** which is responsible for the proper interaction of the antibody with other components of the immune system.

There are five classes of antibodies, IgM, IgD, IgG, IgE, and IgA. The class to which an antibody belongs, and thus the function that it carries out, is

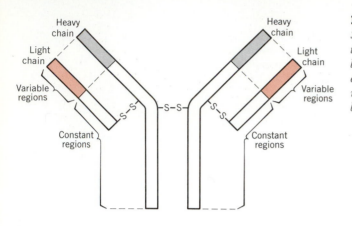

FIGURE 11.24

Schematic diagram of antibody structure. Each antibody is a tetramer composed of four polypeptide chains: two identical light chains and two identical heavy chains. Each chain consists of a variable region (shown shaded) and a constant region. Every antibody has two antigen binding sites, formed by heavy- and light-chain variable regions (see Fig. 11.25).

determined by the structure of its heavy chain constant region (i.e., the structure of its effector function domain). For example, IgD antibodies usually remain bound to the surfaces of the cells in which they are synthesized, whereas IgG antibodies are usually secreted and circulate through the body in the bloodstream. The light chains of antibodies are of two types, **kappa** and **lambda,** with type being determined by the structure of the light chain constant region. As we shall see, antibodies may have the same antigen-binding specificity, as determined by the variable regions of the four chains, but dif-

ferent immunological functions, as determined by the constant regions of the two heavy chains.

Thus, when we examine the structure of antibodies, we see that their diversity resides almost entirely within the variable regions of the molecules. If these polypeptides were synthesized from colinear nucleotide-pair sequences of genes, one gene per polypeptide chain, the genome would have to contain a vast array of genes with highly variable sequences at one end and essentially identical sequences at the other end. This, however, is not the case. Recombinant DNA techniques have made it

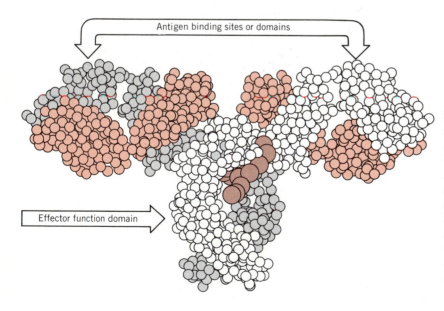

FIGURE 11.25

Space-filling model of antibody structure showing the three functional domains of immunoglobulins. The two heavy chains are shown in white and gray, respectively. The two light chains are shown in color. The dark color components are the associated carbohydrate moieties; their location and structure will vary depending on the immunoglobulin (Ig) class. The structure shown is for a human IgG molecule. (After Silverton, Navia, and Davies, Proc. Natl. Acad. Sci., U.S. *74:5140–5144, 1977.)*

possible to isolate and sequence many of the segments of chromosomal DNA of mice and humans coding for antibody chains. The results of these studies have provided an elegant explanation for the generation of proteins with great diversity in certain regions and constancy in other regions.

Very simply, the **genetic information coding for antibody chains is stored in bits and pieces, and these bits and pieces are put together in the appropriate sequences by genome rearrangements occurring during the development of the antibody-producing cells (called B lymphocytes) of the body.** Each B lymphocyte produces only a single type of antibody. That is, all of the antibodies produced by a given B lymphocyte have the same antigen-binding specificity.

Each antibody chain is synthesized using information stored in several different "genes" or "gene segments." (Some geneticists call these DNA sequences "genes"; others prefer to call them "gene segments." We will call them gene segments in the following discussion.) Note that the classical concept of one gene–one polypeptide is not adequate, at least in its simplest form, to explain gene–antibody relationships.

Synthesis of the kappa light chain is controlled by three different gene segments: (1) a V_κ gene segment, coding for the N-terminal 95 amino acids of the variable region, (2) a J_κ gene segment (**J for joining segment**), coding for the last (constant region-proximal) 13 amino acids of the variable region, and (3) a C_κ gene segment, coding for the C-terminal constant region. A fourth gene segment, the L_κ segment, codes for an N-terminal **hydrophobic leader sequence** 17–20 amino acids long, which is essential for the transport of the antibody chain through the cell membrane. The leader sequence is cleaved off the chain as it passes through the membrane, and thus is not part of the final antibody.

The arrangement of the kappa chain gene segments in germ line cells (in fact, in all cells **not** producing antibodies) is shown in Fig. 11.26. In mice and humans, all of the kappa chain gene segments are located on the same chromosome (chromosome 2 in humans). The same is true for the lambda gene segments (chromosome 22 in humans)

and the heavy chain gene segments (chromosome 14 in humans). **There are a large number, probably about 300, of V_κ gene segments, each with a nearby L_κ gene segment** (Fig. 11.26a). On the other hand, **there is only one C_κ gene segment. Five J_κ gene segments** (one of which is nonfunctional in the mouse) **are located between the V_κ gene segments and the C_κ gene segment.**

In germ line cells, the five J_κ segments are separated from the V_κ segments by a long (we don't yet know how long) noncoding sequence and from the C_κ segment by an approximately 2000 nucleotide-pair-long noncoding sequence. **During the development of a B lymphocyte, the particular kappa light chain gene that will be expressed in that cell is assembled from one L_κ-V_κ segment, one J_κ segment, and the single C_κ segment by a process of somatic recombination** (Fig. 11.26, a and b). This process joins any one of the approximately 300 L_κ-V_κ segments with any one of the five J_κ segments, with the deletion of all intervening DNA (Fig. 11.26b). It yields a fused $V_\kappa J_\kappa$ gene segment coding for the entire variable region of the kappa chain. The noncoding sequence between the J_κ gene-segment cluster and the C_κ gene segment, and the C_κ-proximal J_κ segments, if any, remain between the fused $V_\kappa J_\kappa$ segment and the C_κ segment in the differentiated B lymphocytes (Fig. 11.26b). This entire DNA sequence (L_κ-$V_\kappa J_\kappa$-noncoding-C_κ) is transcribed (Fig. 11.26c) and the noncoding sequences are removed during RNA processing (Fig. 11.26, c and d), just like the noncoding sequences or introns of any other eukaryotic gene.

Lambda light chain genes are also assembled from separate segments during B lymphocyte development. The major difference is that each J_λ gene segment comes with its own C_λ gene segment. That is, the genome rearrangements required for lambda chain synthesis join L_λ-V_λ segments to J_λ-C_λ segments. Mice have only four J_λ-C_λ gene segments, whereas humans have six. This correlates with the fact that only 5 percent of the antibodies of mice are of the lambda type, whereas 40 percent of the antibodies of humans have lambda light chains.

The genetic information coding for antibody heavy chains is organized into L_H-V_H, J_H, and C_H

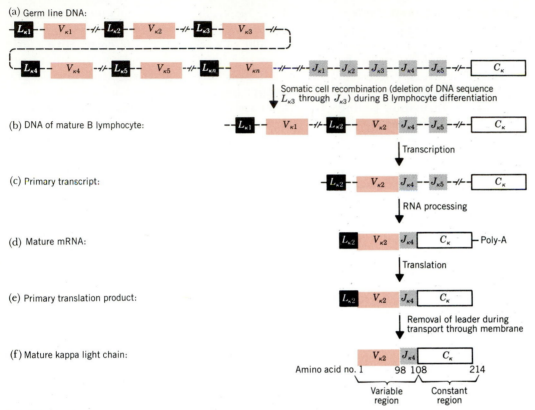

FIGURE 11.26

The genetic control of human antibody kappa light chains. See the text for a description of the processes involved. [After Leder, Sci. Amer. 246(5):102–115, 1982.]

gene segments analogous to those for kappa light chains; but, **there is one additional gene segment, called D for diversity, which codes for 2–13 amino acids of the variable region.** The variable region of the heavy chain is thus encoded in three separate gene segments that must be joined during B lymphocyte development. In addition, there are from one to four C_H gene segments for each Ig class.

In the mouse, there are a total of **eight C_H gene segments,** all functional, arranged on the chromosome in the sequence $C_{H\mu}$, $C_{H\delta}$, $C_{H\gamma3}$, $C_{H\gamma1}$, $C_{H\gamma2b}$, $C_{H\gamma2a}$, $C_{H\varepsilon}$, $C_{H\alpha}$ (Fig. 11.27a). $C_{H\mu}$, $C_{H\delta}$, $C_{H\varepsilon}$, and $C_{H\alpha}$ code for the heavy chain constant regions of IgM, IgD, IgE, and IgA, respectively. Four gene segments $C_{H\gamma3}$, $C_{H\gamma1}$, $C_{H\gamma2b}$, and $C_{H\gamma2a}$ code for IgG heavy chain constant regions.

In humans, there are nine or 10 functional C_H gene segments: $C_{H\mu}$, $C_{H\delta}$, $C_{H\gamma1}$, $C_{H\gamma2}$, $C_{H\gamma3}$, $C_{H\gamma4}$, $C_{H\varepsilon1}$, probably $C_{H\varepsilon2}$, $C_{H\alpha1}$, and $C_{H\alpha2}$. The human C_H gene cluster also contains two nonfunctional "genes," commonly called **pseudogenes,** with very similar structures. Pseudogenes are partial duplicates of structural genes that have incorporated sufficient changes that they are not biologically active and usually are not transcribed. Pseudogenes are turning out to be quite common in eukaryotes.

In mouse germ line cells, there are **about 300 L_H-V_H gene segments,** something like **10–50 D gene segments, 4J_H gene segments,** and the 8 C_H gene segments, arranged on the chromosome in the above order (Fig. 11.27a). During the development of a B lymphocyte from a stem cell (a mitotically

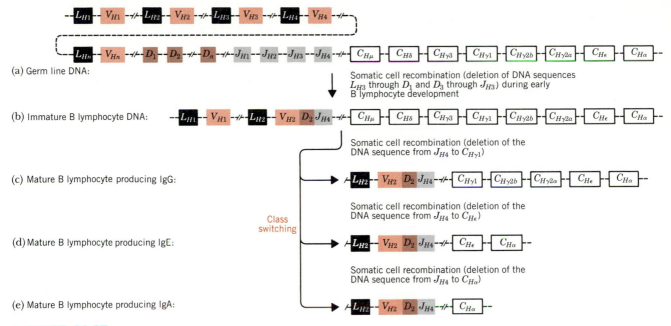

(a) Germ line DNA:

Somatic cell recombination (deletion of DNA sequences L_{H3} through D_1 and D_3 through J_{H3}) during early B lymphocyte development

(b) Immature B lymphocyte DNA:

Somatic cell recombination (deletion of the DNA sequence from J_{H4} to $C_{H\gamma1}$)

(c) Mature B lymphocyte producing IgG:

Somatic cell recombination (deletion of the DNA sequence from J_{H4} to $C_{H\epsilon}$)

Class switching

(d) Mature B lymphocyte producing IgE:

Somatic cell recombination (deletion of the DNA sequence from J_{H4} to $C_{H\alpha}$)

(e) Mature B lymphocyte producing IgA:

FIGURE 11.27

The genetic control of mouse antibody heavy chains. $C_{H\mu}$, $C_{H\delta}$, $C_{H\epsilon}$, and $C_{H\alpha}$ code for the heavy chain constant regions of IgM, IgD, IgE, and IgA, respectively. $C_{H\gamma3}$, $C_{H\gamma1}$, $C_{H\gamma2b}$ and $C_{H\gamma2a}$ code for the heavy chain constant regions of four closely related forms of IgG. See the text for a description of the processes involved. (After Davis, Kim, and Hood, Cell 22:1–2, 1980.)

active somatic cell from which other types of cells "stem" or arise by cell division and differentiation), somatic recombination joins one L_H-V_H gene segment with one D gene segment and one J_H gene segment, deleting the two intervening sequences of DNA, to form one continuous DNA sequence $(V_H D J_H)$ that codes for the entire heavy chain variable region (Fig. 11.27, *a* and *b*).

At the time that antibody synthesis begins in the developing B lymphocyte, all of the C_H gene segments are still present, separated from the newly formed L_H-$V_H D J_H$ gene segment by a short noncoding sequence (Fig. 11.27*b*). At this stage, all antibodies synthesized have IgM heavy chains ($C_{H\mu}$ gene products). If an antigen is recognized and bound to an antibody on the surface of a developing B lymphocyte, however, that cell is stimulated to differen-

tiate into a mature B lymphocyte. During this differentiation, some B lymphocytes will switch from producing antibodies of class IgM to producing antibodies of another class. This phenomonon, called **class switching,** often involves further genome rearrangements during which the C_H gene segments closest to the previously joined L_H-$V_H D J_H$ gene segments are deleted (Fig. 11.27, *c*–*e*). The class of antibodies produced after class switching is determined by which gene is brought into the closest proximity with the L_H-$V_H D J_H$ gene segment, as illustrated in Fig. 11.27, *c*–*e*.

Another type of class switching during B lymphocyte differentiation occurs at the level of RNA processing ("splicing"). Certain mature B lymphocytes produce both IgM and IgD antibodies. It should be emphasized, however, that these antibod-

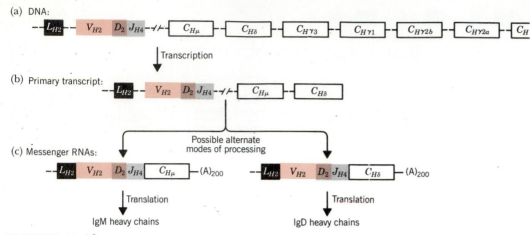

(a) DNA:

(b) Primary transcript:

Transcription

Possible alternate
modes of processing

(c) Messenger RNAs:

Translation   Translation

IgM heavy chains   IgD heavy chains

FIGURE 11.28

*Mechanism for the production of IgM and IgD antibody heavy chains in the same B
lymphocyte as a result of alternate modes of splicing the heavy chain transcript.*

ies differ only in their effector function domains;
they have identical antigen-binding domains, speci-
fied by the same $V_\kappa J_\kappa$ (or $V_\lambda J_\lambda$) and $V_H D J_H$ fused gene
segments. In these cells, a primary transcript that
extends through both the $C_{H\mu}$ and $C_{H\delta}$ gene segments is synthesized. During processing, the $V_H D J_H$
transcript sequence may be spliced to either the $C_{H\mu}$
sequence or the $C_{H\delta}$ sequence, such that both types
of heavy chain are synthesized in the same cell (Fig.
11.28).

A further complexity observed in antibody synthesis is the **sequential production of membrane-
bound and secreted forms of a given antibody.**
The first antibodies to appear in developing B lymphocytes are membrane-bound IgM molecules.
Subsequently, these cells switch to the production of
a secreted form of IgM. These two forms of IgM
differ only in the C-terminal portions of their heavy
chains. The heavy chain of the membrane-bound
form is 21 amino acids longer than that of the
secreted form. The **membrane-bound heavy chain
has a 41 amino acid-long hydrophobic sequence** at
the C-terminus that is probably responsible for an-
choring it to the cell surface. This hydrophobic
sequence is replaced by a **20 amino acid-long hydrophilic sequence** in the **secreted form.**

The coding sequences (exons) of the C_H gene
segments are interrupted by noncoding sequences
(introns) just like those of many other eukaryotic
genes (Chapter 10, pp. 361–365). The C_H gene
segments contain four to six exons and three to five
introns (Fig. 11.29). In membrane-bound antibodies, the heavy chain constant regions are produced
by splicing all six exons together (Fig. 11.29, a and
b). The last two exons code for the hydrophobic tails
of the membrane-bound heavy chains. During synthesis of the membrane-bound form, the fifth C_H
exon is spliced to a site 20 codons from the end of
the fourth exon (Fig. 11.29b), thus changing the
amino acid sequence of this portion of the heavy
chain constant region. In secreted antibodies, the
heavy chain constant regions are the product of the
first four exons (Fig. 11.29c).

The use of alternate pathways of transcription
and RNA processing to synthesize membrane-
bound and secreted forms has been firmly established for the IgM class of antibodies. Recent evidence suggests that similar alternate pathways of
transcription and splicing are responsible for the
production of the membrane-bound and secreted
forms of the other classes of immunoglobulins as
well.

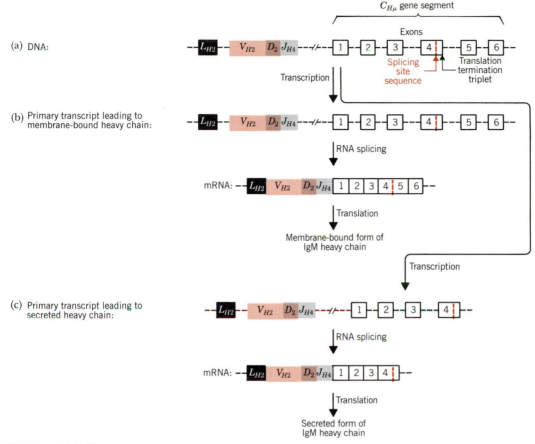

FIGURE 11.29

Genetic control of the membrane-bound and secreted forms of IgM. Only the heavy-chain genes and gene products are shown since both forms have identical light chains. [After Leder, Sci. Amer. *246(5):102–115, 1982.]*

How are the genome rearrangements that occur during B lymphocyte development regulated? What controls the somatic recombination events such that a *V* gene segment is joined to a *J* segment and not to another *V* segment or directly to a *C* segment? Several long segments of chromosomal DNA carrying clusters of *V* gene segments, *D* gene segments, and *J* gene segments of both mice and humans have now been sequenced, and the resulting nucleotide-pair sequences suggest the presence of specific *V-J*, *V-D*, and *D-J* joining signals. The same **signal sequences** are found adjacent to all *V*

gene segments. Similarly, all *J* gene segments have identical signal sequences located adjacent to their coding sequences; however, their signal sequence is different from that adjacent to *V* gene segments. Likewise, *D* and *C* gene segments have their own adjacent signal sequences.

The signal sequences controlling *V-J*, *V-D*, and *D-J* joining contain 7 base-pair (heptamer)- and 9 base-pair (nonamer)-long sequences separated by spacers of different, but specific lengths. For V_κ-J_κ joining, the spacer in the V_κ signal sequence is 12 nucleotide-pairs long, while that in the J_κ signal sequence is 22 nucleotide-pairs long. The heptamer and nonamer sequences located "after" (to the right as drawn in Figs. 11.26 and 11.30) the V_κ gene segments are complementary (with the exception of one base-pair) to those "preceding" (to the left as drawn in Figs. 11.26 and 11.30) the J_κ gene segments. These signal sequences have the potential to form "stem and loop" structures as diagramed in Fig. 11.30, thus bringing the V_κ and J_κ gene segments into juxtaposition for joining. Apparently, joining will occur only when one signal sequence contains a 12 base-pair spacer and the other contains a 22 base-pair spacer. This requirement would supposedly be enforced by the specific protein(s) mediating the joining process. Very similar signal sequences appear to control V_H-D and D-J_H joining, while somewhat different signal sequences mediate class switching [see Leder, *Sci. Amer.* 246(5):102–115, 1982].

One can readily see that a large amount of diversity can be generated by the joining of antibody gene segments as described above. For example, consider the number of different kappa light chains possible in humans: $\cong 300$ V_κ gene segments $\times 5$ J_κ segments $\cong 1500$ fused $V_\kappa J_\kappa$ gene segments. The heavy chain variable region provides even greater diversity because of the multiple D gene segments. If there are 300 V_H gene segments, 25 D gene segments, and 6 J_H gene segments in human germ line cells, 45,000 different heavy chain variable regions could be assembled. Using the above estimates, 67,500,000 different antigen-binding sites could be produced using just kappa light chains. Lambda light chains produce another level of diversity.

Clearly, these antibody gene-segment fusions provide for a vast amount of antibody diversity. We now know, however, that further diversity is generated in two additional ways: (1) somatic mutation, and (2) variability in the sites at which *V-J*, *V-D*, and *D-J* joining events occur. (For a discussion of the evidence for the roles of somatic mutation and variable joining sites in the generation of antibody diversity, see Lewin, *Genes*, 1983, Chapter 39).

Up to this point, we have avoided the question of how an organism initiates the synthesis of antibodies specific to antigens that it has not previously encountered. This is nicely explained by the **clonal selection theory.** Recall that **all of the antibodies produced by a single B lymphocyte have the same antigen-binding specificity.** But **different cells** in a **population** of B lymphocytes will have undergone **different genome rearrangements leading to the production of antibodies with different specificities.** Thus, the population of B lymphocytes in a human or a mouse will be producing a very large variety of antibodies. The clonal selection theory states that the **binding of a particular foreign antigen to an antibody on the surface of a B lymphocyte stimulates that cell to divide,** producing large numbers of this particular B lymphocyte (a "clone" of identical cells) and thus large amounts of the particular antibody that recognizes the foreign antigen.

The immune response in mammals is a very complex process involving a large number of different macromolecules and different cell types. Our discussion has been totally restricted to the genetic control of the synthesis of antibody chains. Many of the other components of the immune response, such as the **transplantation antigens** largely responsible for the rejection of foreign tissues in transplant operations, are controlled by a multigene complex called the **major histocompatibility complex** (*MHC*, on chromosome 6 in humans). A more complete discussion of the immune response is beyond the scope of this book, but we hope your curiosity will stimulate you to pursue this topic further.

Consider one final point before leaving this topic. Each B lymphocyte makes only **one** type of antibody. Why? Mammalian cells are diploid; they carry two sets of genetic information coding for each of the antibody chains. But only one **productive** genome rearrangement occurs in each cell! This phenomenon is called **allelic exclusion** because one of the "alleles" is **excluded from being expressed.** How? Why? At present, we still don't know.

CONTROL OF CELL DIVISION: ONCOGENES AND PROTO-ONCOGENES

Cell division, like all other biological processes, is under genetic control. Certain genes must regulate

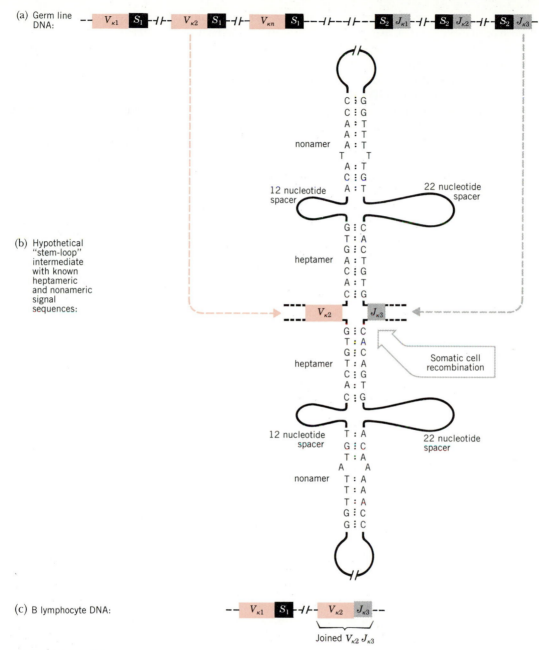

(a) Germ line DNA:

(b) Hypothetical "stem-loop" intermediate with known heptameric and nonameric signal sequences:

(c) B lymphocyte DNA:

Joined $V_{\kappa 2} J_{\kappa 3}$

FIGURE 11.30

Signal sequences and their proposed role in $V_{\kappa} J_{\kappa}$ joining. The germ line and B lymphocyte DNA arrangements shown in (a) and (c), respectively, as well as the heptamer and nonamer sequences and spacer sizes shown in (b), are known. The "stem and loop" structure shown in (b) is a hypothetical intermediate.

the process of cell division in response to environmental signals. These regulatory genes are undoubtedly subject to mutation, like all other genes. Mutations that abolish the function of these regulatory genes would be expected to lead to abnormal cell division — in the extreme, either the inability to divide at all or the inability to stop dividing.

To date, we do not know how cell division is controlled, nor have we identified the genes that regulate the process, in any organism. However, recent studies of viral genes called **oncogenes** (from the Greek *onkos*, meaning "tumor"), which can cause a loss of the normal control of cell division, have led to the identification of a set of homologous genes called **proto-oncogenes** in the genomes of normal animals, including humans. These normal cellular proto-oncogenes can be converted into tumor-causing **cellular oncogenes** by mutation or by becoming associated with new regulatory sequences through recombination processes. These and related observations indicate that the normal cellular function(s) of the proto-oncogenes involves some aspect of the control of cell division. In fact, it now seems likely that breakthroughs in understanding the normal control of cell division may result from studies on the disruption of normal control that occurs in cancer cells.

Cancer is a large class of diverse diseases, all of which exhibit **uncontrolled cell growth and division.** In noncirculatory tissues, such uncontrolled cell growth produces cell masses called **tumors.** Cancerous or **malignant tumors** are those from which cells detach and migrate to other parts of the body giving rise to secondary tumors (a process called **metastasis**). Noncancerous or **benign tumors** do not metastasize.

Human cancers are responsible for an enormous amount of suffering. Thus, a large amount of money and effort have been directed to the study of this disease. While there has been great progress in the detection and treatment of cancers, there has been little progress toward understanding the molecular bases of cancers. Recently, however, two different experimental approaches have provided evidence for the involvement of at least 15 different normal cellular genes, the proto-oncogenes, in the occurrence of certain types of cancer in animals.

One approach involved looking for cellular genes homologous to the oncogenes of animal viruses. The other involved looking directly for cancer-causing genes in the genomes of cancer cells by **transfection experiments,** experiments in which the DNA of tumor cells is isolated and added to normal tissue culture cells to see if it will convert any of them to the cancerous state. Both approaches have been successful and, in some cases, both have resulted in the identification of the same cellular oncogenes.

Most of our information about oncogenes has come from studies of **RNA tumor viruses** or **retroviruses.** The name retrovirus is derived from the fact that these viruses store their genetic information in a single-stranded RNA genome, and then convert it into an homologous double-stranded DNA form after infecting a host cell. Thus, they use a "backward" ("retro") flow of genetic information. Usually, genetic information flows from DNA to RNA during transcription. The retroviruses code for a special enzyme, called reverse transcriptase, which catalyzes the synthesis of homologous DNA sequences using RNA molecules as templates (Chapter 10, p. 356).

The best known of the retroviruses is **Rous sarcoma virus** ("Rous" for the scientist who discovered it; "sarcoma" for the type of cancer induced), which induces cancer in chicken cells. The life cycle of Rous sarcoma virus is diagramed in Fig. 11.31. Shortly after Rous sarcoma virus infects a cell, its RNA genome is replicated to its DNA form by reverse transcriptase, and the viral DNA is integrated into the chromosomal DNA of the host cell. In this integrated state, it is replicated and transcribed by the host cell's metabolic machinery, just like the normal genes of the host cell.

The genome of Rous sarcoma virus contains just four genes: *gag*, which codes for the capsid protein of the virion; *pol*, which codes for reverse transcriptase; *env*, which codes for the protein spikes of the viral envelope; and the oncogene *src* (derived from *sarcoma*), which codes for a membrane-bound protein kinase. The viral genome also carries its own strong promoter; thus, the four viral gene

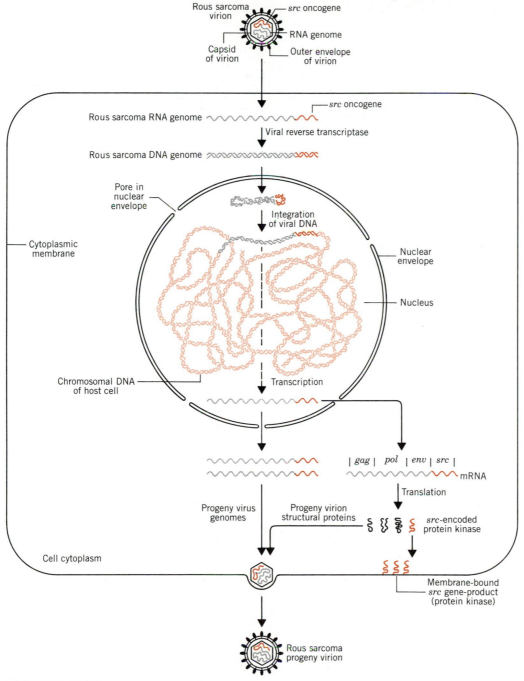

FIGURE 11.31

Life cycle of the RNA tumor virus Rous sarcoma. The ability of this virus to cause cancer resides in the src *oncogene (shown in dark color). [After J. M. Bishop,* Sci. Amer. *246(3):80–92, 1982.]*

products are made in large amounts. The *src* gene is entirely responsible for the ability of Rous sarcoma virus to cause cancer, because the deletion of this one gene yields a virus that infects and replicates just like the *src*-containing virus, but is totally nononcogenic.

Studies of other retroviruses have led to the identification of at least 15 different viral oncogenes, and more undoubtedly remain to be discovered. Recent studies have shown, however, that different retroviruses that induce similar types of cancers often carry the same oncogene. Thus, there may be only a relatively small number of different viral oncogenes.

When viral oncogenes such as *src* are cloned by recombinant DNA techniques and are used as hybridization probes to search for homologous sequences in normal host cells, such sequences are almost always found. These homologous sequences present in the chromosomes of normal cells and normal animals are **not** integrated viral oncogenes, because they differ from the viral oncogenes in having interrupted coding sequences, like most other eukaryotic genes. That is, the cellular oncogenes and proto-oncogenes have multiple exons separated by introns (see Chapter 10, pp. 361–365), whereas the viral oncogenes are single exons. For example, the cellular oncogene homologous to the *src* gene of Rous sarcoma virus contains six introns separating seven coding sequences, whereas the viral *src* gene has a single, uninterrupted coding sequence (Fig. 11.32). The viral *src* gene and its cellular homolog are designated *v-src* and *c-src*, respectively.

The *v-src* and *c-src* genes both code for protein kinases. Moreover, these two protein kinases are the same size and have very similar structures. Both proteins are complexed by antibodies prepared using the viral protein as antigen.

One argument for the importance of the proto-oncogenes, and the protein kinases that they encode, in normal cell growth and/or cell division is that these genes have been highly conserved during evolution. *C-src* genes are found not only in chickens, but in other birds, in mammals (including humans), and even in fish. While we do not yet know what role(s) the protein kinases encoded in proto-oncogenes play in normal cells, it seems very likely that they are somehow involved in the control of cell division.

We also do not know how the protein kinases encoded in viral oncogenes cause tumors, but the mechanism probably relates to the large quantities of these enzymes made in retrovirus-infected cells. There is 100 times as much *v-src* protein kinase per cell in chicken tumors induced with Rous sarcoma virus as there is *c-src* protein kinase in normal chicken cells.

The detection of cellular oncogenes by transfection experiments is based on the ability of the oncogenes to convert noncancerous cells (characterized by controlled cell division) growing in culture to the cancerous state (characterized by uncontrolled cell division). This phenomenon is called **cell transformation** or simply **transformation.** It should not be confused with the recombination process in bacteria called transformation (discussed in Chapters 5 and 7); the two are totally different processes.

Normal (nontransformed) cells growing in culture will stop dividing when they make contact with neighboring cells (a phenomenon called **contact inhibition**); they will thus form a monolayer of cells on the surface of the culture flask or petri dish in which they are growing. Transformed cells do not exhibit contact inhibition. They will keep on dividing despite contact with their neighbors and will form piles of cells or "tumors" on the surface of the culture flask (Fig. 11.33).

When DNA from normal cells is used in transfection experiments, a very low, but detectable, level of cell transformation is observed. When the DNA from the transformed cells is used in a second round of transfection experiments, a higher frequency of transformation is sometimes observed. That is, higher frequencies of transformation are observed using DNA isolated from certain transformed cell clones, but not using DNA isolated from other transformed cell clones. This indicates that genetic changes are responsible for the transformed state in the first group of cell clones, but that **epigenetic changes** (noninherited developmental

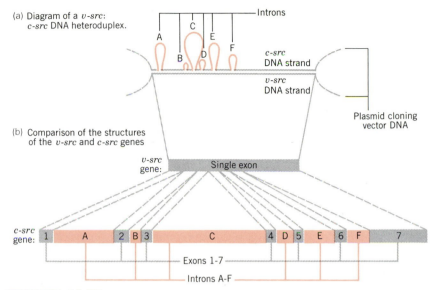

(a) Diagram of a *v-src*:
    c-src DNA heteroduplex.

(b) Comparison of the structures
    of the *v-src* and *c-src* genes

FIGURE 11.32

Structures of the v-src *and* c-src *genes.* (a) *Diagram of a DNA heteroduplex formed by hybridization of one strand carrying the* c-src *gene (top strand) with a (partially) complementary strand carrying the* v-src *gene (bottom strand) as observed by electron microscopy. The intron segments of the* c-src *strand form single-stranded loops (shown in color). The dashed lines represent the DNA strands of the plasmid in which these genes were cloned.* (b) *Schematic comparison of the organization of the coding sequences (exons) of the two genes. The* v-src *gene has a single, uninterrupted coding sequence. The* c-src *gene contains seven exons (numbered 1–7) separated by six introns (designated A–F and shown in color). Exons 1–7 are approximately 100, 270, 120, 100, 120, 180, and 1000 nucleotides, respectively, in length. Introns A–F are approximately 1200, 200, 2300, 450, 1100, and 550 nucleotides, respectively, in length. The dashed lines indicate the approximate correlations between the coding sequences of the two genes. (Based on the data of Parker, Varmus, and Bishop,* Proc. Natl. Acad. Sci., U.S. *78:5842–5846, 1981.)*

changes) are responsible for the transformed state in the second group of cell clones.

Transfection experiments have also been used to demonstrate the presence of cellular oncogenes in cell culture lines derived from various spontaneously occurring and chemically induced animal tumors. Some of the cellular oncogenes detected by transfection experiments have been isolated using recombinant DNA and gene cloning techniques. When these isolated cellular oncogenes were compared with the oncogenes of retroviruses by various procedures (e.g., DNA hybridization, restriction

enzyme analysis, DNA sequencing), some of them were found to be homologous to the viral oncogenes. The oncogene identified by transfection experiments in DNA from human bladder carcinoma cells turned out to be homologous to the oncogene of Harvey sarcoma virus, for example.

Surprisingly, **when the oncogene from human bladder carcinoma cells was sequenced, and its sequence was compared with the sequence of the homologous proto-oncogene, its oncogenicity was found to result from a single base-pair substitution.** That is, a single base-pair difference was shown

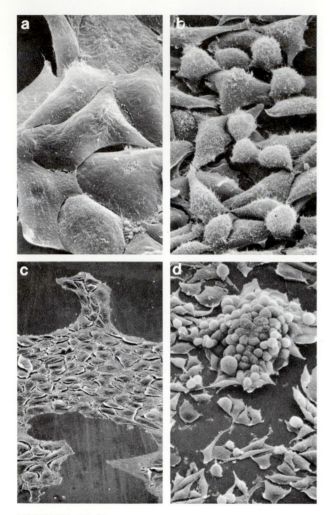

FIGURE 11.33

Scanning electron micrographs showing the altered morphology and cell-cell interaction characteristics of normal (a and c) and transformed (b and d) rat kidney cells growing in culture. Note that the normal cells adhere to the surface of the culture dish forming monolayers of flat, extended cells. The transformed cells, on the other hand, tend to overgrow one another, forming clusters or piles of cells. All cells are of the NRK (normal rat kidney) cell culture line. The NRK cells shown in b and d were transformed by infection with the Prague A strain of avian sarcoma virus. Magnification: (a) ×1100; (b) ×1000; (c) ×200; (d) ×280. (Original photographs courtesy of Mike Atkinson and Chris Frethem, Department of Anatomy, University of Minnesota.)

to correlate with the ability or inability of the two genes to transform "normal" cells growing in culture. The oncogene was apparently produced from its proto-oncogene by a GC → TA transversion. This mutation results in the substitution of valine for the glycine present as the twelfth amino acid (from the amino terminus) in the normal protein kinase.

In contrast to the retroviral oncogenes, the human bladder carcinoma oncogene does not result in the synthesis of abnormally large amounts of its protein kinase gene product. At present, we don't know why such a small change in a proto-oncogene, a normal cellular gene, should produce an oncogene capable of transforming cells to the cancerous state. Recent studies have indicated that both the normal cells and the tumor cells of one patient with a bladder carcinoma may be heterozygous for the oncogene and the proto-oncogene. These results suggest that the oncogene causes a predisposition, rather than an immediate change, to the cancerous state.

For years, cytogeneticists have documented correlations between certain types of cancer and specific changes in chromosome structure. In particular, translocations (the breakage and transfer of parts of chromosomes to nonhomologous chromosomes; see Chapter 12, p. 442) and deletions or deficiencies (the breakage and loss of parts of chromosomes; see Chapter 12, p. 437) involving specific chromosomes and, more importantly, often breakpoints at the same positions on these chromosomes, were repeatedly observed in certain types of cancer cells. The best-known example of this is the so-called "Philadelphia" chromosome, an altered chromosome 22 that has lost a large segment of its long arm. This abnormal chromosome has been found in up to 90 percent of the patients suffering from a specific type of cancer called chronic myelogenous leukemia in various studies.

Now that the chromosomal locations of several human cellular oncogenes and proto-oncogenes have been determined, a striking correlation is evident between their locations and the chromosomal breakpoints of translocations and deficiencies observed in specific types of cancer cells (Fig. 11.34). These results have led to speculation that the observed chromosome breaks and rearrangements

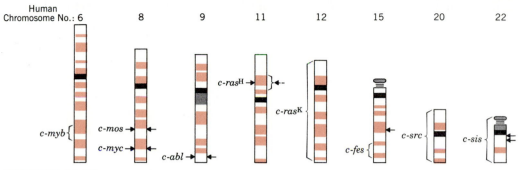

FIGURE 11.34

Chromosomal locations of nine human cellular oncogenes and of breakpoints (arrows to the right of the chromosomes) observed in translocations (solid arrows) and a deficiency (dashed arrow) found associated with specific types of cancer. The symbols for the various cellular oncogenes are derived from the names of the retroviruses that carry the homologous viral oncogenes. Those shown above are homologous to the following viral oncogenes.

Oncogene	Virus
v-myb	*Avian myeloblastosis virus*
v-mos	*Moloney murine sarcoma virus*
v-myc	*Avian myelocytomatosis-29 virus*
v-abl	*Abelson murine leukemia virus*
v-ras^H	*Rat sarcoma virus, Harvey strain*
v-ras^K	*Rat sarcoma virus, Kirsten strain*
v-fes	*Snyder-Theilin feline sarcoma virus*
v-src	*Rous sarcoma virus*
v-sis	*Simian sarcoma virus*

For example, the translocation breakpoints on chromosome 8 at or near the locations of c-mos and c-myc have been found correlated with the occurrence of acute myeloblastic leukemia and Burkitt's lymphoma, respectively. These translocations involve chromosomes 8 and 21 (leukemia) and 8 and 14 (lymphoma). For a summary of the translocations and deficiencies correlated with the other cellular oncogenes and a list of references, see J. D. Rowley, Nature 301:290–291, 1983.

may have caused altered expression of proto-oncogenes or cellular oncogenes.

In the case of mouse and chicken cells, the insertion of "long terminal repeats" (the redundant ends; see Fig. 7.21) of the DNA provirus-forms of RNA tumor viruses next to proto-oncogenes has been indicated in oncogenicity. These retroviral long terminal repeats contain very strong promoters that might cause overproduction of the products of adjacent genes.

Regardless of how oncogenes induce cancers, it now seems quite clear that retroviral oncogenes have evolved from normal cellular genes. All that

needs to occur is for a proto-oncogene's mRNA, produced by splicing the intron sequences out of the primary transcript, to be ligated to the RNA genome of a retrovirus. The viral reverse transcriptase will then convert the mRNA-viral RNA hybrid into homologous DNA for integration into the host genome. What could be of greater value to a virus than to have a new gene that stimulates uncontrolled growth of its host, while its integrated genome goes along for the ride?

Finally, we should emphasize that a large amount of data indicates that the cancerous state is the end-product of a multistep process. The cell lines used in transfection experiments are probably already at some intermediate stage in this pathway, possibly simply due to the selection for the ability to grow under cell culture conditions. The oncogene-induced transformation observed in cell cultures is therefore probably only part of a more complex pathway.

SUMMARY

Gene expression is frequently under the control of **regulator genes.** These regulator genes act **primarily at the level of transcription,** often affecting the ability of **RNA polymerase to bind to promoter sequences.** The effects of regulator gene products are, in turn, controlled by the presence or absence of specific **effector molecules** in the environment.

In prokaryotes, genes with related functions are frequently present in coordinately regulated units called **operons.** Each operon is one unit of transcription; that is, a single mRNA transcript carries the coding sequences for all of the genes in the operon.

Certain regulator genes code for proteins called **repressors,** which function by means of their **sequence-specific binding to DNA.** Some repressors (those for **repressible operons**) bind to DNA only in the presence of specific effector molecules called **co-repressors;** others (those for **inducible operons**) bind to DNA only in the absence of specific effector molecules called **inducers.**

Repressors are **allosteric proteins,** that is, proteins that undergo conformational shifts and correlated changes in activity in response to the binding of specific effector molecules. Repressors bind to their specific DNA sequences in one conformation, but not in the other conformation.

Repressors act by binding to DNA sequences called **operators,** which are located adjacent to the structural genes whose transcription they control. When a repressor, or the complex of repressor and co-repressor, is bound to the operator sequence of an operon, it prevents RNA polymerase from binding to the contiguous promoter sequence and initiating the transcription of the operon.

Operons containing genes that code for enzymes involved in catabolic pathways are also often under **positive control** by the **catabolite activator protein (CAP)** and **cyclic AMP (cAMP).** The binding of the CAP-cAMP complex to a binding site within the promoter region of such an operon is required for the efficient binding of RNA polymerase to the RNA polymerase binding site of the promoter region. High concentrations of glucose result in low intracellular concentrations of cAMP. Thus, such operons cannot be induced in the presence of high concentrations of glucose, a phenomenon known as **catabolite repression.**

Operons controlling enzymes involved in amino acid biosynthetic pathways are frequently controlled by **attenuation.** Attenuation occurs by premature termination of transcription at an **attenuator** site located within the mRNA leader sequence. The attenuator contains a transcription-termination sequence; however, the ability of the attenuator RNA sequence to form the secondary structure that leads to termination of transcription depends on the presence of the amino acid that is the end product of the pathway controlled by the operon in question.

Very little is known about the mechanisms involved in the regulation of gene expression in eukaryotes. In higher eukaryotes, less than 10 percent of the genes present are represented among the mRNA populations in any given cell at any given time. Moreover, mRNA molecules produced from different sets of genes are present in different tissues and different cells. Hormones appear to represent one type of effector molecule that can trigger specific patterns of gene expression in the specific target tissues or target cells of higher eukaryotes. The ecdysone-induced patterns of polytene chromosome "puffing" in *Chironomus* and *Drosophila* indicate that hormones can trigger preprogramed patterns of sequential gene expression.

The existence of **homeotic mutants** in *Drosophila* provides evidence for the role of genetically preprogramed patterns of gene expression in cell determination and differentiation. But what kinds of regulator genes are involved? How do these regulator genes function? How are the patterns of gene expression in the different cells of a multicellular organism coordinated? These and many related questions promise to challenge geneticists for several years to come.

The **genetic information coding for antibody**

chains is stored in **several sets of gene segments,** and the segments are put together in the appropriate sequences by **genome rearrangements occurring during the development of the antibody-producing cells (B lymphocytes).**

Retroviral **oncogenes** code for protein kinases that somehow cause normal cells to undergo **transformation** to the cancerous state. Genes that are homologous to the viral oncogenes are found as normal components of the chromosomes of all higher animals. These genes, called **proto-oncogenes,** can be converted to **cellular oncogenes** by mutation or by becoming associated with new regulatory sequences.

REFERENCES

Baltimore, D. 1981. "Somatic mutation gains its place among the generators of diversity." *Cell* 26:295–296.

Beermann, W. (ed.). 1972. *Developmental studies on giant chromosomes.* Springer-Verlag, New York.

Bishop, J. M. 1981. "Enemies within: the genesis of retrovirus oncogenes." *Cell* 23:5–6.

———. 1982. "Oncogenes." *Sci. Amer.* 246(3):80–92.

Britten, R. J., and E. H. Davidson. 1969. "Gene regulation for higher cells: a theory." *Science* 165:349–357.

Brown, D. D. 1981. "Gene expression in eukaryotes." *Science* 211:667–674.

Cooper, G. M. 1982. "Cellular transforming genes." *Science* 217:801–806.

Davidson, E. H. 1976. *Gene activity in early development.* Academic Press, New York.

Davidson, E. H., and R. J. Britten. 1979. "Regulation of gene expression: possible role of repetitive sequences." *Science* 204:1052–1059.

Davis, M. M., S. K. Kim, and L. Hood. 1980. "Immunoglobulin class switching: developmentally regulated DNA rearrangements during differentiation. *Cell* 22:1–2.

Dickson, R. C., J. Abelson, W. M. Barnes, and W. S. Reznikoff. 1975. "Genetic regulation: the *lac* control region." *Science* 187:27–35.

Early, P., and L. Hood. 1981. "Allelic exclusion and nonproductive immunoglobulin gene rearrangements." *Cell* 24:1–3.

Ehrlich, M., and R. Y.-H. Wang. 1981. "5-Methylcytosine in eukaryotic DNA." *Science* 212:1350–1357.

Freifelder, D. 1983. *Molecular biology, a comprehensive introduction to prokaryotes and eukaryotes.* Science Books International, Boston.

Gilbert, W., and B. Müller-Hill. 1967. "The *lac* operator is DNA." *Proc. Natl. Acad Sci., U.S.* 58:2415–2421.

Groudine, M., and H. Weintraub. 1982. "Propagation of globin DNase I-hypersensitive sites in absence of factors required for induction: a possible mechanism for determination." *Cell* 30:131–139.

Jacob, F., and J. Monod. 1961. "Genetic regulatory mechanisms in the synthesis of proteins." *J. Mol. Biol.* 3:318–356.

Kolter, R., and C. Yanofsky. 1982. "Attenuation in amino acid biosynthetic operons." *Ann. Rev. Genet.* 16:113–134.

Koshland, Jr., D. E. 1973. "Protein shape and biological control." *Sci. Amer.* 229(4):52–64.

Leder, P. 1982. "The genetics of antibody diversity." *Sci. Amer.* 246(5):102–115.

Lewin, B. 1983. *Genes.* John Wiley & Sons, New York.

Ley, T. J., J. DeSimone, N. P. Anagnou, G. H. Keller, R. K. Humphries, P. H. Turner, N. S. Young, P. Heller, and A. W. Nienhuis. 1982. "5-Azacytidine selectively increases γ-globin synthesis in a patient with β^+ thalassemia." *New England J. Medicine* 307:1469–1475.

Lipps, H. J., A. Nordheim, E. M. Lafer, D. Ammermann, B. D. Stollar, and A. Rich. 1983. "Antibodies against Z DNA react with the macronucleus but not the micronucleus of the hypotrichous ciliate *Stylonychia mytilus.*" *Cell* 32:435–441.

Maniatis, T., and M. Ptashne. 1976. "A DNA operator-repressor system." *Sci. Amer.* 234(1):64–76.

McGinnis, W., A. W. Shermoen, J. Heemskerk, and S. K. Beckendorf. 1983. "DNA sequence changes in an upstream DNase I-hypersensitive region are correlated with reduced gene expression." *Proc. Natl. Acad. Sci., U.S.* 80:1063–1067.

Nordheim, A., E. M. Lafer, L. J. Peck, J. C. Wang, B. D. Stollar, and A. Rich. 1982. "Negatively supercoiled plasmids contain left-handed Z-DNA segments as detected by specific antibody binding." *Cell* 31:309–318.

Nordheim, A., M. L. Pardue, E. M. Lafer, A. Möller, B. D. Stollar, and A. Rich. 1981. "Antibodies to left-handed Z-DNA bind to interband regions of *Drosophila* polytene chromosomes." *Nature* 294:417–422.

Nordheim, A., P. Tesser, F. Azorin, Y. H. Kwon, A. Möller, and A. Rich. 1982. "Isolation of *Drosophila* proteins that bind selectively to left-handed Z-DNA." *Proc. Natl. Acad. Sci., U.S.* 79:7729–7733.

O'Malley, B. W., and W. T. Schrader. 1976. "The receptors of steroid hormones." *Sci. Amer.* 234(2):32–43.

Pabo, C. O., and M. Lewis. 1982. "The operator-binding domain of λ repressor: structure and DNA recognition." *Nature* 298:443–447.

Parker, R. C., H. E. Varmus, and J. M. Bishop. 1981. "Cellular homologue (*c-src*) of the transforming gene of Rous sarcoma virus: isolation, mapping, and transcriptional analysis of *c-src* and flanking regions." *Proc. Natl. Acad. Sci., U.S.* 78:5842–5846.

Platt, T. 1981. "Termination of transcription and its regulation in the tryptophan operon of *E. coli*." *Cell* 24:10–23.

Ptashne, M., A. D. Johnson, and C. O. Pabo. 1982. "A genetic switch in a bacterial virus." *Sci. Amer.* 247(5):128–140.

Razin, A., and A. D. Riggs. 1980. "DNA methylation and gene function." *Science* 210:604–610.

Reddy, E. P., R. K. Reynolds, E. Santos, and M. Barbacid. 1982. "A point mutation is responsible for the acquisition of transforming properties by the T24 human bladder carcinoma oncogene." *Nature* 300:149–152.

Stein, G. S., J. S. Stein, and L. J. Kleinsmith. 1975. "Chromosomal proteins and gene regulation." *Sci. Amer.* 232(2):46–57.

Tabin, C. J., S. M. Bradley, C. I. Bargmann, R. A. Weinberg, A. G. Papageorge, E. M. Scolnick, R. Dhar, D. R. Lowy, and E. H. Chang. 1982. "Mechanism of activation of a human oncogene." *Nature* 300:142–149.

Varmus, H. E. 1982. "Form and function of retroviral proviruses." *Science* 216:812–820.

Weinberg, R. A. 1982. "Fewer and fewer oncogenes." *Cell* 30:3–4.

Yanofsky, C. 1981. "Attenuation in the control of expression of bacterial operons." *Nature* 289:751–758.

PROBLEMS AND QUESTIONS

11.1. How can inducible and repressible enzymes of microorganisms be distinguished?

11.2. Distinguish between (a) repression and (b) feedback inhibition caused by the end product of a biosynthetic pathway. How do these two regulatory phenomena complement each other to provide for the efficient regulation of metabolism?

11.3. In the lactose operon of *E. coli* (Fig. 11.3), what is the function of each of the following genes or sites: (a) regulator, (b) operator, (c) promoter, (d) structural gene z, and (e) structural gene y?

11.4. What would be the result of inactivation by mutation of the following genes or sites in the *E. coli* lactose operon: (a) regulator, (b) operator, (c) promoter, (d) structural gene z, and (e) structural gene y?

11.5. Groups of alleles associated with the lactose operon (Fig. 11.3) are as follows (in order of dominance for each allelic series): repressor, i^s (superrepressor), i^+ (inducible), and i^- (constitutive); operator, o^c (constitutive, *cis*-dominant) and o^+ (inducible, *cis*-dominant); structural, z^+ and z^-, y^+ and y^-. (a) Which of the following genotypes will produce β-galactosidase and permease if lactose is present: (1) $i^+\,o^+\,z^+\,y^+$, (2) $i^-\,o^c\,z^+\,y^+$, (3) $i^s\,o^c\,z^+\,y^+$, (4) $i^s\,o^+\,z^+\,y^+$, (5) $i^-\,o^+\,z^+\,y^+$? (b) Which of the above genotypes will produce β-galactosidase and permease if lactose is absent?

11.6. For each of the following partial diploids indicate whether enzyme formation is constitutive or inducible (see Problem 11.5 for dominance relationships):

(a) $\dfrac{i^+\,o^+\,z^+\,y^+}{i^+\,o^+\,z^+\,y^+}$ (b) $\dfrac{i^+\,o^+\,z^+\,y^+}{i^+\,o^c\,z^+\,y^+}$

(c) $\dfrac{i^+\,o^c\,z^+\,y^+}{i^+\,o^c\,z^+\,y^+}$ (d) $\dfrac{i^+\,o^+\,z^+\,y^+}{i^-\,o^+\,z^+\,y^+}$

(e) $\dfrac{i^-\,o^+\,z^+\,y^+}{i^-\,o^+\,z^+\,y^+}$

11.7. Write the partial diploid genotype for a strain that will (a) produce β-galactosidase constitutively and permease inducibly and (b) produce β-galactosidase constitutively but not permease either constitutively or inducibly, even though a y^+ gene is known to be present.

11.8. Constitutive mutations produce elevated enzyme levels at all times; they may be of two types: o^c or i^-. Assume that all other DNA present is wild-type. Outline the way by which the two constitutive mutants can be distinguished with respect to: (a) map position, (b) regulation of enzyme levels in o^c/o^+ versus i^-/i^+ partial diploids, and (c) the position of the structural genes affected by an o^c mutation versus the genes affected by an i^- mutation in a partial diploid.

11.9. How could the tryptophan operon in *S. typhimurium* have developed and been maintained by evolution?

11.10. Of what biological significance is the phenomenon of catabolite repression?

11.11. How might the concentration of glucose in the medium in which an *E. coli* cell is growing regulate the intracellular level of cyclic AMP?

11.12. Is the CAP-cAMP effect on the transcription of the *lac* operon an example of positive or negative regulation? Why?

11.13. Would it be possible to isolate *E. coli* mutants in which the transcription of the *lac* operon is not sensitive to catabolite repression? If so, in what genes might the mutations be located?

11.14. Are operons more common in bacteria or in higher organisms? Why?

11.15. If the operon theory had been established at the time the two classical hypotheses for the mechanics of differentiation were being considered — that is, (1) segregation of nuclear elements and (2) intervention of cytoplasm, which hypothesis would have been most acceptable?

11.16. Using examples, distinguish between regulation based on negative mechanisms and that based on positive regulatory mechanisms.

11.17. Why is the Britten and Davidson model more acceptable than the operon model for explaining regulation in cells of higher animals?

11.18. (a) How do histones regulate gene activity? (b) When in the cell cycle are the histones synthesized? (c) How is the synthesis of the histones themselves regulated?

11.19. How can steroid hormones regulate gene activity?

11.20. Why are salivary chromosomes in the larvae of *Diptera* useful in studying hormonal regulation of gene activity?

11.21. (a) Why are certain organisms such as molds, bacteria, and viruses favorable materials for biochemical genetic studies? (b) What type of experimental material would be most suitable for a study of (1) operons, (2) repressing effects of histones, (3) hormonal control of mRNA synthesis, and (4) chromosome puffs?

11.22. At birth, rabbits of the Himalayan breed are all white, but as they grow older the extremities (paws, nose, ears, and tail) become black. When the white fur is shaved from the spot on the body of the adult and the rabbit is kept in a cool place, the new hair that grows in the shaved spot is black. The temperature of the body proper is about 33°C, but in the extremities it is about 27°C. (a) How may genetic and environmental factors be involved? (b) Formulate an explanation for the difference in pigmentation in different areas of body surface.

11.23. If all cells in a given organism carry the same genes, how can gene expressions that are localized in time and space be explained?

11.24. How are (a) differentiation, (b) organization, and (c) growth involved in the development of an animal?

11.25. The father of two albino children has made widespread inquiries among geneticists and physicians concerning a possible cure for albinism. The steps in pigment production have been elucidated and it seems feasible to him that something might be added to the diet or given by injection that would supply the missing step or steps in melanin production in his children. Evaluate the possibility of such a development.

11.26. Ordinarily, *Drosophila* eye-disc transplants placed in hosts with different genotypes develop according to the genotype of the transplant. For example, transplants from larvae with the genotype for white eye develop white eyes in wild-type hosts, and transplants from wild-type larvae develop wild-type red eyes in hosts with the genotype for white. Beadle and Ephrussi performed transplantation experiments on larvae with vermilion (v) and cinnabar (cn) eyes. The phenotypes for these two mutants are similar. They have bright red color because they lack the brown pigment that is a part of the wild-type red eye. (1) When discs from wild-type larvae were transplanted into vermilion or cinnabar hosts they developed into wild-type, and when v or cn transplants were placed in wild-type hosts they also developed into wild-type. (2) When discs from cn larvae were placed in v hosts, no brown pigment was formed, and the eyes were bright red. From the reciprocal transplant — that is, v in cn hosts — brown pigment was produced and the eyes were wild-type. Formulate an explanation for these results.

11.27. Why are the homeotic mutants of *Drosophila* particularly interesting?

11.28. How can one determine whether a particular gene is being transcribed in the cells of two different tissues or two different cell cultures derived from the same organism?

11.29. A deletion of the regulator (i) gene of the *lac* operon in *E. coli* would be expected to result in the constitutive synthesis of the *lac* operon enzymes. What would the expected phenotype be for a mutant strain of *E. coli* with a deletion of the regulator gene of the *ara* operon? Why?

11.30. Is the genetic information specifying antibody chains stored in germ line cells in the same format as that specifying most other polypeptides?

11.31. How many polypeptide chains are present in each antibody molecule? How many antigen-binding sites are present per antibody? How many different antibodies are produced in each mature B lymphocyte?

11.32. What are three different sources of antibody variability?

11.33. Does class switching during B lymphocyte differentiation occur at the DNA level or the RNA level? By what mechanism does it occur?

11.34. What evidence is there that the proto-oncogenes found in chromosomes of normal cells and normal animals are not simply integrated retroviral oncogenes?

TWELVE

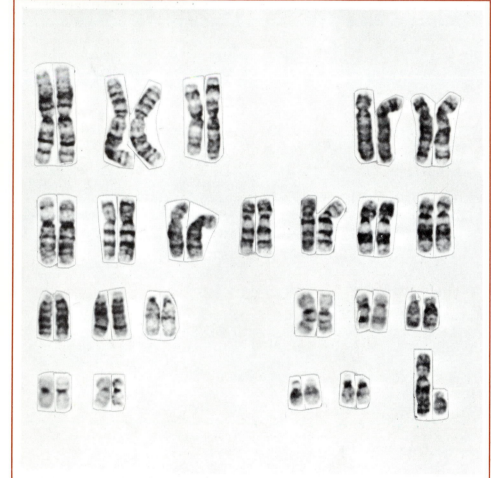

Karyotype of a normal male child. Chromosomes were prepared with G-staining technique. The horizontal bands on the chromosomes afford longitudinal comparisons and thus identify members of pairs. All 23 pairs of human chromosomes in this karyotype can be distinguished. (Courtesy of A. D. Stock.)

CHROMOSOME ABERRATIONS

Structural changes require breaks in the chromosomes. More than one break can occur in a single chromosome or set of chromosomes, and the broken parts may then reunite in new arrangements. Any **broken end may unite with any other broken end,** thus potentially resulting in **new linkage** arrangements. The loss or addition of a chromosome segment may also occur in the process. More than one type of aberration may occur at the same time. For example, a section may be broken off and lost during the formation of an inversion or translocation; this simultaneously produces a deficiency.

Identifying chromosomal aberrations by observation is a major problem for the cytogeneticist because members of a pair usually lack visible markers for identifying different areas along their length. Primary and secondary constrictions and total length are useful, but more markers are needed. Maize chromosomes are exceptions. They are favored materials for microscopic study because meiotic prophase chromosomes have deep-staining bodies, called **heteropyknotic knobs** (Fig. 12.1), distributed along their length. With the aid of these visible markers, many chromosome changes have been detected in maize. The acetocarmine smear method, first applied to other plant materials by J. Belling in 1931, greatly facilitated studies of the maize chromosomes. This technique permits whole chromosomes to be fixed, stained, and spread on a microscope slide in one operation. It provides a way to compare individual chromosomes within a chromosome set or genome, as well as whole sets from different organisms.

The first cytological demonstration of plant chromosomal rearrangements was made in maize by B. McClintock. Working with pachytene and other meiotic prophase stages that present large chromosomes for microscopic observation, she eventually demonstrated that irregular configurations made by chromosome rearrangements in the pairing process led to four different kinds of structural changes: (1) **deficiencies** (parts of chromosomes lost or deleted), (2) **duplications** (parts added or duplicated), (3) **inversions** (sections detached and reunited in reverse order), and (4) **translocations** (parts of chromosomes detached and joined to nonhomologous chromosomes). Comparable demonstrations were later made with the giant polytene chromosomes of *Drosophila*.

GIANT POLYTENE CHROMOSOMES IN DIPTERA

Large coiled bodies about 150–200 times as large as gonad-cell chromosomes were observed in the nuclei of glandular tissues of dipterous larvae as

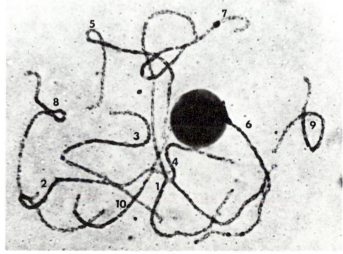

FIGURE 12.1

The ten pairs of chromosomes in maize, shown in pachytene of prophase I of meiosis. The large nucleolus is associated with the nucleolar organizer region near the end of chromosome 6. The positions of the centromeres and some of the more conspicuous knobs are used to identify specific chromosome pairs (bivalents). The 10 bivalents are numbered 1–10; these correspond with the 10 linkage groups of maize shown in Fig. 6.8. (From R. L. Phillips, "Molecular Cytogenetics of the Nucleolus Organizer Region," pp. 711–741, in Maize Breeding and Genetics, *D. B. Walden, ed. Copyright © 1978 by John Wiley & Sons. Original photograph courtesy of R. L. Phillips.)*

early as 1881 by E. G. Balbiani. He described banded structures in the nuclei of cells of larval midges in the genus *Chironomus* but did not attach any genetic significance to the observation. D. Kostoff (in 1930) suggested a relation between the bands of these structures and the linear sequence known to occur among genes. The anatomical significance of the nuclear bodies was further studied by E. Heitz and H. Bauer in the genus *Bibio,* a group of March flies whose larvae feed on the roots of grasses. These authors identified the bodies as **giant chromosomes** that occurred **in pairs.** They described the morphology in detail and discovered the relation between the salivary gland chromosomes and other somatic and germ cell chromosomes. They also demonstrated that comparable elements occurred in the giant chromosomes and in the chromosomes of other cells of the same organism.

It was largely because of the work of T. S. Painter that *Drosophila* salivary gland chromosomes (Fig. 12.2) were first used for the cytological verification of genetic data. Painter related the **bands** on the giant chromosomes to genes, but he was more interested in the morphology of the chromosomes and implications concerning speciation than in associating chromosome sections with particular genes. About 25,000 base pairs are now estimated for each band. C. B. Bridges made extensive and detailed studies of the salivary gland chromosomes and, in the course of his investigations, developed a tool of practical usefulness in **relating genes to chromosomes.** In applying this method to *Drosophila melanogaster,* he prepared a series of cytological (physical) maps to correspond with the genetic linkage maps already available.

FEATURES OF POLYTENE CHROMOSOMES

The unusual size of the salivary gland chromosomes is a product of the type of growth that occurs in larval glandular tissues of dipterous insects. Salivaries and other glands grow by enlargement rather than by duplication of individual cells. This can be demonstrated by cell counts and measurements taken at different stages in the development of a larva. As a larva develops, the synapsed *Drosophila* chromosomes replicate (usually nine times), giving rise to a polytene strand of about 1000 DNA double helix molecules. The cross bands are chromomeres, densely packed with chromatin, and the interband regions are chromonemata. Giant chromosomes are thus **cablelike structures** with many DNA strands (polytene). Giant polytene chromosomes thus correspond in linear structure with other chromosomes of the same species. The difference is that the **duplicate strands** (chromatids) are held together in bundles, and they do not separate out to new cells through cell division.

Another feature that makes the giant salivary chromosomes valuable for study is their continuous state of **somatic synapsis.** If one member of a homologous pair is altered by deficiency, duplication, inversion, or translocation, an irregularity occurs in pairing. Characteristic and observable irregularities make it possible to recognize different kinds of chromosome modifications and to identify their location on the chromosome.

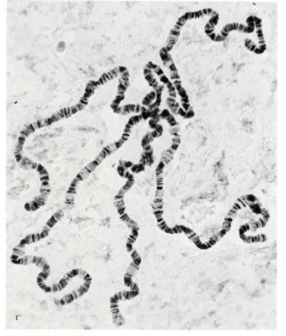

FIGURE 12.2
Salivary gland chromosomes of Drosophila melanogaster. *(Courtesy of B. P. Kaufmann, reprinted with permission from* The Journal of Heredity.)

About 5000 single cross bands have been noted on the four pairs of salivary gland chromosomes in *D. melanogaster.* This number was considered by H. J. Muller to be a minimum approximation of the number of genes in that insect. B. Judd demonstrated by studies of deficiencies that each band corresponds with a unit of genetic function (complementation unit; see Chapter 10). Genes have thus been associated with individual bands.

Linkage data do not correspond exactly with cytological locations, but the linear sequence of genes can be verified from salivary preparations. Not only can the linear sequence be verified, but it agrees generally with the linear sequence determined by cytological mapping. The lack of identical correspondence lies in the relative distances between genes determined by the two methods. Frequency of crossing over is not constant along the length of the chromosome, but is less near the centromere. The main uses of polytene chromosomes are in (1) **locating genes** and (2) **identifying structural changes** in chromosomes.

DEFICIENCIES

A single break near the end of a chromosome would be expected to result in a terminal deficiency. If two breaks occur, a section may be deleted and an **intercalary deficiency** created. Terminal deficiencies might seem less complicated and more likely to occur than those involving two breaks. Instead, the great majority of deficiencies detected thus far are of the intercalary type within the chromosome. The ends of chromosomes have folded-over "hairpins" protecting them from breaks and rejoining. When a deficiency occurs, the chromosome set is left without having the genes carried in the deleted portion, unless the deleted part becomes fused to a chromosome that has a centromere. Without a centromere, a chromosome section cannot move to the pole of the spindle during cell division, but lags in the dividing cell and is excluded from the chromosome group when the nuclear membrane forms around the chromosomes of a daughter cell.

When an intercalary part of a chromosome is missing, a buckling effect may be observed microscopically in the paired salivary gland chromosomes (Fig. 12.3). Large deficiencies are more readily detected than small ones, but with good optical equipment and patience, an investigator may be able to see single bands of the salivary preparations and thus to identify minute heterozygous deficiencies. By identifying the part of the polytene chromosome in which the buckle occurs and then studying the phenotype of flies carrying a recessive gene in the homologous chromosome opposite the deficiency, the gene can be spatially positioned on the chromosome. Expression of a single recessive opposite a deficiency is called pseudodominance. Chromosome deficiencies have greatly facilitated the **checking of linkage maps.** The physical location of many genes is now precisely known in *D. melanogaster* and other species of Diptera because of the effective use of this technique.

A somatic cell that has lost a small chromosome segment may live and produce other cells heterozygous like itself, each with the deleted section of a chromosome. Phenotypic effects sometimes indicate which cells or portions of the body have descended from the originally deficient cell. If the deficient cell is a gamete that is subsequently fertilized by a gamete carrying a nondeficient homolog, all cells of the resulting organism will carry the deficiency in the heterozygous condition. Recessive genes on the nondeficient chromosome in the region of the deficiency may express themselves (pseudodominance). Heterozygous deficiencies usu-

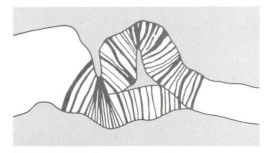

FIGURE 12.3
Appearance in the Drosophila *salivary gland chromosomes of an intercalary deficiency showing a buckling effect. A section of about 45 bands is missing in one X chromosome. This deficiency is associated with a notched phenotype at the tips of the wings. (After H. Slizynska,* Genetics *23:291–299, 1938.)*

ally decrease the general viability of the flies that carry them. Flies carrying deficiencies in the homozygous condition usually die. Some very small homozygous deficiencies, however, are viable in *Drosophila*. Such deficiencies have been identified in the region in which the *w* (white eye) and *fa* (facet eye) genes are located near the end of the X chromosome.

DUPLICATIONS

Duplications represent additions of chromosome parts. Some chromosome segments behave as dominants and some as recessives with respect to certain phenotypes. Others show intermediate inheritance and still others have cumulative effects. Duplications provide a means for determining effects of chromosome segments when three, four, or **more similar sections are present** in individual animals or plants.

The first duplication to be critically examined involved the *B* (bar) locus in the X chromosome of *Drosophila*. The eye of a heterozygous *B/B*⁺ female is somewhat smaller than the normal eye (Fig. 12.4), and the sides are straighter, giving an oblong or bar appearance. In the homozygous or hemizygous condition, the eye is considerably smaller. Bridges and

Muller discovered independently that the bar phenotype was the result of a duplication involving a part of the X chromosome already present in the wild-type flies. Both investigators observed not only the effect of a duplication that produced bar eye, but also a duplication that resulted in an extreme decrease in the size of the eye, which was called "double bar." By examining the salivary polytene chromosome, they identified the segments of the chromosome actually involved in the duplication. The different phenotypes and the corresponding segments of the salivary gland chromosome pairs are illustrated in Fig. 12.4. Polytene chromosome section 16*A* is present in flies with wild-type eyes. When this section is duplicated, it produces the bar phenotype; but when it is represented three times in a single chromosome, the double-bar phenotype results.

Each additional duplicate segment of section 16*A* makes the eye smaller. Other duplications that have since been found in *Drosophila* work in the opposite direction. These suppress the effects of mutant genes and make the fly appear more nearly normal with respect to certain traits. Further studies demonstrated that duplications need not occur in the immediate vicinity of the section duplicated to exert an influence. Chromosome fragments may

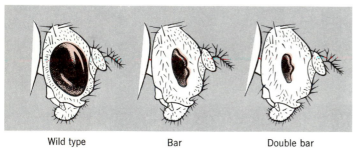

Wild type   Bar   Double bar

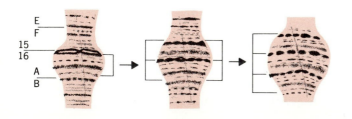

FIGURE 12.4

Effects on eye size of different arrangements of duplicated sections of 16A in the D. melanogaster *X chromosome. The 16A identifies a particular segment on the salivary chromosome following the standard numbering system for* D. melanogaster. *Positions of 16A segment on a particular chromosome are related to size of eye. (After C. B. Bridges,* Science *83:210–211, 1936.)*

become attached to entirely different chromosomes. Through the assortment of such chromosomes in gametes, duplications may be carried to succeeding generations.

INVERSIONS

Inversions occur when parts of chromosomes become detached, turn through 180°, and are reinserted in such a way that the genes are in **reversed order.** Some inversions presumably result from entanglements of the threads during the meiotic prophase and from the chromosome breaks that occur at that time. For example, a certain segment may be broken in two places, and the two breaks may be in close proximity because of a chance loop in the chromosome. When they rejoin, the wrong ends may become connected. The part on one side of the loop connects with a broken end different from the one with which it was formerly connected. This leaves the other two broken ends to become attached. The part within the loop thus becomes turned around or inverted. It is not known whether all inversions occur in this way, but this is a plausible explanation for many chromosome inversions.

Inversions may survive the meiotic process and segregate into viable gametes. As indicated in earlier chapters, chromosome pairing is essential to the production of fertile gametes. The mechanism by which homologous chromosomes heterozygous for inversions accomplish such pairing in the meiotic sequence is remarkable. The part of the uninverted chromosome corresponding to the inversion forms a **loop.** A similar loop is formed by the inverted sections of the homologous chromosome, but in reverse direction. If, for example, the loop of the uninverted section is formed with the gene sequence in a clockwise direction, the inverted part will form in a counterclockwise direction. In this way, corresponding parts come together even though one of the sections is inverted, as illustrated diagrammatically in Fig. 12.5.

Inversions have been associated with the **suppression of crossing over.** Before *Drosophila* chromosomes were studied extensively, investigators had already identified genetic crossover suppressors in this organism. These were first considered to be genes that somehow interfered with crossing over.

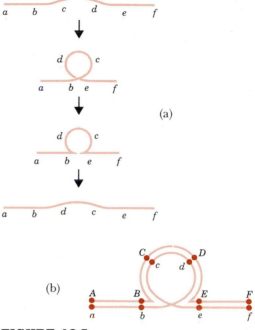

FIGURE 12.5

(a) *Diagram illustrating the mechanism by which some inversions may occur.* (b) *Pairing of an inverted homolog has occurred with alleles* c *and* d *in reverse position. Only two of the four strands are shown. A loop may be observed in salivary gland preparations of* Drosophila *that carry heterozygous inversions.*

It was shown that the locations of inversions and crossover suppressors coincided and that the apparent suppression of crossing over was associated directly with inversions. The main process is not a suppression of crossing over, although the frequency of physical crossing over may be reduced. Crossover gametes that do occur are not recovered; that is, the zygotes die before they can be detected.

The relation between inversions and crossing over can be demonstrated from studies of heterozygous inversions. Heterozygous paracentric inversions (Fig. 12.6), those with the centromere outside the inverted area, and heterozygous pericentric inversions (Fig. 12.7), those with the centromere inside the inverted region, can be used to illustrate the reason that crossover gametes lead to inviability.

Chromosomes that carry **paracentric** inversions cannot cross over within the loop without producing fragments of chromosomes lacking cen-

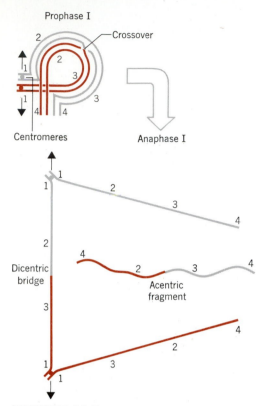

FIGURE 12.6

Meiotic prophase and anaphase, illustrating the mechanism through which a paracentric inversion acts as a crossover "suppressor." Dicentric (1 2 3 1) and acentric (4 3 2 4) chromosomes, which are also unbalanced, are formed by crossover chromatids. Crossover "suppression" results from the unbalanced chromatids and the inability of the chromatids to reach the poles of the cell in division.

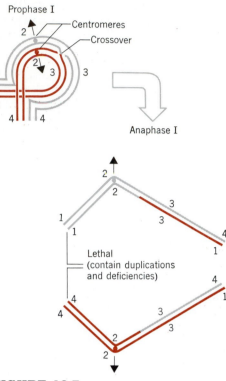

FIGURE 12.7

Diagram illustrating the crossover "suppressor" action of a pericentric inversion. The resulting chromatids carry both duplications and deficiencies and therefore lead to lethality. Crossover "suppression" results from the loss of the gametes containing chromosomes involved in crossovers.

tromeres (**acentric**) and chromosome complexes with two centromeres (**dicentric**). These result either in fragments that lag in the center of the spindle or chromatid bridges that tie together the two homologs involved and interfere with the division process. In either case, the chromosomes do not separate properly to their respective poles. Crossovers within the loops of **pericentric** inversions result in **duplications and deficiencies.** The fate of these cells varies in animals and plants. Following a single crossover within the inversion loop of a maturing plant cell, the gametes receiving crossover chromatids are inviable; thus, crossing over is effectively suppressed. In *Drosophila,* unbalanced zygotes

produced by abnormal gametes die, thus eliminating crossover chromatids from the population. The apparent suppressing effect of inversions on crossing over can thus be explained mostly on the basis of secondary results that follow crossing over in inverted segments.

Polytene chromosomes of *Drosophila* have been especially useful for detecting heterozygous inversions in the flies. A characteristic loop in these chromosomes is illustrated in Fig. 12.8. Loops in giant chromosomes of salivary gland tissues presumably resemble those in meiotic prophase chromosomes, where observation is much more difficult.

Chromosomes with inversions have practical applications in maintaining *Drosophila* stocks. They are used as "balancers," that is, as chromosomes that can be placed opposite homologous chromo-

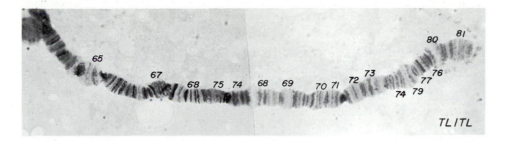

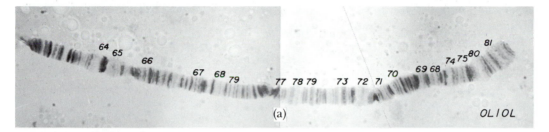

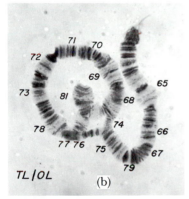

FIGURE 12.8

Salivary gland chromosomes illustrating an inversion. (a) Homozygous chromosomes TL/TL *and* OL/OL. *(b) Heterozygote,* TL/OL, *with inversion loop. (C. D. Kastritsis and D. W. Crumpaker. Reprinted with permission from* The Journal of Heredity.*)*

somes carrying certain genes that are homozygous lethal. Crossing over is suppressed in such chromosomes, and it is possible to maintain a gene heterozygous that would die in the homozygous condition. Laboratory stocks carrying several mutants are kept intact without crossing over. Because recessive genes are not expressed when they are in heterozygous condition, however, frequent checks must be made to ensure that the gene is not lost from the stock. Appropriate outcrosses are conducted occasionally to check for the continued presence of the gene or the genes in the stock. A mechanism through which some laboratory stocks are kept balanced and heterozygous involves "balanced lethal mutations."

INVERSIONS AND BALANCED LETHAL MUTATIONS

Sometimes through the process of mutation, two recessive lethals occur at different loci on each member of the same pair of homologous chromosomes. If the loci involved are near each other, or if chromosomal aberrations "suppress" crossing over between them, an **enforced heterozygous** condition may be established. Since individuals homozygous for any lethal die, only those heterozygous for both lethals survive. The first documented case of a balanced lethal was that described by Muller in *Drosophila*. He had been maintaining in his laboratory the *Bd* (beaded) stock, which produced flies with scalloped wings. Since the beaded phenotype was controlled by a dominant homozygous lethal gene (*Bd*), the progeny from the crosses between beaded flies were two-thirds beaded and one-third normal, as illustrated in Fig. 12.9*a*. All those homozygous for beaded were dead; thus, a **2:1 ratio** replaced the 3:1 ratio, which would otherwise have been expected. This pattern is similar to that described in Chapter 2 for creeper chickens.

Abruptly, with no visible change in the beaded phenotype, all the flies in the culture began to appear beaded. In an attempt to explain the change that had occurred, Muller postulated that a new lethal (ℓ) had been created by mutation in the homologous chromosome opposite *Bd*. An inversion that was associated with the gene for beaded "suppressed" crossing over, and only two kinds of gametes were produced by the beaded flies—one carrying the gene (*Bd*) for beaded, the other carrying the new lethal (ℓ). The cross representing the new arrangement is illustrated in Fig. 12.9*b*. When these chromosomes segregated and fertilization occurred, some zygotes became homozygous for *Bd* and were therefore lethal; others became homozygous for ℓ and they also died. Only those heterozygous for both *Bd* and ℓ survived. Therefore, the heterozygous condition was enforced. This explanation was substantiated, and the term **"balanced lethal"** was used to designate such an arrangement involving two lethals in the same chromosome pair in which recovered crossovers are infrequent or entirely absent.

TRANSLOCATIONS

Sometimes a part of a chromosome becomes detached and **joins** a part of a **nonhomologous chromosome,** thus producing a translocation. In heteromorphic sex chromosomes (homologous chromosomes that differ morphologically), exchanges may occur between nonhomologous parts of the same chromosome pair. Reciprocal translocations occur when parts of chromosomes belonging to members of two different pairs become exchanged. Part of chromosome 1, for example, may be detached and attached to chromosome 2, whereas a section of chromosome 2 becomes attached to chromosome 1. Reciprocal translocations

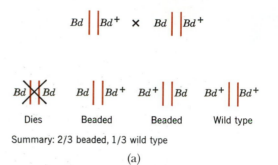

Summary: 2/3 beaded, 1/3 wild type

(a)

FIGURE 12.9a

A cross between two Bd *flies. Since the gene* Bd *is homozygous lethal, all surviving beaded flies are heterozygous and one-fourth of the progeny from beaded parents die before hatching. One-fourth of the total and one-third of the flies that survive are wild-type (not beaded).*

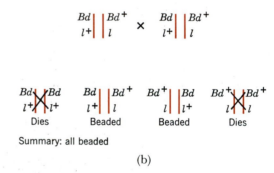

Summary: all beaded

(b)

FIGURE 12.9b

A balanced lethal cross. A cross between two beaded flies after a new lethal mutation (ℓ) had occurred on the homolog of the Bd. *All progeny homozygous for* Bd *and those homozygous for ℓ died. All the survivors expressed the beaded phenotype.*

have been described in a number of plants and are important factors in the evolution of certain plant groups such as *Datura* and *Oenothera*. Translocations may not involve a loss or an addition of chromosome material, but frequently they do become associated with deficiencies and duplications, unbalanced combinations of genetic units.

Translocations can be detected from genetic data by noting altered linkage arrangements brought about by exchanges of parts between different chromosomes. Consider, for example, two nonhomologous autosomes number 2 and number 3 in one organism. In number 2, the sequence of

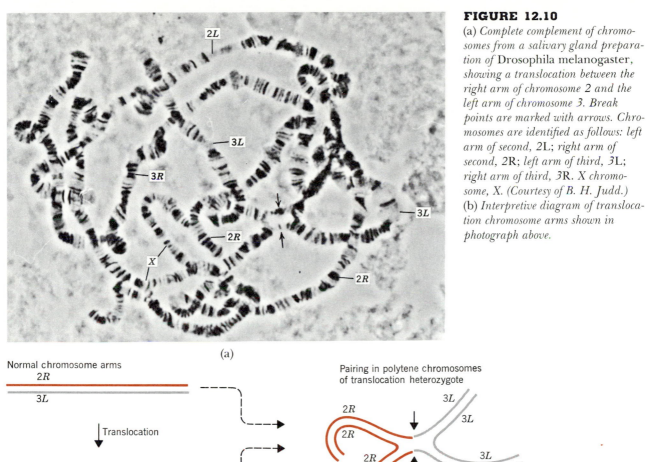

(a)

FIGURE 12.10

(a) *Complete complement of chromosomes from a salivary gland preparation of* Drosophila melanogaster, *showing a translocation between the right arm of chromosome 2 and the left arm of chromosome 3. Break points are marked with arrows. Chromosomes are identified as follows: left arm of second,* 2L; *right arm of second,* 2R; *left arm of third,* 3L; *right arm of third,* 3R. X *chromosome,* X. *(Courtesy of B. H. Judd.)* (b) *Interpretive diagram of translocation chromosome arms shown in photograph above.*

(b)

genes is: *a, b, c, d, e, f, g, h, i,* and the sequence on number 3 is: *r, s, t, u, v, w, x, y, z.* Genes *a–i* are in linkage group 2, and genes *r–z* are in linkage group 3. If appropriate testcrosses should show linkage between *d* and *u,* a translocation between chromosome 2 and 3 would be suggested.

Cytological evidence for translocations can be obtained from microscopic studies of polytene chromosomes in heterozygous *Drosophila* and meiotic prophase stages in plant materials. A characteristic cytological pattern for translocations is marked by arrows in Fig. 12.10. It represents the place

where the right arm of one chromosome 2 homolog has joined the left arm of one chromosome 3 homolog. At this point, the translocation homolog (2 to 3) changes direction, and each part of the homolog is paired with the corresponding part from its homolog. On the other side of the cross, the translocation homolog (3 to 2) changes direction and is paired with its homolog. The diagram with Fig. 12.10 outlines the changes among homologs.

Characteristics of translocations in the germ cells of barley crosses and a method of detecting them are illustrated further from an actual study on barley (Fig. 12.11). The alleles (r^+ and r) for rough and smooth awns and those (s^+ and s) for long and short rachilla hairs are in linkage group 7. Alleles (n^+ and n) for hulled and hull-less and those (l^+ and l) for lax and club head are in group 1. When the progeny of a certain plant homozygous for all four genes (r, s, n, and l) were classified, the results indicated that s was linked with n. This genetic evidence suggested that a translocation joining chromosomes 1 and 7 had occurred. When the plants were observed in the field and studied in the laboratory, at least 30 percent of the pollen was found to be sterile. Further investigations showed that many ovules also were sterile. Cytological studies of pollen mother cells (Fig. 12.12) showed one quadripartite ring and five bivalents instead of the seven bivalents usually observed in barley preparations. This cytological evidence demonstrated that a

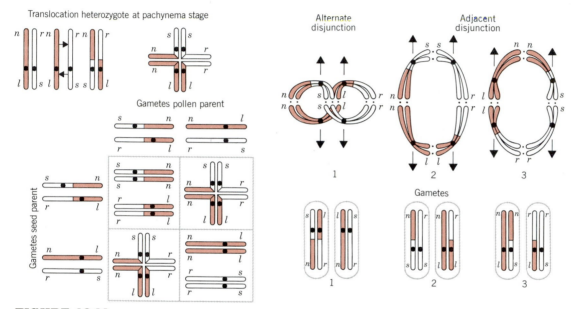

FIGURE 12.11

Explanation to account for consequences of reciprocal translocation in barley germ cells. A translocation had occurred between chromosomes 1 and 7, bringing the genes n *and* s *into the same linkage group. The cross configuration at the top of the diagram represents the pairing of the translocated chromosomes (indicated as single rather than duplicated, for simplicity. Chiasmata necessary to hold homologous chromosomes at this stage are not shown.) Two viable male and two viable female gametes of the selfed plant are represented at the top and left, respectively, of the Punnett square. In the four squares, some zygotes from the selfed plant are illustrated with the chromosomes as they would appear following synapsis. The two zygotes that have crosslike configurations are heterozygous for the translocation and produce plants that are partially sterile — that is, semisteriles — because of the deficiencies and duplications of chromosome parts occurring in the gametes. Meiotic prophase configuration of the heterozygous translocation is shown at left in pachynema: (1) alternating disjunction resulting in balanced viable gametes, (2) adjacent disjunction resulting in deficiency-duplication, inviable gametes, (3) adjacent disjunction resulting in deficiency-duplication, inviable gametes. Arrows at the centromeres indicate the directions of movement of chromosomes in anaphase. (After W. S. Boyle and A. H. Holmgren,* Genetics 40:539–545, 1955.)*

FIGURE 12.12

Photograph of meiotic chromosomes in barley, showing a ring of four chromosomes (at left) in a pollen mother cell that resulted from a reciprocal translocation. The four chromosomes are similar to that for adjacent disjunction shown in Fig. 12.11. The quadrivalent ring in the photograph has been broken and opened in the process of preparation. Along with the quadrivalent ring, five bivalents are shown, making a total of 14 barley chromosomes (4 + 2 + 2 + 2 + 2 + 2).

reciprocal translocation had occurred. The explanation for the **altered linkage** grouping and the mechanism of gamete and zygote formation that involves chromosomes with translocations are illustrated in Fig. 12.11.

A consequence of heterozygous translocation is **semisterility** in pollen or ovules. Some gametes from a plant carrying a heterozygous translocation will be unbalanced and inviable. The imbalance causing the sterility occurs when chromosomes separate to the poles in meiosis. Chance combinations of translocation chromatids leave some gametes deficient in chromosome parts, and some have duplications. Viable and inviable gametes from a plant carrying a translocation heterozygote are shown in Fig. 12.11. Changes in linkage groups were demonstrated by these experiments to be direct consequences of exchanges between nonhomologous chromosomes. Semisterility in barley was shown to result from the irregularities in segregation of heterozygous translocations that developed as deficiencies and duplications.

POSITION EFFECTS

When a chromosome rearrangement involves no change in the amount of genetic material, but only

in the order of genes, the term **position effect** is used to describe any associated phenotypic alteration. Along with gene mutations, position effects represent a source of genetic variation. The extent to which chromosome rearrangements such as **inversions** and **translocations** are associated with new phenotypic variation is, however, open to question. In this regard, it must be remembered that chromosome modifications, particularly inversions, curtail recombination and thus would be expected generally to restrict genetic variation.

Nevertheless, several well-established position effects are on record. The first example, from the studies of Sturtevant and Bridges on the bar-eye duplication in *Drosophila* (Fig. 12.13), also demonstrated a "dosage" effect. These investigators found a relation between the number of chromosome sections (16A) present and the number of facets in the eye. Further critical experiments showed, however, that it is not a strictly proportional relation. The arrangement of the chromosome segments with respect to each other, as well as their presence or absence, influences the size of the eye. The effect of different arrangements was demonstrated by manipulating the chromosomes through appropriate matings and counting the facets in the eyes of the female offspring.

When section 16A was duplicated and the extra segment occurred in homozygous conditions with a total of four segments (Fig. 12.13C), the number of facets in the eyes averaged 68. But when three 16A sections were side by side in one homolog and one section in the other homolog (Fig. 12.13D), the eyes averaged 45 facets. Since the same number of 16A units is present in the eyes represented by Figs. 12.13C and D, the difference depends on the arrangement or position of the genes with respect to each other. This phenomenon was interpreted as a position effect.

Several well-established and many possible position effects have now been described in *Drosophila*. E. B. Lewis has shown that position effects fall into two classes: (1) **stable** and (2) **variegated.** Stable position effects are uniform phenotypic effects that result from changes of specific segments of chromosomes. The bar-eye position effect is an example.

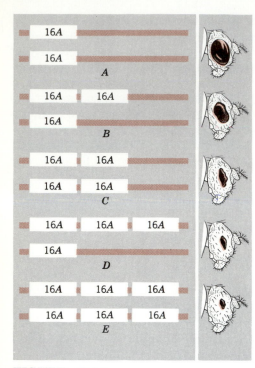

FIGURE 12.13

*Effects on eye size of different arrangements of duplicated sections of 16*A *in the* D. melanogaster X *chromosome. The 16*A *identifies a particular segment on the salivary chromosome following the standard numbering system for* D. melanogaster. *Positions of 16*A *segment on a particular chromosome are related to size of eye. (A) Wild-type female (wild-type male would have same phenotype); (B) bar female, heterozygous; (C) bar female, homozygous (hemizygous bar male would have same phenotype); (D) double-bar female, heterozygous, showing position effect as compared with (C); (E) double-bar homozygous. (After A. H. Sturtevant,* Genetics *10:117–147, 1925.)*

Variegated position effects result in a mosaic pattern of a trait usually evidenced in a particular structure or area of the body. Specks of different colors, for example, may occur in the eyes of *Drosophila* following rearrangements of the *w* (white eye) locus. Inversions or translocations that place *w*+ in heterochromatin may cause white variegation or mosaicism for eye color.

Variegated position effect in *Drosophila* is associated with chromosome structural changes. Action of a gene is depressed when the gene is transferred to a **heterochromatin** region. A position effect usually involves two chromosome breaks, one in the heterochromatin region and one near a euchromatic gene. Regulation of this gene activity occurs at the chromosome rather than the cell level and presumably results from suppression of gene transcription. **Variegated transcription** of linear DNA could result in variegated expression or a position effect. Experiments of H. J. Becker on *Drosophila* eye development showed that mutational disturbances in areas of the chromosomes resulted in the same phenotypes as position effects caused by chromosome structural changes. With X rays he induced mitotic exchanges of eye-color genes at time intervals during development. Because each affected sector was made up of the cell progeny of a single cell carrying an exchange product, it was possible to trace the developmental sequence of the tissue composing the eye. Becker then induced structural changes in chromosomes and found that alterations in the position of genes, with reference to heterochromatin, created effects that appeared the same phenotypically as those of X-ray-induced disturbances.

CHROMOSOME ABERRATIONS AND EVOLUTION

A consequence of chromosomal structural aberration in populations is related to evolutionary change including speciation. Chromosomal aberrations are associated with position effects that may be significant in natural selection. More important for evolution is the genetic isolation that is mostly caused by inversions and translocations. Speciation in the *Drosophila* group of dipterous insects, for example, has been related to chromosome inversions. These structural changes occur in chromosomes of individual flies and are carried homozygous in populations. Populations have developed over periods of time with different chromosome inversions. Each may be isolated because matings of flies from a particular population with those of another population carrying a different inversion result in sterile or inviable hybrids. This strengthens the boundaries around a particular population and prevents the exchange of genes between related populations. Speciation in *Drosophila* has been associated with a series of different inversions that occurred by chance in breeding populations and were eventually recognized in different taxonomic groups. Translocations have been shown to occur in certain plant groups and to cause genetic isolation, thus promoting evolutionary stability in populations.

HUMAN CHROMOSOME TECHNIQUES

The 22 pairs of human autosomes were first classified according to length and position of the centro-

mere into **seven groups,** identified with letters A–G. All autosomes could be placed satisfactorily with a group, but the numbering within the groups was more or less tentative (Fig. 12.14) until 1970.

Chromosome **banding techniques** along with new methods of identification finally distinguished all 46 human chromosomes. Bands are defined as parts of chromosomes that appear lighter or darker than adjacent regions when treated with particular staining methods. Q-staining methods employ quinacrine compounds and produce fluorescent Q-bands along the chromosomes. G (Giemsa)-staining methods (Fig. 12.15) result in G-bands and, with some Giemsa techniques, the R (reverse)-staining

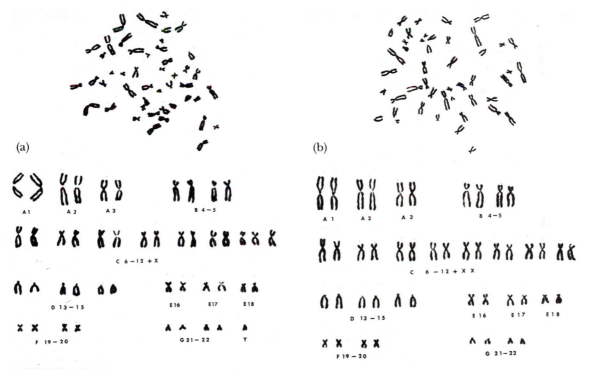

FIGURE 12.14

Human metaphase chromosomes as they appear in microscope preparations and as they are classified (karyotyped) in seven major groups according to criteria established at the Denver (1960), London (1963), and Chicago (1966) conferences. (a) Metaphase and karyotype of a normal male cell, with both X and Y chromosomes present. (b) Metaphase and karyotype of a normal female cell, with two X chromosomes placed among the C group chromosomes. (Courtesy of A. D. Bloom, Division of Genetics, College of Physicians and Surgeons of Columbia University.)

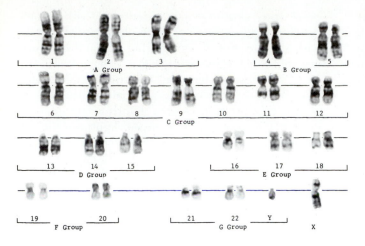

FIGURE 12.15

Karyotype of a normal male child. Chromosomes were prepared with G-staining technique. Bands along the chromosomes identify members of pairs. All 23 pairs of chromosomes can be distinguished by Giemsa-staining methods. (Courtesy of R. M. Fineman, Department of Pediatrics, University of Utah.)

methods result in R-bands. The centric region and other heterochromatin areas are stained by C-banding (Fig. 12.16). Refinements in staining techniques now permit all human chromosomes to be identified (Fig. 12.17). The amount of DNA packed into one G-band is on the order of $10^6 - 10^7$ base pairs of DNA, and clearly is not a gene.

A conference (in 1971) on the standardization of human cytogenetics was held in Paris. Conferees agreed that to describe extra autosomes, the num-

ber of the extra chromosome should be placed after the total number and sex chromosomes with a plus (+) or minus (−) sign before the number of the autosome involved; for example, $47,XX + 21$ is the karyotype of a female with trisomy-21. A male with an extra X chromosome is symbolized $47,XXY$. A plus or minus sign is placed following a chromosome symbol to signify increase or decrease in arm length. The letter q symbolizes the long arm and p the short arm. For example, $46,XY,1q +$ indicates an increase

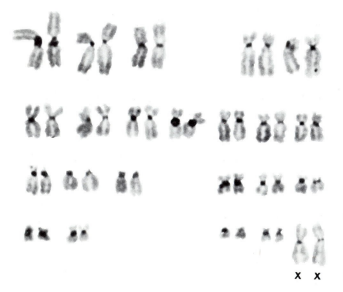

FIGURE 12.16

Karyotype of a normal young woman. Chromosomes were prepared with C-staining technique. Heterochromatin regions of chromosomes are darkly stained. (Courtesy of A. D. Stock, Utah State University.)

in the length of the long arm of chromosome 1. The person represented by this example is a male with 46 chromosomes. The karyotype, 47,XY,+14p+, symbolizes a male with 47 chromosomes, including an additional chromosome (No. 14), with an increase in the length of its short arm. The types of chromosome aberrations are represented by abbreviations for convenience in presenting chromosome formulas: def (deficiency), dup (duplication), r (ring), inv (inversion), and t (translocation).

CHROMOSOME ABERRATIONS IN HUMANS

Chromosome deletions are usually lethal even as heterozygotes, resulting in zygotic loss, stillbirths, or infant deaths. Sometimes infants with small chromosome deficiencies, however, survive long enough to permit observation of some of the abnormal phenotypes they express. Lejeune and his colleagues, for example, discovered a chromosome deficiency in humans that has been associated with the *cri-du-chat* (cat cry) **syndrome** (Fig. 12.18). The name of this syndrome came from a plaintive catlike mewing cry from small weak infants with the disorder. Other characteristics are microcephaly (small head), broad face and saddle nose, widely spaced eyes with epicanthic folds, unique facial features, and physical and mental retardation. IQs of *cri-du-chat* children studied are in the range of 20–40. The chromosome deficiency is in the short arm of chromosome 5 and is designated 5p—. A karyotype (46,XX,5p—) of a newborn is illustrated in Fig. 12.19.

Cri-du-chat patients die in infancy or early childhood and do not transfer the chromosome deletion to offspring. This chromosome deficiency, however, has been shown by Lejeune and others to become involved sometimes in a reciprocal translocation and thus to be transmitted. When the short arm of chromosome 5 became translocated to chromosome 15, the heterozygous translocation was carried in a normal healthy parent. Some gametes, however, carried only the **deficient member** of the translocation pair. Children inheriting the 5p chromosome expressed the *cri-du-chat* syndrome.

Another human disorder that is associated with a chromosome abnormality is chronic myelocytic (myelogenous) leukemia. A deletion of chromosome 22 was first described by Nowell and Hungerford and was called the Philadelphia (Ph') chromosome after the city in which the discovery was made. It was observed consistently in bone marrow preparations of patients with chronic myelocytic leukemia. Later, a translocation discovered by J. Rowley in a leukemia patient provided the correct chromosome rearrangement. A part of the long arm of chromosome 22 was translocated to another chromosome, usually chromosome 9 (46,XX,9q+, 22q—), leaving a **deficiency in the long arm of No. 22.** The karyotype prepared from a bone marrow biopsy is presented in Fig. 12.20.

More chromosomal anomalies are now being associated with **malignancy.** New staining methods permit more precise comparisons of (1) chromosome arms and (2) differentially stained sister chromatids. With banded chromosomes, Burkitt lymphoma was related to a translocation of chromosome 14 [$t(8q-; 14q+)(q24; q32)$]. A deletion in chromosome 13q (band 14.1) is associated with the human embryonic tumor retinoblastoma. Wilms tumor, an embryonic kidney tumor, is associated with a deletion in band 11p13. **Sister chromatid exchanges** (Chapter 3), exchanges occurring between sister chromatids, occur with high frequency in somatic cells of patients with Bloom syndrome, indicating chromosome instability. Most people with this syndrome die with some form of cancer before age 30. Irregular chromosome numbers are also observed in malignant cells, particularly in later stages of cancer.

Like deficiencies, chromosome duplications are usually lethal even as heterozygotes, but sometimes they are sufficiently viable to permit observations of the abnormalities they produce. Like deficiencies, they may be associated with translocations and thus transmitted by normal healthy parents. Remember that a translocation is an exchange of parts between nonhomologous chromosomes or a transfer from one chromosome to a nonhomolog. Broken parts may be further divided and some may be lost or gained in a transfer. Thus, deficiencies and duplications may accompany a translocation carried in a

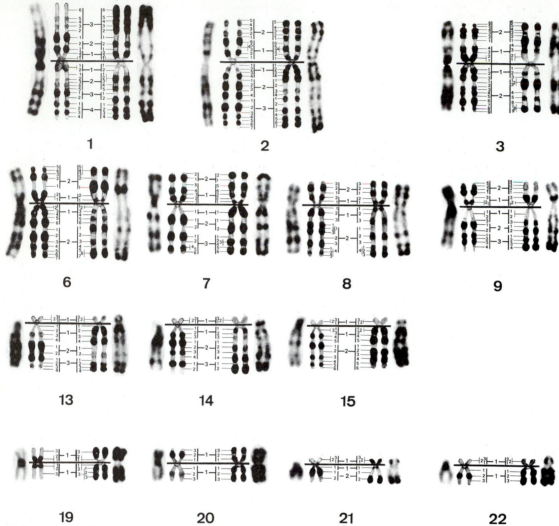

FIGURE 12.17

Photographs of normal human karyotype G and R bands. Autosomes are numbered in order from 1 to 22, and X and Y are included. Short arms of chromosomes (p) *and long arms* (q) *have regions numbered from 1 (next to centromere) to the distal end of each arm, and bands are numbered within the regions from the centromere to the distal end of each arm. (From "The Normal Human Karyotype." Originally reproduced from "An International System for Human Cytogenetic Nomenclature (1978)."* **Birth Defects:** *Original Article Series, XIV:8, 1978, The National Foundation, New York, and Cytogenetics and Cell Genetics 21:6, 1978, S. Karger, Basel, Switzerland.)*

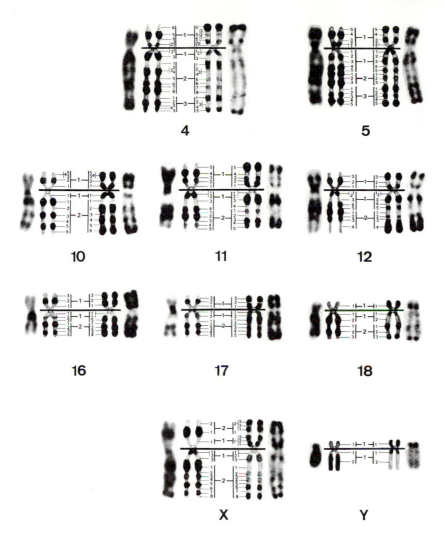

parent. Such a parent may be a translocation heterozygote, with the long arm of one chromosome 21 attached at the centromere to the long arm of chromosome 14 along with normal chromosomes 14 and 21. Some gametes will contain a normal 21 and translocation 14q21q. When such a gamete is fertilized with a gamete from a normal individual, a translocated 21, along with two normal chromosomes 21, results in a viable infant with trisomy-21 and the Down syndrome.

With fluorescent microscope techniques, I. A. Uchida and C. C. Lin discovered a partial trisomy-12 that could not be identified by conventional

methods. A boy (Fig. 12.21) with some clinical features resembling the Down syndrome had been studied with conventional chromosome techniques. Because of continued lack of motor development at seven months of age, further chromosomal investigations were carried out on lymphocyte cultures and slides stained with quinacrine dihydrochloride. With this technique, an additional band was identified in the short arm of chromosome 8. The mother's chromosomes were normal; but in the father, the short arm of No. 12 had become translocated to the short arm of No. 8 [46,XY,(8p+, 12p−)]. The same chromosomal rearrangement was

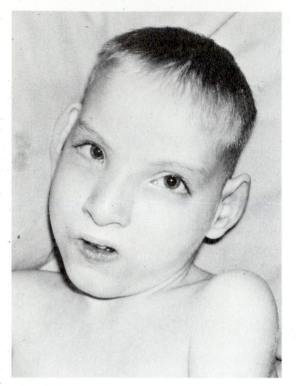

FIGURE 12.18

Patient with cri-du-chat *syndrome, a malady associated with a deletion of part of the short arm of chromosome 5. The syndrome includes microcephaly and a catlike cry. (Courtesy of I. A. Uchida, Department of Pediatrics, McMaster University.)*

found in his daughter, an older sister of the patient. Since both father and daughter were clinically normal, the translocation was presumed to be reciprocal. The patient's No. 8 pair (Fig. 12.22) consisted of one normal and one translocation chromosome, similar to that of his father and sister (Fig. 12.23), but both of his No. 12 chromosomes were normal (Fig. 12.22). Thus, the boy had a duplication of part of the short arm of No. 12 and a deficiency of the tip of No. 8. Many translocations, in addition to those cited above, have been detected in studies of human chromosomes (see Chapter 11, Oncogenes and Proto-oncogenes), but most were apparently reciprocal and produced no phenotypic anomalies. At the Yale-New Haven Hospital, for example, cytological studies were conducted on 4500 infants born consecutively during one year. Lymphocytes from umbilical cord blood of each infant were grown *in vitro* and prepared for microscopic observation of chromosomes. Six translocations were detected. None of these was associated with a distinctive phenotypic anomaly. It is not the translocation process per se that produces abnormalities, but the imbalance of genetic material reflected in chromosome deficiencies and duplications that are produced by segregation of translocated chromosomes.

P. W. Allerdice et al. described a syndrome called the **chromosome 3 duplication-deletion syndrome.** Phenotypically, this syndrome includes a group of morbid symptoms: stillbirths, neonatal

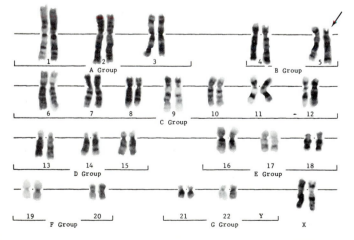

FIGURE 12.19

Karyotype of newborn with the cri-du-chat *syndrome. A heterozygous deficiency is shown in chromosome 5 (see arrow). The chromosome arrangement is 46,XX(5p−). (Courtesy of R. M. Fineman, Department of Pediatrics, University of Utah.)*

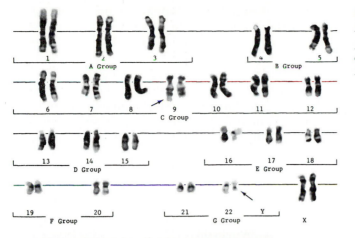

FIGURE 12.20
Karyotype prepared from the bone marrow biopsy of a patient with chronic myelogenous leukemia. The chromosome arrangement is 46,XX,9q+,22q−. (Courtesy of R. M. Fineman, Department of Pediatrics, University of Utah.)

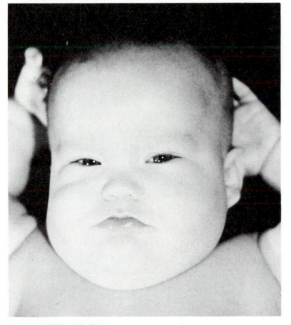

FIGURE 12.21

A boy with some features of the Down syndrome, who was found by quinacrine fluorescence to have partial chromosome trisomy-12 (12p+) and deficiency of the tip of chromosome 8 (8p−). There was no apparent involvement of chromosome 21. Down syndrome features were mental retardation and hypotonia; a rather flat facies with prominent epicanthic folds; Brushfield spots in the irides; flat, broad nasal bridge; and ears in the normal position but with small pinnae. The fingers were somewhat spade shaped with incurving of both fifth digits. Simian creases were present on both palms. (From I. A. Uchida and C. C. Lin, "Identification of Partial 12 Trisomy by Quinacrine Fluorescence," J. Pediatr. 82:269–272, 1973.)

deaths, and spontaneous abortions. Most pregnancies are lost, but two children survived and became probands for the investigation. One of these (Case 1) is shown in Fig. 12.24 at six years of age. He cannot sit up, turn over, or eat solid food. Facial malformation of the living children includes a distorted head shape; thick, low eyebrows; low hairline; long eyelashes; persistent lanugo; distended veins on scalp; hypertelorism; oblique palpebral fissures; a very short nose with a broad, depressed bridge and anteverted nares; protruding maxilla; thin upper lip; micrognathia; low-set ears; and short, webbed neck. Port-wine stains, congenital glaucoma, cloudy corneas, cleft palate, and harelip also occur frequently. Each infant had difficulty sucking and swallowing. Internal physical abnormalities were noted in infants who died neonatally.

Giemsa- and quinacrine-banded karyotypes from a parent of an affected child revealed an inversion [inv(3)($p25q21$)]. Fetal cells cultured for prenatal diagnosis from a subsequent pregnancy of this couple carried a recombinant chromosome 3, with the long arm described as rec(3)del p, dup q, inv(3)($p25q21$). With a banded chromosome 3, the inversion was analyzed. The inverted segment included the centromere (pericentric). Breaks had occurred in the short arm (p) at band 25 and in the long arm (q) at band 21. The central part of the chromosome including the centromere had rotated 180° and become reinserted into chromosome 3. Inv 3($p25q21$) had been carried in the kindred for

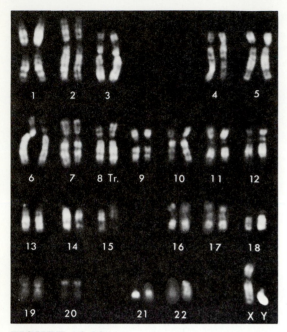

FIGURE 12.22

Karyotype of patient with abnormal chromosome 8 and two normal No. 12 chromosomes. Note the extra band on the second No. 8 chromosome. (From I. A. Uchida and C. C. Lin, "Identification of Partial 12 Trisomy by Quinacrine Fluorescence," J. Pediatr. 82:269–272, 1973.)

at least four generations and 35 kindred members were presumed to carry it. Karyotype of Case 1 is 46,XY,rec(3), dup *q*, inv(3)(*p*25*q*21). He was the second child born to a 23-year-old carrier mother with karyotype 46,XX,inv(3)*p*25*q*21 and a 26-year-old father, karyotype 46,XY.

Crossing over within the inversion loop is presumed to produce the imbalance of genetic material

associated with duplications and deficiencies and to cause the symptoms of the syndrome. Odd numbers of crossovers within the inverted area of a chromosome 3 would be expected to result in genetic imbalance. Earlier in this chapter, crossing over in *Drosophila* inverted chromosomes was shown to create irregular combinations resulting in lethals. This, in turn, resulted in "suppressing" crossing over. In humans, unbalanced zygotes from inverted chromosome crossovers sometimes survive. Some infants carrying unbalanced chromosomes are kept alive, but with varying degrees of birth defect handicaps. The degree of imbalance of genetic material may be the determining factor for life or death and degree of abnormality.

BANDED CHROMOSOMES AND PHYLOGENY

Chromosome-banding techniques have now been applied to nonhuman organisms. Comparisons of chromosome-banding patterns within animal groups such as mammals, birds, reptiles, amphibians, and fish, and also between representatives of particular taxonomic groups, have been accomplished. Compared with other phenotypes employed by taxonomists, chromosome structure has proved to be stable and broadly representative of phylogenetic change. Data from comparisons of chromosome-banding patterns in members of different species therefore have broad phylogenetic significance. Relations among different breeding populations can be estimated by comparing chromosomal characteristics of different taxonomic

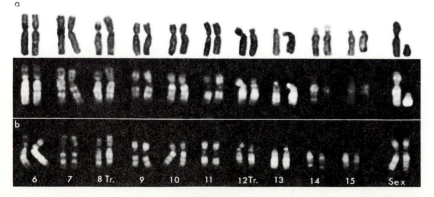

FIGURE 12.23

Partial karyotypes of the father (a) and sister (b) of the patient whose karyotype is shown in Fig. 12.22. Note translocation of chromosomes 8 and 12. (From I. A. Uchida and C. C. Lin, "Identification of Partial 12 Trisomy by Quinacrine Fluorescence," J. Pediatr. 82:269–272, 1973.)

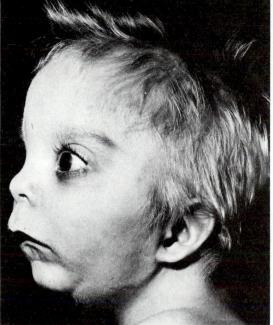

FIGURE 12.24

Six-year-old child with chromosome 3 duplication-deletion syndrome. The karyotype is 46,XY,rec(3),dup q inv(3)(p22q21). He could never suck adequately, and at six years is still spoon or tube fed. All developmental milestones have been grossly delayed. His functioning as tested on the Denver Developmental Screening Test varies between newborn and four-month level. He has had recurrent urinary, respiratory, and eye infections. (Courtesy of P. W. Allderdice. Copyright © 1975 by the American Society of Human Genetics, University of Chicago Press.)

groups. Chimpanzees (*Pan troglodytes*) and human beings (*Homo sapiens*) are placed taxonomically in different genera, *Pan* and *Homo*, respectively. These species are much alike in anatomical and physiological features as well as in DNA and protein composition. Comparisons of chromosomes show **basic similarities** with a few **superimposed structural rearrangements.**

The overall number of chromosomes is comparable (46 for *Homo*, 48 for *Pan*). Homologous pairs can be identified, and general chromosome structure can be matched band for band between the

pairs. Two acrocentrics in *P. troglodytes* have the same bands as one submetacentric in *H. sapiens*, to make the equivalent of 46 in both species.

A. D. Stock compared the banded chromosomes of members of the two species (Fig. 12.25) and noted the following structural differences:

1. Chromosome 2 in *H. sapiens* is long and submetacentric, but is comparable with two acrocentric chromosomes in *P. troglodytes*. When the two *P. troglodytes* chromosomes are fused together near the centromere on the short arm of one chromosome and near the telomere of the short arm of the other, a chromosome similar to *H. sapiens* chromosome 2 is produced.
2. Chromosomes 4, 5, 9, 12, 17, and 18 of the two species each differ by a single pericentric inversion.
3. A probable pericentric inversion distinguishes chromosome 8 of the two species.
4. Chromosomes 1 and 9 of the two species have small, special differences in heterochromatin content.

When chromosome bands are compared and homologies are identified, a close relationship between the two species can be established. In this comparison, at the species level, chromosome morphology is a measure of phylogenetic change. Chromosome number, on the other hand, depends on a more superficial "packaging" process and does not represent a basic criterion for relationship. In this example (Fig. 12.25), one near centric change has made a difference in number, while the basic chromosome units are essentially homologous.

RHESUS AND AFRICAN GREEN MONKEY

Rhesus (*Rhesus macaque*) and African green (*Cercopithecus aethiops*) monkeys are both Old World primates. They are not closely related but they are members of the same family. When Stock and Hsu compared the chromosomes of the two species, differences were observed that could be explained by translocations. Rhesus has a total of $2n = 42$ chromosomes, all biarmed, making a total of 84 arms. The chromosomes, in addition to the sex pair, were arranged in four morphological groups: A: seven pairs of medium to large submetacentrics; B: seven

Pan | Homo

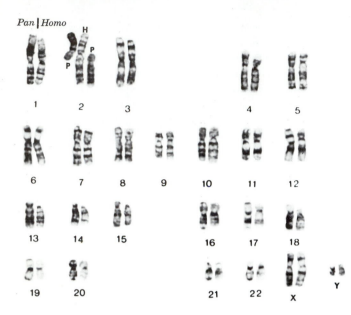

FIGURE 12.25
Chimpanzee chromosomes compared with human chromosomes. Both species have the same total number if two chimpanzee acrocentrics are joined to correspond with Homo *No. 2. Six pericentric inversions are recognized in comparing chromosomes of the two species. (Courtesy of A. D. Stock.)*

pairs of medium submetacentrics with higher arm ratios than those in group A; C: five pairs of small metacentrics and submetacentrics; and D: one pair that carries a pronounced secondary constriction on its short arm. In the sex pair, the X is a large submetacentric, and the Y is the smallest element of the complement. C-banding demonstrated that the constitutive (stationary) heterochromatin of all chromosomes in this species is centromeric.

The African green monkey has a total of $2n = 60$ biarmed chromosomes (120 arms). These chromosomes were arranged in four morphological groups corresponding with those of the Rhesus monkey: A: 11 pairs of subtelocentric; B: 12 pairs of medium to small submetacentrics of varying lengths; C: 5 pairs of small metacentrics and submetacentrics that appeared identical with group C of Rhesus; and D: 1 pair of small subtelocentrics with the secondary constriction in the long arm. The long arm is homologous with the short arm of Rhesus D chromosomes. In gross morphology, the sex chromosomes were similar to those of Rhesus. Constitutive heterochromatin of all members of groups B and C and the X is centromeric. In chromosomes of group A, the short arms of all except one (A2 is euchromatic) are heterochromatic.

When the chromosomes of Rhesus and the African green monkey were compared (Fig. 12.26), groups B and C and sex chromosomes had matching elements in the karyotypes of the African green and Rhesus. The group A chromosomes in the two species were different, but matching parts were found when chromosomes were compared side by side. Some places that did not fit contained constitutive heterochromatin. When excess heterochromatin was trimmed from the African green monkey chromosomes, and the euchromatin parts remaining on acrocentrics were fused, a nearly perfect fit between the two karyotypes was obtained. The main differences between the chromosomes of the two monkey species were (1) amount and distribution of heterochromatin and (2) translocations. Translocations were whole-arm fusions of centromere to telomere that reduced three arms into two. These species are distantly related. Comparisons of banded chromosomes reveal many **homologies** and **differences.**

SUMMARY

Structural aberrations of chromosomes occur in plants and animals. They result in increased variation and reproductive isolation, thus playing a sig-

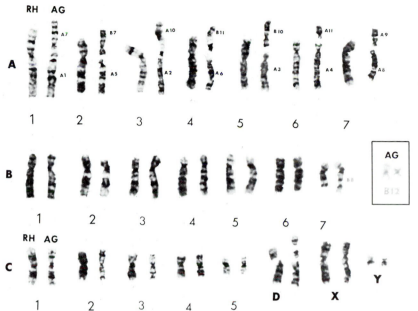

RH AG

A

1   2   3   4   5   6   7

B

1   2   3   4   5   6   7

AG

B12

RH AG

C

1   2   3   4   5   D   X   Y

FIGURE 12.26

Matching between the two monkey species. Rhesus chromosomes, left side of each pair; African green, right side. (From A. D. Stock and T. C. Hsu, Chromosoma *43:211–224, 1973.)*

nificant role in the evolution of some groups of living things. Polytene chromosomes in Diptera and large meiotic prophase chromosomes in maize have been used extensively for identifying and investigating chromosome structural changes. The mechanism of structural aberration consists of the breaking and rejoining of chromosomes, resulting in deficiencies, duplications, inversions, and translocations. Linear linkage relations among genes in chromosomes are altered by structural aberrations. Both stable and variegated position effects are associated with chromosome structural changes.

All 46 human chromosomes can now be identified with chromosome-banding techniques. Descriptive symbols have been standardized for labeling structural and numerical chromosome changes. Chromosome banding has now been applied to non-human as well as human organisms, and comparisons with phylogenetic significance have been established within and among taxonomic groups.

REFERENCES

Allerdice, P. W. 1975. "Chromosome 3 duplication $q21 \rightarrow q$ter deletion $p25 \rightarrow p$ter syndrome in children

of carriers of a pericentric inversion." *Am. J. Hum. Genet.* 27:699–718.

Bergsma, D. (ed.). 1972. "Advances in human genetics and their impact on society." *Birth Defects* 8(4):1–118.

———. 1972. "Paris conference: Standardization in human cytogenetics." *Birth Defects* 8(7):1–46.

Bridges, C. B. 1935. "Salivary chromosome maps with a key to the banding of the chromosomes of *Drosophila melanogaster. J. Hered.* 26:60–64.

Dalla-Favera, R. et al., 1982. "Human *c-myc onc* gene is located on the region of chromosome 8 that is translocated in Burkitt lymphoma cells." *Proc. Natl. Acad. Sci., U.S.* 79:7824–7827.

Darlington, C. D., and L. F. La Cour. 1976. *The handling of chromosomes,* 6th ed. Halsted Press, New York.

Dutrillaux, B., and J. Lejeune. 1975. "New techniques in the study of human chromosomes: methods and applications." *Advances in human genetics,* Vol. 5. H. Harris and K. Hirschhorn (eds.). Plenum Press, New York.

Hsu, T. C. 1972. "Procedures for mammalian chromosome preparation." *Methods in Cell Physiology* 5:1–36.

Judd, B. H., M. W. Shen, and T. C. Kaufman. 1972. "The anatomy and function of a segment of the X chromosome of *Drosophila melanogaster.*" *Genetics* 71:139–156.

Kaiser-McCaw, B., A. L. Epstein, H. S. Kaplan, and F. Hecht. 1977. "Chromosome 14 translocation in African and North American Burkitt's lymphoma." *Int. J. Cancer* 19:482–486.

Klein, G. 1981. "The role of gene dosage and genetic transposition in carcinogenesis." *Nature* 294:313–318.

Klein, G., et al. 1982. "A cellular oncogene is translocated to the Philadelphia chromosome in chronic myelocytic leukemia." *Nature* 300:765–767.

Lewis, E. B. 1950. "The phenomenon of position effect." *Advances in genetics,* Vol. 3. M. Demerec (ed.). Academic Press, New York.

Priest, J. H. 1977. *Medical cytogenetics and cell culture,* 2nd ed. Lea and Febiger, New York.

Rowley, J. 1983. "Human oncogene locations and chromosome aberrations." *Nature* 301:290–291.

———. 1973. "A new consistent chromosomal abnormality in chronic myelogenous leukemia identified by quinacrine fluorescence and Giemsa staining." *Nature* 243:290–293.

Stock, A. D., and T. C. Hsu. 1973. "Evolutionary conservatism in arrangement of genetic material." *Chromosoma* 43:211–224.

Turleau, C., and J. de Grouchy. 1973. "New observations on the human and chimpanzee karyotypes." *Humangenetik* 20:151–157.

Turleau, C., J. de Grouchy, and M. Klein. 1972. "Phylogénie chromosomique de l' homme et des primates hominiens. Essai de reconstition du caryotype de l' ancetre commun." *Ann. Génét.* 15:225–240.

Turpin, R., and J. Lejeune. 1969. *Human afflictions and chromosomal aberrations.* Pergamon Press, Oxford.

Uchida, I. A., and C. C. Lin. 1973. "Identification of partial 12 trisomy by quinacrine fluorescence." *J. Pediat.* 82:269–272.

PROBLEMS AND QUESTIONS

12.1. (a) What genetic evidence first suggested chromosome structural changes? (b) Why did it take so many years to obtain cytological verification for the genetic evidence? (c) How was cytological verification obtained?

12.2. Compare the methods now available for cytogenetic studies of the fruit fly with those available for maize.

12.3. What characteristics of *Drosophila* salivary gland chromosomes make them especially suitable for cytogenetic studies?

12.4. What are the advantages of the acetocarmine smear technique as compared with fixing, sectioning, and staining methods for chromosome studies?

12.5. Formulate a plausible explanation for the origin of the giant salivary gland chromosomes in the developing larva of the fly. What do the cross bands represent?

12.6. What is the difference between a linkage and a cytological chromosome map?

12.7. Describe or illustrate with appropriate sketches how a recessive gene may be expressed through pseudodominance.

12.8. How can the extent of a chromosome deficiency be determined (a) genetically and (b) cytologically?

12.9. If a trait such as "waltzing" in mice, known to depend on a single recessive gene (v), should appear in an animal considered to be heterozygous for the gene, how could it be determined (a) genetically and (b) cytologically whether a mutation had occurred or a deficiency was present in the chromosome opposite v?

12.10. How can (a) paracentric and (b) pericentric inversions act as "crossover suppressors"? Describe or illustrate. (c) Is crossing over really suppressed?

12.11. In *D. melanogaster,* the gene *Bd* is dominant with respect to a wing abnormality but is homozygous lethal. Another homozygous lethal gene (ℓ) is located on the homologous chromosome, and a crossover "suppressor" prevents crossing over between *Bd* and ℓ. What results would be expected from a cross between two flies with the genotype $Bd\ell^+/Bd^+\ell$?

12.12. In barley, a_n for white seedlings and x_c for yellow seedlings are on the same chromosomes with a crossover value of 10 percent. Plants with the genotype $a_n^+x_c^+$ are green. Homozygous a_n and x_c plants die in the seedling stage. A plant with the genotype $a_nx_c^+/a_n^+x_c$ was selfed. Give the expected results.

12.13. Reciprocal crosses were made between flies with striped *sr* bodies and wild-type flies *sr^+*. No differences were found between results of crosses in either the F_1 or the F_2 progeny. Homozygous *sr* females were then mated to males bearing the balanced curly and plum dominant genes on chromosome 2, and stubble and dichaete dominant genes on chromosome 3. Crossing over was suppressed. F_1 progeny showing both curly and stubble were mated. Some progeny had striped bodies and curly wings. (a) Is *sr* autosomal or sex-linked? (b) Which chromosome is it on? (Chromosome 4 is very small and may be eliminated from this consideration.)

12.14. Describe or illustrate with sketches the appearance of the following heterozygous chromosome modifications in salivary preparations: (a) deficiency; (b) duplication; (c) inversion; (d) reciprocal translocation.

12.15. What (a) genetic and (b) cytological evidence would indicate that a translocation was present in a plant material such as barley?

12.16. How is pollen sterility associated with translocations? Illustrate.

12.17. In a *Drosophila* salivary chromosome section, the bands have a sequence of 1 2 3 4 5 6 7 8. The homolog with which this chromosome must synapse has a sequence 1 2 3 6 5 4 7 8. (a) What kind of a chromosome change has occurred? (b) Describe or draw a diagram to illustrate the possible pairing arrangement.

12.18. Other chromosomes have sequences as follows: (a) 1 2 5 6 7 8; (b) 1 2 3 4 4 5 6 7 8; (c) 1 2 3 4 5 8 7 6. What kind of chromosome modification is present in each? Illustrate with diagrams the pairing of these chromosomes with their normal homologs in salivary preparations.

12.19. Chromosome 1 in maize has the sequence A B C D E F, whereas chromosome 2 has the sequence M N O P Q R. A reciprocal translocation resulted in the following arrangements: A B C P Q R and M N O D E F. Diagram the expected pachytene configuration and describe the causes of the pollen sterility that might be expected.

12.20. How could a phenotypic effect, such as the number of facets in expressions of bar eye in *Drosophila*, be demonstrated as a position effect?

12.21. (a) How do variegated position effects originate and (b) how can they be explained?

12.22. What is the (a) theoretical and (b) practical significance of the Philadelphia chromosome?

12.23. Describe with standard nomenclature the following human chromosome complements: (a) male karyotype with all normal chromosomes except one missing, No. 21; (b) female karyotype with a translocation between long arms of a No. 14 and a No. 21 chromosome, with the short arms of No. 14 and No. 21 missing; (c) male karyotype with two translocations, involving interchange of both whole arms of chromosomes 5 and 12.

12.24. What is the significance of chromosome inversions and translocations in evolution?

12.25. Why were critical chromosome studies in humans not carried out before 1956?

12.26. Why was it difficult before 1970 to identify individual human chromosomes?

THIRTEEN

Karyotypes for two species of the cotton rat. (Top) *Sigmodon hispidon* (2n = 52). (Bottom) *Sigmodon arizonae* (2n = 22). (Courtesy of A. D. Stock.)

VARIATIONS IN CHROMOSOME NUMBER

Somatic cells of higher plants and animals usually have chromosomes in pairs ($2n$); that is, two of each kind of chromosome are present in each cell. Mature germ cells, having undergone reduction division, normally have one member of each pair (n). Many individual plants and animals, however, have local areas of somatic tissue characterized by a multiple of the basic chromosome number. A **doubling process** in cell division is the usual explanation for these deviations.

With the exception of sex differences, somatic doubling, and minor variations that occur in natural and experimental populations, all members of a species of plants or animals have the same basic chromosome number. *Ascaris lumbricoides* has only one chromosome pair, at least in early developmental stages. With this exception, the range of reported chromosome numbers in animals extends from 2 pairs in a rhabdocoel, *Gyratrix hermaphroditus,* and some mites, midges, and scale insects, to more than 100 in some butterflies and Crustacea. The crustacean, *Paralithodes camtschatica,* for example, has 208 chromosomes or 104 pairs. The reported range in plants is from 2 pairs in the small composite plant, *Haplopappus gracilis,* to several hundred in some ferns. A species of fernlike plants of the genus *Ophioglossum* is reported to have 768 chromosomes.

Changes in the number of chromosomes may be reflected in high **inviability** and phenotypic **anomalies** in those that survive. This constitutes a useful tool for identifying the influence of individual chromosomes. For example, if phenotypically distinguishable individuals with different chromosome numbers can be identified in natural populations or produced experimentally, it is sometimes possible to determine the effect of adding or removing certain chromosomes. Some plants with increased chromosome numbers have phenotypic changes in morphological or physiological characteristics that are of practical importance to man. Tomato plants, for example, with chromosome numbers above $2n$ are larger and produce more desirable fruit than do corresponding varieties with the usual $2n$ number.

Chromosome changes are classified in terms of additions or eliminations of parts of chromosomes (Chapter 12), whole chromosomes, or whole sets of chromosomes (genomes). Two main classes are **euploidy** and **aneuploidy** (*ploid,* Greek for "unit"; *eu,* "true or even"; and *aneu,* "uneven"). Euploids have chromosome complements consisting of whole sets, or genomes. The basic chromosome number of euploid organisms is represented by the haploid (n). The symbol n represents the **reduced** or half the diploid ($2n$) number in a somatic cell. In polyploid plants, the functional reduced haploid number (n) may not represent the smallest number (x) that can make up a chromosome set or genome. Bread wheat, *Triticum aestivum,* for example, has $2n = 42$ chromosomes, $n = 21$. The 21 chromosomes segregate to each gamete in meiosis. Small seedlings that sometimes occur in a wheat field carry one genome of 21 chromosomes. But emmer wheat, *T. dicoccum,* has $n = 14$, and einkorn, *T. monococcum,* has $n = 7$, where $2n = 14$. Seven is an odd number representing a complete genome in this species. It cannot be divided further into equal sets of whole chromosomes. The symbol x represents the **smallest possible** (monoploid) number in a **genome.** Bread wheat is a hexaploid with $6x = 42$. In garden peas, n (haploid) $= x$ (monoploid) $= 7$, and $2x = 2n = 14$ (diploid). Euploids with more than one genome (monoploid) may be diploid ($2n$), triploid ($3n$), tetraploid ($4n$), and so on.

In humans, additions and deletions of chromosomes, particularly the large chromosomes (groups A, B, and C), almost always result in lethals. Some newborns with extra chromosomes of the smaller groups such as G (e.g., chromosome 21) survive but show multiple physical and mental abnormalities.

TRISOMY IN HUMANS

THE DOWN SYNDROME (47,+21)
The best known and most common chromosome-related disease syndrome, formerly known as "mongolism," is now designated the **Down syndrome,** after Langdon Down, who first described the clinical signs in 1866. Cost of training and maintaining the Down syndrome cases in the United

FIGURE 13.1

Facial features of a child with Down syndrome. Down syndrome children are usually sensitive, happy, and lovable. They are skillful and dependable when properly trained and employed. (Courtesy of A. Rakoczy, Upstate Home for Children, Milford, N.Y.)

States is estimated at **$1 billion per year.** Emotional stress in families with Down syndrome children and adults is also a factor in their care. The need for effective counseling and prevention is readily apparent.

Patients studied by Down, and those with the Down syndrome who have since been observed (Fig. 13.1), are short in stature (about four feet tall); they have an epicanthal fold (thus the earlier name "mongolism"), broad short skulls, wide nostrils, large tongues with distinctive furrowing, stubby hands (particularly the fifth digit) with a simian crease (Fig. 13.2) on the palm and a single crease on the fifth digit, and general loose jointedness, observed particularly in the ankles. They are characterized as low in mentality (Table 13.1), but they can be trained in routine mechanical skills. Through the investigation of J. Lejeune in 1959, the Down syndrome became the first chromosomal disorder to be described in humans.

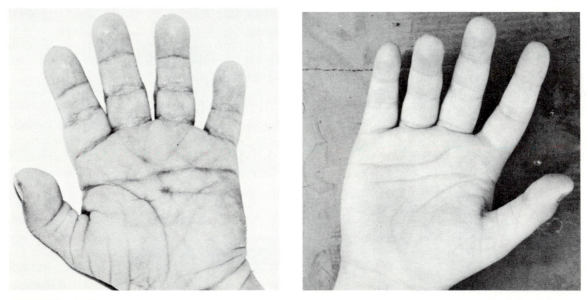

FIGURE 13.2

Hand of a child with Down syndrome, left, showing simian crease on palm and single crease on little finger. (Courtesy of I. A. Uchida.) Right, hand of a normal child. (K. Bendo.)

TABLE 13.1   Aneuploidy Resulting from Nondisjunction in the Human Population

CHROMOSOME NOMENCLATURE	CHROMOSOME FORMULA	CLINICAL SYNDROME	ESTIMATED FREQUENCY AT BIRTH	MAIN PHENOTYPIC CHARACTERISTICS
47,+21	$2n + 1$	Down	$\frac{1}{700}$	Short broad hands with simian-type palmar crease, short stature, hyperflexibility of joints, mental retardation, broad head with round face, open mouth with large tongue, epicanthal fold.
47,+13	$2n + 1$	Trisomy-13	$\frac{1}{20,000}$	Mental deficiency and deafness, minor muscle seizures, cleft lip and/or palate, polydactyly, cardiac anomalies, posterior heel prominence.
47,+18	$2n + 1$	Trisomy-18	$\frac{1}{8000}$	Multiple congenital malformation of many organs; low-set, malformed ears; receding mandible, small mouth and nose with general elfin appearance; mental deficiency; horseshoe or double kidney; short sternum. 90% die in the first 6 months.
45,X	$2n - 1$	Turner	$\frac{1}{2500}$ female births	Female with retarded sexual development, usually sterile, short stature, webbing of skin in neck region, cardiovascular abnormalities, hearing impairment.
47,XXY 48,XXXY 48,XXYY 49,XXXXY 50,XXXXXY	$2n + 1$ $2n + 2$ $2n + 2$ $2n + 3$ $2n + 4$	Klinefelter	$\frac{1}{500}$ male births	Male, subfertile with small testes, developed breasts, feminine pitched voice, long limbs, knock knees, rambling talkativeness.
47,XXX	$2n + 1$	Triple X	$\frac{1}{700}$	Female with usually normal genitalia and limited fertility. Slight mental retardation.

A small autosome, chromosome 21 (Fig.13.3), added to the normal complement (47,+21) causes Down syndrome. This is a **trisomic** for No. **21**; all other chromosomes are "disomes." Chromosome 21 had been difficult to distinguish from No. 22. Chromosome 21 was considered to be slightly larger than No. 22 when the chromosomes were first arranged (on the basis of size) in a karyotype. Critical study has since shown that No. 22 is larger than No. 21, but the Paris Conference (1971) agreed to leave them in their original positions in the standard karyotype. Both No. 21 and No. 22 usually have satellites—distal chromosome segments that are separated from the rest of the chromosome by a chromatic filament that can be observed in good chromosome preparations. Differential staining and photography through a fluorescent microscope have now distinguished between No. 21 and No. 22 (Fig. 13.4). Trisomy of No. 21 is the result of primary nondisjunction, which can occur at either of the two meiotic divisions and in either parent. Paired chromosomes do not separate properly to the poles at anaphase; one gamete will receive two No. 21 chromosomes and the other none.

The Down syndrome occurs once in about 700 live births among European people. Incidence at conception is estimated to be considerably higher (7.3 per 1000), the difference being reflected in fetal loss due to **spontaneous abortion.** About one in six of the Down syndrome children born alive

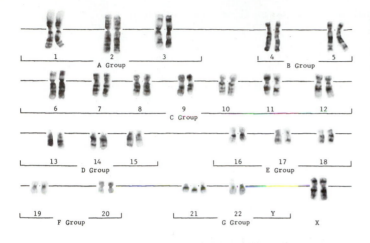

FIGURE 13.3

Karyotype of a child with the Down syndrome, showing trisomy for chromosome 21 (47,XX,+21). (Courtesy of R. M. Fineman.)

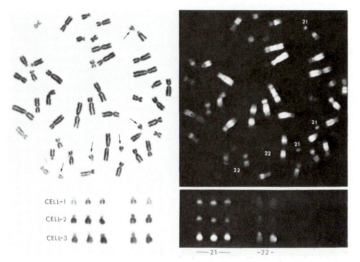

FIGURE 13.4

Photograph showing the difference between chromosomes 21 and 22 in a trisomic with the Down syndrome. The spread on the left is stained with orcein. The one on the right has been stained with quinacrine dihydrochloride and photographed through a fluorescent microscope. This technique shows up the brighter fluorescence of No. 21. The No. 21 and 22 chromosomes at the bottom came from three different cells from the same child with the Down syndrome. (Courtesy of I. A. Uchida and C. C. Lin.)

dies within the first year. The average expectation of life is about 16.2 years. Some patients with Down syndrome have a total of only 46 instead of 47 chromosomes, but in such cases a translocation has joined the long arm of chromosome 21 with another chromosome in the same complement (Fig. 13.5), most frequently No. 14. When an **inherited translocation** is associated with the Down syndrome in an infant, the risk figure of a second occurrence in the same family can be calculated. With assurance that the translocated chromosome t(14q21q) will remain intact and assuming that one parent has a normal chromosome complement, the theoretical risk is one in three that each additional child will have the Down syndrome. Karyotypes of parents who have already produced a child with the inherited translocation Down syndrome are 45,XX, t(14q21q) × 46,XY (normal). Furthermore, gametes from the first parent are expected to be of six different kinds instead of the normal number: (1) normal 14,21, (2) balanced translocation (14q21q), (3) t(14q21q),+21, (4) 14, (5) t(14q21q),+14, (6) 21. If each is fertilized by a normal gamete (14,21) from the second parent, the resulting ratio is: 2 normal to 1 trisomy-21 (Down syndrome, Fig. 13.6). All zygotes with **extra or missing large chromosomes** (trisomics or monosomics) **do not survive.** Only zygotes with the smallest autosomes (No. 21) in

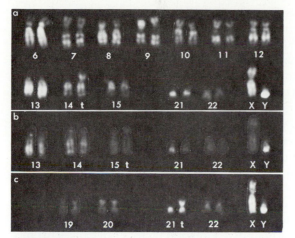

FIGURE 13.5

*Partial karyotypes from three different translocation patients
with the Down syndrome. The top band (a) is from a child with a
translocation of chromosome 21 to a 14 chromosome; the middle
band (b) involves No. 21 and 15; and the bottom band (c) shows
centric fusion of two No. 21 chromosomes. (Courtesy of I. A.
Uchida and C. C. Lin.)*

trisomy can survive, and this combination produces
the Down syndrome. Thus, one-third of the viable
zygotes from the cross, representing possible future
children in the family, are expected to have the

Down syndrome. The risk factor of $\frac{1}{3}$ rests on the
assumption that the three surviving zygotes are
equally viable and that the other three will die.
Apparently, this assumption is not valid since the
observed frequency for the Down syndrome is far
lower, about 11 percent, when the mother is a
balanced translocation carrier; it is 2 percent when
the father is the translocation carrier.

AMNIOCENTESIS FOR DETECTING ANEUPLOIDY

Chromosomal abnormalities are sufficiently well
understood to permit genetic counseling. A fetus
may be checked in early stages of development by
karyotyping the cultured cells obtained by a process
called **amniocentesis** (Figs. 13.7 and 13.8). A sample of fluid is withdrawn with a needle from the
amniotic sac. Fetal cells are cultured and, after a
period of two to three weeks, chromosomes in dividing cells can be stained and observed. If three
No. 21 chromosomes are present, the Down syndrome is confirmed. The risk for mothers less than
25 years of age to have the trisomy is about 1 in 1500
births; at 40 years of age, 1 in 100 births; at 45, 1 in
40 births. If all pregnant women of age 45 were
checked, about 1 in 40 would be expected to be

Cross: 45,XX,t(14q21q) × 46,XY

GAMETE OF SECOND PARENT	GAMETES OF FIRST PARENTS					
	14,21	14q21q	t(14q21q), + 21	14	t(14q21q), + 14	21
(14,21)	14,21 14,21	(14q21q) 14,21	t(14q21q), + 21 14,21	14 14,21	t(14q21q), + 14 14,21	21 14,21
Phenotypes of zygotes	Normal	Normal translocation carrier	Trisomy-21, Down syndrome	Mono-somic-21, lethal	Trisomic-14, lethal	Mono-somic-14, lethal

Summary: 2 normal, 1 with Down syndrome.

FIGURE 13.6

*Diagram of a cross between an individual with an inherited translocation, carrying
chromosome 21, 45,XX,t(14q21q) and a normal person, 46,XY. The theoretical risk of
$\frac{1}{3}$ for a parent carrying the translocation t(14q21q) for Down syndrome is illustrated.
The actual risk is far less, presumably because of inviability of the translocation carrier,
the only zygote capable of producing a Down syndrome child.*

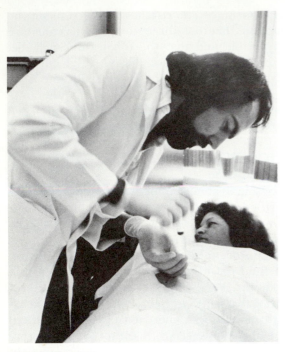

FIGURE 13.7

A physician taking a sample of fluid from the amniotic sac of a pregnant woman for prenatal diagnosis of a chromosomal or biochemical abnormality. (Photo by L. Sloan / Woodfin Camp.)

carrying a child with trisomy-21. Pregnant women over 45 are a special **high-risk** group. Some trisomy-21 cases are not related to maternal age. Ge-

netic factors rather than aging mothers are postulated to control nondisjunction in these cases.

Most, and perhaps all, cases of trisomy in humans can be explained by nondisjunction at meiosis, as illustrated in Fig. 13.9. Older women are more likely than younger women to evidence chromosome damage and meiotic irregularities. An ovary may form about 400,000 eggs, but only about 400 will mature. All the eggs remain in prophase I from the time they are formed in the female fetus until the time of life when they become mature and capable of fertilization and development. The longer the time before fertilization, the greater the chance for damage and irregularity. Some drugs and other environmental factors can cause chromosome anomalies.

TRISOMY-13 and TRISOMY-18

The **trisomy-13 syndrome** (47,+13; Fig. 13.10), described by K. Patau in 1960, occurs in about 1 in 20,000 newborns. It is rare in children and nonexistent in adults because the severe symptoms result in early death. Most of the deaths occur within the first three months after birth, but a few victims have lived for as long as five years. Symptoms include small brain, apparent mental deficiency, deafness, and numerous other external and internal abnormalities (Table 13.1).

The **trisomy-18 syndrome** (47,+18; Fig. 13.11), described by J. H. Edwards and his colleagues in 1960, includes mental deficiency and

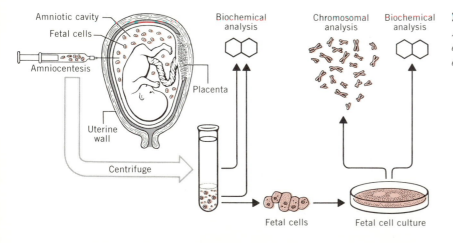

FIGURE 13.8

Amniocentesis and procedure for prenatal diagnosis of biochemical and chromosomal disorders.

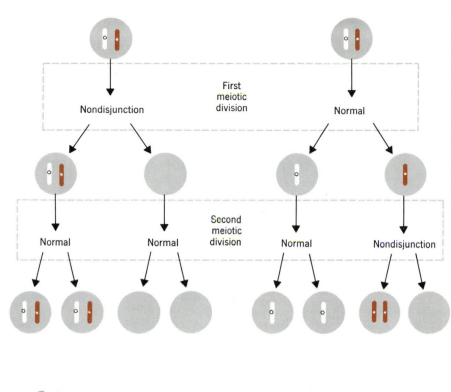

FIGURE 13.9

Consequences of nondisjunction at the first and second meiotic divisions. Nondisjunction at meiosis I produces no normal gametes. Nondisjunction at meiosis II produces two gametes containing (or lacking) two identical chromosomes, both derived from the same member of the homologous pair, and two normal gametes. It is assumed that other chromosome pairs not shown in this diagram behave normally in the first and second meiotic divisions.

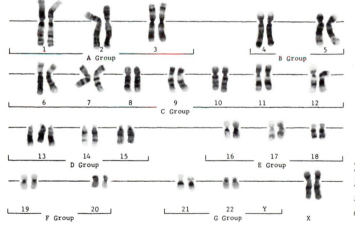

FIGURE 13.10

Karyotype of a newborn with the trisomy-13 syndrome, showing three No. 13 chromosomes: 47,XX,+13. (Courtesy of R. M. Fineman.)

multiple congenital malformations (Figs. 13.12 and 13.13) involving virtually every organ system (Table 13.1). Most infants with this syndrome die at an early age, some 90 percent within their first six months. Nearly all are deceased before they reach one year, but a few have been reported to be alive in their teen years.

The incidence of trisomy-18 is about 1 in 8000. Thus far, only a few cases have been observed, and it is not known whether racial or other population groups differ in incidence. In general, the incidence of this deformity is greater among infants of older women, as expected if the cause of this trisomy is **primary nondisjunction in meiosis.**

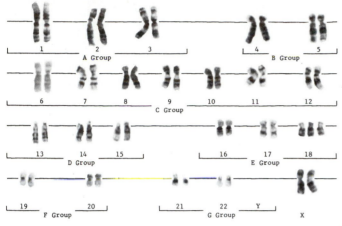

FIGURE 13.11
Karyotype of a child with the trisomy-18 syndrome, showing three No. 18 chromosomes: 47,XX+18. (Courtesy of R. M. Fineman.)

THE TURNER SYNDROME (45,X)

This monosomic has a chromosome complement of 44 autosomes and one X chromosome (Fig. 13.14). The chromosome anomaly is associated with an

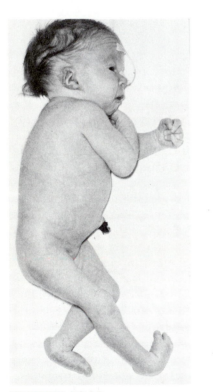

FIGURE 13.12
Infant with trisomy-X and trisomy-18 (48,XXX,+18). Note abnormal ears, receding mandible, flexion deformity of fingers, small hips, left rockerbottom foot, and right club foot characteristic of trisomy-18. (Courtesy of I. A. Uchida.)

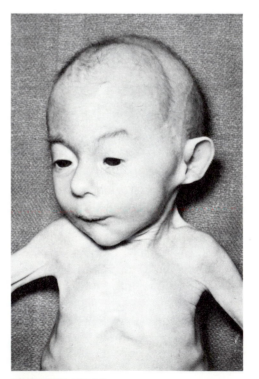

FIGURE 13.13
Facial features associated with trisomy-18. (Courtesy of I. A. Uchida.)

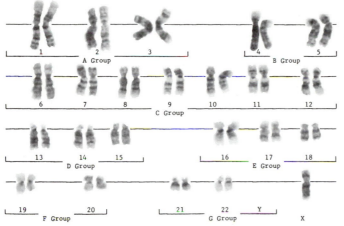

FIGURE 13.14
Karyotype of a child with the Turner syndrome, showing only one sex (X) chromosome (45,X). (Courtesy of R. M. Fineman.)

abnormal female phenotype described in 1938 by H. H. Turner and associates and known as the **Turner syndrome** (Fig. 13.15). It occurs in about 1 per 2500 live female births. More than 90 percent abort spontaneously. A rough estimate for 45,X

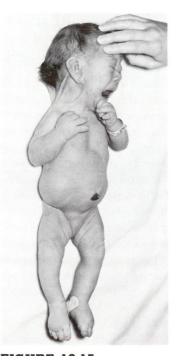

FIGURE 13.15
Infant with Turner syndrome (45,X). Note webbed neck and puffy hands and feet. (Courtesy of I. A. Uchida.)

adults in the general population is 1 in 5000. These adults (Fig. 13.16) have virtually no ovaries, have limited secondary sexual characteristics, and are sterile. Microscopic sections of the ovaries show fibrous streaks of tissue representing remnants of ovaries. Affected females have short stature, low-set ears, webbed neck, and a shieldlike chest (Table 13.1). Mental deficiency is not usually associated with this syndrome. Epithelial cells of 45,X patients are **X chromatin negative** (Chapter 4), as expected when only one X chromosome is present.

X monosomics probably originate from exceptional eggs or sperm with no sex chromosome or from the loss of a sex chromosome in mitosis during early cleavage stages, after an XX or XY zygote has been formed. This latter probability is supported by the high frequency of mosaics that result from postzygotic events in patients with the Turner syndrome. **Mosaics** with X/XX sex chromosomes show symptoms of the Turner syndrome but are usually taller than X and have fewer anomalies than nonmosaic 45,X females. They show more feminization, more normal menstruation, and may be fertile. Many cases of the somatic Turner phenotype without the typical 45,X chromosome constitution are now known. Most of these have one normal X chromosome and a fragment of a second X chromosome. Both arms of the second X chromosome are apparently necessary for normal ovarian differentiation. Individuals with only the long arm of the second X are short in stature and show other so-

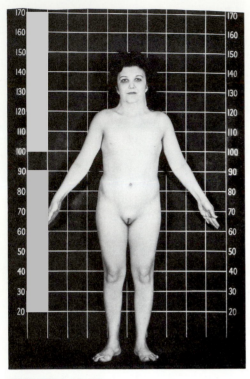

FIGURE 13.16

Female with the Turner syndrome (45,X). She is short in stature, has a short neck, and lacks most female secondary sex characteristics. The breast development shown here has been induced by administering estrogen. Note the broad, shieldlike chest with widely spaced nipples, mild neck webbing, and low-set ears. (Courtesy V. A. McKusick, Johns Hopkins.)

matic symptoms of the Turner syndrome, whereas those with only the short arm of the second X have normal stature and do not show as many signs of the Turner syndrome. This indicates that the Turner phenotype is mostly controlled by **genes on the short arm of the X chromosome.**

Patients with partial deletion of one X chromosome are X chromatin positive and therefore may be misdiagnosed if a buccal smear is the only test used. The deficient X chromosome always forms the X chromatin body. A Y chromosome also occurs in some individuals with the Turner phenotype. These patients are usually mosaic for

45,X/46,XY, with a normal Y. People with one X and a Y fragment, not including the Y short arm, have only streak ovaries but are normal in phenotype. This suggests that **male-determining genes** are in the short arm of the **Y chromosome,** and those that prevent Turner phenotype are in the Y long arm as well as the X short arm. Major features of the Turner phenotype occur in some males as well as in females. The male Turner syndrome is characterized by defective development of the testes, sterility, and limited male secondary sexual characteristics, along with somatic features of the Turner phenotype. These people have normal male karyotypes (46,XY).

THE KLINEFELTER SYNDROME (47,XXY)

An extra X chromosome in addition to the usual male (XY) chromosome complement (47,XXY) has been associated with the abnormal male syndrome described (in 1942) by H. F. Klinefelter and known as the **Klinefelter syndrome** (Fig. 13.17). It is estimated to occur in 1 per 500 live male births. Individuals with this syndrome (Table 13.l) are phenotypically males but with some tendency toward femaleness, particularly in secondary sex characteristics. Such features as enlarged breasts (Fig. 13.18), underdeveloped body hair, small testes, and small prostate glands are a part of the syndrome. Presumably, the XXY constitution originates either by fertilization of an exceptional XX egg by a Y sperm or of an X egg by an exceptional XY sperm. Studies of Klinefelter syndrome and Turner syndrome indicate that the Y chromosome in human beings, unlike that in *Drosophila*, determines male sex.

The most common karyotype (about three-quarters of the cases) for the Klinefelter syndrome is **47,XXY,** but the symptoms of the syndrome will usually occur whenever more than one X chromosome is present along with a Y chromosome. More complex karyotypes associated with the Klinefelter syndrome include: XXYY, XXXY, XXXYY, XXXXY, XXXXYY, and XXXXXY. All patients with the Klinefelter syndrome have one or more X chromatin bodies in their cells. Mental retardation is usually found when there are more than two X chromosomes. The XY/XXY mosaicism in patients

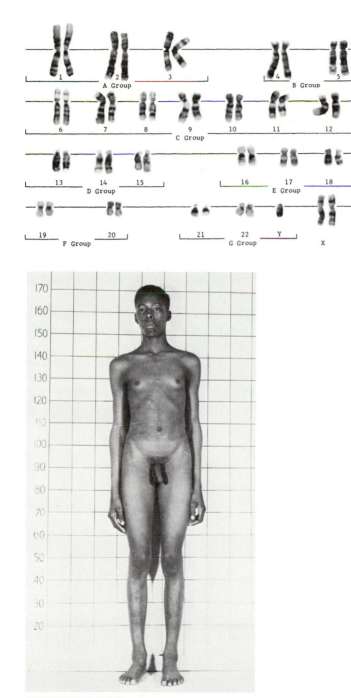

FIGURE 13.18

Male with the Klinefelter syndrome. The syndrome includes small testes and very little body hair. Affected males tend to be long-legged and thin. They often develop breasts (gynecomastia) like females and are sterile. (Courtesy V. A. McKusick, Johns Hopkins.)

with the Klinefelter syndrome is associated with less severe **physical and reproductive anomalies.**

ANEUPLOIDY OF X CHROMOSOMES AND MENTAL DEFICIENCY

Other irregular combinations of X chromosomes have also been recognized among females with X chromosome aberrations. About 1 percent of all mentally defective women in institutions have been shown to have one or more extra X chromosomes. This chromosome abnormality occurs in about 1 in 700 live births in the general population. Individuals with the **"triple X syndrome"** are comparable in some ways with *Drosophila* metafemales (XXX). In *Drosophila*, however, such individuals are usually lethal, and those that survive are strikingly abnormal and sterile. Human XXX individuals are sometimes visibly indistinguishable from normal XX females, but there is considerable range in phenotypic expression. They may be mentally abnormal.

The best-known symptoms in this syndrome are abnormalities associated with functional processes such as menstruation. One patient cited by P. A. Jacobs was a 37-year-old female who reported that the first suggestion of an abnormality was highly irregular menstruation. When the abdominal wall was opened, the ovaries appeared as if they were postmenopausal. Microscopically, they showed deficient ovarian follicle formation. Of 63 cells observed, 51 had 47 chromosomes; the extra chromosome was an X. **Nondisjunction** in the production of the egg from which this woman developed was postulated as the mechanism for the occurrence of the extra chromosome. Buccal smears showed two sex chromatin bodies in the epithelial cells as expected when three X chromosomes are present. Individuals with tetrasomic X chromosomes (48,XXXX) are all mentally defective. The degree of mental deficiency increases with the number of X chromosomes present.

47,XYY AND BEHAVIOR

P. A. Jacobs and her associates reported in 1965 that seven XYY males were detected in a population of 197 male, mentally subnormal inmates of a penal institution in Scotland. The XYY (Fig. 13.19) men were unusually tall, with an average height of 73.1 inches, compared with 64 inches for XY men in the same prison. Numerous other studies, mostly in institutionalized populations, have since confirmed that a high proportion of XYY individuals are tall, subnormal in intelligence (with IQs ranging from 80 to 95), and **antisocial.** The aggressive behavior that brought them into conflict with the law was usually against property rather than people.

XYY trisomy occurs about once in 1000 live male births in the general European population. Only a few of these can be accounted for in the criminal population. Furthermore, most XYY men have been described as perfectly normal in behavior. Hook has shown that only 3.6 percent of all XYY men are institutionalized for any reason.

Environmental factors are presumed to be involved in the development of **aggressiveness.** Since some XYY men are subnormal in intelligence and excessively tall in stature, particular environmental situations in childhood or adulthood may lead to withdrawal from society or aggressive behavior. Unfavorable social conditions such as frustration in personal accomplishment and taunting from associates may encourage physical aggression as a means of adaptation. Males with this sex trisomy have not been found to transmit the extra Y chromosome to their sons. This extra chromosome seems to be weeded out in gametogenesis. A wide range of physical and mental abnormalities has been detected in institutionalized XYY men, but most of these are irregular in occurrence and do not form a syndrome. Tallness of stature and mental dullness,

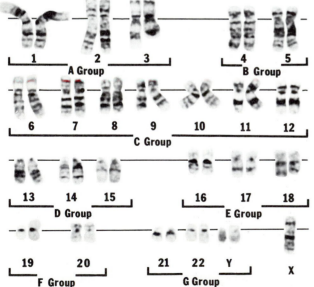

FIGURE 13.19

Karyotype of a boy showing an extra Y chromosome (47,XYY).
(Courtesy of S. W. Rogers.)

however, are fairly constant characteristics among institutionalized XYY men.

CHROMOSOMAL MOSAICS

Individuals who have at least two cell lines, with different karyotypes derived from one zygote, originate from nondisjunction in a cleavage division after fertilization. One daughter cell would thus receive one too many and the other would be one deficient, as illustrated in Fig. 13.20. Each cell would give rise to a cell line with its irregular chromosome number. Proportions of cells representing the different cell lines would vary in different tissues, making the extent of the mosaicism and the effect on the organism difficult to predict.

Many sex chromosome mosaics have been detected in human beings. The main phenotypic characteristic is **extreme variability.** Some sex chromosome mosaics have been reported—X/XX, X/XY, XX/XY, XXY/XX, XX/XXX, XXX/X, XXX/XXXXY—and several other combinations reflecting two or three cell lines. Mild to severe phenotypic symptoms have been associated with these sex chromosome mosaics.

NONDISJUNCTION AND ANEUPLOIDY IN *DROSOPHILA*

Bridges' example of the trisomic X chromosome (Chapter 4) was explained on the basis of primary nondisjunction of the X chromosomes. When fertilized with normal sperm, eggs with two X chromosomes produced two kinds of trisomics, **XXX and XXY.** The metafemales (XXX) were always **sterile** and usually **inviable,** but the XXY combinations produced females phenotypically indistinguishable from wild-type females. Zygotes with normal pairs of autosomes but single X or Y chromosomes were monosomics. Although those with a single X were normal in appearance but were sterile, zygotes with a single Y chromosome were inviable. Later investigations by Bridges showed that chromosome IV, which is very small, could be added or eliminated without seriously affecting the viability of the flies. When, however, the large II and III chromosomes

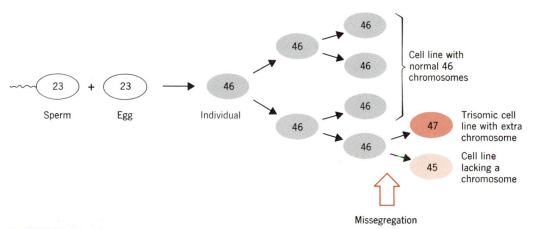

FIGURE 13.20

Mechanism through which chromosomal mosaicism may occur. A normal individual with 46 chromosomes is formed when sperm and egg unite. During an early cell cleavage, nondisjunction gives rise to a cell with 47 chromosomes and a cell with 45 chromosomes. Each of these cells initiates a cell line, thus forming a mosaic individual with cells carrying different numbers of chromosomes.

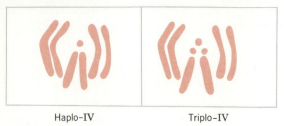

Haplo-IV   Triplo-IV

FIGURE 13.21

Trisomic called "triplo-IV," produced by the addition of a fourth chromosome, and a monosomic called "haplo-IV," which results from the elimination of one fourth chromosome in Drosophila.

were lost or added, the resulting cells were always inviable.

A monosomic $(2n - 1)$ called **haplo-IV** was obtained through the elimination of one IV chromosome. Through the addition of the small fourth chromosome to the wild-type complement (Fig. 13.21), a trisomic $(2n + 1)$ called **triplo-IV** was obtained. Haplo-IV flies are small with slender bristles and deviate in several minor respects from the wild-type. Triplo-IV and diplo-IV flies cannot be separated on the basis of phenotype. Bridges considered triplo-IV to have a darker body than wild type with coarser bristles, smaller eyes, and more elongate wings, but the departure from normal is very slight and the phenotype overlaps with normal.

Chromosome IV genes, such as *ey* for eyeless (a phenotype characterized by small or missing eyes), segregate differently with different chromosome

numbers, as expected. When $2n$ eyeless flies (ey/ey) were crossed with haplo-IV (Fig. 13.22), the F_1 generation consisted of normal $2n$ and eyeless haplo-IV flies. When eyeless flies were crossed with triplo-IV, none of the first-generation progeny were eyeless. About half were triplo-IV, and half were normal (diplo-IV), as expected. When F_1 triplo-IV females were crossed with eyeless (ey/ey) males, as shown in Fig. 13.23, the progeny again consisted of about half triplo-IV and half diplo-IV. Normal and eyeless phenotypes were present in a ratio of about $5:1$, respectively. This **trisomic ratio** is explained on the basis of the extra chromosome in the triplo-IV flies and the dominance of $1ey^+$ over $2ey$.

ANEUPLOID SEGREGATION IN PLANTS

The first critical study of aneuploid plants was initiated in 1924 by A. F. Blakeslee and J. Belling when they discovered a "mutant type" with 25 chromosomes in the common Jimsonweed, *Datura stramonium*, which normally has 24 chromosomes in the somatic cells. At the meiotic metaphase, one of the 12 pairs was found to have an **extra member;** that is, one **trisome** $(2n + 1)$ was present along with 11 disomes. This trisomic plant differed from wild-type plants in several specific ways, particularly in shape and spine characteristics of seed capsules. Because the complement was composed of 12 chromosome pairs differing in the genes they carried, 12 distinguishable trisomics were possible in Jimsonweed. Through experimental breeding, Blakeslee and his associates succeeded in producing **all 12 possible trisomics.** These were grown in Blakeslee's garden, and each was found to have a distinguishable phenotype that was attributed to the extra set of genes contained in one of the 12 chromosomes.

One of the 12 trisomic types, known as poinsettia, had several distinguishing traits, including morphological characteristics of seed capsules, that were attributed to the basic trisomic arrangement. It was also possible to identify some traits that were determined by genes on particular chromosomes by trisomic ratios; that is, $5:1$, $17:1$ (Fig. 13.24), and $35:1$, in contrast to $3:1$ and $1:1$ expected from

Diplo-IV eyeless female   Haplo-IV male

   ey ey   ×   *ey*⁺

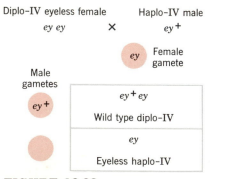

FIGURE 13.22

Cross between a diplo-IV eyeless female and a haplo-IV male.

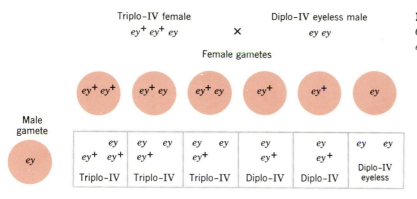

FIGURE 13.23

Cross between a triplo-IV female and a diplo-IV eyeless male.

monohybrid crosses in regular $2n$ plants. The extra chromosome in poinsettia, for example, was found to carry the locus for alleles p^+ and p, for purple or white flowers, respectively. Any one of three chromosome arrangements with the dominant gene p^+ ($p^+ p^+ p^+$, $p^+ p^+ p$, or $p^+ p p$) produced poinsettia plants with purple flowers, whereas only one, the fully recessive trisomic ($p p p$), gave rise to poinsettia plants with white flowers. The $2n$ plants had two chromosome arrangements ($p^+ p^+$ and p^+p) for purple and one ($p p$) for white.

Plant trisomics experience interesting **complications** when undergoing **meiosis**. Two chromosomes go to one pole and one goes to the other in the reduction division of megasporogenesis, thus giving rise to different kinds of gametes, some with two and some with one member of the trisome. Trisomic ratios reflect increased proportions of progeny carrying wild-type dominant alleles as compared with diploid ratios. In the Jimsonweed, developing megaspores tolerate extra chromosomes and form gametes with relatively little loss of viability. When, however, additional chromosomes above the n number enter developing microspores, the resulting spores cannot successfully compete against those with the normal haploid complement. A cross between two purple poinsettia plants $p^+ p^+ p \times p^+ p^+ p$ [seed parent (female) written first] may be constructed as illustrated in Fig. 13.24. **Female gametes** are of four kinds, two haploid and two carrying an extra chromosome, in the proportion: $2p^+$, $2p^+ p$, $1p^+ p^+$, $1p$. Because male gametes receiving extra chromosomes are nonfunctional, only two kinds, p^+ and p, occurring in the proportion $2p^+$ and $1p$, were involved in fertilization. All poinsettia ($2n + 1$) that resulted from the cross carried at least one p^+ gene and had purple flowers. The $2n$ plants occurred in the proportion of 8 purple to 1 white. Complete sets of trisomics have since been discovered for other plants, including rice (Fig. 13.25). The different trisomics may be distinguished by characteristics of the grain as well as other features of the plants.

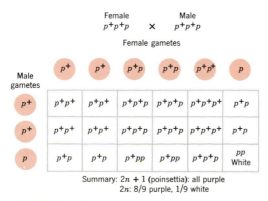

FIGURE 13.24

Cross between two purple trisomic plants of the genus Datura, *illustrating a trisomic ratio of 17:1.*

TETRASOMICS AND NULLISOMICS

Aneuploids more complex than trisomics have been produced in some species, but ordinarily these are highly inviable. **Tetrasomics ($2n + 2$)** have been identified in wheat, but no phenotypic characteris-

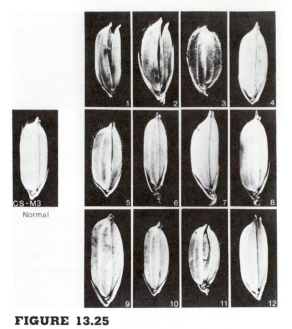

CS-M3
Normal

FIGURE 13.25

A set of 12 primary trisomics of rice, variety CS-M3, showing the grain characteristics of each type along with the normal diploid. The numbers correspond to the extra chromosome in each trisomic in decreasing order of its length. (Courtesy of S. H. Khan and J. N. Rutger, University of California, Davis, and ARS, USDA, Davis, California.)

tics were observed that could distinguish them from trisomics. Another variation that could be expected to occur in an aneuploid series is the complete absence of a certain kind of chromosome ($2n - 2$). Plants in which a chromosome pair is completely missing are called **"nullisomics."** These may be produced occasionally in nature, but seldom survive long enough to be recognized or to perpetuate the chromosome type. E. R. Sears experimentally produced all the 21 possible nullisomics in wheat, *Triticum aestivum*. By associating certain phenotypes with corresponding chromosome arrangements, nullisomics have been used effectively in locating several different genes in wheat.

EUPLOIDY

While aneuploids differ from standard $2n$ chromosome complements in single chromosomes, euploids differ in multiples of n or of x if $n = x$. Monoploids (n) carry one genome. The n or x chromosome number is usual for gametes of diploid animals, but unusual for somatic cells. Monoploidy is seldom observed in animals, but is found in the male honeybee and other insects in which male haploids occur.

By contrast, plants have a gametophyte stage in their cycle that is characterized by the reduced (n) chromosome number. In higher plants, this stage is brief and inconspicuous, but in some lower plant groups it is the major part of the cycle. Occasionally, plants in natural populations or plots can be recognized as monoploids by observation and verified by cytological procedures. These plants are usually frail in structure, with small leaves, low viability, and a high degree of sterility. Sterility is attributed to **irregularities at meiosis.** Obviously, no pairing is possible because only one set of chromosomes is present. Therefore, if the meiotic process succeeds at all, the dispersal of chromosomes to the poles is irregular, and the resulting gametes are highly inviable. Because monoploids undergo no segregation and carry a single set of genes, they are valuable experimental tools when they can be produced successfully. Microorganisms that are propagating monoploids are especially useful in genetics (Chapter 7). Diploid plants with two genomes ($2n$) are most common among euploids. Normal chromosome behavior in animals and plants is based on diploids, which are used in the following examples as standards for comparison.

POLYPLOIDY IN ANIMALS

Organisms with three or more genomes are polyploids. Fully one-half of all known plant genera contain polyploids, and about two-thirds of all the grasses are polyploids, but polyploids are rarely seen in animals. One reason is that the sex balance in animals is much more delicate than that in plants. **Sterility** in animals is virtually always associated with a departure from the diploid number. The few animals (such as the brine shrimp, *Artemia salina*) that show evidence of polyploidy utilize **parthenogenesis** to escape the hazard of anomalous gametes.

An exceptional case of triploidy in salamanders

related to *Ambystoma jeffersonianum* has been reported. Female salamanders of this particular group that have large erythrocytes and erythrocyte nuclei produced some triploid larvae with 42 chromosomes, whereas those with small erythrocytes and erythrocyte nuclei produced diploid larvae with 28 chromosomes. Field observations and laboratory studies indicated that distinct, persisting populations of triploid females have become established in parts of the range occupied by this species complex.

Although animals composed entirely of polyploid cells are rare, many diploid animals have polyploid cells within certain tissues of their bodies. For example, in teleostean embryos, giant nuclei (presumably polyploid) have been observed in many species. Giant polyploid nuclei occur in particular tissues of a wide range of diploid animals (e.g., the liver and kidneys in humans).

POLYPLOIDY IN PLANTS

SOMATIC AND GERM CELL DOUBLING

Two basic irregular processes have been discovered by which polyploids may evolve from diploid plants and become established in nature. (1) With **somatic doubling,** cells sometimes undergo irregularities at mitosis and give rise to meristematic cells that perpetuate these irregularities in new generations of plants. (2) Reproductive cells may have an irregular reductional or equational division in which the sets of chromosomes fail to separate completely to the poles in anaphase. Both sets thus become incorporated in the same **"restitution"** nucleus, which doubles the chromosome number in the gamete. Both of these irregularities occur in nature. Once polyploidy is established, intercrossing among plants with different chromosome numbers may give rise to numerous chromosome combinations. Most of these are sterile, but some may be fertile and come under the influence of natural selection. All degrees of viability are encountered — from lethal combinations to those that compete favorably with diploids in particular environmental situations.

Two main kinds of polyploids, **autopolyploids** and **allopolyploids,** may be distinguished on the basis of their source of chromosomes. Autopolyploids occur when the same genome is duplicated (e.g., $n_1 + n_1$). Apparently, this occurs frequently in single cells of many plants, but these cells usually do not survive. Allopolyploids result when different genomes come together through hybridization (e.g., $n_1 + n_2$). Among the surviving plants, it is usually impossible to determine whether the genomes are alike and therefore whether the polyploids are autopolyploids or allopolyploids unless information from ancestral history is available and detailed chromosome studies are performed.

The presence of varying numbers of quadrivalents (i.e., four homologous chromosomes instead of the usual two pairing with one another in meiosis) suggests autopolyploidy. Unequal segregation of chromosomes in quadrivalents is the main reason why **autopolyploids are sterile** to varying degrees. In the meiotic prophase, chromosomes must pair with one another throughout their entire length. When four similar chromosomes are present, they usually pair with different chromosomes at different places along their length, complicating the meiotic process and resulting in nonviable gametes or zygotes. Unequal chromosome pairing is not, however, the whole basis for sterility in autopolyploids. **Unequal disjunction** occurs and perhaps more important phenomena, none of which is well enough established to permit discussion here. Chromosomes in some autopolyploids appear to pair properly but form bivalents rather than quadrivalents.

Some plant groups have a series of chromosome numbers based on a multiple of a basic number. In the genus *Chrysanthemum*, for example, the basic number is 9, and species are known that have 18, 36, 54, 72, and 90 chromosomes. In *Solanum*, the genus of nightshades including the potato, *S. tuberosum*, the basic number is 12. Members of this genus include species with 24, 36, 48, 60, 72, 96, 108, 120, and 144 chromosomes. In spite of such conspicuous examples where autopolyploidy would appear superficially to be involved in the origin of plants with different chromosome numbers, however, it is doubtful that autopolyploidy alone has played a major role in the evolution of plant groups. Inviability and sterility would seem to preclude the perpetuation of true autopolyploids in nature. Autopolyploidy combined with allopolyploidy, however, produces **"autoallopolyploids"** and has apparently been an important process in the evolution of some plants.

TRIPLOID PLANTS

Triploids ($3n$) with three complete genomes occur when tetraploid ($4n$) or unreduced diploid ($2n$) plants produce viable $2n$ gametes that unite in fertilization with normal n gametes. Reduction sometimes fails in diploid plants, resulting in gametes with more than a single genome. But some $2n$ gametes are produced, fertilized with n gametes, and viable triploids occur in nature. The triploids, however, do not ordinarily reproduce and become established because of irregularity during meiosis that results in low survival and sterility of the survivors. **Synapsis** in meiosis can take place in several ways, but only between two homologous chromosomes in the same region. A $2n + n$ segregation may occur in three ways:

$$\frac{1+2}{3} \quad \frac{1+3}{2} \quad \text{and} \quad \frac{2+3}{1}$$

The possibility of getting either a $2n$ or an n chromosome set in a gamete is remote. The few gametes that are functional become fertilized and are mostly **unbalanced and inviable.** Meiotic segregation of homologs results in gametes with varying members from n to $2n$, with all integral values in between. In this respect, triploids are like other polyploids with odd numbers of genomes ($5n$, $7n$, and so on) in sexually reproducing organisms. They are all sterile.

But there is a way for triploids to be propagated. Like other plants, they can be **propagated vegetatively.** Bananas are triploids that are propagated through asexual cuttings. Winesap, Gravenstein, and Baldwin apples and European pears are perpetuated by grafting and budding and thus maintain their triploid characteristics. Tulips with three sets of chromosomes ($3n$) are also propagated by vegetative bulbs. Triploids occur in grasses, forest trees, vegetables, and flower garden varieties. **Seedless watermelons** and other seedless fruits are often triploids. Among the animals, triploids are very rare, but they have been found in *Drosophila*, salamanders, lizards, and the land isopod, *Triconiscus*. On the whole, triploids have been unsuccessful in establishing themselves in natural populations because the usual consequence of triploidy is sterility. Triploids have, however, provided opportunities for human ingenuity in genetic engineering to develop species with odd ploidy for practical use.

TETRAPLOID PLANTS

Tetraploids ($4n$) with four genomes frequently originate from a **doubling of diploids.** They also may result from the **duplication** of somatic chromosomes following irregularities at mitosis. If the spindle does not develop properly in the mitotic sequence of a diploid, and cell division fails to follow chromosome duplication, a single nuclear membrane may develop around the two sets of chromosomes that ordinarily would produce daughter nuclei, forming a single restitution nucleus. If this tetraploid cell perpetuates itself through normal mitosis, the increased chromosome number may become established in a group of cells or tissues within the organism. When such plants are capable of vegetative reproduction, they may be manipulated to produce whole tetraploid plants. Failure at reduction division in the oöcytes of some plants also results in polyploids. Chromosome irregularity in mature pollen is rare because developing male gametes with irregular chromosome numbers do not compete favorably with normal gametes.

Early in the nineteenth century, some seeds of the American saltmarsh grass (*Spartina alterniflora*) were accidentally transported by ship to Bayonne, France, and to Southampton, England. The American species became established in the same localities where a European saltmarsh grass (*S. maritima*, formerly *S. stricta*) was growing. A new saltmarsh grass, *S. townsendii* (currently known as *S. anglica*), commonly called Townsend's grass, was later identified in these localities. By 1907, it had become common along the coast of southern England and northern France. Townsend's grass was more vigorous and aggressive than either the American or the European species and crowded out the native grasses in many places. It therefore was intentionally introduced into Holland to support the dikes, and it was also imported into other localities for similar purposes. Townsend's grass was considered to be a hybrid between the American and European species; but, unlike most hybrids, it was fertile and true breeding. The European grass had $2n = 60$ chromosomes, the American species $2n = 62$, and Townsend's grass $2n = 122$; some plants had $2n = 120$ and others 124 (Fig. 13.26). These facts suggested that a cross had occurred in which allopolyploidy was involved. Townsend's grass was an "amphidiploid," with the sum of the diploid chromosomes carried by the two species. This evidence, along with the high fertility and intermediate appearance, indicated that the new plant arose from natural hybridization and doubling of chromosomes. The chromosome doubling had presumably given the hybrid its **fertility and ability to survive.**

The low fertility and marked phenotypic variation associated with chromosome irregularity are illustrated in Fig. 13.27. A cross was made between Polish wheat, *T. polonicum*, a tetraploid with large, amber kernels, and Marquis, a hexaploid with hard, red kernels. The entire F_2 from the cross is shown in the illustration. Only a few plants were produced from an extensive experiment, and those observed varied widely in phenotype.

Allopolyploidy has occurred in various plant groups, and some present-day plants have resulted from this kind of hybridization. Polyploid species that are established in nature have genomes that correspond more or less completely to the combined chromosome complements of two different but related diploid plants (Fig. 13.28). These "amphidiploids" or allotetraploids have undergone hybridization somewhere in their ancestral history. Allopolyploidy thus represents a method by which new species may be formed almost immediately, whereas autopolyploidy alone results in meiotic anomalies and "dead ends" with reference to evolution.

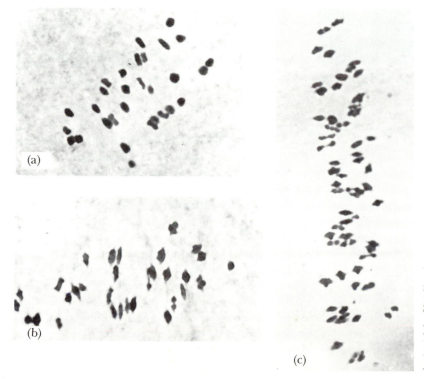

(a)

(b)

(c)

FIGURE 13.26

Metaphase I (paired) chromosomes of two parent species and the amphidiploid hybrid, Townsend's grass. (a) European marshgrass Spartina maritima, *2n = 60; (b) American saltmarshgrass* S. alterniflora, *2n = 62; (c) Townsend's grass* S. townsendii (S. anglica), *2n = 122 (some plants had 2n = 120 and some 2n = 124). (Courtesy of C. J. Marchant.)*

FIGURE 13.27

Heads of wheat resulting from an F_2 cross between Polish with 28 chromosomes and Marquis with 42 chromosomes. These are the only heads produced from an extensive experiment, indicating the low fertility encountered in crosses between plants with different chromosome numbers.

INDUCED POLYPLOIDY

Polyploids have been induced experimentally by several methods in various plants. Anything that interferes with spindle formation during mitosis might result in a doubling of the chromosomes. Induced polyploidy was first demonstrated by subjecting growing plants to a higher than usual temperature. Maize and some other plants responded to such treatment with an increase in the chromosome number of certain cells. Some of these cells gave rise to germinal tissue and whole plants were propagated. Other such cells were cultured artifically and polyploid plants were produced. When, for example, the buds of tomato plants were removed, cells of some shoots developing from the **scar tissue** were **tetraploid.** These were propagated and whole plants with 4*n* chromosomes were produced.

The method of inducing polyploidy in plants that has become most widely used was developed by A. F. Blakeslee, A. G. Avery, and B. R. Nebel. These investigators found that an alkaloid, **colchicine,** extracted from the autumn crocus, *Colchicum autumnale,* could **disturb spindle formation during cell division.** When root tips or other growing plant parts were placed in appropriate concentrations of colchicine, chromosomes of the treated cells duplicated properly, but spindle formation was inhibited and the cytoplasmic phase of cell division did not occur. Instead, restitution nuclei with different numbers of chromosomes were produced in the treated tissue. Some cells had a completely doubled chromosome number. When these cells were propagated, tetraploid plants were produced and tetraploid seed was obtained.

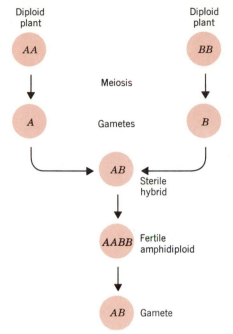

Diploid
plant

AA

Diploid
plant

BB

Meiosis

A

Gametes

B

AB

Sterile
hybrid

AABB Fertile
amphidiploid

AB Gamete

FIGURE 13.28

The production of a fertile amphidiploid (AABB) by doubling of the chromosomes of a sterile hybrid (AB) produced by crossing two normal diploid plants (AA and BB). A represents a haploid set of chromosomes from diploid plant AA, and B represents a haploid set of chromosomes from diploid plant BB. Since A chromosomes are not homologous with B chromosomes, meiosis is highly irregular in hybrid AB, resulting in a high degree of sterility. The fertile amphidiploid (AABB) may be produced by the rare union of diploid (AB) gametes that are produced by the hybrid AB. In the amphidiploid, meiosis is normal—A chromosomes pair with A chromosomes and B chromosomes pair with B chromosomes.

In some plants, growing areas at the stem tips and lateral buds have three distinct cell layers, as illustrated in Fig. 13.29. Cells from each layer are much alike in early stages of development, but later they give rise to separate tissues in stems, leaves, and other organs. The outer layer (I) becomes the epidermis; the middle layer (II) gives rise to the reproductive cells (eggs and pollen); and the inner layer (III) produces the internal parts of stems and leaves. Colchicine placed in the medium of the growing tips may interfere with division and result in transformation of $2n$ cells into $4n$ cells in one or more layers. Cells usually divide vertically in such a way that the number within a given layer is increased. In exam-

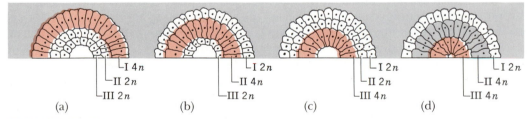

(a) I 4n II 2n III 2n (b) I 2n II 4n III 2n (c) I 2n II 2n III 4n (d) I 2n II 4n III 4n

FIGURE 13.29

Mechanism through which reproductive cells of a plant may have 4n chromosomes and produce tetraploids. Colchicine added to the medium interferes with cell division and transforms 2n cells to 4n cells. (a) The two inner layers are 2n and the outer layer is 4n. (b) Horizontal duplication has resulted in a 4n cell, which gave rise to a 4n middle layer. (c) The inner layer has become 4n. (d) The inner layer and middle layer have become 4n, and 4n reproductive cells may produce 4n (tetraploid) plants under experimental treatment.

ples (a), (b), and (c) in Fig. 13.29, the 4n cells are restricted to the first, second, and third layer, respectively. Sometimes cells divide horizontally and the daughter cells enter a new layer. Thus, a doubled (4n) cell in layer III may give rise to a 4n cell in layer II, as shown in (d). This pattern of irregularity may extend 4n cells to the reproductive tissue and thus perpetuate 4n cells in a new plant. By propagating tetraploid tissue, it is possible to produce **tetraploid plants.**

EXPERIMENTAL PRODUCTION OF POLYPLOIDS

In an early experiment, the Russian cytologist G. D. Karpechenko sythesized a polyploid from crosses between two common vegetables that belong to different genera, the radish, *Raphanus sativus,* and the cabbage, *Brassica oleracea.* Although these plants were only distantly related, they were enough alike to be crossed successfully with the intervention of the experimenter. Both had 9 pairs of chromosomes. The diploid hybrid had 18 chromosomes, 9 from each parent, but it was sterile and could not perpetuate itself, largely because of the failure in pairing between the unlike chromosomes in meiosis. Some unreduced gametes were formed, however, and in the F$_2$ population Karpechenko recovered some tetraploids. When the chromosomes of the F$_1$ hybrid were doubled in this way, a fertile polyploid named *Raphanobrassica* was produced with 18 radish and 18 cabbage chromosomes. Because two sets of chromosomes were now present from each parental variety, pairing was quite regular. Normal gametes were produced and a high degree of fertility was obtained. This experiment had theoretical significance because it demonstrated a method by which fertile interspecific hybrids could be produced. It also suggested the possibility of incorporating desirable genotypes from two different species into a new polyploid species. Seed capsules of the parents, the sterile hybrid, and the tetraploid plants of Karpechenko's experiment are shown in Fig. 13.30. Unfortunately, from the practical standpoint, *Raphanobrassica* had the foliage of a radish and the root of a cabbage.

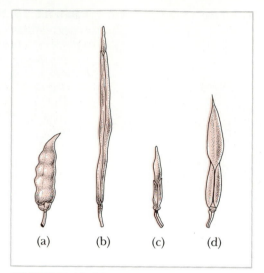

FIGURE 13.30

Seed pods of (a) radish (Raphanus), *with 18R chromosomes; (b) cabbage* (Brassica), *with 18B chromosomes; (c) sterile diploid hybrid with 9R + 9B chromosomes; and (d) tetraploid resulting from chromosome doubling, with 18R + 18B chromosomes. (After Karpechenko.)*

STERILE HYBRID GRASS MADE FERTILE BY CHROMOSOME DOUBLING

Interesting experiments that made use of induced polyploidy and cytological analysis in certain grasses have been conducted by W. S. Boyle, A. H. Holmgren, and their associates at Utah State University. A completely sterile perennial grass was observed in Cache Valley, Utah, and in several other locations. On the basis of its sterility and morphological characteristics, the grass was tentatively identified as a natural, sterile hybrid between two genera in the tribe Hordeae. Examination of the Hordeae species in the vicinity of the hybrids indicated that the two parents were probably *Agropyron trachycaulum* and *Hordeum jubatum* (Fig. 13.31). Cytological studies on the two presumed parents and the hybrid supported this view. Both *A. trachycaulum* and *H. jubatum* were found to be allotetraploids ($2n = 28$). During meiosis of both species, normal pairing of chromosomes occurred to form 14 paired chromosomes (bivalents). The sterile hybrid was also a tet-

FIGURE 13.31
Parent plants representing different genera and a hybrid produced by these plants (a) Hordeum jubatum; (b) F_1 *hybrid;* (c) Agropyron trachycaulum. *(Courtesy of W. S. Boyle.)*

raploid with the same chromosome number ($2n = 28$), but its chromosome behavior during the meiotic process was highly irregular (Fig. 13.32). Fourteen unpaired chromosomes (univalents) and seven bivalents were frequently observed at metaphase in pollen mother cells. Many lagging chromosomes remained in the center of the spindle during anaphase, and numerous small micronuclei, reflecting chromosome irregularity, were observed following division. No viable pollen was produced. These observations indicated that **meiotic irregularity** was a major factor in the sterility.

Sterile hybrids were produced through controlled reciprocal crosses between *A. trachycaulum*

and *H. jubatum,* thus confirming the predicted parentage. The hybrids were then treated with colchicine. Some stalks or clums with double chromosome numbers (octoploids) were produced and set seed. All this seed was viable and produced **fertile** plants considered to be autoalloploids. This name was applied because both autopolyploidy and allopolyploidy had entered into the formation of the fertile octoploid. Chromosome studies indicated that the two parents carried a genome in common. Therefore, the genomic formulas of the parents were represented as *AABB* and *AACC.* The sterile hybrid was *AABC* and the colchicine-induced octoploid was *AAAABBCC* (Fig. 13.33).

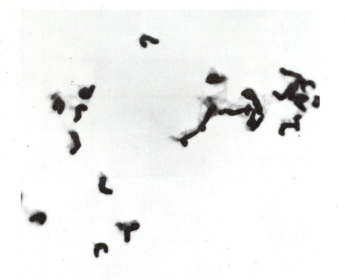

FIGURE 13.32

Chromosomes of the sterile hybrid between A. trachycaulum *and* H. jubatum *during meiosis. Seven bivalents and 14 single chromosomes were frequently found as in this photograph. (Courtesy of W. S. Boyle.)*

APPLICATIONS OF POLYPLOIDY

Among the cultivated varieties of wheat, three different chromosome numbers are represented: 14, 28, and 42 ($x = 7$). For example, the primitive small-grained einkorn type of Europe and Asia, *Triticum monococcum*, has 14 chromosomes in its vegetative cells. Its yield is low and it is of comparatively little value. An emmer wheat (durum), *T. dicoccum*, grown chiefly in southern Europe but also in the United States, has 28 chromosomes. It has thick heads with large hard kernels and is used mainly for macaroni, spaghetti, and stock feed. The **bread wheats**, *T. aestivum*, with 42 chromosomes, were postulated by J. Percival in England to have come from a cross between emmer wheat and goat grass (*Aegilops*), both of which are native to the Babylonian region where bread wheat originated.

When techniques for artificial chromosome doubling became established, investigations of the origin of bread wheat confirmed Percival's theory.

Experimental evidence obtained by E. S. McFadden and E. R. Sears and separately by H. Kihara traced the pathway for the origin of one type of bread wheat, *T. spelta*.

Aegilops squarrosa ($n = 7$) was found to carry a group of major characteristics that distinguish the hexaploid ($n = 21$) *T. spelta* from the tetraploids ($n = 14$) *T. dicoccum* and *T. dicoccoides*. Hybrids between these tetraploid species of wheat and *A. squarrosa* proved to have all of the major taxonomic characters of *T. spelta* but the hybrids were completely or nearly sterile. When the F_1 hybrids of *T. dicoccoides* $\times$ *A. squarrosa* were treated with colchi-

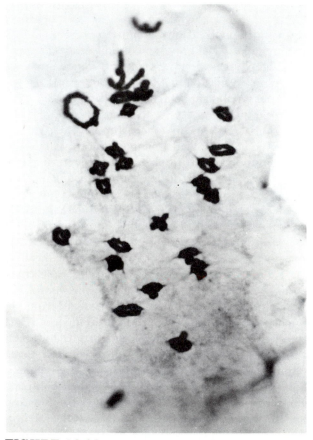

FIGURE 13.33

Diakinesis stage of colchicine-induced octoploid. In this photograph the following chromosome associations are present: 1 ring of 4; 22 rings of 2; 3 rods of 2; and 2 single chromosomes, making a total of 56. (Courtesy of W. S. Boyle.)

cine, highly fertile allopolyploids with 42 chromosomes were obtained. These synthetic hexaploids closely resembled the cultivated *T. spelta,* and they produced highly fertile hybrids with that species and with *T. vulgare,* known to be in the ancestry of the bread wheats. This demonstrated that the genome of the hexaploid wheats corresponded to one chromosome set of *A. squarrosa.* It was postulated that *T. spelta* is the ancestral hexaploid wheat of Europe, having arisen, possibly in fairly recent times, in southeastern Europe or southwestern Asia following chromosome doubling of natural hybrids of *T. dicoccoides* (or its cultivated close relative, *T. dicoccum*) × *A. squarrosa. T. spelta* is believed to have been carried over the northerly route into central and western Europe. Experiments of McFadden, Sears, and Kihara reconstructed the pathway through which a moderately useful wheat and a goat grass **hybridized in nature** and produced forerunners of a most valuable crop, bread wheat.

NEW WORLD COTTON

Crosses can be made between distinct species of cotton, members of the genus *Gossypium.* The hybrids show a wide range of vigor and fertility, making the material favorable for studies of origins. Three cytological groups have been found to correspond with the major world distributional areas. Old World cotton had 13 pairs of large chromosomes. American cotton, which originated in Central or South America, has 13 pairs of small chromosomes. New World cotton (the cultivated long-staple type) has 26 pairs, 13 large and 13 small. Evidently, **hybridization and chromosome duplication** occurred somewhere in the ancestry of the New World cotton.

J. O. Beasley used the colchicine technique and succeeded in doubling the chromosomes of a hybrid between the Old World and American cotton. The resulting hybrids, with four sets of chromosomes (amphidiploids), crossed readily among themselves and produced fertile plants resembling New World cotton. The process by which the valuable **polyploid cotton** may have originated in nature was thus duplicated in the laboratory.

PRIMROSE HYBRIDIZATION

The primrose, *Primula kewensis,* is an allotetraploid with 36 ($2n$) chromosomes. It was derived from a cross between two diploids, *P. floribunda* ($x = 9$) and *P. verticillata* ($x = 9$). Plants from these two **species crossed** readily, producing hybrids with 18 chromosomes in their vegetative cells, 9 from *P. floribunda* and 9 from *P. verticillata,* but the hybrids were **sterile.** Eventually, however, a branch on a hybrid plant developed from a cell in which the chromosome number was doubled (36), so that each chromosome had a homologous partner. This branch was propagated and gave rise to a fertile primrose plant with cells containing 36 chromosomes. In Fig. 13.34, metaphase chromosomes of the two diploid parents, the sterile diploid hybrid, and the **fertile allotetraploid** are shown.

TOBACCO RESISTANCE

Induced polyploidy has been exploited to a great extent. Practical applications may become more common as additional data are accumulated. By artificially induced polyploidy, **disease resistance** and other desirable qualities have been incorporated into some commercial crop plants. Tobacco, *Nicotiana tabacum,* for example, is susceptible to the tobacco mosaic virus (TMV), whereas *N. glutinosa* appeared at first observation to be resistant. Further investigation, however, showed that in *N. glutinosa* the virus killed the cells that were invaded and the virus particles became isolated in the dead cell. The apparent resistance thus was attributable to hypersensitivity. When the two tobacco species were crossed, the hybrid was found to be "resistant" to the virus, but totally sterile. When the chromosomes were doubled, it was possible to secure a fertile polyploid "resistant" to the virus.

POLYPLOID FRUITS, FLOWERS, AND WHEAT

Some varieties of plants that serve human needs more effectively than others have now been identified as polyploids. Many polyploids were selected and cultivated because of their large size, vigor, and ornamental values, before their chromosome num-

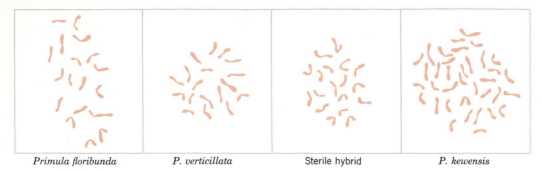

Primula floribunda	*P. verticillata*	Sterile hybrid	*P. kewensis*

FIGURE 13.34

Chromosomes of Primula floribunda, P. verticillata, *the sterile hybrid, and the allotetraploid,* P. kewensis. *The two parents and the hybrid each have 18 chromosomes, whereas the fertile tetraploid* P. kewensis *has 36 chromosomes.*

bers were known. Giant "sports" from twigs of **McIntosh apple** trees that were found to be tetraploid (4*n*) were propagated into whole trees, which produce extra-large fruit. The texture of the giant apples is as fine as that of diploids, but the yield is inferior. Mass selection of seedlings may overcome this difficulty. **Bartlett pears,** several varieties of **grapes,** and **cranberries** have also produced sports with **giant** fruits. some of these show promise of practical usefulness. With colchicine treatment, a number of polyploids have been developed artificially. This technique has provided a way to explore the mechanism involved in **polyploid formation** and to make use of the good qualities of polyploids. Tetraploid (4*n*) maize is more vigorous than the ordinary diploid and produces some 20 percent more vitamin A. Its fertility is somewhat reduced, but this drawback responds to selection. Polyploid watermelons have been developed from colchicine treatment by Kihara and others. The tetraploid with 44 chromosomes is large and has practical value. Triploid watermelons with 33 chromosomes are especially desirable because they are sterile and have no seeds. Among the flower garden varieties, **4*n* marigolds** and snapdragons are widely cultivated.

Polyploid plants respond to artificial selection and hybridization, as do diploid species. The recent history of plant breeding has been characterized by a marked improvement in many polyploid plant crops. The yield of wheat, for example, has increased appreciably. This has been accomplished by developing disease-resistant strains and breeding for increased hardiness and greater efficiency so that crops may survive under the various environmental conditions found in wheat-growing areas.

A constant threat to the wheat crop is rust—a fungus that attacks the stems and leaves of the growing plants and destroys the ripening grain. Spores are borne by wind and, when conditions are right, they spread like fire through wheat fields. The disease can be combated by developing **rust-resistant** strains and by eradicating barberry bushes, which are hosts to the spores during the spring months. But new varieties of rust that destroy previously resistant grain keep evolving, thus perpetuating the job of plant breeders. The advantage of incorporating rust resistance is illustrated in Fig. 13.35. The larger kernels at the left are from a new strain of rust-resistant spring wheat. At the right are shown kernels of wheat, similar in other respects but not resistant, that are dwarfed from infection with stem rust. The number of kernels of grain per plant as well as the size of the kernels is decreased by rust infection. Investigators in agricultural experiment stations are constantly alert for new rusts. When a new one is found, the standard wheat varieties are tested against it. If they are not resistant, breeding programs are initiated immediately to develop **new strains** resistant to that particular rust.

FIGURE 13.35
Wheat kernels illustrating the advantage of rust resistance. Left, rust-resistant wheat; right, rust-susceptible wheat infected in nature, similar in other respects to the strain represented at left. (Courtesy of Utah State Experiment Station.)

SUMMARY

A particular chromosome number is a characteristic of cells of plant and animal species. The normal range in different organisms extends from one pair to several hundred chromosomes. Euploid plants have a multiple of a basic chromosome number (n or x) for the species ($2n$, $3n$, $4n$, and so on). Aneuploids deviate from a base number by one, two or a few chromosomes (e.g., $2n + 1$; $2n - 1$). Human beings have $2n = 46$ chromosomes. Irregularities in number, such as a loss of one or the gain of one or more chromosomes, have been associated with abnormal syndromes. Nondisjunction of chromosomes in meiosis will account for most aneuploid varieties. Irregularities in mitotic division will account for chromosome mosaics in individual organisms. Polyploids, which are multiples of whole sets of chromosomes (e.g., $2n$, $3n$, $4n$), are common among plants but uncommon among animals. In some plant groups, polyploidy has been an important factor in evolution. By using colchicine and some other agents that interfere with cell division, polyploids can be induced. In wheat and cotton, induced poly-ploidy has paralleled what has occurred in nature; induced polyploidy also has practical value in such plants as apples, grapes, and tobacco.

REFERENCES

Blakeslee, A. F. 1941. "Effect of induced polyploidy in plants." *Am. Natur.* 75:117–135.

de Grouchy, J., and C. Turleau. 1977. *Clinical atlas of human chromosomes.* John Wiley & Sons, New York.

McFadden, E. S., and E. R. Sears. 1946. "The origin of *Triticum spelta* and its free-threshing hexaploid relatives." *J. Hered.* 37:107–116.

Raven, P. H., and R. Evert. 1976. *Biology of plants,* 2nd ed. Worth Publishers, New York.

Sears, E. R. 1948. "The cytology and genetics of wheats and their relatives." *Adv. Genet.* 2:240–270. M. Demerec (ed.). Academic Press, New York.

Shepard, J., D. Bidney, T. Barsby, and R. Kemble. 1983. "Genetic transfer in plants through interspecific protoplast fusion." *Science* 219:683–688.

Smith, G. F., and J. M. Berg. 1976. *Down's anomaly.* Churchill Livingstone, New York.

Sparks, R. S., D. E. Comings, and C. F. Fox. 1977. *Molecular human cytogenetics.* Academic Press, New York.

Swanson, C. P., and P. Webster. 1977. *The cell,* 4th ed.

Prentice-Hall, Englewood Cliffs, New Jersey.

Uchida, I. A. 1970. "Epidemiology of mongolism: the Manitoba study." *Annals N. Y. Acad. Sci.* 171:361–369.

Uchida, I. A., and C. C. Lin. 1973. "Identification of partial 12 trisomy by quinacrine fluorescence." *J. Pediat.* 82(2):269–272.

PROBLEMS AND QUESTIONS

13.1. According to the Lyon hypothesis (Chapter 4), all but one X chromosome in multi-X individuals are inactivated and form sex chromatin bodies. How many sex chromatin bodies would be expected to occur in a cell from a person with (a) the Turner syndrome, (b) trisomic X, and (c) the Down syndrome female?

13.2. How can the origin of trisomy in humans be explained?

13.3. If the Down syndrome occurs in 1 out of every 700 births in the general population, (a) what is the chance that two cases will be recorded in a city hospital on the same day? (b) If the number of live births for a given year in a country is 42 million, how many would be expected to have the Down syndrome? (c) If 40 percent of the babies with the Down syndrome are born to mothers over 45 years of age, and mothers in this age group produce 4 percent of all children, what is the chance that a given woman in this age-group would have a baby with the Down syndrome?

13.4. If nondisjunction of chromosome 21 is known to have occurred in the division of the secondary oöcytes in a particular woman, what is the chance that a mature egg arising from this cell division will receive two No. 21 chromosomes?

13.5. If the Down syndrome occurs in about 1 in 700 and the Turner syndrome occurs in about 1 in 5000 in the general population, and each is separately and randomly distributed in the population, what is the chance that a baby will be born with both these abnormalities?

13.6. (a) If X chromosome trisomy occurs in 1 in 1000 of the general population and No. 18 trisomy occurs in 1 in 4000, and each is separately and randomly distributed, what is the chance that a baby, such as one described by Uchida, will be born with both abnormalities? (b) If the mother is over 40 years old and if the increased occurrence of nondisjunction makes mothers in this age-group 10 times more likely to have babies with each of these abnormalities, what is the probability of the two trisomies occurring in the same baby?

13.7. The poinsettia type of *Datura* carries an extra member of the chromosome set $(2n + 1)$ in which the genes for purple (p^+) and white (p) flower color are located. From the following crosses, give the expected proportions of purple and white. (Female parent is always written first. Female gametes may carry either one or two chromosomes of this set, but viable pollen carries only a single chromosome.) (a) $p^+ p^+ p \times p^+ p^+ p$; (b) $p^+ p^+ p \times p^+ p$; (c) $p^+ p \ p \times p^+ p^+ p$; and (d) $p^+ p \ p \ \times p^+ p$. (e) How do trisomic ratios differ from the usual Mendelian ratios?

13.8. Triplo-IV fruit flies have an extra member of the fourth chromosome in which the gene *ey* is located. Give the expected results from a cross between triplo-IV flies of the genotype $ey^+ ey^+ ey$ and diplo-IV, $ey^+ ey$, flies in terms of (a) $2n$ and $2n + 1$, and (b) eyeless and normal eye phenotypes.

13.9. (a) What evidence concerning the influence of the Y chromosome on sex determination in humans can be obtained by comparing the characteristics of X, XXY, and XXX individuals? (b) Compare the influence of the human Y chromosome on sex determination with that of *Drosophila* and *Melandrium* Y chromosomes (see Chapter 4).

13.10. How can aneuploidy be used as a tool to determine the chromosome on which particular genes are located?

13.11. Why are tetrasomics and nullisomics found in nature less frequently than trisomics?

13.12. What values, potential if not realized at present, could monoploids have for genetic studies?

13.13. Polyploidy is rare in animals, yet some tissues in the bodies of certain diploid animals show evidence of polyploidy. Why do numbers above $2n$ persist in somatic tissues when they do not occur in the whole animal?

13.14. Describe two methods by which polyploidy might occur in nature.

13.15. (a) How may autopolyploidy and allopolyploidy originate? (b) Evaluate the significance that each might have in evolution.

13.16. Why do tetraploids behave more regularly in meiosis than triploids, and perpetuate themselves more readily in populations?

13.17. Why is chromosome irregularity associated with low fertility in plants?

13.18. What is the significance of chromosome number in (a) taxonomy and (b) studies of evolution in plants and animals?

13.19. A plant species A, which has seven chromosomes in its gametes, was crossed with a related species B, which has nine. Hybrids were produced, but they were sterile. Microscopic observation of the pollen mother cells of F_1 showed no pairing of chromosomes. A section of the hybrid that grew vigorously was propagated vegetatively, and a plant was produced with 32 chromosomes in its somatic cells. What steps might have been involved?

13.20. A plant species A ($n = 5$) was crossed with a related species B with $n = 7$. Only a few pollen grains were produced by the F_1 hybrid. These were used to fertilize the ovules of species B. A few plants were produced with 19 chromosomes. They were highly sterile but following self-fertilization produced a few plants with 24 chromosomes. These plants were different in phenotype from the original parents, and the progeny were fertile. What steps might have been involved?

13.21. How does colchicine treatment result in chromosome doubling in plants?

13.22. What (a) practical and (b) theoretical significance may be associated with colchicine-induced polyploidy?

13.23. How could polyploidy be a significant factor in the evolution of such plants as cotton and wheat?

13.24. How might new species be produced through a combination of polyploidy and hybridization?

13.25. Give a plausible explanation for the origin of (a) *Triticum spelta;* (b) *Raphanobrassica;* (c) *Spartina townsendii;* (d) *Primula kewensis;* and (e) New World cotton.

13.26. What chromosome arrangement is symbolized by each of the following: (a) n; (b) $2n$; (c) $2n + 1$; (d) $2n - 1$; (e) $2n + 2$; (f) $3n$; and (g) $4n$?

FOURTEEN

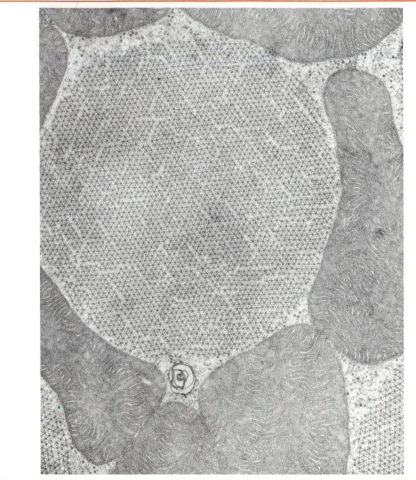

Transverse section of a fibril within the wing-depression muscle of the honey bee worker (magnification ×56,000). Mitochondria are surrounding the muscle bundle. (Courtesy of N. N. Youssef.)

EXTRANUCLEAR INHERITANCE MECHANISMS

Thus far in our treatment of transmission genetics in eukaryotes, we have dealt with nuclear chromosomes and genes. Certainly, nuclear DNA is the most important and very nearly the universal genetic material. Nevertheless, throughout the history of genetics, sporadic reports have indicated that extranuclear or cytoplasmic elements also act as agents for hereditary transmission. But most examples originally attributed to extranuclear inheritance have eventually been explained by nuclear genes. A few cases that appeared at first to depend on cytoplasmic genes and were classified under maternal inheritance were shown by further investigations to be associated with the genes of the mother. Phenotypes were expressed in her progeny. These cases were reclassified as maternal effects. Since the pattern of transmission for maternal effects is similar to that of cytoplasmic inheritance, maternal effects are discussed at the end of this chapter.

What criteria will distinguish extranuclear from nuclear inheritance? Criteria for identifying this heterogeneous group of traits are consequences of the definition of extranuclear or cytoplasmic inheritance and also of the kinds of organisms and mechanisms involved. **Extranuclear inheritance** is defined as **non-Mendelian inheritance,** usually involving DNA in replicating cytoplasmic organelles such as mitochondria and plastids. A few bacteria and viruses are also agents for extranuclear inheritance. General guidelines can be set forth at the outset, and criteria for specific cases can await the context of particular examples. Since the criteria must depend on phenotypes associated with extranuclear DNA, a persistent deviation from the Mendelian pattern will suggest extranuclear inheritance.

CRITERIA FOR EXTRANUCLEAR INHERITANCE

Five major criteria can be used to distinguish between traits controlled by nuclear genes and traits controlled by extranuclear genes. These are summarized below.

1.   Differences in the results of **reciprocal crosses** would suggest a deviation from the pattern of Mendelian autosomal gene transmission. To conduct reciprocal crosses, a female from strain A is mated with a male from strain B and a male from strain A is mated with a female from strain B. If sex linkage were ruled out, differences in results of reciprocal crosses would indicate that one parent (usually maternal) is exerting a greater influence than the other on a particular trait.

2.   The female reproductive cell usually carries more cytoplasm and cytoplasmic organelles than the male cell and would be expected to influence non-Mendelian traits. Organelles and symbionts in the cytoplasm might be isolated and analyzed for more specific evidence concerning maternal transmission in inheritance. When extranuclear DNA can be associated with the transmission of particular traits, the case of extranuclear inheritance is established.

3.   Chromosomal genes occupy particular loci and map in certain places with respect to other genes. This kind of information may rule out chromosomal inheritance and suggest extranuclear inheritance if sufficient data can be obtained.

4.   Lack of Mendelian segregation and characteristic Mendelian ratios that depend on the chromosomal cycle in meiosis would suggest extrachromosomal transmission.

5.   Experimental substitution of nuclei might clarify the relative influence of nucleus and cytoplasm. Transmission of traits without transmission of nuclear genes would suggest extranuclear inheritance. Genes and viruses have much in common. A fine line of distinction may be required to distinguish between persistent infection and the influence of **cytoplasmic DNA.** But phenotypes from either could qualify broadly for extranuclear inheritance.

CYTOPLASMIC ORGANELLES (SYMBIONTS)

Extranuclear inheritance associated with cytoplasmic organelles meets the requirements stated above and deserves an explanation even though organelles represent only a very small proportion of genetic material, perhaps a few hundred genes, based on the amount of functional DNA that is involved. How and why has this independent pocket of extranuclear DNA been preserved in nature? Apparently local, more or less independent cellular control has been advantageous for a few special phenotypes.

It should be noted that **cytoplasmic organelles** are especially significant and basic to the functions and, indeed, to the continued existence of living things. Enzymes for cellular respiration and energy production, for example, are located in mitochondria, and foodstuffs are oxidized to produce adenosine triphosphate (ATP), the fuel for biochemical reactions. Chlorophyll and other plant pigments are synthesized in plastids. It is not likely, however, that many, if any, of the autonomous genes in mitochondrial and plastid DNA are directly associated with these basic and vital phenotypes. The fascinating possibility suggested by several earlier investigators, and recently elaborated by Margulis, is that **mitochondria were once free-living bacteria.** Over long periods of time, they established a hereditary symbiosis with their eukaryote host cells and ultimately evolved into organelles within animal and plant cells. They brought with them from the free-living state their own **DNA** and other equipment for genetic mechanisms. Their processing factories, partially independent from control of nuclear genes, must have been favored in evolution and found worthy of continued existence in eukaryote cells.

Similarly, chloroplasts in green plant cells are postulated to have come from free-living algae that established a symbiotic relation with early eukaryote cells. They had much to contribute to their host cells. Chlorophyll—the essential pigment for photosynthesis, with its synthesizing machinery including specific DNA, mRNA, tRNA, ribosomes, and the assembly plant for chlorophyll production—was already assembled in the free-living algae. In addition, plastids of green algae are presumed to carry other genetic mechanisms such as streptomycin resistance, which was discovered in the alga *Chlamydomonas.*

Symbiont bacteria have been discovered in cytoplasm of the protozoan, *Paramecium aurelia,* where they produce a toxic substance that kills other susceptible paramecia placed in the same culture medium. This symbiont, now dignified with a specific Latin name, *Caedobacter taeniospiralis,* has worked its way into the genetic system of its host, but it can reproduce only in the presence of a particular host genotype.

DNA IN MITOCHONDRIA

Mitochondria in presently living organisms arise from preexisting mitochondria. They are usually small cytoplasmic organelles (Fig. 14.1) with distinct shelflike internal layers or cristae that arise as invaginations from the inner mitochondrial membrane. They are about the same size as bacteria and occur in cells of eukaryotes but not in bacteria and viruses.

Mitochondria provide higher animals and plants with life-sustaining cellular energy through

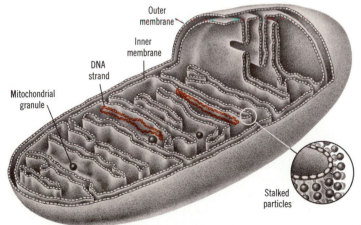

Outer membrane

Inner membrane

DNA strand

Mitochondrial granule

Stalked particles

FIGURE 14.1

A mitochondrion showing a smooth continuous outer membrane and a periodically convoluted inner membrane that forms double membranes called cristae. Mitochondria are the principal energy source in all cells of eukaryotes. They contain a small amount of DNA, RNA polymerase, transfer RNAs, and ribosomes, which are presumably responsible for the extrachromosomal inheritance and protein synthesis in mitochondria.

the oxidative processes of the citric acid and the fatty acid cycles, as well as the coupled processes of oxidative phosphorylation and electron transport. They contain a small amount of unique **DNA** that has remained autonomous outside the nuclear genome throughout the long evolutionary history of animals and plants. The mitochondrial genome is small and codes only for a limited number of structures and functions. Mitochondria contain a distinctive **protein-synthesizing apparatus** with specific ribosomes, tRNAs, aminoacyl-tRNA synthetases, and sensitivity to antibiotics like that of bacteria. Mitochondrial protein-synthetic machinery is significantly different from eukaryote machinery for the same purpose. For example, ribosomes in the protein-synthesizing apparatus of mitochondria are like those found in bacteria and yet very different from those in the cytoplasm of eukaryote cells. **Mitochondrial rRNA** molecules are the same size as those in bacteria and consistently smaller than those in eukaryote cells.

In yeast cells, 10–20 percent of the cellular DNA is localized in a single mitochondrion. Mitochondrial DNA has properties different from those of nuclear DNA in density and proportion of GC and AT base-pairs. One yeast study showed that mitochondrial DNA had a density of 1.683 g/cm³ and a GC content of 21 percent, whereas nuclear DNA had a density of 1.699 g/cm³ and a GC content of 40 percent.

The life cycle of normal baker's yeast, *Saccharomyces cerevisiae*, includes a haploid and diploid phase. Mating normally occurs between vegetative haploid cells of opposite mating type (*A* or *a*). These cells fuse to form vegetative diploid cells that divide by mitosis. Cell division is usually unequal, with a small "daughter" cell budding from a larger "mother." Both cells, however, are identical in nuclear composition. Vegetative diploid cells may undergo the complex process of sporulation in which meiosis occurs. The resulting four ascospores divide to form a clone.

The first mutant found in yeast, a small colony type called "petite," has provided the best evidence that now exists for **mitochondrial mutations.** Petites are defective in their ability to utilize oxygen in the metabolism of carbohydrates. When, for example, glucose is in the medium, petite yeast will grow to only small-sized colonies. Enzyme analysis indicates that the mitochondria lack the respiratory enzyme cytochrome oxidase that is normally associated with mitochondria. Not only does this deficiency produce defective growth, but it prevents petites from producing spores. Petite strains that have been analyzed show only a small proportion of G and C and a preponderance of **repetitive AT base-pairs.** This kind of DNA does not code meaningful biological information. Absence of cytochrome oxidase from mitochondria does not mean that this enzyme is coded for by mitochondrial DNA. But it does indicate that mutational changes in mitochondrial DNA lead to **heritable alterations** in mitochondrial phenotypes.

Mutations other than those causing petites can be induced in yeasts and transmitted by the cytoplasm. For example, resistance to the antibiotics chloramphenicol and erythromycin has been induced. These antibiotics have selective affinity for mitochondrial ribosomal proteins, suggesting that structural genes are present for some ribosomal proteins.

W. L. French has presented evidence that **sterility** in hybrid *Culex* mosquitoes is caused by interaction involving mitochondrial DNA. Several other investigators have shown that mitochondrial DNA is **inherited maternally in frogs.** J. B. David has compared mitochondrial DNA in cell cultures of different mammals, including rats, mice, and human beings; he has also hybridized cells of different mammals in culture. In hybrid mouse and human cells, for example, he has shown that not only homogeneous mouse and homogeneous human mitochondrial DNA can be detected but also heterogeneous **hybrid DNA.** In one series of experiments, 20 percent of each circular DNA unit was mouse and 80 percent was human mitochondrial DNA. Heterogeneous DNA was shown to result from mitochondrial DNA recombination in the hybrids.

Yeast chromosomal genes must specify most enzymes associated with mitochondria. Petite yeast strains with damaged DNA continue to synthesize abnormal mitochondrial DNA. This indicates that the proteins needed for mitochondrial DNA replication are not coded for by mitochondrial DNA. Similarly, petite strains continue to synthesize the enzymes of the Krebs cycle that are located in the mitochondria. Control must come from chromosomal genes.

DNA IN PLASTIDS

Carl Correns (in 1908) observed a difference in the results of reciprocal crosses and was the first to describe deviations from Mendelian heredity. Dif-

ferent shades of color from white (albino) to dark green in the leaves of some plants were investigated. Instead of equal inheritance from the seed and pollen parent, as demonstrated by Mendel in garden peas, Correns showed in studies of four-o'clock, *Mirabilis jalapa*, that inheritance of certain traits came entirely from the seed parent. Color differences were related to cytoplasmic plastids, most important of which are chloroplasts (Fig. 14.2), which carry chlorophyll. Chloroplasts arise from cytoplasmic particles called **proplastids that contain DNA** and duplicate themselves independently of other

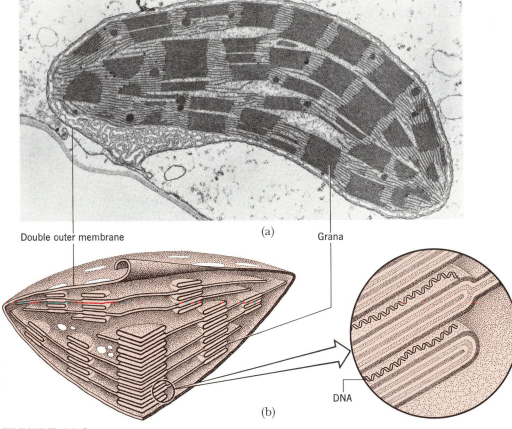

Double outer membrane

(a)

Grana

DNA

(b)

FIGURE 14.2
Chloroplast, cytoplasmic green plant organelle containing DNA. (a) Electron micrograph of a chloroplast; magnification ×15,200. (b) Enlarged diagram of grana— stacks of membrane sacs within chloroplast (magnification ×78,500)—showing position of DNA. (Photo courtesy T. E. Weier. From Weier, Stocking, and Barbour, Botany, *4th ed. John Wiley & Sons, 1970).*

cell parts. They are distributed more or less equally during cell division. Although some proplastids are transmitted in the cytoplasm of the egg, few if any are transmitted in the pollen of most plants. Thus, some chloroplast characteristics are **inherited** from the **seed-parent cytoplasm.**

Many investigators have followed the early lead of Correns, and now almost everything that is known about plastid genetics has come from studies on variegation in seed-bearing plants. Any plant that develops patches of different colors in leaves or other vegetative parts is said to be variegated. Many variegations are not inherited, some are controlled by nuclear genes, and others depend on plastid inheritance. Interactions occur, and it is difficult to distinguish examples that are solely dependent on plastid inheritance. Figurative patterns (marginal bands, diffuse spots, prominent veins and stripes on leaves) that are true-breeding usually represent physiological modifications that influence regulation of normal chloroplast development and not specific gene mutations. The sorting of normal and mutant plastids into color patterns may depend on mutations within plastids.

Ovules, as well as somatic cells of mottled plants (e.g., the four-o'clock), may carry both abnormal nearly **colorless plastids** and normal green chloroplasts in their cytoplasm (Fig. 14.3). The mottled effect is transmitted through the maternal line, generation after generation. Because the pollen of the four-o'clock has little if any cytoplasm, its influence on variegation is negligible. A single plant with

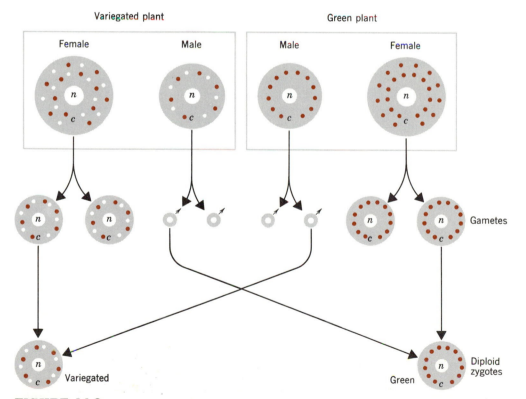

FIGURE 14.3

Diagrammatic illustration of maternal plastid inheritance in a diploid plant Mirabilis, *which has abundant cytoplasm in the seed parent gamete cells but little or no cytoplasm in pollen parent gamete cells:* n, *nucleus; and* c, *cytoplasm.*

green, white, and variegated branches or sectors may produce seed that perpetuates each of the three types. Seeds borne on white branches contain only primordia for colorless plastids, those on green branches only green, and those on variegated branches might contain either colorless or green chloroplasts or a combination of both.

In plants such as the primrose, *P. sinensis*, chimeras (sectors containing different plastid types) are sometimes formed, with only part of the plant containing chlorophyll. The areas with abnormal plastids that lack chlorophyll can rely on the green parts of the plant for the products of photosynthesis and therefore can continue to live. Each part of the chimera may produce reproductive cells and thus transmit its type of plastids through female gametes.

Chloroplasts have now been isolated and found capable of **protein synthesis** in the presence of either adenosine triphosphate or light. The products are identical with authentic chloroplast proteins, demonstrating that isolated chloroplasts have fully functional protein synthesis machinery in which mRNA is translated accurately.

With DNA analysis and the use of restriction endonucleases for DNA fragmentation, much has been learned about plastid DNA. Some 30–60 copies of the chloroplast genome are found in each chloroplast of higher plants; about 100 genome copies occur in each plastid of some algae. Enough unique chloroplast DNA has been discovered to code for about 126 proteins, and about 12 percent of the **plastid DNA** sequence **codes** for plastid components.

CHLOROPLAST DNA AND DRUG RESISTANCE

When Ruth Sager placed *Chlamydomonas* algae cells on a culture medium containing the antibiotic streptomycin, most of the cells were killed, but about one per million survived and multiplied, each to form a **streptomycin-resistant** colony. Mutants with resistance to streptomycin were being selected from the predominantly streptomycin-susceptible alga. About 90 percent of the mutants involved nuclear genes (*sr-1*). Such mutations were merely being demonstrated by the antibiotic challenge. Approximately 10 percent of the mutations (*sr-2*), how-

ever, were uniparental and **nonchromosomal.** Eventually, nonchromosomal mutants were recovered from almost every colony. Nonchromosomal DNA mutations expressed the same phenotypes as chromosomal DNA mutants. These nonchromosomal genes are presumed to be located in the chloroplasts.

Reciprocal crosses (Fig. 14.4) demonstrated that antibiotic resistance, controlled by nonchromosomal genes, was uniparental in inheritance. On the other hand, mating type in this sexual unicellular alga was controlled by chromosomal genes, which were designated by the investigators mt^+ and mt^- or simply plus (+) and minus (−), instead of female and male. All progeny from each reciprocal mating were alike, the plus (+) mating type with respect to relative streptomycin resistance thus demonstrating maternal inheritance. When the plus (+; female) mating type was resistant, all progeny were resistant; when the plus (+) mating type was nonresistant, all progeny were nonresistant. These results of reciprocal crosses demonstrated non-Mendelian inheritance, which involved a single pair of contrasting traits. Nonchromosomal genes, *sr* for streptomycin resistance and *ss* for streptomycin sensitive, were postulated to control these two alternative characteristics.

Another mutant, ac_2, which blocked photosynthetic activity, was induced and a pair of nonchromosomal alleles, ac_1 and ac_2, was thus available for study in the same strain of *Chlamydomonas*. The mutant required acetate in the medium for growth. With two pairs of nonchromosomal genes available, a dihybrid cross could be conducted in the same system to check for evidence of recombination. Crosses of the dihybrid type $ac_1 \, ss \times ac_2 \, sr$ were prepared, and progeny were allowed to grow for a few vegetative multiplications. Each cell was then classified for its segregating markers, both nonchromosomal and chromosomal (i.e., mating type and others known to be chromosomal). Both the ac_1/ac_2 and the sr/ss pairs of alleles were observed to segregate, but not always in the same division. After four or five mitotic doublings, both parentals ($ac_1 \, ss$ and $ac_2 \, sr$) and recombinations ($ac_1 \, sr$ and $ac_2 \, ss$) had been obtained. The results indicated independent assortment, suggesting that

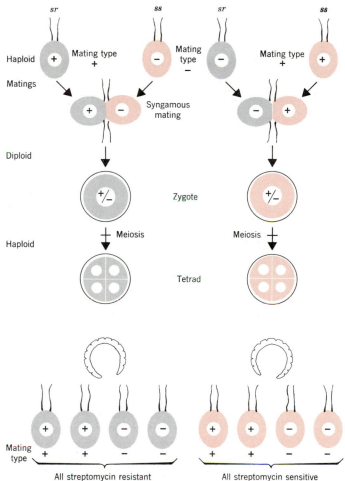

Mating type +

Mating type −

Mating type +

Mating type +

Haploid

Matings

Syngamous mating

Diploid

Zygote

Meiosis

Meiosis

Haploid

Tetrad

Mating type

All streptomycin resistant

All streptomycin sensitive

FIGURE 14.4

Maternal (cytoplasmic) inheritance of resistance to streptomycin and nuclear gene-controlled mating type inheritance. The plus and minus signs refer to mating types, which are inherited as single chromosomal gene differences. Progeny are, with rare exceptions, like the plus (maternal) parent (P) *in their reaction to streptomycin. In reciprocal crosses streptomycin resistance shows maternal inheritance, indicating uniparental and cytoplasmic inheritance. (Based on R. Sager and F. J. Ryan,* Cell Heredity, *John Wiley & Sons, 1961.)*

the two pairs of nonchromosomal genes were carried on different plastids. Three- and four-point crosses and reciprocal crosses have been made with the addition of several mutants, which are presumed to be carried in chloroplasts and mitochondria. A **genetic map of non-Mendelian genes** in *Chlamydomonas* has been constructed, but uncertainty still exists as to whether some "chloroplast" linkage groups are solely in the chloroplast genome.

SYMBIONT BACTERIA IN PARAMECIUM CYTOPLASM

Paramecia are favorable organisms for genetic investigation. They are large, unicellular protozoans that reproduce by both asexual and sexual proc

esses. Asexual reproduction occurs through cell fission to produce clones of genetically identical cells. In the sexual phase, paramecia conjugate periodically and transfer genetic material from one cell to another. Paramecia and other ciliates have two kinds of nuclei: a large vegetative macronucleus and a small micronucleus, which goes through the meiotic sequence and produces haploid gametes. A micronucleus also gives rise to the macronucleus that divides in asexual cell division. It is possible in the laboratory to make sexual crosses through which nuclear DNA is transferred from a donor to a recipient, resulting in heterozygous progeny; that is, *AA* × *aa* → *Aa*. A process of self-fertilization, called **autogamy,** results in the complete homozygosis of

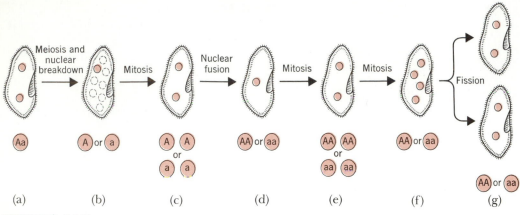

FIGURE 14.5

Autogamy in Paramecium *results in homozygosis through the following steps:* (a *and* b) *Meiosis of micronuclei resulting in one haploid product that divides mitotically, and of micronuclei that disappear;* (c) *the haploid micronucleus has divided; and* (d) *the two micronuclei have fused to form a diploid (2n) cell.* (e *and* f) *Fission of the cell results in two cells, each homozygous as shown in* (g).

the resulting progeny (Fig. 14.5). Following meiosis, the cells are haploid, but through autogamy they become homozygous diploids. This provides a basis for comparing extranuclear and nuclear inheritance — and thus for demonstrating that progeny can differ from wild-type in traits controlled by both nuclear and extranuclear genes.

G. H. Beale discovered that erythromycin resistance in paramecium, like that in yeast, results from non-Mendelian inheritance. A number of additional cytoplasmic and nuclear mutations affecting antibiotic resistance have been studied by both Beale and J. Beisson. These and other investigators made transfers of cytoplasm and also transfers of isolated mitochondria between strains of paramecia and showed that mitochondria (presumably mitochondrial DNA) control the resistance. Studies have also shown that while some mitochondrial traits are determined by the mitochondria themselves, others are dependent on other elements in the protoplasm.

T. M. Sonneborn and others have investigated a persistent extranuclear effect in paramecium. Some strains of *P. aurelia* produce a substance that has a lethal effect on members of other strains of the same species. Paramecia from strains capable of

producing the toxic substance are called **"killers."** When killers are subject to low temperatures, their killing capacity gradually disappears. The toxic effect also decreases after repeated cell divisions. Separate elements in the cytoplasm were postulated for the production of a toxic substance. From mathematical calculation, it was estimated that about **400 particles** are required to make a killer effective. Killers were then observed microscopically and "particles" called "kappa" were observed in the expected numbers. These "particles," shown to be symbiotic bacteria, have been named *Caedobacter taeniospiralis* (the killer bacterium with the spiral ribbon).

A "toxic substance" (paramecin), produced by the killer bacteria, is diffusible in the fluid medium (Fig. 14.6). When killers are allowed to remain in a medium for a time and are then replaced by sensitives, the sensitives are killed. Paramecin, which has no effect on killers, is associated with a particular kind of kappa that occurs in about 20 percent of a kappa population. These kappa bacteria possess a refractile protein-containing "R" body and are called "brights," because they are infected with a virus that controls the synthesis of viral protein. The

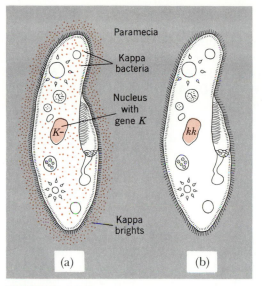

FIGURE 14.6

Kappa particles in Paramecium aurelia. *(a) Killer with kappa particles inside the paramecium and paramecin in the liquid medium outside the organism. Gene* K *is present in the nucleus. (b) Sensitive paramecia with no kappa particles, no paramecin, and genes* kk *in the nucleus. (After T. M. Sonneborn.)*

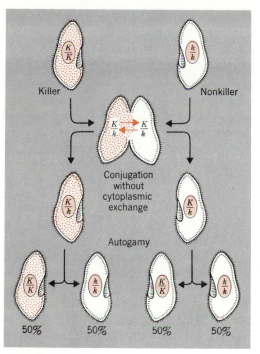

FIGURE 14.7

Conjugation between a killer paramecium (stippled) and a sensitive paramecium without cytoplasmic exchange but with an exchange of K *and* k *genes. The autogamy which follows results in* KK *and* kk *homozygotes, which may give rise to* KK *and* kk *clones. (After T. M. Sonneborn.)*

virus is toxic to sensitive paramecia but is not toxic in "nonbright" bacteria.

Kappa bacteria are perpetuated only in organisms carrying the dominant **nuclear allele K,** which establishes the environment necessary for the bacteria to reproduce. When killers conjugate with sensitives under appropriate conditions (to avoid killing the mate) and no cytoplasmic exchange occurs (Fig. 14.7), two kinds of clones emerge: one from the original killer cell, which contains allele *K* (*Kk*) and kappa bacteria, and the other from the original sensitive cell, which carries the allele *K* (*Kk*) and lacks kappa. Following autogamy, half the progeny of the killers are killers and half are sensitive paramecia. All progeny of sensitives are sensitive. Since no cytoplasm was transferred in this conjugation, only the cells from original killers inherit kappa bacteria. Kappa cannot reproduce in cells unless a *K* allele is present in the nucleus.

Under some conditions, conjugation persists much longer; a larger connection is established between conjugants, and cytoplasm as well as nuclear genes is exchanged (Fig. 14.8). When the conjugants are *KK* and *kk*, alleles *K* and *k* are exchanged and both exconjugants are *Kk*. **Cytoplasmic exchange** has transferred kappa bacteria from the killer to the nonkiller cell. Autogamy produces homozygotes *KK* and *kk* cells, which produce clones of killers or nonkillers, respectively.

PLASMID DNA AND TUMOR TRANSFORMATION

Extrachromosomal DNA molecules that replicate independently and maintain themselves in the cytoplasm of plant cells are called **plasmids** (Chapter 7). They have much in common with chromosomes of mitochondria and plastids, but they are not orga-

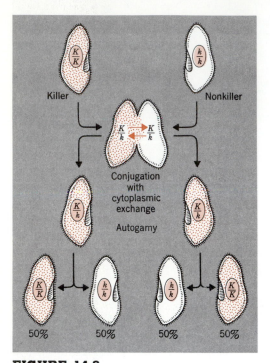

FIGURE 14.8

Conjugation as in Fig. 14.7, but with cytoplasmic exchange followed by autogamy. (After T. M. Sonneborn.)

nized into organelles that are vital to their host cells. Some plasmids are fragments of the bacterial chromosome and some are recombinants of DNA fragments. Most plasmids are not essential to their host cells, but some control a favorable reaction to antibiotics. Because of their ability to replicate independently, to combine with other DNA, and to carry DNA to cell centers of synthetic activity, they are useful in genetic engineering.

A plasmid called Ti (for tumor-inducing) carries a DNA sequence that transforms cells of dicotyledonous plants (tobacco, sunflower, carrot, tomato, etc.) to tumor cells. Tumor transformation is associated with the crown gall disease. This disease, manifested as a bulbous growth or gall, is induced by a bacterium, *Agrobacterium tumefaciens*. The disease is economically important, especially in fruit crops and nursery stocks. It is caused by viable bacteria that enter a wounded surface of a plant, usually at the crown (junction of stem and soil). But the bacteria that initiate the gall disease are not needed to perpetuate the tumor. They can be killed after a few days and the tumor continues to grow. A fragment of the Ti plasmid carried by the bacterium has been combined with a DNA segment of an infected plant cell. Genes carried by the plasmid, now integrated into plant cells, **code for enzymes** that promote continuous and uncontrolled tumor growth, which extend from the bacterial-induced gall.

CYTOPLASMIC MALE STERILITY IN PLANTS

Another example of cytoplasmic inheritance is associated with *pollen failure.* This occurs in many flowering plants and results in male sterility. In maize, wheat, sugar beets, onions, and some other crop plants, fertility is controlled at least in part by cytoplasmic factors. In other plants, however, male sterility is controlled entirely by nuclear genes. Critical observation and tests must be made in individual cases to determine the mechanism of inheritance. Male sterility has practical importance when crosses are made on a large scale to produce hybrid seed. Hybrid plants are produced commercially in maize, cucumbers, onions, sorghum, and other plants to obtain hybrid vigor (Chapter 16).

MALE STERILITY IN A CROSS-POLLINATING PLANT

Classical examples of maternal-inheritance mechanisms that transmit male sterility in maize (corn) were discovered and carefully analyzed by M. M. Rhoades. Pollen was aborted in the anthers of certain corn plants, causing them to be male sterile, but female structures and fertility were normal. Nuclear genes did not control this type of sterility. It was transmitted from generation to generation through the **egg cytoplasm.**

A particular male-sterile variety produced only male-sterile progeny when fertilized with pollen from normal maize plants. The male-sterile seed-parent plants were then backcrossed repeatedly with pollen-fertile lines until all chromosomes from the male-sterile line had been exchanged for those of the male-fertile line (Fig. 14.9). In the genetically restored sterile line, male sterility persisted, demonstrating that inheritance was maternal and not controlled by chromosomal genes. As the investigations progressed, a small amount of pollen was obtained from the male-sterile line, making reciprocal crosses possible. These crosses produced progeny from the male-sterile seed plant line that were male-fertile. Thus, inheritance of **male sterility was maternal,** regardless of the direction in which the cross was made. Male sterility in this example was attributed to cytoplasmic genes (plasmagenes) that were transmitted by female gametes.

However, the cytoplasmic effect is not the only factor in male sterility. Specific nuclear genes are now known to suppress maternally inherited sterility in maize. A single dominant chromosomal gene, for example, can restore pollen fertility in the presence of cytoplasm that ordinarily would ensure sterility. In one experiment, pollen abortion occurred only when a specific kind of cytoplasm was present along with a dominant gene for male sterility. The homozygous recessive allele was present at a suppressor locus.

Large-scale use of male-sterile maize for seed production brought disaster to the United States corn crop in 1970. Because of the advantages of uniformity in corn and the great advantage of male sterility in seed production, a single source of cytoplasm, known as Texas (T) male-sterile cytoplasm, had been used in producing seed for most of the corn hybrids planted that year.

DANGER OF UNIFORMITY

What caused disaster to the corn crop? A new mutant of the fungus *Helminthosporium maydis* (Nisikado and Miyake) became a virulent pathogen on a particular kind of hybrid corn. It was especially destructive on corn with (T) male-sterile cytoplasm. Pathologists and plant breeders met the epidemic by searching for corn varieties that were resistant to the fungus. Because of a previous, less serious yellow leaf blight, some 1970 seed production had been shifted to corn without T cytoplasm. This corn required manual detasseling but was used for winter planting in many 1971 fields. It also produced some resistant seed for immediate general farm use. Some growers preferred the predictable 20–30 percent loss of yield to the high risk of much larger losses from growing susceptible hybrids.

Most of the 1971 seed production was, therefore, accomplished without the use of male sterility and T cytoplasm. The T race of *H. maydis* was not serious in 1972. Still another race of *H. maydis* may appear, or one of the other corn diseases could become a threat to the highly uniform hybrid corn with T male-sterile cytoplasm. Several varieties of

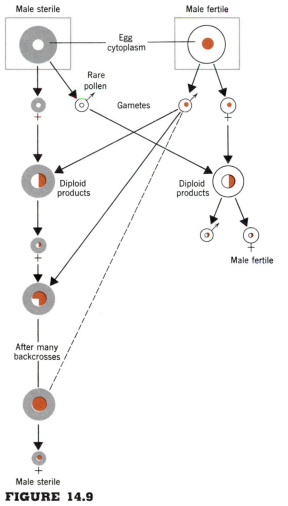

FIGURE 14.9

Maternal inheritance of male sterility in maize. (After M. M. Rhoades.)

corn that are resistant to the existing T race of *H. maydis* have now been identified and are available for seed production.

This example illustrates the danger of uniformity in germ plasm for a crop grown on a large scale. It also indicates that sustained research programs are essential in protecting food supplies from potential losses of catastrophic magnitude.

MATERNAL EFFECTS

Eggs and embryos are expected to be influenced by the maternal environment in which they develop. Even those removed from the body of the mother at an early stage receive cytoplasm and nutrients in the egg from the mother, and special influences on gene action may have already taken effect. Certain potentialities of the egg are known to be determined before fertilization, and, in some cases, these have been influenced by the surrounding maternal environment. Such predetermination by genes of the mother, rather than those of the progeny, is called **maternal effect.** Existence of a maternal effect is commonly substantiated or disproved by reciprocal crosses. If a maternal effect is involved, results from reciprocal crosses will be different from each other, with genes of the mother being expressed.

MATERNAL EFFECT IN SNAIL SHELL COILING

One of the earliest and best-known examples of a maternal effect is that of the direction of coiling in snail shells, *Limnaea peregra.* Some strains of this species have dextral shells, which coil to the right; others have sinistral shells, which coil to the left. This characteristic is determined by the **genotype of the mother** (not her phenotype) rather than by the genes of the developing snail. Allele s^+ for right-handed coiling is dominant over allele s for coiling to the left.

When crosses (Fig. 14.10) were made between females coiled to the right and males coiled left, the F_1 snails were all coiled to the right. The usual 3 : 1 ratio was not obtained in the F_2 because the phenotype of ss was not expressed. Instead, the pattern determined by the mother's (P) genes (s^+s^+) was expressed in the F_1, and the F_1 mother's genotype

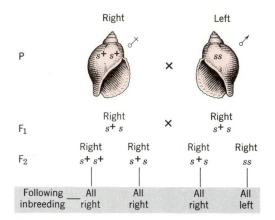

FIGURE 14.10

A cross illustrating a maternal effect. The coiling pattern is controlled by the genes of the mother. Following inbreeding of snails with ss *genotypes,* ss *mothers produced progeny that coiled to the left. (Data and interpretation from A. E. Boycott et al. and A. H. Sturtevant.)*

(s^+s) was expressed in the F_2. When ss individuals were inbred, only progeny that coiled to the left were produced. When the s^+s^+ or s^+s snails were inbred, however, they produced offspring that all coiled to the right. From the reciprocal cross between left-coiling females and right-coiling males (Fig. 14.11), all F_1 progeny were coiled to the left. The F_2 all coiled to the right; but, when each F_2 snail was inbred, those with the genotype ss produced progeny that coiled to the left.

Further investigation of coiling in snails has shown that the spindle formed in the metaphase of the first cleavage division influences the direction of coiling. The spindle of potential "dextral" snails is tipped to the right, but that of "sinistral" snails is tipped to the left. This difference in the arrangement of the spindle is controlled by the genes of the mother. They determine the orientation of the spindle, which in turn influences further cell division and results in the adult pattern of coiling. The actual phenotypic characteristic, therefore, is influenced directly by the mother, with no immediate relation to the genes in the egg, sperm, or progeny. However, most other snail traits do not show the maternal effect pattern. The striping color pattern, for example, is also determined in the early embryo, but it is controlled directly by chromosomal genes of

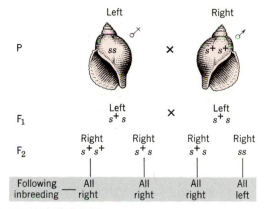

FIGURE 14.11

Cross reciprocal to that of Fig. 14.10, illustrating a maternal effect. The ss (P) mother produced only left coiling progeny, and the s⁺s (F₁) mother produced only right coiling progeny. Following inbreeding, the ss (F₂) mother produced only left coiling progeny. (Data and interpretation from A. E. Boycott et al. and A. H. Sturtevant.)

both parents. In this example, comparable color patterns are obtained from the results of reciprocal crosses.

MATERNAL EFFECT IN *DROSOPHILA*

Abnormal growth in the head region of *Drosophila melanogaster* appeared sporadically at the University of Texas in a sample from a wild population collected at Acahuizotla, Mexico. At the University of Utah, these flies were inbred and selected for the abnormal head growths over a period of several years. The proportion of flies expressing the trait, named "tumorous head" (tu-h, Fig. 14.12), was increased to about 76 percent at 22°C when the flies were raised on cornmeal, molasses medium. When reciprocal crosses were conducted, a maternal effect was indicated.

Tu-h females mated separately with three wild-type males and males from 11 laboratory stocks produced from 14 to 52 percent (average of 30 percent) of abnormal flies in the first generation.

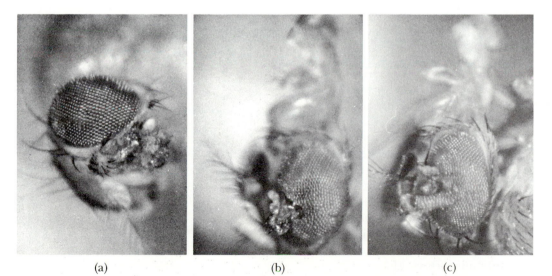

(a) (b) (c)

FIGURE 14.12

Drosophila melanogaster of the tumorous head strain, showing abnormal growths. A sex-linked gene homozygous in the mother controls the proportion of female and male progeny that express the abnormal growths. An autosomal gene in the third chromosome is required in the progeny for the expression, but this gene alone produces the phenotype in only one percent or less of the flies. With the maternal effect, 30 to 80 percent of the flies express the trait, depending on the combinations of genes in mothers and progeny.

From the reciprocal cross between tu-h males and the same three wild-type and the same 11 laboratory stock females, 0–1 percent (average of less than 1 percent) of tumorous-head flies were obtained. Further studies demonstrated the maternal effect. Genes of the mother were exerting an influence in the direction of abnormal growths on the heads of adult progeny during the first 22 hours of development. Two major genes were found to control the tumorous head trait: (1) a sex-linked gene at 64.5 map units on the X chromosome controlling the maternal effect and (2) a structural gene at 58 map units on the third chromosome controlling the tumorous head phenotype.

SUMMARY

Most heritable traits are controlled by nuclear chromosomal genes, but some depend on DNA in cytoplasmic organelles. Mitochondria and chloroplasts carry small amounts of unique DNA that behave independently with respect to nuclear genes. These cytoplasmic organelles have presumably evolved from free-living bacteria and algae, respectively, that entered into symbiotic relationships with eukaryotic cells. Streptomycin resistance in some present-day algae is dependent on plastids carrying DNA. In *paramecium*, symbionts with their own DNA are established in the cytoplasm, but they can reproduce only in the presence of a particular nuclear genotype *(K)*. Cytoplasmic organelles carrying DNA and having developed from prokaryote symbionts have become established in evolution and have retained a limited genetic operation more or less independent of nuclear genes. Plasmids are DNA molecules in the cytoplasm that may transform normal plant cells to tumor cells and produce enzymes that produce tumor growth. Male sterility in maize and some other crop and flower garden plants is controlled by cytoplasmic factors. Maternal effects are controlled by nuclear **genes** of the **mother** and therefore are not examples of extranuclear inheritance. Extranuclear contents of the egg, however, reflect the influence of the mother's genotype, and the pattern of inheritance is like that of extranuclear inheritance.

REFERENCES

Ashwell, M., and T. W. Work. 1970. "The biogenesis of mitochondria." *Ann. Rev. Biochem.* E. E. Snell (ed.), 39:251–290. Annual Reviews, Inc., Stanford University Press, Palo Alto, California.

Birky, C. W., Jr., P. S. Perlman, and T. J. Byers (eds.). 1974. *Genetics and biogenesis of mitochondria and chloroplast.* Ohio State University Press, Columbus.

Duvick, D. N. 1965. "Cytoplasmic pollen sterility in corn." *Adv. Genet.* E. W. Caspari and J. M. Thoday (eds.), 13:2–56. Academic Press, New York.

Gibson, I. 1970. "Interacting genetic systems in *Paramecium*." *Adv. Morphogen.* M. Abercrombin, J. Bracker, and T. J. King (eds.), 9:159–208. Academic Press, New York.

Goodenough, U. W., and R. P. Levine. 1970. "The genetic activity of mitochondria and chloroplasts." *Sci. Amer.* 223:22–29.

Kirk, J. T. O., and R. A. E. Tilney Bassett. 1978. *The plastids: their chemistry, structure, growth and inheritance,* 2nd ed. Elsevier/North-Holland, New York.

Kroon, A. M., and C. Saccone. 1974. *The biogenesis of mitochondria.* Academic Press, New York.

Margulis, L. 1971. "Symbiosis and evolution." *Sci. Amer.* 225:49–57.

Nester, E. W., and A. Montoya. 1979. "Crown gall: a natural case of genetic engineering." *ASM News* 45:283–287.

Preer, J. P., Jr. 1971. "Extrachromosomal inheritance: hereditary symbionts, mitochondria, chloroplasts." H. L. Roman (ed.), *Ann. Rev. Genet.* 5:361–406.

———. 1969. "Genetics of the protozoa." T. T. Chen (ed.), *Res. Protozool.* 31:130–278.

Sager, R., and Z. Raminis. 1970. "A genetic map of non-Mendelian genes in *Chlamydomonas*." *Proc. Natl. Acad. Sci., U.S.* 65:593–600.

Swift, H., and D. R. Wolstenholme. 1970. "Mitochondria and chloroplasts: nucleic acids and problems of biogenesis." A Lima de Faria (ed.), *Handbook Mol. Cytol.* 972–1046.

Tatum, L. A. 1971. "The southern corn leaf blight epidemic." *Science* 171:1113–1116.

PROBLEMS AND QUESTIONS

14.1. If a particular trait in a plant could be shown to be inherited solely through mitochondrial DNA, would it be classified as a case of maternal inheritance or maternal effect? Why?

14.2. In most animals, a larger amount of cytoplasm is

carried by the egg than by the sperm. Similarly, the egg in plants carries more cytoplasm than the pollen. How could this difference affect the expression of inherited traits (a) dependent on nuclear chromosomal genes and (b) dependent on extranuclear genes?

14.3. Reciprocal crosses with experimental animals or plants sometimes give different results in the F_1. This may be due to (a) sex-linked inheritance, (b) cytoplasmic inheritance, or (c) maternal effects. If such a result were obtained, how could the investigator determine experimentally which category was involved?

14.4. Explain how single plants such as four-o'clocks could have green, pale green, and variegated sectors. If such sectors reached sexual maturity, what color characteristics would each type be expected to transmit through male or female gametes?

14.5. What practical applications could be made with male-sterile lines of maize?

14.6. How could kappa particles have become established in their host organism, *Paramecium,* in evolution?

14.7. O. Renner carried out reciprocal crosses between two types of the evening primrose, *Oenothera hookeri* and *O. muricata,* known to have the same chromosome constitution. When the seed parent was *O. hookeri,* the plastids of the progeny were yellow; but when the seed parent was *O. muricata,* the plastids of the progeny were green. How might this difference in the results of reciprocal crosses be explained?

14.8. In snails of the genus *Limnaea,* coiling is transmitted as a maternal effect. (a) Give the phenotypes that could be associated with the following genotypes in individual snails, and give the reason for each answer: s^+s^+, ss^+, and ss. (b) What might be said about the female and male parents and grandparents of snails represented by each of the three genotypes?

14.9. Diagram a cross between a female snail with dextral coiling and the genotype s^+s^+ and an inbred ss male with sinistral coiling. Carry the cross to the F_2, and represent the expected results from inbreeding each of the F_2 snails. Explain the results.

14.10. In the beach hopper of the genus *Gammarus,* pigment of the eyes is influenced in early stages by the genotype of the mother, but is later influenced by the genes of the individual hopper. Give the expected results of the following crosses in young and adult stages: (a) dark females *(AA)* × light males *(aa)* and (b) light females *(aa)* × dark males *(AA).* (c) Give a plausible explanation for the change that sometimes occurs from light eyes in young organisms to dark eyes in later stages.

14.11. When ovaries from light-colored *(aa)* flour moths of the genus *Ephestia* are implanted into dark *(AA)* females, which are then mated to *aa* males, the *aa* progeny have dark eyes when first hatched, but the eyes gradually become lighter. Give a plausible explanation for such a change in eye color.

14.12. A female fruit fly known to be heterozygous for *y* (for yellow body color) had patches of yellow on the thorax. Is this expression more likely to be the result of (a) nonhereditary environmental modification, (b) maternal effect, or (c) nondisjunction of chromosomes, resulting in a gynandromorph?

FIFTEEN

Contrasting sizes of "normal" (unselected) flour beetles *(Tribolium castaneum)* and "giant" flour beetles produced by over 120 generations of selection for increased body size (pupa weight). The beetles in the selected population attained an average body weight of 6000 micrograms — more than twice the weight of the largest beetles in the unselected control population. This type of response to selection is expected when a trait is affected by a large number of genes each with a relatively small effect on the trait. Such long-term selection experiments have provided important information about the nature of the genetic variation observed with quantitative traits. (Original photograph courtesy of Franklin D. Enfield, Department of Genetics and Cell Biology, University of Minnesota).

MULTIPLE GENE INHERITANCE

In 1760, Joseph Kölreuter reported the results of crosses between tall and dwarf varieties of tobacco, genus *Nicotiana*. The F_1 plants were intermediate in size between the two parent varieties; the F_2 progeny showed a continuous gradation from the size of the dwarf to that of the tall parent. Kölreuter did not use curves as a device to represent his data, but a curve may now be drawn to illustrate diagrammatically the difference between continuous variation and discontinuous variation as described by Mendel (Fig. 15.1). Kölreuter was a careful experimenter and a good biologist, but he could not explain his results from crosses because the basic principles of Mendelian genetics had not yet been established. Only after Mendel's work was discovered, more than a hundred years after Kölreuter's experiments were completed, was an explanation for **discontinuous variation** (falling into distinct classes) formulated. [It should be noted that Mendel also described continuous variation. When he crossed white-flowered and purple-red-flowered beans, an intermediate flower color (pink) was obtained in the F_1 progeny and a spread from white to purple-red was found in the F_2.]

Mendel's results from garden peas could be analyzed in terms of simple frequencies that require only simple arithmetic, mostly counting peas. Bateson supported Mendel and strengthened the case for discontinuous variation. Later investigators reported simple ratios from crosses that involved many traits in a wide variety of plants and animals. The more elusive and problematical results, which were not readily explained by Mendelian segregation, were sometimes pigeonholed or discarded. Several years elapsed after the discovery of Mendel's work before progress was made in the analysis of continuous variation.

When Mendel crossed tall and dwarf varieties of peas, the F_1 progeny were all tall. Some of the F_2 plants were tall and some were dwarf, in the proportion of about 3 : 1. From these results, pairs of particulate genetic elements were postulated, with one member of each pair being dominant over its allele. Distinct and clear-cut contrasting traits were observed and readily classified under such headings as tall or dwarf, yellow or green, and red or white.

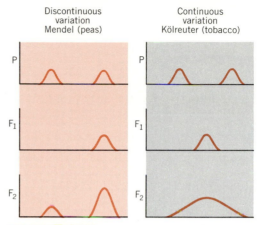

FIGURE 15.1

Curves representing the results of Mendel's experiments compared with those of Kölreuter. (Left) P, F_1, and F_2 from Mendel's crosses with garden peas illustrating discontinuous variation; (right) P, F_1, and F_2 from Kölreuter's crosses on tobacco plants illustrating continuous variation. The ordinate represents the number of plants and the abscissa the range in height.

Kölreuter's results, on the other hand, showed continuous variation with no distinct class boundaries. The F_1 hybrids were intermediate between the parents, and the F_2 generation covered the entire range between the heights of the parents. Each gene alone had a slight effect on height.

QUANTITATIVE TRAITS

Between 1900 and 1910, many geneticists thought continuous variation reflected an entirely different mechanism of inheritance from that of discontinuous variation. A few keen investigators, however, began to envision a common basis for the results of Mendel and those of Kölreuter. Alleles that had small but **cumulative effects** with semidominance rather than complete dominance were postulated to behave in a Mendelian fashion. Several to many nonallelic genes were postulated to have cumulative effects on particular traits. An explanation for continuous variation thus emerged in the form of the **multiple-gene hypothesis.** Experimental results and interpretations substantiating this hypothesis were obtained from the classical investigations of

P	Red	White		
	$AABB$ $\times$	$A'B'A'B'$		
	AB	$A'B'$		
F₁	Medium	Medium		
	$AA'BB'$ $\times$	$AA'BB'$		
F₂	AB	AB'	$A'B$	$A'B'$

	AB	AB'	$A'B$	$A'B'$
AB	$AABB$ Red	$AABB'$ Dark	$AA'BB$ Dark	$AA'BB'$ Medium
AB'	$AABB'$ Dark	$AAB'B'$ Medium	$AA'BB'$ Medium	$AA'B'B'$ Light
$A'B$	$AA'BB$ Dark	$AA'BB'$ Medium	$A'A'BB$ Medium	$A'A'BB'$ Light
$A'B'$	$AA'BB'$ Medium	$AA'B'B'$ Light	$A'A'BB'$ Light	$A'A'B'B'$ White

Summary: $\frac{1}{16}$ red, $\frac{4}{16}$ dark, $\frac{6}{16}$ medium, $\frac{4}{16}$ light, $\frac{1}{16}$ white

FIGURE 15.2

Cross between a wheat variety with red kernels and another variety with white kernels. This cross illustrates quantitative inheritance, which is dependent on multiple genes.

H. Nilsson-Ehle in Sweden and E. M. East in the United States during the period 1910–1913.

One of these studies was based on crosses between two varieties of wheat, one producing red kernels and the other white kernels (Fig. 15.2). The F₁ seeds were intermediate in color between those of the two parents. They were lighter than those of the red parent but distinctly more colored than those of the white-parent variety. When the F₂ seeds were classified according to intensity of color, a continuous gradation was observed from red to white, and classes were more or less arbitrary. About $\frac{1}{16}$ of the F₂ seeds were as red as those of the red parent, and about $\frac{1}{16}$ were white. About $\frac{14}{16}$ were in between, ranging from near the color of the red parent (P) to that with only slight pigment near the white original parent.

When the $\frac{14}{16}$ of the F₂ seeds were classified further, on the basis of color intensity, it was shown that about $\frac{4}{16}$ had more color than the F₁ intermediates, about $\frac{6}{16}$ were intermediate like the F₁ seeds, and about $\frac{4}{16}$ were lighter than the F₁ seeds. This result suggested a segregation of two allele pairs and

was explained on the basis of two different loci acting on the same trait and producing a cumulative effect. When the results of this cross were examined critically, they were found to resemble those that Kölreuter had obtained many years before.

A second cross between red-kernel and white-kernel varieties of wheat was carried to the F₂; this time about $\frac{1}{64}$ of the F₂ progeny produced white kernels and about $\frac{1}{64}$ produced red kernels. Some $\frac{62}{64}$ were between the parents in color intensity, from near red to near white. Again, continuous gradations were recognized between the extremes of the parents. The results of this cross resembled those of a Mendelian trihybrid cross in that three independent allele pairs were postulated to explain the results, in contrast to the two pairs for the previous cross. Evidently, one pair of alleles, which was segregating in the second cross, had been homozygous in both parents in the first cross.

The concept of multiple genes for quantitative inheritance is now one of the most important principles of genetics. It has been strengthened greatly by the use of statistical methods devised by R. A. Fisher (Fig. 15.3) in England, Sewall Wright in the United States, and others. The explanation is based on the action of many genes (polygenes), usually segregating independently but influencing the same phenotype in a cumulative fashion, with a large environmental influence.

POLYGENE CONCEPT

A **polygene** is defined as a gene that individually exerts a slight effect on a phenotype but, in conjunction with a few or many other genes, controls a quantitative trait, such as height. Polygenic inheritance differs from the classical Mendelian pattern in that a **graded series** extends from one parental extreme to the other. Only averages and variances of populations are considered and not discrete values for individuals. Such factors as epistasis, cytoplasmic influences, interactions among genes and gene products, and interactions with the environment are reflected in the averages and variances. Polygenic inheritance is a Mendelian concept that is studied by statistical methods.

Because most characteristics of domestic plants

FIGURE 15.3

Ronald A. Fisher (1890–1962), British statistician and scholar in biometrical and population genetics. He made biometry a most useful tool for analysis and interpretation of quantitative data. (Courtesy of Everett Thorpe.)

and animals that have practical significance (including height, weight, time required to reach maturity, and qualities relevant to nutrition) depend on polygenic inheritance, much attention has centered around this principle. If all the practical genetic experimental projects that are now in progress at the various experiment stations throughout the world could be listed and classified, the results would probably indicate that some 80–90 percent involve polygenic or quantitative inheritance. Some human characteristics of interest and significance such as intelligence depend on **multiple genes.**

The inheritance of polygenic traits depends on the cumulative or additive action of several or many genes, each of which produces a small proportion of the total effect. This is in marked contrast to the inheritance of major gene traits, which is an all-or-none phenomenon dependent on one gene or a few interacting genes. **Polygenic traits can be measured,** and these measurements can be subjected to statistical treatment. Environmental factors also influence the final result of these traits (e.g., height, weight, and color intensity). The genetic component or heritability of the trait must therefore be separated from the environmental effect. Variation in such traits as kernel color in wheat and skin color in humans is mostly hereditary under the conditions of the usual observation.

MULTIPLE-GENE HYPOTHESIS

A classical study of quantitative inheritance, which did much to establish the multiple-gene hypothesis, was made by R. A. Emerson and E. M. East on the inheritance of ear length in maize. A variety with ears averaging 6.6 cm in length was crossed with a variety that averaged 16.8 cm. The F_1 progeny were intermediate, averaging 12.1 cm and ranging from 9 to 15 cm. The F_2 represented a wider spread of variation than the F_1, with some of the ears as extreme in size as those of the original parents. This is the characteristic pattern when only a few polygenes are segregating. Homozygosity of alleles may occur in the F_2. The histograms (Fig. 15.4) represent the maize data and illustrate graphically the characteristic pattern of multiple-gene inheritance. Since the parental lines were inbred and presumably homozygous, the variation within these lines and that shown for the fully heterozygous F_1 plants are environmental. Results of genetic segregation of genes from the two parents are illustrated by the wider variation in the F_2.

As in other examples involving polygenic inheritance, **environmental influences** are confusing factors for analysis. The environment can produce results similar to those of genes with respect to size differences between large and small varieties. Plants raised in unfavorable environments — that is, without sufficient water, sunlight, or soil nutrients — will be smaller than others of the same genotype that have a more favorable environment. On the other hand, under ideal environmental conditions, organisms with inferior genotypes (e.g., many mutational variants) may develop phenotypes equivalent or superior to those with better genotypes but infe-

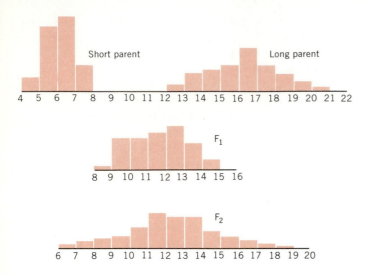

FIGURE 15.4

Histograms illustrating maize data from a cross involving multiple gene inheritance. (Data from Emerson and East.)

rior environments. Environmental influence is especially important in the analysis of polygenic traits. Environmental influences must be identified and controlled as completely as possible with experimental design or statistical adjustment.

In the maize example, let us assume that two pairs of alleles are active and that the variation is all hereditary. If each gene produces an equal effect on the size of the ear (beyond that of the residual genotype, 6.6 cm), the individual contribution would be

$$\frac{\text{maximum height} - \text{minimum height}}{\text{number of alleles}}$$

$$= \text{relative contribution of each allele}$$

$$= \frac{16.8 - 6.6}{4} = 2.55 \text{ cm per allele}$$

Each active allele thus would produce 2.55 cm in addition to the 6.6 cm inherent in the residual genotype for the small variety. When the F_2 plants were classified according to phenotype, a ratio of $1:4:6:4:1$ was obtained. This is a modification of the $1:2:1:2:4:2:1:2:1$ genotype ratio, which may be changed to $9:3:3:1$ by dominance. The analysis (Fig. 15.5) represents a model for these data based on the multiple-gene hypothesis.

In other plant crosses that involve several thousand F_2 progeny, none even approached the parental phenotypes. Many pairs of alleles were therefore

presumed to be involved. As many as 200 pairs of alleles were postulated to segregate in some experiments. E. M. East, for example, studied the corolla length in a cross between two varieties of tobacco, *Nicotiana longiflora,* and found that none of the sev-

Individual Contribution of Alleles (*A* or *B*): 2.55 cm
Size Produced by Residual Genotype: 6.6 cm
Genotypes of Parents:
Black Mexican   Tom Thumb
   AABB   ×   *A'A'B'B'*

F_1 *AA'BB'* (12.1 cm)

F_2	Geno-type	Fre-quency	Pheno-type (cm)	Pheno-typic Ratio
	AABB	1	16.8	1
	AA'BB	2	14.2	
	AABB'	2	14.2	} 4
	AA'BB'	4	11.7	
	A'A'BB	1	11.7	} 6
	AAB'B'	1	11.7	
	A'A'BB'	2	9.1	
	AA'B'B'	2	9.1	} 4
	A'A'B'B'	1	6.6	1

FIGURE 15.5

Analysis of Emerson's maize data, which is based on the multiple gene hypothesis with two pairs of alleles, each contributing equally to the phenotype.

eral hundred F_2 plants resembled a parent (P). Many pairs of alleles were then postulated.

ESTIMATING THE NUMBER OF GENE DIFFERENCES

The contributions of individual genes to a particular trait can be roughly evaluated for some crosses. In the maize data, as indicated, each extreme of the parents occurred in the proportion of about 1 in 16 in the F_2. Two pairs of alleles were assumed to be segregating. As noted, environmental as well as genetic variations are represented by the same measurements. A further complication is that the genes may not be equal in their influence on the phenotype. Models devised to estimate the number of genes are oversimplified if based on the assumption of equal effects, but this is the assumption of choice when the degree of influence of individual genes is not known. A rough estimate of genes involved has been made by determining the frequency of occurrence in the F_2 population of the **extremes** that represent the parental phenotypes (Table 15.1). More complicated mathematical procedures are employed when sufficient F_1 and F_2 data are available for comparison.

TRANSGRESSIVE VARIATION

Extremes in some F_2 results have exceeded the corresponding parental (P) values. This pattern is called **transgressive variation;** its explanation is based on the hypothesis that the parents did not represent the extremes possible from the combined genotypes. If the genotype of the large parent did not include all the alleles for large size and the genotype of the small parent did not include all the alleles for small size, an F_2 individual might receive a combination of genes, producing a larger or smaller size than that seen in either parent. For example, in a cross between a large Hamburgh chicken and a small Sebright bantam, Punnett found that the F_1 chickens were intermediate between the two parents. While the F_2 progeny included some birds larger and some smaller than the parental varieties, most were intermediate between the original parents. Punnett explained the result on the basis of a four-factor difference between the two parents. The Hamburghs were postulated to have the genotype $AABBCCD'D'$ and the bantams $A'A'B'B'C'C'DD$. The F_1 birds would be uniformly heterozygous $AA'BB'CC'DD'$, which accounted for the intermediate weight. Some F_2 individuals might have the genotype $AABBCCDD$ and be heavier than the original Hamburgh parent, whereas others could have the genotype $A'A'B'B'C'C'D'D'$ and be smaller than the bantam parent.

Genes responsible for polygenic inheritance are considered to behave in the typical **Mendelian fashion.** Only their phenotypic results differ from the familiar ratios. Thus, a $9:3:3:1$ ratio expected from a dihybrid cross involving complete dominance is modified to a $1:4:6:4:1$ ratio with continuous variation and no dominance (i.e., $AAB'B' = AA'BB'$ phenotypically). With no dominance, each gene is presumed to have the same effect as any other gene, and a normal distribution results for the phenotypes. If three pairs of independent alleles were involved in a cross, the ratio of $1:6:15:20:15:6:1$ would replace the familiar $27:9:9:9:3:3:3:1$ ratio based on discontinuous variation and complete dominance. The **more polygenes** that are segregating, the **more continuous** is the phenotypic variation.

MECHANISM OF QUANTITATIVE INHERITANCE

Quantitative traits such as body weight of an animal must depend on a number of systems, at both the gross and biochemical levels. The digestive system,

TABLE 15.1 Probability that F_2 Individuals Will Be as Extreme as Either Parent

PAIRS OF SEGREGATING ALLELES	FRACTION OF F_2 AS EXTREME AS EITHER PARENT
1	$\frac{1}{4}$
2	$\frac{1}{16}$
3	$\frac{1}{64}$
4	$\frac{1}{256}$
5	$\frac{1}{1024}$

including the liver and pancreas, and the vast array of digestive and other enzymes, the fat storage tissues, the nervous system in controlling appetite, endocrine glands controlling metabolic rates, and numerous other organs and systems must all influence body weight. These systems are established from the basic DNA "blueprints" carried in the zygote for the organism. Thus, it would be expected that a large number of genes acting cumulatively would influence body weight. This same quantitative mechanism would explain height, skin color, intelligence, and many other quantitative traits.

Similarly, yield in cabbage plants would depend on many inherited systems (as well as noninherited environmental conditions). An efficient root complex, transport systems for water and nutrients, chlorophyll synthesis, disease resistance, and many other factors would be involved in yield. Polygenes representing the stable genetic system could maintain the cumulative record of the many interrelated systems. Parent plants could transmit the DNA "blueprints" received in the seeds that produced them, along with any mutational changes that might have occurred, to their generation of cabbage plants.

Modifier genes can influence size difference in animals and plants, but they do not always follow a uniform pattern. Some act as enhancers and some as inhibitors of a particular effect, and the end result is the average effect of many genes. Certain pleiotropic genes may influence both qualitative and quantitative traits. Furthermore, the effect of a particular gene substitution may vary with different genetic backgrounds.

The same phenotype may be produced by different genetic systems. Size in animals and plants, for example, is generally controlled by a polygenic system, but a single major gene may produce a dwarf animal or a dwarf plant, such as among Mendel's peas (Chapter 2), and may thus accomplish the same end result as that produced by multiple genes. In cases where only one or a few genes are involved, it is possible to develop homozygous, pure-breeding types in a few generations of inbreeding. This is not possible in systems of polygenes where many genes are involved; the genes cannot be individually recognized, and analytical techniques ordinarily applied to Mendelian inheritance are inadequate. New techniques making use of computers have been applied to various models based on different hypothetical patterns. It has been possible to demonstrate results comparable with those of natural systems of polygenes using the model of Mendelian segregation. These studies support the basic premise that polygenic inheritance is Mendelian. R. D. Milkman has studied a series of polygenes that controls the crossvein-making ability in the *Drosophila* wing. Some of these genes have been located by linkage studies in the three major chromosomes of *Drosophila melanogaster.*

Although there are many unresolved problems concerning the nature and action of polygenes associated with quantitative traits, the multiple-gene concept is a good working hypothesis. Because it is impossible at this time to identify more than a few particular polygenes or to obtain a detailed assessment of their properites, polygenic systems must be considered in **statistical terms.** Statistical studies provide comprehensive averages in terms of **quantitative values.** In natural selection, the balance within the operative polygenic system is more important than the effect of individual genes. Polygenic inheritance fits the Darwinian pattern of gradual and continuous changes, which are "selected" by natural environments.

HERITABILITY

Effectiveness of selecting for a particular characteristic depends on the relative importance of heredity and environment in the development of that trait. The extent of inheritance is a significant consideration, particularly in animal breeding. Breeding animals are selected on the basis of visible or measurable traits (phenotypes) that are displayed by themselves or their relatives. Agreement between phenotype and genotype is measured by the coefficient of heritability h^2. Heritability of a trait (in a narrow sense) is the ratio of its cumulative genetic variance to total variance, as shown by the following expression:

$$h^2 = \frac{V(G)}{V(P)}$$

where $V(G)$ is the variance of genetic effects and $V(P)$ is the total phenotypic variance. **Heritability is a measure of the degree to which a phenotype is genetically influenced and, therefore, the degree to which it can be modified by phenotypic selection.**

The simplest method of selection used by animal breeders is **phenotypic mass selection.** This is a systematic choosing of the part of the population in which the desired qualities are most strongly developed and then using the individuals chosen as parents of the next generation. In farm animals, this procedure involves evaluation of production records or other evidence of desired traits. Rate of progress depends on the ability of the breeder to choose individuals that are not only phenotypically but also genotypically superior. Mass selection is most efficient when heritability is high. However, progeny tests, family selection, and pedigrees can be used to supplement information pertaining to the specific animal when heritability of the trait is low. It follows that the animal's phenotype is not an accurate indicator of its genotype when heritability is low.

Three different approaches have been used for estimating heritability. The first is the time-honored practice of trial and error combined with good husbandmanship that began in prehistoric time—the actual gain or success that is realized through selection. The estimate of heritability derived by this method is the degree of progress related to the effort expended in selection as measured by the selection differential. This is the difference in mean phenotype between the selected parents and the entire parental population.

The second approach is based on how much more alike in phenotype are related individuals having somewhat similar genotypes than are unrelated or less closely related individuals. In isogenic stocks made up of individuals with exactly the same gene component, all variation is environmental, as shown by the relation:

$$h^2 = \frac{V(P) - V(E)}{V(P)}$$

where $V(P)$ is the variance among individuals in the general population, which is made up of genetic and environmental differences, and $V(E)$ (environmental variance) is the variance among individuals within the same isogenic line. Comparisons of variation in populations that are otherwise similar to isogenic stocks but undergo random breeding rather than inbreeding can provide estimates of **heritability.** This method can be readily utilized with fruit flies and other laboratory animals, but among farm animals, the only isogenic individuals are identical twins. These are infrequent and sometimes difficult to identify.

Another approach that is suitable for some laboratory organisms but not practical for farm animals is to select in both directions and thus develop high and low lines with reference to a particular trait. The extent to which the lines differ in response to selection provides an estimate of heritability. Such an experiment has been performed in honeybees, as illustrated in Fig. 17.3 in Chapter 17.

The most widely used procedure for estimating heritability for farm animals is to compare the resemblance between relatives. Greater average resemblance between parents and progeny of half or full sib groups is associated with higher heritability. When little correlation exists between the degree of expression of a trait by members within these related groups, the greater part of the observed variation in that trait is environmental.

When heritability of a trait is known, the progress to be expected from a generation of selection can be predicted. The coefficient of heritability is not a constant, but only indicates the proportion of variance caused by differences in additive gene effects in a particular population at a particular time. Some examples of heritability coefficients are listed in Table 15.2. Egg production in chickens has a low heritability coefficient, hence little progress would be expected per generation of mass selection.

Why do modern chickens have low heritability for egg production? They differ significantly from the ancestral jungle fowl that laid only a few eggs per year for genetically influenced reproduction. The change has been accomplished mostly by environmental control that has converted the chicken into an egg factory to meet human needs. Nutrition, housing, temperature control, and extending the day with artificial light have all been contrived to improve egg production. The hereditary compo-

TABLE 15.2 Heritability Coefficients

TRAIT AND ANIMAL	h^2
Egg production in chickens	.05 – .15
Back fat in swine	.50 – .70
Milk yield in cattle	.2 – .4
Daily weight gain in cattle	.3 – .5
Fleece weight in sheep	.3 – .6

nent may represent little more than the jungle fowl level for egg production. Selection on some basis other than mass selection or the use of breeding systems that increase genetic variation may result in slight further improvement in egg production. Rate of weight gain in cattle, on the other hand, has a moderately high heritability coefficient, and considerable progress is possible and, indeed, has been made through mass selection.

Assume that in a cattle population the average daily gain is 2.4 pounds, and the bulls selected to be sires of the next generation average 3.4 pounds and the cows 3.0, giving a selection differential of $\frac{(3.4 + 3.0)}{2} - 2.4 = 0.8$ pounds. If heritability is 0.4 or 40 percent, the expected gain will be 40 percent of the selection differential or 40 percent of 0.8 pounds, giving 0.32 pounds. The offspring would then be expected to average 2.4 + 0.32 = 2.72 pounds as compared to 2.4 pounds for the entire population from which the parents were selected. If heritability had been only 10 percent, as it is for egg production, the gain would have then been 0.08 (10 percent of 0.8) and the average for the next generation would be only 2.48. The generation interval in beef cattle is $4\frac{1}{2}$ to 5 years, so the expected progress of 0.32 pounds per generation would represent an annual increase of about 0.06 to 0.07 pounds. This points out the value of reducing the generation interval in livestock.

HERITABILITY IN POLYGENIC TRAITS
The heritability of a trait represents the cumulative effect of all loci affecting the trait. It says little or nothing about the mode of inheritance of a quantitative trait, but it is useful in predicting response to selection. Calculations are usually made from results of selection, but the meaning of heritability is greater than implied by this method. For a single locus, heritability is dependent on a complex of interrelated values. The number of segregating genes for a given polygenic trait usually is not determined. Sometimes it can be determined but only with elaborate and specifically designed experiments. For a quantitative trait determined by 20 loci, for example, 60 quantities would be involved in the calculation of heritability: 20 allele frequencies, 20 values for G, and 20 values for P in the formula for heritability. Heritability is measured as a single number that must include the combined effect of all 60 quantities, but it says nothing about any of them.

When more than 20 loci affect the trait or when multiple alleles are involved, the problems become more complex. The assumption thus far has been that all loci affecting the trait act independently with respect to each other. When linkage and gene interaction are included, the quantitation of the problem is even more complex. Furthermore, the task of interpreting heritability must depend on the range of environments involved. Nevertheless, the concept of heritability is indispensable to the plant and animal breeder. Values are usually estimated from the degree of selection that is possible and used by "rule of thumb." Although the heritability value is real, it is virtually impossible to quantitate a precise estimate for polygenic inheritance.

MEASUREMENT DATA
Data are numerous and complex in studies of quantitative inheritance because many genes each produce a small segment of the same phenotype. Unit results are additive or cumulative, providing for a sum, average, or other statistical device. The data require analysis and organization before their significance can be appreciated. Large numbers of individuals usually typify populations to be studied and compared; therefore, sampling methods are used to facilitate comparison. One requirement that must be observed in sampling techniques is **randomness.** A sufficiently **large sample,** taken at random without bias or favoritism, should adequately

provide an estimate for a complete population that cannot be measured in its entirety. Most biological populations are so large that they are assumed to be infinite. In other situations, laboratory populations do not actually exist but are only theoretical.

It is important to distinguish clearly between estimates based on samples and the actual values that would be obtained if it were possible to measure the entire population. Estimates are called **statistics,** and the true values based on entire populations are called **parameters.** In biological investigations, parameters are seldom known. If they were known, direct comparisons could be made between populations. For example, species A could compared directly with species B in terms of the actual mean or degree of variability. When parameters are not known, statistics based on samples are used for comparison. Measurements especially useful to the geneticist are the **mean** $\left(\bar{x} = \dfrac{\Sigma X}{n} \right)$ and the **variance** $\left(s^2 = \dfrac{\Sigma (X - \bar{x})^2}{n - 1} \right)$. The square root of the variance is the familiar statistic called the **standard deviation** $(s = \sqrt{s^2})$.

The first statistic to be considered here is the mean ($\bar{x}$), which is an estimate of magnitude. It can be defined as the sum of measurements divided by the total number of measurements. The variance (s^2) is an estimate of variation.

The variance in each of two inbred populations in the same environment would be similar because the variations in both would be almost entirely environmental in origin. If homozygous plants from two populations (P) were crossed, the variance of the F_1 fully heterozygous population would be similar to those of the two parent populations, because the variation again would be environmental. When F_1 heterozygotes are crossed, segregation and independent assortment result in genetic recombination. Genotype differences plus environmental variation result in a higher variance than that of the P and F_1 populations. Variance in the F_3 populations would differ according to the number of segregating alleles. The variance thus measures the genotype and environmental variation and is useful for analyzing the genetic structure of populations. Variances of gene-controlled phenotypic differences

with adaptive significance are useful in studies of evolution.

The difficulty with using variance as a measure of variation is that it is given in terms of the units of measurements squared. Extracting the square root of the variance converts it to the same units in which the measurements were taken, and the statistic thus obtained is called the standard deviation, or s.

The symmetrical bell-shaped curve presented in Fig. 15.6 represents a normal distribution of individual measurements. If one population standard deviation (σ) was plotted on either side of the mean, the arc would include about two-thirds (68.26 percent) of the population represented. The shape of the curve is determined by the amount of variation, as indicated by the standard deviation. If the same area is maintained, a small s is associated with a high curve and a large s with a flat curve. Furthermore, if s is small and the curve is high, most of the observations are clustered around the mean. On the other hand, if s is large and the curve is flat, the observations are spread away from the mean.

Another useful statistical measurement is the standard deviation of means or the "standard error of the mean." This is a measure of how reliable the sample mean $\bar{x}$ is as an estimate of the population mean (μ). As suggested by the name, the standard error is an estimate of the standard deviation of means of samples drawn from a single population. Means, which represent values for groups of individuals, are expected to vary less than would indi-

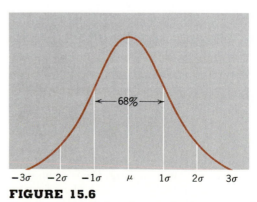

FIGURE 15.6

A symmetrical bell-shaped curve showing a normal distribution of individuals in a population on either side of the mean (μ).

viduals drawn at random from the population. It is not necessary to actually go through the process of taking several samples from the same population, since the standard error is inversely proportional to the square root of the sample size n.

Two factors are involved in an evaluation of a sample's **reliability:** (1) the **size** of the sample and (2) the amount of **variation** in the population sampled. If great variation is present, a large sample will be required to represent the population adequately. If individuals are fairly uniform, a smaller sample may be satisfactory. These two factors are taken into consideration in the following formula for the standard error $s_{\bar{x}}$ of the sampling mean:

$$s_{\bar{x}} = \sqrt{\frac{s^2}{n}} = \frac{s}{\sqrt{n}}$$

The standard error of the mean indicates the reliability with which the **sample mean $\bar{x}$ estimates the population mean** μ The smaller the standard error, the more reliable the estimate. It can be seen from the formula that, as the sample size n increases, the value of the standard error tends to decrease, and vice versa. This relation illustrates the advantage of large samples for obtaining the best possible estimates of parameters.

The application of statistical methods to genetics has provided indirect as well as direct benefits. Statistical requirements have made it necessary for geneticists to carefully design each experiment before undertaking actual investigations. Thus, they develop a critical attitude toward methods and interpretations of experimental results. A number of variations formerly attributed to genetic mechanisms have now been explained on the basis of sampling errors.

CONTROL OF ENVIRONMENTAL VARIATION

Some genetic patterns were not recognized in the past or were confused with the effects of environmental factors. The modern geneticist controls all possible aspects of a particular study and allows for the element of chance in sampling. Development of effective statistical methods has helped foster this trend. The significance of mean, variance, and standard error of the mean as basic statistical tools can be demonstrated by applying them directly to a

problem. For example, in cereal crops, the time required for the plants to mature depends to some extent on inheritance. Since environmental factors such as temperature are also involved, however, the plants to be compared must be grown in a single, controlled environment, with adequate checks for uncontrollable environmental factors.

In a particular experiment, seeds representing each of four types of wheat were planted in randomized field plots. The four types of wheat were two inbred parent varieties (PA and PB) and F_1 and F_2 progeny from a cross between PA and PB. The seedlings were all raised in the same season, and the randomized plots were designed to equalize minor environmental variations in soil, moisture, and other factors. The time required for maturing was recorded as days from the time of planting to the time when the heads were fully formed. A sample of 40 plants was taken at random from each population. The data representing the four samples of 40 individuals are given as follows:

PA

75	74	72	72	73	71	72	71
76	73	72	72	72	70	71	72
71	73	74	73	73	72	71	72
72	74	73	72	71	72	73	72
74	71	72	73	75	70	72	76

PB

58	55	56	56	53	55	55	57
54	55	56	55	58	57	55	56
55	57	55	57	56	57	55	55
56	57	55	54	59	57	55	55
58	56	57	54	53	56	58	56

F1PAPB

60	65	63	61	65	50	62	63
61	60	63	64	64	61	62	63
65	62	64	62	60	59	61	62
61	60	63	62	60	63	60	65
64	61	62	64	64	61	62	64

F2PAPB

69	66	62	60	63	67	72	64
61	63	62	63	60	59	64	63
56	62	63	65	64	73	60	65
57	64	63	70	68	62	71	63
65	66	64	58	61	65	62	64

TABLE 15.3 Data and Calculations from Samples of the Two Parents, F₁ and F₂, of Wheat Populations, Tabulated According to Time in Days Required for Maturity; 40 Plants Are Included in Each Sample

	DAYS																								$\bar{x}$	$s_{\bar{x}}$	s^2	s
	53	54	55	56	57	58	59	60	61	62	63	64	65	66	67	68	69	70	71	72	73	74	75	76				
PA																		2	7	15	8	4	2	2	72.47±	.23	2.05	1.43
PB	2	3	13	9	8	4	1																		55.85±	.22	1.92	1.39
F₁							1	7	7	8	6	7	4												62.20±	.27	2.88	1.70
F₂				1	1	1	1	3	2	5	7	6	4	2	1	1	1	1	1	1	1				63.72±	.60	14.26	3.78

The data from each sample were classified and are summarized in Table 15.3. The means, variances, standard deviations, and standard errors of the means are also included in the table. Several facts can be derived from the data and calculations. First, each parental class represents a different population in regard to the length of time required for maturity. One sample (PA) has a mean $(\bar{x}_A)$ of 72.47 and the other (PB) has a mean $(\bar{x}_B)$ of 55.85 (Fig. 15.7). What is the mode of inheritance? The F₁ plants were intermediate between the two parents, and the F₂ were also generally intermediate, but with a wider range than the F₁. This pattern indicates polygenic inheritance with few segregating genes. Why are the variances for the PA, PB, and F₁ about equal, whereas the F₂ variance is much greater? The inbred parents were highly homozygous. If they were completely homozygous, the variation would be entirely environmental. If PA and PB plants were completely uniform, F₁ plants from homozygous parents would be uniformly heterozygous. Again, all variation would be environmental. Segregation expected in the F₂ provides for genetic variation in addition to that directly associated with the environment.

The standard error of the F₂ mean (used as a measure of reliability) was larger than those of the parents and of the F₁ because of the increased variation in the F₂ and the comparable sample size (Table 15.3). A larger sample would provide a better estimate of the parameter (mean). For a sample of 80 with $s = 3.78$, the $s_{\bar{x}}$ would be .425. A sample of 200 with the same s would give $s_{\bar{x}} = .267$.

CONTINUOUS VARIATION

Most differences among normal animals and plants follow the pattern of continuous variation. For example, if 5000 men were arranged in order of height, each would differ only slightly from his neighbors on either side. Many polygenic traits could be identified by appropriate comparisons among individuals in a population. For most traits, however, only part of the continuous variation is inherited; the other part is due to environmental factors. In the example of stature in men, barring gross malnutrition or other severe environmental

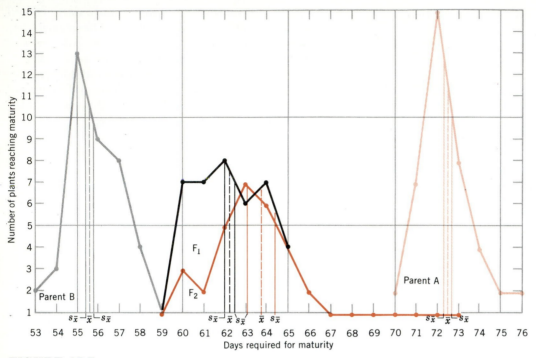

FIGURE 15.7

Graphic presentation of data in Table 15.3 for experiment on time required for maturity in wheat. PB: curve for parent B sample; PA: parent A sample; F_1: first filial generation sample; and F_2: second filial generation sample (n = 40 for each sample). Calculations for the mean ($\bar{x}$) and ±one standard deviation ($s_{\bar{x}}$) of the mean are illustrated for each sample. As a measure of reliability, the ±$s_{\bar{x}}$ of .60 for the F_2 sample indicates that the true population mean (parameter) is expected to fall within the area of .60 on either side of the sample mean statistic in 68 percent of similar trials.

influence, about 90 percent of the variability would be hereditary. Measurements for other traits such as arterial pressures and pulse rates show a smaller genetic component. Hereditary continuous variation depends on the combined action of **many genes** (polygenes), each contributing to the same trait.

An important consideration in assessing relative contributions of heredity and environment is the requirement of one time and one place for all observations. When environmental conditions change, the relative importance of heredity must usually also change. For example, susceptibility to pulmonary tuberculosis depends to some extent on hereditary predisposition. If the incidence of the

disease decreases appreciably, the relative importance of the hereditary constitution will not remain the same. On the one hand, certain environmental conditions may offset high constitutional susceptibility. In such a case, the relative importance of heredity is increased among those who still suffer from the disease. Or, on the other hand, genetic susceptibility may be overcome as environmental improvements take place, in which event the relative importance of heredity decreases.

Another equally important consideration is that each person, irrespective of kinship, must have an equal chance of experiencing the various kinds of environmental factors that could mimic the effect of

FIGURE 15.8
Identical or monozygotic twins. Since monozygotic twins develop from a single zygote, they are genetically identical. Thus, all differences between them are effects of environmental differences. (K. Bendo.)

heredity and thus increase the measured resemblance. At least in theory, the genes that make relatives alike and the shared environmental factors that can also make them alike can be distinguished. Genes held in common decrease rapidly as relationships become more remote. Environmental similarities fall off less rapidly. A **regression** (correlation between organisms and offspring or other relatives as a measure of inheritance) for uncles and aunts or for cousins that was higher than expected in proportion to that for sibs or parents would indicate shared **nongenetic factors.** Some data give indications of this, but the practical difficulties of making observations and measurements on any but the closest relatives are formidable. Better evidence for the relative influence of heredity and environmental factors comes from studies of twins. Usefulness of twins in genetic studies depends on comparisons between two kinds: (1) one-egg, identical and (2) two-egg, fraternal twins. Since monozygotic (MZ) twins are from the division of the same zygote, identical genetic material goes to each member of the pair (Fig. 15.8). All phenotypic differences **except those resulting from rare mutations** are environmental.

Fraternal or dizygotic (DZ) twins arise from two separate eggs produced simultaneously in the mother and fertilized by separate sperm from the same father. They are no more alike genetically than ordinary sibs. Their differences depend on genetic segregation and environmental influences. Comparisons between MZ and DZ twins have provided the best data to indicate whether genetic variation is present and how much of the total variation is genetic. MZ and DZ twins were compared in a study of cleft lip and/or cleft palate [CL(P)] and cleft lip (CL) (Table 15.4). A much higher concordance was observed for the MZ as compared with the DZ twins.

Monozygotic twins vary less than DZ twins and the differences are environmental. It is possible from comparisons to determine how much heritability occurs. But the question of environmental differences from immediate postnatal environmental influences for children raised together still leaves the environment uncontrolled. The ideal situation would be to have a sufficient number of both MZ and DZ twins separated at birth and reared apart.

The frequency of twinning in the white population in the United States is about 1 twin pair per 100

TABLE 15.4 Facial Clefts — Twin Studies (1922 – 1979)

	MONOZYGOTIC TWINS				DIZYGOTIC TWINS			
	CL(P)		CP		CL(P)		CP	
	C	D	C	D	C	D	C	D
	29	53	5	14	7	149	2	32
Concordance	35%		26%		4.5%		5.8%	

C indicates concordance, D indicates discordance. (Courtesy of Robert J. Gorlin. Based on Gorlin, 1970, together with cases of Blake and Wreakes, Schweckendiek, Shields et al., and sundry other authors.)

births. Dizygotic twins are about twice as frequent as monozygotic twins at birth. In the United States black population, about 1 twin pair occurs in 73 births. The white and black groups differ only in the frequencies of fraternal twins. Identical pairs have about the same frequency in all populations studied.

DISTINGUISHING MONOZYGOTIC AND DIZYGOTIC TWINS

The mechanism that divides zygotes is apparently a random and irregular event that occurs with about equal frequency in all human populations. The cause is unknown. No hereditary implication is associated with this form of twinning. Except for rare mutations, MZ twins are identical in genotype and essentially alike in phenotype. In contrast, DZ twins that arise from separate zygotes are no more alike genetically than two sibs from the same parents. Since, however, they have had the same prenatal environment and more or less the same postnatal environment, they are expected to have phenotypic similarities.

Separation of MZ and DZ twin pairs depends in one way or another on confirming identical phenotypes in the two MZ twins. If a single reliable difference can be detected, a pair is classified as DZ. Only when no basic difference can be established is a pair classified as MZ. In making the determination, observable or measurable characteristics are compared systematically. Sex is the simplest genetic trait to detect. Twins of different sex must arise from different zygotes. Other genetic traits useful in detecting zygosity are blood groups; plasma protein types; antigen similarity; chromosome polymorphism; biochemical identity; eye color; fingerprints; characteristics of teeth, nostrils, and ears; and many aspects of physiology.

The obstetrician may make a preliminary determination by observing the fetal and maternal tissue in the afterbirth, which includes the amnion, chorion, and placenta. Undifferentiated cells form an amnion membrane early in development that surrounds an embryo of each monozygotic or dizygotic twin. The placenta of dizygotic twins may be fused, thus making it unreliable for distinguishing the two kinds of twins. Monozygotic twins may have a single chorion, but dizygotic twins never do. A single chorion may thus identify monozygotic twins. But obstetricians are usually very busy at the time when appropriate observations might be made; zygosity is not a top priority for them. Unless the twins are unlike in sex, the determinations made at birth are subject to error.

In the past, many errors were apparently made in determining zygosity. One study showed that about one-fourth of the twins classified as MZ considered themselves to be DZ, whereas one-fifth of those classified as DZ considered themselves to be MZ. Permanent acceptance by the body of a skin graft is the most rigorous test for zygosity. But this requires a surgical procedure and is seldom carried out unless the need is great. Twin graft acceptance may be tested when a twin is injured.

Sir Francis Galton in the nineteenth century

apparently made many mistakes in his initial demonstrations of twin studies for investigating quantitative inheritance, but he opened the way for others who followed at a time when better distinguishing features were available. Galton, a pioneer in the application of statistics to biology, introduced the field of **biometry.** He attempted to measure everything related to human structure and behavior. Walking through the streets of London, he would carry a counter or two in his pocket, counting people with one trait after another. The data would be calculated and analyzed when he arrived back at his desk. His data on MZ twins were obtained from conversations with other people and letters from colleagues and friends who had twins in their families or knew about others who knew about twin pairs. Some 35 of these pairs were probably identical and 20 were probably fraternal. He recognized the value of identical and fraternal twins for distinguishing between genetic and environmental components for quantitative traits.

In spite of the limitations in numbers of MZ twin pairs and difficulties in obtaining suitable matched pairs of DZ twins for comparison, twin studies have supported statistical analysis for investigations of multiple genes and quantitative inheritance. They have been used successfully in investigating hereditary and environmental factors in human variation. Three groups of twins have been used on different quantitative human traits in studies conducted with identical twins reared apart: James Shields in Britain with 44 pairs, Horatio Neuman in the United States with 19 pairs, and Niels Juel-Nielsen with 12 pairs. Extensive studies on twins were carried out over the years by Gedda in Rome, Kallman in New York, and Rife in Columbus, Ohio.

SUMMARY

Polygenic inheritance is a Mendelian concept. Its study makes use of statistics. Many genes with cumulative effects in population aggregates are associated with metric traits. A consequence of the action of polygenes is continuous rather than discontinuous variation. Measurements are required for sampling the variations, but the Mendelian foundation forms the basis for the genetic mechanisms involved. If a small number of active genes is involved in the making of a trait, a discontinuous genotype may be reflected in a continuous phenotype. Environmental variation tends to smooth over the Mendelian discontinuity. Heritability is a measure of the degree to which a phenotype is genetically influenced. It is difficult to measure precisely for quantitative traits, but it is indispensable as a concept used by plant and animal breeders. Parameters of populations may be estimated from population samples in statistical terms. Mechanisms for polygenic inheritance in specific examples are not defined because it is difficult to distinguish hereditary and environmental influence. In continuous variation, numbers and locations of polygenes are not readily established, effects of individual polygenes are usually not recognizable, and populations rather than individuals must be considered in statistical terms. Nevertheless, polygenic inheritance can be inferred from experimental data, and it represents a highly significant and practical mode of inheritance.

REFERENCES

Bodmer, W. F., and L. L. Cavalli-Sforza. 1976. *Genetics, evolution, and man,* Chapter 14. W. H. Freeman, San Francisco.

Dunn, L. C., and D. R. Charles. 1937. "Studies on spotting patterns." *Genetics* 22:14–42.

East, E. M. 1910. "A Mendelian interpretation of variation that is apparently continuous." *Am. Nat.* 44:65–82.

———. 1916. "Studies on size inheritance in *Nicotiana.*" *Genetics* 1:164–176.

Elseth, G. D., and K. D. Baumgardner. 1981. *Population biology.* D. Van Nostrand, New York.

Emerson, R. A., and E. M. East. 1913. "The inheritance of quantitative characters in maize." *Nebraska Agric. Exp. Res. Bull.* (120 pp.).

Falconer D. S. 1967. "The inheritance of liability to disease with variable age of onset, with particular reference to diabetes mellitus." *Ann. Hum. Genet.* 31:1–20.

Hartl, D. L. 1980. *Principles of population genetics.* Sinauer Associates, Inc., Sunderland, Massachusetts.

Holt, S. B. 1968. *The genetics of dermal ridges.* Charles C. Thomas, Springfield, Illinois.

Mather, K., and J. L. Jinks. 1971. *Biometrical genetics.* Cornell University Press, Ithaca, New York.

Mather, K. 1974. *Genetic structures of population.* Halsted Press, New York.

Stern, C. 1970. "Model estimates of the number of gene pairs involved in pigmentation variability of the Negro-American." *Hum. Hered.* 20:165–168.

PROBLEMS AND QUESTIONS

15.1. Using the forked-line method, diagram the cross between a wheat variety with red kernels (*AABB*) and a variety with white kernels (*A'A'B'B'*) in which the two pairs of genes have a cumulative effect. Classify the F_2 progeny under the headings red, dark, medium, light, and white, and summarize the expected results.

15.2. From another cross between red and white kernel varieties, $\frac{1}{64}$ of the F_2 plants had kernels as deeply colored as the red parent and $\frac{1}{64}$ had white kernels. About $\frac{62}{64}$ were between the extremes of the parents. How can the difference in F_2 results in Problems 15.1 and 15.2 be explained?

15.3. Different F_2 plants from the cross in Problem 15.2 were crossed with the white parent (*A'A'B'B'C'C'*). Give the genotypes of the individual plants from which the following backcross results could have been obtained: (a) 1 colored, 1 white, (b) 3 colored, 1 white, and (c) 7 colored, 1 white.

15.4. Different F_2 plants from the cross in Problem 15.2 were selfed. Give the genotypes of the parents that could have produced the following results: (a) all white, (b) all colored, (c) 3 colored, 1 white, (d) 15 colored, 1 white, and (e) 63 colored, 1 white.

15.5. Different F_2 plants from the cross in Problem 15.2, all producing kernels with some red color, were crossed with each other. Give the genotypes of parents that could have produced the following results: (a) 7 colored, 1 white, (b) 31 colored, 1 white, and (c) 63 colored, 1 white.

15.6. Assume that two pairs of genes are involved in the inheritance of skin pigmentation. For the purpose of this problem, the genotype of a black person may be symbolized *AABB* and that of the white *A'A'B'B'*. What color might be expected in the F_1 from the crosses between black and white people?

15.7. If mulattoes (*AA'BB'*) mated with other mulattoes of the same genotype, what results might be expected with reference to the intensity of skin pigmentation?

15.8. (a) If people with various degrees of skin pigmentation married only people known to have the genotype *A'A'B'B'*, could they have "black" babies? Explain. (b) Could a "white" couple with black ancestry have a "black" baby? Explain.

15.9. If the number of gene pairs involved in skin pigmentation should actually be 4, 5, or 6, (a) how would the expected results of crosses between white and black differ from those considered in the above problem? (b) What criteria and data would be necessary to determine the number of genes actually involved?

15.10. Two pairs of genes (*AABB*) with equal and additive effects are postulated to influence the size of corn in certain varieties. A tall variety averaging 6 feet was crossed with a dwarf averaging 2 feet. (a) If the size of the dwarf is attributed to the residual genotype (*A'A'B'B'*), what is the effect of each gene that increases the size above 2 feet? (b) Diagram a cross between a large and small variety, and classify the expected F_2 phenotypes.

15.11. The size of rabbits is presumably determined by genes with an equal and additive effect. From a total of 2012 F_2 progeny from crosses between large and small varieties, eight were as small as the average of the small-parent variety and about eight were as large as the large-parent variety. How many allele pairs were operating?

15.12. A sample of 20 plants from a certain population was measured in inches as follows: 18, 21, 20, 23, 20, 21, 20, 22, 19, 20, 17, 21, 20, 22, 20, 21, 20, 22, 19, and 23. Calculate (a) the mean, (b) the standard deviation, and (c) the standard error of the mean.

15.13. What is measured by (a) the mean, (b) the standard deviation, and (c) the standard error of the mean?

15.14. If the population sampled in Problem 15.12 were an F_2 involving parents from varieties averaging 7 inches and 33 inches, respectively, would you conclude that few or many gene pairs were involved? Why?

15.15. A sample of 20 plants from a certain population was measured in inches as follows: 7, 10, 12, 9, 10, 12, 10, 9, 10, 11, 8, 12, 10, 10, 9, 11, 10, 9, 10, and 11. Calculate (a) the mean, (b) the standard deviation, (c) the standard error of the mean.

15.16. If the population sampled in Problem 15.15 were an F_2 from parents averaging 7 inches and 12 inches, respectively, would you conclude that few or many gene pairs were involved? Why?

15.17. The width or spread of 122 guayule plants representing a random sample were measured 107 days after planting, in inches, as follows:

13	12	11	13	13	13	12	11	13	11
13	13	12	11	11	12	12	12	13	10
11	10	9	10	10	10	14	12	11	11
11	12	12	14	13	14	13	16	13	13
10	11	11	11	10	11	12	13	12	12
10	10	10	13	11	10	13	14	12	9
12	10	10	13	11	11	13	10	12	12
12	14	11	12	11	12	12	13	12	13
11	11	11	11	12	10	12	10	10	11
9	11	12	13	11	11	12	10	9	14
11	14	12	13	15	12	11	11	14	11
12	12	13	12	10	13	10	12	10	11
9	10								

Calculate (a) the mean, (b) the standard deviation, (c) the standard error of the mean.

15.18. Seeds from four different types of wheat were planted and samples of each type were measured as to the time required for maturity. This is the number of days elapsing from the time of planting the seed until the heads of the grain appear. The following statistics were computed from the samples of different sizes:

	MEAN	STANDARD DEVIATION	STANDARD ERROR OF THE MEAN
Strain A	72.37	1.31	.221
Strain B	55.85	.93	.179
$F_1(A \times B)$	61.40	1.35	.350
$F_2(A \times B)$	63.84	3.45	.330

Several of the F_2 plants were selfed. The F_3 seeds were planted in randomized plots and raised in the same season. The progeny thus produced were classified according to the time required for maturity. A sample of 24 plants from the seed of a single F_2 was recorded as follows:

56	55	54	56	57	56
55	56	57	56	57	56
55	57	56	55	56	58
56	55	57	56	59	58

Compute (a) the mean, (b) the standard deviation, (c) the standard error of the mean.

15.19. (a) Discuss briefly the biometrical significance of the sample in Problem 15.18. (b) From which part of the F_2 population was the F_2 parent likely obtained? (c) What suggestions can be made concerning the genotype of the F_2 parent?

15.20. Make a similar analysis of the following samples, each of which represents a random sample from the F_3 derived from the seed of a single F_2 plant:

(a)
73	72	70	74	76	71
72	74	74	74	71	76
73	72	77	73	78	75
76	70	74	70	71	70

(b)
67	65	64	66	65	66
68	64	65	66	69	71
70	63	62	61	60	64
63	65	64	63	68	67

(c)
65	64	66	67	65	64
64	68	65	64	63	65
65	64	66	65	67	66
65	68	65	64	66	65

15.21. (a) What does it mean to be 68 percent confident? (b) What does $s_{\bar{x}} = \pm 1.02$ cm tell the investigator? (c) What is the value of knowing the variance of a set of data?

15.22. Can heritability be estimated for polygenic traits from studies of twins? If so, how can it be done?

SIXTEEN

A population of zebras. (Ira Kirschenbaum/Stock, Boston)

POPULATION GENETICS AND EVOLUTION

We have considered mechanisms and consequences of Mendelian inheritance among individuals. Now we are concerned with alleles in a Mendelian population. Such a population must be a natural evolving population. It could be an entire species with all members capable of exchanging genes with each other. But natural species populations are usually not continuous in a geographical area. Physical barriers, distance, and other factors make it impossible for random mating to occur among members of an entire species. Individuals tend to mate with members of their own subpopulation, which may be called a local population, deme, or Mendelian population.

A **Mendelian population** is an interbreeding group of organisms sharing a common **gene pool.** A gene pool is the total genetic information possessed by reproductive members in a population of sexually reproducing organisms. Alleles in the pool have **dynamic relations** with other alleles and with the environment in which the organisms reside. Environmental factors that collectively provide selection tend to alter allele frequencies and thus to cause evolutionary changes in local populations.

In 1908, an English mathematician, G. H. Hardy, and a German physician, W. Weinberg, independently discovered the principle concerned with the frequency of alleles in a population, called the **Hardy–Weinberg equilibrium principle.** It states that at equilibrium, both allele and genotype frequencies remain constant from generation to generation. This occurs among diploid, sexually reproducing organisms with nonoverlapping generations and in large interbreeding populations in which mating is random and no selection or other factors are present for changing the allele frequencies.

Mendelian segregation is represented mathematically by the binomial expansion of $(a + b)^n$, where a is the probability that an event will occur and b is the probability that it will not occur. The familiar ratio of $1:2:1$, representing the segregation of a single pair of alleles (Aa), in a monohybrid cross may be represented by the simple expansion of $(a + b)^n = (A + a)^2 = 1AA + 2Aa + 1aa$. To ex-

TABLE 16.1 Combinations of Sperm and Eggs from Gamete Pools for an Entire Population, Illustrating the Allele Frequency Relations that Form the Basis for the Hardy–Weinberg Equilibrium

EGGS	SPERM	
	$A(p)$	$a(q)$
$A(p)$	$AA(p^2)$	$Aa(pq)$
$a(q)$	$Aa(pq)$	$aa(qz^2)$

Summary: $p^2(AA) + 2pq(Aa) + q^2(aa) = 1$

press the relation in more general terms that will apply to any frequency of alleles in a population, the symbols p and q are introduced. At equilibrium, the frequencies of the genotypic classes are **$p^2(AA)$, $2pq(Aa)$,** and **$q^2(aa)$.** A **frequency** is the ratio of the actual number of individuals falling in a single class to the total number of individuals; **a probability** represents the likelihood of occurrence of any particular form of an event. Possible combinations of sperm and eggs from gamete pools for an entire population where the genotypes can be any distribution of AA, Aa, and aa and p and q can take on any values so long as they sum to 1, as shown in Table 16.1. In a large population including AA, Aa, and aa genotypes, an equilibrium is established for a single pair of alleles after one generation of random (panmictic) mating. The genetic proportions of an equilibrium population are entirely determined by its allele frequencies, as illustrated in Table 16.2.

When only two alleles are involved, $p + q = 1$. Since $p + q = 1$, $p = 1 - q$. Now, if $1 - q$ is substituted for p, all the relationships in the formula can be represented in terms of q as follows: $(1 - q)^2 + 2q(1 - q) + q^2 = 1$. (This alteration provides a means for solving a problem with a single unknown, q.) If allele A has a frequency of $1 - q$ and allele a has a frequency of q, the expected distribution of these alleles under conditions of random mating in succeeding generations may be calculated with the Hardy–Weinberg formula; it is applied to any pair of alleles if the frequency of one member of the pair in the population can be determined.

TABLE 16.2 Algebraic Proof of Genetic Equilibrium in a Randomly Mating Population for Any Two Alleles (i.e., $p + q = 1$) in the Population[a]

Parental matings	AA × AA	2(AA × Aa)	2(AA × aa)	Aa × Aa	2(Aa × aa)	aa × aa	Summations
Parental mating frequencies	$p^2 \times p^2$	$2(p^2 \times 2pq)$	$2(p^2 \times q^2)$	$2pq \times 2pq$	$2(2pq \times q^2)$	$q^2 \times q^2$	
Offspring frequencies AA	p^4	$2p^3q$		p^2q^2			$\Sigma_{AA} = p^2$
Aa		$2p^3q$	$2p^2q^2$	$2p^2q^2$	$2pq^3$		$\Sigma_{Aa} = 2pq$
aa				p^2q^2	$2pq^3$	q^4	$\Sigma_{aa} = q^2$
Total offspring frequencies	p^4	$+ 4p^3q$	$+ 2p^2q^2$	$+ 4p^2q^2$	$+ 4pq^3$	$+ q^4$	$= \sum_{\text{Total}} = 1$

$$(p^4 + 2p^3q + p^2q^2) + (2p^3q + 4p^2q^2 + 2pq^3) + (p^2q^2 + 2pq^3 + q^4)$$
$$p^2(p^2 + 2pq + q^2) + 2pq(p^2 + 2pq + q^2) + q^2(p^2 + 2pq + q^2) \quad = p^2 \; = 2pq \; = q^2$$
$$p^2 + 2pq + q^2$$
$$(p + q)^2 \quad = 1$$

[a] One generation of random mating is required to reach genetic equilibrium.

EQUILIBRIUM

Genotypes in a population tend to establish an equilibrium with reference to each other, expressed as $p^2 : 2pq : q^2$. The absolute frequency of each genotype is thus seen to depend on the values of p and q. For example, if two alleles should occur in equal proportion in a large, isolated breeding population and neither has an advantage over the other, they would be expected to remain in equal proportion generation after generation. This would be a special case because alleles in natural populations seldom if ever occur in equal frequency. They may, however, be expected to maintain their relative frequency, whatever it is, subject only to change by such factors as **natural selection, differential mutation rates, migration, genetic drift, and meiotic drive,** all of which may alter the level of the allele frequencies. An equilibrium in genotype frequencies is maintained through random mating, with absolute frequencies of genotypes being determined by the allele frequency. As long as the allele frequency does not change and matings are at random, genotypic proportions remain constant.

ALLELE FREQUENCY FOR CODOMINANT INHERITANCE

The M and N blood antigens provide a model by which the segregation of a single pair of autosomal alleles having codominant inheritance can be studied in a human population. To be consistent with symbols for alleles, the same letter should be used to represent the different alleles at the same locus. L^M and L^N will represent the two alleles in this discussion. These alleles are especially suitable for a beginning study of allele frequency because dominance is not expressed. The M, N, and MN phenotypes are detectable by serological tests. Table 16.3 shows the relationships between genotypes and phenotypes with the specific reactions of the blood cells to anti-M and anti-N sera. (Although no significant selective value has been associated with these blood traits, several investigators have suggested that a type of immunization is possible that may give these antigens significance in selection.) Numerous samples of blood representing different populations have been tested, and the fre-

TABLE 16.3   Detection of M, N, and MN Phenotypes

PHENOTYPE	GENOTYPE	ANTI-M[a]	ANTI-N[a]
M	$L^M L^M$	+	○
MN	$L^M L^N$	+	+
N	$L^N L^N$	○	+

[a] The + sign means that the red blood cells of the individual react with the antiserum. The ○ means that no reaction occurs.

quencies of the L^M and L^N alleles have been calculated from the data obtained. Here are the proportions of the different phenotypes based on a sample of 6129 Caucasian people in the United States:

M	MN	N
1787	3039	1303

When the data are interpreted in terms of the frequency of genotypes ($L^M L^M + L^M L^N + L^N L^N$), as shown at the bottom of this page, the total numbers and frequencies of alleles are obtained, as shown in Table 16.4.

Allele frequencies represented in decimal fractions, as in this example, can be used directly as probabilities. Thus, the probability is .5395 that a member of the particular pair will carry L^M, and the probability is .4605 that a member of the pair will

carry L^N. From such information a determination of genetic equilibrium can be made for the population.

If in the present example only allele frequencies (.5395 for L^M and .4605 for L^N) had been given, the following proportions of genotypes predicted for the population at equilibrium could be calculated, as shown at the bottom of this page. These proportions based on probability may now be related to the population based on the total of 6129. When the χ^2 test for goodness of fit is applied, the differences between the predicted numbers based on allele frequencies and the actual number of M, MN, and N blood groups in the population are not significantly different ($P = .01$). The genotypic frequencies fit the equilibrium formula:

$$p^2 + 2pq + q^2 = 1$$

indicating that the $L^M L^M$, $L^M L^N$, and $L^N L^N$ genotypes are at equilibrium in the population sampled.

ALLELE FREQUENCY WHEN DOMINANCE IS INVOLVED

Dominance and recessiveness of alleles do not directly influence allele frequency; that is, dominance alone does not make an allele occur more frequently in the population. The same type of equilibrium is maintained for autosomal alleles showing dominance or recessiveness, as is maintained for those with codominant inheritance, such as the L^M and L^N alleles just described. It is true, however, that dominant alleles express themselves in heterozygous combination and therefore are expressed more frequently than recessives when equal numbers of dominant and recessive alleles are present. If one phenotype has a selective advantage over another, dominance could indirectly influence allele frequency.

An interesting pair of contrasting traits, which

TABLE 16.4   Frequency of L^M and L^N Alleles

ALLELE	PROPORTION	ALLELE FREQUENCY
$L^M = 2 \times 1787$ $+ 3039 = 6613$	6613/12,258	.5395
$L^N = 2 \times 1303$ $+ 3039 = 5645$	5645/12,258	.4605

	PROPORTION OF 6129 TOTAL	χ^2 OF OBSERVED/ CALCULATED
$L^M L^M = .5395 \times .5395 = .2911$	1784.15	.0045
$L^M L^N = .5395 \times .4605 = .2484$	1522.44	.0115
$L^N L^M = .4605 \times .5395 = .2484$	1522.44	.0115
$L^N L^N = .4605 \times .4605 = .2121$	1299.96	.0071
		.0346

has been detected in human populations and has no known selective value, is the ability or inability to taste phenylthiocarbamide (PTC). The differences in the ability of people to taste this chemical was discovered accidentally by investigators in a university laboratory. A hereditary basis for the mechanism was postulated to account for tasters and nontasters. Inability to taste the chemical ("taste blindness") was found to be dependent on a single recessive allele *(t)*.

From a group of 228 university students who were invited to taste the chemical, the following results were obtained: 160 tasters and 68 nontasters. What are the relative allele frequencies of *T* and *t* in the population? Since *T* is dominant over *t*, the 160 tasters include genotypes *TT* and *Tt*. The 68 nontasters must have the genotype *tt*. Assuming equilibrium, the best approach to the Hardy–Weinberg formula is through the nontasters, all of whom must have the same genotype. The 68 *tt* individuals represent .30 (q^2) of the population, $\hat{q} = \sqrt{.30} = .55$. (The "hat" over a *p* or a *q* symbol indicates an equilibrium frequency.) Thus, the frequency of *T* in the sample is .45 and the frequency of *t* is .55. These allele frequencies are comparable to those obtained from larger samples of the Caucasian population in the United States. Predicted genotype frequencies based on random mating in the population are

$$\text{Frequency of } TT = p^2 = .45 \times .45 = .20$$
$$\text{Frequency of } Tt = pq = .45 \times .55 = .25$$
$$\text{Frequency of } tT = qp = .55 \times .45 = .25$$
$$\text{Frequency of } tt = q^2 = .55 \times .55 = .30$$
$$p^2 = .20 \qquad 2pq = .50 \qquad q^2 = .30$$

When the allele frequencies are known, it is possible to predict the likelihood of occurrence of genotypes and expressions of corresponding traits in populations. The probability of an expression of a trait dependent on a recessive allele may thus be calculated, even in family groups or other populations where no previous expression has occurred and no evidence is available to indicate which individuals carry the recessive allele in question.

FREQUENCIES FOR MULTIPLE ALLELES

The equation $p + q = 1$ applies when only two autosomal alleles in the population occur at a given locus. If the system includes more alleles, more symbols must be added to the equation. The four human ABO blood types, for example, are controlled by three alleles, I^A, I^B, and i. Both I^A and I^B are dominant over i, but neither is dominant over the other; that is, both anti-A and anti-B sera will coagulate the cells from an individual with I^AI^B genotype. For substitution into the formula, the frequencies of I^A, I^B, and i may be symbolized *p*, *q*, and *r*, respectively. Since these three alleles represent all the alleles involved, it follows that $p + q + r = 1$. Multiple alleles establish an equilibrium in the same way as the single pairs of alleles previously described.

Blood types of 173 students in genetics laboratory classes were determined as follows: O, 78; A, 71; B, 17; and AB, 7. Genotypes and probabilities in the population sampled are summarized in Table 16.5. From these data, the frequency of the various alleles may be calculated. The gene pool includes all the alleles in this sample as well as all others in the population. Since r^2 (O) = 78/173 = .45, $r = \sqrt{.45} = .67$. The Hardy–Weinberg formula may now be applied to two of the three alleles. The proportion of the A *(p)* and O *(r)* individuals in the population is represented by the equation:

$$\text{the frequency of (O + A)}$$
$$= r^2 + 2pr + p^2$$
$$= (r + p)^2.$$

Therefore,

$$r + p = \text{the } \sqrt{\text{frequency of O + A}}$$
$$= \sqrt{78/173 + 71/173}$$
$$= \sqrt{.45 + .41} = \sqrt{.86} = .93$$

and $\quad p = .93 - .67 = .26.$

The frequencies of *p* and *r* have now been obtained and the next step is to calculate the frequency of *q*. Since $p + q + r = 1$, $.26 + .67 = 1 - q$, or $q = 1 - (.67 + .26) = .07$. Now summing the allele frequencies calculated from the sample of 173 students, we have .26 *(p)* + .07 *(q)* + .67 *(r)* = 1. The

TABLE 16.5 Phenotypes, Frequencies, and Genotypes Represented by a Sample of 173 Genetics Students

PHENOTYPES	PHENOTYPE FREQUENCIES OF STUDENTS	GENOTYPES	GENOTYPIC FREQUENCY	SUM OF FREQUENCIES OF GENOTYPES WITH SIMILAR PHENOTYPES
O	.4509	ii	r^2	r^2
A	.4104	$I^A I^A$	p^2	$p^2 + 2pr$
		$I^A i$	$2pr$	
B	.0983	$I^B I^B$	q^2	$q^2 + 2qr$
		$I^B i$	$2qr$	
AB	.0405	$I^A I^B$	$2pq$	$2pq$

value of 1 represents all the alleles considered in this example. Because of dominance, allele distribution in the gene pool differs from the distribution of phenotypes that result from combinations of alleles. Results from this sample of 173 students are comparable to those from larger samples in the general populations. The A and B from the AB blood group were not used during the calculations because the frequencies of alleles and not phenotypes were being determined. All three segregating alleles were included in the calculations.

FREQUENCY FOR SEX-LINKED ALLELES

Alleles in the sex chromosomes occur in a different frequency from those in autosomes because of the arrangements of sex chromosomes in the two sexes. In organisms such as *Drosophila* and humans, with one heterogametic sex, there are five possible genotypes: *AA, Aa, aa* for the female and *a, A* for the male. In organisms with male heterogametic sex determination, the genotypic values at equilibrium for sex-linked genes are as follows:

1. For females: $p^2 + 2pq + q^2$
    AA   Aa   aa
2. For males: $p + q$
    A   a

In females, X chromosomes are paired and behave like autosomes. Hemizygous males with only one X

chromosome follow the crisscross sex-linkage pattern described in Chapter 4. In human populations, the X chromosome is transmitted from a father through his daughter(s) to half of her sons (her father's grandsons). Hardy–Weinberg equilibrium for sex-linked genes requires that allele frequencies be equal in the two sexes. If equilibrium is not already established, the attainment of allele-frequency equality is not accomplished in a single generation of random mating. It occurs gradually over several generations of random mating and is governed by two principles resulting from the mechanism of sex-linked inheritance.

If alleles *A* and *a* are at one locus in the X chromosome, let m_n and f_n represent the frequency of the *A* allele in males (m) and females (f) in generation *n*. (1) The crisscross mechanism of inheritance dictates that the allele frequency in males in any generation equals the frequency in females in the previous generation; that is: $m_n = f_{n-1}$, because a male receives his X chromosome from his mother. (2) The frequency in females in any generation equals the average allele frequency in the previous generation, because females receive one X chromosome from each parent: $f_n = (\frac{1}{2})(m_{n-1} + f_{n-1})$.

If, for example, the initial (n_0) frequency for *A* is .8 in males and .5 in females, the first generation (n_1) frequency will be the average generation frequency (3) for males and females in $n_0 : (\frac{1}{2})$

$(m_0 + f_0) = \dfrac{.3}{2} = .15$; $.15 + .5 = .65$ for females in generation n_1 and $.5$ for males in generation n_1. In generation n_2: $\dfrac{.15}{2} + .5 = .575$ for females and $.5 + \dfrac{3}{2} = .65$ for males. With continued random mating, about 10 generations would be required to reach equilibrium with $f = \hat{f}$ and $m = \hat{m}$. When $f = \hat{f}$ and $m = \hat{m}$, f and m remain constant thereafter. The equilibrium frequencies would be:

$$A = \hat{f} = \hat{m} = \frac{2 f_0 + m_0}{3} = \frac{2(.5) + .8}{3}$$

$$= \frac{1.8}{3} = .6, \quad a = .4$$

EQUILIBRIUM AT MORE THAN ONE LOCUS

When two allele-pair differences (e.g., Aa and Bb) segregate independently and are considered simultaneously, the number of possible genotypes is 3^2 (i.e., $AABB$, $AABb$, $AaBB$, $AaBb$, $aaBB$, $aaBb$, $AAbb$, $Aabb$, and $aabb$). Attainment of equilibrium at both loci together requires more than one generation of random mating. More terms are involved and a multinomial rather than a binomial expansion is required to obtain equilibrium ratios. Let p, q, r, and s represent the allele frequencies of A, a, B, and b, respectively. The equilibrium formula depends on the terms pr, ps, qr, and qs, which are the equilibrium frequencies of the gametes AB, Ab, aB, and ab, respectively. The equilibrium ratios of the genotypes are: $(pr + ps + qr + qs)^2$ or $p^2 r^2 AABB$, $2 p^2 rs AABb$, $2 pqr^2 AaBB$, $4 pqrs AaBb$, $q^2 r^2 aaBB$, $2 q^2 rs aaBb$, $p^2 s^2 AAbb$, $2 pqs^2 Aabb$, and $q^2 s^2 aabb$. When the gametic equilibrium frequencies have been reached, the genotypic frequencies will also be at equilibrium.

How long will it take for gametic frequencies to reach equilibrium? In a population of heterozygotes ($AaBb \times AaBb$) in which all alleles have the same frequency, that is, $p = q = r = s = .5$, equilibrium should already have been attained and gametic frequencies would be at equilibrium in the next gener-

ation. But this example is an extreme, hypothetical case unlikely to occur in nature. Allele frequencies can take any value as long as $p + q = 1$ and $r + s = 1$. Equilibrium will require several to many generations for any initial population not already at equilibrium. Linkage delays the attainment of equilibrium. Alleles of different gene pairs located in the same member of a chromosome pair do not assort independently. Many generations might be required for chance recombination to place them in different members of the chromosome pairs and thus permit segregation and independent assortment.

At equilibrium, frequencies of unlinked alleles in gametes would be the same in coupling (AB and ab) as in repulsion (Ab and aB; i.e., $pr \times qs = ps \times qr$). If coupling and repulsion products are not equal in the initial population, the difference (d) may be used to calculate the time required to attain equilibrium. One-half the difference from equilibrium is reduced in each generation of random mating. If, for example, the frequencies of A and B are each $.6$ and the frequencies of a and b are each $.4$, the equilibrium frequencies $= ps \times qr = (.24)(.24) = pr \times qs = (.36)(.16) = .0576$. The difference in coupling and repulsion products represents the change in gametic frequencies required to attain equilibrium. If the initial genotypic frequencies of $AABB$, $AAbb$, and $aaBB$ are all $.3$ and the frequency of $aabb = .1$, the initial frequencies of gametes are $.3$ for AB, Ab, and aB, and $.1$ for ab; $d =$ frequency of $Ab \times$ frequency of Ab for repulsion $-$ frequency of $AB \times$ frequency of ab for coupling $= (.3)(.3) - (.3)(.1) = .06$. Calculations to determine the number of generations of random mating required to reach equilibrium, with given initial gene frequencies for unlinked genes and genotype frequencies, are summarized in Table 16.6.

NONRANDOM MATING

Hardy–Weinberg equilibrium of genotype and phenotype frequencies is based on random mating. Two nonrandom systems of mating are encountered in nature and in humanly controlled popula-

TABLE 16.6 Calculation for Approach to Equilibrium of Gametes in a Population with Two Separate Unlinked Allele Pairs *Aa* and *Bb*, with Initial Allele Frequencies of .6 for *A* and *B* and .4 for *a* and *b* and Initial Genotypic Frequencies of .3 for *AABB*, *AAbb*, and *aaBB*, and .1 for *aabb*[a]

GENERA-TION	PROPORTION OF $d = .06$[b]; ADD TO *AB* & *ab*; SUBTRACT FROM *Ab* & *aB*	GAMETES			
		AB	*Ab*	*aB*	*ab*
1.		.3	.3	.3	.1
2.	$.5d$	.33	.27	.27	.13
. . . 5.	$.9375d$	.35625	.24375	.24375	.15625
. . . 11.	$.9990234375d$	.35994140625	.24005859375	.24005859375	.15994140625
Equilibrium	d	.36	.24	.24	.16

[a] After M. Strickberger.
[b] The difference (d) between repulsion and coupling products is: d = frequency of *Ab* $\times$ frequency of *aB* − frequency of *AB* $\times$ frequency of *ab* = [(.3)(.3) − (.3)(.1)] = .06. Generation 11 approaches equilibrium.

tions: (1) **inbreeding** and **outbreeding** and (2) positive and negative assortative mating. Assortative mating is accomplished by human intervention in plant and animal breeding. It is positive when mates are more alike than would be expected by chance and negative when mates are more dissimilar than expected by chance.

INBREEDING

Parthenogenesis in animals such as *Drosophila* is the most intense form of inbreeding possible. Self-fertilizing plants typify extreme inbreeding, whereas cross-fertilizing plants and animals can evidence varying degrees of inbreeding and outbreeding, depending on such factors as compatibility, motility, and proximity.

A model (Table 16.7) involving a very small population of mice (two families) can illustrate the effects of a particular mating system on allele fre-

quency. In family I, one parent is white *(cc)* and the other parent is black *(CC)*. Their progeny, four females and four males, are all black and heterozygous *(Cc)*. In family II, both parents and all eight (four female and four male) progeny are black *(CC)*. If only the F$_1$ of the two families is considered as the population, there are $8c$ and $24C$.

When the entire population is considered, the same proportion of c and C alleles is maintained whether inbreeding or outbreeding is practiced. The system of mating in itself does not influence the proportion of alleles. A population with a frequency of .9 for allele A and .1 for a will maintain these frequencies, whether close relatives or unrelated individuals are mated, unless some other factor changes the proportions.

The main difference between the results of inbreeding and outbreeding is that more recessive alleles express themselves when the mating is be-

TABLE 16.7   Expected Distribution of Alleles in Families of Normal Mice of the Same Strain from Inbreeding and Outbreeding When the Initial Proportion of Alleles Is the Same in the Parents and All Progenies Are Equal in Size (4)

	FAMILY	SECOND GENERATION	THIRD GENERATION
Inbreeding	I	$Cc \times Cc$	$1CC\ 2Cc\ 1cc$
	I	$Cc \times Cc$	$1CC\ 2Cc\ 1cc$
	II	$CC \times CC$	$4CC$
	II	$CC \times CC$	$4CC$
	Total genotypes		$10CC\ 4Cc\ 2cc$
Outbreeding	I × II	$Cc \times CC$	$2CC\ 2Cc$
	I × II	$Cc \times CC$	$2CC\ 2Cc$
	II × I	$CC \times Cc$	$2CC\ 2Cc$
	II × I	$CC \times Cc$	$2CC\ 2Cc$
	Total genotypes		$8CC\ 8Cc\ 0cc$

tween related individuals. **Although the total allele frequency remained the same in the model, the proportion of phenotypes was different.** Two of the 16 progeny resulting from inbreeding expressed the trait (white) controlled by the recessive allele, but none of those resulting from outbreeding were white. Also, more homozygous blacks *(CC)* resulted from inbreeding.

His choice of self-fertilizing plants such as peas and beans was a real advantage to Mendel in his studies. It meant pure lines of peas were available to hybridize, and it also facilitated his experimental procedure (Chapter 2). In 1903, Johannsen recognized the uniformity that characterized self-fertilizing plants grown in the same environment. On this basis, he postulated **pure lines:** populations that breed true without appreciable genetic variation. Johannsen considered the pure lines that resulted from inbreeding to depend on the similarity of alleles in the various individuals making up the group. Completely pure lines, in which all allelic pairs are homozygous, would represent the ultimate result of inbreeding. It is doubtful that complete homozy-

gosity ever actually occurs, because of new mutations and the tendency for segregation to maintain a few heterozygotes in the system; but for all practical purposes, lines subjected to inbreeding over long periods of time are pure.

How do pure lines develop in nature? If tall heterozygous *(Dd)* garden peas and their descendants were allowed to self-fertilize over a period of time, and neither tall nor dwarf was favored by selection, what would be expected in five or 10 generations? In the first generation, the proportion would be $1DD:2Dd:1dd$. In terms of fractions of the total population, the proportion would be $\frac{1}{4}DD:\frac{1}{2}Dd:\frac{1}{4}dd$. The heterozygotes represent only 50 percent of the population instead of 100 percent as in the beginning. If, for simplicity, the population at this point is considered to consist of four, and each plant produces four progeny in the next generation, the *DD* would produce four *DD* and the *dd* would produce four *dd*, but the two *Dd* plants would each produce four distributed as $1DD:2Dd:1dd$. The total from all four plants would be $6DD:4Dd:6dd$. Now 25 percent are heterozygotes

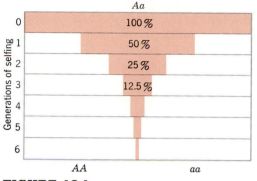

FIGURE 16.1

Consequences of self-fertilization in the various generations of selfing. In each generation of selfing, the frequency of heterozygous individuals is reduced by one-half.

and 37.5 percent represent each of the two homozygotes (Fig. 16.1). This trend would go on generation after generation, as illustrated in Table 16.8. In the fifth generation, the proportion in whole numbers would be about 48 percent *DD*, 4 percent *Dd*, and 48 percent *dd*. In the tenth generation, only about 1 in 1000 would be heterozygous. Most of the plants would eventually be *DD* or *dd*, with a smaller and smaller proportion of heterozygotes. The population of peas would eventually consist of two types, tall and dwarf, in essentially equal proportion. In contrast, if the plants had been cross-fertilized and mated at random, the proportion of $1DD : 2Dd : 1dd$ would have continued generation after generation.

Mendel worked this out with a model that included 1600 plants. He asked what would be the proportion of the three genotypes in five or ten generations if the original population consisted of 1600 heterozygous (*Dd*) plants and each produced only one offspring. In the first generation, the proportion would be $1DD : 2Dd : 1dd$, or $400DD : 800Dd : 400dd$. In the second generation, there would be $600DD : 400Dd : 600dd$, and so on. The heterozygotes would be decreased by half in each generation.

Now suppose that in the beginning the hybrid peas had been heterozygous for flower color alleles (*Rr*) as well as plant height. About half the plants over a period of time would be tall and half dwarf. Similarly, they would be half red and half white. The following types would be expected to occur in about equal proportions: *DDRR, DDrr, ddRR,* and *ddrr*. Four "pure" phenotypic lines—tall, red; tall, white; dwarf, red; and dwarf, white—would have developed. When other allelic pairs are included,

TABLE 16.8   Expected Progeny in Generations Following Self-fertilization of an Annual Plant with Genotype *Dd*, for Progenies of Equal Size (4)[a]

GENERATION	PROGENY REPRESENTING DIFFERENT GENOTYPES BASED ON FAMILIES OF 4			PERCENT OF EACH GENOTYPE		
	DD	*Dd*	*dd*	*DD*	*Dd*	*dd*
I	1	2	1	25	50	25
II	6	4	6	37.50	25	37.50
III	28	8	28	43.75	12.50	43.75
IV	120	16	120	46.875	6.25	46.875
V	496	32	496	48.4375	3.125	48.4375

[a] After Gregor Mendel.

they would behave in the same way, and a greater variety of pure lines would be expected. The entire population would eventually be made up of distinct types or races. This has actually occurred in many self-fertilizing plants in nature.

HOMOZYGOSITY UNDER DIFFERENT DEGREES OF INBREEDING

Self-fertilization occurs only among a limited group of plants and is nonexistent among the higher animals. In experimental animals such as mice, therefore, homozygosis can be approached most rapidly and conveniently by brother and sister matings. Backcrosses between progeny and one parent are efficient, especially when sex-linked alleles are involved; but technical factors, such as the age difference between parents and progeny, make this system less practical than matings between litter mates. Another simple method of inbreeding in mice involves double first cousins; that is, the mother and father of child 1 are brother and sister of the parents of child 2. S. Wright has contributed greatly to the theoretical aspects of homozygosis with his studies on the effect of inbreeding in guinea pigs. His series of papers on systems of mating (see References at the end of this chapter) represents a landmark in the field of population genetics.

Wright's coefficient of inbreeding (F) can measure the proportion by which inbreeding has reduced heterozygosity in a given situation. F is the probability that two homozygous alleles united in a zygote have descended from an ancestor common to both parents. When $F = 0$, no inbreeding is occurring, and the genotype frequencies are those of the Hardy–Weinberg formula ($p^2 + 2pq + q^2$). When $F = 1$, inbreeding is complete and the entire population is homozygous ($p + 0 + q$). By inbreeding, the proportion of heterozygotes is reduced from $2pq$ to $2pq(1 - F)$.

In a population with an inbreeding coefficient of F, individuals may be homozygous for a particular allele for one of two reasons: (1) Because of inbreeding, they may receive the identical allele from the same common ancestor and become homozygous for p or q. The probability of receiving the identical allele from a common ancestor is F;

hence the terms pF and qF. (2) Individuals may be homozygous for the usual reason, namely, that they received the same allele from each parent, but the alleles are not identical by descent. The probability that the alleles are not identical by descent is $1 - F$, and the probability that both are the same is p^2 or q^2; hence the terms $p^2(1 - F)$ and $q^2(1 - F)$. Table 16.9 lists the frequencies of the three genotypes that may occur in a population. F is the probability that two alleles are identical by descent.

The problem is to determine the value of F in a given situation. The F values may be determined from pedigree studies. In the pedigree shown in Fig. 16.2, for example, representing a marriage between first cousins, the probability that any allele (e.g., a_1 or a_2) carried by either of the common ancestors would be present in both parents of the child would be the **product** of the probabilities that this particular allele was transmitted through each step in the pathway. The probability that an allele such as a_2 came from an ancestor common to the father and also to the mother of the child would be $\frac{1}{2}$ for each step in a pathway. The number of individuals is the

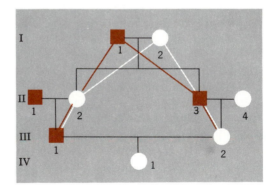

FIGURE 16.2

Pedigree of a man (III-1) and a woman (III-2) from common ancestry, illustrating a first-cousin marriage. The probability that any allele carried by either the common grandfather (I-1) or common grandmother (I-2) would go to both III-1 and III-2 would be the sum of the probabilities that the man and the woman would receive this gene from each grandparent. The probability for each parent of the child (IV-1) would be the product of the separate probabilities (each $\frac{1}{2}$) that the particular allele and not the alternative allele would be transmitted through each of the five steps in the pathway.

TABLE 16.9   Frequencies of Three Genotypes Represented by One Pair of Alleles under Random Mating, Complete Inbreeding, and Inbreeding with Coefficient of Inbreeding Equal to F

GENOTYPE	FREQUENCY		
	$F = 0$ (RANDOM MATING)	$F = 1$ (COMPLETE INBREEDING)	INBREEDING $= F$
a_1/a_1	p^2	p	$p^2(1 - F) + pF$
a_1/a_2	$2pq$	0	$2pq(1 - F)$
a_2/a_2	q^2	q	$q^2(1 - F) + qF$

exponent of $\frac{1}{2}$ for that path. When there is more than one path, the probabilities are added.

In Fig. 16.2, there are two paths, each with five individuals. Path 1 is III-1, II-2, I-1, II-3, III-2; and path 2 is III-1, II-2, I-2, II-3, III-2. Therefore, $F = (\frac{1}{2})^5 + (\frac{1}{2})^5 = \frac{1}{16}$.

If, in the general population, a particular trait such as albinism dependent on a recessive allele (a_2) in homozygous condition occurs in about one person in 40,000, how much is the probability of expression enhanced if the parents are cousins?

$$q^2 = 1/40,000 \qquad q = \sqrt{\frac{1}{40,000}} = 1/200$$

The proportion of homozygous recessives in the population when $F = \frac{1}{16}$ (first cousins) would be:

h (after inbreeding with $F = \frac{1}{16}$)

$= q^2 (1 - F) + qF$

$= \left(\frac{1}{40,000} \times \frac{15}{16}\right) + \left(\frac{1}{200} \times \frac{1}{16}\right) = \frac{13.4}{40,000}$.

Therefore, the probability of an expression of this trait would be about 13.4 times higher if the parents were cousins than if the parents were unrelated.

Percentages of homozygosis compared with the original degree of heterozygosis have been calculated by S. Wright and others for successive generations under different systems of inbreeding. Results for simple combinations are shown graphically in Fig. 16.3. Not shown in the graph is the case of parthenogenesis in *Drosophila*, in which homozy-

gosis occurs in one generation. Of the examples in the graph, self-fertilization produces homozygosis most rapidly. The 90 percent mark is passed in the third generation and the 95 percent mark in the fourth. After eight generations, nearly all the genes that were heterozygous before inbreeding began would theoretically be homozygous. Brother and sister matings are somewhat less efficient than selfing in producing homozygosis, but these and parent–offspring matings are the most intense form of inbreeding possible among nonparthenogenetic or nonself-fertilizing organisms. Under this system, 95 percent of the genes that were originally heterozygous would become homozygous in 11 generations of inbreeding, and 5 percent would remain hetero-

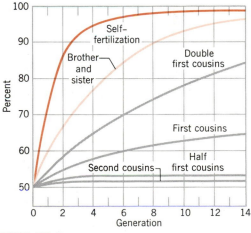

FIGURE 16.3

Graph representing percentage of homozygotes in successive generations under different systems of inbreeding. (After S. Wright.)

zygous. Following continued brother and sister matings, homozygosity is closely approached in 20 generations. Double first cousins would yield about 85 percent homozygosity in the fourteenth generation. It might be expected, on the basis of superficial observations, that any system of inbreeding followed consistently would lead to complete homozygosis of all the originally heterozygous pairs. This is not true for matings between individuals in a large population who are less related than first cousins. Continued matings between half first cousins will result in a rise of only two percent (from 50 to 52 percent) after an infinite number of generations. Continued matings of second cousins would cause a rise in homozygosis of only one percent (from 50 to 51 percent).

INBRED MICE AND CANCER INHERITANCE

In the early part of the twentieth century, C. C. Little recognized a practical application of inbreeding in connection with studies on the genetics of cancer. He and others had been studying the inheritance of tumors in ordinary laboratory mice, with little success. The possibilities of greater success with homozygous strains led to an extensive inbreeding program that continued for several years. After many generations of inbreeding and selection, strains highly susceptible to specific types of cancer were produced. The inbred lines have lived up to expectations and proved to be a most valuable material for investigating the genetics of different kinds of tumors as well as other aspects of genetics. Some inbred mouse strains had a high incidence of mammary tumors; others were high in the incidence of lung tumors or leukemia; and still others were relatively free of all types of tumors. The cancer incidence in some well-known inbred strains is summarized in Table 16.10.

OUTBREEDING

In contrast to inbreeding, which increases homozygosity, matings among nonrelated individuals increase **heterozygosity.** Outbreeding augments, temporarily at least, the vigor of the individuals. By keeping allelic pairs heterozygous, recessives are not expressed and are thus protected from the forces of selection. At the same time that recessives are being protected from selection, increased heterozygosity is providing greater variation on which selection can

TABLE 16.10   Incidence of Mammary Tumors, Lung Tumors, and Leukemia in Several Inbred Strains of Mice[a]

STRAIN	MAMMARY TUMOR (PERCENT)	LUNG TUMOR (PERCENT)	LEUKEMIA (PERCENT)
dba	55–75	low	30–40
A	70–85	80–90	low
C$_3$H	75–100	5–10	low
C57 black	low	low	20
C57 brown	low	low	Most common neoplasm in strain
C57 leaden	low	low	low
C58	low	low	90
Ak	low	low	60–80

[a] After W. E. Heston.

operate. Heterozygosity has practical application in the **heterosis (hybrid vigor)** of several animals and plants, particularly maize (corn). Both inbreeding and outbreeding are involved in the production of seed for hybrid corn.

HETEROSIS IN MAIZE

In the second half of the nineteenth century, it was shown by W. J. Beal and others that the products of varietal crosses were superior to the products of inbred varieties of maize. G. H. Shull, working at Cold Spring Harbor (1908), and later D. F. Jones, H. K. Hayes, and others, continued the investigation and found that hybrid corn had more vigor than the inbred lines or the open-pollinated varieties then in use. This observation has been elaborated and tremendous increases in production have been obtained. Field-corn acreage in the United States has decreased from slightly over 100 million acres in 1929 to less than 80 million acres in 1983, but the yield has increased by one-third. In 1929, practically none of the acreage was in hybrid corn; 10 years later, about 23 percent was planted to hybrid seed. Now 99 percent of the field-corn acreage in the United States is planted with hybrid seed. Hybrid corn is one of the most practical successes of genetics thus far; but, paradoxically, the early contributors were not located in the Corn Belt and apparently were not motivated by the prospect of a practical contribution.

Hybrid seed is produced by crossing inbred strains, which are developed by controlled self-pollination. This is done on a small scale by placing paper bags over the ears and tassels of the corn plants when these parts first develop. Self-pollination is accomplished by transferring the pollen from the tassel to the silk of the same plant and protecting the silk from foreign pollen. Only the most desirable lines are kept during the several generations of inbreeding. To make the hybrid cross on a larger scale, a few rows of one type of inbred corn are planted near a few rows of another variety in the direction of the prevailing wind. Tassels are removed from the seed-parent plants, and the silks are fertilized by the wind-borne pollen of the other inbred line (pollen parent). Male-sterile corn is used

on a very large scale for hybrid seed production. The F_1 plants are larger and more productive than those from selfing, as illustrated diagrammatically in Fig. 16.4. In this figure, experimental results for the different generations are illustrated by sketches drawn to scale.

Vigorous hybrids from two properly matched F_1 plants produce seed more efficiently and economically than do the weak inbred plants used in the single cross. Therefore, the double cross (Fig. 16.5) has been developed for hybrid seed production in corn. This is accomplished by developing four inbred varieties, making parallel crosses between two pairs in the same season, and, in the next season, crossing the hybrids. Inbred strains A and B, for example, are crossed together, and other inbred strains, C and D, are crossed with each other. F_1 plants from AB are then crossed with F_1 plants from CD. Seed from this cross is sold to the farmer. Double crosses do not actually improve the hybrid vigor above that of the single cross; they mainly provide large, uniform, vigorous plants for seed production and thus reduce the cost of commercial seed. The uniformity of the crop in height, yield, and ear characteristics can also be improved. Plant

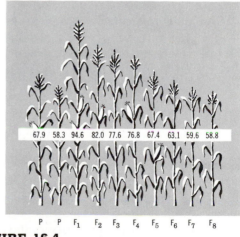

67.9 58.3 94.6 82.0 77.6 76.8 67.4 63.1 59.6 58.8

P   P F_1 F_2 F_3 F_4 F_5 F_6 F_7 F_8

FIGURE 16.4

Heterosis in corn. The figures in the center represent the average height in inches for the different generations identified at the bottom. (Data from D. F. Jones.)

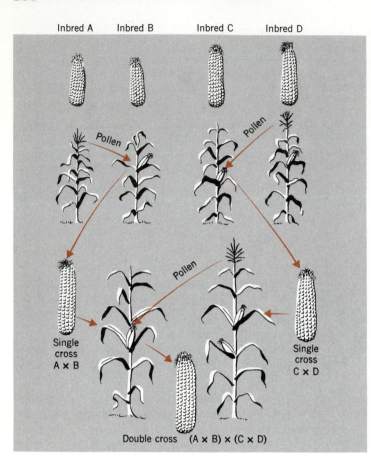

Inbred A   Inbred B   Inbred C   Inbred D

Pollen   Pollen

Single
cross
A × B

Pollen

Single
cross
C × D

Double cross   (A × B) × (C × D)

FIGURE 16.5

Double-cross method for producing hybrid corn. (From T. Dobzhansky, Evolution, Genetics, and Man, *John Wiley & Sons, New York, 1955.)*

size and vigor, and particularly the size of the cob, determine the amount of seed that can be produced. A weak inbred plant must be the cob producer for F_1 and, therefore, the amount of seed is limited. On the other hand, if an F_1 plant can be used as a seed parent, a much larger cob and greater plant vigor are available.

One theory postulated to explain heterosis is based on the use of dominant alleles to produce increased vigor. It requires that these alleles accumulate in inbred lines by random mutations and selection. Vigor depends on a large and efficient root system, well-developed leaves with a good supply of chlorophyll, firm supporting tissue, and other properties. Alleles for vigor are brought together in hybrids and, because of their dominance, produce maximum expression in the F_1. When all the many

avenues through which plants could become better fitted to their environment are considered, different inbred lines may be expected to accumulate many different alleles for increased vigor. Dominant alleles postulated in this theory would become homozygous through continued inbreeding. Crosses between inbred lines would result in the heterozygous F_1 plants, which would express the favorable characteristics from both inbred lines. Under this system, maximum vigor would be expected to occur in individuals having the loci with dominant favorable alleles.

Let us consider a simple model with only two pairs of genes involved. One inbred line will be considered to have become better able to meet the conditions of the environment through an improved root system, controlled by dominant allele

A. Another inbred line is also improved because it has a better chlorophyll system, controlled by dominant allele *B*. Through inbreeding, each line has become homozygous for its dominant allele; that is, *Ab/Ab* and *aB/aB*, respectively. A cross between these inbred lines brings the good qualities from both together in the hybrid *Ab/aB*. The weak alleles *a* and *b* are also present, but they are recessive and do not influence the efficiency of the hybrid. If the inbred parents were *AB/AB* and *ab/ab*, the hybrid would be no better than the more vigorous parent. Complete homozygosity would be rare if many genes are involved. The theory is dependent on dominant alleles controlling in separate ways the vigor of progeny. Assuming that such alleles for vigor are distributed between the two inbred parents, the hybrids should be better than either parent.

What happens when hybrids with a high level of vigor are inbred? If the hybrids illustrated in the oversimplified model presented above were self-fertilized—that is, *Ab/aB* × *Ab/aB*—nine combinations would be expected, but only those with *A* and *B* (i.e., *AABB, AABb, AaBB,* and *AaBb*) would be as vigorous as the F_1. Continued inbreeding over many generations would result in more or less purer lines than those crossed to produce the hybrids. If fully homozygous dominant plants could be distinguished from the others, these should be as vigorous, but no more so, than the hybrid. The population as a whole would decrease in vigor and the plants would lose the uniformity that was conspicuous in the original hybrids. By the eighth generation, when the lines would be relatively pure, further inbreeding would have little effect.

On the basis of this explanation, it would seem to be possible eventually to develop inbred lines carrying the genes for increased vigor. This has been attempted many times without success. A possible explanation is that dominant alleles for vigor are linked with recessives possessing undesirable characteristics, which also accumulate through inbreeding and become homozygous. The deleterious recessives, which on this assumption come together in later generations, more than counteract the good effects observed in the F_1. The genetic basis of heterosis is probably more complex than that expressed in any single explanation offered thus far. Such genetic principles as (1) complementary gene action, (2) epistasis masking deleterious recessives, (3) effects of multiple alleles, and (4) overdominance (selective advantage of heterozygote over both homozygous types) may be involved in the process. It now seems evident that only certain genes produce heterosis when they are in heterozygous condition. Favorable chemical combinations controlled by particular alleles may be responsible for the increased vigor.

FACTORS INFLUENCING ALLELE FREQUENCY

The Hardy and Weinberg explanation for equilibrium in a population requires five assumptions: (1) individuals of each genotype must be as reproductively fit as those of any other genotype in the population; (2) the population must consist of an infinitely large number of individuals; (3) random mating must occur throughout the population; (4) individuals must not migrate into or out of the population; and (5) there must be mutation equilibrium. This is just what Hardy and Weinberg intended, because their formula described the **statics** of a Mendelian population. Something more was required to formulate a mathematical explanation of naturally occurring population change or **dynamics** in terms of allele frequencies. This need was filled by R. A. Fisher, S. Wright, and J. B. S. Haldane, who conceived additional theoretical models and imposed mechanisms for change in allele frequencies using the Hardy–Weinberg equilibrium. Population statics thus became population dynamics.

Now that mathematical tools have been established, the dynamic relations that exist between individuals representing particular genotypes and their variable environments can be determined. Population genetics seeks an explanation for the genetic architectures of different populations. This discipline is concerned with the origins of adaptive norms or arrays of genotypes as they become consonant with the demands of environments. The genetic architecture of a population is the manner in which the genetic material is arranged and the frequency with which each unit occurs. In ascertaining the genetic architecture of a population, questions must be answered concerning the allele frequency, the frequency of chromosome structural changes such as inversions and translocations, and chromosome numbers. Changes in allele frequency depend on selection, mutation, chance (genetic drift), differential migration, and meiotic drive.

SELECTION

Selection is the nonrandom differential reproduction of genotypes. Under selection, frequencies established by alleles in a population at equilibrium are subject to change. If, for example, allele A makes the organism more efficient in reproduction (fit) than a, A is expected to increase generation after generation at the expense of a. Continued selection of this kind would tend to decrease the proportion of a in favor of A.

Selection can alter the genetic composition of a population only to the extent that preexisting genetic variation occurs in that population. Natural populations are rich storehouses of genetic variation. All individuals, except identical twins, in most populations are different; they carry both genetic and environmental variation. Population geneticists are determining how much genetic variation exists and explaining the variation in terms of its origin, maintenance, and evolutionary importance. The most useful current procedure for detecting variation is **electrophoresis,** a laboratory procedure for separating out chemical differences in **an electric field** from tissue or blood samples for individuals within a population. If an enzyme in the sample has an identifiable amino acid substitution, a mutant may be detected. When two kinds of polypeptide chains appear, heterozygous alleles may be detected. Enzymes differing in electrophoretic mobility as a result of allelic differences at a single locus are called allozymes. They are associated with genetic variation that is common in natural populations.

Observations made in natural populations, such as one involving moths in England, have pro-

vided examples of selection occurring in nature. In 1850, surveys indicated that most moths in various nonindustrialized communities were light in color, whereas a very few, less than 1 percent, were dark. As the cities became industrialized and factory smoke darkened the buildings and countryside, a parallel change occurred among the moths. Later surveys showed that 80–90 percent of the moths in certain industrial areas were dark. L. Doncaster, R. Goldschmidt, A. Kühn, and E. B. Ford showed, from experimental breeding, that dark color was inherited and controlled by one or two dominant genes. Presumably, those moths that matched their environment best lived, reproduced, and transmitted their alleles for dark color to their progeny, whereas the more conspicuous light-colored moths on a dark background (Fig. 16.6) were often caught by predators. The alternative explanation, that in some way the direct influence of the environment changed the genetic color-producing mechanism, was considered but not substantiated. Observations have now borne out the explanation that the change is based on selection favoring moths that most closely blend with their surroundings.

In a particular study conducted by H. B. D. Kettlewell, 447 black moths, *Biston betularia*, were collected elsewhere and released in smoky Birmingham. At the same time, 137 white moths of a type not already present in the vicinity were also released. After a period of time, moths in the area were trapped and classified. Only 18 (13 percent) of the original 137 white moths and 123 (27.5 percent) of the blacks could be found. The collections were extensive and it was presumed that virtually all the released moths that were still alive were captured. The blacks were favored 2 : 1 over the whites in survival. In Dorset, which is not industrialized, 398 black and 376 white moths were released and retrapped in a similar way. This time, 14.6 percent of the whites but only 4.7 percent of the blacks survived. The moths that blended with their surroundings best were concealed from predators and therefore survived, whereas the conspicuous dark moths were more readily seen and devoured by birds.

The implication in the theory of natural selection is that certain genotypes in a species endow

FIGURE 16.6

Dark and light forms of the peppered moth photographed on the trunk of an oak tree blackened by the polluted air of the English industrial city of Birmingham. The light form Biston betularia *is clearly visible; the dark form* (carbonaria) *is well camouflaged.* (H. B. D. Kettlewell, Sci. Amer. 200:48–53, 1956. *From the experiments of H. B. D. Kettlewell, University of Oxford.)*

their carriers with advantages in survival and reproduction. In dealing with "fitness" traits that are less conspicuous than traits like color in moths, it is difficult to evaluate the effects of individual alleles on survival and reproduction because it is whole organisms with their full complements of genes that survive or fail to survive. Usually, only slight differences in survival and fitness characterize different genotypes. Nevertheless, single gene pairs contribute to total fitness in particular environments, and it is appropriate to approach a complex problem through a consideration of its component parts.

The process of change in allele frequency through selection can be illustrated with a large, random-mating population that has a single pair of alleles, A and a, segregating. Let $1 - q$ represent the frequency of A, which in this example is completely dominant, and q the frequency of the recessive allele a. A constant selective advantage is assumed for the dominant phenotype over the others. Reproductive rates for genotypes AA, Aa, and aa would be $1 : 1 : 1 - s$, respectively. The **selection coefficient,** s, is a measure of the disadvantage of an organism, relative to a particular reference genotype under selection, and thus represents the **intensity of selection.** It expresses the fitness of one genotype compared with the other. It has a positive value if selection favors A over aa. If selection is complete and in favor of A (aa is lethal), $s = 1$. Change per generation in allele frequency can be determined when q and s are known.

For example, in a population with three-fourths (.75) of the individuals expressing the phenotype of A (dominant allele) and one-fourth (.25) the phenotype of aa, the relations may be summarized as shown in Table 16.11. Out of the total gene frequency of $1 - sq^2$, the AA parents would contribute proportionality $\dfrac{(1 - q)^2}{(1 - sq^2)}$ A gametes to the gene pool, and the Aa parents would contribute $\dfrac{[q(1 - q)]}{(1 - sq^2)}$ A gametes and $[q(1 - q)](1 - sq^2)$ a gametes. The aa individuals would contribute proportionally $\dfrac{q^2(1 - s)}{(1 - sq^2)}$ a gametes. Therefore, the frequency of a after one generation would be:

$$\frac{q(1 - q) + q^2(1 - s)}{(1 - sq^2)} \quad \text{or} \quad \frac{(q - sq^2)}{(1 - sq^2)}$$

$$= q_1 = \text{frequency of } a \text{ after the first generation of selection}$$

Under selection pressure, the change (Δ) in the frequency of a (q) from one generation to the next would be:

$$\Delta q_s = q_1 - q \quad \text{or} \quad \frac{q - sq^2}{1 - sq^2} - q$$

$$= -\frac{sq^2(1 - q)}{1 - sq^2}$$

TABLE 16.11   Relations among Genotypes after Selection

| | GENOTYPES | | | |
	AA	Aa	aa	TOTAL
Relative proportion before selection	$(1-q)^2$	$2q(1-q)$	q^2	$= 1$
Genotypic frequencies	.25	.50	.25	$= 1$
Selective value	1	1	$1-s$	
Relative proportion after selection	$(1-q)^2$	$2q(1-q)$	$q^2(1-s)$	$= 1 - sq^2$

This measures the amount of **change in q per generation of selection.** If s is very small, as is usually the case in nature, Δq_s can be approximated as $-sq^2(1-q)$, since $1 - sq^2 \simeq 1$.

Now that q_1 (frequency of a in the first generation of selection) has been derived, the following example may be completed. Given: Initial $q = 0.5$ and $s = 1$. What will be the frequency of a in the first generation?

$$q_1 = \frac{(q - sq^2)}{(1 - sq^2)} = \frac{(.5 - .25)}{(1 - .25)} = \frac{.25}{.75} = \frac{1}{3}$$

The frequency of allele a has decreased from .5 to .33 in one generation. The new genotype frequencies are:

$$q_1^2 = \text{frequency of } aa$$
$$= (\tfrac{1}{3})^2 = .11$$
$$2q_1(1-q_1) = \text{frequency of } Aa$$
$$= 2(\tfrac{1}{3})(\tfrac{2}{3}) = .44$$
$$(1-q_1)^2 = \text{frequency of } AA$$
$$= (\tfrac{2}{3})^2 = .44$$

The genotypes are not at equilibrium. They will change in the next generation and in all subsequent generations until selection is removed or all a alleles are removed from the population. Genotypic and phenotypic frequencies of A- have been favored under the condition of complete selection against aa.

The efficiency of selection against a recessive trait depends on the allele frequency. Selection efficiency increases with increasing allele frequency, but decreases with decreasing allele frequency, as shown in Table 16.12. Progress of selection is rapid at first, when the allele is frequent enough to produce a significant number of recessive homozygotes, but decreases as allele frequency declines. When q approaches 0, no rapid change occurs because most recessive alleles are in heterozygotes and

TABLE 16.12   Allele Frequency and Efficiency of Selection against a Recessive Trait[a]

GENERATION	FREQUENCY
0	.500
1	.333
2	.250
3	.200
4	.167
8	.100
10	.083
20	.045
30	.031
40	.024
50	.020
100	.010
200	.005
1000	.001

[a] A recessive allele for an undesirable trait is present in the initial population with a frequency of .5, and homozygous recessives are eliminated in every generation ($s = 1$). Allele frequencies are given for successive generations.

TABLE 16.13 Frequencies *p* of Allele A after One Generation from an Initial Population Where Alleles A and α Are Equally Frequent (*p* = *q* = .5), Fitness of the Dominants (AA and Aα) Is 1, and That of the Recessives (αα) is 0 (α Recessive Lethal), .4 (α Semilethal), .9, .99 (Subvital)

Fitness	0	.4	.9	.99
Selection coefficient (s)	1.0	.6	.1	.01
Frequency p after one generation of selection	.67	.59	.5128	.5012
Increment of gene frequency (Δp)	+.17	+.09	+.0128	+.0012

therefore are not exposed to selection. Recessive alleles cannot be entirely eliminated by selection. When q is small enough to allow homozygotes to occur very infrequently, recessive alleles are still carried in heterozygotes.

Frequencies of alleles under different intensities of selection (s) will change according to the severity of selection. Frequencies of allele A after one generation of selection at s for aa = 1.0, .6, .1, and .01, with a corresponding advantage for AA and Aa of 0, .4, .9, and .99, respectively, are given in Table 16.13. As genotype aa decreases, those of AA and Aa with fitness of 1 increase.

Thalassemia (or Cooley's anemia) provides an example of selection in human populations. It occurs mostly in children and is nearly 100 percent fatal. The disease is controlled by an allele (c) that, in homozygous condition (cc), produces the severe thalassemia major. The same allele in heterozygous condition results in a mild form of the disease called thalassemia minor or microcythemia. People with the heterozygous combination may have no outward symptoms, but they can be identified with a blood test. The importance of the disease from a public health standpoint and the ease with which carriers can be detected have led to many allele frequency surveys in areas where the disease is prevalent.

The thalassemia allele is widespread, mostly in equatorial areas (as shown in Fig. 16.7). It is rarely seen in the United States, but is especially frequent in the Mediterranean region, particularly in Italy, Greece, and Syria. Silvestroni and Bianco (1975) reported as high as 27–30 percent carriers in

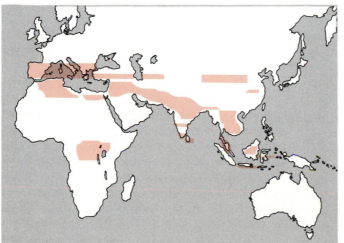

FIGURE 16.7
World distribution of thalassemia. This blood disease is concentrated mostly in equatorial regions.

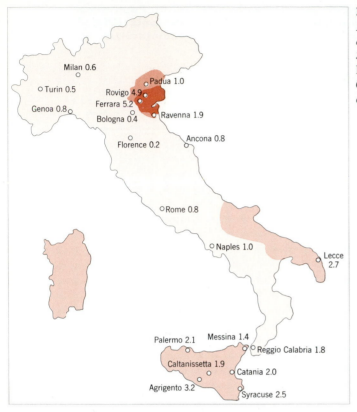

FIGURE 16.8
Heterozygous carriers of the thalassemia allele (c), expressed mostly as microcythemia in Italy, Sicily, and Sardinia. (From E. Silvestroni and I. Bianco, Am. J. Hum. Genet. 27:198–212, 1975. The University of Chicago Press. Copyright © 1975 by the American Society of Human Genetics.)

southern Sardinia (Fig. 16.8). In areas with 5 percent carriers, the chance of the severe form (which requires the homozygous condition) occurring by random mating is about 1 in 1600. Since the homozygous form is lethal and heterozygotes can be identified early in life, the allele frequency could be greatly reduced or even removed from the population if heterozygotes would not have children.

In a similar situation of complete dominance of allele *A* but with complete selection against *AA* and *Aa* genotypes (*A* phenotype) in a particular environment, *A* could be completely eliminated in one generation. This allele would be lost from the population and would be restored only by new mutation.

Examples involving complete dominance and complete selection against *aa* and *A* phenotypes, respectively, are dramatic but infrequent in nature. When they do occur, however, the allele that is selected against is eventually lost from the popula-

tion. Normally, effectiveness of selection is graded. Under one set of conditions, selection may be strong, producing significant changes in allele frequency within only a few generations. Under other conditions, selection might be weak, resulting in only slight changes over long periods of time. Apparently, in most common situations in nature, one genotype is only slightly less efficient than another.

Mechanisms for selection have been identified in many higher forms, but questions have arisen concerning the effectiveness of selection in viruses. S. Spiegelman and his associates have demonstrated that nucleic acids produce phenotypes that are subject to selection and can evolve to lesser or greater complexity. Qβ—a single-stranded RNA bacteriophage—can replicate in an *in vitro* reaction mixture that consists of an RNA template, precursors for RNA synthesis, and the replicase enzyme. In working with Qβ RNA, the investigators selected molecules with a high rate of replication for this *in vitro* system and showed that considerable variation can occur in a population of RNA molecules. Progress was made in a series of experiments that worked toward producing a molecule with a replication rate that was 15-fold greater than the normal rate. This corresponded with the elimination of 83 percent of the original genome, a significant change in base composition, and an increased efficiency of interaction of the genome with replicase. Further studies, with suboptimal concentrations of ribonucleoside triphosphates and inhibitory base analogs, showed that selection can occur among mutant viruses in a selective environment.

A large number of certain polymer sequences (composed of few nucleotides) occur in both strands of the Qβ replicative RNA molecule. Complements of these sequences also appear in both strands. Polymer sequences common to both strands pair with their complementary parts. Studies of sequences and of the chemical properties of the Qβ RNA molecule indicated extensive base-pairing between complementary regions of the same strand, as well as a resultant extensive secondary and tertiary structure to the RNA molecule. This allowed a distinction between the genotype (primary structure) and phenotype (secondary and tertiary structure) of a simple molecular species. Results indicated that when a molecule exists as a highly evolved self-replicating structure, the structural "phenotype" of the molecule may be acted upon by a selective agent. The argument here is indirect. No evidence is presented that molecules with different structures differ in the number of progeny viruses produced. The reasoning is that the regular structural features observed must have been selected as a result of the evolution of the virus.

These experiments, along with those of J. H. Campbell and others, showed that selection may occur at the molecular level in viruses as well as in higher forms. Selection depends on interactions between genetically determined phenotypes and environmental agents. In this way, the selection process resembles that in higher forms. Selection is a mechanism for changing allele frequencies in all forms of life.

MUTATION

Recurrent or **"one-way" mutations** that change one allele to another, such as A to a, alter the relative frequency of the two alleles by increasing the proportion of a at the expense of A. If mutations occur only in one direction — that is, A to a — eventually, only a alleles will be expected in the population. Because mutation rates for particular genes are low, a long period of time is required for any appreciable change in the relative frequency of alleles. In fact, mutation rates for most genes in higher organisms are so low that it is doubtful that mutation alone is an important factor for allele frequency changes. In addition to the direct effect of mutations in changing the relative proportion of alleles, however, mutations have a more general and indirect influence on allele frequency changes in populations. They represent an original **source for variation and provide alternative alleles with their corresponding traits.** This variation is basic to evolution as it is expressed through segregation, recombination, and selection.

Assume a population that is very large so that sampling errors may be ignored. Consider a single genetic locus at which two alleles, A and a, are segregating. The population will be composed of three genotypes, AA, Aa, and aa. Suppose that all of these reproduce with equal frequency (i.e., there is no differential selection). Let the experiment begin by counting the number of individuals of each of the three genotypes in the population. From these counts the following computation is made:

Frequency of $A = p_0$ (in the initial generation)
Frequency of $a = 1 - p_0$
$\qquad\qquad\quad = q_0$ (in the initial generation)

In each generation, a certain fraction, μ, of the A alleles will change by mutation to a alleles. If, for example, the initial frequency of $A = p_0 = .9$, and one in each million A alleles mutates to a in each generation, what will be the frequency of A after 1000 generations? The calculations are complex, but it can be shown that after 1000 generations p is reduced from .9 to .899. The influence of a recurring mutation in altering the frequency of a particular allele in a population of organisms is called **mutation pressure.** This example shows that mutation pressure acting alone over 1000 generations does not make much difference in allele frequency. Over a very long range, however, one-way mutation pressure may result in the near elimination of a recessive allele.

Mutations are usually reversible as well as recurrent (Chapter 9). If the mutation of A to a is reversible (from a to A), the proportion of A will increase and the relative rates of the two kinds of mutations will determine the proportion of alleles at any given time. Equilibrium $\hat{q}$ is independent of the initial q (Fig. 16.9). If an allele starts at a high frequency (q_1 greater than $\hat{q}$) or with a low frequency (q_2 less than $\hat{q}$), over time it approaches $\hat{q}$. Usually, two kinds of mutations are not equally frequent in occurrence, and a mutation trend is established in one direction. If mutations A to a occur at rate u, and a to A mutations occur at rate v, the forward and reverse mutations may be presented according to the following scheme:

$$A \underset{v}{\overset{u}{\rightleftharpoons}} a.$$

The frequency of A and a will be influenced by the relative rates u and v. If the initial frequency of A is $1 - q$ and the initial frequency of q is a, the population allele frequencies will be stable with reference to these alleles when $(1 - q)u = qv$. Solving for q, the equilibrium value of $\hat{q}$ is

$$\hat{q} = \frac{u}{u + v}$$

This expression is independent of the initial allele frequency q. The value of $\hat{q}$ is approached but never reached, as shown in Fig. 16.9.

Recombination of alleles already present in the population is much more significant for changing the gene frequency of higher organisms than original spontaneous mutations. In such organisms, recombination (Chapter 6) provides the immediate source of variety upon which natural selection can act.

Observations in nature indicate that mutation,

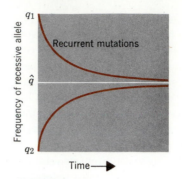

FIGURE 16.9

Time scale for change in allele frequency of a recessive allele that is subject only to mutation pressure: q_1 (an initial allele frequency greater than $\hat{q}$) falls exponentially toward $\hat{q}$ but never reaches equilibrium while q_2 (an initial allele frequency less than $\hat{q}$) rises exponentially toward $\hat{q}$ but never reaches equilibrium.

selection, and especially recombination operate in natural populations, and they must be recognized in defining population trends. In a natural environment, mutant organisms usually are at a disadvantage when compared with wild-type forms. A few mutations might be inconsequential or neutral, but the chance of making an already well-adjusted organism better able to meet the conditions of an environment by random change is remote, indeed. How often would the engine of an automobile, or any other well-adjusted machine, be improved by random changes? The random origin of newness that maintains population fitness is purchased at the cost of a genetic burden or "genetic load." **Genetic load** is the average number of lethal genes or equivalents per individual in a population. It may be defined in terms of reduced fitness:

$$\frac{W_{\max} - \overline{W}}{W_{\max}}$$

where $W_{\max}$ is the fitness of the best genotype and $\overline{W}$ is the average fitness of the entire population. When H. J. Muller introduced the concept of the genetic load in 1950, he was especially concerned with the mutations caused in human beings by radiation.

When such a mutation occurs, it almost always confers a lower fitness, at least in the homozygous condition. Lower fitnesses of the mutations are too often produced by hereditary diseases that cripple and prematurely kill the affected individuals. But this is only part of the story. The genetic load is also based on genetic differences that may not harm the individuals in overt, physical ways, but may produce morbid effects in their progeny.

In organisms such as fungi and bacteria, which reproduce primarily by rapid, asexual means, mutations associated with selection may abruptly change the trend in the population. If, for example, a single bacterium, *Staphylococcus aureus*, should undergo a spontaneous mutation in an environment where penicillin is present and become resistant to penicillin, it might give rise to an entire population of penicillin-resistant bacteria in a short period of time. In this example, penicillin is not a mutagen but a selective agent. Animals as far up the phylogenetic ladder as scale insects in California and rabbits in Australia have possessed chance mutations that have made them more fit for their current environment. These mutants may rapidly give rise to whole populations of individuals favored in that environment. Organisms becoming resistant to a particular insecticide, for example, would be favored in an orchard sprayed with that insecticide. They and their descendants could take immediate advantage of the imbalanced environment from which competing insects have been eliminated. In the citrus groves of California, for example, whole populations of resistant scale insects have developed within a single season. Again, the insecticide is not a mutagen but a selective agent.

Mutations occur as frequently in higher, sexually reproducing organisms as in microorganisms, but the numbers in a population are too few to allow many mutations to become established, even in long periods of geologic time. This is mainly because the individuals in the population are too few in number. A mutation with a frequency of 1 per 100 million (10^8) germ cells might occur a few times in the population of human beings on the earth. A virus population of 10^8 units, on the other hand, may occur in a single plate culture. The F_2 virus, for example, may produce 20,000–40,000 viral particles in a single bacterial cell.

Under most circumstances of recurrent mutation and selection, a dynamic equilibrium is reached with regard to genotype frequencies. As indicated previously, recombination (through meiosis and fertilization) of genes already present in a population of higher organisms provides a more effective source of variation than mutation. Recombination also occurs through crossing over (Chapter 6), with the rate depending on the strength of linkage. In sexually reproducing organisms, when random matings occur in a population, the genotype of each individual incorporates genetic contributions from many preexisting members of the species. As time goes on, mutant alleles thus become widely dispersed throughout the entire population.

GENETIC DRIFT

Random fluctuation in allele frequencies, called **genetic drift,** also occurs in breeding populations. The effect of genetic drift is negligible in large populations; but in a small effective breeding population, the limited numbers of progeny might all become the same type with respect to certain allelic pairs because of genetic drift. This is fixation or homozygosity at that locus. Fixation is defined in terms of the allele frequency reaching $p = 1.00$ or $q = 1.00$. Genetic drift may or may not lead to fixation, but fixation is much more likely in very small than in large effective breeding populations. The effective breeding populations may be much smaller than the actual population because of nonreproducing individuals, isolation into breeding units (demes), and limited sampling time. Distribution of gene frequencies in populations of different sizes is given in Fig. 16.10. Genetic drift is a mathematical necessity because population samples are finite.

If a pair of alleles, A and a, is present in all members of a small breeding population, normally one-fourth of the progeny would be AA, one-half Aa, and one-fourth aa. If chance decrees that all progeny are either AA or aa, fixation would have occurred. Without mutation, no further genetic fluctuation would be expected at that locus. In small breeding populations, genetic drift will act on all

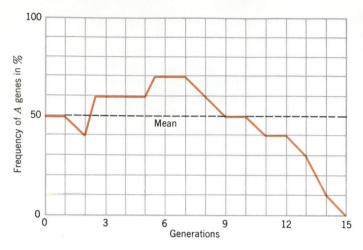

FIGURE 16.10
Random variation in frequency of an allele beginning at the midpoint where A = *50 percent or .5, as shown in simulation experiment on genetic drift. (Data in Table 16.14.)*

loci represented by two or more alleles, but the direction and magnitude of the effect may be different at each locus. The chance factor accounts for an appreciable amount of variation in small populations of cross-mating organisms.

L. L. Cavalli-Sforza and associates have designed and conducted sampling experiments to simulate random genetic drift. In simulation studies, the procedure of random sampling is imitated by using random numbers. These are numbers of one or more digits in which each digit has the same probability of occurring, and no correlation exists between successive digits or numbers. For example, a series of 10 random numbers simulating *A* and *a* alleles were taken from a population of random numbers (Table 16.14). The first step was to rule that random digits between 0 and 4, inclusive, are *A* "alleles" and those between 5 and 9, inclusive, are *a* "alleles." In the first generation, 5 (50 percent) are *A* and 5 (50 percent) are *a*. The proportion of first-generation "progeny" is the probability of *A* alleles in the parents for the second generation. From 5*A* : 5*a* "parents" the ruling again is that digits 0, 1, 2, 3, 4 are *A* alleles and that 5, 6, 7, 8, 9 are *a* alleles. In the second generation, 40 percent are *A*. This procedure is followed until the fifteenth generation (Table 16.14), when all 10 alleles were *a*, and *A* was lost.

The results of the simulation experiment outlined above are shown graphically in Fig. 16.10.

Allele frequencies may fluctuate about their mean from generation to generation. If an effective population is large, the numerical fluctuations are small and random sampling of parents—and thus gametes from the parental gene pool—has little or no effect on succeeding generations. If, on the other hand, the effective population is small, random fluctuations may lead to complete fixation of one allele or another.

Cavalli-Sforza and his associates have investigated actual gene samplings with elaborate computerized experiments. In one series of experiments, some 20 alleles of blood groups, particularly ABO, Rh, and MN, were studied in different world populations. These computerized studies have indicated that random genetic drift has been an important factor in the variations that have developed in different populations. The establishment of small but relatively stable agricultural communities has in the past provided a favorable situation for diversity among communities.

As a further example of probable genetic drift, an Irish colony in Liverpool, which originated from a few founders, was observed to have a different frequency for blood-group alleles than the Irish in Ireland. Again, some small Swiss isolates have a high proportion of A- and others a high proportion of O-type blood; yet all these populations came from the same ancestry. Chance fluctuations in allele frequency presumably caused these changes.

TABLE 16.14   Simulation Experiment for Random Genetic Drift with Random Numbers[a]

GENERATION	RANDOM NUMBER	RULE FOR TRANSFORMING RANDOM NUMBERS INTO GENES	CORRESPONDING GENES	PERCENTAGE OF A GENES
0	—	—	—	50
1	9452274358	$0,1,2,3,4 = A$; $5,6,7,8,9 = a$	aAaAAaAAaa	50
2	4262686819	$0,1,2,3,4 = A$; $5,6,7,8,9 = a$	AAaAaaaaAa	40
3	1605133763	$0,1,2,3 = A$; $4,5,6,7,8,9 = a$	AaAaAAAaaA	60
4	0824427647	$0,1,2,3,4,5 = A$; $6,7,8,9 = a$	AaAAAAaaAa	60
5	5949704392	$0,1,2,3,4,5 = A$; $6,7,8,9 = a$	AaAaaAAAaA	60
6	9715513428	$0,1,2,3,4,5 = A$; $6,7,8,9 = a$	aaAAAAAAAa	70
7	9840966162	$0,1,2,3,4,5,6 = A$; $7,8,9 = a$	aaAAaAAAAA	70
8	4547684882	$0,1,2,3,4,5,6 = A$; $7,8,9 = a$	AAAaAaAaaA	60
9	8930069700	$0,1,2,3,4,5 = A$; $6,7,8,9 = a$	aaAAAaaaAA	50
10	5005195137	$0,1,2,3,4 = A$; $5,6,7,8,9 = a$	aAAaAaaAAa	50
11	3175385178	$0,1,2,3,4 = A$; $5,6,7,8,9 = a$	AAaaAaaAaa	40
12	7915253829	$0,1,2,3 = A$; $4,5,6,7,8,9 = a$	aaAaAaAaAa	40
13	4456038750	$0,1,2,3 = A$; $4,5,6,7,8,9 = a$	aaaaAAaaaA	30
14	6832883378	$0,1,2 = A$; $3,4,5,6,7,8,9 = a$	aaaAaaaaaa	10
15	4693938689	$0 = A$; $1,2,3,4,5,6,7,8,9 = a$	aaaaaaaaaa	0

[a] After L. L. Cavalli-Sforza.

MIGRATION

Allele frequencies may be altered in local populations of a species by an exchange of alleles with other demes. This exchange effectively modifies allele frequencies if the breeding populations have been partially or completely separated for enough time to have developed markedly different frequencies for the same alleles. Physical features in the environment such as rivers and islands can provide models to explain isolation, which makes later migration more effective. The significance of migration in changing allele frequencies depends on the degree of previous isolation of the subpopulations involved. Migration is a source of variation similar to muta-

tion in that alleles can be recurrently added and lost from a population, but migration may be more effective in changing allele frequency.

Together with natural selection, the swiftest way by which gene frequencies can conceivably be altered is by introducing into the population groups of genetically different individuals. Migration may enhance the effect of natural selection or may blur the effect of selection by replacing genes removed by selection. Let population a of butterflies, containing a frequency q_a of a certain allele for white color, receive some fraction m of its individuals in the next generation from a second population b with a frequency q_b of the same allele (Fig. 16.11). The

(a) (b)

FIGURE 16.11

Evolution by gene flow. Population (a) of butterflies, with a frequency of (q_a) of alleles for white color, receives a fraction (m) of its individuals each generation from population (b), which has a frequency of q_b of alleles for white.

frequency (q_a) of the white allele in population a is altered to the frequency of the allele in the nonmigrant part of the population of a, times the portion of individuals that are not migrants $(1 - m)$, plus the frequency of the same allele among the immigrants q_b, times the proportion of individuals in the population that are new immigrants (m). The new frequency (q_a') is thus

$$q_a' = (1 - m)q_a + mq_b$$

and the amount of change in one generation is

$$\Delta q = q_a' - q_a = -m(q_a - q_b)$$

The fractional change m in the size of the population depends on M, the number of migrants,

$$m = \frac{M}{M + N}$$

where N is the number of individuals already in the isolated population. A population isolated for a prolonged period in the same environment has become specialized and has achieved optimum fitness. With the onset of migration, the population could become extinct because of the disruption of optimum fitness.

Only a small difference in allele frequencies (of the magnitude that often separates populations), together with a moderate migration coefficient (m), is needed to effect a significant evolutionary change. The phenomenon is referred to as **gene flow** or **migration pressure.** Two categories can be distinguished; intraspecific gene flow between geographically separate populations of the same species and **interspecific hybridization.** The former occurs constantly within many plant and animal species and is a major determinant of the patterns of geographic variation. Interspecific hybridization occurs during breakdowns of normal species-isolating barriers. Although much less common than gene flow within species, it has a greater effect because of the larger number of gene differences that normally separate species.

THE FOUNDER PRINCIPLE

New populations are often started by small numbers of individuals, which carry only a fraction of the genetic variability of the parental population and hence may be unrepresentative. If chance operates in the selection of the founder individuals (and it usually does to some extent), new populations will differ from the parent population and from each other. The founder principle (or founder effect) is of potential importance in the origin of species.

An isolated population such as an island population with restricted variability and genotype-that migrates to another area occupied by related populations will most likely become extinct because of the broken population boundary that was established by selection pressure. If it does not become extinct, genetic drift will cause the small remaining migrant population to become divergent from the original population. Divergence is further enhanced because the different evolutionary pressures in the different areas occupied by the parent and daughter populations will be operating on different gene pools. This may be the beginning of race formation.

MEIOTIC DRIVE

Another factor that may alter allele frequencies in a population is any **irregularity** in the mechanics of the meiotic divisions, called **meiotic drive.** Ordinarily, heterozygous Aa individuals produce A and a gametes in equal proportions and these gametes have equal probabilities of fertilization and development. For many years, cases of preferential segregation have been reported, but most of these have been sporadic and nonhereditary. Now genetically based examples of systematic deviations from Mendelian ratios are on record from both male and female parents.

The SD (segregation distorter) complex in chromosome II of *Drosophila melanogaster,* for example, has at least the Sd and Rsp loci. Sd is thought to be the locus required for distortion; that is, it makes the product, and Rsp makes the target site for distortion. Thus, an SD/SD^+ heterozygote is really $Sd\ Rsp^{ins}/Sd^+Rsp^{sens}$. In the presence of the wild-type allele (Sd^+), the No. 2 chromosomes segregate normally. Heterozygous males that carry the mutant (Sd) allele, however, show a marked departure from

the 1 : 1 ratio. The mutant allele interacts with its wild-type homolog (Sd^+) associated with the Rsp^{sens} target site. Distortion occurs in the form of irregular behavior during spermatogenesis. As a result, only a few sperm carry the wild-type (normal) allele. In SD, the drive is thus due to an induced dysfunction of the sperm bearing the Sd^+Rsp^{sens} chromosome; this involves a failure of normal chromatin condensation during sperm formation.

In the presence of the homozygous wild-type allele, the chromosomes segregate normally. Heterozygous males that carry the mutant allele show a marked departure from the 1 : 1 ratio. The mutant allele apparently interacts with its wild-type homolog, causing it to behave irregularly during spermatogenesis. As a result, only a few sperm contain the normal allele.

Similarly, *Drosophila* females that have a pair of structurally different chromosomes may undergo preferential segregation. One member of the pair may be systematically retained in the egg and the other segregated to the nonfunctional polar body. If the two homologs are of unequal length, the shorter one is usually included in the egg nucleus. Chromosomes that have undergone structural changes are often extruded. Any persistent factor disrupting a particular chromosome would result in preferential rather than random segregation.

Sex ratios are expected to be altered by irregularities in the segregation of X and Y chromosomes. Examples have been reported in certain races of *Drosophila pseudoobscura* where abnormal spermatogenesis resulted in failure of the X and Y chromosomes to pair. The Y degenerates and the X undergoes an extra division. As a result, all sperm carry X chromosomes. When these males, which carry only X chromosomes in their sperm, are mated, only female progeny are produced. In other examples, all X chromosomes have systematically degenerated, and only males have been observed in the next generation.

Meiotic drive could be a significant factor in evolution. Even the most favorable genes would not be perpetuated if the chromosomes in which they were carried were systematically excluded from gametes.

GENETIC VARIATION AT THE MOLECULAR LEVEL

Genetic variation can be evaluated at the molecular level by (1) direct nucleotide sequencing of the genes and complete genomes of specific organisms (e.g., as for phages MS2 and ΦX174, Chapter 10), (2) direct amino acid sequencing of the primary gene products (the polypeptides) of specific genes in different individuals (as with the hemoglobin variants, Chapter 9), (3) measuring the amount and kind of variation within a species and between closely related and more distantly related species, and (4) estimating gene changes over time by comparing nucleotide sequences of specific genes or amino acid sequences of specific gene products in species related by evolution.

Base-pair substitutions of the type giving rise to the hemoglobin variants (Chapter 9) undoubtedly account for a significant amount of the natural variation in different species. The accumulation of such mutations over long periods of time might well account for much of the divergence of different species. In fact, comparisons of the amino acid sequences of conserved proteins such as the hemoglobins and cytochromes in different species show a high correlation between the divergence of individual polypeptides and the divergence estimated by other taxonomic criteria of phylogenetic relatedness. Similar comparisons have been made for different polypeptides of the same organism in cases where the gene products are believed to have evolved from a common progenitor, such as for the different hemoglobin chains.

HEMOGLOBINS

All of the hemoglobin chains have similar amino acid sequences and conserved features, which suggests that they evolved from a common progenitor, probably the muscle protein myoglobin (Fig. 16.12) or an ancestral form of myoglobin. Although the alpha and beta polypeptides of human hemoglobin A apparently evolved from lineages that separated early in the evolution of chordates, they still exhibit many similarities (Figs. 16.12 and 16.13). The beta chains of adult hemoglobin and the gamma chains of fetal hemoglobin presumably diverged much more recently since they are identical at 71 percent of the amino acid positions (Fig. 16.12). All hemoglobins that have been characterized have two histidine residues, spaced 29 amino acids apart, that

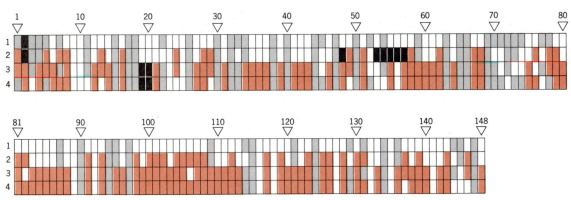

FIGURE 16.12

Amino acid replacements in human globin series: (1) myoglobin, (2) alpha hemoglobin, (3) beta hemoglobin, (4) gamma hemoglobin chains. Light gray areas indicate similarities; white areas indicate differences; colored areas indicate identity of alpha, beta, and gamma hemoglobin sites with each other; black areas indicate gaps. (From T. H. Jukes, Molecules and Evolution, *New York: Columbia University Press, 1966, reprinted by permission of the publisher.)*

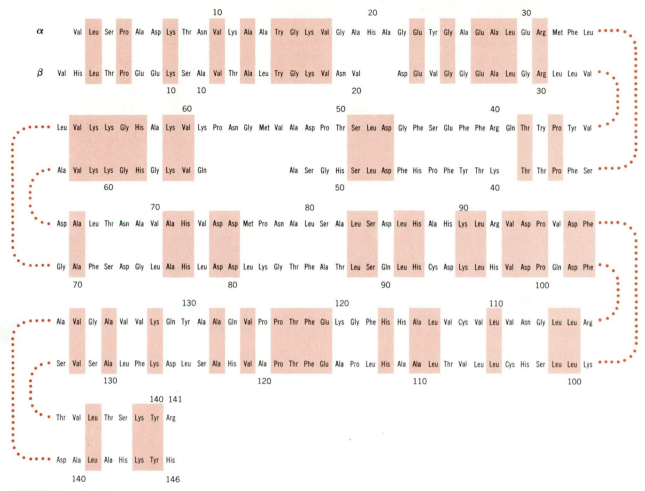

FIGURE 16.13

The amino acid sequences of the alpha (α) and beta (β) polypeptides of normal adult hemoglobin (hemoglobin A). The amino acids enclosed in boxes are identical and occupy corresponding positions along the polypeptides. The amino acids are numbered sequentially from the amino terminus. (After Ingram.)

bind the heme group to the polypeptide chain. This feature of the hemoglobin molecule thus appears to have been universally conserved.

Numerous amino acid changes (and, thus, base-pair substitutions in the structural genes) have occurred in different species. The number of differences observed between two species usually correlates well with their phylogenetic divergence. Closely related species have very similar hemoglobins, whereas those that are farther apart on the evolutionary scale exhibit a greater number of amino acid differences. The alpha and beta polypeptides of the same species have more differences than the alpha chains or the beta chains of different species. For example, human alpha chains and horse alpha chains have fewer differences than human alpha chains and human beta chains.

The hemoglobin molecule, with its heme groups bound by four globin chains, apparently is sufficiently efficient in oxygen transport that its basic structure has been conserved in the course of evolution.

CYTOCHROME C

Cytochrome is widely distributed in animal and plant tissues, where it is involved in oxidative phosphorylation. Cytochromes A, B, and C are distinguished by their absorption spectra. Cytochrome C is a protein electron carrier that has been identified in many different animals. Some 35 different amino acid sequences of mammalian-type cytochrome C have been compared and are remarkably similar. Similarities and differences in such animals as lamprey, tuna, chicken, rabbit, cow, horse, rhesus monkey, chimpanzee, and human have been detected. Cytochrome C taken from human, horse, chicken, tuna, and yeast tissues are compared in Fig. 16.14.

Most cytochromes have 104 amino acids in their polypeptide chains. Some amino acids are in the same positions in all species studied. Glycine, for example, occurs at eight sites (1, 6, 29, 34, 41, 45, 77, and 84), lysine at six sites, and from one to three representatives of all other amino acids (represented by 70–80 residues) have been found in all samples. These series of amino acids in present-day animals must have descended from ancestral proteins. The probability of their occurring by chance is extremely remote. However, differences as well as similarities are noted. In the horse, cow, and rabbit, for example, alanine is in position 15; but humans have serine in this position. Such changes have arisen through mutation.

The findings of certain residues in the same place in all samples suggest that they must have important and specific functions. For example, arginine is located in positions 39 and 91 in all samples. The amino acid sequences in cytochrome C are identical in humans and chimpanzees; the human and the rhesus monkey differ in only one residue. There are 12–13 amino acid differences between tuna and human tissues, and cytochrome C in humans differs from that of yeast in 46–48 residues. Assuming that each difference is the result of a single mutational change somewhere in the ancestral history, time estimates are possible for divergences. A phylogenetic tree (Fig. 16.15) based on the cytochrome C gene alone resembles those based on studies from other areas of biology. Spontaneous mutations arising from single base substitutions in DNA, reflected in changes in nucleotides in the RNA, have brought about profound changes in hemoglobin, cytochrome, and other proteins over long periods of time. Some studies of other proteins do not yield phylogenetic trees that are as informative.

GENETIC POLYMORPHISM

Genetic polymorphism is the existence of two or more genetically different phenotypes in the same interbreeding population. In a constant environment, the frequencies of two alleles are carried to some intermediate equilibrium point between zero and one by the countervailing forces of selection, gene flow, and mutation pressure (with meiotic drive as a possible fourth factor). Even in the absence of a balance between the primary evolution-

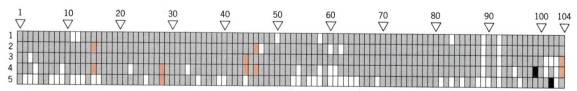

FIGURE 16.14

Amino acid sequences of cytochrome C from: (1) human, (2) horse, (3) chicken, (4) tuna, and (5) yeast. Gray areas indicate similarities; white areas indicate amino acid differences; colored areas indicate differences from (1) but identity of other chains with each other; and black areas indicate gaps. (From T. H. Jukes, Molecules and Evolution, *New York: Columbia University Press, 1966, reprinted by permission of the publisher.)*

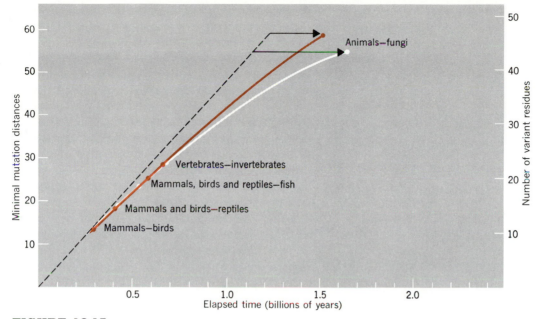

FIGURE 16.15

Relationship between the number of variant residues among cytochromes C of different classes and phyla of organisms and the time elapsed since the divergence of the corresponding lines of evolutionary descent (dashed line and curve indicated by white circles, ordinate to the right); similar relationship between the number of variant nucleotide positions in the corresponding structural genes (dashed line and curve indicated by colored circles; ordinate to the left). The straight dashed line applies to both residue and nucleotide variations and is calculated on the basis of a value of 280 million years for the time elapsed since the divergence of the avian and mammalian lines of descent. The solid curves show the relationship corrected for probable multiple changes in single residue (white circle) or nucleotide (colored circle) positions, assuming a total of 76 variable residues. (Reproduced with permission from "Comparative Aspects of Primary Structures of Proteins," Annual Review of Biochemistry, *Vol. 37, p. 740. Copyright © 1968 by Annual Reviews, Inc.)*

ary agents, two or more alleles can be maintained together in the same panmictic population for indefinite periods of time. This condition is called **balanced polymorphism.** It is the preservation of genetic variability through selection. Such a balance can be achieved by several means. It can occur, for example, through frequency-dependent selection, when the fitnesses of the genotypes are not constant but change with their frequency. If one genotype has a lower fitness than the other while it is at higher frequencies, but gains the advantage when its frequency descends to a certain level, the allele frequencies will tend to stabilize.

A second condition that causes balanced polymorphisms is **heterozygote superiority** or **heterosis** (see p. 537). If a heterozygote Aa is superior to both homozygotes, AA and aa, neither allele will be eliminated. The frequency of a (q) and the frequency of A (p) are expected to stabilize at some intermediate frequency between zero and one. The

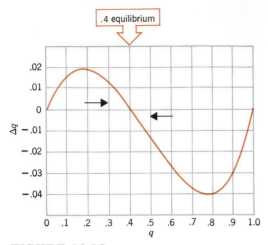

FIGURE 16.16

Selection favoring heterozygotes. Change in the frequency of a (Δq) when the relative fitnesses are AA = .80, Aa = 1.00, aa = .70, and population size is infinite. These values provide a stable, balanced polymorphism (Δq̄ = 0) at q̄ = .4. That is, Δq is positive (q increases) if q is less than .4 and negative (q decreases) if q is more than .4. If one allele is accidentally eliminated (i.e., q = 0 and p = 1), Δq is 0, but polymorphism is lost. (After Crow, Li, and Strickberger.)

equilibrium value $\hat{q}$ (at which selection against the two homozygotes is in balance) and $\Delta\hat{q} = 0$ (Fig. 16.16) is obtained from the following mathematical relation:

$$\hat{q} = \frac{S_A}{S_A + S_a}$$

where $\hat{q}$ is the equilibrium frequency of a; S_A, selection value of A; and S_a, selection value of a.

An example of balanced polymorphism is the sickle-cell trait (Chapter 9) that is common in the human population in Africa and parts of the Middle East. Sickle-cell anemia apparently became established in human populations while the way of life in parts of Africa was changing from hunting and gathering to a more stable agricultural pattern. The environment in agricultural areas was favorable for mosquitoes and some mosquitoes carried malaria. Both sickle-cell anemia and one type of malaria, that caused by *Plasmodium falciparum*, are diseases of human red blood corpuscles. Patients with the sickle-cell disease have collapsed, sickle-shaped cor-

puscles that clog the capillaries, thus interfering with circulation and depriving the cells in the body of oxygen. Heterozygotes for the allele for sickle-cell anemia are more resistant to falciparum malaria.

The allele Hb_β^S controls the sickle-cell trait in which the red blood cells assume a sickle-like shape when they are exposed to low oxygen tension. Hb_β^A designates the wild allele for normal red corpuscles. People in the heterozygous condition (Hb_β^S/Hb_β^A) show the trait in less than 1 percent of their red corpuscles. The hemoglobin is slightly abnormal, and the disease symptoms are slight but usually sufficient to make possible the recognition of heterozygous individuals. A simple blood test can verify carriers. About 40 percent of the people in some tribes in parts of Africa carry the allele. Homozygotes (Hb_β^S/Hb_β^S) display the trait in a large percentage of their red blood cells and they suffer from sickle-cell anemia, which usually proves fatal in childhood.

Malaria is common in the same areas where sickle-cell anemia is prevalent, but the greatest death rate from malaria occurs among natives with normal hemoglobin in their red corpuscles. Those homozygous for the sickle-cell trait (Hb_β^S/Hb_β^S) usually die early from anemia ($Hb_\beta^S/Hb_\beta^S = 1 - s_a \therefore W = 0$), while those homozygous for normal hemoglobin (Hb_β^A/Hb_β^A) suffer from malaria ($Hb_\beta^A/Hb_\beta^A = 1 - s_A$). The heterozygotes ($Hb_\beta^S/Hb_\beta^A$) do not contract malaria and do not suffer seriously from anemia ($W = 1$), but they perpetuate the sickle-cell allele in their progeny.

Selective forces in a malaria-plagued environment and among people whose agricultural mode of life favors exposure to the malaria-bearing mosquito have apparently resulted in three different genotypes being maintained in the population. By effective control of mosquitoes and other public health measures, malaria is being eradicated from many parts of the world. With this change in the environment, the polymorphism is expected to disappear and the selection factor favoring sickle-cell anemia will be removed. The population in previously malarial areas is expected to become homozygous for Hb_β^A except for rare mutations from Hb_β^A to Hb_β^S. Therefore, the relative fitness of the heterozygote will no longer equal one, and p and q will no longer be stable. The balanced polymorphism will no longer exist.

GEOGRAPHIC POLYMORHISM

Another major type of polymorphism is geographic polymorphism in which allele frequencies differ in different geographical regions. The ABO blood alleles are present in different frequencies in populations originating in different geographical areas of the world. When the hereditary antigens that help define the different blood profiles are studied individually in various populations, it is apparent that their distribution varies among different races. Thus, traits that have a known mode of inheritance allow us to determine how the frequency of the given allele in a given population compares with the frequency of that allele in another population. This objective approach to racial classification offers a number of advantages over morphological charac-

teristics that are much more difficult to analyze genetically.

Among American Indians, the Utes from Utah, the Navajos from New Mexico, and most of the tribes in North, Central, and South America, there is a high incidence of O, comparatively little A, and no B or AB. On the other hand, Blackfeet Indians (Fig. 16.17) of Montana have less O and a high proportion of A (76.5 percent in the sample cited). The Polynesians sampled have 60.8 percent A and very little B and AB. Australian aborigines have a high A (57.4 percent in the sample cited) and no B or AB. Basques, who are believed to be similar to the ancestors of present-day European populations, have a high incidence of A, moderate B, and virtually no AB. Siamese and Japanese, on the other hand, have a high proportion of A and B. Western European peoples and Caucasians of the United States have a fairly high proportion of A and some B and AB. In none of the extensive samples, however,

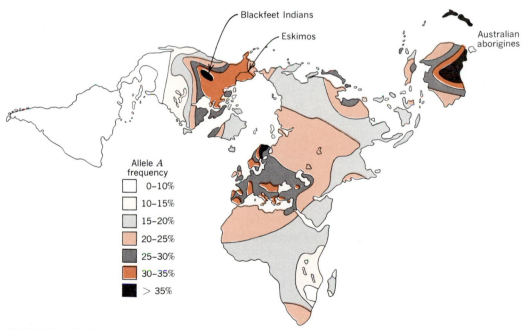

FIGURE 16.17

World map showing percentage frequencies of blood-group allele I^A. High frequencies occur among the Blackfeet Indians and Australian aborigines. (Data from Mourant, Kopeč, Domaniewska-Sobczak, Boyd, and Matson.)

was the frequency of A as high as among the Blackfeet and related tribes of Indians.

CHROMOSOMAL POLYMORPHISM

Different types of chromosomes as well as different types of alleles may become established in populations. Chromosome polymorphism is the inclusion in the population of two or more alternative structural forms of chromosomes. The structurally changed chromosomes are the results of rearrangements (Chapter 12). On the basis of limited observations, chromosome inversions would seem to be more common in animals, whereas translocations are more common in plants. The rearrangements may be fixed as homozygotes and heterozygotes in a certain percentage of the population; this is called **balanced chromosomal polymorphism.** Chromosomal structural types then fluctuate around the mean in roughly constant proportions from generation to generation and possess adaptive value. They behave as homologous chromosomes in their ability to mutually replace the alternative structural forms in the karyotype (see Chapter 12). Chromosomal polymorphism may be eliminated if one of the structural variants is superior under prevailing environmental conditions. On the other hand, its proliferation may be strongly enhanced by natural selection. A classical investigation of chromosomal polymorphism has been conducted by Dobzhansky and his associates in *Drosophila* strains inhabiting the mountains of southern California.

POPULATION GENETICS AND EVOLUTION

Plant and animal populations tend to evolve in nature. The biological principle of evolution covers this progressive change in populations. Evolution has, therefore, become a central principle of biology. In the words of the distinguished biologist, T. Dobzhansky, "Nothing in biology makes sense except in the light of evolution." Genetic variation in plant and animal populations is basic to evolution. It occurs continuously over time in natural populations because changes occur in gene frequency. Factors causing changes in gene frequency, and thus major forces in evolution, are natural selection, mutation, migration, random genetic drift, and, to an unknown extent, meiotic drive. Two frequent questions about these forces are (1) what roles do the various forces play in the overall evolutionary process and (2) how important are these processes relative to one another at the various levels of evolution? Neither of these questions has been answered satisfactorily. Two broad views have been expressed: (1) gradualism and (2) punctuationalism, along with many intermediate shades of opinion. The most far-reaching hypothesis is the **shifting balance theory of evolution** developed by Sewall Wright (1977, 1978). It encompasses polymorphism, pleiotropy, the relation between genotype and fitness, and population structure (particularly size of the breeding population).

CHARLES DARWIN AND GRADUALISM

The pioneer scientist in the field of evolution was Charles Darwin (1809–1882, Fig. 16.18). In his *On the Origin of Species by Natural Selection* (1859) and other writings, he had two objectives: (1) to establish the fact of organic evolution and (2) to explain the cause of evolution. In the first objective, he succeeded very well by bringing together the data available from geology, paleontology, biology, and other disciplines to show that evolutionary change had occurred in plant and animal populations. The second objective, on the mechanics of evolutionary change, could not be resolved at Darwin's time, and has not been resolved now. But Darwin did more than others of his time by establishing a theory called **natural selection** as an explanation to be tested as further data became available. Natural selection acts on phenotypes, not on genotypes; and it acts on whole phenotypes (organisms) as determined by many loci and countless environmental factors. Darwin was mainly concerned with **microevolution,** that is, evolutionary change from increased adaptation of organisms to their environments that occurs in local populations. Darwin was also interested in the origin of biological units of

FIGURE 16.18

Charles Darwin (1809–1882), pioneer in the study of evolution by natural selection. (Portrait by Everett Thorpe.)

living things called species: one or more populations, the individuals of which can interbreed, but which in nature cannot exchange genes with members of other species. Species are thus groups of actual or potentially interbreeding natural populations that are reproductively isolated from other such groups. Organisms that are actually or potentially capable of exchanging genes share a common gene pool.

Natural selection is Darwin's theory for the cause of microevolution and speciation that is called Darwinism. It is based on three premises and one conclusion: (1) all species have more offspring than can survive and reproduce, (2) organisms vary in their ability to survive and reproduce, and (3) part of the variation in ability to survive and reproduce is hereditary. The conclusion from the three premises is the occurrence of natural selection, the process by which genotypes with greater fitness leave more offspring than do less fit genotypes. Favorable alleles that promote higher fitness will be represented disproportionately in further generations. Types and frequencies of alleles in the population will gradually change and promote greater adaptation to the particular environment.

Another consequence of natural selection and random mutation has been postulated for the basic mechanism of sexual reproduction required in a Mendelian population. Female mating behavior depends on genetic mechanisms and the presence of males in a random mating population. Hampton Carson and associates (1982) have found experimental support for this aspect of Darwinism.

A selected laboratory strain of *Drosophila mercatorum* has existed for 20 years without males and therefore without natural selection operating to maintain the genetic basis of female mating behavior. The unisexual (all female) laboratory strains of *D. mercatorum* were established by selection from normal wild bisexual ancestors. This was accomplished by selecting for enhancement of a naturally present, low-grade parthenogenesis (development of individuals from an egg without fertilization). Females in the new strains are diploids produced by duplication of single haploid nuclei. They reproduce genetically identical females without fertilization.

The females in this strain have recently undergone a genetic impairment of mating capacity. This observation exemplifies the mode of evolution of vestigial traits and supports Muller's (1949) theory that random mutations tend to destroy the genetic basis of a character from which selection has been removed.

Darwinism remains a theory and is known to be an inadequate explanation for some aspects of evolution. "Darwinism" and "evolution" are sometimes used synonymously and considered to mean the same thing, but they are entirely different. Evolution is change, unfolding or evolving of animal and plant populations, in nature. It is established scientifically and recognized by informed people. Darwinism is one theory, natural selection, presented to explain how part of evolution occurs. Natural selection, based on small variations in reproductive capacity that are influenced by the environment, is not adequate to account for major evolutionary changes, or macroevolution, occurring in the higher taxonomic categories. Macroevolution includes the processes that lead to the formation of new species, genera, families, orders, and higher taxonomic categories. It can involve phenomena quite different from those involved in microevolution, such as hybridization and polyploidy in plants and major chromosome changes in plants and animals. Whether microevolutionary events that occur gradually over many generations are involved or macroevolution results from entirely new principles is not clear. Biologists are attempting to supplement natural selection with new observations and theories to account for all aspects of evolution.

SUCCESSION AND THE FOSSIL RECORD

The fact that life has evolved, with occasional interruptions but with increased variety and complexity, has been established. Successive populations have arisen as bacteria, sponges, fish, amphibia, reptiles (including dinosaurs), birds, and mammals. Fossil records are known to be sporadic and incomplete, but they confirm in a general way the succession of living forms on the earth. They are deficient, however, in some important places. They do not demonstrate that the different major groups of living things were actually linked to each other. Large gaps in the fossil record that are associated with profound physiological changes in organisms seem impossible to demonstrate with fossils and to explain by natural selection. How did bats evolve wings from their flightless shrew-like ancestors? How did whales, presumed to have come from land mammals, get from land to sea? No intermediate links appear in the fossil record. Darwin, himself, was at a loss even to attempt to explain the origin and perfect construction of the human eye. Darwin expected that the major groups of animals — the fish, amphibians, reptiles, birds, and mammals — that appeared in succession in the fossil record would be **linked together, but he did not know how such transitions could occur.** No evidence for such connections was available in the fossil record when Darwin published *The Origin of Species* in 1859, and virtually none is available now. The fossil record, however, is better known and much more complete. When Darwin published the first edition of the

Origin, only distinctive fossil faunas of various ages were recognized. No well-established succession of fossils for any one family of animals was known.

WORLDWIDE UPHEAVALS AND EVOLUTION

In the geological history of the earth, an interesting correlation has occurred with the origin of major taxonomic groups such as fish, amphibia, reptiles, and mammals. Major changes in phylogeny have coincided with massive global upheavals called **geological revolutions.** These upheavals could confuse the geological record and make whatever links might be expected to have been established between phyla difficult or impossible to decipher. Another more basic relationship could be associated with a non-Darwinian, comparatively rapid mechanism for original major changes to occur in animal and plant populations. Thus, global upheavals might provide a setting necessary for unique methods for major changes at the family and phylum taxonomic level.

In the 1940s, Richard Goldschmidt (1878– 1958, Fig. 16.19), a renowned German geneticist who moved to the United States in 1936, was theorizing on the causes for major evolutionary change. He called himself a neutralist and macroevolutionist in contrast to the contemporary Darwinian selectionists and microevolutionists, a group that included T. Dobzhansky and most of the other scholars in the field of evolution. One of Goldschmidt's extreme presentations was a possible explanation for the transition from water to land animals — that is, amphibia to reptiles. In the wake of a global disaster, Goldschmidt imagined, a large number of amphibious creatures could be thrown up from a water habitat to a mostly land environment and become stranded in moist areas. Most of the disturbed creatures would die, he theorized, but a very few surviving mothers in entirely new surroundings might somehow manage to live and continue reproduction. Successive environmental changes over comparatively short geological time periods could provide new challenges and opportunities for relatively unspecialized organisms to survive. A swim bladder and moist outer skin might provide for sufficient respiration in air. A strange

FIGURE 16.19

Richard A. Goldschmidt (1878–1958), German and American biologist who developed the theory of macro-evolution. (Portrait by Everett Thorpe.)

diet, exposure to the elements, and population pressure might be accommodated by a few of the many competitors for space and sustenance. A very few animals in some local areas might have the resiliency and stamina to adapt, survive, and reproduce in their new environmental situations.

If the strange, changing environment included radioactive material, gene mutations and chromosome changes might occur. These would tend to decrease viability and interfere with reproduction, but some organisms with substantial chromosomal alterations might survive and reproduce markedly changed progeny. Professor Goldschmidt called these "hopeful monsters" that might occur in different locations. Some, he thought, might happen to fit the new environment well enough to survive and reproduce more changed progeny. Again, most progeny would be inviable, but many natural experiments could be in progress at the same time. Once in a while a grossly changed chromosome combina-

tion might be perpetuated in a living fertile female. According to Goldschmidt, a single female could theoretically carry the makings of an entirely new living form. After generations of struggle and change associated (by chance) with a favorable environmental trend, a few of the different kinds of organisms might become established. With the new, changed environment now becoming stabilized, and little or no competition for the very few new organisms, rapid increase in numbers of the changed populations could occur.

When Professor Goldschmidt presented his theories and possible applications at his own university in the United States and at scientific meetings throughout the world, many questions were raised. How could major chromosome changes occur without killing the organism? How could fertility be maintained? How could the numerous changes needed, for example, to transform a water animal into a land animal be accomplished? Goldschmidt, like Darwin, could always theorize, but he had few answers for technical questions. He would reply, "I am not a chemist or a detail man; I am a theoretical biologist." Sometimes he would reply with a broad generalization such as (1) mathematical laws suggest that large-scale transformations may occur abruptly or (2) the environment does more than merely weed out the unfit genes as expected from natural selection. The environment may determine the pace of evolution; large-scale evolution may occur when conditions are right for change.

Goldschmidt was an extremist. He gave too much weight to structural and numerical changes in chromosomes and perhaps not enough consideration to gene changes in bringing about wide variations, but leaving the changed organisms viable and fertile. He was ahead of his time in evaluating the gene as a chemical determiner and he was one of the first to develop a model for large-scale macroevolution in contrast to Darwinian natural selection. Goldschmidt's model was in the direction of a current "punctuational" model for large-scale abrupt changes in evolution as compared with a modern view of the traditional Darwinian model now called the "gradualistic" model.

Both Darwin and Goldschmidt would be

pleased with the treatment that has been given to their theories and the current models for two aspects of evolution because progress in science occurs by discussing, challenging, and testing alternate hypotheses. Understanding of the mechanisms of evolution has come a long way and each model has its place in the complete picture of population change. The evolution theory is getting better. Fossil records are much improved and are now scaled against absolute time rather than unmeasured geological periods. Life spans of some species have been studied. Some species have come into being rather rapidly, survived for a million generations with little evolutionary change, and become extinct. On the other hand, rapid change in some species has resulted from frequent budding off from ancestral stocks and production of new species.

GRADUALISTIC AND PUNCTUATIONAL MODELS

Darwin was right in recognizing genetic variation as the basis for evolutionary change, but he did not know the cause, now known to include such factors as mutation, recombination of genes, and chromosome structural and numerical changes. He resorted to a Lamarckian (inheritance of acquired characteristics) model for an explanation for variation. Selection of genes for reproductive fitness and isolation of populations in the development of unique populations were well considered by Darwin as major factors for speciation. Random genetic drift operating in small breeding populations and the founder principle have been emphasized by others since Darwin's time. The Darwinian pattern of small changes accumulating in response to natural selection associated with environmental change resulted in a slow, gradual process of evolutionary change. This widely accepted gradualistic pattern of evolution may now be compared with the more rapid, macroevolution or sporadic punctuational evolution.

Punctuational change involves all major elements included in gradual change. It differs only in its emphasis on certain elements and is applied mostly to large-scale evolution. Random genetic drift in small breeding populations plays a greater role in punctuational evolution than in evolution of large breeding populations. Selection is important, but it is a coarse-grained kind of selection, operating in very small populations where unusual new features loom large. Small size in a population has two major effects: (1) random mating spreads new genes and their traits throughout members of the population and (2) chance (genetic drift) is more effective in producing change.

MOLECULAR DRIVE IN EVOLUTION

As indicated earlier in this chapter, the origin of species and mechanisms for other aspects of evolution have been investigated for a long time. Theories have been established and applications have been tested, but relatively few experimental proofs for genetic mechanisms for evolution have arisen. Natural selection and genetic drift have been widely applied to explain the consequences of evolutionary novelty.

With the understanding of molecular genetics and studies of eukaryotic genomes with multiple-copy families of genes and noncoding sequences, more problems have been added to the complex. Solutions to the problems have revealed a new level for investigation. Many gene families exhibit unexpected sequence homogeneity within and between individuals of a species. A gene family that is shared between several species, however, may reveal substantial heterogeneity between the species.

Gene family heterogeneity could be accomplished by several molecular mechanisms. These ensure that existing members of a family are replaced with new variant members. Modes of operation of such mechanisms have unusual genetic consequences through which variation is distributed in nature. Fixation of variants in a population through such processes is defined as **molecular drive.** Three known mechanisms through which such fluctuations can occur are (1) unequal chromatid (or chromosome) exchange, (2) gene conversion, a situation in which products of meiosis from AA' individuals are $3A:1A'$ or $1A:3A'$ instead of $2A:2A'$ as expected [An A gene appears to have been converted to an A' gene (or vice versa); a possible cause is a switch in templates during DNA synthesis: an A

gene is copied twice and the *A'* gene not at all (or vice versa)], and (3) a gain or loss of DNA through transposition, movement of a segment of DNA from one position in a chromosome to another position or even to another chromosome or genome.

Thus, families of genes and noncoding sequences within a population may have a fixation of mutations as a consequence of molecular mechanisms of turnover within the genome. Such mechanisms may be both random and directional in activity. Unusual patterns of fixation permit the origin of biological novelty and species discontinuity not accomplished by selection or genetic drift.

SUMMARY

The Hardy–Weinberg theorem describes mathematically the equilibrium established among genotypes in a population. Equilibrium is maintained through random mating. Nonrandom mating such as inbreeding does not alter frequencies of alleles but alters the proportion of homozygotes. Outbreeding mixes genotypes and keeps infrequent recessives hidden by their dominant alleles. It provides for new combinations that may be tried out in selection. Factors influencing relative allele **frequency** are selection, mutation, genetic drift, meiotic drive, and migration. Change in allele frequencies is the basis for evolutionary change. When a breeding population is isolated for a prolonged period of time, its gene pool becomes distinctive in terms of allele frequencies as compared with that of other isolated populations. Subunits may thus be formed within breeding populations. Genetic load in a human population is based on the number of deleterious genes per individual. Polymorphism may be balanced, geographic, or chromosomal. Population genetics forms the basis for evolution. Charles Darwin (1) established the fact of evolution and (2) presented the theory of natural selection to explain the cause of evolution. The theory was based on natural variation selected by the environment. Gradual Darwinian evolution does not adequately explain major rapid changes. Geological revolutions and chance in small breeding populations are associated with rapid, major changes for the punctuational model.

REFERENCES

Ayala, F. J., 1977. "Nothing in biology makes sense except in the light of evolution." Theodosius Dobzhansky: 1900–1975. *J. Heredity* 68:3–10.

Bodmer, W. F., and L. L. Cavalli-Sforza. 1976. *Genetics, evolution, and man.* W. H. Freeman, San Francisco.

Bryant, S. V., V. French, and P. J. Bryant. 1981. Distal regeneration and symmetry. *Science* 212:993–1002.

Carson, H. L., L. S. Chang, and T. W. Lyttle. 1982. "Decay of female sexual behavior under parthenogenesis." *Science* 218:68–70.

Cavalli-Sforza, L. L. 1977. *Elements of human genetics*, 2nd ed. Benjamin Co., New York.

Crow, J. F. 1976. *Genetic notes*, 7th ed. Burgess Publishing Co., Minneapolis.

Darwin, C. 1951. *The origin of species by means of natural selection or the preservation of favoured races in the struggle for life.* Philosophical Library, New York. (Reprint of first edition published November 24, 1859.)

Dobzhansky, T. 1951. *Genetics and the origin of species,* 3rd ed. Columbia University Press, New York.

Hardy, G. 1908. "Mendelian proportions in a mixed population." *Science* 28:49–50.

Hartl, D. L. 1980. *Principles of population genetics.* Sinauer Associates, Inc., Sunderland, Massachusetts.

Hayes, H. K. 1963. *A professor's study of hybrid corn.* Burgess Publishing Co., Minneapolis.

Kettlewell, H. B. D. 1959. "Darwin's missing evidence." *Sci. Amer.* 200(3):48–53.

Lewontin, R. C. 1974. *The genetic basis of evolutionary change.* Columbia University Press, New York.

Matson, G. A., H. E. Sutton, J. Swanson, A. R. Robinson, and A. Santiana. 1966. "Distribution of hereditary blood groups among Indians in South America." *Am. J. Phys. Anthro.* 24:51–70.

Mueller, L. D., and F. J. Ayala. 1981. Fitness and density-dependent population growth in *Drosophila melanogaster. Genetics* 97:667–677.

Mueller, L. D., and F. J. Ayala. 1981. Trade-off between *r* selection and *k* selection in *Drosophila melanogaster. Proc. Natl. Acad. Sci. U.S.* 78:1303–1305.

Mourant, A. E., A. C. Kopeč, and K. Domaniewska-Sobczak. 1976. *The distribution of human blood groups*, 2nd ed. Oxford University Press, New York.

Peacock, W. J., and G. L. G. Miklos. 1973. "Meiotic drive in *Drosophila*: new interpretations of the segregation distorter and sex chromosome systems." *Adv. Genet.* 17:361–409.

Silvestroni, E., and I. Bianco. 1975. "Screening for microcythemia in Italy: analysis of data collected in the past 30 years." *Amer. J. Hum. Genet.* 27:198–212.

Stanley, S. M. 1981. *The new evolutionary timetable.* Basic Books, New York.

Stebbins, G. L., and F. J. Ayala. 1981. Is a new evolutionary synthesis necessary? *Science* 213:967–971.

Stern, C. 1962. "Wilhelm Weinberg." *Genetics* 47:1–5.

Wallace, B. 1970. *Genetic load, its biological and conceptual aspects.* Prentice-Hall, Englewood Cliffs, New Jersey.

Wright, S. 1978. *Evolution and genetics of populations.* Vol. 4. *Variability within and among natural populations.* University of Chicago Press, Chicago.

———. 1977. *Evolution and the genetics of populations,* Vol. 3. *Experimental results and evolutionary deductions.* University of Chicago Press, Chicago.

———. 1969. *Theory of gene frequencies.* University of Chicago Press, Chicago.

———. 1934. "On genetics of subnormal development of the head (octocephaly) in the guinea pig, an analysis of variability in number of digits in an inbred strain of guinea pigs, the results of crosses between inbred strains of guinea pigs, differing in number of digits." *Genetics* 19:474–551.

PROBLEMS AND QUESTIONS

16.1. (a) Do dominant alleles spread more rapidly in a population than recessives? (b) What factors influence the relative frequency of alleles in a population? (c) When the relative frequency of alleles in a population is altered, how can equilibrium be maintained?

16.2. The frequency of children homozygous for a recessive lethal allele is about 1 in 25,000. What is the proportion of carriers (heterozygotes)?

16.3. The following MN blood types were obtained from the entire populations of an isolated North American Indian village and an isolated village of Central American Indians.

GROUP	POPULA-TION SIZE	M	MN	N
Central American Indian	86	53	29	4
North American Indian	278	78	61	139

Calculate the allele frequencies of L^M and L^N for (a) the Central American Indian and (b) the North American Indian populations. List the results in tabular form. (c) Discuss possible reasons for the differences in allele frequency.

16.4. (a) Among 205 American Negroes, the frequencies of the L^M and L^N alleles were .78 and .22, respectively. Calculate the percentage of individuals with M-, MN-, and N-type blood. (b) A group of 212 college students were invited to taste PTC. There were 149 tasters and 63 nontasters. Calculate the allele frequencies of T and t.

16.5. Among 798 students, 70.2 percent were tasters. (a) What proportion of the students were TT, Tt, and tt? (b) What proportion of the tasters who marry nontasters might expect only taster children in their families? (c) What proportion might expect some nontaster children?

16.6. Among 11,335 people, the following blood types were obtained: 5150 O, 4791 A, 1032 B, and 362 AB. Calculate the allele frequencies for i, I^A, and I^B.

16.7. Among 237 Indians, the allele frequencies of i, I^A, and I^B blood alleles were .96, .03, and .01, respectively. Calculate the percentage of individuals with O-, A-, B-, and AB-type blood.

16.8. Blood samples from 999 (883 male and 116 female) students were typed as follows:

	O	A	B	AB
Male	419	371	68	25
Female	68	38	7	3

(a) Calculate the allele frequencies for males and females separately. (b) Determine the proportion of heterozygotes ($I^A i$, $I^B i$, and $I^A I^B$) and the proportion of homozygotes (ii, $I^A I^A$, $I^B I^B$) for males and females in the sample. (c) Give the genotypic frequencies (r^2, p^2, $2pr$, q^2, $2qr$ and $2pq$) for the entire sample (males and females combined).

16.9. When the blood samples from 999 students were tested with anti-Rh serum, 74.9 percent were positive and 25.1 percent were negative. Assuming a single pair of alleles R and r (for this problem only), what proportion of the students would be expected to be RR, Rr, and rr?

16.10. In a species of animals with heterogametic (XY) males, alleles A and a may occur in the X chromosome and no allele is in the Y chromosome. Females have genotypes AA, Aa, and aa. (a) Give the genotypic and allele equilibrium frequencies for A and a for males and females. (b) Calculate the progress toward equilibrium in six generations of random mating if the initial frequency for A is .8 in males and .5 in females (assume number of males equals number of females).

16.11. In a large population of random-mating animals, .84 of the individuals express the phenotype of the dominant allele (A-) and .16 express the phenotype of the recessive (*aa*). Calculate the amount of change in gene frequency in the first generation under .05 selection pressure against the *aa* phenotype.

16.12. In a large population of random-mating animals, .84 of the individuals express the phenotype of the dominant allele (*A*-) and .16 express the phenotype of the recessive (*aa*). Under complete selection ($s = 1$) against the *A*- phenotype, what proportions of *AA*, *Aa*, and *aa* would be expected in the next generation?

16.13. In a large random-mating population in which *A* is completely dominant, the gene frequency of *A* $(1 - q) = .7$ and *a* $(q) = .3$. Calculate the rate of change of *q* per generation if selection is against *aa* with $s = .005$.

16.14. In a large random-mating population, if the mutation rate of *A* to *a* is .00001 and the reverse mutation rate *a* to *A* is .000001 and neither *A* nor *a* has a selective advantage, at what gene frequency level would *A* and *a* be at equilibrium?

16.15. In a large random-mating population, if an unfavorable recessive recurrent mutation of *A* to *a* should occur at a net rate (*u*) of .000001 and selection against the phenotype *aa* is $s = .01$, what would be the frequency of *a* at equilibrium?

16.16. In a large random-mating population, an unfavorable recessive, recurrent mutation of *A* to *a* occurs at a rate of $u = .000001$ and $s = .01$. What will be the frequency of *A* at equilibrium?

16.17. Why is random genetic drift effective in altering gene frequencies only in small breeding populations?

16.18. Why are the mechanisms of meiotic drive different in males and females?

16.19. Why has the polymorphism between Hb_β^A and Hb_β^S (for sickle-cell anemia) been maintained in African populations?

16.20. In a small African village, the frequency (*q*) of the sickle-cell allele (Hb_β^S) is .10. What does this fact suggest about the incidence of falciparum malaria in this area?

16.21. If the frequency of the Hb_β^S allele is .10 and there is no more malaria ($S_A = 0$), what will the frequency of Hb_β^S be in two generations?

16.22. On the oceanic island of St. Helena in the South Atlantic, where winds are strong and nearly constant, a number of species of wingless insects have become established. (a) How might the wingless insects have developed? Teissier carried out the following experiment with *Drosophila* in a windy part of the Mediterranean region: Equal numbers of flies with normal wings and those with vestigial wings were placed in open but protected dishes well supplied with food on the roof of a building. At the end of several generations, most of the flies in the dishes had vestigial wings. (b) How could

this change in proportion of long-wing and vestigial-winged flies be explained?

16.23. Discuss why two linked genes do not attain equilibrium as rapidly as two unlinked genes.

16.24. Explain how the onset of migration into a previously isolated population can cause extinction of that population.

16.25. How does inbreeding affect (a) gene frequency and (b) heterozygosity? (c) What are the general effects of inbreeding and outbreeding among plants or animals that are ordinarily cross-fertilized?

16.26. (a) Why do self-fertilizing plants such as peas and beans not lose vigor through continued inbreeding? (b) If it were known that self-fertilization resulted in decreased vigor in a certain type of plant but the natural method of breeding was not known, what speculation might be drawn as to the natural system of mating?

16.27. Why are pure lines valuable for (a) hybridization experiments and (b) experiments designed to compare genetic and environmental variation?

16.28. (a) Garden peas heterozygous for three genes, *g* for seed color (yellow or green), *y* for pod color (green or yellow), and *c* for pod shape (inflated or constricted), are allowed to self-fertilize, and their descendants are self-fertilized for many generations. What phenotypic pure lines would be expected to develop? (b) When pure lines are obtained, would the self-fertilizing individuals be expected to produce anything but homozygous progeny? Explain.

16.29. If 2000 peas heterozygous (*Dd*) for height were self-fertilized for five generations, about how many plants would be expected to be heterozygous? (Assume that there is no differential adaptive value and that each parent produced one offspring, so that 2000 plants will be present in each generation.)

16.30. A large number of fruit flies, all heterozygous for gene *a*, were allowed to mate at random and their descendants also mated at random for 10 generations in a population cage. What proportion of *AA*, *Aa*, and *aa* would be expected, assuming no differential adaptive value?

16.31. Evaluate the following systems of mating with reference to their relative efficiency in producing homozygosis: self-fertilization, brother and sister mating, double first cousins, first cousins, second cousins, and random mating in a large population.

16.32. In some herds of beef cattle, dwarf calves occur. A recessive gene is responsible for the abnormality. (a) How could this condition have developed? (b) What measures might be taken to avoid the financial losses that come from the production of dwarfs?

16.33. If genes *A*, *B*, *C*, and *D* produce vigor in maize, evaluate the efficiency with which superior hybrid seed could be produced from the following crosses and give reasons for evaluation: (a) *abCD/abCD* × *ABcd/ABcd*, (b) F₁ × F₁, and (c) F₁ from *aBCD/aBCD* × *AbCd/AbCD* × F₁ from *ABcD/ABcD* × *ABCd/ABCd*?

16.34. In certain breeding stocks of maize, assume that gene *A* for increased vigor is closely linked with a recessive gene *w*, which makes the plant weak, and that gene *B* for another quality also results in increased vigor, but is closely linked with a recessive gene *l* for low viability. (a) What combination would be most efficient for production of vigor? (b) What kind of mating would produce the desired combination? (c) What would be the likelihood of obtaining a pure breeding strain with superior vigor?

16.35. A trait dependent on a recessive gene (*a*) occurs in the general population at the rate of 1 in 10,000 people. Note the following pedigree:

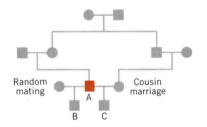

(a) What would be the chance of this trait occurring in A's child if he married at random and if the trait was not known to occur in his family? (b) What is the inbreeding coefficient for C? (c) What is the chance that C will express the trait? (d) If A's grandmother was known from family history to be a carrier for *a* and A's grandfather was known not to carry the gene, what is the chance that C will express the trait? (e) If A's grandmother was known to have expressed the trait and his grandfather was known not to carry the gene, what is the chance that C will express the trait?

16.36. Distinguish between (1) Darwinism and evolution, (2) views of Darwin and Goldschmidt, and (3) random genetic drift and molecular drive.

16.37. Assume that pattern baldness in humans is controlled by an allele, *B*, of a single gene. Further assume that this allele is dominant in males and recessive in females, that is, is "sex-influenced." If a bald man and a nonbald woman in a randomly mating population in which 51 percent of the men are bald have a child, what is the probability that the child will eventually become bald?

SEVENTEEN

A peacock demonstrating his plumage. (Frederick D. Bodin/Stock, Boston)

GENETICS OF BEHAVIOR

Genetic mechanisms for structural and numerical traits have been investigated more successfully than those for behavioral characteristics. The **complexity of behavioral traits** and the difficulty of studying them at the molecular level are the main reasons for the difference. Furthermore, behavioral characteristics of any animal develop under the joint, tightly entwined effects of heredity and environment. DNA in the genome determines the individual's physiological, structural, and behavioral potentials, but not all of the potentials are realized in the developing individual. **Behavior genetics is concerned with the effects of genotype on behavior and with the role that genetic differences play in the determination of behavioral differences in a population.**

A fundamental question in the study of the relationship between genes and behavior is whether heredity directly affects behavior or merely defines the stage on which behavioral patterns may be molded by environmental factors. Biologists and psychologists have taken opposite views on this issue, but now the two groups recognize that both heredity and environment are relevant for virtually all behavior patterns. The question of direct versus indirect effects of heredity can be approached by recognizing the sequential levels of organization of behavior patterns in developing animals. **Environmental factors are interwoven with inheritance** mechanisms at every point in development. The problem is to recognize the role of each and to evaluate their relative importance in specific situations.

In some cases, genetic programing simply states that animals have specific learning capabilities at certain developmental periods. Zebra finches, for example, have inherited physical equipment for the learning process, but when they are isolated from species members early in life, they can never learn to distinguish males and females of their own species. Other aspects of behavior, such as the calls of some birds, are less learned and more inherited. An incubator-raised chicken, for example, having never heard the sound of a hen, will still mature and produce notes typical of other chickens. In other cases, it has been shown that the environment in which an animal matures can drastically affect certain aspects of its adult behavior. H. F. Harlow and his associates, for example, have shown that female Rhesus monkeys separated at birth from their own mothers and deprived of early interaction with other monkeys were deficient in basic patterns of maternal behavior, social play, and sexual activity. Again, structural and physiological characteristics that provide the tangible framework for these behavior patterns must follow the DNA "blueprints" for the organism, but an important component of the behavior itself must be learned through contact with the environment.

Basically, then, there are **ranges of modifiability** that are inherited, with the segregation of genes and the forces of natural selection accounting for observable individual differences in behavior patterns within a given population.

GENETIC MECHANISMS

In contrast to the more common complex meshing of environment and heredity, the genetic mechanisms of some behavior traits have proved surprisingly simple. Some specific examples depend on a few genes and respond to a limited range of environmental stimuli.

W. C. Rothenbuhler, for example, has found evidence that an interesting behavior pattern in honeybees is controlled by two pairs of recessive alleles in simple Mendelian fashion. Two different races of bees differ in their "hygienic" behavior. Worker bees from the Brown line, a hygienic race, open compartments in the hive that contain pupae dead from American foulbrood and remove the dead (Fig. 17.1a). Those from the Van Scoy line, a nonhygienic race, leave the dead pupae in the closed compartments and thus allow the infectious agent for American foulbrood, *Bacillus larvae,* to spread in the colony (Fig. 17.1b).

Rothenbuhler crossed the two races and obtained F_1 worker bees, all of which were nonhygienic. When F_1 drones (from unfertilized F_1 gametes) were backcrossed to hygienic queens, four kinds of backcross colonies were obtained in about equal proportions: (1) hygienic bees, (2) bees that

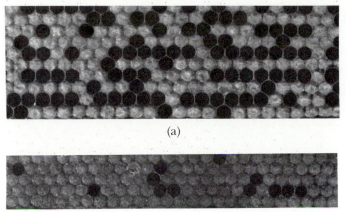

(a)

(b)

FIGURE 17.1

Combs of brood from (a) *Brown, hygienic colony and* (b) *Van Scoy, nonhygienic colony. In the Brown colony about two days before brood emergence, many individuals were missing from the spore-inoculated rows, but all brood remaining in the comb were found to be alive. In the Van Scoy colony, most of the brood were present but, when the cells were uncapped in the laboratory, many individuals in the spore-inoculated rows were dead of American foulbrood. (From* Am. Zool. *4:111–123, 1964. Photos courtesy of W. C. Rothenbuhler.)*

opened cells but did not remove the dead pupae, (3) bees that did not open cells but removed dead pupae when cells were opened by the beekeepers, and (4) nonhygienic bees. If a single recessive gene *u* controls the behavior pattern of uncapping cells and another single recessive gene *r* controls the behavior pattern for removal of dead pupae, the results obtained by Rothenbuhler can be explained by Mendelian independent assortment. The dihybrid cross is reconstructed in Fig. 17.2. Although many genes and environmental influences may be assorted with the complicated neuronal mechanism underlying the behavior patterns of uncapping and removing, the response **threshold** is determined primarily by the **single alleles.**

In this example, two colony-behavior characteristics that appeared together in parental lines separated in the backcross colonies. This manifestation of genetic segregation is expected among individuals but not among colonies. It can occur only when members of a colony are genetically similar. Colonial bees can be genetically similar when the mother (the queen) is highly inbred and has mated with a single haploid drone to produce the colony. An inbred queen–single drone mating performed by artificial insemination facilitates the study of colony behavior.

In the course of studying the 63 colonies of honeybees in their hygienic behavior experiment, Rothenbuhler and his associates made observations on several other behavioral characteristics of bees. Most obvious and impressive was stinging behavior. In the course of 98 visits to seven Van Scoy colonies, the beekeepers were stung only once, whereas the same number of visits to seven Brown colonies brought 143 stings. The first proposed explanation was that keeping the brood nest free of dead larvae and defending the colony against beekeepers were manifestations of the same general characteristics —a high level of vigor in the worker bees. It was suggested that the same genes might explain the two behavior patterns. If this were the explanation, the hygienic colonies among the backcross colonies would also be the stingers. This was not the case. Only a very few of the bees in the hygienic backcross colonies were stingers. The stinging behavior in all of the 29 backcross colonies indicated that more than two pairs of genes were involved.

The genetic aspects of such observable elements of behavior are believed to be determined basically by the same mechanisms that function for the more tangible physical traits that Mendel and others have described. However, the genetic bases for behavior patterns in animals are characteristically difficult to confirm through experimental procedures.

INHERITANCE AND LEARNING IN BEES

W. P. Nye and O. Mackensen have shown that honeybee preference for alfalfa pollen, *Medicago*

P	Hygienic queens (uncapping and removing)			Nonhygienic drones
P	Genotypes $uurr$	$\times$		u^+r^+
P	Gametes ur			u^+r^+
F_1	Genotype	u^+ur^+r		
F_1	Phenotype	Nonhygienic		
F_1	Gametes (drones) u^+r^+, ur, u^+r, ur^+ used to inseminate $uurr$ queens			

From backcross:

Females	$uurr$	uur^+r	u^+urr	u^+ur^+r
Phenotypes	Hygienic	Uncapping, no removal	No uncapping, removal	Non-hygienic
Ratio	1 colony	1 colony	1 colony	1 colony

FIGURE 17.2

Cross between hygienic and nonhygienic bees and a backcross of four kinds of F_1 drones (from unfertilized haploid F_1 gametes) with uurr *queens; u is the gene for uncapping, r is for removing. The bees (Hymenoptera) used in the P cross are colonial and inbred. Many hygienic worker females and several queens with the same genotype (uurr) were produced. Sex is determined by haploidy (nonfertilization) or diploidy (fertilization). By a timing process, queens can determine which eggs are fertilized and thus become females (either workers or queens depending on diet), and which are unfertilized and thus become drones. The nonhygienic drones in the P cross came from a different race. Fully heterozygous (u$^+$ur$^+$r) F_1 females were produced by artificial insemination. The four kinds of gametes (u$^+$r$^+$, ur, u$^+$r, ur$^+$) produced by F_1 queens that were not fertilized became the drones that were used to inseminate P queens (uurr). Four kinds of backcross female progeny were produced in equal proportion—1 uurr: 1 uurr$^+$: 1 u$^+$urr: 1 u$^+$ur$^+$r, all phenotypically different in a behavior characteristics. (From experiments by Rothenbuhler.)*

sativa, depends to a large extent on genetic determinants that respond to **selective breeding.** These investigators observed that some colonies of honeybees collected a much higher percentage of alfalfa pollen than did others. Separate inbred lines were developed from colonies that showed either a high or a low preference for collecting alfalfa pollen. At the end of the eighth generation of inbreeding, the high and low lines had been completely separated (Fig. 17.3). Subsequent hybridization of the bees from the two lines produced bees that were intermediate between the high and low lines. Because the preference for alfalfa pollen can be changed markedly by selection, and the trait follows a predictable pattern based on Mendelian inheritance, preference for alfalfa pollen is presumed to have a high hereditary component.

In later experiments, Nye took high and low alfalfa pollen–collection lines of bees to Donnelly, Idaho, where alsike clover was flowering. Nearly all the collection of all lines was clover pollen, and no significant difference occurred among lines. Inherited behavior for collecting alfalfa pollen, in contrast, was found to be specific to alfalfa pollen.

Both inheritance and learning are involved in the complex social patterns of several Hymenoptera, such as the honeybee, *Apis mellifera,* that make use of chemical, optical, and aural signals. Effectiveness of the activities of bees in and around a hive, as well as in foraging areas, depends on the exchange of information among individual bees. Communication symbols for **distance and direction of food source are mostly learned.** Even so, in some groups, this behavior is so stereotyped and well

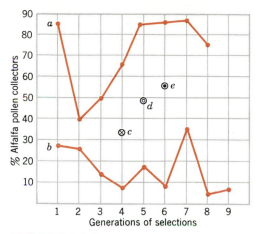

FIGURE 17.3

Results of selection experiments for high alfalfa pollen collectors (HAPC), low alfalfa pollen collectors (LAPC), and results from outcrosses and backcrosses. (a) HAPC queens and drones selected in each of 8 generations. (b) LAPC queens and drones selected in each of 9 generations. (c) Hybrid colonies resulting from crosses between LAPC queens and HAPC drones from generation 3. (d) Results from backcrosses between queens from c (hybrid colonies) and drones from generation 4 HAPC queens. (e) Results from outcrosses between HAPC queens from generation 5 and drones from a commercial stock. The dip from 85 percent to 40 percent between generation 1 and 2 in a is presumed to be the result of heterozygosity of the genes involved in the parents selected to initiate the HAPC line. Both HAPC and LAPC lines are expected to become more homozygous as inbreeding continues. Environmental as well as genetic factors are known to influence pollen collection. In the spring of the seventh year, frosts destroyed most of the early plants on which LAPC bees usually forage, and the bees were forced to move to more abundant alfalfa. (Data from W. P. Nye.)

established that it can serve as well as a morphological characteristic for distinguishing species.

Although the basic communication system is the same for all *Apis mellifera*, different "dialects" have developed in different races. Members of an Italian race, for example, have a slower dancing rhythm than those of an Austrian race. When bees of these two races are mixed, they misunderstand each other. An Austrian bee receiving information from an Italian bee about food 100 meters from the nest will fly 120 meters because she interprets the "Italian dialect" on the basis of her Austrian knowledge. Conversely, the Italian bee will fly 80 meters when given the information for 100 meters by the Austrian forager. Nevertheless, when M. Lindauer compared the communication systems of three different species of *Apis*, he observed wider differences in behavior as well as in structural characteristics between the species than between races.

EXPERIMENTAL BEHAVIOR GENETICS

Behavior genetics is a challenging field for experimental work. Several investigators, including the late T. Dobzhansky (Fig. 17.4), S. Benzer, J. Adler, C. Kung, M. Nirenberg, G. Stent, M. Levinthal, S. Brenner, and C. M. Woolf, who have already distinguished themselves in other aspects of experimental genetics, have moved into the area of behavior ge-

FIGURE 17.4

Theodosius Dobzhansky, eminent researcher, writer, and teacher in the fields of genetics and evolution. He has made basic contributions to population genetics as well as behavior genetics through his extensive studies on Drosophila pseudoobscura.

netics. Each investigator has an "ideal" experimental material with properties such as large size, rapid reproduction, few neurons, and wide behavior patterns. Organisms such as *E. coli,* phycomycetes, paramecia, nematodes, rotifers, fresh-water crustacea (*Daphnia*), leeches, *Drosophila,* and mice have been chosen for basic investigations on behavior. Examples from *E. coli* and *Drosophila* will be cited here.

E. COLI CHEMOTAXIS

In the bacterium *E. coli,* behavior patterns are comparatively simple. **Response (taxis)** to chemicals, light, gravity, and temperature have been studied by different investigators. J. Adler, for example, investigating chemotaxis, has shown that the mechanics include detection of the chemical (attractant) by a chemoreceptor and transmission to an effector that produces a flagellar response, the swimming of the organism (Fig. 17.5). By isolating mutants, genes resposible for each step in the process have been postulated. Three *che* mutants, for example, render the bacteria nonchemotactic to specific chemicals: galactose, aspartate, and serine. Eight *fla* (nonflagellar) genes interfere with swimming activity.

Movements of the bacteria are followed with a tracking microscope and the data are fed into the computer. When no chemical gradient is present in the medium, the **movement is haphazard.** A bacterium moves in one direction (run) for 2–4 seconds, tumbles around (twiddles), makes another short run, twiddles, and so on, without appreciably changing location (Fig. 17.6). If an **attractant** is present, the runs are longer in the **direction of** the **gradient.** At the end of a run, the organism twiddles, takes a

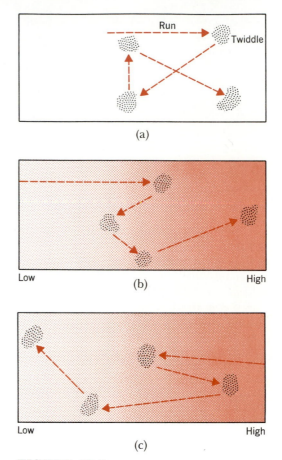

FIGURE 17.6
Movement of E. coli *in a liquid medium.* (a) *No chemical gradient in medium;* (b) *gradient for attractant;* (c) *gradient for repellent.*

short run, and repeats the activity until the orientation is again in the direction of the gradient. At this time a long run occurs. If a **repellent** instead of an attractant is present, long runs, twiddles, and short runs occur, but the **direction** is toward the **low concentration** of the chemical. Several theories, not always exclusive of each other, have been advanced to explain the cause of twiddling: (1) a diffusible substance, (2) a change in membrane potential, (3) a change in membrane configuration, and (4) a gene-controlled enzyme that influences chemotaxis. Such mutations have now been obtained. Some mutations

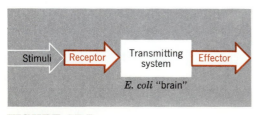

FIGURE 17.5
Receptor-effector system in E. coli.

block the twiddle and others make bacteria twiddle all the time.

DROSOPHILA BEHAVIOR GENETICS

Drosophila adults are much more complex than *E. coli.* Their brain contains some 10^5 neurons that are arranged in nerve tracts. For such complex animals, the *Drosophila* are small; they also reproduce within a few days, live on simple food in the laboratory, and are well-known genetically. Their chromosomes have already been extensively mapped, and behavior genes have been detected in particular areas of the chromosomes. On the X chromosome of *D. melanogaster,* for example, S. Benzer and associates have mapped several behavior loci, including visual receptors, stress receptors, and "wings-up" mutants. Developmental studies by these investigators have related cells in the blastocyst stage with imaginal disks that give rise to eyes, antennae, legs, and wings in the adult flies. Mosaics (gynandromorphs, see Chapter 4) are being used to relate structural parts of the fly to cells in the blastoderm and to genes in the zygote. The wings-up mutation, for example, has been related to a developmental abnormality of the muscles. The "drop-dead" mutation is related to a defect in the brain. Sex and learning behavior are analyzed in component parts through mutant blocks and are then related to **sequential steps in gene activity.**

RESPONSE TO LIGHT AND GRAVITY IN *DROSOPHILA*

C. M. Woolf has carried out hybridization studies with strains of *D. pseudoobscura* that differ in phototactic (i.e., response to light) and geotactic (response to gravity) behavior. These behavior traits were measured by running virgin females and males separately through Hirsch–Hadler classification mazes (Fig. 17.7). Matings were then made among the flies that earned particular classifications, to obtain evidence with respect to **genetic mechanisms.**

Hirsch–Hadler mazes provide 15 downward or upward choices, that is, toward or away from light. Eventually, a fly enters one of 16 different collecting tubes. The No. 1 collecting tube is en-

tered by flies that make 15 choices upward or away from light. The No. 16 collecting tube will be entered if the 15 opposite choices are made. If an equal number of downward and upward choices are made, the fly enters No. 8 or 9 collecting tube. The number of flies reaching each collecting tube can be used to calculate the reaction of particular flies to light. A completely neutral strain has an expected mean of 8.5. The strongest possible positive strain would have an expected mean of 16. The strongest possible negative strain would have an expected mean of 1.

The strains of flies used by Woolf had been strengthened through selection as part of the extensive research program of T. Dobzhansky and his associates. Beginning with strains that were phototactically and geotactically neutral, the Dobzhansky group had selected flies with positive and negative response to light and gravity. Strain 25, for example, had become strongly negative. Observations of this kind provided evidence for genetic variance in these strains of *D. pseudoobscura.* The rate of divergence under artificial selection indicated that the genetic component consisted of polygenes. When selection was relaxed, the positive and negative strains reverted to the neutral state, suggesting that the genes responsible for the positive and negative phototactic behavior were maintained in the heterozygous state in natural and laboratory populations by natural selection.

Woolf crossed females from strain 25 with males from strain 27 and also crossed females from strain 27 with males from strain 25. As indicated in Fig. 17.8, the 311 female flies from strain 25 had a mean of 12.28. The 321 male flies from strain 27 chosen for this same experiment had a mean of 4.88. The means of the F_1 progeny and of the F_2 (that were obtained from mating F_1 females with F_1 males) were intermediate between the parental strains. The particular arrangement of parental chromosomes in the F_1 and F_2 male and female progeny are indicated in Fig. 17.8, along with the histograms. **Phototactic behavior** scores are characterized as being relatively consistent from replication to replication and relatively similar in males and females of the F_1 and F_2 generations.

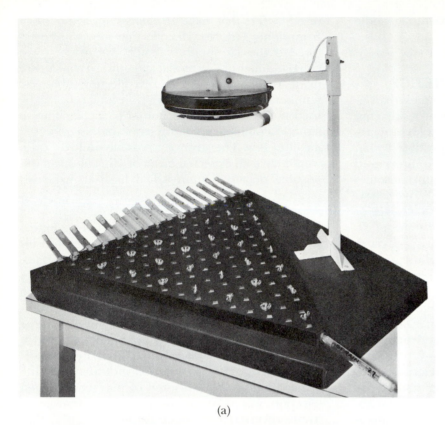

(a)

FIGURE 17.7

Hirsch-Hadler classification mazes designed to select **Drosophila** *for positive and negative geotactic and phototactic behavior. (a) Phototactic behavior maze. (b) Geotactic behavior maze. In each generation, virgin females and males are introduced separately at the maze entrance. The flies have a number of choices, indicating positive or negative behavior in the maze. Those showing each response are collected at the exit and used for progenitors of the next generation. Beginning with strains that were phototactic and geotactic and neutral, it was possible to obtain positive and negative strains after several generations of selection. (Courtesy of C. M. Woolf, Arizona State University.)*

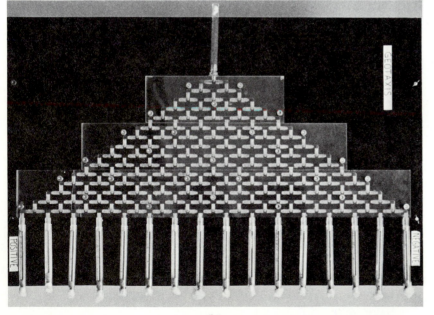

(b)

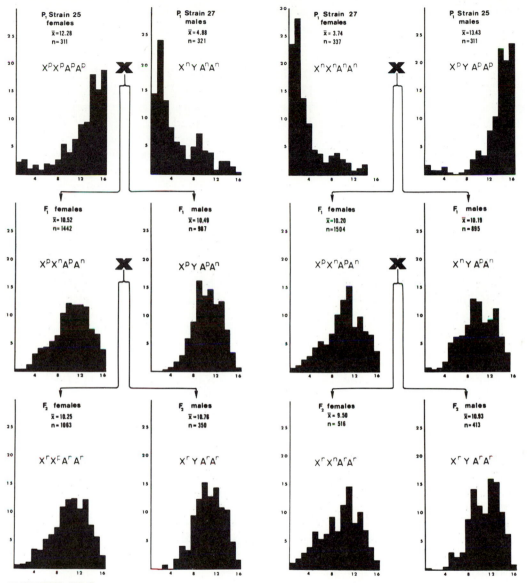

FIGURE 17.8

Distribution of phototactic scores (pooled data) in the P_1, F_1, and F_2 generations for the parental mating between (left) female strain 25 (positive) $\times$ male strain 27 (negative), and (right) female strain 27 (negative) $\times$ male strain 25 (positive). In this series of histograms, the area of each rectangle represents the percentage of flies occurring in each of the 16 collecting tubes of the Hirsch-Hadler classification maze. The X chromosome is symbolized by X, the Y chromosome by Y, and the set of autosomes by A. The p and n indicate whether the X chromosome and autosomes are from the positive or negative strain, respectively. The r designates that the X chromosome and autosomes are recombinant chromosomes as a result of crossing over a segregation. (Courtesy of C. M. Woolf.)

The position of the F_1 means between the values of the parental means indicates that **polygenic inheritance** (groups of genes controlling quantitative traits, see Chapter 15) is involved. A consistent shift, however, was observed in the F_1 and F_2 means toward that of the parental positive strain (strain 25), suggesting that some type of nonadditive influence is present. The similarity of the means of F_1 males and F_1 females indicated that the X chromosome had little or no effect on phototactic behavior. Therefore, the responsible genes are largely located in the autosomes. The variances of the F_1 and F_2 scores are similar. This exception to classical polygenic inheritance, in which a wider spread is usually observed in the F_2 as compared with the F_1, apparently results from the **low heritability** (Chapter 15) of phototactic behavior in *D. pseudoobscura*, the lack of homozygosity of the parental strains, and perhaps some unknown interaction between genotype and environment.

In other comparable experiments carried out in other types of mazes and designed to determine geotactic behavior, Woolf's results were quite different. As in the studies of phototactic behavior, the consistency of the replications indicated the importance of the genotype for these types of behavior under controlled laboratory conditions in spite of the low heritability. Female progeny in the F_1 generation were essentially intermediate between the two parents, but the F_1 males were strongly positive for **geotactic behavior.** This is the pattern of crisscross inheritance expected when the **X chromosome** carries the gene or genes involved in the transmission of the trait. This same pattern was held through the F_2, F_3, and backcross results, giving evidence that the X chromosome is strongly involved in the transmission of geotactic behavior. The evidence indicated further that the autosomes carried some genes for this trait, but that the great proportion of genes for positive geotactic behavior were located in a region of the X chromosome not readily divisible by crossing over. Females homozygous or males hemizygous for this region tended to express positive geotactic behavior. Although geotactic behavior in *Drosophila* is not controlled by single major genes, like those associated with hygienic behavior in bees,

particular regions of DNA are involved in the mechanism of this behavior characteristic.

GENETIC AND ENVIRONMENTAL INTERACTIONS IN DOGS

Dogs have personality differences even within a breed. Some are timid and others are confident; some are gentle and others are aggressive. Those that are "socialized" (allowed to interact with people) early in life function as friendly and understanding companions of humans, whereas others of the same breed (or even of the same litter) that are not given similar experiences while young may become fearful or even hostile toward people.

J. P. Scott and J. L. Fuller have made extensive observations on **genetic and environmental factors** involved in the **building** of **behavior patterns** in dogs. In some of their work, daily observations were continued until the dogs were 16 weeks of age. Every effort was made to observe the earliest manifestation of hereditary difference before the effects of experience were noted. Of course, during the first few days after birth, there was very little behavior to observe in the pups.

As soon as recognizable behavior became apparent, interaction between hereditary and environmental influences was already present. The puppies changed markedly in reactions from day to day in response to learning and increasing physical and mental maturity. The environment was optimal for learning, and innate faculties became active at **progressive developmental stages. Genetic** control was **acting on** a very **different animal** with each new developmental stage, and environmental influence was apparent.

Results obtained by Scott and Fuller were quite unexpected and significant. During the very early stages of development when behavior is minimal, genetic differences had few opportunities to be expressed. When behavior patterns did appear, however, genetically determined differences did not appear all at once early in development, to be modified by later experience. Instead, they developed under the influence of environmental factors. Scott and Fuller concluded that the raising and lowering of response thresholds is one of the most important ways behavior in dogs is affected by heredity. These studies showed that while **heredity** is an **important** factor in dog behavior, details—and sometimes the actual appearance of specific patterns—depend to varying degrees on the **individual's experience.** Futhermore, they demonstrated that at least some genetic behavior difference can be measured and compared as readily as hereditary physical differences.

Dog breeds generally have managed to retain a great deal of genetic flexibility, despite man's intensive selection over time. This was borne out by the further studies of Scott and Fuller in which 50 traits were examined in five pure breeds of dogs. Almost all the traits were significantly different among the various breeds. But a very few of them were found to breed true, as would be expected if they were controlled by single homozygous pairs of alleles. In addition, lack of correlation was observed between behavior and phenotypic "type" within each breed. In their crossbreeding work with cocker spaniels and dogs of the Basenji breed, the Basenji personality was often seen in spaniel-appearing dogs, and vice versa. However, selection has apparently produced near homozygosity for certain traits in particular breeds. Fighting behavior, for example, is almost nonexistent in the hound but is well developed in the terrier. But such instances of near homozygosity are rare.

Through selection of genetic qualities and training, remarkable behavior patterns have been made available in some breeds of dogs. The Basenji for example, commonly known as the African non-barking hound, is by nature a "scent" hunter. This dog is used in Africa to find and drive wild game. He is basically intelligent and can be taught such feats as the advanced American Kennel Club obedience program, which includes retrieving a dumbbell over a high jump. In this demonstration, the Basenji's inherent intelligence is put to a relatively artificial use.

Retrieving dogs have been selected and trained for more than 100 years in England by enthusiasts of the waterfowling sport. Several breeds exhibited some of the characteristics necessary for an excellent retriever, such as strength, moderate size, endurance, enthusiasm, aquatic ability, keen scenting ability, courage, favorable temperament, and trainability. By intercrossing the most favorable breeds —Newfoundland, setter, and spaniel—and selecting progeny, a litter of four puppies with a favorable combination of traits was obtained in 1868. From this beginning the golden retriever stock was developed. These dogs (Fig. 17.9) are light yellow in coat color. They also have aquatic ability, pleasing temperament, keen nose, tracking ability, and tenacity to retrieve under severe conditions. It is natural for a pup from this stock to want to retrieve. Training merely perfects and polishes the performance.

Breed differences in behavior are both **real** and **important.** The great variability available in

FIGURE 17.9
Golden retriever. This breed of dogs has been selected and trained for more than 100 years to produce the characteristics that make good retrievers. (Leonard Lee Rue III / National Audubon Collection—Photo Researchers.)

dogs means that it is possible to **modify** a breed markedly within a few generations **by careful selection.** Through crossbreeding, entirely new and unique combinations of behavioral traits can be created and studied.

SOCIOBIOLOGY

Groups of animals have behavior characteristics in common that involve biological mechanisms responsive to selection. Adherence to a particular social order has value when it provides that the group may act as one in behalf of the individual. On the other hand, without a certain amount of social disorder or freedom, the strengths inherent in diversity are lost. A dynamic **balance** between **order** and **disorder** is a healthy condition within living populations.

One aspect of "order" requires respect for one another's "distance" needs. Different species have different **individual distance** requirements and differ in their responses to crowding. Black-headed gulls resting in a row will space themselves at about one foot apart. Flamingos maintain about twice that much space between neighbors, and swallows re-

quire about half as much space as the gulls (Fig. 17.10). Some tortoises and hedgehogs will crowd together and make an animal pile. Inherent species differences control such behavior patterns. Individual distance or "personal space" requirements, however, may vary with individuals and with seasons. For example, virtually all bird species tend to gather in tight flocks during the winter, but as soon as the breeding season comes in the spring they disperse and strenuously defend their individual territories. **Social distance** is the farthest point an animal will stray from the group. A baboon troop is dispersed widely while feeding in the daytime, but at night the group will come together and sleep in a few adjacent trees. In this case, the acceptable social distance varies with environmental conditions.

Overcrowding, whether among human beings or other animals, can lead to a breakdown in social structure and open the gates to overaggressiveness. **Innate aggressiveness,** which may have been necessary throughout evolution to help some species to survive, can take a destructive turn under conditions of **overcrowding.** In some cases, aggression is directed toward others, but it can also become manifest in ulcers, nervous disorders, and various psychosomatic maladies.

FIGURE 17.10

Normal spacing of swallows on power lines. (Photograph by Allan D. Cruickshank/National Audubon Society Collection—Photo Researchers.)

Fortunately, aggression, like many other behavior patterns, is subject to modification—though only within genetic bounds. Aggression can be increased or decreased by purposeful education. If mice or chickens are paired with others of their kind, fighting ensues and one becomes dominant and the other subordinate. If these encounters are arranged so that a particular animal always wins, that animal becomes more and more emphatically dominant over his fellows. In psychological terms, winning reinforces further aggression. On the other hand, repeated losses make an animal submissive.

One can also train animals to be nonaggressive in other ways. Scott prevented puppies from making playful attacks on their handlers by picking them up frequently, thus rendering them helpless with their feet off the ground. This process of **passive inhibition** produced nonaggressive adults. He obtained similar results with young male mice by repeatedly stroking them at an early age. Aggressiveness is a product of inheritance, maturation, various endogenous factors, and experience. However, manifestations of aggression among nonhumans depend on the presentation of proper external stimuli, usually specific sign stimuli from other individuals of the same species.

THE EVOLUTIONARY APPROACH

In trying to trace the evolution of behavior in humans, researchers use the same general approaches that have proved productive with other species. It has been found (predictably) that some behavior patterns that serve the function of food intake are phylogenetically quite old—and the human infant shares them with many other mammals. One example is the rhythmic searching for the nipple.

The grasping reflex is also characteristic, perhaps originally serving to keep the baby attached to the mother. In premature infants, it is especially strong, and they are able to cling to an outstretched cord. Climbing movements can also be elicited in premature infants.

Observation of congenitally deaf-blind children (in effect deprivation experiments) is giving new insights into situations where human behavior

patterns are truly unlearned. Expressions of anger (stamping of feet, facial contortions), high-intensity laughter, and the rejection of strangers seem to be innate. Similarly, components of facial expressions—such as raising the eyebrows when greeting someone—are common to all human cultures that have been studied.

Inborn behavior mechanisms also seem to account for a remarkably universal desire for cover and unobstructed vision, as evidenced in modern-day life by a preference for corner and wall tables in restaurants. Culture imposes specific limits on individual distance requirements, but the basic need for space seems inborn and probably evolved.

The tendency to seek membership in some sort of group and to accept the group-dictated exclusion of others also seem to be universal human behaviorisms and are therefore likely evolutionary phenomena.

Obviously human beings have many kinds of behavior in common, despite a heavy veneer of individually learned, culturally dictated modifications. It has also been proved (e.g., as in Scott and Fuller's work with puppies) that experiments with other animals can provide valid insights into human behavior. So it seems reasonable to expect that, as more is learned about behavior as it exists today and also about how it got that way, that the future evolution of humankind may be modified in the direction of true humaneness.

SUMMARY

DNA in the genome determines the behavioral potentials of individuals, but environmental factors, which are interwoven with inheritance mechanisms, create the setting in which individuals will realize their potentials. Ranges of modifiability are inherited. Segregating genes, forces of natural and artificial selection, and direct environmental influences account for individual differences in behavior patterns within a given population. Although the genetics of most behavior patterns is complex and cannot at this time be reduced to simple Mendelian ratios, the stability and the evolutionary background of these patterns imply genetic mechanisms.

Behavior of an animal **develops** in response to interactions between inherited limitations and environmental factors.

REFERENCES

Ardrey, R. 1976. *The hunting hypothesis: a personal conclusion concerning the evolutionary nature of man.* Atheneum, New York.

———. 1974. *The social contact,* 2nd ed. Atheneum, New York.

Benzer, S. 1973. "Genetic dissection of behavior." *Sci. Amer.* 229(6):24–37.

Dobzhansky, T., B. Spassky, and F. Sved. 1969. "Effects of selection and migration on geotactic and phototactic behavior of *Drosophila.* II." *Proc. Roy. Soc. of London B.* 173:191–207.

Ehrman, L., and P. A. Parsons. 1976. *The genetics of behavior.* Sineuar Assoc., New York.

Eibl-Eibesfeldt, I. 1975. *Ethology, biology of behavior,* 2nd ed. Holt, Rinehart and Winston, New York.

Hirsch, J. (ed.). 1969. *Behavior-genetic analysis.* W. C. Brown, New York.

McClearn, G. E., and J. C. DeVries. 1973. *Introduction to behavioral genetics.* W. H. Freeman, San Francisco.

McGill, T. (ed.). 1977. *Readings in animal behavior,* 3rd ed. Holt, Rinehart and Winston, New York.

Nye, W. P. 1971. "Pollen collection from alsike clover by high and low alfalfa pollen collecting lines and by a commercial line of honeybees." *J. Apicultural Res.* 10:115–118.

Nye, W. P., and O. Mackensen. 1970. "Selective breeding of honeybees for alfalfa pollen collection: with tests on high and low alfalfa collection regions." *J. Apicultural Res.* 9:61–64.

Rothenbuhler, W. C. 1968. "Bee genetics." *Ann. Rev. Genet.* 2:413–437.

Scott, J. P., and J. L. Fuller (eds.). 1974. *Dog behavior: the genetic basis.* University of Chicago Press, Chicago.

PROBLEMS AND QUESTIONS

17.1. In general, how is animal behavior related to genetics and environmental factors?

17.2. Why has the genetics of behavior patterns developed more slowly than the genetics of other characteristics, such as size and color patterns?

17.3. Why do certain studies, such as Rothenbuhler's study of hygienic and nonhygienic bees, have particular significance in behavior genetics?

17.4. Why are comparative studies of behavior patterns especially useful for investigating the evolution of behavior?

17.5. What evidence suggests a genetic basis for the preference of some bees for alfalfa pollen?

17.6. (a) How may hormones influence behavior? (b) How and to what extent is hormone production under genetic control?

17.7. What conclusions may be drawn from the studies of Scott and Fuller on dogs concerning the relative influence of heredity and environment on behavior?

17.8. Evaluate the extent of influence of (a) genetic factors and (b) training on retrieving dogs.

17.9. Why are dogs highly reactive to selection?

17.10. Why was the dog one of the first animals to be domesticated and to come into close association with humans?

17.11. How could natural or human selection account for nonbarking in Basenji dogs even though they are capable of barking?

17.12. What are the potential difficulties and dangers in trying to transfer behavioral activities in dogs that have developed in laboratory or wild populations to individual domesticated animals?

17.13. If a behavioral trait has demonstrable survival value, is it more likely to be genetically controlled or a learned phenomenon? Why?

17.14. What characteristics should be considered when choosing suitable material for the following kinds of studies on genetics of behavior: (a) study of simple response to environmental stimuli; (b) study of location of a gene that affects a behavior mechanism.

17.15. If two strains of rats seemed to differ in their ability to solve maze-running problems, what sorts of environmental factors should be considered before concluding that the difference was genetically controlled (whether through general physiology or brain capacity)?

17.16. How may aggression be reduced in animal populations?

17.17. Why are *Drosophila* more suitable than mice as experimental material for maze studies that are designed to determine genetic mechanisms?

17.18. How could it be determined whether sex-linked polygenes control geotactic behavior in *Drosophila*?

17.19. A fully heterozygous (u^+ur^+r) nonhygienic queen bee was mated with a hygienic drone (ur). (Remember that queen bees are diploid and drones are haploid.) What genotypes and phenotypes (uncapping of cells and removal of dead pupae) might be expected in the first generation and in what proportions?

GLOSSARY

This glossary provides an introduction to some basic and recurring terms in the text. Names of chemical compounds, definitions of specialized terms, and variants of basic names have been omitted from the Glossary but are given in the Index, where they are identified by page number; the Index specifies where a term first appears and where it is defined in the book. For definitions not in the Glossary, please locate the term by referring to the Index.

Acentric chromosome. Chromosome fragment lacking a centromere.

Acrocentric. A modifying term for a chromosome or chromatid that has its centromere near the end.

Acrosome. An apical organelle in the head of the sperm.

Adaptation. Adjustment of an organism or a population to an environment.

Additive factors. Polygenes affecting the same trait, with each enhancing the phenotype.

Adenine. A purine base found in RNA and DNA.

Agglutinin. An antibody in blood plasma that brings about clumping (agglutination) of blood cells that carry an incompatible agglutinogen.

Agglutinogen. An antigen carried by red blood cells that reacts with a specific agglutinin in the plasma and causes clumping of the cells. When a specific antigen is injected into an animal body, it stimulates the production of a corresponding antibody.

Albinism. Absence of pigment in skin, hair, and eyes of an animal. Absence of chlorophyll in plants.

Alcaptonuria. An inherited metabolic disorder. Alcaptonurics excrete excessive amounts of homogentisic acid (alcapton) in the urine.

Allele (allelomorph; *adj.,* **allelic, allelomorphic).** One of a pair, or series, of alternative genes that occur at a given locus in a chromosome: one contrasting form of a gene. Alleles are symbolized with the same basic symbol (e.g., D for tall peas and d for dwarf). (See **Multiple alleles.**)

Allopolyploid. A polyploid having chromosome sets from different sources, such as different species. A polyploid containing genetically different chromosome sets derived from two or more species.

Allosteric effect. Reversible interaction of a small molecule with a protein molecule, causing a change in shape of the protein and a consequent alteration of the interaction of that protein with a third molecule.

Allotetraploid. An organism with four genomes derived from hybridization of different species. Usually, in forms that become established, two of the four genomes are from one species and two are from another species.

Amino acid. Any one of a class of organic compounds containing the amino (NH_2) group and the carboxyl (COOH) group. Amino acids are building blocks of proteins. Alanine, proline, threonine, histidine, lysine, glutamine, phenylalanine, tryptophan, valine, arginine, tyrosine, and leucine are among the common amino acids.

Amniocentesis. A procedure for obtaining amniotic fluid from a pregnant woman. Chemical contents of the fluid are studied directly for the diagnosis of some diseases. Cells are cultured, and metaphase chromosomes are studied for detection of chromosomal irregularities (e.g., the Down syndrome).

Amnion. The thin membrane that lines the fluid-filled sac in which the embryo develops in higher vertebrates.

Amniotic fluid. Liquid contents of the amniotic sac of higher vertebrates containing cells of the embryo (not of the mother). Both fluid and cells are used for diagnosis of genetic abnormalities of the embryo or fetus.

Amphidiploid. A species or type of plant derived from doubling the chromosomes in the F_1 hybrid of two species; an allopolyploid. In an amphidiploid the two species are known, whereas in other allopolyploids they may not be known.

Anaphase. The stage of mitosis or meiosis during which the daughter chromosomes pass from the equatorial plate to opposite poles of the cell (toward the ends of the spindle). Anaphase follows metaphase and precedes telophase.

Androgen. A male hormone that controls sexual activity in vertebrate animals.

Anemia. Abnormal condition characterized by pallor, weakness, and breathlessness, resulting from a deficiency of hemoglobin or a reduced number of red blood cells.

Aneuploid (heteroploid). An organism or cell having a chromosome number that is not an exact multiple of the monoploid (n) with one genome, that is, hyperploid, higher (e.g., $4n + 1$); or hypoploid, lower (e.g., $4n - 1$).

Antibody. Substance in a tissue or fluid of the body that acts in antagonism to a foreign substance (antigen).

Anticodon. Three bases in a transfer RNA molecule that are complementary to the three bases of a specific codon in messenger RNA.

Antigen. A substance, usually a protein, introduced into a living organism that elicits antibody formation.

Antihemophilic globulin. Blood globulin that reduces the clotting time of hemophilic blood.

Ascospore. One of the spores contained in the ascus of certain fungi such as *Neurospora.*

Ascus (*pl.,* **asci**). Reproductive sac in the sexual stage of a type of fungi (Ascomycetes) in which ascospores are produced.

Asexual reproduction. Any process of reproduction that does not involve the formation and union of gametes from the two sexes.

Asynapsis. The failure or partial failure in the pairing of homologous chromosomes during the meiotic prophase.

Atavism. Reappearance of an ancestral trait after several generations. A gene comes to expression after a period of nonexpression because of recessiveness or other masking effects. (See **Reversion.**)

ATP. Adenosine triphosphate: an energy-rich compound that promotes certain activities in the cell.

Autogamy. A process of self-fertilization within one undivided cell, resulting in homozygosity. This occurs in *Paramecium.*

Autopolyploid. A polyploid that has multiple and identical or nearly identical sets of chromosomes (genomes). A polyploid species with genomes derived from the same original species.

Autoradiograph. A record or photograph prepared by labeling a substance such as DNA with a radioactive material such as tritiated thymidine and allowing the image produced by decay radiations to develop on a film over a period of time.

Autosome. Any chromosome that is not a sex chromosome.

Auxotroph. A mutant organism (bacterium) that will not grow on a minimal medium but requires the addition of some growth factor.

Backcross. The cross of an F_1 hybrid to one of the parental types. The offspring of such a cross are referred to as the backcross generation or backcross progeny. (See **Testcross.**)

Bacteriophage. Virus that attacks bacteria. Such viruses are called bacteriophages because they destroy their bacterial host.

Balanced lethal. Lethal genes on the same pair of chromosomes that remain in repulsion because of close linkage or crossover suppression. Only heterozygotes for the gene pairs survive.

Balanced polymorphism. Two or more types of individuals maintained in the same breeding population.

Basal body. Small granule to which a cilium or flagellum is attached.

Base analog. Unnatural purine or pyrimidine base, differing slightly from the normal base, that can be incorporated into nucleic acid.

Binomial expansion. Exponential multiplication of an expression consisting of two terms connected by a plus ($+$) or minus ($-$) sign, such as $(a + b)^n$.

Biometry. Application of statistical methods to the study of biological problems.

Biotype. Distinct physiological race or strain within morphological species. A population of individuals with identical genetic constitution. A biotype may be made up of homozygotes or of heterozygotes, of which only the former would be expected to breed true.

Bivalent. A pair of synapsed or associated homologous chromosomes that may or may not have undergone the duplication process to form a group of four chromatids.

Blastomere. Any one of the cells formed from the first few cleavages in animal embryology.

Blastula. A form of early animal embryo following the morula stage; typically, a single-layered, hollow ball stage.

Carcinogen. A substance capable of inducing cancer in an organism.

Carrier. An individual who carries a recessive gene that is not expressed (i.e., obscured by a dominant allele).

Catabolite repression. Decreased rates of transcription of certain operons that specify enzymes involved in catabolic pathways (such as the *lac* operon) in the presence of glucose.

Centriole. Central granule in many animal cells that appears to be the active principle for spindle formation in mitosis.

Centromere. Spindle-fiber attachment region of a chromosome.

Character (*contraction of the word* **characteristic**). One of the many details of structure, form, substance, or function that make up an individual organism. The Mendelian characters represent the end products of development, such as tall garden peas.

Chemotaxis. Attraction or repulsion of organisms by a diffusing substance.

Chiasma (*pl.,* **chiasmata**). A visible change of partners or crossover in two of a group of four chromatids during the first meiotic prophase. In the diplotene stage of meiosis, the four chromatids of a bivalent are associated in pairs, but in such a way that one part of two chromatids is exchanged. This point of "change of partner" is the chiasma.

Chimera (animal). Individual derived from two embryos by experimental intervention.

Chimera (plant). Part of a plant with a genetically different constitution as compared with other parts of the same plant. It may result from different zygotes that grow together or artificial fusion (grafting); it may also be periclinal, with parallel layers of genetically different tissues, or sectorial.

Chloroplastid. Green structure in plant cytoplasm that contains chlorophyll and in which starch is synthesized. A mode of cytoplasmic inheritance, independent of nuclear genes, has been associated with these cytoplasmic structures.

Chromatid. One of the two identical strands resulting from self-duplication of a chromosome during mitosis or meiosis.

Chromatin. The deoxyribonucleohistone in chromosomes; originally named because of the readiness with which it stains with certain dyes (chromaticity).

Chromatography. A method for separating and identifying the components from mixtures of molecules having similar chemical and physical properties.

Chromocenter. Body produced by fusion of the heterochromatin regions of the autosomes and the Y chromosome in salivary gland preparations of certain *Diptera.*

Chromomeres. Small bodies, described by J. Belling, that are identified by their characteristic size and linear arrangement on the chromosome thread.

Chromonema (*pl.,* **chromonemata).** An optically single thread forming an axial structure within each chromosome.

Chromosome aberration. Abnormal structure or number of chromosomes; includes deficiency, duplication, inversion, translocation, aneuploid, polyploid, or any change from normal pattern.

Chromosome banding. Staining of chromosomes in such a way that light and dark areas occur along the length of the chromosomes. Lateral comparisons identify pairs. Each human chromosome can be identified by its banding pattern.

Chromosomes. Nucleoprotein bodies, which are dark-staining with basic dyes, microscopically observable in the cell during cell division. They carry the genes that are arranged in linear order. Each species has a characteristic chromosome number.

Cilium (*pl.,* **cilia;** *adj.,* **ciliate).** Hairlike locomotor structure on certain cells; a locomotor structure on a ciliate protozoan.

cis-**arrangement.** See **Coupling.**

Cistron. A unit of function in DNA. One DNA cistron specifies one polypeptide chain in protein synthesis.

Clone. All the individuals derived by vegetative propagation from a single original individual.

Codominant genes. Alleles, each of which produces an independent effect when heterozygous.

Codon. A set of three adjacent nucleotides that will code one amino acid (or chain termination).

Coefficient. A number expressing the amount of some change or effect under certain conditions (e.g., coefficient of inbreeding).

Coenzyme. A substance necessary for the activity of an enzyme.

Coincidence. The ratio of observed double crossovers to expected doubles calculated on the basis of probability for independent occurrence and expressed as a decimal fraction.

Colchicine. An alkaloid derived from the autumn crocus that is used as an agent to arrest spindle formation and interrupt mitosis.

Colinearity. The state in which the sequence of nucleotides in a cistron corresponds with the order of the amino acids in the polypeptide it specifies.

Competence. Ability of a bacterial cell to incorporate DNA and become genetically transformed.

Complementation test (*cis-trans* **test).** Introduction of two mutants into the same cell to determine whether the two mutations occurred in the same gene. If the mutations are nonallelic, the genotype of the hybrid may be symbolized (ab^+/a^+b). The wild phenotype will be expressed, since each mutant "makes up for" or complements the defect in the other.

Concordance. Identity of matched pairs or groups for a given trait; for example, identical twins both expressing the same genetic syndrome.

Conidium (*pl.,* **conidia).** An asexual spore produced by a specialized hypha in certain fungi.

Conjugation. Union of sex cells (gametes) or unicellular organisms during fertilization; in *Escherichia coli,* a one-way transfer of genetic material from a donor ("male") to a recipient ("female").

Consanguinity. Related by descent from a common ancestor.

Constitutive enzyme. An enzyme that is produced in fixed quantities regardless of need (cf. **Inducible enzyme** and **Repressible enzyme).**

Continuous variation. Variation not represented by distinct classes. Individuals grade into each other and measurement data are required for analysis (cf. **Discontinuous variation).** Multiple genes or polygenes are usually responsible for this type of variation.

Coordinate repression. Control of structural genes in an operon by a single operator gene.

Copolymers. Mixtures consisting of more than one polymer. Example: Polymers of two kinds of organic bases such as uracil and cytosine (poly-UC) have been combined for studies of genetic coding.

Coupling (*cis*-arrangement). The condition in linked inheritance in which an individual heterozygous for two pairs of genes received the two dominant members from one parent and the two recessives from the other parent (e.g., $AABB \times aabb$). (See **Repulsion**.)

Covalent bond. A bond in which an electron pair is equally shared by protons in two adjacent atoms.

Crossbreeding. Mating between members of different races or species.

Crossing over. A process inferred genetically by the new association of linked genes and demonstrated cytologically from new associations of parts of chromosomes. It results in an exchange of genes and therefore produces combinations differing from those characteristic of the parents. The term "genetic crossover" may be applied to the new gene combinations. (See **Recombination**.)

Crossover unit. A frequency of exchange of one percent between two pairs of linked genes; 1 percent of crossing over is equal to 1 unit on a linkage map.

Cytogenetics. Area of biology concerned with chromosomes and their implications in genetics.

Cytokinesis. Cytoplasmic division and other changes exclusive of nuclear division that are a part of mitosis or meiosis.

Cytology. The study of the structure and function of the cell.

Cytoplasm. The protoplasm of a cell outside the nucleus in which cell organelles (mitochondria, plastids, etc.) are located. All living parts of the cell except the nucleus.

Cytoplasmic inheritance. Hereditary transmission dependent on the cytoplasm or structures in the cytoplasm rather than the nuclear genes; extrachromosomal inheritance. Example: Plastid characteristics in plants may be inherited by a mechanism independent of nuclear genes.

Cytosine. A pyrimidine base found in RNA and DNA.

Deficiency (deletion). Absence of a segment or band of a chromosome, reducing the number of loci.

Degenerate code. The same amino acid has more than one code word. The genetic code is degenerate because more than one nucleotide triplet codes for the same amino acid.

Denaturation. Loss of native configuration of a macromolecule, usually accompanied by loss of biological activity. Denatured proteins often unfold their polypeptide chains and express changed properties of solubility.

de novo. Arising anew, afresh, once more.

Determination. Process by which embryonic parts become capable of developing into only one kind of adult tissue or organ.

Deviation. As used in statistics, a variation from an expected number.

Diakinesis. A stage of meiosis just before metaphase I in which the bivalents are shortened and thickened.

Dicentric chromosome. One chromosome having two centromeres.

Differentiation. Modification of different parts of the body for particular functions during development of the organism. Continual restriction of types of transcription that each cell can undertake.

Dihybrid. An individual that is heterozygous for two pairs of alleles. The progeny of a cross between homozygous parents differing in two respects.

Dimer. A compound having the same percentage composition as another but twice the molecular weight; one formed by polymerization.

Dimorphism. Two different forms in a group as determined by such characteristics as sex, size, or coloration.

Diploid. An organism or cell with two sets of chromosomes ($2n$) or two genomes. Somatic tissues of higher plants and animals are ordinarily diploid in chromosome constitution in contrast with the haploid (monoploid) gametes.

Diplonema (*adj.*, **diplotene**). That stage in prophase of meiosis following the pachytene stage, but preceding diakinesis, in which the chromosomes are visibly double; stage characterized by centromere repulsion of bivalents resulting in loops.

Discontinuous variation. Distinct classes such as red versus white, tall versus dwarf (cf. **Continuous variation**).

Discordant. Twins are discordant if one shows a genetic trait and the other does not.

Disjunction. Separation of homologous chromosomes during anaphase of mitotic or meiotic divisions. (See **Nondisjunction**.)

Disome. See **Monosomic**.

Ditype. A tetrad that contains two kinds of meiotic products (spores) (e.g., $4AB$ and $4ab$).

Dizygotic twins. Two-egg or fraternal twins.

DNA. Deoxyribonucleic acid; the information-carrying genetic material or genes.

DNase. Any enzyme that hydrolyzes DNA.

Dominance. Applied to one member of an allelic pair that has the ability to manifest itself at the exclusion of the expression of the other allele.

Drift. See **Random genetic drift**.

Duplication. The occurrence of a segment more than once in the same chromosome or genome. Multiplication of cells.

Egg (ovum). A germ cell produced by a female organism.

Electrophoresis. The migration of suspended particles in an electric field.

Embryo. An organism in the early stages of development; in humans, the first two months in the uterus.

Embryo sac. A large thin-walled space within the ovule of the seed plant in which the egg and, after fertilization, the embryo develop; the mature female gametophyte in higher plants.

Endogenote. The part of the bacterial chromosome that is homologous to a genome fragment (exogenote) transferred from the donor to the recipient cell in the formation of a merozygote.

Endomitosis. Duplication of chromosomes without division of the nucleus, resulting in increased chromosome number within a cell or endopolyploidy. Chromosome strands separate but the cell does not divide.

Endonuclease. An enzyme that breaks strands of DNA at internal positions; some are involved in recombination of DNA.

Endoplasmic reticulum. Network of membranes in the cytoplasm to which ribosomes adhere.

Endopolyploidy. Occurrence of cells in diploid organisms containing multiples of the $2n$ genomes (i.e., $4n$, $8n$, etc.).

Endosperm. Nutritive tissue that develops in the embryo sac of most angiosperms. It usually follows the fertilization of the two fused primary endosperm nuclei of the embryo sac by one of the two male gamete nuclei. In most diploid plants, the endosperm is triploid ($3n$), but in some (e.g., the lily) the endosperm is $5n$.

Enhancer. A substance or object that increases a chemical activity or a physiological process; a major or modifier gene that increases a physiological process.

Environment. The aggregate of all the external conditions and influences affecting the life and development of the organism.

Enzyme. A protein that accelerates a specific chemical reaction in a living system.

Episome. A genetic element that

may be present or absent in different cells and that is associated with a chromosome or independent in the cytoplasm. Example: Fertility factor (F) in *Escherichia coli*.

Epistasis. Interactions between products of nonallelic genes. Genes suppressed are said to be hypostatic. Dominance is associated with members of allelic pairs, whereas epistasis is interaction among products of nonalleles.

Equational (homotypic) division. Mitotic-type division that is usually the second division in the meiotic sequence; somatic mitosis and the nonreductional division of meiosis.

Equatorial plate. The figure formed by the chromosomes in the center (equatorial plane) of the spindle in mitosis.

Estrogen. Female hormone or estrus-producing compound.

Euchromatin. Genetic material not stained by common dyes during interphase and normally transcribed into mRNA (cf. **Heterochromatin**).

Eugenics. The science of improving the qualities of the human race; the application of the principles of genetics to the improvement of humankind.

Eukaryote. A member of the large group of higher organisms composed of cells with true nuclei that are enclosed in nuclear membranes. These organisms undergo meiosis (cf. **Prokaryote**).

Euploid. An organism or cell having a chromosome number that is an exact multiple of the monoploid (n) or haploid number. Terms used to identify different levels in an euploid series are diploid, triploid, tetraploid, and so on (cf. **Heteroploid** and **Aneuploid**).

Exogenote. Chromosomal fragment homologous to an endogenote and donated to a merozygote.

Exons. The segments of eukaryotic genes that code for amino acid

sequences or RNA sequences of the final gene products; often separated by noncoding introns.

Exonuclease. An enzyme that digests DNA or RNA, beginning at the ends of strands.

Expressivity. Degree of expression of a trait controlled by a gene. A particular gene may produce varying degrees of expression in different individuals.

Extrachromosomal. Structures that are not a part of the chromosomes; DNA units in the cytoplasm that control cytoplasmic inheritance.

F_1. The first filial generation. The first generation of descent from a given mating.

F_2. The second filial generation produced by crossing *inter se* or by self-pollinating the F_1. The inbred "grandchildren" of a given mating. The term is loosely used to indicate any second-generation progeny from a given mating, but in controlled genetic experimentation, self-fertilization of the F_1 (or equivalent) is implied.

F factor. Fertility factor in a bacterium. In the presence of F^+, a bacterial cell functions as a male.

Feedback inhibition. Accumulated end product stops synthesis of that product. A late metabolite of a synthetic pathway regulates synthesis at an earlier step of that pathway.

Fertilization. The fusion of a male gamete (sperm) with a female gamete (egg) to form a zygote.

Fetus. Prenatal stage of a viviparous animal between the embryonic stage and the time of birth. In humans, the final seven months before birth.

Filial. See **F_1** and **F_2**.

Fitness. The number of offspring left by an individual as compared with the average of the population or compared to individuals of different genotype. Reproductive value determined by a genotype in a population.

Fixation. Chance fluctuation of

members of a pair of alleles around the mean that may result in homozygosity of one or the other allele (i.e., *AA* or *aa*) and that henceforth will prevent segregation between the different alleles.

Flagellum (*pl.*, **flagella;** *adj.,* **flagellate**). A whiplike organelle of locomotion in certain cells; locomotor structures in flagellate protozoa.

Founder principle. The premise that a new, small, isolated population diverges from the parent population because of sampling error. Parent and daughter populations come to operate on different gene pools.

Gamete. A mature male or female reproductive cell (sperm or egg).

Gametogenesis. The formation of gametes.

Gametophyte. That phase of the plant life cycle that bears the gametes; cells have *n* chromosomes.

Gastrula. An early animal embryo consisting of two layers of cells; an embryological stage following the blastula.

Gene. A hereditary determiner specifying a biological function; a unit of inheritance (DNA) located in a fixed place on the chromosome.

Gene conversion. Nonreciprocal recombination resulting, for example, in tetrad ratios of other than 1:1 (6:2 or 5:3) for a given pair of alleles. Process by which one gene is replicated in excess of expectation.

Gene flow. The spread of genes from one breeding population to another by migration, which may result in changes in gene frequency.

Gene frequency. The proportion of one allele relative to all alleles at a locus in a particular population.

Gene pool. Sum total of all different alleles in the breeding members of a population at a given time.

Genetic drift. See **Random genetic drift.**

Genetic equilibrium. Condition in a group of interbreeding organisms in which particular gene frequencies remain constant throughout succeeding generations.

Genetic load. The proportion by which the fitness of the optimum genotype is decreased by deleterious genes, as expressed in lethal equivalents or "genetic deaths." Average genetic load of the human population was estimated to be four to six lethal equivalents.

Genetics. The science of heredity and variation.

Genome. A complete set (*n*) of chromosomes (hence, of genes) inherited as a unit from one parent.

Genotype. The genetic constitution (gene makeup) of an organism (i.e., *Dd* or *dd*). Individuals of the same genotype breed alike (cf. **Phenotype**).

Germ cell. A reproductive cell capable when mature of being fertilized and reproducing an entire organism (cf. **Soma cell**).

Germ plasm. The hereditary material transmitted to the offspring through the germ cells.

Globulins. Common proteins in the blood that are insoluble in water and soluble in salt solutions. Alpha, beta, and gamma globulins can be distinguished in human blood serum. Gamma globulins are important in developing immunity to diseases.

Gonad. A sexual gland (i.e., ovary or testis) that produces gametes.

Guanine. A purine base found in DNA and RNA.

Gynandromorph. An individual in which one part of the body is female and another part is male; a sex mosaic.

Haploid (monoploid). An organism or cell having only one complete set (*n*) of chromosomes or one genome.

Haptoglobin. A serum protein, alpha globulin in the blood.

Hardy–Weinberg equilibrium. Mathematical relationship between the allele frequencies within populations. At equilibrium, frequencies of genotypic classes are $p^2(AA)$, $2pq(Aa)$, and $q^2(aa)$.

Helix. Any structure with a spiral shape. The Watson and Crick model of DNA is in the form of a double helix.

Hemizygous. The condition in which only one allele of a pair is present, as in sex linkage or as a result of deletion.

Hemoglobin. Conjugated protein compound containing iron, located in erythrocytes of vertebrates; important in the transportation of oxygen to the cells of the body.

Hemolymph. The mixture of blood and other fluids in the body cavity of an invertebrate.

Hemophilia. A bleeder's disease; tendency to bleed freely from even a slight wound; hereditary condition dependent on a sex-linked recessive gene.

Heredity. Resemblance among individuals related by descent; transmission of traits from parents to offspring.

Heritability. Degree to which a given trait is controlled by inheritance.

Hermaphrodite. An individual with both male and female reproductive organs.

Heterochromatin. Chromatin staining darkly even during interphase. In eukaryotes, repetitive DNA that is rarely transcribed. In *Drosophila* salivary preparations, the heterochromatin including most of the Y chromosome is in the chromocenter. The nucleolar organizing region, the site of ribosomal RNA synthesis, is flanked on either side by heterochromatin.

Heteroduplex. A double-stranded nucleic acid containing one or more mismatched (noncomplementary) base pairs.

Heterogametic sex. Producing unlike gametes with regard to the sex chromosomes. The XY male is heterogametic, the XX female is homogametic.

Heterokaryon. A cell containing two or more different nuclei.

Heteroploid (aneuploid). An organism characterized by a chromosome number other than the true haploid (monoploid) or diploid number ($2n + 1$ or $2n - 1$) (cf. **Euploid**).

Heteropyknosis (*adj.*, **heteropyknotic**). Property of certain chromosomes, or of their parts, to remain more dense and to stain more intensely than other chromosomes or parts during the nuclear cycle.

Heterosis. Superiority of heterozygous genotypes in respect to one or more traits in comparison with corresponding homozygotes.

Heterozygote (*adj.*, **heterozygous**). An organism with unlike members of any given pair or series of alleles that consequently produces unlike gametes.

Hfr. High-frequency recombination strain of *Escherichia coli;* the F episome is integrated into the bacterial chromosome.

Histones. Group of proteins rich in basic amino acids. They function in the coiling of DNA in chromosomes and in regulation of gene activity.

"Holandric" gene. A gene carried on the Y chromosome and therefore transmitted father to son.

Homogametic sex. Producing like gametes (cf. **Heterogametic sex**).

Homologous chromosomes. Chromosomes that occur in pairs and are generally similar in size and shape—one having come from the male and one from the female parent. They contain the same gene loci and form bivalents at meiosis.

Hormone. An organic product of cells of one part of the body that is transported by the body fluids to another part where it influences activity or serves as a coordinating agent.

Hybrid. An offspring of homozygous parents differing in one or more genes.

Hybridization. Interbreeding of species, races, varieties, and so on, among plants or animals. A process of forming a hybrid by cross-pollination of plants or by mating animals of different types.

Hybrid vigor (heterosis). Unusual growth, strength, and health of heterozygous hybrids from two less vigorous homozygous parents.

Hypostasis. See **Epistasis**.

Imaginal disc. Thickenings of epidermis containing mesodermal cells in the larvae of *Drosophila* and other holometabolous insects. They give rise to adult organs such as antennae, eyes, and wings.

Immunoglobulin. See **Globulin**.

Inbreeding. Matings among related individuals.

Incomplete dominance. Expression of heterozygous alleles, different from those of the parents, producing distinguishable hybrids.

Independent assortment. The random distribution of alleles to the gametes that occurs when genes are located in different chromosomes. The distribution of one pair of alleles is independent of other genes located in nonhomologous chromosomes.

Inducer. A substance of low molecular weight that combines with a repressor and thus decreases the repression of protein synthesis.

Inducible enzyme. An enzyme that is synthesized only in the presence of the substratum that acts as an inducer.

Inhibitor. Any substance or object that retards a chemical reaction; a major or modifier gene that interferes with a reaction.

Interference. Crossing over at one point that reduces the chance of another crossover in adjacent regions; detected by studying crossovers of three or more linked genes.

Interphase. The stage in the cell cycle when the cell is not dividing; the metabolic stage; the stage following telophase of one division and extending to the beginning of prophase in the next division, during which DNA replication occurs.

Intersex. An organism displaying secondary sexual characters intermediate between male and female; a type that shows some phenotypic characteristics of both males and females.

Introns. Intervening sequences of DNA bases within eukaryotic genes; they are not present in mRNA. Thus, they must be spliced out of eukaryotic primary transcripts to obtain continuous coding sequences like those in prokaryotes.

Inversion. A rearrangement of a linear array of genes in a chromosome in such a way that their order in the chromosome is reversed.

in vitro. Biological processes made to occur experimentally outside the organism in a test tube or other container.

in vivo. Within the living organism.

IS element (insertion sequence). A short (800–1400 nucleotide pairs) sequence capable of mediating transposition of itself and of other (attached) genetic elements from one position to a new position in the genome.

Isoagglutinogen. An antigen, such as A or B blood type factor, that occurs normally (i.e., in an individual, without artificial stimulation).

Isochromosome. A chromosome with two identical arms and identical genes. The arms are mirror images of one another.

Isogenic stocks. Strains of organisms that are genetically uniform; completely homozygous.

Kappa particles. DNA-containing, self-reproducing cytoplasmic particles in certain strains of *Paramecium aurelia.* They control a toxic substance, *paramecin,* that is released into the culture medium and kills sensitive paramecia. Nuclear gene K is required for maintenance of kappa in the cytoplasm of killers.

Karyotype. The chromosome constitution of a cell or an individual; chromosomes arranged in order of length and according to position of centromere; also the abbreviated formula for chromosome constitution, such as $47,XX+21$ for trisomy-21.

Kinetosome. Granular body at the base of a flagellum or a cilium.

Lamella. A double-membrane structure, plate, or vesicle that is formed by two membranes lying parallel to each other.

Lampbrush chromosomes. Large diplotene chromosomes present in oöcyte nuclei, particularly conspicuous in amphibians. These chromosomes have extended regions called loops, which are active sites of transcription.

Leptonema (*adj.,* **leptotene**). Stage in meiosis immediately preceding synapsis in which the chromosomes appear as single fine threadlike structures (but they are really double because DNA replication has taken place).

Lethal gene. An allele that renders inviable an organism or a cell possessing it in proper arrangement for expression.

Ligase. An enzyme that joins the ends of two strands of nucleic acid.

Linkage. Linear arrangements of genes in the same chromosome, forming a group that may be inherited together.

Linkage map. Genes of a given species listed in linear order to show their relative positions on the chromosomes.

Locus (*pl.,* **loci**). A fixed position on a chromosome that is occupied by a given gene or one of its alleles.

Lysis. Bursting of a bacterial cell by the destruction of the cell membrane following infection by bacteriophage.

Lysogenic bacteria. Those harboring temperate bacteriophages.

Macromolecule. A large molecule; term used to identify molecules of proteins and nucleic acids.

Map unit. One percent of crossing over represents one unit on a linkage map.

Maternal effect. Trait controlled by a gene of the mother but expressed in the progeny.

Maternal inheritance. Inheritance controlled by extrachromosomal (e.g., cytoplasmic) factors; phenotypic differences found between individuals of identical genotype, which is the result of maternal influence.

Maturation. The formation of gametes or spores.

Mean. The arithmmetic average; the sum of all measurements or values of a group of objects divided by the number of objects.

Meiosis. The process by which the chromosome number of a reproductive cell becomes reduced to half the diploid ($2n$) or somatic number; results in the formation of gametes in animals or of spores in plants; important source of variability through recombination.

Meiotic drive. Any mechanism that results in the unequal frequencies of gametes that are produced by a heterozygote.

Melanin. Brown or black pigment.

Mendelian population. A natural interbreeding unit of sexually reproducing plants or animals sharing a common gene pool.

Merozygote. Partial zygote produced by a process of partial genetic exchange, such as transformation in bacteria. An exogenote may be introduced into a bacterial cell in the formation of a merozygote.

Mesoderm. The middle germ layer that forms in the early animal embryo and gives rise to such parts as bone and connective tissue.

Messenger RNA (mRNA). RNA that carries information necessary for protein synthesis from the DNA to the ribosomes.

Metabolic cell. A cell that is not dividing.

Metabolism. Sum total of all chemical processes in living cells by which energy is provided and used.

Metacentric chromosome. A chromosome with the centromere near the middle and two arms of about equal length.

Metafemale (superfemale). In *Drosophila,* abnormal female, usually sterile, with an overbalance of X chromosomes compared with sets of autosomes.

Metaphase. That stage of cell division in which the chromosomes are most discrete and arranged in an equatorial plate; stage following prophase and preceding anaphase.

Microtubules. Hollow filaments in the cytoplasm making up a part of the locomotor apparatus of any motile cells; component of the mitotic spindle.

Mitochondria. DNA-containing bodies in the cytoplasm of plant and animal cells; site of reactions producing ATP.

Mitosis. Chromosome duplication followed by migration of chromosomes to the ends of the spindle and division of cytoplasm.

Modifier (modifying gene). A gene that affects the expression of a nonallelic gene.

Monohybrid. An offspring of two homozygous parents that differ from one another by the alleles present at only one gene locus.

Monohybrid cross. A cross between parents differing in only one trait or in which only one trait is being considered.

Monomer. A simple molecule of a compound of relatively low molecular weight.

Monoploid (haploid). Organism or cell having a single set of chromosomes or one genome (n).

Monosomic. A diploid organism lacking one chromosome of its proper complement ($2n - 1$); an aneuploid. Monosome refers to the single chromosome, disome to two chromosomes of a kind, trisome to three chromosomes of a kind.

Monozygotic twins. One-egg or identical twins.

Morphology. Study of the form of an organism. Developmental history of visible structures and the comparative relation of similar structures in different organisms.

Mosaic. An organism or part of an organism that is composed of cells of different genotypes.

Multiple alleles. Three or more alternative alleles that represent the same locus in a given pair of chromosomes.

Mutable genes. Those with an unusually high mutation rate.

Mutagen. An environmental agent, either physical or chemical, that is capable of inducing mutations.

Mutant. A cell or individual organism that shows a change brought about by a mutation; a changed gene.

Mutation. A change in the DNA at a particular locus in an organism. The term is used loosely to include point mutations involving a single gene change as well as a chromosomal change. Recurrent mutation: one-way mutation; continuous changes from one allele to another particular allele, such as $A \rightarrow a$.

Mutation pressure. A constant mutation rate that adds mutant genes to a population; repeated occurrences of mutations in a population.

Mycelium (*pl.,* **mycelia**). Thread-like filament making up the vegetative portion of thallus fungi.

Natural selection. Differential fertility in nature favoring individuals that are better adapted and tending to eliminate those unfitted to their environment.

Nondisjunction. Failure of disjunction or separation of homologous chromosomes in mitosis or meiosis, resulting in too many chromosomes in some daughter cells and too few in others. In meiosis, one pair goes to the same pole and the other pole receives neither of them. In mitosis, sister chromatids go to the same pole.

Nucleic acid. An acid composed of phosphoric acid, pentose sugar, and organic bases. DNA and RNA are nucleic acids.

Nucleolar organizer (NO). A chromosomal segment with genes that control the synthesis of ribosomal RNA, located at the secondary constriction of some metaphase chromosomes.

Nucleolus. An RNA-rich, spherical sack in the nucleus of metabolic cells; associated with the nucleolar organizer; storage package for ribosomes and ribosome precursors.

Nucleoprotein. Conjugated protein composed of nucleic acid and protein; the material of which the chromosomes are made.

Nucleosome (Nu body). Spherical subunits, or "beads-on-a-string" of chromatin, in eukaryotes that measure approximately 100 Å and are composed of a core particle (octamer of histones and 146 nucleotide pairs).

Nucleotide. A unit of the DNA molecule containing a phosphate, a sugar, and an organic base.

Nucleus. Part of a cell containing genetic material and surrounded by cytoplasm.

Nullisomic. An otherwise diploid cell or organism lacking both members of a chromosome pair (chromosome formula: $2n - 2$).

Octoploid. Cell or organism with eight genomes or sets of chromosomes ($8n$).

Oncogene. A gene that can cause cell transformation in animal cells growing in culture and tumor formation in animals.

Oöcyte. The egg-mother cell; the cell that undergoes two meiotic divisions (oögenesis) to form the egg cell. Primary oöcyte—before completion of the first meiotic division; secondary oöcyte—after completion of the first meiotic division.

Oögenesis. The formation of the egg or ovum in animals.

Oögonium (*pl.,* **oögonia**). A germ cell of the female animal before meiosis begins.

Operator gene. A part of an operon that controls the activity of one or more structural genes.

Operon. A group of genes making up a regulatory or control unit. The unit includes an operator, a promoter, and structural genes.

Organelle. Specialized part of a cell with a particular function or functions (e.g., cilium of a protozoan).

Organizer. An inductor; a chemical substance in a living system that determines the fate in development of certain cells or groups of cells.

Otocephaly. Abnormal development of the head of a mammalian fetus.

Outbreeding. Mating of unrelated individuals.

Ovary. The swollen part of the pistil of a plant flower that contains the ovules. The female reproductive gland or gonad in animals.

Ovule. The macrosporangium of a flowering plant that becomes the seed. It includes the nucellus and the integuments.

P. Symbolizes the parental generation or parents of a given individual.

Pachynema (*adj.,* **pachytene**). A midprophase stage in meiosis immediately following zygonema and preceding diplonema. In favorable microscopic preparations, the chromosomes are visible as long, paired threads. Rarely, four chromatids are detectable.

Palindrome. A segment of DNA in which the base-pair sequence reads the same in both directions from a point of symmetry.

Panmictic population. A population in which mating occurs at random.

Panmixis. Random mating in a population in contrast to assortative mating.

Paracentric inversion. An inversion that is entirely within one arm of a chromosome and does not include the centromere.

Parameter. A value or constant based on an entire population (cf. **Statistic**).

Parthenogenesis. The development of a new individual from an egg without fertilization.

Paternal. Pertaining to the father.

Pathogen. An organism that causes a disease.

Pedigree. A table, chart, or diagram representing the ancestral history of an individual.

Penetrance. The appearance of a trait exppressed as the percentage of individuals (of those known to carry the gene responsible) that actually express the gene and exhibit the phenotype.

Peptide. A compound containing amino acids. A breakdown or buildup unit in protein metabolism.

Peptide bond. A chemical bond holding amino acid subunits together in proteins.

Pericentric inversion. An inversion including the centromere, hence involving both arms of a chromosome.

Phage. See **Bacteriophage.**

Pharmacogenetics. Biochemical genetics associated with genetically controlled variations in responses to drugs.

Phenocopy. An organism whose phenotype (but not genotype) has been changed by the environment to resemble the phenotype of a different (mutant) organism.

Phenotype. Characteristics of an individual observed or discernible by other means (i.e., tallnes in garden peas; color blindness or blood type in humans).

Phenylalanine. See **Amino acid.**

Phenylketonuria. Metabolic disorder resulting in mental retardation; transmitted as a Mendelian recessive and treated in early childhood by special diet.

Phylogeny. Evolutionary history of a population of related organisms. History of an animal or vegetable type.

Plaque. Clear area on an otherwise opaque culture plate of bacteria where the bacteria have been killed by a virus.

Plasmid. An extrachromosomal hereditary determinant that exists in an autonomous state and is transferred independently of chromosomes. Some plasmids, such as F, are also episomal.

Plastid. A cytoplasmic body found in the cells of plants and some protozoa. Chloroplastids produce chlorophyll that is involved in photosynthesis.

Pleiotropy (*adj.,* **pleiotropic**). Condition in which a single gene influences more than one trait.

Polar bodies. In female animals, the smaller cells produced at meiosis that do not develop into egg cells. The first polar body is produced at division I and may not go through division II. The second polar body is produced at division II.

Polarity mutation. Gene mutation that influences the functioning of other more distal genes in the same operon.

Polydactyly. The occurrence of more than the usual number of fingers or toes.

Polygene. One of a series of multiple genes involved in quantitative inheritance.

Polymer. A compound composed of many smaller subunits; results from the process of polymerization.

Polymerase (p). An enzyme that catalyzes the formation of DNA or RNA. DNA p—specific enzyme that catalyzes the synthesis of DNA. RNA p—specific enzyme that catalyzes the formation of ribonucleic acid.

Polymerization. Chemical union of two or more molecules of the same kind to form a new compound having the same elements in the same porportions but a higher molecular weight and different physical properties.

Polymorphism. Two or more kinds of individuals maintained in a breeding population.

Polynucleotide. A linear sequence of joined nucleotides in DNA or RNA.

Polypeptide. A compound containing two or more amino acids and one or more peptide groups. They are called dipeptides, tripeptides, and so on, according to the number of amino acids contained.

Polyploid. An organism with more than two sets of chromosomes ($2n$ diploid) or genomes [e.g., triploid ($3n$), tetraploid ($4n$), pentaploid ($5n$), hexaploid ($6n$), heptaploid ($7n$), octoploid ($8n$)].

Polysaccharide capsules. Carbohydrate coverings with antigenic specificity on some types of bacteria.

Polytene chromosomes. Giant chromosomes produced by interphase replication without division and consisting of many identical chromatids; cablelike formation.

Populations. Entire group of organisms of one kind: an interbreeding group of plants or animals. The extensive group from which a sample might be taken.

Population (effective). Breeding members of the population.

Population genetics. The branch of genetics that deals with frequencies of alleles and genotypes in breeding populations.

Position effect. A difference in phenotype that is dependent on the position of a gene or group of genes or heterochromatin in relation to genes.

Primary oöcyte. See **Oöcyte.**

Prokaryote. A member of a large group of lower organisms (including viruses, bacteria, and blue-green algae) that do not go through meiosis and lack nuclear membranes.

Promotor. A part of an operon; a nucleotide sequence to which RNA polymerase binds and initiates transcription. Also, a chemical substance that enhances the transformation of benign cells to cancerous cells.

Prophase. The stage of mitosis between the interphase and the metaphase. During this phase, the centriole divides and the two daughter centrioles move apart. Each sister DNA strand from interphase replication becomes coiled and the chromosome is longitudinally double except in the region of the centromere. Each partially separated chromosome is called a chromatid. The two chromatids of a chromosome are sister chromatids.

Proto-oncogene. A normal cellular gene that can be changed to an oncogene by mutation.

Pseudogene. An inactive but stable component of a genome resembling a gene; apparently derived from active genes by mutation.

Quantitative inheritance. Inheritance of quantitative traits (height, weight, color intensity) that depend on cumulative action of many genes, each producing a small effect on the same trait.

Radioactive isotope. An unstable isotope (form of an atom) that emits ionizing radiation.

Random genetic drift. Changes in gene frequency in small breeding populations due to chance fluctuations.

Recessive. Applied to one member of an allelic pair lacking the ability to manifest itself when the other or dominant member is present.

Reciprocal crosses. A second cross involving the same strains but carried by sexes opposite to those in the first cross—for example, a female from strain A $\times$ a male from strain B and a male from A $\times$ a female from B are reciprocal crosses.

Recombination. The observed new combinations of traits different from those combinations exhibited by the parents. Percentage of recombination equals percentage of crossing over only when the genes are relatively close together. Cytological chiasma refers to the observed change of partners among chromatids, whereas recombination or genetic crossing over refers to the observed genetic result.

Reduction division. Phase of meiosis in which the maternal and paternal elements of the bivalent separate [cf. **Equational (homotypic) division**].

Regulator gene. A gene that controls the rate of expression of another gene or genes. Example: The operon involved in lactose production of *Escherichia coli* has a regulator, an operator, and structural genes.

Replication. A duplication process that is accomplished by copying from a template (e.g., reproduction at the level of DNA).

Replicon. A unit of replication. In bacteria, replicons are associated with segments of the cell membrane that control replication and coordinate it with cell division.

Repressible enzyme. Enzyme produced in a cell when its end product is not present. It is repressed as the end product increases.

Repulsion (*trans*-arrangement). The condition in linked inheritance in which an individual heterozygous for two pairs of linked genes receives the dominant member of one pair and the recessive member of the other pair from one parent and the reverse arrangement from the other parent (e.g., *AAbb* $\times$ *aaBB*). (See **Coupling.**)

Resistance factor. A plasmid that confers antibiotic resistance to the recipient bacterium.

Restitution nucleus. A nucleus with unreduced or doubled chromosome number that results from the failure of a meiotic or mitotic division.

Restriction enzyme. An endonuclease that recognizes a specific short sequence in DNA and cleaves the molecule; many cleave DNA molecules in a site-specific manner.

Reticulocyte. A young red blood cell.

Reversion. Appearance of a trait expressed by a remote ancestor; a throwback; atavism.

Rh factor. Antigen in the red blood corpuscles of certain people. A pregnant Rh negative woman carrying an Rh positive child may develop anitbodies, causing the child to develop a hemolytic disease.

Ribonucleic acid. See **RNA.**

Ribosome. Cytoplasmic structure on which proteins are synthesized.

RNA. Ribonucleic acid. The information-carrying material in some viruses. Certain kinds of RNA are invloved in the transcription of genetic information from DNA (i.e., mRNA); in transfer of amino acids to the ribosomes for incorporation into proteins (i.e., tRNA); and the makeup of the ribosome (i.e., rRNA).

Roentgen (r). Unit of ionizing radiation.

Secondary oöcyte. See **Oöcyte.**

Secondary spermatocyte. See **Spermatocyte.**

Secretor. A person with a water-soluble form of antigen A or B. In such a person, the antigen may be detected in body fluids (e.g., saliva) as well as on the erythrocytes.

Segregation. The separation of the paternal from maternal chromosomes at meiosis and the consequent separation of alleles and their phenotypic differences as observed in the offspring. Mendel's first principle of inheritance.

Segregation distorter (*SD*). A gene or other factor that alters the ratio of segregating alleles: *SD*$^+$ cells are eliminated during spermiogenesis.

Selection. Differential reproduction of different genotypes. The most important of factors that change the frequencies of alleles and genotypes in a large population and thus influence evolutionary change.

Selection pressure. Effectiveness of the environment in changing the frequency of alleles in a population of individuals.

Self-fertilization. The process by which pollen of a given plant fertilizes the ovules of the same plant. Plants fertilized in this way are said to be selfed.

Semisterility. A condition of only partial fertility in plant zygotes (e.g., maize); usually associated with translocation.

Serology (*adj.*, **serological**). The study of interactions between antigens and antibodies.

Sex chromosomes. Chromosomes that are connected with the determination of sex.

Sexduction. The incorporation of bacterial genes by sex factors and their subsequent transfer by conjugation to a recipient cell.

Sex factor. In bacteria, an episome (F$^+$ in *E. coli*) that enables a bacterial cell to be a donor of genetic material. The episome may be transferred in the cytoplasm during conjugation, or it may be integrated into the bacterial chromosome.

Sex-influenced dominance. A dominant expression that depends on the sex of the individual (e.g., horns in sheep are dominant in males and recessive in females).

Sex-limited. Expression of a trait in only one sex. Examples: milk production in mammals; horns in Rambouillet sheep; egg production in chickens.

Sex linkage. Association or linkage of a hereditary trait with sex; the gene is in a sex chromosome, usually the X.

Sex mosaic. See **Gynandromorph.**

Sexual reproduction. Reproduction involving the formation of mature germ cells (i.e., eggs and sperm).

Sib-mating (crossing of siblings). Matings involving two or more individuals of the same parentage; brother–sister matings.

Soma cell. A cell that is a component of the body, in contrast with a germ cell that is capable, when fertilized, of reproducing the organism.

Somatic cells. Referring to body tissues; having two sets of chromosomes—one set normally coming from the female parent and one from the male (as contrasted with germinal tissue that will give rise to germ cells).

Species. Interbreeding, natural populations that are reproductively isolated from other such groups.

Sperm (*abbreviation of spermatozoön;* *pl.,* **spermatozoa**). A mature male germ cell.

Spermatids. The four cells formed by the meiotic divisions in spermatogenesis. Spermatids become mature spermatozoa or sperm.

Spermatocyte (sperm mother cell). The cell that undergoes two meiotic divisions (spermatogenesis) to form four spermatids; the *primary* spermatocyte before completion of the first meiotic division; the *secondary* spermato-

cyte after completion of the first meiotic division.

Spermatogenesis. The process by which maturation of the gametes (sperm) of the male takes place.

Spermatogonium (*pl.,* **spermatogonia**). Primordial male germ cell that may divide by mitosis to produce more spermatogonia. A spermatogonium may enter a growth phase and give rise to a primary spermatocyte.

Spermiogenesis. Formation of sperm from spermatids; the part of spermatogenesis that follows the meiotic divisions of spermatocytes.

Standard deviation. A measure of variability in a population of individuals.

Standard error. A measure of variation of a population of means; used to indicate how well samples represent populations or parameters.

Stem cell. A mitotically active somatic cell from which other cells arise by differentiation.

Statistic. A value based on a sample or samples of a population from which estimates of a population value or parameter may be obtained.

Sterility. Inability to produce offspring.

Structural gene. A gene that controls actual protein production by determining the amino acid sequence (cf. **Operator gene** and **Regulator gene**).

Sublethal gene. A lethal gene with delayed effect. The gene in proper combinations kills its possessor in infancy, childhood, or adulthood.

Symbiont. An organism living in intimate association with another, dissimilar organism.

Synapsis. The pairing of homologous chromosomes in the meiotic prophase

Synaptinemal complex. A ribbonlike structure formed between synapsed homologs at the end of the first meiotic prophase, binding the chromatids along their length and facilitating chromatid exchange.

Syndrome. A group of symptoms that occur together and represent a particular disease.

Synkaryon. A nucleus formed by the fusion of nuclei from two different somatic cells during somatic cell hybridization.

Synteny. The occurrence of two loci on the same chromosome, independent of distance that separates them.

Telophase. The last stage in each mitotic or meiotic division in which the chromosomes are assembled at the poles of the division spindle.

Temperate phage. A phage (virus) that invades but may not destroy (lyse) the host (bacterial cell) (cf. **Virulent phage**). It may continue to the lytic cycle.

Template. A pattern or mold. DNA stores coded information and acts as a model or template from which information is taken by messenger RNA.

Terminalization. Repelling movement of the centromeres of bivalents in the diplotene stages of the meiotic prophase that tends to move the visible chiasmata toward the end of the bivalents. The point where the exchange of chromatids or the chiasma occurs is believed to be the point of a genetic crossover, but the chiasmata appear to slip toward the ends of the bivalents, and, therefore, after the diplotene looping begins, there is no longer a relationship between the chiasma and the point of crossing over.

Testcross. Backcross to the recessive parental type, or a cross between genetically unknown individuals with a fully recessive tester to determine whether an individual in question is heterozygous or homozygous for a certain allele. Also used as a test for linkage.

Tetrad. The four cells arising from the second meiotic division in plants (i.e., pollen tetrads). The term is also used to identify the quadruple group of chromatids that is formed by the association of split homologous chromosomes during meiosis, but the term *bivalent* is preferred in this usage.

Tetraploid. An organism whose cells contain four haploid ($4n$) sets of chromosomes or genomes.

Tetrasomic (*noun*, **tetrasome**). Pertaining to a nucleus or an organism with four members of one of its chromosomes whereas the remainder of its chromosome complement is diploid. (Chromosome formula: $2n + 2$.)

Thymine. A pyrimidine base found in DNA. The other three organic bases—adenine, cytosine, and guanine—are found in both RNA and DNA, but in RNA, thymine is replaced by uracil.

Totipotent cell (or **nucleus**). An undifferentiated cell (or nucleus) such as a blastomere that when isolated or suitably transplanted can develop into a complete embryo.

trans-**arrangement.** See **Repulsion.**

Transcription. Process through which messenger RNA is formed along a DNA template. The enzyme RNA polymerase catalyzes the formation of RNA from ribonucleotide triphosphates.

Transduction (t). Genetic recombination in bacteria mediated by bacteriophage. Abortive t: Bacterial DNA is injected by a phage into a bacterium but it does not replicate. Generalized t: Any bacterial gene may be transferred by a phage to a recipient bacterium. Restricted t: Transfer of bacterial DNA by a temperate phage is restricted to only one site on the bacterial chromosome.

Transferrin. Blood serum protein, beta globulin. (See **Globulin.**)

Transfer RNA (tRNA). RNA that transports amino acids to the ribosome, where they are assembled into proteins.

Transfection. The uptake of DNA by eukaryotic cells, followed by the incorporation of genetic markers present in this DNA into the cell's genome.

Transformation (bacteria). Genetic recombination in bacteria brought about by adding foreign DNA to a bacterial culture.

Transformation (eukaryotic cells). The conversion of eukaryotic cells growing in culture to a state of uncontrolled cell growth (similar to tumor cell growth).

Transgressive variation. The appearance in the F_2 (or later) generation of individuals, showing a more extreme development of a trait than shown in either original parent.

Transition. A mutation caused by the substitution of one purine by another purine or one pyrimidine by another pyrimidine in DNA or RNA.

Translation. Protein synthesis directed by a specific messenger RNA occurring on a ribosome.

Translocation. Change in position of a segment of a chromosome to another part of the same chromosome or to a different chromosome.

Transposons. DNA elements that can move ("transpose themselves") from one DNA molecule or chromosome to another.

Transversion. A mutation caused by the substitution of a purine for a pyrimidine or a pyrimidine for a purine in DNA or RNA.

Trihybrid. The offspring from homozygous parent differing in three pairs of genes.

Trisomic. An otherwise diploid cell or organism that has an extra chromosome of one pair. (Chromosome formula: $2n + 1$.)

Tryptophan. See **Amino acids.**

Unipartite structures (chromosomes). Single units.

Univalent. A chromosome unpaired at meiosis.

Uracil. A pyrimidine base found in RNA but not in DNA. In DNA, uracil is replaced by thymine.

Variance. The square of the standard deviation. A measure of variation.

Variation. In biology, the occurrence of differences among the individuals of the same species.

Viability. Degree of capability to live and develop normally

Virulent phage. A phage (virus) that destroys the host (bacterial) cell (cf. **Temperate phage**).

Wild type. The customary phenotype or standard for comparison.

Wobble hypothesis. Hypothesis to explain how one tRNA may recognize two codons. The first two bases of the mRNA codon and anticodon pair properly, but the third base in the anticodon has some play (or wobble) that permits it to pair with more than one base, which occupies the third position of a different codon.

X chromosome. A chromosome associated with sex determination. In most animals, the female has two and the male has one X chromosome.

Xenia. Hereditary influence from the pollen parent to the endosperm in the seed in certain plants.

Y chromosome. The mate to the X chromosome in the male of most animal species. In *Drosophila* it is composed mostly of heterochromatin and carries few genes, but does not influence maleness. In man, the Y chromosome carries genes that influence maleness.

Zygonema (*adj.,* **zygotene**). Stage in meiosis during which synapsis occurs; after the leptotene stage and before the pachytene stage in the meiotic prophase.

Zygote. The cell produced by the union of two mature sex cells (gametes) in reproduction; also used in genetics to designate the individual developing from such a cell.

ANSWERS TO PROBLEMS

CHAPTER 2

2.1. (a) All tall; (b) $\frac{3}{4}$ tall, $\frac{1}{4}$ dwarf; (c) all tall; (d) $\frac{1}{2}$ tall, $\frac{1}{2}$ dwarf.

2.2. (a) $WW \times ww$; (b) $\textcircled{W}$ and $\textcircled{w}$; (c) Ww; (d) $Ww \times Ww$; (e) $\textcircled{W}$, $\textcircled{w}$ and $\textcircled{W}$, $\textcircled{w}$; (f) see the following table:

PHENOTYPES	GENOTYPES	GENOTYPIC FREQUENCY	PHENO-TYPIC RATIO
Round	WW	1	3
	Ww	2	
Wrinkled	ww	1	1

2.3. (a) The $3:1$ ratio suggests a single pair of genes, with the gene for color dominant over that for white.

(b)
CC	$\times$	cc	P
$\textcircled{C}$		$\textcircled{c}$	Gametes
	Cc		F_1
Cc	$\times$	Cc	$F_1 \times F_1$
$\textcircled{C}\textcircled{c}$		$\textcircled{C}\textcircled{c}$	F_1 gametes
$1CC : 2Cc : 1cc$			F_2

Phenotypes:	Observed	Calculated	Deviation
Colored	198	202.5	-4.5
White	72	67.5	4.5

2.4. (a) Woman, Pp; her father, Pp; her mother, pp. (b) Half of her children are expected to have ptosis.

2.5. $CCBB \times ccbb$ P
$\textcircled{CB}$ $\textcircled{cb}$ Gametes
$CcBb \times CcBb$ $F_1 \times F_1$

Gametes	$\textcircled{CB}$	$\textcircled{Cb}$	$\textcircled{cB}$	$\textcircled{cb}$
$\textcircled{CB}$	$CCBB$	$CCBb$	$CcBB$	$CcBb$
$\textcircled{Cb}$	$CCBb$	$CCbb$	$CcBb$	$Ccbb$
$\textcircled{cB}$	$CcBB$	$CcBb$	$ccBB$	$ccBb$
$\textcircled{cb}$	$CcBb$	$Ccbb$	$ccBb$	$ccbb$

Summary of F_2:

PHENOTYPES	GENO-TYPES	GENOTYPIC FREQUENCY	PHENO-TYPIC RATIO
Checkered red	$CCBB$	1	9
	$CCBb$	2	
	$CcBB$	2	
	$CcBb$	4	
Checkered brown	$CCbb$	1	3
	$Ccbb$	2	
Plain red	$ccBB$	1	3
	$ccBb$	2	
Plain brown	$ccbb$	1	1

2.6. (a) $CCVv \times ccVv$; (b) $CcVv \times CcVv$; (c) $CcVv \times ccvv$.

2.7. (a) $BBRR \times bbrr$ P
$\textcircled{BR}$ $\textcircled{br}$ Gametes
$BbRr$ F_1
$BbRr \times BbRr$ $F_1 \times F_1$

Gametes	$\textcircled{BR}$	$\textcircled{Br}$	$\textcircled{bR}$	$\textcircled{br}$
$\textcircled{BR}$	$BBRR$	$BBRr$	$BbRR$	$BbRr$
$\textcircled{Br}$	$BBRr$	$BBrr$	$BbRr$	$Bbrr$
$\textcircled{bR}$	$BbRR$	$BbRr$	$bbRR$	$bbRr$
$\textcircled{br}$	$BbRr$	$Bbrr$	$bbRr$	$bbrr$

Summary of F_2 phenotypes: 9 black, long: 3 black, rex: 3 brown, long: 1 brown, rex.

(b) $\frac{1}{9}$.

(c) $BbRr \times bbrr$.

Gametes	BR	Br	bR	br
br	$BbRr$	$Bbrr$	$bbRr$	$bbrr$

Summary of backcross results: 1 black, long: 1 black, rex: 1 brown, long: 1 brown, rex.

2.8. (a) All red; (b) $\frac{1}{2}$ red: $\frac{1}{2}$ roan; (c) all roan; (d) $\frac{1}{4}$ red: $\frac{1}{2}$ roan: $\frac{1}{4}$ white; (e) $\frac{1}{2}$ roan: $\frac{1}{2}$ white; (f) all white: (g) red. Mate red × red and all progeny will be red. If roan animals are mated together, red and white as well as roan will be produced.

2.9. (a) $\frac{1}{2}$; (b) $\frac{1}{2}$; (c) $\frac{3}{8}$.

2.10. (a) $\frac{81}{256}$; (b) $\frac{108}{256}$; (c) $\frac{54}{256}$; (d) $\frac{12}{256}$.

2.11. (a)

	Dd × *Dd*		P
	DD 2*Dd* *dd*		Progeny
	dies dichaete wild-type		

Summary: 2 dichaete, 1 wild.

(b)

	Dd × *dd*		P
	Dd *dd*		Progeny

Summary: 1 dichaete, 1 wild.

2.12. (a) $\frac{1}{4}$; (b) $\frac{1}{4}$; (c) $\frac{1}{4}$; (d) $\frac{1}{4}$.

2.13. (a) Single recessive gene. (b) 1. Parents, *aa* × *Aa*; progeny, *Aa, aa, Aa, aa.* 2. Parents, *Aa* × *Aa*; progeny *Aa* or *AA, aa, Aa* or *AA, aa.* 3. Parents, *Aa* × *Aa*; progeny, *aa, Aa* or *AA, Aa* or *AA, Aa* or *AA, aa.* 4. Parents, *aa* × *aa*; progeny, *aa, aa, aa, aa.*

2.14. (a)

DDggWW × *ddGGww*	P
DgW dGw	Gametes
DdGgWw	F₁
DGW DGw DgW Dgw dGW dGw dgW dgw	F₁ gametes

(b) *DdGgWw* × *DdGgWw*

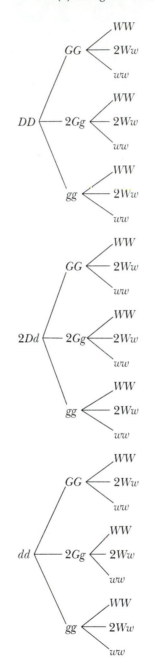

Summary of F₂ phenotypes:

27 tall, yellow, round
9 tall, yellow, wrinkled
9 tall, green, round
9 dwarf, yellow, round
3 tall, green, wrinkled
3 dwarf, yellow, wrinkled
3 dwarf, green, round
1 dwarf, green, wrinkled

(c) *DdGgWw* × *ddggww*
Summary of backcross results:

		PHENOTYPES	GENOTYPES	GENOTYPIC RATIO	PHENOTYPIC RATIO
Dd → *Gg* → *Ww*		Tall, yellow, round	*DdGgWw*	1	1
Dd → *Gg* → *ww*		Tall, yellow, wrinkled	*DdGgww*	1	1
Dd → *gg* → *Ww*		Tall, green, round	*DdggWw*	1	1
Dd → *gg* → *ww*		Tall, green, wrinkled	*Ddggww*	1	1
dd → *Gg* → *Ww*		Dwarf, yellow, round	*ddGgWw*	1	1
dd → *Gg* → *ww*		Dwarf, yellow, wrinkled	*ddGgww*	1	1
dd → *gg* → *Ww*		Dwarf, green, round	*ddggWw*	1	1
dd → *gg* → *ww*		Dwarf, green, wrinkled	*ddggww*	1	1

2.15.

P CROSS	$AA \times aa$	$AABB \times aabb$	$AABBCC \times aabbcc$	GENERAL FORMULA
F_1 gametes	2	4	8	2^{n}[a]
F_2 genotypes	3	9	27	3^{n}
F_2 phenotypes[b]	2	4	8	2^{n}

[a] n = number of segregating pairs of alleles.
[b] Under complete dominance of *A*, *B*, and *C*.

2.16. The Punnett square method is illustrated as follows:

$$RRLL \times R'R'L'L' \qquad P$$

ⓇⓁ   Ⓡ'Ⓛ'   Gametes

$$RR'LL' \qquad F_1$$
$$RR'LL' \times RR'LL' \qquad F_1 \times F_1$$

Gametes	ⓇⓁ	ⓇⓁ'	Ⓡ'Ⓛ	Ⓡ'Ⓛ'
ⓇⓁ	RRLL	RRLL'	RR'LL	RR'LL'
ⓇⓁ'	RRLL'	RRL'L'	RR'LL'	RR'L'L'
Ⓡ'Ⓛ	RR'LL	RR'LL'	R'R'LL	R'R'LL'
Ⓡ'Ⓛ'	RR'LL'	RR'L'L'	R'R'LL'	R'R'L'L'

PHENOTYPES	GENOTYPES	GENOTYPIC FREQUENCY	PHENO-TYPIC RATIO
Red, long	RRLL	1	1
Red, oval	RRLL'	2	2
Red, round	RRL'L'	1	1
Purple, long	RR'LL	2	2
Purple, oval	RR'LL'	4	4
Purple, round	RR'L'L'	2	2
White, long	R'R'LL	1	1
White, oval	R'R'L'L	2	2
White, round	R'R'L'L'	1	1

2.17. (a) 3 walnut: 1 rose; (b) 1 walnut: 1 pea; (c) 3 walnut: 3 rose: 1 pea: 1 single; (d) 1 rose: 1 single.

2.18. $Rrpp \times RrPp$.

2.19. 12 white: 3 yellow: 1 green.

2.20. 13 white: 3 colored.

2.21. (a) 3:1; (b) 3:1; (c) 1:1:2; (d) 2:1:1; (e) 1:1:1:1.

2.22. (a) $\chi^2 = .75$, P = .20 − .50, not significant; (b) $\chi^2 = 8.39$; P = .004, significant.

2.23. (a) $\chi^2 = .246$, P = .80 − .50, not significant; (b) $\chi^2 = .390$, P = .80 − .50, not significant; (c) $\chi^2 = .450$, P = .80 − .50, not significant; (d) $\chi^2 = .618$, P = .95 − .80, not significant; (e) $\chi^2 = .530$, P = .95 − .80, not significant.

2.24. (a) $\frac{1}{16}$; (b) $\frac{1}{16}$; (c) $\frac{4}{16}$; (d) $\frac{4}{16}$; (e) $\frac{6}{16}$.

2.25. (a) $\frac{6}{16}$ or $\frac{3}{8}$; (b) $\frac{1}{16}$; (c) 2 boys and 2 girls, because more combinations of the four independent events result in 2 and 2 than in any other arrangement; (d) $\frac{1}{2}$.

2.26. (a) $\frac{6}{64}$; (b) $\frac{20}{64}$; (c) $\frac{1}{64}$.

2.27. (a) 0; (b) $\frac{1}{2}$.

2.28. (a) $\frac{1}{2} \times 1 \times \frac{1}{4} = \frac{1}{8}$; (b) $\frac{1}{16}$; (c) $\frac{1}{6}$; (d) $\frac{1}{24}$.

2.29. (a) $\frac{243}{1024}$; (b) $\frac{405}{1024}$; (c) $\frac{270}{1024}$; (d) $\frac{90}{1024}$; (e) $\frac{15}{1024}$; (f) $\frac{1}{1024}$.

2.30. 9 black: 3 grey: 52 white.

2.31. 9 black: 39 grey: 16 white.

2.32. (a) $(\frac{3}{4})^3(\frac{1}{4}) = \frac{27}{256}$; (b) $(\frac{3}{4})^4 + (\frac{3}{4})^2(\frac{1}{4}) = \frac{117}{256}$; (c) $\frac{256}{256} - \frac{144}{256} = \frac{112}{256}$

CHAPTER 3

3.1. (a) 0; (b) 0; (c) 0; (d) 0; (e) 0; (f) +; (g) +.

3.2. (a) 23; (b) 23; (c) 23; (d) 23; (e) 46; (f) 46; (g) 23; (h) 46.

3.3. (a) 200; (b) 50.

3.4.

$$\frac{M}{m} \times \frac{m}{m} \qquad P$$

Ⓜ ⓜ   ⓜ   Gametes

$$\frac{M}{m} \qquad \frac{m}{m} \qquad$$ F_1, half of the progeny are expected to have myopia.

3.5. Model in text, Fig. 3.6, right.

3.6. The chromosome mechanism is similar in animals and plants. Division of the cytoplasmic part of the flexible animal cell is accomplished by constriction (cytokinesis), whereas the rigid plant cell forms a partition or cell plate.

3.7. Meiosis includes a pairing (synapsis) of corresponding maternal and paternal chromosomes. In the cell division that follows, the chromosomes that have previously paired separate. This results in a reduction of chromosome number from $2n$ (diploid) to n (haploid).

3.8. (a) Many plants have male and female parts on the same plant or in the same flower. Unlike animals, plants have a gametophyte stage that consists (in higher plants) of a few cell divisions. (b) The chromosome mechanism is essentially the same in the gamete formation of plants and animals.

3.9. An egg and an endosperm nucleus are developed in the ovule. Two haploid nuclei are introduced by the pollen tube. One nucleus fuses with the egg and the other with the endosperm nucleus. The fertilized egg is the zygote that develops into an embryo. The endosperm forms the nutrient material that supports the developing embryo.

3.10. Model for diagram, Fig. 2.11. About $\frac{1}{4}$ of all children would be expected to have only the gene for intestinal polyposis; $\frac{1}{4}$ to have only the gene for Huntington's chorea; $\frac{1}{4}$ to have neither; and about $\frac{1}{4}$ to have both.

3.11. Model in text, Fig. 3.6, right.

3.12. Model in text, Fig. 3.6, left.

3.13. $AaBb$.

3.14. Bisexual organisms. Asexual reproduction provides for no genetic variation, except that of rare mutations. Self-fertilization tends toward homozygosity or pure lines. Bisexual reproduction in higher organisms is associated with great hereditary variation through recombination.

3.15. (a) Early primary oöcyte; (b) prophase, first meiotic division; (c) suspended prophase; (d) first meiotic division is completed just before ovulation of each egg.

CHAPTER 4

4.1. Male Y and female X bearing sperm.

4.2. (a) Female (tetraploid); (b) intersex; (c) intersex; (d) metamale; (e) female (diploid); (f) male (sterile if no Y chromosome is present in *Drosophila*).

4.3. Female gametes would be (2X 2A); (2X A); (X 2A); and (X A). Zygotes and sex would be 3X 3A female (triploid); 3X 2A metafemale; 2X 3A intersex; 2X 2A female (diploid); 2XY 3A intersex; 2XY 2A female; XY 3A metamale; XY 2A male.

4.4. (a) Male; (b) female; (c) bisex; (d) female.

4.5. (a) Male; (b) female; (c) male; (d) female.

4.6. The single gene (*ba*) removes the female part of the monoecious plant and makes the stalk only staminate. Another gene (*ts*) transforms the tassel into a pistillate structure. A plant of the genotype *ba ba ts ts* would be only pistillate (female), whereas a plant with the genotype *ba ba ts⁺ ts⁺* would be only staminate (male).

4.7. (a) 0; (b) 1; (c) 1; (d) 2; (e) 3; (f) 0.

4.8. (a) $\frac{3}{4}$; (b) $\frac{1}{4}$.

4.9. (a) 3 hen-feathered: 1 cock-feathered; (b) all hen-feathered.

4.10. 2 rods; 1 hook and 1 rod; 1 knob and 1 rod; 1 hook and 1 knob.

4.11. (a) $\frac{1}{2}$ $X^{bb}X^{bb}$ bobbed females, $\frac{1}{2}$ $X^{bb}Y^{bb+}$ wild males; (b) $\frac{1}{2}$ $X^{bb}X^{bb+}$ wild females, $\frac{1}{2}$ $X^{bb}Y^{bb}$ bobbed males; (c) $\frac{1}{2}$ $X^{bb+}X^{bb+}$ and $X^{bb}X^{bb+}$ wild females, $\frac{1}{4}$ $X^{bb+}Y^{bb}$ wild males, $\frac{1}{4}$ $X^{bb}Y^{bb}$ bobbed males; (d) $\frac{1}{4}$ $X^{bb+}X^{bb}$ wild females, $\frac{1}{4}$ $X^{bb}X^{bb}$ bobbed females, $\frac{1}{2}$ $X^{bb+}Y^{bb+}$ and $X^{bb}Y^{bb+}$ wild males.

4.12. (a) Neither. Both the dominant and recessive would be expressed with equal frequency in males. (b) The dominant would be expressed more frequently in females.

4.13. (a) If it breeds true in succeeding generations, it is probably hereditary and may be assumed to have arisen through mutation. (b) The crisscross pattern of inheritance is evidence for sex linkage. (c) Females whose father had white eyes could be crossed with white-eyed males. Half of the females would be expected to have white eyes. This would conform to genetic theory based on sex linkage.

4.14.

4.15.

(1) $\dfrac{g^+}{\longrightarrow}$ (2) $\dfrac{g^+}{g}$ (3) $\dfrac{g}{\longrightarrow}$ (4) $\dfrac{g^+}{g}$ (5) Probability is $\frac{63}{64}$ for $\dfrac{g^+}{g^+}$ and $\frac{1}{64}$ for $\dfrac{g^+}{g}$ (6) $\dfrac{g^+}{g}$ (7) $\dfrac{g}{\longrightarrow}$ (8) $\dfrac{g^+}{\longrightarrow}$ (9) $\dfrac{g^+}{\longrightarrow}$

4.16. Half of their sons would be color-blind, and their daughters would be normal.

4.17. No. A son receives his X chromosome from his mother. The father contributes a Y chromosome.

4.18.

$\dfrac{g^+}{g} \times \dfrac{g}{\longrightarrow}$ P

$\dfrac{g^+}{}\quad\dfrac{g}{}\quad\dfrac{g}{}\longrightarrow$ Gametes

$\dfrac{g^+}{g}\quad\dfrac{g}{g}\quad\dfrac{g^+}{\longrightarrow}\quad\dfrac{g}{\longrightarrow}$ Sons and daughters: $\frac{1}{2}$ normal, $\frac{1}{2}$ color-blind

4.19.

(a) $\dfrac{h}{h} \times \dfrac{h^+}{\longrightarrow}$ P

$\dfrac{h}{}\quad\dfrac{h^+}{}\longrightarrow$ Gametes

$\dfrac{h}{h^+}\quad\dfrac{h}{\longrightarrow}$ F$_1$: normal daughters, hemophiliac sons

(b) $\dfrac{h^+}{h} \times \dfrac{h}{\longrightarrow}$ P

$\dfrac{h^+}{}\quad\dfrac{h}{}\quad\dfrac{h}{}\longrightarrow$ Gametes

$\dfrac{h^+}{h}\quad\dfrac{h}{h}\quad\dfrac{h^+}{\longrightarrow}\quad\dfrac{h}{\longrightarrow}$ F$_1$ daughters and sons: $\frac{1}{2}$ normal, $\frac{1}{2}$ hemophiliac

(c) $\dfrac{h^+}{h^+} \times \dfrac{h}{\longrightarrow}\quad\dfrac{h^+}{}\quad\dfrac{h}{}\longrightarrow$ P and gametes

$\dfrac{h}{h^+}\quad\dfrac{h^+}{\longrightarrow}$ F$_1$ all normal

4.20. $\frac{1}{2}$ chance for each son or $\frac{1}{4}$ for each child ($\frac{1}{2}$ for male $\times \frac{1}{2}$ affected).

4.21. (a) $\frac{1}{2}$; (b) $\frac{1}{2}$.

CHAPTER 5

5.1. (a) Griffith's *in vivo* experiments demonstrated the occurrence of transformation in pneumococcus. They provided no indication as to the molecular basis of the transformation phenomenon. Avery and colleagues carried out *in vitro* experiments, employing biochemical analyses to demonstrate that transformation was mediated by DNA. (b) Griffith showed that a transforming substance existed; Avery et al. defined it as DNA. (c) Griffith's experiments did not include any attempt to characterize the substance responsible for transformation. Avery et al. isolated DNA in "pure" form and demonstrated that it could mediate transformation.

5.2. (a) No effect; (b) no effect; (c) DNase will destroy the capacity of the extract to transform Type IIR cells to Type IIIS by degrading the DNA in the extract. Protease and RNase will degrade the proteins and the RNA, respectively, in the extract. They will have no effect, since the proteins and RNA are not involved in transformation.

5.3. Purified DNA from Type III cells was shown to be sufficient to transform Type II cells. This occurred in the absence of any dead Type III cells.

5.4. About $\frac{1}{2}$ protein, $\frac{1}{2}$ DNA. A single long molecule of DNA is enclosed within a complex "coat" that is composed of many proteins.

5.5. DNA contains phosphorus (normally ^{31}P), but no sulfur; it can be labeled with ^{32}P. Proteins contain sulfur (normally ^{32}S), but usually no phosphorus; they can be labeled with ^{35}S.

5.6. (a) The objective was to determine whether the genetic material was DNA or protein. (b) By labeling phosphorus, a constituent of DNA, and sulfur, a constituent of protein in a virus, it was possible to demonstrate that only the labeled phosphorus was introduced into the host cell during the viral reproductive cycle. This was enough to produce new phages. (c) Therefore DNA, not protein, is the genetic material.

5.7. (a) The ladderlike pattern was known from X-ray diffraction studies. Chemical analyses had shown that a 1:1 relationship existed between the organic bases adenine and thymine and between cytosine and guanine. Physical data concerning the length of each spiral and the stacking of bases were also available. (b) Watson and Crick developed the model of a double helix, with the rigid strands of sugar and phosphorus forming spirals around an axis, and hydrogen bonds connecting the complementary bases in base-pairs.

5.8. (a) A multistranded, spiral structure was suggested by the X-ray diffraction patterns. A double-stranded helix with specific base-pairing nicely fits the 1:1 stoichiometry observed for A:T and G:C in DNA. (b) Use of the known hydrogen-bonding potential of the bases provided a means of holding the two complementary strands in a stable configuration in such a double helix.

5.9. (a) 400,000; (b) 20,000; (c) 400,000; (d) 680,000 Å.

5.10. 3'-CAGTACTG-5'.

5.11. (a) DNA has one atom less of oxygen than RNA in the sugar part of the molecule. In DNA, thymine replaces the uracil that is present in RNA. (In certain bacteriophages, DNA-containing uracil is present.) DNA is most frequently double-stranded, but bacteriophages such as ΦX174 contain single-stranded DNA. RNA is most frequently single-stranded. Some viruses, such as the Reoviruses, however, contain double-stranded RNA chromosomes.

5.12. 13%.

5.13. No. TMV RNA is single-stranded. Thus, the base-pair stoichiometry of DNA does not apply.

5.14. (a) False; (b) false; (c) true; (d) true; (e) true; (f) true; (g) false; (h) true; (i) true; (j) false; (k) true; (l) false; (m) true.

5.15. (a) (1) One-half of the DNA molecules with ^{15}N in both strands and $\frac{1}{2}$ with ^{14}N in both strands; (2) all DNA molecules with one strand containing ^{15}N and the complementary strand containing ^{14}N; (3) all DNA molecules with both strands containing roughly equal amounts of ^{15}N and ^{14}N. (b) (1) $\frac{1}{4}$ of the DNA molecules with ^{15}N in both strands and $\frac{3}{4}$ with ^{14}N in both strands; (2) $\frac{1}{2}$ of the DNA molecules with one strand containing ^{15}N and the complementary strand containing ^{14}N and the other $\frac{1}{2}$ with ^{14}N in both strands; (3) all DNA molecules with both strands containing about $\frac{1}{4}$ ^{15}N and $\frac{3}{4}$ ^{14}N. See Fig. 5.13.

5.16. One-half of the DNA molecules fully heavy (^{15}N in both strands); the other half of the molecules "hybrid" (^{15}N in one strand, ^{14}N in the complementary strand).

5.17.

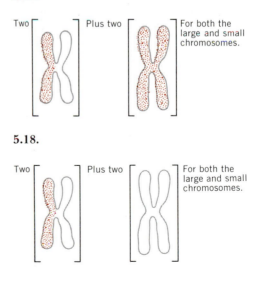

Two [ ] Plus two [ ] For both the large and small chromosomes.

5.18.

Two [ ] Plus two [ ] For both the large and small chromosomes.

5.19. (a) Both $3' \rightarrow 5'$ and $5' \rightarrow 3'$ exonuclease activities. (b) The $3' \rightarrow 5'$ exonuclease "proofreads" the nascent DNA strand during its synthesis. It a mismatched base-pair occurs at the 3'-OH end of the primer, the $3' \rightarrow 5'$ exonuclease removes the incorrect terminal nucleotide before polymerization proceeds again. The $5' \rightarrow 3'$ exonuclease is responsible for the removal of RNA primers during DNA replication and functions in pathways involved in the repair of damaged DNA (see Chapter 9). (c) Yes, both exonuclease activities appear to be very important. Without the $3' \rightarrow 5'$ proofreading activity during replication, an intolerable mutation frequency would probably occur. The $5' \rightarrow 3'$ exonuclease activity is essential to the survival of the cell. Conditional mutations (see Chapter 9) that affect the $5' \rightarrow 3'$ exonuclease activity of DNA polymerase I are lethal to the cell under conditions where the exonuclease is nonfunctional.

5.20. (a)

$$3' \text{ (P)-T G C G A A T T A G C G A C A T-(P) } 5'$$
$$5' \text{ (P)-A T C G G T A C G Ä Č Ğ Č T̈ Ä Ä T̈ Č Ğ Č T̈ Ğ Ä-OH } 3';$$

note that DNA synthesis will *not* occur on the left end since the 3'-terminus of the potential primer strand is blocked with a phosphate group—all DNA polymerases require a free 3'-OH terminus. (b) The first step will be the removal of the mismatched C (exiting as dCMP) from the 3'-OH primer terminus by the $3' \rightarrow 5'$ exonuclease ("proofreading") activity.

5.21. The satellite DNA fragments would renature much more rapidly than the main-band DNA fragments. In *D. virilus* satellite DNAs, all three have repeating heptanucleotide-pair sequences. Thus, essentially every 40 nucleotide-long (average) single-stranded fragment from one strand will have a sequence complementary (in part) with every single-stranded fragment from the complementary strand. Many of the nucleotide-pair sequences in main-band DNA will be unique sequences (present only once in the genome).

5.22. It indicates that highly repetitive DNA sequences do not contrain structural genes specifying RNA and polypeptide gene products.

5.23. (1) The nucleosome level; the core containing an octamer of histones plus 146 nucleotide pairs of DNA arranged as $1\frac{3}{4}$ turns of a supercoil (see Fig. 5.43), yielding an approximately 100 Å diameter spherical body; or juxtaposed, a roughly 100 Å diameter fiber. (2) The 300 Å fiber observed in condensed mitotic and meiotic chromosomes; structure unknown, apparently a second level of folding or coiling. (3) The highly condensed mitotic and meiotic chromosomes (e.g., metaphase chromosomes); the tight folding or coiling maintained by a "scaffold" composed of nonhistone chromosomal proteins (see Fig. 5.50).

5.24. If nascent DNA is labeled by exposure to ^{3}H-thymidine for very short periods of time, continuous replication predicts that the label would be incorporated into chromosome-sized DNA molecules, whereas discontinuous replication predicts that the label would first appear in small pieces of nascent DNA (prior to covalent joining, catalyzed by polynucleotide ligase).

5.25. That DNA replication was unidirectional rather than bidirectional. As the intracellular pools of radioactive ^{3}H-thymidine are gradually diluted after transfer to nonradioactive medium, less and less ^{3}H-thymidine will be incorporated into DNA at each replicating fork. This will produce autoradiograms with tails of decreasing grain density at each growing point. Since such tails appear at only one end of each track, replication must be unidirectional. Bidirectional replication would produce such tails at both ends of an autoradiographic track (see Fig. 5.58).

5.26. In the diploid nucleus of *D. melanogaster,* 10^6 nucleosomes would be present; these would contain 2×10^6 molecules of each histone, H2a, H2b, H3, and H4.

5.27.

PROTEINS	FUNCTIONS
1. DNA polymerase III	1. (a) Catalyzes polymerization (covalent extension) of new DNA chains. (b) The $3' \rightarrow 5'$ exonuclease activity "proofreads" the product, removing any mismatched base-pairs at the $3'$-end of the primer strand.
2. DNA polymerase I	2. Removes the RNA primers ($5' \rightarrow 3'$ exonuclease activity) and replaces them with new DNA strands ($5' \rightarrow 3'$ polymerase activity).
3. DNA ligase	3. Catalyzes covalent joining of "Okazaki fragments."
4. Primase (*dnaG* protein)	4. Catalyzes RNA primer synthesis.
5. DNA gyrase	5. Catalyzes the formation of negative supercoils; facilitates unwinding?
6. DNA helicase	6. Catalyzes unwinding.
7. DNA single-strand binding protein	7. Maintains an "extended" single-stranded template; aids unwinding?
8. *dnaB* protein	8. Required for initiation of replication.
9. proteins i, n, n', *dnaC* protein	9. "Prepriming"—required prior to initiation of primer synthesis.
10. *dnaI*, *dnaJ*, etc., proteins	10. Required; but functions unknown.

CHAPTER 6

6.1. (a) Prepare a testcross and compare results with those expected from the hypothesis of independent assortment. If they do not fit, linkage may be the next hypothesis. (b) The parental combinations should occur in greater proportion in the progeny. If the parental combinations are not known and only progeny data are available, determine whether the expressions controlled by the two dominants or those controlled by a dominant and recessive are in greater proportion. (c) With most materials, a testcross is easily prepared, and the results can be compared with the simple $1:1:1:1$ ratio rather than with the more complex $9:3:3:1$ F_2 ratio.

6.2. The one class that is represented by $\frac{351}{1000}$ is out of proportion for independent assortment. At least two of the three gene pairs must be in the same chromosome.

6.3. The three classes given fit the hypothesis of independent assortment, suggesting that the gene pairs are on three different chromosome pairs.

6.4. 20%.

6.5. (a) $\dfrac{a^+b^+}{a^+b^+} \times \dfrac{a\ b}{a\ b}$ P

$\underline{(a^+b^+)}$ $\underline{(a\ b)}$ Gametes

$\dfrac{a^+b^+}{a\ b}$ F_1

(b) $\dfrac{(a^+b^+)}{}$ 40%

$\dfrac{(a^+b)}{}$ 10%

$\dfrac{(a\ b^+)}{}$ 10%

$\dfrac{(a\ b)}{}$ 40%

(c) $\dfrac{a^+b^+}{a\ b}$ 40%

$\dfrac{a^+b}{a\ b}$ 10%

$\dfrac{a\ b^+}{a\ b}$ 10%

$\dfrac{a\ b}{a\ b}$ 40%

(d) Coupling.

6.6. (a) $\dfrac{a^+b}{a\ b^+}$ F_1

(b) $\dfrac{(a^+b)}{}$ 40%

$\dfrac{(a^+b^+)}{}$ 10%

$\dfrac{(a\ b)}{}$ 10%

$\dfrac{(a\ b^+)}{}$ 40%

(c) $\dfrac{a^+b}{a\ b}$ 40%

$\dfrac{a^+b^+}{a\ b}$ 10%

$\dfrac{a\ b}{a\ b}$ 10%

$\dfrac{a\ b^+}{a\ b}$ 40%

(d) Repulsion.

6.7. The parental gametes would be in the proportion of 30 percent each and the recombinants in the proportion of 20 percent each. The zygotes from the testcross would be in the same proportion as the F_1 gametes.

6.8. (a)

	(a)	(b)
Two dominant expressions $(a^+\text{-}b^+\text{-})$	66%	51%
One dominant and one recessive $(a^+\text{-}bb)$	9%	24%
Other dominant and other recessive $(aab^+\text{-})$	9%	24%
Two recessives $(aabb)$	16%	1%
	100%	100%

6.9. (a) No; (b) -.
(c) *Parentals, frequency = .496*

$\dfrac{b^+}{b}\ \dfrac{ts^+}{ts}$; red, normal; $\frac{124}{498} = .249$

$\dfrac{b}{b}\ \dfrac{ts}{ts}$; green, tassel; $\frac{123}{498} = .247$

Recombinants, frequency = .504

$\dfrac{b^+}{b}\ \dfrac{ts}{ts}$; red, tassel; $\frac{126}{498} = .253$

$\dfrac{b}{b}\ \dfrac{ts^+}{ts}$; green, normal; $\frac{125}{498} = .251$

(d) $\dfrac{b}{b^+}\ \dfrac{ts}{ts^+} \times \dfrac{b}{b}\ \dfrac{ts}{ts}$

6.10. (a) Yes; (b) 16.7%; (c) b and vg are linked; see (b).

(d) $\dfrac{b^+vg^+}{b\ vg} \times \dfrac{b\ vg}{b\ vg}$

6.11. (a) Yes; (b) 16.3%. (c) $\dfrac{b^+vg}{b\ vg^+} \times \dfrac{b\ vg}{b\ vg}$

6.12. (a) Yes; (b) 34%. (c) $\dfrac{c^+b}{c^+b} \times \dfrac{c\ b^+}{c\ b^+}$

6.13. (a) $\dfrac{d^+p^+}{d\ p}$ (b) $\dfrac{d^+p}{d\ p^+}$

(c) 54% $d^+\text{-}p^+\text{-}$: 21% $d^+\text{-}pp$: 21% $ddp^+\text{-}$: 4% $ddpp$.

6.14.

(a) $\dfrac{a^+b^+}{a^+b^+}\ \dfrac{c^+d^+}{c^+d^+} \times \dfrac{a\ b}{a\ b}\ \dfrac{c\ d}{c\ d}$ P

(b) $\dfrac{a^+b^+}{a\ b}\ \dfrac{c^+d^+}{c\ d}$ F_1

(c)

12% $\dfrac{a^+b^+}{}\ \dfrac{c^+d^+}{}$	8% $\dfrac{a^+b^+}{}\ \dfrac{c^+d}{}$
12% $\dfrac{a^+b^+}{}\ \dfrac{c\ d}{}$	8% $\dfrac{a^+b^+}{}\ \dfrac{c\ d^+}{}$
12% $\dfrac{a\ b}{}\ \dfrac{c^+d^+}{}$	8% $\dfrac{a\ b}{}\ \dfrac{c^+d}{}$
12% $\dfrac{a\ b}{}\ \dfrac{c\ d}{}$	8% $\dfrac{a\ b}{}\ \dfrac{c\ d^+}{}$
3% $\dfrac{a^+b}{}\ \dfrac{c^+d^+}{}$	2% $\dfrac{a^+b}{}\ \dfrac{c^+d}{}$
3% $\dfrac{a^+b}{}\ \dfrac{c\ d}{}$	2% $\dfrac{a\ b^+}{}\ \dfrac{c\ d^+}{}$
3% $\dfrac{a\ b^+}{}\ \dfrac{c^+d^+}{}$	2% $\dfrac{a^+b}{}\ \dfrac{c\ d^+}{}$
3% $\dfrac{a\ b^+}{}\ \dfrac{c\ d}{}$	2% $\dfrac{a\ b^+}{}\ \dfrac{c^+d}{}$

6.15. (a) Either cataract or polydactyly. The genes would be in repulsion. In the separation at meiosis, each gamete would get one or the other. A crossover would be required to produce a gamete with both or neither.

6.16. (a) 46% $\dfrac{sr\ e^+}{}$

46% $\dfrac{sr^+e}{}$

4% $\dfrac{sr^+e^+}{}$

4% $\dfrac{sr\ e}{}$

(b) 46% striped, gray
46% not striped, ebony
4% not striped, gray
4% striped, ebony

6.17. (a) Four kinds in proportion of 5%; 45%; 45%; 5%. (Repulsion.) (b) Wild-type: 5%; vestigial, red: 45%; long wing, cinnabar: 45%; vestigial, cinnabar: 5%.

6.18. Map distances may be converted to probabilities as follows: the chance that crossing over will occur between *st* and *ss* is .14 and the chance that crossing over will not occur is .86; the chance that crossing over will occur between *ss* and *e* is .12 and the chance that it will not occur is .88. By applying the multiplication theorem, the probabilities for the different combinations expected to be represented in the gametes can be calculated. Only one kind of gamete is produced by the male, so the proportion of zygotes will be the same as the proportion of gametes. Phenotypes may be expected in the following percentages:

Red eyes, normal bristles, gray body;
$$\frac{(.12 \times .86 \times 100)}{2} = 5.16\%$$

Red, normal, ebony:
$$\frac{(.86 \times .88 \times 100)}{2} = 37.84\%$$

Red, spineless, gray:
$$\frac{(.14 \times .88 \times 100)}{2} = 6.16\%$$

Red, spineless, ebony:
$$\frac{(.14 \times .12 \times 100)}{2} = .84\%$$

Scarlet, normal, gray:
$$\frac{(.14 \times .12 \times 100)}{2} = .84\%$$

Scarlet, normal, ebony:
$$\frac{(.14 \times .88 \times 100)}{2} = 6.16\%$$

Scarlet, spineless, gray:
$$\frac{(.86 \times .88 \times 100)}{2} = 37.84\%$$

Scarlet, spineless, ebony:
$$\frac{(.12 \times .86 \times 100)}{2} = 5.16\%$$

6.19.
Purple, salmon silk, pigmy:
$$\frac{.81}{2} = .405 = 40.5\%$$

Green, yellow silk, normal:
$$\frac{.81}{2} = .405 = 40.5\%$$

Purple, yellow silk, normal:
$$\frac{.09}{2} = .045 = 4.5\%$$

Green, salmon silk, pigmy:
$$\frac{.09}{2} = .045 = 4.5\%$$

Purple, salmon silk, normal:
$$\frac{.09}{2} = .045 = 4.5\%$$

Green, yellow silk, pigmy:
$$\frac{.09}{2} = .045 = 4.5\%$$

Purple, yellow silk, pigmy:
$$\frac{.01}{2} = .005 = .5\%$$

Green, salmon silk, normal:
$$\frac{.01}{2} = .005 = .5\%$$

These predictions are based entirely on probability, assuming equal crossing over in all areas along the chromosome. Interference could reduce the frequency of the double crossover classes.

6.20.
(*Tu* is a dominant mutant; Tu^+ is recessive.)

$$Tu^+ j_2 gl_3 : \frac{.893}{2} = .4465 = 44.65\%$$

$$Tu j_2^+ gl_3^+ : \frac{.893}{2} = .4465 = 44.65\%$$

$$Tu j_2 gl_3 : \frac{.047}{2} = .00235 = 2.35\%$$

$$Tu^+ j_2^+ gl_3^+ : \frac{.047}{2} = .0235 = 2.35\%$$

$$Tu^+ j_2 gl_3^+ : \frac{.057}{2} = .0285 = 2.85\%$$

$$Tu j_2^+ gl_3 : \frac{.057}{2} = .0285 = 2.85\%$$

$$Tuj_2gl_3^+: \frac{.0030}{2} = .0015 = .15\%$$

$$Tu^+j_2^+gl_3: \frac{.0030}{2} = .0015 = .15\%$$

6.21. The classes with the smallest numbers (+ + w and y ec +) must be double crossovers. The gene in these classes that differs from the parentals must be in the center, that is, + w + and y + ec. With the sequence established, single crossovers can be identified and percentages calculated.

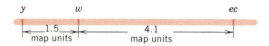

Or with y arbitrarily set at position 0 on the linkage map,

6.22. The smallest class is the double crossover class, and the v locus must be in the center. The double crossover value must be added to each single crossover value to obtain each map distance.

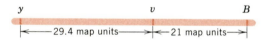

Or with y arbitrarily set at position 0 on the linkage map,

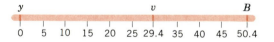

6.23. (a) Yes. The testcross results do not follow the 1:1:1:1:1:1:1:1 pattern expected from a trihybrid cross with independent assortment.

(b)

(c) Parental cross: $\dfrac{s^+ \; ss \; e}{s^+ \; ss \; e} \times \dfrac{s \; ss^+ \; e^+}{s \; ss^+ \; e^+}$

Testcross: $\dfrac{s^+ \; ss \; e}{s \; ss^+ \; e^+} \times \dfrac{s \; ss \; e}{s \; ss \; e}$

(d) 30.3 units.

(e) 14.0 units.

(f) $\dfrac{18}{.14 \times .163 \,(1000)} = .79$

6.24. 80% (30% 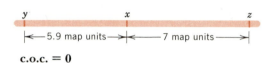 + 30% + 20%); only the 20% $\dfrac{w^+ \; dor^+}{}$ recombinants will exhibit a wild-type phenotype.

6.25.

c.o.c. = 0

Note that these genes are sex-linked; thus male progeny are hemizygous for X chromosome markers that are transmitted through the eggs (from their female parents). All female progeny will necessarily carry the wild-type alleles of all three genes, namely, wild-type alleles transmitted through the sperm (from their male parents). Therefore, linkage relationships can be determined only by analyzing the male progeny of the cross. Only six genotypes are represented among the progeny; thus, no double crossover types were present among the progeny analyzed. However, the missing double crossover genotypes can be determined by elimination.

6.26. $\dfrac{P^2}{4}$.

6.27. $\dfrac{(.5)(.02)(1000)}{2} = 5$

6.28.

6.29. ; 7.2 map units; 13.2 map units; 20.4 map units; c.o.c. = $\dfrac{.006}{(.072)(.132)} \cong .63$.

6.30. 2A-:1aa.

6.31. From the frequency of the homozygous recessive genotype aabb; this genotype can be formed only when an ab ovule is fertilized by an ab pollen. The frequency of the ab gamete type is therefore equal to the square root of the frequency of the double homozygous recessive genotype. In our example, the frequency of the aabb genotype equals $\frac{2}{200}$ or .01. Thus, the frequency of the ab gamete equals $\sqrt{.01}$ or .1. Since the cross was done in repulsion, this is a recombinant gamete type. Therefore, the recombination frequency equals 2 (.1) or .2, and the A and B loci are approximately 20 map units apart.

6.32. 21 map units.

6.33. It gives definitive evidence that crossing over occurs in the postreplication four-strand or tetrad stage rather than in the prereplication two-strand stage.

6.34.

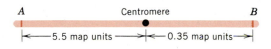

6.35.

0 map units *thi* to centromere.

0.17 map units *arg* to centromere; from $\dfrac{\frac{1}{2}(1)}{300} \times 100$.

11 map units *arg* to *leu*; from $\dfrac{\frac{1}{2}(44)}{200} \times 100$.

6.36. All loci are unlinked; the parental ditype to nonparental ditype ratios are 52:46, 50:40, and 46:46 for gene pairs *xy*, *xz*, and *yz*, respectively. Distances from genes to centromeres cannot be determined from unordered tetrad data by the simple procedure discussed in this text. In some cases, these distances can be calculated by a more complex procedure (see Whitehouse, *J. Genetics* 55:348, 1957).

6.37. The two genes are unlinked, but each is very tightly linked to its respective centromere.

6.38. The ascus pattern shown is most simply explained by a four-strand double crossover, with one crossover between *x* and *y* and the second between *y* and *z*.

6.39. $kar\text{-}ry^{402}\text{-}ry^8$

6.40. Chromosome 17.

CHAPTER 7

7.1. Recombination has occurred, producing wild-type or prototrophic bacteria.

7.2. Perform two experiments: (1) determine whether the process is sensitive to DNase, and (2) determine whether cell contact is required for the process to take place. The cell contact requirement can be tested by a U-tube experiment (see Fig. 7.1). If the process is sensitive to DNase, it is similar to transformation. If cell contact is required, it is similar to conjugation. If it is neither sensitive to DNase nor requires cell contact, it is similar to transduction.

7.3.

RECOMBINATION PROCESS	AGENT MEDIATING DNA TRANSFER	SIZE OF DNA UNITS TRANSFERRED	STATE OF DONOR DNA IN RECOMBINANT CELL
Transformation	Active uptake of free DNA	Small (about $\frac{1}{200}$ to $\frac{1}{100}$ of a chromosome)	Single strand integrated (see Fig. 7.19)
Transduction	Bacteriophage	Small (about $\frac{1}{100}$ to $\frac{1}{50}$ of a chromosome)	Integrated into host chromosome (except autonomous in abortive transduction)
Sexduction	F factor	Variable (see Fig. 7.16)	Initially added to the host cell as separate plasmid; may undergo recombination with host chromosome to yield stable sexductant

7.4. F$^-$ cells, no F factor present; F$^+$ cells, autonomous F factor: Hfr cells, integrated F factor (see Fig. 7.9). (b) F$^+$ and Hfr cells have F pili; F$^-$ cells do not (see Fig. 7.8). (c) F$^-$ cells are converted to F$^+$ cells by the conjugative transfer of F factors from F$^+$ cells; Hfr cells are formed when F factors in F$^+$ cells become integrated into the chromosomes.

7.5. F$'$ factors are useful for genetic mapping and, particularly, for *cis-trans* or complementation tests in bacteria. (b) F$'$ factors are formed by abnormal excision of F factors from Hfr chromosomes (see Fig. 7.15). (c) By the conjugative transfer of an F$'$ factor from a donor cell to a recipient (F$^-$) cell.

7.6. Generalized transduction: (1) transducting particles often contain only host DNA; (2) transducing particles may carry any segment of the host chromosome. Thus, all host genes are transduced. Specialized transduction: (1) transducing particles carry a recombinant chromosome, which contains both phage DNA and host DNA; (2) only host genes that are adjacent to the prophage integration site are transduced.

7.7. (a) A prophage is a phage chromosome that has become integrated into the host chromosome. The prophage is dormant in the sense that the phage genes involved in lytic development (replication and maturation) are repressed (see Chapter 11). The prophage replicates during host chromosome replication just as any other segment of the host chromosome. (b) No. (c) The mature (packaged) lambda chromosome and the lambda prophage are circular permutations of one another (see Fig. 7.4).

7.8. IS elements (or insertion sequences) are short (800–1400 nucleotide pairs) DNA sequences found in chromosomes (extensively studied so far only in bacteria and bacterial viruses) that are transposable—that is, capable of moving from one position in a chromosome to another position or from one chromosome to another chromosome. (b) IS elements appear to mediate recombination between nonhomologous DNA molecules or sequences (so-called "illegitimate recombination"). (c) IS elements are implicated in the evolution of R plasmids that carry multiple resistance to drugs and antibiotics (see Fig. 7.22).

7.9. By interrupting conjugation at various times after the donor and recipient cells are mixed (using a blender or other form of agitation), one can determine the length of time required to transfer a given genetic marker from an Hfr to an F$^-$. Since the chromosome is transferred in a linear sequence, the positions of genetic markers can be ordered relative to each other.

7.10. Cotransduction refers to the simultaneous transduction of two different genetic markers to a single recipient cell. Since bacteriophage particles can package only $\frac{1}{100}$ to $\frac{1}{50}$ of the total bacterial chromosome, only markers that are relatively closely linked can be cotransduced. The frequency of cotransduction of any two markers will be an inverse function of the distance between them on the chromosome. As such, this frequency can be used as an estimate of the linkage distance. Specific cotransduction-linkage functions must be prepared for each phage–host system studied.

7.11.

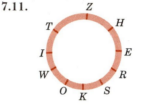

7.12. *lacY* — *lacZ* — *proC*.

7.13. *anth*-A34-A223-A46.

7.14. *anth*-A487-A223-A58.

7.15. *am*A453-*nrd*11-*nd*28-*am*M69-*am*N54.

CHAPTER 8

8.1. (a) RNA contains the sugar ribose, which has an hydroxyl (OH) group on the 2-carbon; DNA contains the sugar 2-deoxyribose, with only hydrogens on the 2-carbon. RNA usually contains the base uracil at positions where thymine is present in DNA. However, some DNAs contain uracil, and some RNAs contain thymine. DNA exists most frequently as a double helix (double-stranded molecule); RNA exists more frequently as a single-stranded molecule; but some DNAs are single-stranded, and some RNAs are double-standed. (b) The main function of DNA is to store genetic information and to transmit that information from cell to cell and from generation to generation. RNA stores and transmits genetic information in some viruses that contain no DNA. In cells with both DNA and RNA, (1) mRNA acts as in intermediary in protein synthesis, carrying the information from DNA in the chromosomes to the ribosomes (sites at which proteins are synthesized), (2) tRNAs carry amino acids to the ribosomes and function in codon recognition during the synthesis of polypeptides, and (3) rRNA molecules are essential components of the ribosomes. (c) DNA is located primarily in the chromosomes (with some in cytoplasmic organelles, such as mitochondria and chloroplasts), whereas RNA is located throughout cells.

8.2. 3'-ACGUCUGU-5'.

8.3. 3'-GACTA-5'.

8.4. (a) Environmental factors may influence developing phenotypes in the same way that genes do. If so, studies of phenocopies might suggest mechanisms of gene action. (b) Phenocopies provide one approach to the study of physiological genetics. Such studies are indirect and can only provide suggestions about the mechanisms of gene action. (c) Mutations are inherited; phenocopies are not inherited.

8.5. Phenocopies in humans are often very difficult to distinguish from inherited abnormalities in which the mutant gene has variable expressivity, such as variable age of onset.

8.6. 70%.

8.7. The genetic information of cells is stored in DNA, which is located predominantly in the chromosomes. The gene products (polypeptides) are synthesized primarily in the cytoplasm on ribosomes. Some intermediate must therefore carry the genetic information from the chromosomes to the ribosomes. RNA molecules (mRNAs) were shown to perform this function by means of RNA pulse-labeling and pulse-chase experiments combined with autoradiography (see Fig. 8.9). The enzyme RNA polymerase was subsequently shown to catalyze the synthesis of mRNA using chromosomal DNA as a template. Finally, the mRNA molecules synthesized by RNA polymerase were shown to faithfully direct the synthesis of specific polypeptides when used in *in vitro* protein synthesis systems.

8.8. Proteins are long chainlike molecules made up of amino acids linked together by peptide bonds. Proteins are composed of carbon, hydrogen, nitrogen, oxygen, and usually sulfur. They provide the enzymatic capacity and much of the structure of living organisms. DNA is composed of phosphate, the pentose sugar 2-deoxyribose, and four nitrogen-containing organic bases (adenine, cytosine, guanine, and thymine). DNA stores and transmits the genetic information in most living organisms. Protein synthesis is of particular interest to geneticists because proteins are the primary gene products—the key intermediates through which genes control the phenotypes of living organisms.

8.9. Protein synthesis occurs on ribosomes. In eukaryotes, most of the ribosomes are located in the cytoplasm and are attached to the extensive membranous network of endoplasmic reticulum (see Fig. 8.12). Some protein synthesis also occurs in cytoplasmic organelles such as chloroplasts and mitochondria.

8.10. Ribosomes are from 100 to 200 Å in diameter. They are located primarily in the cytoplasm of cells. In bacteria, they are largely free in the cytoplasm. In eukaryotes, many of the ribosomes are attached to the endoplasmic reticulum. Ribosomes are complex structures composed of about 50 different polypeptides and 3–5 different RNA molecules.

8.11. (a) The nucleus, specifically the nucleoli. (b) The cytoplasm.

8.12. Messenger RNA molecules carry genetic information from the DNA in the chromosomes (where the information is stored) to the ribosomes in the cytoplasm (where the information is expressed during protein synthesis). The linear sequence of triplet codons in an mRNA molecule specifies the linear sequence of amino acids in the polypeptide(s) produced during translation of that mRNA. Transfer RNA molecules are small (about 80 nucleotides) molecules that carry amino acids to the ribosomes and provide the codon-recognition specificity during translation. Ribosomal RNA molecules provide part of the structure and function of ribosomes; they represent an important part of the machinery required for the synthesis of polypeptides.

8.13. (a) Polysomes are formed when two or more ribosomes are simultaneously translating the same mRNA molecule. Ribosomes are usually spaced about 90 nucleotides apart on an mRNA molecule. Thus, polysome size is determined by mRNA size. (b) A ribosome, containing its three rRNA molecules, can participate in the synthesis of any polypeptide specified by the ribosome-associated mRNA. In that sense, rRNA is *nonspecific*. Messenger RNAs and tRNAs, in contrast, are *specific*, in directing the synthesis of a particular polypeptide or set of polypeptides (mRNA) or in attaching to a particular amino acid (tRNA). (c) Transfer RNA molecules are much smaller (about 80 nucleotides) than DNA or mRNA molecules. They are single-stranded molecules but have complex secondary structures because of the base-pairing between different segments of the molecules (see Figs. 8.15 and 8.16).

8.14. A specific aminoacyl-tRNA synthetase catalyzes the formation of an amino acid-AMP complex from the appropriate amino acid and ATP (with the release of pyrophosphate). The same enzyme then catalyzes the formation of the aminoacyl-tRNA complex, with the release of AMP. The amino acid-AMP and aminoacyl-tRNA linkages are both high-energy phosphate bonds.

8.15. (a) Synthetic RNA molecules (polyuridylic acid molecules) containing only the base uracil were prepared. When these synthetic molecules were used to activate *in* *vitro* protein synthesis systems, small polypeptides containing only the amino acid phenylalanine (polyphenylalanine molecules) were synthesized. Codons composed only of uracil were therefore shown to specify phenylalanine. Similar experiments were carried out using synthetic RNA molecules with different base compositions. (b) Better *in vitro* systems activated with synthetic RNA molecules with known repeating base sequences were developed. Ultimately, *in vitro* systems in which specific aminoacyl-tRNAs where shown to bind to ribosomes activated with specific mini-mRNAs, which were trinucleotides of known base sequence, were developed and used in codon identification.

8.16. (a) The genetic code is degenerate in that all but two of the 20 amino acids are specified by two or more codons. Some amino acids are specified by six different codons. The degeneracy occurs largely at the third or 3′ base of the codons. "Partial degeneracy" occurs where the third base of the codon may be either of the two purines or either of the two pyrimidines and the codon still specifies the same amino acid. "Complete degeneracy" occurs where the third base of the codon may be any one of the four bases and the codon still specifies the same amino acid. (b) The code appears to be almost completely universal. Known exceptions to universality include strains carrying suppressor mutations that alter the reading of certain codons (with low efficiencies in most cases) and the use of UGA as a tryptophan codon in yeast and human mitochondria.

8.17. Colinearity received strong support from studies that showed a direct correlation between the linear sequence of mutational sites in a gene (established by genetic mapping experiments) and the linear sequence of mutational defects (amino acid substitutions or chain terminations) in the polypeptide gene product (established by purification and amino acid sequencing, etc., of mutant polypeptides). See Figs. 8.28 and 10.19.

8.18. Colinearity is an important prediction of our present concepts of transcription and translation. Whether noncoding "spacers" or "intervening sequences" in eukaryotic genes will require modifications of our concept of colinearity is still uncertain.

8.19. Blueprints transcribed into building instructions and translated into structures composed of boards, bricks, and mortar by skilled craftsmen and craftswomen may be likened to DNA, mRNA, and tRNA functions in the assembly of amino acids into polypeptides by ribosomes and other required factors.

8.20. (a) Met → Val. This substitution occurs as a result of a transition. All other amino acid substitutions listed would require transversions.

8.21. His → Arg results from a transition; His → Pro would require a transversion (not induced by 5-bromouracil).

8.22. (a) By RNA polymerase binding to *promoter sequences* (see Chapter 11). (b) By specific *terminator sequences* (apparently AT-rich sequences of DNA) that are recognized by RNA polymerase or the transcription termination protein *rho*. (c) By a complex reaction involving mRNA, ribosomes, initiation factors (IF-1, IF-2, and IF-3), GTP, the initiator codon AUG, and a special initiator tRNA ($tRNA_f^{Met}$). It probably also involves a base-pairing interaction between a base sequence near the 3′-end of the 16S rRNA and a base sequence in the "leader sequence" of the mRNA. (d) By recognition of one or more of the chain-termination codons (UAG, UAA, and UGA) by the appropriate release factors (RF-1 and/or RF-2).

8.23. At least 813 nucleotides.

8.24. By changing the anticodon or by changing the sequence recognized by the aminoacyl-tRNA synthetase. Yes. Mutations in genes coding for aminoacyl-tRNA synthetases. Possibly also mutations in genes coding for certain components of the ribosome.

8.25. 3′-TACACC-5′ (transcribed strand)
5′-ÄTGTGG-3′ (nontranscribed strand)

8.26. Incorporation of alanine into polypeptide chains.

8.27. *Amber* (UAG). This is the only nonsense codon that is related to tryptophan, serine, tyrosine, leucine, glutamic acid, glutamine, and lysine codons by a single base-pair substitution in each case.

8.28. (1) The promoters of eukaryotic and prokaryotic organisms are different. Thus, the eukaryotic genes will probably have to be joined to bacterial promoters. (2) Eukaryotic genes often contain introns. Bacteria probably do not have the enzymatic machinery needed to remove introns. Thus, unless the introns are "self-splicing," they will probably have to be removed from the eukaryotic genes before their insertion into bacterial plasmids in order to obtain expression.

CHAPTER 9

9.1. (a) Transition; (b) transition; (c) transversion; (d) transversion; (e) frameshift; (f) transition.

9.2. (a) C*l*B method; (b) attached-X method (see Figs. 9.16 and 9.17).

9.3. Bacteria treated with a mutagen or expected to carry mutations may be introduced into media with particular drugs in appropriate concentrations. Colonies that appear have originated from cells carrying preexisting mutations for resistance. This may be verified by the replica-plating technique (see Fig. 9.1). The frequency of mutation of wild-type (drug-sensitive) cells to drug resistance can be measured in the presence or the absence of the drug.

9.4. Probably not. A human is larger than a bacterium, with more cells and a longer life span. If mutation frequencies are converted to cell generations, the rates for human cells and bacterial cells are similar.

9.5. A dominant mutation presumably occurred in the woman in whom the condition was first known.

9.6. The sex-linked gene is carried by mothers, and the disease is expressed in half of their sons. Such a disease is difficult to follow in pedigree studies because of the recessive nature of the gene, the tendency for the expression to skip generations in a family line, and the loss of the males who carry the gene. One explanation for the sporadic occurrence and tendency for the gene to persist is that, by mutation, new defective genes are constantly being added to the load already present in the population.

9.7. Plants can be propagated vegetatively, but no such methods are available for widespread use in animals.

9.8. The sheep with short legs could be mated to unrelated animals with long legs. If the trait is expressed in the

9.9. Enzymes may discriminate among the different nucleotides that are being incorporated. Mutator enzymes may utilize a higher proportion of incorrect nucleotides, whereas antimutator enzymes may select fewer incorrect bases in DNA replication. In the case of the phage T4 DNA polymerase, the relative efficiencies of polymerization and proofreading by the polymerase's 3′ → 5′ exonuclease activity play key roles in determining the mutation rate.

9.10. If both mutators and antimutators operate in the same living system, an optimum mutation rate for a particular organism in a given environment may result from natural selection.

9.11. *Dt* is a mutator gene that induces somatic mutations in developing kernels.

9.12. (a) Yes. (b) A block would result in the accumulation of phenylalanine and a decrease in the amount of tyrosine, which would be expected to result in several different phenotypic expressions.

9.13. These hemoglobins can be distinguished by mobility of molecules in an electric field (electrophoretic mobility) and by determining the amino acid sequences of their β polypeptides.

9.14.

Amino acid	mRNA	DNA
Glutamic acid	—GAA→	—GAA→ ←CTT— ← Transcribed strand ↓ Mutation
Valine	—GUA→	—GTA→ ←CAT— Mutation
Lysine	—AAA→	—AAA→ ←TTT—

first generation, it could be presumed to be inherited and to depend on a dominant gene. On the other hand, if it does not appear in the first generation, F_1 sheep could be crossed back to the short-legged parent. If the trait is expressed in one-half of the backcross progeny, it might be presumed to be inherited as a simple recessive. If two short-legged sheep of different sex could be obtained, they could be mated repeatedly to test the hypothesis of dominance. In the event that the trait is not transmitted to the progeny that result from these matings, it might be considered to be environmental or dependent on some complex genetic mechanism that could not be identified by the simple test used in the experiments.

9.15. The label "molecular disease" became common in speaking of sickle-cell anemia because its molecular basis (the substitution of a valine residue for the glutamic acid residue at amino acid position number 6 in the β chain) was recognized quite early during the emergence of the science of molecular biology. Actually, most if not all inherited diseases probably have very similar molecular bases. We just don't know what the molecular defects are in most instances.

9.16. Mutations: transitions, transversions, frameshifts.

9.17. Irradiate the nonresistant strain and plate the irradiated organisms on a medium containing streptomycin. Those that survive and produce colonies are resistant.

They could then be transferred to a medium without streptomycin. Those that survive would be of the first type; those that can live with streptomycin but not without it would be the second type.

9.18. 3%; 4%; 6%.

9.19. Each quantum of energy from the X rays that is absorbed in a cell has a certain probability of hitting and breaking a chromosome. Hence, the greater the number of quanta of energy or dosage, the more likely breaks are to occur. The rate at which this dosage is delivered does not change the probability of each quantum inducing a break.

9.20. The person receiving a total of 100 r would be expected to have twice as many mutations as the one receiving 50 r.

9.21. During the replicating process, ultraviolet light produces mispairing alterations mostly in pyrimidines (e.g., cytosine to thymine transitions). Thymine may be altered to cytosine, which pairs with guanine. A reverse mutation may occur when cytosine is changed to thymine, which pairs with adenine. The T-A base-pair may thus be changed to a C-G, and the reverse mutation may occur from C-G to T-A.

9.22. Nitrous acid brings about a substitution of an OH group for an NH$_2$ group in those bases (A, C, and G) having NH$_2$ side groups. In so doing, adenine is converted to hypoxanthine, which base-pairs with cytosine, and cytosine is converted to uracil, which base-pairs with adenine. The net effects are GC $\rightleftharpoons$ AT base-pair substitutions (see Fig. 9.27).

9.23. Transitions.

9.24. Nitrous acid acts as a mutagen on nonreplicating DNA (resting DNA) and produces transitions from A to G or C to T, whereas 5-bromouracil does not affect resting DNA but acts during the replication process causing GC $\rightleftharpoons$ AT transitions.

9.25. Large numbers of seeds have been given massive X-ray doses and have been screened to see if induced mutations have improved any of them for particular envi-

ronments. In one study of oats, a few seeds produced plants with improved disease resistance.

9.26. The immediate effects are surface and internal burning. Irradiation is known to induce leukemia and, in general, to shorten the life span—an effect that is difficult to evaluate. Genetic effects are insidious; small doses produce mutations and have a cumulative effect. The number of mutations generally is proportional to total dosage. This is a potential danger to future generations.

9.27. Mutations induced by acridine dyes are primarily insertions or deletions of single base-pairs. Such mutations alter the reading frame (triplets specifying mRNA codons) for that portion of the gene distal (relative to the direction of transcription and translation) to the mutation (see Fig. 9.14).

9.28. Proline and serine.

9.29. No. Leucine $\rightarrow$ proline would occur more frequently. Leu (CUA) $\xrightarrow{5\text{-Bu}}$ Pro (CCA) occurs by a single base-pair transition, whereas Leu (CUA) $\xrightarrow{5\text{-Bu}}$ ser (UCA) requires two base-pair transitions. Recall that 5-bromouracil (5-BU) induces only transitions (see Figs. 9.25 and 9.26).

9.30. No. 5-Bromouracil is mutagenic only to replicating nucleic acids.

9.31.

Yes. DNA: $\leftarrow$ GGX $-$ $\xrightarrow{HNO_2}$ $\leftarrow$ GGX $-$ or $\leftarrow$ GGX $-$ or $\leftarrow$ GGX

$-$ CCX′ $\rightarrow$ $-$ UCX′ $\rightarrow$ $-$ CUX′ $\rightarrow$ $-$ UUX′ $\rightarrow$

mRNA: GGX   AGX   GAX   AAX

Polypeptide: Gly   Ser or Arg   Asp or Glu (depending on X)   Asn or Lys

Note: The X at the third position in each codon in mRNA and in each triplet of base-pairs in DNA refers to the fact that there is complete degeneracy at the third base in the glycine codon. Any base may be present in the codon, and it will still specify glycine.

9.32. No. The glycine codon is GGX, where X can be any one of the four bases. Because of this complete degeneracy at the third position of the glycine codon, changing X to any other base will have no effect (i.e., the codon will still specify glycine). Nitrous acid deaminates guanine (G) to xanthine, but xanthine still base-pairs with cytosine. Thus guanine is not a target for mutagenesis by nitrous acid.

9.33. Tyr $\rightarrow$ Cys substitutions; Tyr to Cys requires a transition, which is induced by nitrous acid. Tyr to Ser would require a transverion, and nitrous acid is not expected to induce transversions.

9.34. (b) Met → Thr. 5-Bromouracil induces transitions, not transversions. All other changes listed require transversions.

9.35. 5'-AUGCCCUUUGGG**G**AAAGG

UUUCCCUAA-3'

CHAPTER 10

10.1. Prior to 1940, the gene was considered a "bead-on-a-string," not subdivisible by recombination or mutation. Today, the gene is considered to be the unit of genetic material that codes for one polypeptide. The unit of structure, not subdivisible by recombination or mutation, is known to be the single nucleotide pair.

10.2. The recombination observed between lz^s and lz^g, two functionally allelic mutations at the *lozenge* locus of *Drosophila*.

10.3. The *cis-trans* test, which defines the unit of genetic material specifying the amino acid sequence of one polypeptide.

10.4. They provide powerful selective sieves for identifying rare recombinants. This is accomplished by using the restrictive environmental conditions to select wild-type recombinant progeny from crosses between pairs of conditional lethal mutants.

10.5. 7-5-6-4-1-2-3.

10.6. Two genes; mutations 1, 2, 3, 4, 5, 6, and 8 are in one gene; mutation 7 is in a second gene.

10.7. Four genes; mutations 1 and 2 in one gene; mutations 3 and 4 in a second gene; mutations 5 and 6 in a third gene; mutations 7 and 8 in a fourth gene.

10.8. The size of the gene (assuming that all nucleotide pairs in the gene are capable of undergoing base-pair substitutions, as seems highly probable). Dominant lethal alleles and recessive lethal alleles in haploids will (under normal conditions) exist only transiently, of course.

10.9. Homoalleles are structurally and functionally allelic; they are *not* separable by recombination. Heteroalleles are functionally allelic (based on *cis-trans* tests), but are structurally nonallelic (based on recombination tests). Heteroalleles thus result from mutations occurring at different sites within a gene.

10.10. (1) Cross the two white-flowered varieties. The F_1 plants will be *trans*-heterozygotes. If the F_1 plants have white flowers, the two varieties probably carry mutations in the same gene, causing white flowers. (2) Cross white-flowered varieties with red-flowered varieties and self-pollinate or intercross the F_1 plants. If alleles of a single gene are involved, monohybrid F_2 ratios should be observed in all cases.

10.11. In most cases, the positions of loci on chromosomes can be determined only by genetic mapping experiments. Such experiments require at least one mutant allele of a gene, so that the segregation of the pairs of alleles can be followed in crosses. The recombinant DNA, cloning, restriction endonuclease mapping, and nucleic acid sequencing techniques developed in the last few years now allow some wild-type genes to be analyzed and localized in the chromosome without genetic mapping data.

10.12. (a) By independent mutations at various sites within an ancestral gene. Such mutations may accumulate over time, resulting in the divergence of different alleles. (b) The same basic symbol is used conventionally for all members of a group of multiple alleles, with superscripts to identify particular alleles. (c) Most series of multiple alleles are associated with gradations in the same phenotype. Some produce quite different end results, such as legs in the place of antennae in *Drosophila*.

10.13. (a) All colored; (b) 3 colored:1 albino; (c) 3 colored:1 chinchilla; (d) 1 chinchilla:1 albino; (e) 3 colored:1 himalayan; (f) 1 himalayan:1 albino.

10.14. (a) Yellow and yellow; 2 yellow:1 agouti light belly ($A^Y A^Y$ is lethal); (b) yellow and agouti light belly; 2 yellow:1 agouti light belly:1 black and tan; (c) black and tan and yellow; 2 yellow:1 black and tan:1 black; (d) agouti light belly and agouti light belly; all agouti light belly; (e) agouti light belly and yellow; 1 yellow:1 agouti light belly; (f) agouti and black and tan; 1 agouti:1 black and tan; (g) black and tan and black; 1 black and tan:1 black; (h) yellow and agouti; 1 yellow:1 agouti light belly; (i) yellow and yellow; 2 yellow:1 agouti.

10.15. (a) 1; (b) 2; (c) 2; (d) 10.

10.16. 30.

10.17. (a) 28; (b) by crossing fish that carry the different alleles, intercrossing the F_1 progeny, and checking for monohybrid F_2 ratios, and by looking for possible complementation in *trans*-heterozygotes that carry pairs of the putative alleles.

10.18. (a) $S_1 S_3$ and $S_2 S_3$; (b) $S_1 S_3$, $S_1 S_4$, $S_2 S_3$, and $S_2 S_4$; (c) none; (d) $S_3 S_5$, $S_3 S_6$, $S_4 S_5$, and $S_4 S_6$.

10.19. (a) all AB; (b) 1 A:1 B; (c) 1 AB:1 A:1 B:1 O-type; (d) 1 A:1 O-type.

10.20. It is **extremely unlikely** (approximately 10^{-5} to 10^{-6}) that the man with AB-type blood was the father of the child with O-type blood. This could happen only if a new mutation had occurred in one of the A or B blood-type alleles of the man's reproductive cells.

10.21. It is **extremely unlikely** (see answer to Problem 10.20 above) that the baby with O-type blood was the daughter of the woman with AB-type blood.

10.22. (a) All children would be heterozygous (*Rr*) and Rh positive. The mother became immunized during the first pregnancy and the next child was affected. (b) All future children would be expected to be affected.

10.23. Half of the children carried by the *rr* mother would be *Rr* (Rh positive) and half would be *rr* (Rh negative), like the mother. Because the woman was sensitized at the second pregnancy, the *Rr* children would be expected to be erythroblastotic and the *rr* children normal.

10.24. (a) The amino acid sequences of polypeptides that are specified by overlapping genes or "genes-within-genes" would not be able to evolve independently. However, the degeneracy of the code (see Chapter 8) will allow for some independence. In addition, a single mutation within a shared nucleotide sequence could result in non-functional products for both genes. (b) 3. (c) 6.

10.25. (a) Both introduce new genetic variability into the cell. In both cases, only one gene or a small segment of DNA representing a small fraction of the total genome is changed or added to the genome. The vast majority of the genes of the organism remain the same. (b) The introduction of recombinant DNA molecules, if introducing genes from a very different species, is more likely to result in a novel, functional gene product in the cell, **if the introduced gene (or genes) is capable of being expressed in the foreign protoplasm.** The introduction of recombinant DNA molecules is more analogous to duplication mutations (see Chapter 12) than to other types of mutations.

10.26. This is a wide-open question at present! There is much speculation (see p. 364), but little hard evidence. In one yeast mitochondrial gene, the introns contain open reading frames that appear to code for "maturases" that splice out these introns—a neat negative feedback control. Other introns may be merely relics of evolution.

10.27.

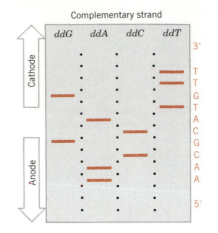

10.28.

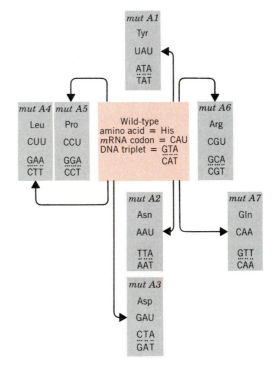

Deductions

(1) Wild-type His codon must be CAU based on DNA sequence of gene given.

(2) The codons for the seven amino acids found at posi-

tion 6 in the mutant polypeptides must be connected to CAU by a single base change since the mutants were all derived from wild-type by a single nucleotide-pair substitution. Thus, the degeneracy of the genetic code is not a factor in deducing specific codon assignments.

(3)   Because of the nature of the genetic code—specifically the degeneracy at the third (3′) position in each codon—there are three possible amino acid substitutions due to single base substitutions at each of the first two positions (the 5′ base and the middle base), but only one possible amino acid change due to a single base change at position 3 (the 3′ base in the codon).

Note: For ease of discussion, the three nucleotide-pair positions in the triplet of nucleotide pairs in the gene specifying the CAU codon in mRNA and the amino acid His in the wild-type polypeptide will be referred to as position 1 (corresponding to the 5′ base in the codon), position 2 (the middle nucleotide pair), and position 3 (corresponding to the 3′ base in the mRNA codon).

(4)   Since *A1, A2,* and *A3* do not recombine with each other, they must all result from base-pair substitutions at the same position in the triplet, at either position 1 or position 2. The same is true for *A4, A5,* and *A6.* Since *A7* recombines with each of the other six mutant alleles, it must result from the one possible substitution at position 3 causing an amino acid change.

(5)   The only amino acid with a codon connected to the His codon CAU by a single base change at position 3 is Gln (codon CAA). Thus, the *mut A7* polypeptide must have glutamine as the sixth amino acid.

(6)   Since *mut A7* (the third position substitution) yields about twice as many wild-type recombinants in crosses with *mut A1* as in crosses with *mut A6,* the *A1* substitution must be at position 1 and the *mut A6* substitution must be at position 2. Combined with (4) above, this places the *A2* and *A3* substitutions at position 1 and the *A4* and *A5* substitutions at position 2.

(7)   Since *mut A1* and *mut A6* are induced to revert to wild-type by 5-BU, they must be connected to the His triplet of nucleotide pairs by transition mutations, that is

$$(mut\ A1)\ \frac{ATA}{TAT} \xleftrightarrow{\text{5-BU}} \frac{GTA}{CAT} \xleftrightarrow{\text{5-BU}} \frac{GCA}{CGT}\ (mut\ A6)$$

(8)   Since *mut A3* and *mut A5* are induced to mutate to *mut A2* and *mut A4,* respectively, by hydroxylamine, *A3* must be connected to *A2* and *A5* to *A4* specifically by GC → AT transitions, i.e.,

$$(mut\ A3)\ \frac{CTA}{GAT} \xrightarrow{\text{HA}} \frac{TTA}{AAT}\ (mut\ A2)$$

and

$$(mut\ A5)\frac{GGA}{CCT} \xrightarrow{\text{HA}} \frac{GGA}{CTT}\ (mut\ A4).$$

CHAPTER 11

11.1.   By studying the synthesis or lack of synthesis of the enzyme in cells grown on chemically defined media. If the enzyme is synthesized only in the presence of a certain metabolite or a particular set of metabolites, it is probably inducible. If it is synthesized in the absence but not in the presence of a particular metabolite or group of metabolites, it is probably repressible.

11.2.   Repression occurs at the level of transcription during enzyme synthesis. The end product, or a derivative of the end product, of a repressible system acts as an effector molecule that usually, if not always, combines with the product of one or more regulator genes to turn off the **synthesis** of the enzymes in the biosynthetic pathway. Feedback inhibition occurs at the level of enzyme **activity;** it usually involves the first enzyme of the biosynthetic pathway. Feedback inhibition thus brings about an immediate arrest of the biosynthesis of the end product. Together, feedback inhibition and repression rapidly and efficiently turn off the synthesis of both the enzymes and the end products that no longer need to be synthesized by the cell.

11.3.

Gene or Regulatory Element	Function
(1) Regulator gene	Codes for repressor
(2) Operator	Binding site of repressor
(3) Promoter	Binding site of RNA polymerase and CAP-cAMP complex
(4) Structural gene *z*	Codes for β-galactosidase
(5) Structural gene *y*	Codes for β-galactoside permease

11.4.   (1) Constitutive synthesis of the *lac* enzymes. (2) Constitutive synthesis of the *lac* enzymes. (3) Uninducibility of the *lac* enzymes. (4) No β-galactosidase activity. (5) No β-galactoside permease activity.

11.5. (a) 1, 2, 3*, and 5; (b) 2, 3*, and 5; *3 may be either noninducible or constitutive, depending on whether the specific o^c mutation eliminates binding of the i^s "superrepressor."

11.6. (a) Inducible, (b) constitutive, (c) constitutive, (d) inducible, (e) constitutive.

11.7. (a) $\dfrac{i^+ \ o^c \ z^+ \ y^-}{i^+ \ o^+ \ z^- \ y^+}$ (b) $\dfrac{i^s \ o^c \ z^+ \ y^-}{i^s \ o^+ \ z^- \ y^+}$

11.8. (a) The o^c mutations map very close to the z structural gene; i^- mutations map slightly farther from the z structural gene (but still very close by; see Fig. 11.7). (b) An $\dfrac{i^+ \ o^c \ z^+ \ y^+}{i^+ \ o^+ \ z^+ \ y^!}$ partial diploid would exhibit constitutive synthesis of β-galactosidase and β-galactoside permease; an $\dfrac{i^+ \ o^+ \ z^+ \ y^+}{i^- \ o^+ \ z^+ \ y^+}$ partial diploid would be inducible for the synthesis of these enzymes. (c) The o^c mutation is *cis*-dominant; i^- is *trans*-recessive.

11.9. The system could have developed from a series of tandem duplications of a single ancestral gene. Mutational changes that make the system more efficient and, therefore, favored in selection could have brought the system to its present level of efficiency.

11.10. Catabolite repression has apparently evolved to assure the use of glucose as a carbon source when this carbohydrate is available, rather than less efficient energy sources.

11.11. Possibly by directly or indirectly inhibiting the enzyme adenylcyclase, which catalyzes the synthesis of cyclic AMP from ATP.

11.12. Positive regulation; the CAP-cAMP complex has a "positive" effect on the expression of the *lac* operon. It functions in turning on the transcription of the structural genes in the operon.

11.13. Yes. In the gene coding for CAP; some mutations in this gene might result in a CAP that binds to the promoter in the absence of cAMP. Also, mutations in the gene (or genes) coding for the protein (or proteins) that regulates the cAMP level as a function of glucose concentration.

11.14. Operons are common in prokaryotes. Gene clusters that resemble operons, if they are not in fact operons, are known to exist in eukaryotic microorganisms such as yeasts and fungi. No operons are known in higher eukaryotes. Why operons are common in prokaryotes but are rare, if present at all, in higher eukaryotes is unknown. More knowledge about the organization of genes and modes of regulation of gene expression in eukaryotes will be required before this question can be answered.

11.15. Probably the hypothesis most acceptable would be (a) the intervention of cytoplasm. The cytoplasmic repressor or activator substances involved in the process of differentiation according to this hypothesis fit rather nicely with the operon concept.

11.16. Negative regulatory mechanisms such as that involving the repressor in the lactose operon block the transcription of the structural genes of the operon, whereas positive mechanisms such as the activator in the arabinose operon or the CAP-cAMP effect on the *lac* operon promote the transcription of the structural genes of the operon.

11.17. In prokaryotes, the structural genes specifying the enzymes in a metabolic pathway are often arranged as clusters of contiguous genes. This facilitates regulation by the operon mechanism. In higher animals, such genes are usually not in clusters, often being unlinked. The complex patterns of gene expression during development in higher animals almost certainly require complex integrated controls that can govern gene expression. Genomes of higher animals contain single-copy DNA sequences (structural genes, in some cases at least) that are interspersed with middle-repetitive sequences. The Britten and Davidson model is consistent with these observations.

11.18. (a) We really don't know yet. Histones are thought to be nonspecific repressors or inhibitors of transcription. This is true *in vitro* where the rate of RNA synthesis from chromatin (native or reconstituted) is inversely related to the amount of histones present, given a constant amount of DNA. Whether this is true *in vivo* is not known. Some researchers believe that histone modifications, such as phosphorylations, acetylations, and so on, play an important role in the regulation of gene expression in eukaryotes. (b) Histones are synthesized along with DNA during the S phase of the cell cycle. (c) Again, we don't know how the synthesis of histones is regulated. However, nonhistone chromosomal proteins apparently play a key role.

11.19. We do not know the details of the mechanism(s) by which steroid hormones regulate gene expression. These hormones form complexes with specific receptor proteins in target cells. In some way, possibly involving nonhistone chromosomal proteins, these steroid hormone-receptor protein complexes trigger the transcription of the appropriate genes in target cell nuclei (see Fig. 11.14).

11.20. When salivary chromosomes are followed in their developmental sequence, puffs that are controlled by a hormone occur in specific regions in a regular pattern. Puffs have been interpreted as regions of RNA synthesis and have been related to particular gene loci. The chromosomes are large and easily observed. Staining techniques are available to identify areas of RNA synthesis, and the gene sequence along the chromosomes is known and identifiable with phenotypes in the flies. *In situ* DNA-RNA hybridization and autoradiography can be used to determine the chromosomal location of genes transcribed in response to a particular hormone.

11.21. (a) These organisms can be cultured readily and investigated in the laboratory. Techniques are available for studies of biochemical pathways. (b) 1. The material most suitable for experiments would be organisms such as *E. coli* and other bacteria that have a single linkage group, many biochemical mutants, and, for the most part, organisms that reproduce asexually; these organisms would be most likely to develop clusters of coordinately regulated genes. 2. The effects of histones might be studied using *in vitro* chromatin-reconstitution (fractionation and reassociation) experiments in combination with *in vitro* transcription systems. These *in vitro* systems might be derived from cells such as chicken erythrocytes, which have large amounts of histones. 3. Insects or higher animals provide excellent systems with which to study the effects of steroid hormones on gene expression. These organisms use many different steroid hormones as signals that control the metabolic patterns in different tissues. 4. *Drosophila* and other *Diptera* exhibit puffing of bands in their giant salivary chromosomes.

11.22. (a) Genes controlling pigment production are present in all cells of the rabbit. The chemical reaction required for black pigment can proceed at a temperature of 27°C, but it is blocked at higher temperatures. (b) While the rabbits are developing in the body of the mother, the temperature is about 33°C, and after birth the main part of the body is maintained at that temperature. Extremities and treated areas in this experiment were cooler as the reaction proceeded.

11.23. Although the genomes remain largely the same in all cells of an organism, the molecular composition of the cytoplasms of cells is constantly changing in response to both the genome and the environment. Many of the metabolic processes occurring in cells take place in the cytoplasm. Thus changes in the cytoplasm may induce or repress the expression of genes. In some cases, these changing patterns of metabolism appear to be genetically preprogrammed. In other cases, they occur in response to environmental stimuli.

11.24. Differentiation provides the necessary cell forms and functions; organization results in appropriate groupings of cells and tissues; and growth, through cell division, determines the ultimate size and shape of the whole organism as well as its various parts.

11.25. At present, there is no known artificial method for producing and supplying to the cells the necessary enzymes for pigment production. Enzymes that operate in cells must be synthesized in these cells. In the future, it is likely that techniques will be developed by which enzymes, mRNAs, or DNA molecules (synthetic or purified natural genes) may be introduced into mutant cells or tissues.

11.26. (a) Wild-type host tissue apparently can compensate for the chemical blocks induced by *v* and *cn* genes. (b) Two steps in the production of wild-type pigment are involved. Discs with *v* can synthesize a material needed for pigment production when supplemented with a substance from the *cn* host. The *cn* discs, however, had their pigment production blocked at a later state and required a substance that could not be supplied by the *v* host.

11.27. Homeotic mutants provide direct evidence that patterns of gene expression occurring during differentiation are under genetic control (at least in part). We can only hope that future studies of these mutants, and of the genes in which the mutations have occurred to produce these mutant phenotypes, will elucidate the mechanism(s) by which choices between different patterns of gene expression are made.

11.28. If purified mRNA transcribed from the gene can be obtained, it can be used to synthesize a radioactive cDNA probe (a complementary strand of DNA), using the enzyme "reverse transcriptase" (RNA-dependent DNA polymerase). The radioactive cDNA can then be allowed to hybridize with total RNA extracted from each of the two tissues. After hybridization, all single-stranded molecules can be degraded with a single-strand-specific nuclease. The amount of radioactivity remaining in macromolecular form (precipitable by trichloroacetic acid) is a measure of the amount of gene-specific mRNA present in the two tissues. This is the most direct approach. If the gene is known to be the structural gene for a specific enzyme or polypeptide, the synthesis of this enzyme or polypeptide can be used as an indirect assay for the expression of the gene. To be certain that synthesis of the protein accurately reflects what is occurring at the transcription level, one would have to make sure that it is not synthesized from a stable mRNA.

11.29. Uninducible, but with a higher level of baseline synthesis of the arabinose enzymes; or, stated differently, a low level of constitutive synthesis of the enzymes. The product of the arabinose regulator gene is required for induction; in its absence, induction could not occur. However, in the absence of arabinose, the regulator gene product represses the level of *ara* operon transcription. This effect would also be eliminated in such a deletion mutant, resulting in a higher baseline level of synthesis (or a low level of constitutive synthesis).

11.30. The genetic information specifying antibody chains is stored in sequences of nucleotide pairs coding for sequences of amino acids, just like the genetic information specifying other polypeptides. However, the information specifying an antibody chain is stored in bits and pieces that are assembled into functional genes specifying antibody chains by genome rearrangements (somatic cell recombination events) occurring during the development of the B lymphocytes (the antibody-producing cells). See Figs. 11.26 and 11.27.

11.31. 4; 2; 1.

11.32. (1) The joining of different *V, D,* and *J* gene segments by somatic recombination during B lymphocyte development (see Figs. 11.26 and 11.27); (2) variability in the exact location of the joining reaction during *V-D-J* joining events; and (3) somatic mutation.

11.33. At the DNA level; class switching occurs by somatic recombination during B lymphocyte differentiation (see Fig. 11.27).

11.34. The proto-oncogenes and retroviral oncogenes have very different structures. The viral oncogenes contain a single uninterrupted coding sequence or exon, whereas the proto-oncogenes contain several coding sequences interrupted by several introns (see Fig. 11.32).

CHAPTER 12

12.1. (a) A recessive gene presumed to be carried in heterozygous condition unexpectedly came to expression. Bridges postulated that a section in the homologous chromosome carrying the wild-type allele was missing (i.e., there was a chromosome deficiency). On another occasion, Bridges found that a recessive gene presumed to be homozygous did not come to expression. He postulated that a gene acting as a dominant allele must be present elsewhere in the chromosome set (i.e., there was a duplication). (b) It was impossible to distinguish microscopically between the structural parts of homologous chromosomes at the time these genetic results were obtained. (c) The discovery of attached-X chromosomes in *Drosophila*, meiotic configurations in maize, and salivary chromosomes in *Drosophila* provided tools for cytological verification.

12.2. Salivary preparations in *Drosophila* provide larger chromosomes to work with than those of meiotic stages in maize. Chromosome parts can be identified with the aid

of Bridges' chromosome maps. Salivary chromosome studies are made from somatic cells, whereas maize studies are made from germ cells.

12.3. Enlarged size, somatic pairing, identifiable bands, and distinguishable anatomical features along the length of the different chromosomes make salivary gland chromosomes especially useful.

12.4. Chromosomes can be spread on a slide, fixed, and stained in a single operation with a single solution. When they are well spread out on the slide, the linear sequence of an individual chromosome can be followed. On the other hand, it would be necessary to reconstruct the chromosomes if sections were employed for study.

12.5. The chromonemata duplicate themselves many times, but the chromosomes and cells do not divide. Bundles are thus developed. The cross bands represent groups of identical chromomeres.

12.6. Linkage maps are constructed by placing the relative gene positions calculated from crossover data along a line representing a chromosome. Cytological maps are constructed by microscopic observations of chromosomes from actual cell preparations. The salivary chromosome maps prepared by Bridges are linkage maps superimposed on cytological maps.

12.7. When a section of a chromosome carrying a dominant gene becomes deleted, a recessive allele carried in the homologous chromosome may come to expression.

12.8. (a) The extent of a deficiency can be determined by testing the genes on either side of the point known to be in the deficiency to see if they behave as pseudodominants. (b) The determination can be made cytologically by observing microscopically the extent in a suitable chromosome preparation (e.g., a salivary chromosome preparation with chromosomes heterozygous for the deficiency).

12.9. (a) This is not an easy determination. New mutants of a particular kind may occur infrequently at a particular locus and produce phenotypic changes like those associated with a deficiency. A mutation from v^+ to v in this case would appear unlikely, but, if it did occur, the pair of recessive genes would behave in a regular Mendelian pattern. Testcross or F_2 results could be predicted and checked. In animals such as *Drosophila* from which more complete data are available, deficiences are usually homozygous lethal, resulting in modified F_2 ratios. If other recessives in the same chromosome, presumed to be heterozygous, came to expression, the case for a deficiency would be strengthened. (b) Cytological determination in mice is also difficult. There are no enlarged salivary or meiotic chromosomes available for study. If a chromosome could be shown to be shorter or structurally different from its homolog, the deficiency hypothesis would gain support. Banded chromosomes could now be used for identifying a deficiency in a mouse chromosome.

12.10. (a) When crossing over occurs in the area of a paracentric inversion, acentric and dicentric chromosomes occur that do not separate properly to the poles in division. Gametes carrying crossover chromatids are abnormal and inviable. (b) Crossovers within pericentric inversions result in unbalanced chromosome arrangements that make the crossover gametes or zygotes inviable. (c) Crossing over is reduced to some extent, but the main "suppression" results from inviable gametes or zygotes.

12.11. All flies would be beaded with the genotype Bd $\ell^+/Bd^+\ \ell$.

12.12. The seedlings would be approximately 1 white : 2 green : 1 yellow; white and yellow seedlings would die, and all mature plants would be green.

12.13. (a) Autosomal; (b) chromosome III.

12.14. (a) A loop is formed or a buckling occurs in the unpaired normal chromosome segment, which corresponds in size with the deficiency. (b) A loop similar in appearance to that described for a deficient is formed. (c) A loop is formed by the synapsed chromosomes; one member of the pair is reversed, making it possible for the corresponding segments of the inverted and univerted parts to pair. (d) A cross is formed by the pairing chromosomes (see Figs. 12.3, 12.5, 12.6, 12.7, 12.10, and 12.11).

12.15. (a) Altered linkage groups, pollen or ovule sterility, and position effects suggest the presence of translocations. (b) Cross configurations, and rings and chains of chromosomes in the meiotic prophase that can be seen microscopically represent cytological evidence.

12.16. Unbalanced chromosome arrangements occur in gamete formation, making pollen inviable (see Fig. 12.11).

12.17. (a) An inversion has occurred involving section 456. (b) A loop would be formed with paired chromosomes, and the elements of the inverted segment would be in reverse order (i.e., 654).

12.18. (a) Segment 34 has been deleted, (b) segment 4 duplicated, (c) segment 876 inverted (see Fig. 12.5 for model).

12.19. See Fig. 12.11 for model.

12.20. If a structural alteration in a chromosome can be definitively related to a phenotypic change, a position effect would be indicated. The phenotypic effect is caused by the change in position of genetic materials rather than by addition or deletion.

12.21. (a) When a section of euchromatin is moved by structural change to a location in or near heterochromatin, variegated position effects may occur. (b) Apparently genes moved to a different chromosome location behave differently from those remaining in the genetic environment to which they are adjusted.

12.22. (a) The Philadelphia chromosome is a fragment of a No. 22 chromosome that is associated with a particular disease. (b) It is a diagnostic factor for chronic myelocytic leukemia.

12.23. (a) 45,XY,-21; (b) 45,XX,$-14,-21$, $+$t$(14q21q)$; (c) 46,XY,$-5,-12,+$t$(5p12p),+$t$(5q12q)$.

12.24. Inversions and translocations act as isolating factors among individuals in populations, thus preventing normal gene exchange and promoting speciation.

12.25. Techniques for preparing slides for critical microscopic observation were not effectively applied to human chromosome studies until 1956. Better sources of human material from surgical procedures and tissue-culture techniques are now available.

12.26. Human chromosomes are small, quite uniform, and, using older techniques, difficult to distinguish. They could be separated into seven groups following the Denver conference criteria, but enough overlap occurred among members of the same group (partly because of variations in techniques) to make it impossible to consistently identify individual chromosomes.

CHAPTER 13

13.1. (a) 0; (b) 2; (c) 1.

13.2. Nondisjunction of chromosomes in the production of gametes (eggs) seems to be the explanation for most, if not all, trisomy.

13.3. (a) $\dfrac{1}{490,000}$; (b) 60,000; (c) $\dfrac{1}{70}$.

13.4. $\dfrac{1}{2}$.

13.5. $\dfrac{1}{3,500,000}$.

13.6. (a) $\dfrac{1}{4,000,000}$; (b) $\dfrac{1}{40,000}$.

13.7. (a) 17 purple : 1 white; (b) 11 purple : 1 white; (c) 15 purple : 3 white reduced to 5 purple : 1 white; (d) 3 purple : 1 white. (e) The mutant trait occurs less frequently than expected in the results of crosses because of the presence of an extra chromosome, usually carrying a wild-type allele.

13.8. (a) $\dfrac{1}{2}(2n)$ and $\dfrac{1}{2}(2n+1)$; (b) 11 normal : 1 eyeless.

13.9. (a) XO is basically female; XXY male; and XXX metafemale. The human Y carries the male-determining capacity. (b) *Drosophila* Y has no influence on sex determination. *Melandrium* Y carries male determiners.

13.10. Some plants and animals with a particular extra chromosome can be recognized phenotypically. Genes that give trisomic ratios (e.g., genes for purple or white in *Datura*) are located in the particular chromosome that makes the trisome. Monosomics are useful because recessive genes on a chromosome that has no homolog (for example, a monosome) express themselves and can thus be associated with a particular chromosome.

13.11. Tetrasomics and nullisomics apparently occur in nature, but they are less viable and usually die before they are detected.

13.12. Monoploids have only one set of genes and, therefore, no gene segregation. They could be used for experimental work where it is desirable to relate genes with phenotypes or traits. In molds such as *Neurospora*, monoploids represent the most important part of the life cycle and they are used extensively for genetic studies.

13.13. Polyploid tissues grow through cell division that can occur regularly among polyploid cells. Sexual reproduction of whole animals requires gamete formation, fertilization, and sex determination. Irregularities associated with polyploidy nearly always result in inviability and sterility.

13.14. Polyploidy might occur through a doubling of chromosomes in the somatic cells or the failure of the reduction division, resulting in unreduced gametes.

13.15. (a) Autopolyploidy is the doubling of a $2n$ complement, resulting in four similar genomes. Allopolyploidy occurs through hybridization, in which two $2n$ complements are involved. (b) Because autopolyploidy results in four duplicate sets of chromosomes, pairing is irregular; parts of all four similar chromosomes may be paired in different places, thus interfering with normal reduction division. This and more subtle genetic factors make autopolyploids sterile and incapable of perpetuating themselves. On the other hand, allopolyploidy provides for two pairs of chromosome sets. The chromosomes can pair properly and produce gametes. Apparently this has been important in the evolution of many plant groups.

13.16. Tetraploids have an even number of chromosome sets or genomes. If they have arisen through hybridization and the four genomes are represented as two pairs, synapsis—an essential part of meiosis—can be accomplished with regularity. Triploids with three genomes have inherent difficulties in pairing and gamete formation, and they usually do not penetrate themselves through normal sexual reproduction. They may arise through fusion of unreduced ($2n$) and reduced (n) gametes, but usually they do not reproduce triploids unless apomixis or some other deviation from sexual reproduction occurs.

13.17. Poor pairing, the most obvious cause of sterility, is associated with chromosome irregularity. More subtle genetic factors are also involved.

13.18. (a) A valuable criterion available to the taxonomist is chromosome number. It is at least as significant as any well-defined morphological characteristic. (b) Numerical as well as structural differences in chromosomes have been significant in the mechanics of evolution. Analysis and comparison of chromosome numbers within and between taxonomic groups have aided in solving problems of evolution.

13.19. (1) Doubling of chromosomes in the section of the hybrid that grew vigorously. (2) Propagation of cells from the polyploid area to reproduce a plant with two sets of chromosomes from each parent species.

13.20. (1) A few cells of the hybrid were apparently polyploid, with 24 chromosomes. (2) These gave rise to pollen grains with 12 chromosomes. (3) When these pollen grains were used to fertilize species-*B* eggs, a few 19-chromosome plants were produced. (4) These produced gametes with different numbers of chromosomes, some with 12. (5) Gametes with 12 chromosomes fused to form 24-chromosome plants with the characteristics of a new polyploid species.

13.21. Colchicine treatment interferes with spindle formation in cell division and some chromosome mechanisms. Irregular numbers of chromosomes are included in the daughter cells. Sometimes the membrane forms around the entire two sets of chromosomes that normally would separate to two cells, and the number in the single cell is doubled.

13.22. (a) Colchicine-induced polyploidy provides a means of accomplishing in the laboratory a process that has apparently occurred naturally in many important plant groups. Some polyploids have qualities that give them practical advantages. New polyploids are being produced and tested against established commercial varieties of grapes, tomatoes, and other plants, and they have been found to have valuable properties. (b) Induced polyploidy has theoretical significance as a tool for discovering the mechanism that has occurred in the evolution of some plant groups. The origin of modern varieties of cotton, wheat, and other valuable and interesting polyploids presents a challenge for present and future investigations.

13.23. Polyploidy that is associated with hybridization could account for the differences among some known strains of cotton and wheat. The processes through which some modern polyploids may have developed in nature have been reconstructed in the laboratory (see text for details).

13.24. Progeny of interspecific crosses are usually sterile. If hybridization is combined with chromosome doubling, fertile hybrids may occur and perpetuate their new chromosome arrangement. They do not cross with the diploids and, if they are perpetuated in nature, they might eventually develop into a new species.

13.25. (a) *Triticum spelta* has been reproduced experimentally by doubling the chromosomes of emmer wheat and goat grass and crossing the two polyploids. (b) *Raphanobrassica* was developed experimentally by crossing the radish and the cabbage. The F_1 progeny were sterile, but some unreduced gametes were obtained, and the F_2 tetraploid (*Raphanobrassica*) was produced from them. (c) *Spartina townsendii* apparently occurred in nature from the crossing of *Spartina alterniflora* with *S. stricta* and the doubling of the chromosome number of the hybrid. (d) *Primula kewensis* was produced by the crossing of *Primula floribunda* and *P. verticillata* and the doubling of the chromosome number of the hybrid. (e) The origin of New World cotton has been reconstructed experimentally by crossing Old World cotton with upland cotton and doubling the chromosome number of the hybrid.

13.26. (a) Haploid; (b) diploid; (c) trisomic; (d) monosomic; (e) tetrasomic, (f) triploid; (g) tetraploid.

CHAPTER 14

14.1. Maternal inheritance. It is inherited through the maternal line and does not involve nuclear genes of the mother.

14.2. (a) The cytoplasm provides the environment in which the genes act. Therefore, the mother would be expected to influence secondary actions of certain genes more than the father. (b) Nonchromosomal genes would be inherited through the maternal line.

14.3. (a) Sex-linked genes are located in the X chromosome, and characteristic crisscross inheritance could be detected from appropriate crosses if sex linkage were involved. (b) Cytoplasmic inheritance would be transmitted through the maternal line because most of the cytoplasm of the zygote comes from the egg. A series of backcrosses could be made from F_1 males and females to the appropriate females and males of the two parent types. If the trait is transmitted repeatedly for several generations from the maternal parent to her progeny and not through the paternal parent, it may be cytoplasmic. (c) If the trait was transmitted from mother to progeny but did not persist in the maternal line, it might be attributed to the influence of the mother's genes on the developing egg or embryo, that is, a maternal effect. In this case, nuclear genes would be involved, but they would be the genes of the mother rather than those of the individual itself.

14.4. Inherited changes in the proplastids and the multiplication of cells in which the changes occurred resulted in sectors that showed different color characteristics, as in the four-o'clocks. If the plastid characteristics are determined solely by eggs, only female gametes would determine green, pale green, or variegated plants. In examples from other species, a small amount of cytoplasm is carried with sperm, which also results in variegated plants.

14.5. Male sterility facilitates crosses involving plants that are ordinarily self-fertilized. Large-scale crossing for obtaining hybrid vigor is accomplished more economically if the plants are male sterile.

14.6. Kappa particles are bacteria that have developed an intimate symbiotic relationship with paramecia of a particular genotype.

14.7. Maternal inheritance. *Oenothera hookeri* must have yellow plastids, and *O. muricata* must have green plastids.

14.8. (a) Because this trait is transmitted as a maternal effect, all three genotypes could have dextral or sinistral coiling, depending on the genes of the mother in question. (b) The mother and grandmother might be expected to determine the coiling characteristics in their immediate progeny. An *ss* mother would produce sinistral progeny. Male parents have no immediate effect on this trait.

14.9.

$s^+s^+ \times ss$		P
(s^+) (s)		Gametes
s^+s		F_1
$1s^+s^+ : 2s^+s : 1ss$		F_2 genotypes; all F_2 snails would coil right because the F_1 mother was s^+s.
right, right, left		Phenotypic results from inbreeding the F_2 snails.

Phenotype depends on the mother's genotype. All F_2 snails would coil to the right because the F_1 mothers were s^+s. Following inbreeding, progeny of *ss* snails would coil to the left.

14.10. (a) Dark young, dark adult; (b) light young, dark adult. (c) The pigment condition is at first influenced by the cells of the mother. When the hoppers grow older, the condition determined by the genes of the individual snail replaces the one that originated from the mother (if there is a difference in genotype between mother and young).

14.11. The eyes became lighter as the kynurenine that diffused from the *AA* host into the egg was metabolized and broken down by the *aa* individuals that were unable to manufacture more kynurenine.

14.12. (c) A gynandromorph.

CHAPTER 15

15.1. $AABB \times A'A'B'B'$ P
                  $AA'BB'$ F_1
          $AA'BB' \times AA'BB'$ $F_1 \times F_1$

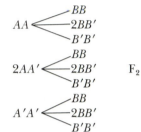

Summary of F_2: $\frac{1}{16}$ red, $\frac{4}{16}$ dark, $\frac{6}{16}$ medium, $\frac{4}{16}$ light, $\frac{1}{16}$ white.

15.2. Three pairs of genes were segregating in Problem 15.2, and only two pairs in Problem 15.1.

15.3. (a) $AA'B'B'C'C'$ or $A'A'BB'C'C'$ or $A'A'B'B'CC'$; (b) $AA'BB'C'C'$ or $AA'B'B'C'C$ or $A'A'BB'CC'$; (c) $AA'BB'CC'$.

15.4. (a) $A'A'B'B'C'C'$; (b) $AAB'B'C'C'$ or $A'A'BBC'C'$ or $A'A'B'B'CC$; (c) $AA'B'B'B'C'C'$ or $A'A'BB'C'C'$ or $A'A'B'B'CC'$; (d) $AA'BB'C'C'$ or $AA'B'B'CC'$ or $A'A'BB'CC'$; (e) $AA'BB'CC'$.

15.5. (a) $AA'BB'C'C' \times AA'B'B'C'C'$ or any other combination with two capital-letter (not prime) genes in one parent and one (of the same genes) in the other. (b) $AA'BB'CC' \times AA'BB'C'C'$ or $AA'BB'CC' \times AA'B'B'CC'$ or $AA'BB'CC' \times A'A'BB'CC'$; (c) $AA'BB'CC' \times AA'BB'CC'$.

15.6. $AA'BB'$ genotype, medium or mulatto phenotype.

15.7. $\frac{1}{16}$ black ($AABB$), $\frac{1}{16}$ white ($A'A'B'B'$), and $\frac{14}{16}$ with pigmentation intensity between black and white ($\frac{4}{16}$ dark, $\frac{6}{16}$ intermediate, $\frac{4}{16}$ light).

15.8. (a) No. The greatest degree of pigmentation possible would be medium ($AA'BB'$). (b) If the parents were genetically $A'A'B'B'$, they could not have a baby with the *A* or *B* pigment-producing genes.

15.9. (a) There would be more gradations in pigmentation, and more than five classes would be required to classify the phenotypes. Individuals homozygous for all

pigment-producing genes or all genes for white would be considerably more infrequent than the model based on two pairs suggests. (b) Better methods of classifying phenotypes, and more data showing the proportions of different color classes in mixed populations, will be required to determine more precisely the number of genes involved. The actual results of matings between large numbers of people with only black ancestry and those with only white ancestry, as well as those of matings involving other particular genotypes, would be useful in making such a determination.

15.10. (a) 1 ft; (b) see Fig. 15.5 for model. Summary of F_2 phenotypes: one plant, 6 ft; four plants, 5 ft; six plants, 4 ft; four plants, 3 ft; and one plant, 2 ft.

15.11. $\frac{2012}{8} = 251$, or 1 extreme in about 256; four pairs of genes were involved.

15.12. (a) $\bar{x} = 20.45$; (b) $s = 1.54$; (c) $s_{\bar{x}} = .34$.

15.13. (a) The mean measures the magnitude or average of the sample in the units of measurement. (b) The standard deviation measures the variation within the sample. (c) The standard error of the mean measures the reliability of the sample in terms of sample size and variability in population sample.

15.14. Many. In a sample of 20, the sizes of the parents were not approached.

15.15. (a) $\bar{x} = 10$; (b) $s = 1.3$; (c) $s_{\bar{x}} = .29$.

15.16. Few, perhaps two pairs. One in 20 represented the size range of the small parent and 3 in 20 that of the large parent. One in 16 would be expected if two pairs of equally effective genes were operating.

15.17. (a) $\bar{x} = 11.65$; (b) $s = 1.37$; (c) $s_{\bar{x}} = .124$.

15.18. (a) $\bar{x} = 56.2$; (b) $s = 1.14$; (c) $s_{\bar{x}} = .23$.

15.19. (a) The mean was near that of the parent that required a short time for maturity. The small standard deviation shows that there is not much variation, and the small standard error indicates that a sample of 24 plants was sufficiently large to properly sample the population. (b) The end of the curve represented the shorter time required for maturity. (c) The F_2 parent of the sample recorded was probably homozygous.

15.20. (a) $\bar{x} = 73.2$; $s = 2.3$; $s_{\bar{x}} = .48$. The mean was slightly larger than that of the original parent with the long time for maturity. The larger standard deviation indicates more variation, probably because of more heterozygosity in parents. The larger standard error indicates that a sample larger than 24 plants would have better sampled the population. (b) $\bar{x} = 65.25$; $s = 2.72$; $s_{\bar{x}} = .56$. The mean is intermediate between those of the original parents (P) and somewhat larger than those of the F_1 and F_2. The parent was taken from the end of the curve between the mean and the end representing long time for maturity. The larger standard deviation indicates more heterozygosity than that found in the other F_3 samples, but not as much as shown by the F_2. The larger standard error indicates that a larger sample would be desirable. (c) $\bar{x} = 65.25$; $s = 1.29$; $s_{\bar{x}} = .26$. The mean is the same as that of sample (b), but the standard deviation and standard error are smaller, indicating less heterozygosity in the parent and a more adequate sample to represent this population.

15.21. (a) The parameter in question will fall within "plus or minus" limits calculated in 68 percent of similar trials. (b) The population mean (parameter) will fall within plus or minus 1.02 cm of the sample mean in about 68 percent (plus or minus one standard deviation) of

similar trials. (c) The variance estimates the amount of variation in the population sampled.

15.22. MZ and DZ twins expressing a particular polygenic trait, such as cleft lip and cleft palate, can be compared for concordance and discordance.

CHAPTER 16

16.1. (a) No, unless they are associated with phenotypes that are favored in selection; (b) mutation, selection, migration, chance (i.e., random genetic drift), and meiotic drive; (c) random mating.

16.2. The proportion is .0126 (about 13 per 1000), assuming the Hardy–Weinberg equilibrium.

16.3.

Group	L^M	L^N
(a) Central American Indians	.78	.22
(b) North American Indians	.39	.61

(c) Assuming a common origin for the two presently isolated American Indian populations and the small size of the populations, the founder principle and random genetic drift could have been major factors in the divergence of allele frequencies.

16.4. (a) $L^M L^M = 61\%$, $L^M L^N = 34\%$, $L^N L^N = 5\%$; (b) $T = .455$ and $t = .545$, assuming equilibrium.

16.5. (a) $TT = .21$, $Tt = .49$, $tt = .30$; (b) about $\frac{2}{7}$; (c) about $\frac{5}{7}$ (all answers assume equilibrium).

16.6. $I^A = .262$; $I^B = .064$; $i = .674$ (all answers assume equilibrium).

16.7. O-type blood, 92.16%; A, 5.85%; B, 1.93%, AB, .06% (all results assume equilibrium).

16.8.

	$r(i)$	$p(I^A)$	$q(I^B)$
Males:	.69	.25	.06
Females:	$r(i)$	$p(I^A)$	$q(I^B)$
	.77	.18	.05

(assuming equilibrium)

16.9. $RR = .25$; $Rr = .50$; $rr = .25$.

16.10. $A = \dfrac{.8 + 2(.5)}{3} = \dfrac{1.8}{3} = .6$; $a = .4$

Males $= .6\ A$; $.4\ a$

Females $= .36\ AA$; $.48\ Aa$; $.16\ aa$ (genotype frequencies)
$= .6\ A$; $.4\ a$ (allele frequencies)

Generations

	0	1	2	3	4	5	6
Males:	$A = .8$	.5	.65	.575	.6125	.59375	.603125
Females:	$A = .5$	.65	.575	.6125	.59375	.603125	.5984375

Equilibrium is approached in the sixth generation, but more generations of random mating would be required to stabilize equilibrium at $.6\ A$ and $.4\ a$.

16.11. $\Delta q \sim -.0048 =$ change of q per generation. Since s is small, the value is approximate.

16.12. No AA or Aa; 100% aa.

16.13. $-.00032 =$ change of q per generation.

16.14. $A = .09$; $a = .91$.

16.15. .01.

16.16. .99.

16.17. Random fluctuations in gene frequencies that result from sampling errors are most effective in very small populations. They become less effective as population size increases. Random genetic drift is a direct result of the sampling process. Since in this process the variance is $\dfrac{pq}{N}$ and the standard deviation is $\sqrt{qp/N}$ (where N is the population size), as N becomes smaller the standard deviation becomes a significant factor.

16.18. The X and Y chromosomes have a differential effect on preferential segregation in meiosis.

16.19. People carrying the gene Hb_β^S for sickle-cell anemia are immune to the falciparum malaria parasite and are thus favored by selection in areas where malaria is prevalent.

16.20. Falciparum malaria is a selective agent, which takes a heavy toll of Hb^A_β/Hb^A_β people. Most Hb^S_β/Hb^S_β people die in childhood from the sickle-cell disease. Those heterozygous (Hb^A_β/Hb^S_β) are immune to malaria and not seriously affected by the sickle-cell disease. To maintain this balance, a high incidence of malaria is required.

$$\hat{q}_a = \frac{S_A}{S_A + S_a}$$

where $S_a =$ selection coefficient of Hb^S_β/Hb^S_β and where $S_A =$ selection coefficient of Hb^A_β/Hb^A_β

$$.10 = \frac{S_A}{S_A + 1}$$

$$S_A = .11$$

16.21.
$$Hb^S_\beta = .10$$
$$Hb^A_\beta = .90$$
$$S_a = 1$$

$$q_1(Hb^S_\beta) = \frac{(q - sq^2)}{(1 - sq^2)} = \frac{(.1 - .01)}{(1 - .01)}$$
$$= \frac{.09}{.99} = .091$$

$$q_2 = \frac{.09 - .0083}{1 - .0083} = \frac{.0817}{.9917} = .0823$$

= frequency of Hb^S_β in two generations

16.22. (a) Wingless insects had a selective advantage on an island where those with wings were in danger of being blown out to sea. (b) Those without wings remained in their protected dishes and produced more wingless flies, whereas those with wings were blown away.

16.23. If two genes are linked, the proportions of all types of possible dihybrid gametes depend on the frequency of crossing over. The amount of time needed for the frequency of coupling-type gametes to equal the frequency of repulsion-type gametes is inversely proportional to the linkage distance between the two genes.

16.24. With the introduction of new genes into this isolated gene pool, the fitness of the individuals will become drastically reduced. This disruption of the genome and subsequent loss of fitness may lead to extinction of the population.

16.25. (a) Inbreeding as such has no effect on gene frequency. (b) It increases homozygosity and decreases heterozygosity. (c) Inbreeding tends to stabilize the type, but decreases the vigor of individuals. Outbreeding results in less constancy in the population, but in an increase in the vigor of individuals.

16.26. (a) They have already lost essentially all of the deleterious recessives from the breeding population through past inbreeding and selection. (b) If self-fertilization resulted in a loss of vigor, the plant must be cross-fertilized in nature; otherwise, the loss would already have been sustained and no further loss would be expected.

16.27. (a) Pure lines are completely homozygous. Results of various crosses can be predicted and experiments can be readily performed to verify the prediction as Mendel did in his garden pea experiments. (b) All variation within the pure line is environmental.

16.28. (a) The expected pure lines would have the following phenotypes: yellow seed, green pod, inflated pod; yellow seed, green pod, constricted pod; yellow seed, yellow pod, inflated pod; yellow seed, yellow pod, constricted pod; green seed, green pod, inflated pod; green seed, green pod, constricted pod; green seed, yellow pod, inflated pod; green seed, yellow pod, constricted pod. (b) Even in pure lines mutations occasionally would create new heterozygotes.

16.29. About 62 of the 2000 plants would be heterozygous at the end of five generations.

16.30. $1:2:1$.

16.31. The systems are less efficient in the order as listed (see Fig. 16.3).

16.32. (a) The abnormality apparently arose through mutation and was maintained through continued inbreeding. (b) The losses can be avoided by detecting carriers and eliminating them from the breeding stock.

16.33. (a) Maximum vigor would be transmitted in the seeds from this cross and would be manifest in the F_1 progeny. (b) The seed parent would be vigorous and would produce a good quantity of seed, but the quality, with reference to hybrid vigor, would be poor. The F_2 plants would be variable with only a small proportion as vigorous as the F_1 plants. (c) If A, B, C, and D are dominant, this cross should produce as much hybrid vigor as (a). If they are not dominant, the hybrid vigor would be slightly less than that of (a). In any case, the parents would be strong hybrid plants through which an abundance of good quality seed could be produced. The F_1 plants raised from the seed would be as vigorous or nearly as vigorous and uniform as those from (a).

16.34. (a) The heterozygous combination would be most efficient for production of vigor. (b) Matings between plants homozygous for A and those homozygous for B would result in the greatest vigor. (c) A pure breeding strain with increased vigor is not likely, because the linked recessives in homozygous condition would offset the advantage of heterosis. If the situation in nature were as simple as that indicated in this problem, the recessives w and l might be removed by crossing over; but in actual populations many recessives for lack of vigor seem to be linked with the dominants postulated to explain hybrid vigor.

16.35. (a) $\frac{1}{10,000}$; (b) $\frac{1}{16}$; (c) $\frac{7.2}{10,000}$ (d) $\frac{1}{64}$; (e) $\frac{1}{16}$.

16.36. (a) The theory of the slow gradual process of evolution by natural selection and the actual procedure of evolving or unfolding in plant and animal populations. (b) Natural selection and rapid major changes in populations associated with global geological upheavals. (c) Fixation of variants in a population by such factors as unequal chromatid exchange, gene conversion, or a gain or loss of DNA through transposition.

16.37.

$\delta\delta$		$\varphi\varphi$	
BB		BB	bald
BB′	} bald	BB′	} nonbald
B′B′	nonbald	B′B′	

Total population

p = freq. B allele
q = freq. B′ allele
$p^2 + 2pq = 0.51$
$q^2 = 0.49$
$q = .7;\ p = .3$

Mating subpopulation

bald men $\begin{cases} \text{BB:} & \dfrac{p^2}{p_2 + 2pq} = \dfrac{.09}{.51} = .176 \\[2mm] \text{BB′:} & \dfrac{2pq}{p^2 + 2pq} = \dfrac{.42}{.51} = .824 \end{cases}$

nonbald women $\begin{cases} \text{BB′:} & \dfrac{2pq}{2pq + q^2} = \dfrac{.42}{.91} = .462 \\[2mm] \text{B′B′:} & \dfrac{q^2}{2pq + q^2} = \dfrac{.49}{.91} = .538 \end{cases}$

		δ gametes	
		p_m 0.588	q_m 0.412
♀ gametes	p_f 0.231	0.1358	0.0952
	q_f 0.769	0.4522	0.3168

$p_{\text{male}} = .176 + \dfrac{.824}{2} = .588$

$q_{\text{male}} = \dfrac{.824}{2} = .412$

$p_{\text{female}} = \dfrac{.462}{2} = .231$

$q_{\text{female}} = \dfrac{.462}{2} + .538 = .769$

Probability bald child $= p_f p_m + \dfrac{p_f q_m}{2} + \dfrac{p_m q_f}{2}$

$= 0.1358 + 0.0476 + 0.2261 \cong 0.41$

CHAPTER 17

17.1. Animal behavior, the sum total of the animal's responses to its environment, has a hereditary basis. The structural and physiological characteristics on which behavior depends are inherited. Now some behavioral patterns are shown to have a genetic basis, but most of these are also strongly influenced by environmental factors.

17.2. Behavior is more difficult to explain in terms of specific genetic mechanisms than tangible, structural characteristics because behavior traits are complex and the genetic control is intimately interwoven with environmental influences. It is difficult to determine exactly what is inherited, but the general genetic basis as indicated by response to selection is well established.

17.3. Such studies indicate that simple Mendelian inheritance can be applied to behavior traits. They suggest that other behavior patterns that now seem complex and involved with environmental factors may be reduced to common denominators when the mechanics are disentangled.

17.4. Such studies identify differences in the behavior of different species and provide specific areas where hypotheses might be developed and experiments designed to investigate mechanisms that have evolved in these species.

17.5. High and low lines have been selected for preference to alfalfa pollen. A trait that responds to selection and becomes established in a line of bees must have a genetic basis.

17.6. (a) Hormones provide chemical coordination and thus influence the general health and temperament as well as the mating behavior of an animal. (b) Glands that produce hormones develop under genetic control, and hormone production in mature animals can be altered by genetic factors.

17.7. Scott and Fuller showed that genetically determined behavior differences develop under the influence of environmental factors.

17.8. The dogs are innately intelligent, and they have been selected for desired physical and mental traits. Training guides and perfects the particular skills for which the dogs are genetically qualified.

17.9. Dogs have a great amount of variability, even within breeds, that responds readily to selection.

17.10. The dog is an intelligent and highly adaptive animal. Man and the dog apparently adopted each other at an early period in their known cultural history and have grown up together. Humans and dogs work together in many ways for mutual advantage. Dogs live in the same shelters with people, feed on their leftovers, and adapt to the requirements for human companionship.

17.11. In their African environment, noisy dogs would call attention to themselves and their masters and either humans or nature (in the form of carnivores) would tend to remove the conspicuous.

17.12. A large component of behavioral patterns in a population having undergone long periods of selection is genetic. Basic differences in motivation, cultural overlay, hormonal response, and physical variations make it difficult and unsettling to transfer into a population behavioral insights that have not developed in that population.

17.13. A trait with survival value is likely to be predominantly and basically genetic. Selection in nature tends to preserve individuals and populations with traits that have survival value.

17.14. (a) Size, simplicity of nervous system, mutability, generation time, response to culture methods, defined behavior traits, and observability; (b) size, generation time, culture requirements, defined simple behavior traits, and abundant linkage data.

17.15. To be sure that the experiments were properly controlled and environmental factors were ruled out, questions such as the following must be considered: Are diets exactly alike? Do mothers differ in treatment of young? Does early environment provide more diversity for one strain than the other? Are both strains run through the same maze? Does time of day when tested alter results? Are temperature and light and possible disturbing factors controlled?

17.16. Aggression can be decreased in animal populations by training, by reducing stimuli that arouse aggressive tendencies, and by avoiding overcrowding.

17.17. *Drosophila* are better known genetically than mice. All chromosomes have been mapped and numerous genes have been located on chromosomes. It is possible, not only to detect behavior traits, but to determine some basic genetic mechanisms that are involved.

17.18. From the results of reciprocal crosses between geotactic positive and geotactic negative strains of *Drosophila*, it can be shown whether the F_1 and F_2 plotted pattern is that of quantitative inheritance (polygenes) and whether a crisscross pattern (from mother to son, sex linkage) can be detected.

17.19.

Genotype	u^+ur^+r	u^+urr	uur^+r	$uurr$
Phenotype	nonhygienic	no uncapping, but removal	uncapping, no removal	hygienic
Proportion	1	1	1	1

INDEX

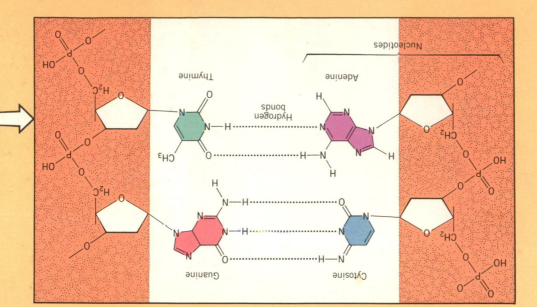

Chemical structures of base-pairs and sugar-phosphate chains

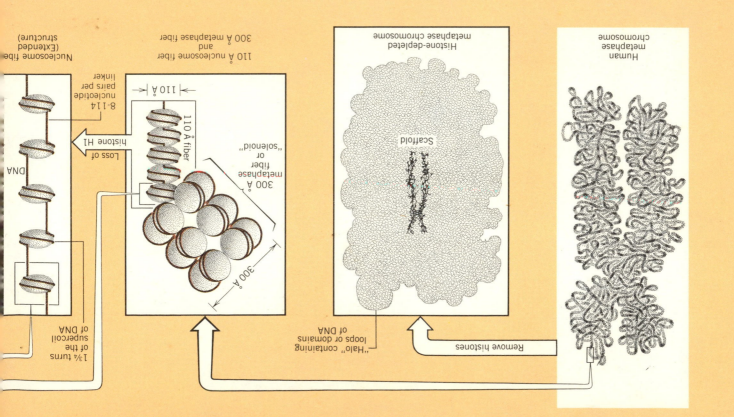